Lecture Notes in Computer Science   6648

Commenced Publication in 1973
Founding and Former Series Editors:
Gerhard Goos, Juris Hartmanis, and Jan va

Mitsunori Ogihara   Jun Tarui (Eds.)

Theory and Applications of Models of Computation

8th Annual Conference, TAMC 2011
Tokyo, Japan, May 23-25, 2011
Proceedings

 Springer

Volume Editors

Mitsunori Ogihara
University of Miami
Department of Computer Science
1365 Memorial Drive, Coral Gables, FL 33146, USA
E-mail: ogihara@cs.miami.edu

Jun Tarui
University of Electro-Communications
Department of Information and Communication Engineering
Chofugaoga 1-5-1, Chofu, Tokyo, 182-8585, Japan
E-mail: tarui@ice.uec.ac.jp

ISSN 0302-9743   e-ISSN 1611-3349
ISBN 978-3-642-20876-8   e-ISBN 978-3-642-20877-5
DOI 10.1007/978-3-642-20877-5
Springer Heidelberg Dordrecht London New York

Library of Congress Control Number: 2011926299

CR Subject Classification (1998): F.2, F.3, F.4, G.2.2, H.1.1, E.1, G.4, I.1

LNCS Sublibrary: SL 1 – Theoretical Computer Science and General Issues

Typesetting: Camera-ready by author, data conversion by Scientific Publishing Services, Chennai, India

Printed on acid-free paper

Springer is part of Springer Science+Business Media (www.springer.com)

Preface

Theory and Applications of Models of Computation 2011 (TAMC 2011) was held May 23–25 in Chofu, Tokyo. After six years of successful meetings in China, the seventh TAMC was held in Prague, Czech Republic, in 2010. This eighth meeting was the second to be held outside China. The conference has a strong focus on computational models. In 2011, there were 136 submissions, out of which 51 papers were selected by the Program Committee. The conference had invited talks by two world-renowned scholars, Tetsuo Asano of the Japan Advanced Institute for Science and Technology and Richard Lipton of the Georgia Institute of Technology.

May 2011

Mitsunori Ogihara
Jun Tarui

Organization

Program Committee

Olaf Beyersdorff	University of Hannover, Germany
Cristian Calude	University of Auckland, New Zealand
Amit Chakrabarti	Dartmouth College, USA
Danny Z. Chen	University of Notre Dame, USA
Zhi-Zhong Chen	Tokyo Denki University, Japan
Marek Chrobak	University of California – Riverside, USA
Pierluigi Crescenzi	University of Florence, Italy
William Gasarch	University of Maryland – College Park, USA
Tero Harju	University of Turku, Finland
Miki Hermann	École Polytechnique, France
Sanjay Jain	National University of Singapore, Singapore
Ming-Yang Kao	Northwestern University, USA
S. Rao Kosaraju	The Johns Hopkins University, USA
Carlos Martín Vide	Universitat Rovira i Virgili, Spain
Peter Bro Miltersen	Aarhus University, Denmark
Mitsunori Ogihara, Chair	University of Miami, USA
Rüdiger Reischuk	University of Lübeck, Germany
Christian Sohler	Technische Universität Dortmund, Germany
Jun Tarui, Co-chair	University of Electro-Communications, Japan
Takeshi Tokuyama	Tohoku University, Japan
Chee-Keng Yap	New York University, USA

Local Organizing Committee

Takaaki Goto	University of Electro-Communications, Japan
Satoshi Kobayashi	University of Electro-Communications, Japan
Tetsuro Nishino	University of Electro-Communications, Japan
Yasuhiko Takenaga	University of Electro-Communications, Japan
Jun Tarui, Chair	University of Electro-Communications, Japan
Seinosuke Toda	Nihon University, Japan
Mitsuo Wakatsuki	University of Electro-Communications, Japan
Osamu Watanabe	Tokyo Institute of Technology, Japan
Katsuhisa Yamanaka	Iwate University, Japan

Organizational Assistance

Yajie Hu (Proceedings)	University of Miami, USA
Seiko Kurakami (Web Design)	Atelier Prefull, Japan

Conference Sponsorship

University of Electro-Communications, Japan

Steering Committee

Manindra Agrawal	Indian Institute of Technology – Kanpur, India
Jin-Yi Cai	University of Wisconsin – Madison, USA
S. Barry Cooper	University of Leeds, UK
John Hopcroft	Cornell University, USA
Angsheng Li	Institute of Software, Chinese Academy of Sciences, China

Referees

Vicente Acuña
Isolde Adler
Luis Antunes
Vikraman Arvind
Chandrajit Bajaj
Yair Bartal
Hervé Baumann
Manuel Bodirsky
Beate Bollig
Vincenzo Bonifaci
Tiziana Calamoneri
Jianer Chen
Miriam Di Ianni
Michael Elberfeld
Bruno Escoffier
Cristina Fernandes
Carlos Ferreira
Alexandre Freire
Naveen Garg
Lee-Ad Gottlieb
Andre Gronemeier
Roberto Grossi
Mohammad Hajiaghayi
Xin He
Frank Hellweg
Susan Hohenberger
Stefan Hougardy
Rahul Jain
Akitoshi Kawamura
Rolf Klein

Hüseyin Koçak
Johannes Köbler
Ranganath Kondapally
Oliver Kullmann
Christiane Lammersen
Massimo Lauria
Peter Lohmann
Dierk Luedemann
Andrew Lyons
Andrea Marino
Dániel Marx
Arne Meier
Daniel Meister
Jochen Messner
Alexander Munteanu
Sebastian Müller
Kiyohito Nagano
Jesper Buus Nielsen
Zeev Nutov
Mikhail Ostrovsky
Francesco Pasquale
Paolo Penna
Rajeev Raman
Burton Rosenberg
Gianluca Rossi
Marie-France Sagot
Dilip Sarkar
Jayalal Sarma M.N.
Gido Scharfenberger-Fabian
Melanie Schmidt

Chris Schwiegelshohn
Akiyoshi Shioura
Somnath Sikdar
Frank Stephan
Dirk Sudholt
Jukka Suomela
Pushkar Tripathi
Mathieu Turuani
Christopher Umans
Jan Vahrenhold
Vinodchandran Variyam
Stéphane Vialette

Paulo Vieira Milreu
Heribert Vollmer
Yoshiko Wakabayashi
Matthias Westermann
Peter Winkler
Maximilian Witek
Philipp Woelfel
Li Yan
Christine Zarges
Shengyu Zhang
Marius Zimand

Table of Contents

Session 4B: Logic and Formal Language Theory

Session 5A: Graph Algorithms II

Session 5B: Approximation II

Session 6A: Games and Learning Theory

Session 6B: Cryptography and Communication Complexity

Session 7A: Optimization II

Session 7B: Complexity II

Session 8A: Graph Algorithms III

Session 8B: Complexity III

Designing Algorithms with Limited Work Space

Tetsuo Asano

Japan Advanced Institute of Science and Technology (JAIST)
1-1 Asahidai, Nomi, Ishikawa 923-1292, Japan
t-asano@jaist.ac.jp

Abstract. Recent progress in computer systems has provided programmers with virtually unlimited amount of work storage for their programs. This leads to space-inefficient programs that use too much storage and become too slow if sufficiently large memory is not available. Thus, I believe that space-efficient algorithms or **memory-constrained algorithms** deserve more attention.

Constant-work-space algorithms have been extensively studied under a different name, log-space algorithms. Input data are given on a read-only array of n elements, each having $O(\log n)$ bits, and work space is limited to $O(\log n)$ bits, in other words, a constant number of pointers and counters, each of $O(\log n)$ bits. This memory constraint in the log-space algorithms may be too severe for practical applications. For problems related to an image with n pixels, for example, it is quite reasonable to use $O(\sqrt{n})$ work space, which amounts to a constant number of rows and columns.

I will start my talk with a simple algorithm for detecting a cycle in a graph using only some constant amount of work space (more exactly, $O(\log n)$ bits in total) and then its applications. Then, I will introduce some paradigms for designing such memory-constrained algorithms and their applications to interesting problems including those in computational geometry and computer vision.

M. Ogihara and J. Tarui (Eds.): TAMC 2011, LNCS 6648, p. 1, 2011.
© Springer-Verlag Berlin Heidelberg 2011

Group-Theoretic Lower Bounds for the Complexity of Matrix Multiplication

Alexey Pospelov*

Saarland University, Computer Science Department
Campus E1.3, 66123 Saarbrücken, Germany
pospelov@cs.uni-saarland.de

Abstract. The complexity of multiplication in group algebras is closely related to the complexity of matrix multiplication. Inspired by the recent group-theoretic approach by Cohn and Umans [10] and the algorithms by Cohn et al. [9] for matrix multiplication, we present conditional group-theoretic lower bounds for the complexity of matrix multiplication. These bounds depend on the complexity of multiplication in group algebras.

Using Bläser's lower bounds for the rank of associative algebras we characterize all semisimple group algebras of minimal bilinear complexity and show improved lower bounds for other group algebras. We also improve the best previously known bound for the bilinear complexity of group algebras by Atkinson. Our bounds depend on the complexity of matrix multiplication. In the special if the exponent of the matrix multiplication equals two, we achieve almost linear bounds.

Keywords: Bilinear complexity, group algebras, complexity of matrix multiplication, lower bounds.

1   Introduction

The complexity of multiplication in algebras is an important problem in the algebraic complexity theory and computer algebra. The goal is to compute the coefficients of the product of two vectors with the minimal number of algebraic operations. The required coefficients are bilinear forms of the coefficients of the input vectors. Naturally, the model and the complexity measure for this problem which received particular attention in the literature are the *bilinear model* and the *bilinear complexity* resp., the latter also called the *rank* of multiplication [16], [7, Chap. 4]. In a bilinear computation, all multiplications other than by constants are of the form $\ell(a) \cdot \ell'(b)$, where a and b are the input vectors, and ℓ, ℓ' are some linear forms. An $O\big((\dim A)^2\big)$ upper bound is straightforward to prove, and all currently known lower bounds for the general case are $\Omega(\dim A)$ [7, Chap. 17].

This research is motivated by the recent group-theoretic approach [10] and the following group-theoretic algorithms [9] for matrix multiplication. It was

---

* This work was supported through funds provided by the Cluster of Excellence "Multimodal Computing and Interaction" at Saarland University.

M. Ogihara and J. Tarui (Eds.): TAMC 2011, LNCS 6648, pp. 2–13, 2011.

shown that finite groups with special properties can be used for the design of fast matrix multiplication algorithms. Our goal is to explore the structure of the group algebras and investigate further the complexity relations between the group algebras and the matrix algebra. We put the group-theoretic approach into a different light by showing that the group algebras of the most promising groups for the group-theoretic approach have roughly the same bilinear complexity as the matrix multiplication itself. For the bilinear complexity of a wide class of group algebras the lower bounds depend on the exponent of the matrix multiplication denoted in the literature by ω [7, Introduction].

Another motivation for this work was the search for the algebras of high bilinear complexity. Over the complex field there exist algebras of arbitrarily high dimensions with bilinear complexity higher than $\frac{(\text{dimension of the algebra})^2}{27}$ [6], [7, Ex. 17.20]. However, no explicit example is known. This situation is similar to a classical problem in the logical synthesis theory. It is known that the circuit complexity in a full basis B of almost all Boolean functions of n variables is $c\frac{2^n}{n}(1 + o(1))$ [14] where the constant c depends solely on B. For $B = \{f_1, \ldots, f_n\}$, where each f_ν is of m_ν variables (with no fictitious dependencies) and has weight w_ν,

$$c = \min_{\substack{1 \le \nu \le n, \\ m_\nu \ge 2}} \frac{w_\nu}{m_\nu - 1}.$$

E.g., for $B = \{\vee, \&, \neg\}$, with unit weights, $c = 1$. However, no explicit function of n variables is known to have a superlinear lower bound on the number of gates in a full finite functional circuit basis.

Algebraic preliminaries. In what follows k will always denote a field, and *algebra* (or *k-algebra*) will always stand for a finite dimensional associative algebra with unity 1 over k. A *basis* of an algebra is a basis of the underlying vector space. The *dimension* of the algebra A, dim A is the dimension of the underlying vector space. We call a basis $\{e_i\}_{i=1}^n$ of a k-algebra A a *group basis* if the vectors e_i form a group with respect to the multiplication in algebra. If A contains a group basis, it is called a *group algebra*. Given a finite group $G = \{g_1, \ldots, g_n\}$, and k we can always build a group algebra $k[G]$ as an n-dimensional vector space over k with a basis $\{g_i\}_{i=1}^n$ and the multiplication in $k[G]$ defined as $(\sum_{i=1}^n \alpha_i g_i) \cdot (\sum_{j=1}^n \beta_j g_j) = \sum_{\ell=1}^n (\sum_{g_i g_j = g_\ell} \alpha_i \beta_j) g_\ell$.

The *direct product* of algebras A and B over k is the algebra $A \times B$ over k which consists of all pairs (a, b), $a \in A$, $b \in B$, and where all operations are performed component-wise: $(a_1, b_1) \circ (a_2, b_2) = (a_1 \circ a_2, b_1 \circ b_2)$, $\circ \in \{+, -, \cdot\}$, and $\lambda \cdot (a, b) = (\lambda a, \lambda b)$, where $\lambda \in k$, and $a_i \in A$, $b_i \in B$, for $i = 1, 2$.

$B \subseteq A$ is called a *subalgebra* of A, if it is a linear subspace of A, which is closed under the multiplication in A. A subalgebra I of A is called a *left* (a *right*) *ideal* of A, if for all $a \in A$, $x \in I$, the product $ax \in I$ ($xa \in I$ resp.) A left ideal which is at the same time a right ideal is called a *two-sided ideal*. A two-sided ideal is called *maximal* if it is not contained in any other proper two-sided ideal of the algebra. An ideal I is called *nilpotent* if $I^m = \{0\}$ for some $m > 0$.[1] The smallest m with this property is called the *nilpotence index* of I. The sum of all nilpotent left ideals of the algebra A is called the *radical* of A and is denoted by $\operatorname{rad} A$. The intersection of all maximal two-sided ideals of A is called the *Jacobson radical* of A and is denoted by $J(A)$. In an algebra, $\operatorname{rad} A = J(A)$, see [17, §98].

A is called a *division algebra* if every $a \in A$ has a multiplicative inverse $a^{-1} \in A$. A is called *semisimple* if $\operatorname{rad} A = 0$, and *simple* if it does not contain any proper twosided ideals except $\{0\}$. The structure of the semisimple and simple algebras is described by Wedderburn's theorem [17]. It states that every finite dimensional semisimple k-algebra is isomorphic to a finite direct product of simple algebras; every finite dimensional simple k-algebra is isomorphic to an algebra $D^{n \times n}$ for an integer $n \geq 1$ and a k-division algebra D, where the integer n and the algebra D are uniquely determined by A (D—up to isomorphism).

Model of computation. Let A be a k-algebra. A *bilinear algorithm* for the multiplication in A is a sequence $\phi = (u_1, v_1, w_1; \ldots; u_r, v_r, w_r)$, where $u_\rho, v_\rho \in A^*$, $w_\rho \in A$, such that for all $x, y \in A$, $x \cdot y = \sum_{\rho=1}^r u_\rho(x) v_\rho(y) w_\rho$. r is called the *length* of ϕ, and the *minimal length of all bilinear algorithms for the multiplication in A* is called the *rank* or the *bilinear complexity* of A, and is denoted by $\operatorname{rk} A$. Trivially, $\operatorname{rk} A \leq (\dim A)^2$, and $\operatorname{rk} A \times B \leq \operatorname{rk} A + \operatorname{rk} B$. However, it is not known if the converse of the latter holds, see [7, p. 360, Strassen's Direct Sum Conjecture].

Let $A = \{A_1, A_2, \ldots\}$ be a family of algebras over k. We define the *rank-exponent of A*, $\omega_A = \inf\{\tau : \operatorname{rk} A_n = O((\dim A_n)^\tau) \text{ for all } n \geq 1\}$. Trivially, $1 \leq \omega_A \leq 2$. Note that this definition makes only sense if A contains algebras of arbitrarily high dimensions. This notion is very similar to the well-known *exponent of the matrix multiplication* $\omega_k = \inf\{\tau : \operatorname{rk} k^{n \times n} = O(n^\tau)\}$.[2] However, ω_k is defined with respect to n and $\dim k^{n \times n} = n^2$. It can be easily verified that the exponent of the matrix multiplication equals twice its rank-exponent.

The rank-exponent is a rather crude estimate. For example, $\operatorname{rk} A_n = \dim A_n$, or $\operatorname{rk} A_n = (\dim A_n) \cdot 2^{\lceil \sqrt{\log \dim A_n} \rceil}$, both imply $\omega_A = 1$. On the other hand, for the complexity of matrix multiplication, the exponent (or twice the rank-exponent) is only known to be within $2 \leq \omega < 2.376$ [7, 15.13 Notes] and the exact value is a long-standing open problem.

Structure of group algebras. For a finite group G, the group algebra $k[G]$ is semisimple iff $\operatorname{char} k \nmid \sharp G$. In this case

$$k[G] \cong D_1^{n_1 \times n_1} \times \cdots \times D_t^{n_t \times n_t}, \tag{1}$$

where D_τ is a division algebra over k, $\dim D_\tau = d_\tau$ for $1 \leq \tau \leq t$. Each $D_\tau^{n_\tau \times n_\tau}$ is called an *irreducible representation* of G over k. $n_1, \ldots, n_t$ are called the *character degrees* of G. If k is algebraically closed then all $D_\tau \cong k$ and $d_\tau = 1$.

---

[1] For a set S with multiplication and a positive integer r, S^r denotes the set of all possible products of r elements of S: $\{s_1 \cdots s_r : s_\rho \in S, 1 \leq \rho \leq r\}$.

[2] As usually, we will write just ω when k is clear from the context.

If char $k = p$, and $\sharp G = np^s$, for $p \nmid n$, then $k[G]$ has a nontrivial radical. We have $k[G]/J(k[G]) \cong D_1^{n_1 \times n_1} \times \cdots \times D_t^{n_t \times n_t}$, and $\sum_{\tau=1}^{t} n_\tau^2 d_\tau + \dim J(k[G]) = \sharp G$. If a Sylow p-subgroup of G is normal (and therefore is *the* Sylow p-subgroup of G), then $\dim J(k[G]) = n(p^s - 1)$. If the Sylow p-subgroups of G are not normal, it is known that $J(k[G])$ contains all ideals generated by all $J(k[H])$, where H is a normal p-subgroup of G. In particular, this holds when H is the intersection of all the p-Sylow subgroups of G.

For more on the representation theory of finite groups we refer to [18].

Lower bounds for the bilinear complexity. The only general lower bound for the bilinear complexity of associative algebras is based on the following fact: if A and B are associative algebras over k and $t(A)$ is the number of maximal twosided ideals of A, then [1]

$$\operatorname{rk} A \times B \geq 2 \dim A - t(A) + \operatorname{rk} B. \qquad (2)$$

In particular, $\operatorname{rk} A \geq 2 \dim A - t(A)$. Algebras for which the latter holds as an equality are called the *algebras of minimal rank*. Their structure was completely defined in [5]. We mention here that a semisimple algebra is not of minimal rank if it contains at least one simple factor of format $m \times m$, for $m \geq 3$.

The next two lower bounds are from [3]:

Theorem 1. *Let A be a k-algebra, $A/\operatorname{rad} A \cong A_1 \times \cdots \times A_t$ with $A_\tau = D_\tau^{n_\tau \times n_\tau}$ for all τ, where D_τ is a k-division algebra. Assume that each A_τ is noncommutative, that is, $n_\tau \geq 2$ or D_τ is noncommutative. Then*

$$\operatorname{rk} A \geq \frac{5}{2} \dim A - 3 \sum_{\tau=1}^{t} n_\tau.$$

Theorem 2. *Let A be a finite dimensional k-algebra. For all $m, n \geq 1$,*

$$\operatorname{rk} A \geq \dim A - \dim\big((\operatorname{rad} A)^{m+n-1}\big) + \dim\big((\operatorname{rad} A)^m\big) + \dim\big((\operatorname{rad} A)^n\big). \qquad (3)$$

Our results. We first show an $O(N^{\frac{\omega}{2}+\epsilon})$ upper bound for the rank of a group algebra of order N, for any $\epsilon > 0$. This improves the best previously known upper bound of $O(N^{\frac{3}{2}})$ [2]. We show how one can do better with an extra knowledge about the group.

Using Bläser's classification of all algebras of minimal rank [5] we give a criterion for a semisimple group algebra over an algebraically closed field to be an algebra of minimal rank. Under different assumptions we prove $\frac{5}{2}$·dimension- and 3·dimension-lower bounds for the rank of group algebras.

For groups having not "too many" different irreducible representations, we deduce from Schönhage's τ-theorem [7, Asymptotic Sum Inequality (15.11)] a lower bound for the rank of group algebras, which is superlinear if $\omega \neq 2$. We show that this lower bound holds for the group algebras of full symmetric groups and finite general linear groups. We prove therefore that the group algebras of some of the promising groups for the group-theoretic approach for matrix multiplication have essentially the same complexity as the matrix multiplication itself.

2   Lower Bounds for the Complexity of Matrix Multiplication

The complexity of the multiplication in group algebras is closely related to the complexity of matrix multiplication (1). We will use the following simple fact.

Proposition 1. *Let* $n_1, \ldots, n_t > 0$ *and* $\alpha \geq 1$. *Then* $\sum_{\tau=1}^{t} n_\tau^\alpha \leq \left(\sum_{\tau=1}^{t} n_\tau\right)^\alpha$.

Proof. The statement follows from the convexity of x^α for $x \geq 0$ and $\alpha \geq 1$.   □

For monotonically growing functions $f(n)$ and $g(n)$ we will write $f(n) \lesssim g(n)$, if for every $\delta > 1$, $f(n) = O\big((g(n))^\delta\big)$. G and G_i will always stand for finite groups, and $\sharp G$ will denote the order of G.

Theorem 3. *Let* $\mathcal{G} = \{G_1, G_2, \ldots\}$ *be a family of finite groups,* $\sharp G_i < \sharp G_{i+1}$. *Assume that* k *is algebraically closed, and* $\operatorname{char} k = 0$, *or* $\operatorname{char} k \nmid \sharp G_i$, *for each* $i \geq 1$. *Then for* $G \in \mathcal{G}$,

$$\operatorname{rk} k[G] \lesssim (\sharp G)^{\frac{\omega}{2}}, \tag{4}$$

where ω *is the exponent of matrix multiplication over* k, *and*

$$\omega \geq 2 \limsup_{n \to \infty} \frac{\log \operatorname{rk} k[G_n]}{\log \sharp G_n}. \tag{5}$$

Proof. We have (1) for $k[G]$, with all $D_\tau \cong k$, and $\operatorname{rk} k[G] \leq \sum_{\tau=1}^{t} \operatorname{rk} k^{n_\tau \times n_\tau}$. Furthermore, $\operatorname{rk} k^{n_\tau \times n_\tau} \lesssim n_\tau^\omega$, and by Proposition 1,

$$\operatorname{rk} k[G] \lesssim \sum_{\tau=1}^{t} n_\tau^\omega = \sum_{\tau=1}^{t} (n_\tau^2)^{\frac{\omega}{2}} \leq \left(\sum_{\tau=1}^{t} n_\tau^2\right)^{\frac{\omega}{2}} = (\sharp G)^{\frac{\omega}{2}},$$

which proves (4).

From (4) we have that for every $\epsilon > 0$, there exists a constant $C = C(\epsilon)$, such that $\operatorname{rk} k[G_n] < C \cdot (\sharp G_n)^{(1+\epsilon)\frac{\omega}{2}}$. This implies $\omega > 2\frac{\log \operatorname{rk} k[G_n]}{\log \sharp G_n} - 2\frac{\log C}{\log \sharp G_n} - 2\epsilon$, for all n. Since $\sharp G_n \to \infty$, it also implies that $\omega > 2\limsup_{n\to\infty} \frac{\log \operatorname{rk} k[G_n]}{\log \sharp G_n} - 2\epsilon$. The latter holds for every $\epsilon > 0$, and (5) follows.   □

Corollary 1. *If for all group algebras* $k[G_n]$ *from the family* $\{k[G_1], k[G_2], \ldots\}$, $\operatorname{rk} k[G_n] = \Omega\big((\sharp G_n)^{1+\epsilon}\big)$ *for arbitrary but fixed* $\epsilon > 0$, *then the exponent of matrix multiplication* $\omega > 2$.

Theorem 4. *Let* $\mathcal{G} = \{G_1, G_2, \ldots\}$ *be a family of finite groups,* $\sharp G_i < \sharp G_{i+1}$, *and* k *is algebraically closed with* $\operatorname{char} k = 0$ *or* $\operatorname{char} k \nmid G_i$ *for every* i. *Let* $f(N)$ *be an integer function such that for each* $G \in \mathcal{G}$, *all character degrees of* G *over* k *are less or equal than* $f(\sharp G)$. *Then for* $G \in \mathcal{G}$,

$$\operatorname{rk} k[G] \lesssim \sharp G \left(f(\sharp G)\right)^{\omega - 2 + \frac{4}{\omega + 2}} = \sharp G \left(f(\sharp G)\right)^{\frac{\omega^2}{\omega + 2}} \leq \sharp G \left(f(\sharp G)\right)^{\omega - 1}, \tag{6}$$

where ω is the exponent of matrix multiplication. If, additionally, for all n, $\operatorname{rk} k[G_n] > \sharp G_n$ and $\limsup_{n \to \infty} f(\sharp G_n) = \infty$, then for $F_n = \frac{\log \operatorname{rk} k[G_n] - \log \sharp G_n}{2}$,

$$\omega \geq \limsup_{n \to \infty} \left(F_n + \sqrt{F_n^2 + 4F_n \log f(\sharp G_n)} \right), \tag{7}$$

Proof. Let $n_1 \geq \cdots \geq n_t$ be the irreducible character degrees of G over k and $h(N) > 0$ be an arbitrary integer function. Let $j(N)$ be the number of n_τ greater than $h(N)$. Then $\sharp G = \sum_{\tau=1}^t n_\tau^2 \geq j(\sharp G)(h(\sharp G))^2$ and $j(\sharp G) \leq \frac{\sharp G}{(h(\sharp G))^2}$. This implies:

$$\operatorname{rk} k[G] \lesssim \left(j(\sharp G)(f(\sharp G))^\omega + \sum_{\tau=j(\sharp G)+1}^t n_\tau^\omega \right) \leq \sharp G \frac{(f(\sharp G))^\omega}{(h(\sharp G))^2} + \sharp G(h(\sharp G))^\omega.$$

We finally obtain (6) by setting $h(N) = (f(N))^{1 - \frac{2}{\omega + 2}} < f(N)$. The last inequality in (6) follows from the fact that $\omega \geq 2$.

(7) is proved by a bit more involved but similar argument as in the proof of (5) in Theorem 3.   □

Corollary 2. *1. If $f(\sharp G) = O(1)$, then $\operatorname{rk} k[G]$ is linear. If for every $\epsilon > 0$, $f(\sharp G) = o((\sharp G)^\epsilon)$, then $\omega_{k[\mathcal{G}]} = 1$.*

2. If for some $\{k[G_1], k[G_2], \dots\}$, $\limsup_{n \to \infty} \frac{\operatorname{rk} k[G_n]}{\sharp G_n \cdot f(\sharp G_n)} > 1$, then the exponent of matrix multiplication $\omega > 2$.

Remark 1. The upper bound of (6) is better than of (4), and the lower bound of (7) is better than the lower bound of (5), if $f(\sharp G_n) = o\big((\sharp G_n)^{\frac{1}{2} - \frac{2}{\omega^2}}\big)$. According to the best known upper bound $\omega < 2.376$ [7, Notes 15.13], this is currently the case when $f(\sharp G_n) = o\big((\sharp G_n)^{0.1457}\big)$.

Theorem 3 can be generalized to the case of a not algebraically closed fields.

Theorem 5. *Let $\mathcal{G} = \{G_1, G_2, \dots\}$. Assume that $\operatorname{char} k = 0$, or k is finite. Then for all $G \in \mathcal{G}$, $\operatorname{rk} k[G] \gtrsim (\sharp G)^{\frac{\omega}{2}}$, where ω is the exponent of matrix multiplication over k. In the same way, $\omega \geq 2 \limsup_{n \to \infty} \frac{\log \operatorname{rk} k[G_n]}{\log \sharp G_n}$.*

Proof. Assume first that $\operatorname{char} k = 0$. In this case $k[G]$ is semisimple, (1) holds, and $\mathbb{Q} \subseteq k$ is the prime subfield. Let K be the algebraically closed extension of k. It is known [12, Theorem 11.4, Chap. XVIII] that every representation of G over K is definable over $\mathbb{Q}(\zeta_m)$ where m is the exponent of G, *i.e.*, the minimal μ such that $g^\mu = 1$ for every $g \in G$, and $\zeta_m \in K$ is a primitive m-th root of unity. Therefore, it is definable over $k(\zeta_m)$. Any irreducible representation of G over k is a simple $k[G]$-module by Maschke's Theorem [12, Theorem 1.2, Chap. XVIII]. Therefore, it is isomorphic to $D^{n \times n}$, where D is a k-division algebra. ζ_m is algebraic over D since it is algebraic over $k \subseteq D$ and $D \cong D' \subseteq k(\zeta_m)$. The latter holds since all irreducible representations of G

over $k(\zeta_m)$ are isomorphic to matrix algebras over $k(\zeta_m)$. Thus, D is a subalgebra of $k(\zeta_m)$, and $D \cong k(\zeta_\ell)$ for some $\ell \mid m$, and for all τ, $\operatorname{rk} D_\tau \le 2d_\tau - 1$ since k is infinite. Therefore, we obtain

$$\operatorname{rk} k[G] \gtrsim \sum_{\tau=1}^{t} n_\tau^\omega (2d_\tau - 1) < 2 \sum_{\tau=1}^{t} n_\tau^\omega d_\tau \le 2 \left(\sum_{\tau=1}^{t} n_\tau^2 d_\tau \right)^{\frac{\omega}{2}} = 2 (\sharp G)^{\frac{\omega}{2}}.$$

The lower bound for ω is proved exactly as in the proof of Theorem 3.

Note that the proof remains valid for $\operatorname{char} k > 0$, whenever the division algebras in (1) of $k[G_i]$ have linear ranks. By Wedderburn's Little Theorem [13, Theorem 2.55], a finite dimensional division algebra over a finite field k is an extension field of k. Its rank is linear due to Chudnovskys' algorithm, see [8,15], [7, Theorem 18.20], and the statement holds also for finite k.   □

3   Lower Bounds for the Bilinear Complexity of Group Algebras

Let $\mathcal{G} = \{G_1, G_2, \dots\}$ be a family of finite groups so that $\sharp G_i < \sharp G_{i+1}$. If $\operatorname{char} k = 0$, or $\operatorname{char} k = p$, and for every $i \ge 1$, $p \nmid \sharp G_i$, $\mathcal{G}$ is called a *semisimple* (over k) family of finite groups. If $\operatorname{char} k = p$ and for some $i \ge 1$, $p \mid \sharp G_i$ then $\mathcal{G}$ is called a *modular* family.

From the Hölder's inequality we have:

Lemma 1. *Let* $n_1, \dots, n_t \ge 0$ *and* $\delta \ge 1$. *Then* $\sum_{\tau=1}^{t} n_\tau \le t^{1-\frac{1}{\delta}} \left(\sum_{\tau=1}^{t} n_\tau^\delta \right)^{\frac{1}{\delta}}$.

We will denote by $t_i(G)$ the number of the irreducible characters of G over k whose degree equals to i. Obviously, $t_i(G) = 0$ if $i > \sqrt{\sharp G}$. We will further denote $T_i(G) = \sum_{j=i}^{\infty} t_j(G)$. By definition, $T_i(G) \ge T_j(G)$, if $i < j$ and $t_i(G) = T_i(G) - T_{i+1}(G)$. Note that the number of the maximal twosided ideals of $k[G]$ is exactly $T_1(G) = t$, where t is the number of factors in (1).

Theorem 6. *Let* G *be a finite group and let* k *be an algebraically closed field of* $\operatorname{char} k = 0$ *or* $\operatorname{char} k \nmid \sharp G$. *Let* t *be as in* (1).

1. $T_3(G) = 0$ *iff* $k[G]$ *is of minimal rank. In this case* $\operatorname{rk} k[G] = t_1(G) + 7t_2(G)$.
2. *If* $T_3(G) > 0$, *then* $\operatorname{rk} k[G] \ge 2\sharp G - t + \max\left(\frac{9}{2}T_7(G), 1\right)$.
3. *Let* $\mathcal{G} = \{G_1, G_2, \dots\}$ *be a semisimple family of groups. Assume that the number of irreducible representations of each* $G \in \mathcal{G}$ *over* k *is* $o(\sharp G)$.[3] *Then for all* $G \in \mathcal{G}$, $\operatorname{rk} k[G] \ge \frac{5}{2}\sharp G - o(\sharp G)$.

Proof. Consider the decomposition (1) for $k[G]$ (here all $D_\tau \cong k$). Assume w.l.o.g. that $n_1 \le \cdots \le n_t$. Let A be the direct product of all matrix algebras from (1) of order 1×1 or 2×2, and let B be such that $k[G] = A \times B$. Then $\dim A = t_1(G) + 4t_2(G) = T_1(G) + 3T_2(G) - 4T_3(G)$,

$$\operatorname{rk} A = t_1(G) + 7t_2(G) = 2 \dim A - (t_1(G) + t_2(G)), \tag{8}$$

---

[3] $o(\sharp G)$ means that for any $\epsilon > 0$, there exists such $N = N(\epsilon) > 0$, that if $G \in \mathcal{G}$ and $\sharp G > N$, then the number of irreducible representations of G over k is less than $\epsilon \cdot \sharp G$.

since $\mathrm{rk}\, k^{2\times 2} = 7$, see [7, (17.10)]. The number of the maximal twosided ideals in A is $t_1(G) + t_2(G)$, and (8) follows.

1. If $T_3(G) = 0$, then $t = t_1(G) + t_2(G)$, and the statement follows from (8).
2. If $T_3(G) > 0$, then $\mathrm{rk}\, k[G] = 2\dim A - (T_1(G) - T_3(G)) + \mathrm{rk}\, B \geq 2\sharp G - t + 1$, by (2). $k[G]$ is not of minimal rank [5]. Let $B = B_1 \times B_2$, so that B_1 consists of all matrix algebras from (1) of order ≤ 6. The number of maximal twosided ideals in B_1 is $t_3(G) + \cdots + t_6(G) = T_3(G) - T_7(G)$. By (2) we have $\mathrm{rk}\, B \geq 2\dim B_1 - (T_3(G) - T_7(G)) + \mathrm{rk}\, B_2$. If B_2 is not empty, then $n_{t-T_7(G)+1} \geq 7$, since $n_1 \leq \cdots \leq n_t$. From Theorem 1 we obtain

$$\mathrm{rk}\, B_2 \geq \frac{5}{2} \sum_{\tau=t-T_7(G)+1}^{t} n_\tau^2 - 3 \sum_{\tau=t-T_7(G)+1}^{t} n_\tau \geq 2\dim B_2 + \frac{7}{2}T_7(G),$$

Gathering it all together, we come up to

$$\mathrm{rk}\, k[G] \geq 2\dim A + 2\dim B_1 + 2\dim B_2 - T_1(G) + \tfrac{9}{2}T_7(G) = 2\sharp G - t + \tfrac{9}{2}T_7(G).$$

which proves the second claim of the theorem.

3. Let $k[G] = k^{t_1(G)} \times C$, $\dim C = n_{t-T_2(G)+1}^2 + \cdots + n_t^2$. By (2) and Theorem 1, $\mathrm{rk}\, k[G] = \mathrm{rk}\, k^{t_1(G)} + \mathrm{rk}\, C \geq t_1(G) + \frac{5}{2}\dim C - 3\sum_{\tau=t-T_2(G)+1}^{t} n_\tau$. By using Lemma 1 for the dimensions of the factors of C, and setting $\delta = \frac{1}{2}$, we obtain $\sum_{\tau=t-T_2(G)+1}^{t} n_\tau \leq \sqrt{T_2(G)\dim C} \leq \sqrt{t\sharp G} = o(\sharp G)$. On the other hand, $t_1(G) \leq t = o(\sharp G)$. Thus, $\dim C = \sharp G - t_1(G) = \sharp G - o(\sharp G)$, and $\mathrm{rk}\, k[G] \geq \frac{5}{2}\sharp G - o(\sharp G)$. □

Remark 2. The lower bound in case 2 can be improved for some groups using the lower bound due to Bläser: for $n \geq 3$, $\mathrm{rk}\, k^{n\times n} \geq 2n^2 + n - 2$ [4]. The best we can do in this case is to employ the Alder-Strassen lower bounds for all matrix algebras in (1) except for one of the highest dimension, for which we will use the Bläser's lower bound: if $n_1 \leq \cdots \leq n_t$, and $n_t \geq 3$, then $\mathrm{rk}\, k[G] \geq 2\sharp G + n_t - t - 1$.

Corollary 3. *Let k be algebraically closed of char $k = 0$, p be an arbitrary fixed prime, and p_n be the n-th prime number. Let $\mathcal{G} = \{G_n\}_{n\geq 1}$, where G_n is*

1. *S_n, full symmetric group of order $n!$, or*
2. *$GL(2, p^n)$, general linear group of nonsingular 2×2 matrices over $\mathbb{F}_{p^n}$, or*
3. *$SL(2, p^n)$, special linear group of 2×2 matrices over $\mathbb{F}_{p^n}$ whose determinant equals 1, or*
4. *F_n, a Frobenius group of order $p_n(p_n - 1)$, generated by $a, b \in F_n$, such that $a^{p_n} = b^{p_n-1} = 1$, $b^{-1}ab = a^u$, where u is a primitive element of $\mathbb{Z}_{p_n}^*$, or*
5. *A non-abelian p_n-group with an abelian subgroup of index p_n.*

Then $\mathrm{rk}\, k[G_n] \geq \frac{5}{2}\sharp G_n - o(\sharp G_n)$.

The statements follow from Theorem 6 and the well-known upper bounds on the number of different irreducible representations of the mentioned groups.

Schönhage's τ-theorem [7, (15.11)] in its weakened form states that if the rank of $k^{n_1 \times n_1} \times \cdots \times k^{n_t \times n_t}$ is at most r, and $n_1 \cdots n_t > 1$, then $n_1^\omega + \cdots + n_t^\omega \leq r$, where ω is the exponent of matrix multiplication over k. Note that Strassen's Direct Sum Conjecture implies that $\mathrm{rk}\, k[G] = \mathrm{rk}\, k^{n_1 \times n_1} + \cdots + \mathrm{rk}\, k^{n_t \times n_t}$, in case of an algebraically closed k. It turns out that an insignificantly weaker version of the corresponding lower bound can be proved independently of the validity of the Direct Sum Conjecture.

Theorem 7. *Let $\mathcal{G} = \{G_1, G_2, \ldots\}$ be a semisimple family of finite nonabelian groups over an algebraically closed field k.*

1. *Let $f(\sharp G)$ be a lower bound for the largest irreducible character degree of $G \in \mathcal{G}$. Then $\mathrm{rk}\, k[G] \geq (f(\sharp G))^\omega$, where ω is the exponent of matrix multiplication. If for all n, $f(\sharp G_n) > 1$, then*

$$\omega \leq \limsup_{n \to \infty} \frac{\log k[G_n]}{\log f(\sharp G_n)}.$$

2. *Let $t(\sharp G)$ be an upper bound for the number of different irreducible representations of $G \in \mathcal{G}$. Then $\mathrm{rk}\, k[G] \geq \dfrac{(\sharp G)^{\frac{\omega}{2}}}{(t(\sharp G))^{\frac{\omega^2}{4} - \frac{\omega}{2}}}$. If for all n, $t(\sharp G_n) > 1$, then for $Q_n = 1 + \frac{\log \sharp G_n}{\log t(\sharp G_n)}$,*

$$\omega \geq \limsup_{n \to \infty} \left(Q_n + \sqrt{Q_n^2 - \frac{\log \mathrm{rk}\, k[G_n]}{\log t(\sharp G_n)}} \right) \tag{9}$$

Proof. $\mathrm{rk}\, k[G] \geq f(\sharp G)^\omega$ follows trivially from (1) and the observation that for any algebras A, B over one field $\mathrm{rk}\, A \times B \geq \max\{\mathrm{rk}\, A, \mathrm{rk}\, B\}$. The related lower bound for ω is obtained by taking logarithms of both sides of the inequality.

Since G is not abelian, at least one $n_\tau > 1$ in (1). By Schönhage's τ-theorem, $n_1^\omega + \cdots + n_t^\omega \leq \mathrm{rk}\, k[G]$. On the other hand, by Lemma 1,

$$\sum_{\tau=1}^{t} n_\tau^\omega = \sum_{\tau=1}^{t} (n_\tau^2)^{\frac{\omega}{2}} \geq \left(t^{1 - \frac{\omega}{2}} \cdot \sum_{\tau=1}^{t} n_\tau^2 \right)^{\frac{\omega}{2}} \geq \frac{(\sharp G)^{\frac{\omega}{2}}}{(t(\sharp G))^{\frac{\omega^2}{4} - \frac{\omega}{2}}}.$$

It follows, that $\omega^2 - 2 Q_n \omega + \frac{\log \mathrm{rk}\, k[G_n]}{\log t(\sharp G_n)} > 0$. Since $\omega > 0$, the solution is $\omega \geq Q_n + \sqrt{Q_n^2 - \frac{\log \mathrm{rk}\, k[G_n]}{\log t(\sharp G_n)}}$. This holds for all n, therefore (9) follows. $\square$

Corollary 4. 1. *If $t(\sharp G_n) = o((\sharp G_n)^\epsilon)$ for every $\epsilon > 0$, then $\omega_{k[\mathcal{G}]} = \frac{\omega}{2}$.[4] If additionally $\omega > 2$, then $\omega_{k[\mathcal{G}]} > 1$.*

2. *If $\omega > 2$ and $\sharp G_n = o((f(\sharp G_n))^\omega)$, then $\omega_{k[\mathcal{G}]} > 1$. One promising family of finite groups which can help to achieve $\omega = 2$ in [10] has $f(\sharp G_n) = (\sharp G_n)^{\frac{1}{2} - \epsilon}$ for some fixed $\epsilon > 0$. It follows from Theorem 7, that one should look for $\epsilon \geq \frac{1}{2} - \frac{1}{\omega}$, since otherwise the lower bound is superlinear unless $\omega = 2$, and for $\epsilon > 0.079$ to improve the Coppersmith-Winograd upper bound for ω.*

---

[4] For a family of finite groups $\mathcal{G} = \{G_1, G_2, \ldots\}$ we denote by $k[\mathcal{G}]$ the family of group algebras $\{k[G_1], k[G_2], \ldots\}$.

3. If $\omega > 2$ and $t(\sharp G_n) = o\big((\sharp G_n)^{\frac{2}{\omega}}\big)$, then $\omega_{k[\mathcal{G}]} > 1$. In particular, this holds if $t(\sharp G_n) = o\big((\sharp G_n)^{0.841}\big)$.

4. Let k be an algebraically closed field of characteristic 0, q be a fixed prime power, $\mathcal{G}_1 = \{S_n\}_{n \geq 1}$, $\mathcal{G}_2 = \{GL(n, q)\}_{n \geq 1}$. Then $\omega_{k[\mathcal{G}_i]} = \frac{\omega}{2}$, for $i = 1, 2$.

Proof. Items 1–3 are trivial. Item 4 for $\mathcal{G}_1$ follows from Corollary 3, and for $\mathcal{G}_2$— from the fact that $GL(n, q)$ contains an irreducible representation of dimension $(\sharp GL(n, q))^{\frac{1}{2} - o(1)}$ [11]. $\qquad\square$

We conclude with an extension of the lower bounds to the more difficult case of non-semisimple group algebras. Let k be an algebraically closed field of characteristic p, and let G be a finite group of order $N = np^d$, where $p \nmid n$. We will assume that G has *the* normal Sylow p-subgroup P of order p^d. In this case $\operatorname{rad} k[G]$ is generated by all vectors $\sum_{h \in P} x_h h$, with $\sum x_h = 0$, and $\dim \operatorname{rad} k[G] = p^d(n-1)$. We will further assume that P is abelian and is isomorphic to a direct product of cyclic p-groups:

$$P = \mathbb{Z}_{p^{s_1}} \times \cdots \times \mathbb{Z}_{p^{s_r}}, \qquad s_1 \leq \cdots \leq s_r, \qquad d = s_1 + \cdots + s_r. \tag{10}$$

Theorem 8. *Let k be a field of characteristic p, and $\mathcal{G} = \{G_1, G_2, \dots\}$ be a modular family of groups. Let $G \in \mathcal{G}$, and $\sharp G = np^d$, where $p \nmid n$, and d is as in (10). Assume that*

- $P = Z(G)^5$ *is the Sylow p-subgroup of G;*
- *For any $D > 0$ there exists $G \in \mathcal{G}$ with $d > D$;*
- *For any $\epsilon > 0$ and for all $G \in \mathcal{G}$ with $\sharp G > N_0 = N_0(\epsilon)$, $s_r - s_1 \leq \frac{1}{2}\log_p \epsilon r$.*

Then $\operatorname{rk} k[G] \geq \big(2 + \frac{1}{n}\big)\sharp G - o(\sharp G)$.

Proof. We will denote the elements of P by $h_{i_1, \dots, i_r}$, $0 \leq i_\rho < p^{s_\rho}$, for all $1 \leq \rho \leq r$, so that

$$h_{i_1, \dots, i_r} \cdot h_{j_1, \dots, j_r} = h_{(i_1 + j_1) \bmod p^{s_1}, \dots, (i_r + j_r) \bmod p^{s_r}}.$$

For $0 \leq \rho \leq r$, let

$$x_\rho = h_{u_1, \dots, u_r} - h_{0, \dots, 0},$$

where $u_\rho = 1$ and $u_\kappa = 0$ for $\kappa \neq \rho$. $R := \operatorname{rad} k[G]$ is generated by $x_1, \dots, x_r$. It is easy to check that $x_\rho^{p^{s_\rho}} = 0$ and for $m \geq 1$, the system

$$\big\{ x_1^{i_1} \cdots x_r^{i_r} \mid i_1 + \cdots + i_r \geq m, \ 0 \leq i_\rho < p^{s_\rho} \big\}$$

is linearly independent and generates R^m. For

$$a_{m-1} = \sharp\big\{(i_1, \dots, i_r) : i_1 + \cdots + i_r \leq m - 1, \ 0 \leq i_\rho < p^{s_\rho}\big\},$$

$\dim R^m = n(p^d - a_{m-1})$.

---

[5] $Z(G)$ is the center of G, *i.e.*, the set of elements of G that commute with all the elements of G.

Let $\mathbb{E}\xi$ be the expectation of a discrete random variable ξ which can assume only finite number of values. If ξ takes value $a_i \in \mathbb{R}$ with probability $p_i \geq 0$ for $1 \leq i \leq n$, $\sum_{i=1}^{n} p_i = 1$, then $\mathbb{E}\xi = \sum_{i=1}^{n} a_i p_i$. $\mathbb{D}\xi = \mathbb{E}(\xi - \mathbb{E}\xi)^2$ denotes the variance of ξ.

The exponent of P is $p^{s_r} = o(\sharp P)$ by the third condition of the theorem. Thus the parameter r is unbounded among all groups from $\mathcal{G}$. According to (3),

$$\operatorname{rk} k[G] \geq \sharp G + \dim R^m + \dim R^m - \dim R^{2m} = \left(2 + \frac{a_{2m-1} - 2a_{m-1}}{np^d} \right) \sharp G.$$

We will choose m in such a way that $\frac{a_{2m-1}}{p^d} \to 1$, $\frac{a_{m-1}}{p^d} \to 0$ when $r \to \infty$.

This following elegant argument is due to B. Chokayev. Consider the indices $\{i_\rho\}_{\rho=1}^{r}$ as independent random variables, i_ρ taking each value in $[0, p^{s_\rho} - 1]$ with probability $\frac{1}{p^{s_\rho}}$. Then $\mathbb{E}i_\rho = \frac{p^{s_\rho} - 1}{2}$, $\mathbb{D}i_\rho = \frac{p^{2s_\rho} - 1}{12}$, and denoting $\xi_r = i_1 + \cdots + i_r$,

$$\mathbb{E}\xi_r = \frac{1}{2} \sum_{\rho=1}^{r} p^{s_\rho} - \frac{r}{2}, \qquad \mathbb{D}\xi_r = \frac{1}{12} \sum_{\rho=1}^{r} p^{2s_\rho} - \frac{r}{12}.$$

Note that ξ_r takes each value in $[0, \sum_{\sigma=1}^{r} p^{s_\rho} - r]$ with probability $\frac{a_m - a_{m-1}}{p^d}$. Now consider $m = \frac{2}{3}\mathbb{E}\xi_r$ as a function of r. By Chebyshev's inequality,

$$\frac{a_{m-1}}{p^d} = \mathbb{P}(\xi_r \leq m - 1) \leq \mathbb{P}(|\xi_r - \mathbb{E}\xi_r| \geq \mathbb{E}\xi_r - m + 1)$$

$$\leq \frac{\mathbb{D}\xi_r}{(\mathbb{E}\xi_r - m + 1)^2} \leq \frac{3p^{2s_r - 2s_1}}{4r} \xrightarrow[r\to\infty]{} 0,$$

$$\frac{a_{2m-1}}{p^d} = \mathbb{P}(\xi_r \leq 2m - 1) \geq \mathbb{P}(|\xi_r - \mathbb{E}\xi_r| \leq 2m - 1 - \mathbb{E}\xi_r)$$

$$\geq 1 - \frac{\mathbb{D}\xi_r}{(2m - 1 - \mathbb{E}\xi_r)^2} \geq 1 - \frac{3p^{2s_r - 2s_1}}{4r} \xrightarrow[r\to\infty]{} 1.$$

$\square$

Corollary 5. *Let k be a field of characteristic p and $\mathcal{G} = \{G_1, G_2, \ldots\}$ be a family of finite groups such that*

1. *For all $i \geq 1$, $\sharp G_i < \sharp G_{i+1}$,*
2. *For every $i \geq 1$, the only Sylow p-subroup of G_i coincides with $Z(G_i)$, and $Z(G_i) = \mathbb{Z}_{n_i} \times \cdots \times \mathbb{Z}_{n_i}$,*
3. *$\limsup_{n\to\infty} \sharp Z(G_i)/(\sharp G_i) > 0$.*

Then there exists such N that the family of group algebras $k[\mathcal{G}]$ does not contain algebras of minimal rank of dimension greater than N.

4   Conclusion

We extended the group-theoretic approach and algorithms for matrix multiplication by showing that there are tight relations between the lower bounds for the rank of matrix multiplication and the rank of group algebras.

The first possible improvement would be a further generalization of Theorem 5 for the case of arbitrary semisimple group algebras. This would link the ranks of group algebras and of the matrix algebra over more fields. On the other hand, relaxing the conditions of Theorem 8 may shed more light on the possible complexity issues that arise in the group algebras with radical which are not related to the complexity of matrix multiplication.

Acknowledgements. I would like to thank M. Bläser, the anonymous reviewers for many helpful comments and suggestions, and V. Alekseyev for introducing me into this topic. Many thoughtful remarks came from D. Khovratovich, M. Fouz and R. Rao.

References

1. Alder, A., Strassen, V.: On the algorithmic complexity of associative algebras. Theoret. Comput. Sci. 15, 201–211 (1981)
2. Atkinson, M.D.: The complexity of group algebra computations. Theoret. Comput. Sci. 5(2), 205–209 (1977)
3. Bläser, M.: Lower bounds for the bilinear complexity of associative algebras. Comput. Complexity 9(2), 73–112 (2000)
4. Bläser, M.: On the complexity of the multiplication of matrices of small formats. J. Complexity 19(1), 43–60 (2003)
5. Bläser, M.: A complete characterization of the algebras of minimal bilinear complexity. SIAM J. Comput. 34(2), 277–298 (2004)
6. Büchi, W.: Über eine Klasse von Algebren minimalen Ranges. PhD thesis, Zürich University (1984)
7. Bürgisser, P., Clausen, M., Shokrollahi, M.A.: Algebraic Complexity Theory. A Series of Comprehensive Studies in Mathematics, vol. 315. Springer, Heidelberg (1997)
8. Chudnovsky, D.V., Chudnovsky, G.V.: Algebraic complexities and algebraic curves over finite fields. J. Complexity 4(4), 285–316 (1988)
9. Cohn, H., Kleinberg, R.D., Szegedy, B., Umans, C.: Group-theoretic algorithms for matrix multiplication. In: 46th FOCS, pp. 379–388. IEEE Comp. Soc., Los Alamitos (2005)
10. Cohn, H., Umans, C.: A group-theoretic approach to fast matrix multiplication. In: 44th FOCS, pp. 438–449. IEEE Comp. Soc., Los Alamitos (2003)
11. Gelfand, S.I.: Representations of the full linear group over a finite field. Math. USSR, Sb. 12, 1339 (1970)
12. Lang, S.: Algebra, 3rd revised edn. Springer, Heidelberg (2005)
13. Lidl, R., Niederreiter, H.: Finite Fields. Encyclopedia of Mathematics and its Applications, vol. 20. Cambridge University Press, Cambridge (2008)
14. Lupanov, O.B.: A method of circuit synthesis. Izvesitya VUZ, Radiofizika 1, 120–140 (1958)
15. Shparlinski, I.E., Tsfasman, M.A., Vladut, S.G.: Curves with many points and multiplication in finite fields. In: Stichtenoth, H., Tsfasman, M.A. (eds.) International Workshop on Coding Theory and Algebraic Geometry. Lecture Notes in Mathematics, vol. 1518, pp. 145–169. Springer, Berlin (1992)
16. Strassen, V.: Vermeidung von Divisionen. J. Reine Angew. Math. 264, 182–202 (1973)
17. van der Waerden, B.L.: Algebra II, 5th edn. Springer, Heidelberg (1967)
18. Wientraub, S.H.: Representation Theory of Finite Groups: Algebra and Arithmetic. Graduate Series in Mathematics, vol. 59. AMS, Providence (2003)

A Real Elementary Approach to the Master Recurrence and Generalizations*

Chee Yap

[1] Courant Institute of Mathematical Sciences
New York University
New York, NY 10012, USA
yap@cs.nyu.edu
[2] Korea Institute of Advanced Study
Seoul, Korea

Abstract. The master theorem provides a solution to a well-known divide-and-conquer recurrence, called here the master recurrence. This paper proves two cook-book style generalizations of this master theorem. The first extends the treated class of driving functions to the natural class of exponential-logarithmic (EL) functions. The second extends the result to the multiterm master recurrence. The power and simplicity of our approach comes from re-interpreting integer recurrences as real recurrences, with emphasis on elementary techniques and real induction.

1 Introduction

Techniques for solving recurrences are among the standard repertoire of algorithmic textbooks [13,5,16,4,1]. A proto-typical recurrence arising in the analysis of efficient recursive algorithms is

$$T(n) = aT(n/b) + d(n) \tag{1}$$

where $a > 0$ and $b > 1$ are arbitrary real numbers, and $d(n) \geq 1$ is the **driving function**. We call (1) the **master recurrence** since theorems providing its solution are widely known as "master theorems". The solutions depend on the nature of $d(n)$. The case where $d(n)$ is multiplicative is treated in [1, p. 301]. In an influential note, Bentley, Haken and Saxe [3, Table 1, p.39] proved a master theorem under a fairly general hypothesis on $d(n)$. Recurrence (1) generalizes to

$$T(n) = \sum_{i=1}^{k} a_i T(n/b_i) + d(n) \tag{2}$$

where $a_i > 0$ and $b_i > 1$ are arbitrary real constants ($k \geq 2$). We call (2) the **multiterm master recurrence**. E.g., the 2-term recurrences $T(n) = T(n/b_1) + T(n/b_2) + n$ and $T(n) = T(n/2) + T(n/4) + \log n$ arise (respectively)

---

* This work is supported by an National Science Foundation Grants #CCF-0728977 and #CCF-0917093, and also with KIAS support.

M. Ogihara and J. Tarui (Eds.): TAMC 2011, LNCS 6648, pp. 14–26, 2011.

in fast median algorithms [4, p. 240] and in conjugate search tree analysis in Computational Geometry [6].

To discuss the literature, it is useful to begin with the "standard" master theorem for (1). This is Proposition 1 in the next Section. It has two kinds of generalizations: (A) The first kind, as in Verma [19], extends the class of driving functions $d(n)$ that are captured by the master theorem. Verma's main result [19, Theorem 13] provided integral bounds on solutions when the driving functions $d(n)$ satisfy some growth properties. (B) The second kind comes from extending the master recurrence itself. Wang and Fu [20, Theorem 3.5] gave integral bounds for a parametric form of (1) where a, b are now functions of n. Of course, the multiterm recurrence (2) is also a generalization of the second kind. An early treatment of multiterm recurrences is found in Purdom and Brown [16]. Multiterm master theorems are given by Kao [11], Akra and Bazzi [2], and Roura [18, Theorem 2.3]. Leighton [14] provides an exposition of [2]. We remark that obtaining generalized bounds in the form of integrals, by itself, is not satisfactory: our goal is to achieve "cookbook style" theorems [12] as exemplified by the master theorem.

¶1. *Contributions and Overview.* Our main contribution is two cookbook style generalizations of the master theorem, **Theorems A** and **B**. They are natural extensions, and completions, of known results. They serve to unify many complexity analysis of individual algorithms: thus, no previous master theorems capture the analysis of Schönhage-Strassen's multiplication algorithm [13], but this is now an application of Theorem A. Similarly, the conjugate tree analysis of Edelsbrunner and Welzl [6] is a consequence of Theorem B. Furthermore, Theorem B shows that the conjugate tree exponent, $\alpha = \lg(\phi - 1) \sim 0.695$ where $\phi = 1.618\ldots$ is the golden ratio, can be systematically obtained, and that this bound is tight to Θ-order. Our second contribution is the introduction of rigorous elementary techniques for these derivations. In particular, we provide summation formulas for Exponential-Logarithmic (EL) functions. Elementary techniques are possible because we exploit bounds which are tight to (only) Θ-order.

Section 2 will review the master theorem and extensions. Section 3 states our two main results: Theorem A is a master theorem that allows the driving function $d(n)$ to be any EL-function. Theorem B is a multiterm master theorem. Section 4 introduces elementary summation techniques. Section 5 addresses elementary sums and proves Theorem A. Section 6 introduces real induction and proves Theorem B. We conclude in Section 7.

¶2. *Approach of Paper.* Our approach has two emphases. The first is on *real recurrences*: in the recurrences (1) and (2), we treat n as a real variable, $T(n)$ as a real function and all constants a, b, a_i, b_i are real. In contrast, most of the literature regards n as an integer variable. E.g., Kao [11] treats this multiterm recurrence:

$$T(n) = \begin{cases} c \cdot n^\alpha \cdot \log^\beta n + \sum_{i=1}^{k} a_i T(\lceil b_i n \rceil) & \text{for } n \geq n_0 \\ c_n & \text{for } n < n_0 \end{cases} \tag{3}$$

where n and k are positive integers, c, c_n, a_i are positive constants, α, β are non-negative constants, $b_i \in (0,1)$ and $n_0 \geq \max_{i=1}^{k} \frac{1}{1-b_i}$. Similar viewpoints are seen in Wang-Fu, Akra-Bazzi and Roura. But most driving functions such as $d(n) = \sqrt{n}$ and $d(n) = n \log n$ are naturally real functions. Hence our real extension remains well-defined if we simply omit the (troublesome) integer-valued functions such as ceiling or floor. A standard approach to avoid ceiling/floor functions is "domain restriction". E.g., restricting the domain of $T(n)$ in recurrence (1) to positive powers of b ([4, p. 145, Problem 4.44] or [19]) and requiring b to be integer. Finally, to restore n to range over all integers, we need special arguments (e.g., [5, pp. 81–84]) or smoothness assumptions on $T(n)$. Although the idea of real recurrences is nascent in several of the papers (e.g., [2,19,18]), it seldom takes on a full-blown form. In this paper, we develop basic tools to rigorously treat real recurrences.

Our second emphasis is the use of elementary methods. Here "elementary" means the avoidance of calculus [10], not that the results are trivial or easy to come by. It is conventional wisdom in algorithmics to solve $T(n)$ up to Θ-order because it yields robust conclusions about complexity (e.g., [3]). But it is seldom noted that Θ-order analysis lends itself to elementary techniques. E.g., below we give elementary Θ-bounds on sums that are usually treated by the Euler-Maclaurin formula [8, p. 217]. Authors also fail to exploit problem simplifications from Θ-order analysis [19,20,18]. For instance, up to Θ-order, most solutions are insensitive to the initial conditions. *So we need not explicitly specify initial conditions.* Instead, this paper assumes the following **default initial condition** (DIC):

$$T(n) = C, \qquad (n \leq n_0) \tag{4}$$

for some constant $C \geq 0$ and real n_0; the recurrence equation is assumed to be operative for $n > n_0$. Usually $C = 0$ is simplest and easily justified. Thus, using real recurrences under DIC, Kao's recurrence (3) greatly simplifies to $T(n) = n^\alpha \log^\beta n + \sum_{i=1}^{k} a_i T(b_i n)$. Roura [18] and Leighton [14] also discuss robustness issues. The pedagogical advantage of avoiding calculus for computer science students is obvious. Also our driving function $d(n)$ need not be differentiable (Lipschitz type bounds suffice).

These two emphases (real and elementary) explain the title of this paper. The simplicity and power of the real approach will hopefully be evident.

2   On Master Theorems

The "standard" master theorem provides the motif for generalizations. Relative to the master recurrence (1), we define a **watershed constant**

$$\alpha := \log_b a \tag{5}$$

and an associated **watershed function** $w(n) := n^\alpha$. The master theorem is a trichotomy based on a comparison between $d(n)$ and $w(n)$:

Proposition 1 (Standard Master Theorem). *The solution to (1) is*

$$T(n) = \Theta \begin{cases} n^\alpha & \text{if } d(n) = O(w(n)n^{-\varepsilon}) \text{ for some } \varepsilon > 0 \ [\text{CASE } (-)] \\ n^\alpha \log n & \text{if } d(n) = \Theta(w(n)) \qquad\qquad\qquad [\text{CASE } (0)] \\ d(n) & \text{if } ``d(n) = \Omega(w(n)n^\varepsilon)\text{''} \text{ for some } \varepsilon > 0 \ [\text{CASE } (+)]. \end{cases} \tag{6}$$

This is taken from Cormen et al [5, p. 73], except n is now a real variable and $a > 0$ (not $a \geq 1$). The trichotomy amounts to $d(n)$ being (resp.) polynomially-slower than, Θ-order of, and polynomially-faster than $w(n)$. The condition for polynomially-faster [CASE (+)] in (6) is written in quotes because the original Ω-notation in [3, p. 39] was non-standard. This was replaced in [5] by the weaker **regularity condition**: for some $C > 1$,

$$d(n) \geq C \cdot a \cdot d(n/b) \text{ (ev. } n) \tag{7}$$

where the qualification "(ev. n)" reads as "eventually n", meaning that the statement holds for large enough n. Our real approach affords a "two-line proof" of Prop. 1: by induction, $T(n) = a^{i+1}T\left(n/b^{i+1}\right) + \sum_{j=0}^{i} a^j \cdot d\left(\frac{n}{b^j}\right)$, for $i = 0, 1, \ldots$. Setting $i = m := \lceil \log_b n \rceil$, and using DIC (with $C = 0$) in (4), we obtain

$$T(n) = \sum_{j=0}^{m} a^j d(n/b^j). \tag{8}$$

The 3 cases follow by plugging in the corresponding bounds for $d(n)$. Q.E.D.

¶3. *Extended Master Theorem.* It is well-known that Prop. 1 does not cover many useful driving functions such as $d(n) = w(n) \log^\delta n$ ($\delta \neq 0$). By applying the general techniques of **domain** and **range transformations** [4, pp. 130-137], we get:

Proposition 2 (Extended Master Theorem). *The solution to (1) is*

$$T(n) = \Theta \begin{cases} d(n) & \text{if } d(n) \text{ satisfies the reg. cond. (7)} & [\text{CASE } (+)] , \\ d(n) \log n & \text{if } d(n) = \Theta(n^\alpha \log^\delta n) \text{ for some } \delta > -1 & [\text{CASE } (0)] , \\ d(n) \log n \log \log n & \text{if } d(n) = \Theta(n^\alpha \log^\delta n) \text{ where } \delta = -1 & [\text{CASE } (1)] , \\ n^\alpha & \text{if } d(n) = O(n^\alpha \log^\delta n) \text{ for some } \delta < -1 & [\text{CASE } (-)] . \end{cases}$$

Prop. 2 generalizes the master theorem (6) since the original CASEs $(-)\&(0)$ are subsumed by the new ones; CASE $(+)$ is unchanged but CASE (1) is new. Prop. 2 is from Brassard and Bratley [4, p. 145] (cf. [5, p.84, Ex.4.4-2]), slightly sharpened here: we state CASE $(+)$ in terms of the regularity condition. Further [4] assumes $n = n_0 b^i$ for integers $n_0 \geq 1$ and $b \geq 2$. Wang and Fu's version of Prop. 2 is in [20, §4.3 and Table 1]. Roura's version [17] missed CASE (1). Case 3 in Verma's version [19, Theorem 1] is weaker than CASE (0) as he assumes $\delta \geq 0$. Still, our Prop. 2 is silent when the driving functions are, for example:

$$\left. \begin{aligned} d_0(n) &:= n^\alpha \log n \log \log n \\ d_1(n) &:= n^\alpha (\log \log n)^r \\ d_2(n) &:= n^\alpha \frac{(\log \log \log n)^s}{\log n \log \log n} \end{aligned} \right\} \tag{9}$$

for non-zero r, s. Note that $d_0(n)$ arise in the Schönhage-Strassen's algorithm [13] with $\alpha = 1$. It turns out that the solutions for $T(n)$ under the driving functions (9) are (resp.)

$$\Theta(n^\alpha \cdot \log^2 n \log \log n), \quad \Theta(n^\alpha \cdot \log n (\log \log n)^r), \quad \Theta(n^\alpha (\log \log \log n)^{s+1}). \quad (10)$$

The last case assumes $s > -1$; different solutions arise if $s = -1$ or if $s < -1$. Theorem A in the next section will provide these solutions, and much more.

3   Two Generalized Master Theorems

This section will state our two main results: Theorems A and B. Both are extensions of Prop. 2. We begin with Theorem B since Theorem A requires a bit more development to formulate.

¶4. We need the multiterm analogues of (5) and (7): the **watershed constant** for (2) is the unique α satisfying the characteristic equation $\sum_{i=1}^{k} \frac{a_i}{b_i^\alpha} = 1$ (see [11,2]). Say $d(n)$ satisfies the **regularity condition** of (2) if, for some $0 < c < 1$,

$$\sum_{i=1}^{k} a_i d\left(\frac{n}{b_i}\right) \leq c \cdot d(n). \quad (11)$$

Theorem B – Multiterm Master Theorem
The solution to (2) satisfies

$$T(n) = \Theta \begin{cases} d(n) & \text{if } d(n) \text{ satisfies the reg. cond. (11)} \quad [\text{CASE (+)}], \\ d(n) \lg n & \text{if } d(n) = \Theta(n^\alpha \lg^\delta n) \text{ for some } \delta > -1 \text{ [CASE (0)],} \\ d(n) \lg n \lg \lg n & \text{if } d(n) = \Theta(n^\alpha \lg^\delta n) \text{ where } \delta = -1 \quad [\text{CASE (1)}], \\ n^\alpha & \text{if } d(n) = O(n^\alpha \lg^\delta n) \text{ for some } \delta < -1 \text{ [CASE (−)].} \end{cases}$$

All previous versions of Theorem B have 3 cases, as in Prop. 1. Our CASE (1) is new, and in some sense it completes this line of analysis. Kao [11] gave an inductive proof for the case $k = 2$ only. Roura [18, Theorem 2.3] treats more general driving functions; like Kao, the treatment is for integer recurrences. Akra and Bazzi [2] deduced their result from a general integral bound, which Leighton [14] simplified.

¶5. *On EL-functions.* We now introduce the family[1] of "EL-functions" which will serve as driving functions for Theorem A. The **iterated logarithm function** (for $k \in \mathbb{N}$) is defined as $\ell\ell g_k(x) := \underbrace{\lg(\lg(\cdots (\lg(x)) \cdots))}_{k \text{ times}}$ where $\lg := \log_2$ is the "computer science logarithm". E.g., $\ell\ell g_0(x) = x$, $\ell\ell g_1(x) = \lg x$, $\ell\ell g_2(x) = \lg \lg x$. We may extend the index k to all integers where, for $k \in \mathbb{N}$, $\ell\ell g_{-(k+1)}(x)$

---

[1] EL is mnemonic for **E**xponential-**L**ogarithmic.

$:= 2^{\ell\ell g_{-k}(x)}$. Thus $\ell\ell g_{-1}(x) = 2^x$ and $\ell\ell g_{-2}(x) = 2^{2^x}$. An **exponent se-quence** is a function $\mathbf{e} : \mathbb{Z} \to \mathbb{R}$ with finite support, i.e., $\mathbf{e}(i) = 0$ for all but finitely many i's. We normally write e_i for $\mathbf{e}(i)$. Call $\mathbf{e}$ **trivial** if $e_i = 0$ for all $i \in \mathbb{Z}$. For nontrivial $\mathbf{e}$, the smallest index i such that $e_i \neq 0$ is called its **order**, and e_i its **leading power**; these are denoted $Ord(\mathbf{e})$ and $Pow(\mathbf{e})$ (resp.). Given $k, \ell \in \mathbb{N}$, if $e_i = 0$ for all $i < -k$ and $i > \ell$, we may repre-sent $\mathbf{e}$ (non-uniquely) as the sequence $\mathbf{e} = (e_{-k}, \ldots, e_{-1}, e_0; e_1, \ldots, e_\ell)$ where the semi-colon ";" between e_0 and e_1 indicates the "origin" of the bidirec-tional sequence. An **elementary function** is a partial function $f(x)$ of the form $\mathrm{EL}^{\mathbf{e}}(x) := \prod_{i \in \mathbb{Z}} \ell\ell g_i^{e_i}(x)$ where $\ell\ell g_k^a(x) := (\ell\ell g_k(x))^a$ is the a-**th power** of $\ell\ell g_k(x)$. E.g., $\mathrm{EL}^{(-2;0,\pi)}(x) = x^{-2}(\lg \lg(x))^\pi$.

We need one more concept to state Theorem A. Any driving function $d(n)$ of the form $\mathrm{EL}^{\mathbf{e}}(n)$ where $Ord(\mathbf{e}) = 0$ and $Pow(\mathbf{e}) = \alpha$ (watershed constant) is just "at the cusp" between CASES $(+)$ and $(-)$, and so we call such $\mathbf{e}$ a **cusp exponent**. The **cusp order** of any $\mathbf{e}$ is the largest index $h \geq 1$ such that $\mathbf{e}(i) = -1$ for $i = 1, 2, \ldots, h-1$; also call $\mathbf{e}(h)$ the **cusp power**. Cusp exponents have form $\mathbf{e} = (\alpha; -1, -1, \ldots, -1, \beta, \ldots)$ where $\beta \neq -1$ is the cusp power.

Theorem A – Generalized Master Theorem

Let the driving function be $d(n) = \mathrm{EL}^{\mathbf{e}}(n)$ with $k = Ord(\mathbf{e})$ and $c = Pow(\mathbf{e})$. Also let the cusp order and cusp power of $\mathbf{e}$ be h and β respectively. The solution $T(n)$ to the master recurrence (1) with watershed constant $\alpha = \log_b a$ satisfies

$$T(n) = \Theta \begin{cases} d(n) & \text{if } (k < 0 \wedge c > 0) \text{ or } (k \geq 0 \wedge \mathbf{e}(0) > \alpha), \text{[CASE }(+)] \\ d(n)LL_h(n) & \text{if } (k = 0 \wedge \mathbf{e}(0) = \alpha \wedge \beta > -1), & \text{[CASE }(h-1)] \\ n^\alpha & \text{otherwise} & \text{[CASE }(-)] \end{cases}$$

(12)

where $LL_h(n) := \prod_{i=1}^{h} \ell\ell g_i(n) = \lg n \cdot \lg \lg n \cdot \ell\ell g_3(n) \cdots \ell\ell g_h(n)$.

This theorem has infinitely many cases, one for each $h \geq 1$. For $h = 1$ and 2, we reproduce CASEs (0) and (1) of Prop. 2. Verma [19, Theorem 13] has driving functions not covered here. To make Theorem A fully comparable to Prop. 1, we could re-formulate CASE $(+)$ using the regularity condition. An interesting corollary of Theorem A is this: *when the driving function is an EL-function, the solution to the master recurrence is, up to Θ-order, another EL-function.*

To see Theorem A in action, recall the driving functions $d_0(n), d_1(n), d_2(n)$ in (9). The exponent sequence for them are (resp.)

$$(\alpha; 1, 1), \quad (\alpha; 0, r), \quad (\alpha; -1, -1, s)$$

For $d_0(n)$ and $d_1(n)$, their cusp order h and cusp power β are (resp.) $(h, \beta) = (1, 1)$ and $(1, 0)$. As these cusp powers are > -1, they are both fall under CASE (0) which has solution $T(n) = \Theta(d(n)LL_1(h))$. This yields the first two solutions in (10). For $d_2(n)$, we have three possibilities: If $s > -1$, then $(h, \beta) = (3, s)$ and the solution falls under CASE (2) with solution $T(n) = \Theta(d(n)LL_3(n))$ as given by the third bound in (10). If $s = -1$, then $(h, \beta) = (4, 0)$ and it falls under CASE (3) with solution $T(n) = \Theta(d(n)LL_4(n)) = \Theta(n^\alpha \ell\ell g_4(n))$. If $s < -1$, then $(h, \beta) = (3, s)$ but it falls under CASE $(-)$ with solution $T(n) = \Theta(n^\alpha)$.

4   Elementary Summation Techniques

A **complexity function** is a partial function $f : \mathbb{R} \to \mathbb{R}$ where $f(x)$ is defined for x large enough. Standard asymptotic notations (big-Oh, big-Omega, Theta, etc) can be extended to partial functions [21]. Define two kinds of sums on f-values between real limits $a, b \in \mathbb{R}$:

$$\left.\begin{array}{l} \sum_{x \geq a}^{b} f(x) := f(b) + f(b-1) + \cdots + f(b - \lfloor b - a \rfloor), \text{ (descending)} \\ \sum_{x=a}^{b} f(x) := f(a) + f(a+1) + \cdots + f(a + \lfloor b - a \rfloor) \text{ (ascending)} \end{array}\right\} \quad (13)$$

Both sums are 0 for $b < a$; else the descending (ascending) sum will include the term $f(b)$ $(f(a))$. The two versions are distinguished by way we write their lower limits: "$\sum_{x \geq a}$" versus "$\sum_{x=a}$". Such sums are always well-defined as *any undefined summand $f(x)$ is replaced by* 0. Our manipulations below exploits:

$$\sum_{x \geq 0}^{b} f(x) \equiv \sum_{x=0}^{b} f(b - x). \quad (14)$$

Our main focus will be descending sums of the form $S^f(n) := \sum_{x \geq 1}^{n} f(x)$ for real values of n. This sum is traditionally bounded with the Euler-Maclaurin formula. But we now provide elementary method based on "growth-types":

• f is **polynomial-type** if $f \geq 0$, f is non-decreasing, and for some $K > 0$, $f(x) \leq K f(x/2)$ (ev.).
• f **increases exponentially** if $f > 0$ and for some $C > 1$ and $k > 0$, $f(x) \geq C \cdot f(x - k)$ (ev.).
• f **decreases exponentially** if $f > 0$ and for some $0 < c < 1$ and $k > 0$, $f(x) \leq c \cdot f(x - k)$ (ev.).

We say that f is **exponential-type** if it increases or decreases exponentially. Polynomial-type functions corresponds to Verma's "slowly growing functions" [19]. These growth-types are non-exhaustive: for instance, it can be shown that the function $x^{\ln x}$ is not captured. Our next result is relatively easy but useful because it reduces estimating $S^f(n)$ to the easier problem of determining the growth type of f.

Theorem 1 (Summation Rules)

$$S^f(n) = \Theta \begin{cases} n f(\Theta(n)) & \text{if } f \text{ is polynomial-type,} \\ f(n) & \text{if } f \text{ increases exponentially,} \\ 1 & \text{if } f \text{ decreases exponentially.} \end{cases}$$

To determine the growth-type of f, we can exploit simple closure properties of growth types (e.g., each type is closed under addition, multiplication, raising to a positive power, etc). Moreover, an EL-functions f (say $f(x) = \mathrm{EL}^{\mathbf{e}}(()x)$) is exponential-type if $Ord(\mathbf{e}) < 0$; otherwise, either f or f^{-1} is polynomial-type. We exploit such properties in our proofs.

5   Elementary Sums and Proof of Theorem A

Our goal is to bound **elementary sums**, i.e., $S^f(n)$ where f is an EL-function. Such sums may be denoted by $S^{\mathbf{e}}(n) := \sum_{x \geq 1}^n \text{EL}^{\mathbf{e}}(x)$.

¶6. *Error Notation.* We write "$x = y \pm z$" to mean that $x = y + \theta z$ for some θ where $|\theta| \leq 1$. The general convention [21] is that in any numerical expression, each occurrence of the symbol "$\pm$" stands for a sequence "$+\theta$" where θ is an anonymous variable satisfying $|\theta| \leq 1$. Like the big-Oh notation, this is a very useful variable hiding device. Thus, the following holds for any continuous f:

$$\sum_{i=1}^n f(n \pm c) = nf(n \pm c). \tag{15}$$

We need 3 operators on $\mathbf{e}$. The **shift operator** σ is: $\sigma(\mathbf{e})(i) = \mathbf{e}(i+1)$ for all i. E.g., $\text{EL}^{\sigma(\mathbf{e})}(n) = \text{EL}^{\mathbf{e}}(2^n)$. For $c \in \mathbb{R}$, let $\mathbf{e}'$ (resp., $\mathbf{e} + c$) denote the exponent sequence where we zero out (resp., add c to) the component $\mathbf{e}(0)$: $\mathbf{e}'(0) = 0$ and $(\mathbf{e} + c)(0) = c + \mathbf{e}(0)$, and $\mathbf{e}'(i) = (\mathbf{e} + c)(i) = \mathbf{e}(i)$ ($i \neq 0$). Usually, $c = 1$. Another result we need is this: if $Ord(\mathbf{e}) \geq 0$ and $c \in \mathbb{R}$,

$$\text{EL}^{\mathbf{e}}(2^{n \pm c}) = \text{EL}^{\sigma(\mathbf{e})}(n \pm c) = \Theta(\text{EL}^{\sigma(\mathbf{e})}(n)) = \Theta(\text{EL}^{\mathbf{e}}(2^n)). \tag{16}$$

The next transformation of elementary sums is the key.

Lemma 1 (Key Transformation). *If $Ord(\mathbf{e}) \geq 0$, $S^{\mathbf{e}}(n) = \Theta(S^{\sigma(\mathbf{e}+1)}(\lg n))$.*

Up to Θ-order, we will show $S^{\mathbf{e}}(n) = \Theta(f(n))$ for some elementary function f. The goal (next Theorem) is to determine the exponent of f. Note that all asymptotic notations assume a fixed $\mathbf{e}$. We need a variant notion of cusp order from Section 3: for any $\mathbf{e}$, its **augmented cusp order** is 0 if $\mathbf{e}(0) \neq -1$; else it is the cusp order of $\mathbf{e}$. Also $\mathbf{e}(h)$ is the **augmented cusp power** if h is the augmented cusp order. If $\mathbf{e}(0) = -1$, then the augmented concepts agree with the original ones.

Theorem 2 (Elementary Sums). *Let $k := Ord(\mathbf{e})$, $c := Pow(\mathbf{e})$. Also, let the augmented cusp order and power of $\mathbf{e}$ be h and β, respectively. Then*

$$S^{\mathbf{e}}(n) = \Theta \begin{cases} \text{EL}^{\mathbf{e}}(n) & \text{if } (k \leq -1 \wedge c > 0), \text{ [CASE (+)]} \\ \text{EL}^{\mathbf{e}}(n)LL_{h+1}(2^n) & \text{if } (k \geq 0 \wedge \beta > -1), \text{ [CASE (h)]} \\ 1 & \text{else} \qquad\qquad\qquad\; \text{[CASE (−)]} \end{cases} \tag{17}$$

The proof uses repeated application of the key transformation, Lemma 1. To see the power of Thm. 2, note that it implies $S^f(n) = \Theta(\ell\ell g_h(n))$ when $f(x) = 1/LL_h(2^x)$ (for any $h \in \mathbb{N}$). Goursat [9, p. 349] has the calculus analogue of this.

¶7. *Proof of Theorem A.* Let $n = b^m$. From (8), we have

$$T(n) = \sum_{i=0}^{m} a^i d(n/b^i) \quad = \sum_{i\geq0}^{m} a^{m-i} d(b^i) \qquad \text{(by (14))}$$
$$= n^\alpha \sum_{i\geq0}^{m} a^{-i} d(b^i) = n^\alpha \sum_{i\geq0}^{m} a^{-i} \mathrm{EL}^{\mathbf{e}}(b^i). \qquad (18)$$

If $k \leq -1$, the function $F(i) := a^{-i}\mathrm{EL}^{\mathbf{e}}(b^i)$ is increasing (decreasing) exponentially when $c > 0$ $(c < 0)$. Applying our summation rules (Thm. 1) to (18),

$$T(n) = n^\alpha \cdot \Theta \begin{cases} a^{-m}\mathrm{EL}^{\mathbf{e}}(b^m) \\ 1 \end{cases} = \Theta \begin{cases} \mathrm{EL}^{\mathbf{e}}(n) & \text{if } c > 0, \\ n^\alpha & \text{if } c < 0. \end{cases} \qquad (19)$$

This proves our theorem for $k \leq -1$. Next assume $k \geq 0$. Writing $e_0 = \mathbf{e}(0)$,

$$T(n) = n^\alpha \sum_{i\geq0}^{m} a^{-i}\mathrm{EL}^{\mathbf{e}}(b^i) = n^\alpha \sum_{i\geq0}^{m}(b^{e_0}/a)^i \cdot \mathrm{EL}^{\mathbf{e'}}(b^i). \qquad (20)$$

But (b^{e_0}/a) is > 1 $(= 1, < 1)$ depending on whether $e_0 > \alpha$ $(= \alpha, < \alpha)$. So the sum (20) is exponential-type (polynomial-type) when $e_0 \neq \alpha$, $(e_0 = \alpha)$. So:

$$T(n) = n^\alpha \cdot \Theta \begin{cases} a^{-m}\mathrm{EL}^{\mathbf{e}}(b^m), \\ \sum_{i\geq0}^{m} \mathrm{EL}^{\mathbf{e'}}(b^i), \\ 1 \end{cases} = \Theta \begin{cases} \mathrm{EL}^{\mathbf{e}}(n) & \text{if } e_0 > \alpha, \\ n^\alpha \cdot \sum_{i\geq0}^{m} \mathrm{EL}^{\sigma(\mathbf{e'})}(i \lg b) & \text{if } e_0 = \alpha \\ n^\alpha & \text{if } e_0 < \alpha. \end{cases} \qquad (21)$$

We are done with the case $k \geq 0$ and $e_0 \neq \alpha$. For $k = 0$ and $e_0 = \alpha$, (21) gives

$$T(n) = \Theta(n^\alpha \cdot \sum_{i\geq0}^{m} \mathrm{EL}^{\sigma(\mathbf{e'})}(i \lg b))$$
$$= \Theta(n^\alpha \cdot \sum_{i\geq0}^{m} \mathrm{EL}^{\sigma(\mathbf{e'})}(i)) \qquad (\lg b \text{ is const. in a poly.-type sum})$$
$$= \Theta(n^\alpha \cdot S^{\sigma(\mathbf{e'})}(m)) \qquad (\text{definition of } S^{\sigma(\mathbf{e'})})$$
$$= \Theta \begin{cases} n^\alpha \cdot \mathrm{EL}^{\sigma(\mathbf{e'})}(m) LL_h(2^m) & \text{if } \beta > -1 \\ n^\alpha & \text{if } \beta < -1 \end{cases} \qquad (\text{by Thm. 2})$$

In applying Thm. 2, we use the fact that $Ord(\sigma(\mathbf{e'})) \geq 0$ and the augmented cusp order of $\sigma(\mathbf{e'})$ is equal to $h-1$. The case $\beta < -1$ falls under CASE $(-)$. Case $\beta > -1$ falls under CASE (h) because $n^\alpha \cdot \mathrm{EL}^{\sigma(\mathbf{e'})}(m) LL_h(2^m) = \Theta(\mathrm{EL}^{\mathbf{e}}(n) LL_h(n))$ because $m = \Theta(\lg n)$. This proves Theorem A.

6   Real Induction and Proof of Theorem B

The principle of natural induction, or induction on $\mathbb{N}$, is well-known. To prove Theorem B, we need induction on $\mathbb{R}$, or **real induction**. Real induction is rarely discussed in the literature although it is needed in areas such as automatic correctness proofs of programs involving real numbers, timing logic [15], and in the programming language Real PCF [7]. We give a simple formulation here:

PRINCIPLE OF (ARCHIMEDEAN) REAL INDUCTION
Let $P(x)$ be a real predicate. Then $P(x)$ is valid provided there exist real numbers x_1 (cutoff constant) and $\gamma > 0$ (gap constant) such that:

(RB) Real Basis: *For all $x < x_1$, $P(x)$ holds.*
(RI) Real Induction: *For all $y \geq x_1$, if $(\forall x \leq y - \gamma)P(x)$ then $P(y)$.*

The derived predicate "$P^+(y) \equiv (\forall x \leq y - \gamma)P(x)$" in (RI) is called the **real induction hypothesis** (RIH). Thus (RI) says $P^+(y) \Rightarrow P(y)$. This principle is "Archimedean" because it exploits the Archimedean property of the reals: for any $x \in \mathbb{R}$, there is a smallest $n(x) \in \mathbb{N}$ such that $x \leq x_1 + n(x)\gamma$. Our principle is easily justified by a strong induction on the natural number $n(x)$. As an application of real induction, we can prove that the multiterm regularity condition (11) implies $d(n) = \Omega(n^{\alpha+\varepsilon})$ for some $\varepsilon > 0$.

¶8. *Proof of Theorem B.* We use the Principle of Real Induction. First we prove the real induction (RI) part of each case in our theorem:

CASE (+): This is the easiest case. The lower bound $T(n) = \Omega(d(n))$ is trivial. For the upper bound, we will show $T(n) \leq D_1 d(n)$ (ev.), for some D_1:

$$
\begin{aligned}
T(n) = d(n) + \sum_{i=1}^{k} a_i T\left(\tfrac{n}{b_i}\right) &\leq d(n) + \sum_{i=1}^{k} a_i D_1 d(n/b_i) && \text{(by RIH)} \\
&\leq d(n) + D_1 c d(n) && \text{(by regularity cond. (11))} \\
&\leq D_1 d(n) && \text{(choosing } D_1 \geq 1/(1-c))
\end{aligned}
$$

CASE (0): Assume that $d(n) = n^\alpha \lg^\delta n$ for some $\delta > -1$. We first show $T(n) \leq D_1 d(n) \lg n$. We have, eventually,

$$
\begin{aligned}
T(n) = d(n) + \sum_{i=1}^{k} a_i T\left(\tfrac{n}{b_i}\right) & \\
\leq n^\alpha \lg^\delta n + \sum_{i=1}^{k} a_i D_1 \left(\tfrac{n}{b_i}\right)^\alpha \lg^{\delta+1}\left(\tfrac{n}{b_i}\right) && \text{(by RIH)} \\
= n^\alpha \lg^\delta n + D_1 n^\alpha \lg^{\delta+1} n \left[\sum_{i=1}^{k} \tfrac{a_i}{b_i^\alpha}\left(1 - \tfrac{\lg b_i}{\lg n}\right)^{\delta+1}\right] \\
= D_1 n^\alpha \lg^{\delta+1} n \left[\tfrac{1}{D_1 \lg n} + \sum_{i=1}^{k} \tfrac{a_i}{b_i^\alpha}\left\{1 - (\delta+1)\tfrac{\lg b_i}{\lg n}(1 + o(1))\right\}\right] \\
= D_1 n^\alpha \lg^{\delta+1} n \left[1 + \tfrac{1}{\lg n}\left\{\tfrac{1}{D_1} - (\delta+1)\sum_{i=1}^{k} \tfrac{a_i \lg b_i}{b_i^\alpha}(1 + o(1))\right\}\right] \\
\leq D_1 n^\alpha \lg^{\delta+1} n
\end{aligned}
$$

provided D_1 is sufficiently large to verify $\tfrac{1}{D_1} < (\delta+1)\sum_{i=1}^{k} \tfrac{a_i \lg b_i}{b_i^\alpha}$. Here the condition $\delta > -1$ is necessary. Similarly, we show the lower bound $T(n) \geq D_2 d(n) \lg n$ using the same derivation above, but with reversed inequalities. The provision is that D_2 is small enough to verify $\tfrac{1}{D_2} > (\delta+1)\sum_{i=1}^{k} \tfrac{a_i \lg b_i}{b_i^\alpha}$.

CASE (1): Assume that $d(n) = n^\alpha / \lg n$. We first show $T(n) \leq D_1 d(n) \lg \lg n$.

$$T(n) = d(n) + \sum_{i=1}^{k} a_i T\left(\frac{n}{b_i}\right)$$

$$\leq \frac{n^\alpha}{\lg n} + \sum_{i=1}^{k} a_i D_1 \left(\frac{n}{b_i}\right)^\alpha \lg\lg\left(\frac{n}{b_i}\right) \qquad \text{(by RIH)}$$

$$= D_1 n^\alpha \left[\frac{1}{D_1 \lg n} + \sum_{i=1}^{k} \frac{a_i}{b_i^\alpha} \lg\left\{(\lg n)\left(1 - \frac{\lg b_i}{\lg n}\right)\right\}\right]$$

$$= D_1 n^\alpha \left[\frac{1}{D_1 \lg n} + \sum_{i=1}^{k} \frac{a_i}{b_i^\alpha}\left\{\lg\lg n + \lg\left(1 - \frac{\lg b_i}{\lg n}\right)\right\}\right]$$

$$= D_1 n^\alpha \left[\lg\lg n + \frac{1}{D_1 \lg n} + \sum_{i=1}^{k} \frac{a_i}{b_i^\alpha} \lg\left(1 - \frac{\lg b_i}{\lg n}\right)\right]$$

$$= D_1 n^\alpha \left[\lg\lg n + \frac{1}{D_1 \lg n} - \sum_{i=1}^{k} \frac{a_i}{b_i^\alpha} \frac{\lg b_i}{\lg n}(1 + o(1))\right]$$

$$= D_1 n^\alpha \left[\lg\lg n + \frac{1}{\lg n}\left\{\frac{1}{D_1} - \sum_{i=1}^{k} \frac{a_i \lg b_i}{b_i^\alpha}(1 + o(1))\right\}\right]$$

$$\leq D_1 n^\alpha \lg\lg n$$

provided D_1 is large enough to verify $\frac{1}{D_1} < \sum_{i=1}^{k} \frac{a_i \lg b_i}{b_i^\alpha}$. Similarly, the lower bound $T(n) \geq D_2 n^\alpha \lg\lg n$ uses the above derivation with inequalities reversed, and D_2 small enough to verify $\frac{1}{D_2} > \sum_{i=1}^{k} \frac{a_i \lg b_i}{b_i^\alpha}$.

CASE (-1): This is the trickiest. By assumption, $0 \leq d(n) \leq n^\alpha \lg^\delta n$ (ev.) for some $\delta < -1$. To show $T(n) = O(n^\alpha)$, the hypothesis $T(n) \leq D_1 n^\alpha$ will not do. Instead, use the stronger hypothesis $T(n) \leq D_1 n^\alpha \left[1 - K \lg^{\delta+1} n\right]$ (ev.) for some $D_1, K > 0$. Eventually,

$$T(n) = d(n) + \sum_{i=1}^{k} a_i T\left(\frac{n}{b_i}\right)$$

$$\leq n^\alpha \lg^\delta n + \sum_{i=1}^{k} a_i D_1 \left(\frac{n}{b_i}\right)^\alpha \left[1 - K \lg^{\delta+1}\left(\frac{n}{b_i}\right)\right] \qquad \text{(by RIH)}$$

$$= D_1 n^\alpha \left[\frac{\lg^\delta n}{D_1} + 1 - K \lg^{\delta+1} n \sum_{i=1}^{k} \frac{a_i}{b_i^\alpha}\left(1 - \frac{\lg b_i}{\lg n}\right)^{\delta+1}\right]$$

$$= D_1 n^\alpha \left[1 - K \lg^{\delta+1} n \left\{-\frac{1}{KD_1 \lg n} + \sum_{i=1}^{k} \frac{a_i}{b_i^\alpha}\left(1 - (\delta+1)\frac{\lg b_i}{\lg n}\right)(1 + o(1))\right\}\right]$$

$$= D_1 n^\alpha \left[1 - K \lg^{\delta+1} n \left\{1 - \frac{1}{\lg n}\left(\frac{1}{KD_1} + (\delta+1)\sum_{i=1}^{k} \frac{a_i \lg b_i}{b_i^\alpha}(1 + o(1))\right)\right\}\right]$$

$$\leq D_1 n^\alpha \left[1 - K \lg^{\delta+1} n\right]$$

provided KD_1 is small (sic) enough to verify $\frac{1}{KD_1} > -(\delta+1)\sum_{i=1}^{k} \frac{a_i \lg b_i}{b_i^\alpha}$ (recall $\delta < -1$). The introduction of K is crucial. For the lower bound, we also use a strengthened hypothesis, $T(n) \geq D_2 n^\alpha (1 + \lg^{\delta+1} n)$. The derivations is essentially the same, except inequalities are reversed. This completes the four cases.

We now provide the real bases (RB) for each of the above cases: first choose n_0 so that $d(n)$ is defined and the recurrence (2) for $T(n)$ holds nonvacuously ($\forall n \geq n_0$). Choose $\gamma = \gamma(n_0)$ as shown in the Appendix. Ensure the cutoff n_1 is $\geq n_0/\gamma$, so that RIH holds nonvacuously.

CASE $(+)$: Choose $n_1 = n_0/\gamma$ and ensure $D_1 \geq T(n)/d(n)$ for all $n \in [n_0, n_1]$. CASEs (0) and (1) are omitted in this abstract. CASE $(-)$: For upper bound, we first choose the product KD_1 to equal the reciprocal of $-(\delta+1)\sum_{i=1}^{k} \frac{3a_i \lg b_i}{2b_i^\alpha}$. Choose $n_1 \geq n_0/\gamma$ to be large enough so that the $o(1)$ term has absolute value

$< 1/2$, and for $n \geq n_1$, the function $f(n) = n^\alpha - (KD_1)\lg^{\delta+1} n$ is increasing and ≥ 1. Finally, choose D_1 as $\sup_{n_0 \leq n \leq n_1}\{T(n)/f(n)\}$. Note that $f(n) \leq D_1 n^\alpha(1 - K\lg^{\delta+1} n)$ and hence $T(n) \leq f(n) \leq D_1 n^\alpha(1 - K\lg^{\delta+1} n)$ for $n \in [n_0, n_1]$.

The proof of Theorem B is complete.

7 Conclusion

Cormen et al [5, p. 90] noted that some generalized master theorems are not easy to use. This echoes Karp's wish for "cookbook theorems" to recurrences [12]. That is the appeal of the standard master theorem. Our Theorem B has similar qualities. Although Theorem A is also cookbook, the generality of its driving function calls for some unavoidable deciphering of the notations. Further generalizations of Theorems A and B are possible: for instance, one could extend Theorem B to driving functions that are general EL functions. Another direction is to treat robustness issues of such solutions – we address this in the full paper.

Features that detract from cookbook property include bounds left in an integral form, tedious details involving integrality assumptions, and tracking of (essentially) arbitrary initial conditions. We have shown that much of this can be removed if we exploit Θ-robustness and embrace real recurrences wholeheartedly. Real induction is another useful tool that ought to be used more widely in this context. We feel our ideas are pedagogically sound. For instance, the summation rules for the various growth-types are easily taught in introductory algorithms. Indeed, our perspectives have developed out of classroom teaching.

References

1. Aho, A.V., Hopcroft, J.E., Ullman, J.D.: Data Structures and Algorithms. Addison-Wesley, Reading (1983)
2. Akra, M., Bazzi, L.: On the solution of linear recurrences. Computational Optimizations and Applications 10(2), 195–210 (1998)
3. Bentley, J.L., Haken, D., Saxe, J.B.: A general method for solving divide-and-conquer recurrences. ACM SIGACT News 12(3), 36–44 (1980)
4. Brassard, G., Bratley, P.: Fundamentals of Algorithms. Prentice-Hall, Englewood Cliffs (1996)
5. Corman, T.H., Leiserson, C.E., Rivest, R.L., Stein, C.: Introduction to Algorithms, 2nd edn. MIT Press & McGraw-Hill Book Co. (2001)
6. Edelsbrunner, H., Welzl, E.: Halfplanar range search in linear space and $O(n^{0.695})$ query time. Info. Processing Letters 23, 289–293 (1986)
7. Escardó, M.H., Streicher, T.: Induction and recursion on the partial real line with applications to Real PCF. Theor. Computer Sci. 210(1), 121–157 (1999)
8. Gonnet, G.H.: Handbook of Algorithms and Data Structures. Addison-Wesley Pub. Co., London (1984)
9. Goursat, É.: A Course in Mathematical Analysis, vol. 1. Ginn & Co., Boston (1904); Trans. by Earle Raymond Hedrick. Available from Google books
10. Greene, D.H., Knuth, D.E.: Mathematics for the Analysis of Algorithms, 2nd edn. Birkhäuser, Basel (1982)

11. Kao, M.: Multiple-size divide-and-conquer recurrences. SIGACT News 28(2), 67–69 (1997); also Proc. 1996 Intl. Conf. on Algorithms, Natl. Sun Yat-Sen U., Taiwan, pp. 159–161
12. Karp, R.M.: Probabilistic recurrence relations. J. ACM 41(6), 1136–1150 (1994)
13. Knuth, D.E.: The Art of Computer Programming: Sorting and Searching, vol. 3. Addison-Wesley, Boston (1972)
14. Leighton, T.: Notes on better master theorems for divide-and-conquer recurrences (1996) (class notes)
15. Mahony, B.P., Hayes, I.J.: Using continuous real functions to model timed histories. In: 6th Australian Software Eng. Conf (ASWEC), pp. 257–270 (1991)
16. Paul Walton Purdom, J., Brown, C.A.: The Analysis of Algorithms. Holt, Rinehart and Winston, New York (1985)
17. Roura, S.: An improved master theorem for divide-and-conquer recurrences. In: Degano, P., Gorrieri, R., Marchetti-Spaccamela, A. (eds.) ICALP 1997. LNCS, vol. 1256, pp. 449–459. Springer, Heidelberg (1997)
18. Roura, S.: Improved master theorems for divide-and-conquer recurrences. J. ACM 48(2), 170–205 (2001)
19. Verma, R.M.: A general method and a master theorem for divide-and-conquer recurrences with applications. J. Algorithms 16, 67–79 (1994)
20. Wang, X., Fu, Q.: A frame for general divide-and-conquer recurrences. Info. Processing Letters 59, 45–51 (1996)
21. Yap, C.K.: Theory of real computation according to EGC. In: Hertling, P., Hoffmann, C.M., Luther, W., Revol, N. (eds.) Real Number Algorithms. LNCS, vol. 5045, pp. 193–237. Springer, Heidelberg (2008)

Multiprocessor Speed Scaling for Jobs with Arbitrary Sizes and Deadlines[*]

Paul C. Bell[1] and Prudence W.H. Wong[2]

[1] Department of Computer Science, Loughborough University
P.Bell@lboro.ac.uk
[2] Department of Computer Science, University of Liverpool
pwong@liverpool.ac.uk

Abstract. In this paper we study energy efficient deadline scheduling on multiprocessors in which the processors consumes power at a rate of s^α when running at speed s, where $\alpha \geq 2$. The problem is to dispatch jobs to processors and determine the speed and jobs to run for each processor so as to complete all jobs by their deadlines using the minimum energy. The problem has been well studied for the single processor case. For the multiprocessor setting, constant competitive online algorithms for special cases of unit size jobs or arbitrary size jobs with agreeable deadlines have been proposed [4]. A randomized algorithm has been proposed for jobs of arbitrary sizes and arbitrary deadlines [13]. We propose a deterministic online algorithm for the general setting and show that it is $O(\log^\alpha P)$-competitive, where P is the ratio of the maximum and minimum job size.

1 Introduction

Energy efficient deadline scheduling. Energy consumption has become an important concern in the design of modern processors, not only for battery-operated mobile devices with single processors but also for server farms or laptops with multi-core processors. A popular technology to reduce energy usage is *dynamic speed scaling* (see e.g., [4, 7, 8, 21]) where the processor can vary its speed dynamically. The power consumption is modelled by s^α when the processor runs at speed s, where α is typically 2 or 3 [11, 20]. Running a job slower saves energy, yet it takes longer to finish the job. The challenge arises from the conflicting objectives of providing good "quality of service" (QoS) and conserving energy. Deadline feasibility is a common QoS measure for job scheduling. Jobs with arbitrary sizes and deadlines arrive at unpredictable times and they are to be run on some processor. Preemption is allowed with no penalty.

The study of speed scaling was initiated by Yao et al. [21]. They studied deadline scheduling on a single processor in which jobs with arbitrary sizes and deadlines arrive online and the aim is to finish all jobs by their deadlines using the minimum amount of energy. The decision at any time is to determine which job

---

[*] This work is partially supported by EPSRC Grant EP/E028276/1.

M. Ogihara and J. Tarui (Eds.): TAMC 2011, LNCS 6648, pp. 27–36, 2011.

to run and at what speed. They gave an optimal offline algorithm and a simple online algorithm AVR which is $2^{\alpha-1}\alpha^\alpha$-competitive and they also proposed an online algorithm OA. Bansal, Kimbrel and Pruhs [8] later showed that OA is α^α-competitive. They also gave a $2(\alpha/(\alpha-1))^\alpha e^\alpha$-competitive algorithm, which is called the BKP algorithm and is better than OA when $\alpha > 5$. The result is further improved to $4^\alpha/(2\sqrt{e\alpha})$-competitive by the qOA algorithm [7].

The problem of energy efficient scheduling has also been studied for other QoS measures. The problem of minimizing flow time and energy has attracted a lot of attention [3, 5, 6, 9, 13, 18, 19]. Energy efficient scheduling has also been extended to the setting with sleep states [14, 16, 17]. The literature also contains results on other aspects of energy efficient scheduling, see [1, 2, 15].

Energy efficient multiprocessor scheduling. The problem of energy efficient deadline scheduling becomes NP-hard in the multiprocessor setting, even when all the jobs have the same arrival times and deadlines. In the multiprocessor setting, in addition to determining processor speeds, a job dispatching algorithm is required to assign jobs to processors. Albers et al. [4] have extended the study to the multiprocessor setting and they study the special cases of unit-size jobs or jobs with agreeable deadlines (jobs arriving earlier have earlier deadlines). If jobs have unit-size and agreeable deadlines, Round Robin (RR) is optimal. For the case of unit-sized jobs with arbitrary deadlines or arbitrary-sized jobs with agreeable deadlines, they gave an $\alpha^\alpha 2^{4\alpha}$-competitive algorithm. Their algorithm, called Classified Round Robin (CRR) first classifies jobs according to the density of the job (the ratio of the job size to the duration between arrival and deadline), and then schedules jobs in each class independently using RR. All jobs (of different classes) dispatched on a processor are run at a speed determined by AVR. The case for jobs of arbitrary sizes and arbitrary deadlines is left as an open question.

Recently, Greiner, Nonner and Souza [13] have shown that any β-competitive algorithm for a single processor yields a randomized βB_α-competitive algorithm, where B_α is the αth Bell number [10] and this result holds for jobs of arbitrary size and arbitrary deadlines. This means that the existing algorithms [7, 8, 21] for single processors lead to randomized online algorithms in the multiprocessor setting. Yet it is still an open question to have a competitive deterministic algorithm for the general case of jobs with arbitrary sizes and arbitrary deadlines.

Our contribution. In this paper we study the generalized problems in the multiprocessor setting where jobs have arbitrary sizes and arbitrary deadlines and give a deterministic online algorithm. We first show that the Classified Round Robin algorithm (CRR) [4] does not scale well when jobs have arbitrary sizes and deadlines. The competitive ratio is at least $m^{\alpha-1}$, where m is the number of processors. We then consider a natural extension of CRR and propose a non-migratory deterministic job dispatching algorithm, called *Dual-Classified Round Robin* (DCRR), which classifies jobs in terms of both density and sizes. We show that DCRR coupled with AVR is $2^{4\alpha}(\log^\alpha P + \alpha^\alpha 2^{\alpha-1})$-competitive where P is the ratio between the maximum and minimum job size. Note that the competitive ratio is independent of m and holds even against an optimal migratory offline algorithm.

Roughly speaking, to analyze the performance of DCRR, we round the density and size of a job to the boundaries that define the classes, and show that the performance on the general set is no more than a constant factor of that on such a "nice" job set. This idea is similar to the proof in [4], which rounds only the density of the jobs. We further show that for a nice job set, the classification of DCRR means that the jobs in the same class satisfy the property of agreeable deadlines, making the analysis easier. We are then able to show that the competitive ratio of DCRR depends on the number of classes, which is related to $\log P$.

Organization of the paper. The rest of the paper is organized as follows. In Section 2, we define the problem and give some preliminary results. In Section 3, we review an existing algorithm CRR and show that it does not work well for jobs of arbitrary sizes and deadlines. In Section 4, we describe and analyze our algorithm DCRR. Finally, we conclude in Section 5.

2 Preliminaries

We are to schedule a set of jobs onto m processors M_0, M_1, $\cdots$, M_{m-1}. Preemption is allowed without penalty. The speed of each processor can be varied. When running at speed s, a processor processes s units of work and consumes s^α units of energy in each time unit, where $\alpha \geq 2$.

We denote the release time, deadline and size of a job j as $r(j)$, $d(j)$, and $w(j)$, respectively. The *span* of job j is $\text{span}(j) = d(j) - r(j)$ and the *density* $\text{den}(j) = \frac{w(j)}{d(j)-r(j)}$. A job j is called *active* at time t if $r(j) \leq t \leq d(j)$.

The problem is to dispatch the jobs to processors, and for each processor, to determine which job and at what speed to run at any time. The objective is to complete all jobs by their deadlines using the minimum energy.

Consider any job set $\mathcal{J}$. For any algorithm A, we overload the symbol $A(\mathcal{J})$ to mean both the schedule of A on $\mathcal{J}$ and the energy required by the schedule. Let OPT_1 and OPT_m denote the optimal schedule on a single processor and m processors, respectively. In [4], it has been shown that $\text{OPT}_1(\mathcal{J})/m^{\alpha-1} \leq \text{OPT}_m(\mathcal{J})$. We further lower bound the value $\text{OPT}_m(\mathcal{J})$. At any time t, the speed of AVR on a processor is the sum of the densities of all active jobs at t scheduled on this processor. It has been shown in [21] that $\text{AVR}_1(\mathcal{J}) \leq \alpha^\alpha 2^{\alpha-1} \text{OPT}_1(\mathcal{J})$, implying $\text{AVR}_1(\mathcal{J}) \leq \alpha^\alpha 2^{\alpha-1} m^{\alpha-1} \text{OPT}_m(\mathcal{J})$. Let $\text{MIN}(\mathcal{J})$ be the minimum energy to run each job of $\mathcal{J}$ independently of other jobs, i.e., $\text{MIN}(\mathcal{J}) = \sum_{j \in \mathcal{J}} (\text{den}(j))^\alpha \text{span}(j)$. Then, we have $\text{MIN}(\mathcal{J}) \leq \text{OPT}_m(\mathcal{J})$. We summarize these bounds on $\text{OPT}_m(\mathcal{J})$ in the following lemma.

Lemma 1 ([4]). *Consider any job set $\mathcal{J}$. (a) $\text{OPT}_m(\mathcal{J}) \geq \text{OPT}_1(\mathcal{J})/m^{\alpha-1}$.* *(b) (i) $\text{MIN}(\mathcal{J}) \leq \text{OPT}_m(\mathcal{J})$; (ii) $\text{AVR}_1(\mathcal{J}) \leq \alpha^\alpha 2^{\alpha-1} m^{\alpha-1} \text{OPT}_m(\mathcal{J})$.*

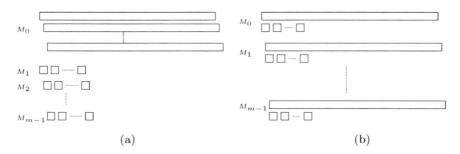

Fig. 1. In the adversary, all jobs have density 1. The span and size of the m large jobs is k and the $m^2 - m$ small jobs is ϵ. (a) CRR schedules all the large jobs to processor M_0 and m small jobs to each of $M_1, \cdots, M_{m-1}$. (b) The optimal schedule dispatches one large job and $m - 1$ small jobs to each processor.

3   Classified Round Robin (CRR)

In this section, we review the algorithm CRR which is $\alpha^\alpha 2^{4\alpha}$-competitive for the special case in which jobs are of unit size, or jobs are of arbitrary sizes but agreeable deadlines [4]. We show that CRR is no longer constant competitive when the jobs have arbitrary sizes and arbitrary deadlines.

Let Δ be the maximum density of the jobs in $\mathcal{J}$. CRR classifies jobs with density Δ into density-class-0, and jobs with density in $[\Delta/2^k, \Delta/2^{k-1})$ into density-class-k, for some positive integer k. Jobs within each class are dispatched to processors by round-robin independently. For each processor, the speed is the sum of the densities of the unfinished jobs dispatched to that processor (i.e., AVR) and the processor processes these jobs by splitting the speed equally among them.

The following theorem shows a lower bound for CRR when jobs are of arbitrary sizes and deadlines. Figure 1 shows the CRR schedule and the optimal schedule for the adversary.

Theorem 1. *For arbitrary size jobs with arbitrary deadlines, CRR has a competitive ratio of at least $m^{\alpha-1}$.*

Proof. Let $\epsilon > 0$ be a small positive value and $k > 0$ be an arbitrary large value. Given m processors, define a job set $\mathcal{J}$ of m^2 jobs such that for any $1 \leq i \leq m^2$, the release time of job j_i is $i\epsilon$. For all jobs j_i with $i \bmod m \neq 0$, we set the span of the job to be ϵ. For all jobs j_i with $i \bmod m = 0$, we set the span of the jobs to be k. We further set the sizes of all jobs to be the same as their span, in other words, all jobs have density 1.

Algorithm CRR classifies all m^2 jobs into the same class C_0 since they have the same density and dispatches jobs according to round robin by their release time. Thus the first processor receives the m jobs of large span k and large size k. The energy used by the first processor is therefore km^α as $\epsilon \to 0$ and the energy of the remaining processors approaches 0.

On the other hand, we can dispatch one large span job and $m - 1$ small span jobs to each processor. As ϵ tends to 0, the energy used by each processor is k and the total energy of the schedule is km. Therefore, the competitive ratio of CRR is at least $m^{\alpha-1}$. $\square$

We note that even if we classify jobs according to their sizes, such a classification plus round robin still does not perform well. We give a similar adversary with m^2 jobs of the same size, m of them having a small span (thus large density) and the rest with very large span. One job of small span arrives followed by $m - 1$ large span jobs and this repeats for m times. Then CRR assigns all the small span jobs to the same processor, dominating the energy used by the algorithm. The optimal offline algorithm can dispatch one small span job to each processor, distributing the energy used much better and thus the same lower bound can be obtained.

4   Dual-Classified Round Robin (DCRR)

4.1   The Algorithm

We now describe our algorithm DCRR (Dual-Classified Round Robin). In addition to classifying jobs into density classes, DCRR also classifies jobs according to sizes. Let Γ be the maximum job size of a job set $\mathcal{J}$. Jobs with size in $(\Gamma/2^{h+1}, \Gamma/2^h]$ are classified into size-class-h, for some integer $h \geq 0$ (note the difference from the definition of density-classes). We then define the set $C_{k,h}$ to be the set of jobs in density-class-k and size-class-h. For simplicity, we assume that Δ and Γ are known in advance[1]. With the definition of $C_{k,h}$, DCRR dispatches jobs in the same $C_{k,h}$ in a round robin manner, independent of other classes. Then all jobs (of different classes) dispatched to the same processor are run using a speed determined by AVR (see Algorithm 1).

4.2   Framework of the Analysis and Nice Job Sets

To analyze the performance of DCRR, we transform job set $\mathcal{J}$ to a *nice* job set $\mathcal{J}^*$ (to be defined) and show that such a transformation only increases the energy usage modestly. Furthermore, we show that for a nice job set $\mathcal{J}^*$, we can bound $\mathrm{DCRR}(\mathcal{J}^*)$ by $\mathrm{OPT}_m(\mathcal{J}^*)$ and in turn by $\mathrm{OPT}_m(\mathcal{J})$. Then we can establish the competitive ratio of DCRR.

A job set $\mathcal{J}^*$ is said to be a *nice job set* if every job j^* in $\mathcal{J}^*$ satisfies the following properties.

---

[1] If Δ and Γ are not known in advance, the class definition could be modified slightly. Specifically, the first job which arrives will define the initial density and size classes Δ' and Γ'. New jobs may have larger sizes or density than these Δ' and Γ' and thus we may have classes with a negative index, but the analysis can be seen to still hold and increasing the competitive ratio by at most a factor of 2, see [4] for further details.

Algorithm 1. Algorithm DCRR

Let Δ and Γ be (respectively) the maximum density and maximum size of all jobs.

Classification: A job is classified into $C_{k,h}$ if its density is in $[\Delta/2^k, \Delta/2^{k-1})$ and its size is in $(\Gamma/2^{h+1}, \Gamma/2^h]$.

Job dispatching: Jobs of the same class $C_{k,h}$ are dispatched (upon their arrival) to the m processors using a round-robin strategy, i.e., the i-th job in $C_{k,h}$ is dispatched to processor-$(i \bmod m)$, and different classes are handled independently.

Speed running: The speed of each processor is determined by AVR on the jobs dispatched to that processor and the speed is split equally among these jobs (note that this gives a feasible schedule).

- The density $\mathrm{den}(j^*) = \Delta/2^k$, for some positive integer k.
- The size $w(j^*) = \Gamma/2^h$, for some positive integer h.

Given a job set $\mathcal{J}$, we transform each job $j \in \mathcal{J}$ into a job j^* as follows. Suppose j is in class $C_{k,h}$.

- We set the release time of j^* to be the same as j, i.e., $r(j^*) = r(j)$.
- We round up the size of j to the maximum in the class $C_{k,h}$, i.e., $w(j^*) = \Gamma/2^h$. Then, we have $w(j) \leq w(j^*) \leq 2w(j)$.
- We round down the density of j to the minimum in the class $C_{k,h}$, i.e., $\mathrm{den}(j^*) = \Delta/2^k$. Then, we have $\mathrm{den}(j)/2 \leq \mathrm{den}(j^*) \leq \mathrm{den}(j)$.
- Effectively, we set the deadline $d(j^*) = r(j^*) + (\frac{\Gamma}{2^h} \cdot \frac{2^k}{\Delta})$.

In other words, job densities only decrease and sizes only increase. The following lemma relates the optimal schedule for $\mathcal{J}$ and $\mathcal{J}^*$, as well as the DCRR schedule for $\mathcal{J}$ and $\mathcal{J}^*$. The implication of the lemma is that we can focus on analyzing the performance of DCRR on nice job set $\mathcal{J}^*$.

Lemma 2. *For any job set $\mathcal{J}$ and its corresponding nice job set $\mathcal{J}^*$, we have (a) $2^\alpha \mathrm{OPT}_m(\mathcal{J}) \geq \mathrm{OPT}_m(\mathcal{J}^*)$; (b) $\mathrm{DCRR}(\mathcal{J}) \leq 2^\alpha \mathrm{DCRR}(\mathcal{J}^*)$.*

Proof. (a) We construct from $\mathrm{OPT}_m(\mathcal{J})$ a feasible schedule S for $\mathcal{J}^*$, and show that this increases the energy slightly. The dispatching of S follows the dispatching of $\mathrm{OPT}_m(\mathcal{J})$. For any processor, at any time t, S runs at double the speed that $\mathrm{OPT}_m(\mathcal{J})$ does. S is feasible for $\mathcal{J}^*$ because $w(j^*) \leq 2w(j)$ and $\mathrm{span}(j) \leq \mathrm{span}(j^*)$, the latter implies that whenever j is run, it is within the span of j^*. Because of the double speed, $S = 2^\alpha \mathrm{OPT}_m(\mathcal{J})$. As S is a feasible schedule for $\mathcal{J}^*$, we have $S \geq \mathrm{OPT}_m(\mathcal{J}^*)$. Then the statement follows.

(b) First we notice that a job j and its corresponding j^* belong to the same class. The release time of j^* is also kept the same as j. Therefore, j^* will be dispatched to the same processor as j. In the schedule of $\mathrm{DCRR}(\mathcal{J}^*)$, at any time when the job j^* is active, it contributes $\mathrm{den}(j^*)$ to the speed of that processor. If we consider a schedule S' that runs double the speed at any time and on any processor as $\mathrm{AVR}(\mathcal{J}^*)$ does, the job j^* contributes $2 \times \mathrm{den}(j^*)$ to the speed.

As for energy usage, $S' = 2^\alpha \text{DCRR}(\mathcal{J}^*)$. On the other hand, in $\text{DCRR}(\mathcal{J})$, at any time that j is active, it contributes $\text{den}(j)$ to the speed of that processor. Since $2\,\text{den}(j^*) \geq \text{den}(j)$ and $\text{span}(j^*) > \text{span}(j)$, for any processor, the schedule S' runs at least the same speed as $\text{DCRR}(\mathcal{J})$, and probably higher. Therefore, $S' \geq \text{DCRR}(\mathcal{J})$, and the statement follows.   □

4.3   Analysis of DCRR

With Lemma 2, the analysis of DCRR on a general job set $\mathcal{J}$ can be done via the analysis of DCRR on $\mathcal{J}^*$. Recall that $\text{MIN}(\mathcal{J})$ is the minimum energy to run each job of $\mathcal{J}$ independently of other jobs, i.e., $\text{MIN}(\mathcal{J}) = \sum_{j \in \mathcal{J}}(\text{den}(j))^\alpha \text{span}(j)$. First, we show in Lemma 3 a property about how DCRR dispatches jobs in a class to the m processors. Then, in Lemma 4, we relate the sum of energy usage of AVR on jobs DCRR dispatched to each machine with $\text{MIN}(\mathcal{J})$ and $\text{AVR}_1(\mathcal{J})$. Finally, together with Lemma 2, we can then conclude in Theorem 2 the competitive ratio of DCRR.

The following is a modification to a lemma from [4]. Since all spans within a class are identical, they have agreeable deadlines and the same proof follows as is shown in [4].

Lemma 3. *For any time t, DCRR assigns to each processor at most $\lceil C_{k,h}(t)/m \rceil$ jobs from $\mathcal{J}^*$, where $C_{k,h}(t)$ is the set of jobs from $C_{k,h}$ active at time t.*

Let $\mathcal{J}_i^*$ be the subset of $\mathcal{J}^*$ that is dispatched to processor i by DCRR. Then $\text{DCRR}(\mathcal{J}^*) = \sum_{1 \leq i \leq m} \text{AVR}_1(\mathcal{J}_i^*)$. We now relate $\sum_{1 \leq i \leq m} \text{AVR}_1(\mathcal{J}_i^*)$ with $\text{MIN}(\mathcal{J}^*)$ and $\text{AVR}_1(\mathcal{J}^*)$.

Lemma 4. *For any nice job set $\mathcal{J}^*$, the following inequality holds*

$$\sum_{1 \leq i \leq m} \text{AVR}_1(\mathcal{J}_i^*) \leq 2^{2\alpha}((\log^\alpha P^*)\,\text{MIN}(\mathcal{J}^*) + \text{AVR}_1(\mathcal{J}^*)/m^{\alpha-1})$$

where $P^* = \frac{max\ \{w(j)|j \in \mathcal{J}^*\}}{min\ \{w(j)|j \in \mathcal{J}^*\}}$.

Proof. We adapt the proof of CRR in [4]. Let $C_{k,h,i}(t)$ for $1 \leq i \leq m$ be the set of jobs from class $C_{k,h}$ *assigned to processor i* active at time t dispatched by DCRR. Let $s_i(t)$ denote the speed of the average rate AVR algorithm on processor i at time t. Since the speed of AVR is the sum of densities of all active jobs at each time point, we see that:

$$s_i(t) = \sum_{k \geq 0} \sum_{h \geq 0} |C_{k,h,i}(t)| \frac{\Delta}{2^k}. \tag{1}$$

Running jobs according to the *Earliest Deadline First* policy yields a feasible schedule. Let $s(t)$ denote the speed of AVR for the whole job set $\mathcal{J}^*$. Then $s(t) = \sum_{k \geq 0} \sum_{h \geq 0} |C_{k,h}(t)| \Delta/2^k$.

Fix a time $t \geq 0$ and a processor $1 \leq i \leq m$. Let K_1 be the set of job class indices (k, h) such that $|C_{k,h,i}(t)| = 1$ and K_2 be the indices (k, h) such that $|C_{k,h,i}(t)| \geq 2$. Define $k_1 = \min\{k | (k, h) \in K_1\}$ for some $h \geq 0$ and $P^* = \frac{\max\ \{w(j)|j \in \mathcal{J}^*\}}{\min\ \{w(j)|j \in \mathcal{J}^*\}}$. Using Equation (1) and Lemma 3, we see that

$$
\begin{aligned}
s_i(t) &= \sum_{(k,h) \in K_1} \frac{\Delta}{2^k} + \sum_{(k,h) \in K_2} |C_{k,h,i}(t)| \frac{\Delta}{2^k} \\
&\leq (\log P^*) \frac{\Delta}{2^{k_1 - 1}} + \sum_{(k,h) \in K_2} \left\lceil \frac{|C_{k,h}(t)|}{m} \right\rceil \frac{\Delta}{2^k} \qquad \text{(by Lemma 3)} \\
&\leq (\log P^*) \frac{\Delta}{2^{k_1 - 1}} + \sum_{(k,h) \in K_2} \frac{2|C_{k,h}(t)|}{m} \frac{\Delta}{2^k} \\
&\leq 4 \cdot \max \left\{ (\log P^*) \frac{\Delta}{2^{k_1}}, \frac{s(t)}{m} \right\}
\end{aligned}
\qquad (2)
$$

We shall integrate $s_i(t)^\alpha$ first over all t when the first term of Equation (2) is dominating to give an upper bound on required energy of:

$$
(4 \log(P^*))^\alpha \sum_{k \geq 0} \sum_{h \geq 0} |C_{k,h} \cap \mathcal{J}_i^*| \left(\frac{\Delta}{2^k} \right)^\alpha \left(2^{k-h} \frac{\Gamma}{\Delta} \right)
$$

Integrating $s_i(t)^\alpha$ when the second term of Equation (2) is dominating gives $\left(\frac{4}{m} \right)^\alpha \mathrm{AVR}_1(\mathcal{J}^*)$. Summing over $1 \leq i \leq m$ shows that $\sum_{i=1}^m \mathrm{AVR}_1(\mathcal{J}_i^*) \leq 4^\alpha ((\log^\alpha P^*)\mathrm{MIN}(\mathcal{J}^*) + m^{1-\alpha} \mathrm{AVR}_1(\mathcal{J}^*))$ as required. $\qquad \square$

Together with Lemma 2, we can conclude the competitive ratio of DCRR in the following theorem.

Theorem 2. *For an arbitrary job set $\mathcal{J}$, the competitive ratio of algorithm DCRR is at most $2^{4\alpha}(\log^\alpha P + \alpha^\alpha 2^{\alpha-1})$, where P is the ratio between the maximum and minimum job size.*

Proof. By Lemma 4, we know that:

$$
\sum_{1 \leq i \leq m} \mathrm{AVR}_1(\mathcal{J}_i^*) \leq 2^{2\alpha}((\log^\alpha P^*)\,\mathrm{MIN}(\mathcal{J}^*) + \mathrm{AVR}_1(\mathcal{J}^*)/m^{\alpha-1}).
$$

Since $\mathrm{MIN}(\mathcal{J}^*) \leq \mathrm{OPT}_m(\mathcal{J}^*)$ and $\mathrm{AVR}_1(\mathcal{J}^*) \leq \alpha^\alpha 2^{\alpha-1} m^{\alpha-1} \mathrm{OPT}_m(\mathcal{J}^*)$ by Lemma 1 (b) (ii), we therefore conclude that

$$
\mathrm{DCRR}(\mathcal{J}^*) \leq \sum_{1 \leq j \leq m} \mathrm{AVR}_1(\mathcal{J}_j^*) \leq 2^{2\alpha} \mathrm{OPT}_m(\mathcal{J}^*)((\log^\alpha P^*) + \alpha^\alpha 2^{\alpha-1}).
$$

By Lemma 2 (a) and (b) above, $\mathrm{DCRR}(\mathcal{J}) \leq 2^\alpha \mathrm{DCRR}(\mathcal{J}^*)$ and $\mathrm{OPT}_m(\mathcal{J}^*) \leq 2^\alpha \mathrm{OPT}_m(\mathcal{J})$. Then, we have

$$
\mathrm{DCRR}(\mathcal{J}) \leq 2^{4\alpha} \mathrm{OPT}_m(\mathcal{J})((\log^\alpha P^*) + \alpha^\alpha 2^{\alpha-1}).
$$

Note that from the proof of Lemma 4, $\log P^*$ is essentially the number of size classes used by DCRR which does not change under $\mathcal{J}$ or $\mathcal{J}^*$, therefore $\log P$ and $\log P^*$ can be taken to be equal and the theorem holds. $\qquad \square$

5   Conclusion

We extend the study of energy efficient deadline scheduling on multiprocessor to jobs with arbitrary sizes and deadlines. We analyze the performance of the deterministic algorithm DCRR. In the proof of Theorem 2, the $\log P$ factor comes in the case K_1, yet we note that this bound is rather loose and we believe that this can be improved. On the other hand, one may consider how DCRR can be coupled with OA instead of AVR to improve the results. Another open question is to consider speed bounded processors [12], in which case, not all the jobs can be completed by their deadlines. The concern becomes to maximize the throughput (number of jobs completed by their deadlines) and to minimize the energy used to achieve this throughput. The problem has been considered in the single processor setting [5, 12]. It would be interesting to derive algorithms that are competitive both in throughput and energy in the multiprocessor setting.

References

1. Albers, S.: Algorithms for energy saving. In: Albers, S., Alt, H., Näher, S. (eds.) Efficient Algorithms. LNCS, vol. 5760, pp. 173–186. Springer, Heidelberg (2009)
2. Albers, S.: Energy-efficient algorithms. Communication ACM 53(5), 86–96 (2010)
3. Albers, S., Fujiwara, H.: Energy-efficient algorithms for flow time minimization. ACM Transactions on Algorithms 3(4), 49 (2007)
4. Albers, S., Muller, F., Schmelzer, S.: Speed scaling on parallel processors. In: Proceedings of ACM Symposium on Parallelism in Algorithms and Architectures (SPAA), pp. 289–298 (2007)
5. Bansal, N., Chan, H.L., Lam, T.W., Lee, L.K.: Scheduling for speed bounded processors. In: Aceto, L., Damgård, I., Goldberg, L.A., Halldórsson, M.M., Ingólfsdóttir, A., Walukiewicz, I. (eds.) ICALP 2008, Part I. LNCS, vol. 5125, pp. 409–420. Springer, Heidelberg (2008)
6. Bansal, N., Chan, H.L., Pruhs, K.: Speed scaling with an arbitrary power function. In: Proceedings of ACM-SIAM Symposium on Discrete Algorithms (SODA), pp. 693–701 (2009)
7. Bansal, N., Chan, H.L., Pruhs, K., Rogozhnikov-Katz, D.: Improved bounds for speed scaling in devices obeying the cube-root rule. In: Albers, S., Marchetti-Spaccamela, A., Matias, Y., Nikoletseas, S., Thomas, W. (eds.) ICALP 2009. LNCS, vol. 5555, pp. 144–155. Springer, Heidelberg (2009)
8. Bansal, N., Kimbrel, T., Pruhs, K.: Speed scaling to manage energy and temperature. Journal of the ACM 54(1), 3 (2007)
9. Bansal, N., Pruhs, K., Stein, C.: Speed scaling for weighted flow time. In: Proceedings of ACM-SIAM Symposium on Discrete Algorithms (SODA), pp. 805–813 (2007)
10. Becker, H.W., Riordan, J.: The arithmetic of Bell and Stirling numbers. American Journal of Mathematics 70, 385–394 (1948)
11. Brooks, D.M., Bose, P., Schuster, S.E., Jacobson, H., Kudva, P.N., Buyuktosunoglu, A., Wellman, J.D., Zyuban, V., Gupta, M., Cook, P.W.: Power-aware microarchitecture: Design and modeling challenges for next-generation microprocessors. IEEE Micro 20(6), 26–44 (2000)

12. Chan, H.L., Chan, W.T., Lam, T.W., Lee, L.K., Mak, K.S., Wong, P.W.H.: Optimizing throughput and energy in online deadline scheduling. ACM Transactions on Algorithms 6(1), 10 (2009); Preliminary version appeared in Proceedings of Symposium on Discrete Algorithms SODA, pp. 795–804 (2007)
13. Greiner, G., Nonner, T., Souza, A.: The bell is ringing in speed-scaled multiprocessor scheduling. In: Proceedings of ACM Symposium on Parallelism in Algorithms and Architectures (SPAA), pp. 11–18 (2009)
14. Han, X., Lam, T.-W., Lee, L.-K., To, I.K.-K., Wong, P.W.-H.: Deadline scheduling and power management for speed bounded processors. Theoretical Computer Science 411(40-42), 3587–3600 (2010)
15. Irani, S., Pruhs, K.: Algorithmic problems in power management. ACM SIGACT News 32(2), 63–76 (2005)
16. Irani, S., Shukla, S., Gupta, R.K.: Algorithms for power savings. ACM Transactions on Algorithms 3(4), 41 (2007)
17. Lam, T.W., Lee, L.K., Ting, H.F., To, I.K.K., Wong, P.W.H.: Sleep with guilt and work faster to minimize flow plus energy. In: Albers, S., Marchetti-Spaccamela, A., Matias, Y., Nikoletseas, S., Thomas, W. (eds.) ICALP 2009. LNCS, vol. 5555, pp. 665–676. Springer, Heidelberg (2009)
18. Lam, T.W., Lee, L.K., To, I.K.K., Wong, P.W.H.: Improved multi-processor scheduling for flow time and energy. Journal of Scheduling (to appear), http://dx.doi.org/10.1016/j.tcs.2010.05.035; Preliminary version appeared in Proceedings of ACM Symposium on Parallelism in Algorithms and Architectures, pp. 256–264 (2008)
19. Lam, T.W., Lee, L.K., To, I.K.K., Wong, P.W.H.: Speed scaling functions for flow time scheduling based on active job count. In: Halperin, D., Mehlhorn, K. (eds.) ESA 2008. LNCS, vol. 5193, pp. 647–659. Springer, Heidelberg (2008)
20. Mudge, T.: Power: A first-class architectural design constraint. Computer 34(4), 52–58 (2001)
21. Yao, F., Demers, A., Shenker, S.: A scheduling model for reduced CPU energy. In: Proceedings of IEEE Symposium on Foundations of Computer Science (FOCS), pp. 374–382 (1995)

Approximating Edge Dominating Set in Dense Graphs

Richard Schmied* and Claus Viehmann**

Department of Computer Science, University of Bonn
{schmied,viehmann}@cs.uni-bonn.de

Abstract. We study the approximation complexity of the *Minimum Edge Dominating Set* problem in everywhere ϵ-dense and average $\bar{\epsilon}$-dense graphs. More precisely, we consider the computational complexity of approximating a generalization of the Minimum Edge Dominating Set problem, the so called *Minimum Subset Edge Dominating Set* problem. As a direct result, we obtain for the special case of the Minimum Edge Dominating Set problem in everywhere ϵ-dense and average $\bar{\epsilon}$-dense graphs by using the techniques of Karpinski and Zelikovsky, the approximation ratios of $\min\{2, 3/(1+2\epsilon)\}$ and of $\min\{2, 3/(3-2\sqrt{1-\bar{\epsilon}})\}$, respectively. On the other hand, we show that it is UGC-hard to approximate the Minimum Edge Dominating Set problem in everywhere ϵ-dense graphs with a ratio better than $2/(1+\epsilon)$ with $\epsilon > 1/3$ and $2/(2-\sqrt{1-\bar{\epsilon}})$ with $\bar{\epsilon} > 5/9$ in average $\bar{\epsilon}$-dense graphs.

Keywords: Edge Dominating Set, Minimum Maximal Matching, Dense Instances, Approximation Algorithms, Approximation Lower Bounds.

1 Introduction

In this paper, we consider the computational complexity of approximating the *Minimum Subset Edge Dominating Set* problem which generalizes the Minimum Edge Dominating Set problem. As a direct result, we obtain improved upper bounds for the Minimum Edge Dominating Set problem in everywhere and average dense graphs, i.e. graphs with bounded minimum and average vertex degree, respectively.

1.1 Problem Statement

An *edge dominating set* (for short EDS) of a finite undirected graph $G = (V, E)$ is a subset $M \subseteq E$ of edges such that each edge in E shares an endpoint with some edges in M. The *Minimum Edge Dominating Set* problem (for short MEDS problem) asks to find an edge dominating set of minimum cardinality $|M|$.

For given graph $G = (V, E)$, the *Minimum Maximal Matching* problem (for short MMM problem) asks for a subset $M \subseteq E$ of non adjacent edges with

---

* Work supported by Hausdorff Doctoral Fellowship.
** Work partially supported by Hausdorff Center for Mathematics, Bonn.

M. Ogihara and J. Tarui (Eds.): TAMC 2011, LNCS 6648, pp. 37–47, 2011.

minimal cardinality such that each edge in E shares an endpoint with some edge in M.

It has been noted that the Minimum Edge Dominating Set and the Minimum Maximal Matching problem admit optimal solutions of the same size and that an optimal solution of the MEDS problem can be transformed in polynomial time into an optimal solution of the MMM problem (cf. [20]), and vice versa.

The *Minimum Subset Edge Dominating Set* problem (for short MSED problem) is a generalization of the MEDS problem and is defined as follows: given a graph $G = (V, E)$ and a subset $S \subseteq V$, find a minimum cardinality EDS M of G with the property $S \subseteq \bigcup_{e \in M} e$.

For some $\epsilon, \bar{\epsilon} > 0$, we call a graph $G = (V, E)$ *everywhere ϵ-dense* if any vertex in G has at least $\epsilon |V|$ neighbors, and we call a graph $G = (V, E)$ *average $\bar{\epsilon}$-dense* if the average degree of a vertex in G is at least $\bar{\epsilon}|V|$, i.e. $(\sum_{v \in V} deg(v))/|V| \geq \bar{\epsilon}|V|$.

1.2   Related Work

The MEDS problem is already referred to in Garey and Johnson [11]. Even for planar or bipartite graphs of maximum degree 3 the MEDS problem remains NP-hard [20] in the exact setting. An inapproximability result was obtained by Chlebík and Chlebíková ([7]), who claimed that it is NP-hard to approximate the MEDS problem within any factor better than $7/6$. They further showed that the MEDS problem is NP-hard to approximate within any constant less than $(7 + \epsilon)/(6 + 2\epsilon)$, in graphs with minimum degree at least ϵn. In the unweighted case, finding an arbitrary maximal matching M provides 2-approximation for the MEDS problem, since each edge in the optimal solution can cover at most two edges of M. The first nontrivial approximation algorithm is due to Gotthilf et al. ([12]) and achieves an approximation ratio of $2 - c \log(n)/n$, where c is an arbitrary positive constant and n is the number of vertices in the graph.

Density parameters such as the number of edges $\bar{\epsilon}$ and the minimum degree ϵ have been used in approximation ratios for various optimization problems (see [15] for a detailed survey, [17,14,2,3] for covering and related problems).

Currently, the best parameterized ratios for the Vertex Cover problem with parameters $\bar{\epsilon}$ and ϵ are $2/(2 - \sqrt{1 - \bar{\epsilon}})$ and $2/(1 + \epsilon)$, respectively ([17]). Imamura and Iwama ([14]) later improved the former result, by generalizing it to depend on both $\bar{\epsilon}$ and $\Delta := \max_{v \in V}\{deg(v)\}$.

As for lower bounds, Clementi and Trevisan ([8]) as well as Karpinski and Zelikovsky ([16]) proved that the Vertex Cover problem restricted to everywhere and average dense graphs remains APX-hard. Later, Eremeev ([9]) showed that it is NP-hard to approximate the Vertex Cover problem in everywhere ϵ-dense graphs within a factor less than $(7 + \epsilon)/(6 + 2\epsilon)$. Finally, Bar-Yehuda et al. ([2]) prove that if the Vertex Cover problem cannot be approximated within a factor strictly smaller than 2 on arbitrary graphs, then it cannot be approximated within factors smaller than $2/(2 - \sqrt{1 - \bar{\epsilon}}) - o(1)$ and $2/(1 + \epsilon) - o(1)$, respectively, on average and everywhere dense graphs.

For the MEDS problem, Cardinal et al. achieved the first upper bound smaller than 2 for sufficiently dense graphs. More precisely, the obtained approximation ratio is asymptotic to $\min\{2, 1/\epsilon\}$ in everywhere ϵ-dense graphs and to $\min\{2, 1/(1 - \sqrt{1 - \bar{\epsilon}})\}$ in average $\bar{\epsilon}$-dense graphs ([4]). More recently, Cardinal, Langerman, and Levy provided an improved bound on the approximation ratio for the MEDS problem in average dense graphs. This bound is asymptotic to $1/(1 - \sqrt{(1 - \epsilon)/2})$, which is smaller than 2 when ϵ is greater than $1/2$ ([5]).

1.3   Our Contributions

This work is the first best to our knowledge studying the approximation complexity of the MSED problem. We give an approximation algorithm that achieves the approximation ratio at most $\min\{2, 3/(1 + 2|S|/|V|)\}$. For the special case of the MEDS problem in dense graphs, it yields by using the techniques of Karpinski and Zelikovsky for the dense Minimum Vertex Cover problem ([17]) an approximation ratio of $\min\{2, 3/(1 + 2\epsilon)\}$ for everywhere ϵ-dense graphs and $\min\{2, 3/(3 - 2\sqrt{1 - \bar{\epsilon}})\}$ for average $\bar{\epsilon}$-dense graphs, respectively.

On the other hand, we use an approximation preserving reduction due to Karpinski and Zelikovsky ([16]) from the Minimum Vertex Cover problem to the Minimum Vertex Cover problem in dense graphs to obtain hardness result for the MEDS problem in dense graphs. We show that it is UGC-hard (cf. [18]) to approximate the MEDS problem in everywhere ϵ-dense graphs with a ratio better than $2/(1 + \epsilon)$ with $\epsilon > 1/3$ and $2/(2 - \sqrt{1 - \bar{\epsilon}})$ with $\bar{\epsilon} > 5/9$ in average $\bar{\epsilon}$-dense graphs. The same reduction shows that the MSED problem is UGC-hard to approximate within any constant better than $2/(1 + |S|/|V|)$ with $3|S| > |V|$.

2   Subset Edge Dominating Set Problem

We start by introducing some basic notations and tools which are used in our algorithms. Afterwards we state our approximation algorithm for the MSED problem and prove the claimed result.

2.1   Definitions and Notations

Given a finite graph $G = (V, E)$ and a subset $S \subseteq V$, the induced subgraph $G[S]$ is defined as $(S, \{e \in E \mid e \subseteq S\})$. For a given set $M \subseteq E$ we introduce the notation $V(M) := \bigcup_{e \in M} e$.

The maximal matching heuristic is a standard algorithm that provides a 2-approximation for the Minimum Edge Dominating Set problem. It is perhaps one of the simplest and best-known approximation algorithm. It consists in finding a collection of disjoint edges (a matching) that is maximal (with respect to edge inclusion) by iteratively removing adjacent vertices until no more edges are left in the graph.

In the Maximum Subset Matching problem (for short MSM problem), which generalizes the Maximum Matching problem, we are given a graph $G = (V, E)$

and $S \subset V$. The goal is to determine the maximum number of vertices of S that can be matched in a matching of G. Alon and Yuster considered this problem and introduced a randomized algorithm in [1]. The Maximum Subset Matching problem can be reduced to the Maximum Weighted Matching problem. Just assign to every vertex with both endpoints in S weight 2, and edges from S to $V \backslash S$ weight 1. The currently fastest algorithm for maximum weighted matchings in general graphs is the algorithm of Gabow and Tarjan (see [10]).

In our setting, it runs in $\tilde{O}(\sqrt{|V|}(|E| + |S|^2))$ time. For a given graph $G = (V, E)$, $S \subseteq V$ and $U \subseteq V \backslash S$, let us denote by $MSM(G, S, U)$ the set of edges of a maximum subset matching in the graph $G[S \cup U]$ and S.

An important theorem for many problems related to the Minimum Vertex Cover problem was proven by Nemhauser and Trotter (cf. [19]). It enables us to reduce the problem to instances in which the value of a minimum vertex cover is at least $|V|/2$ together with other nice properties. Here, we use a generalized version of the NT-Theorem given by Chlebík and Chlebíková.

Theorem 1. (Optimal Version of the NT-Theorem [6])
There exists a polynomial time algorithm that partitions the vertex set V of any graph G into three subsets $(V_0, V_1, V_{1/2})$ with no edges between V_0 and $V_{1/2}$ or within V_0 such that

1. *for any vertex cover VC of $G[V_{1/2}]$ it holds $|VC| \geq |V_{1/2}|/2$*
2. *every minimum vertex cover C for G satisfies $V_1 \subseteq C \subseteq V_1 \cup V_{1/2}$ and $C \cap V_{1/2}$ is a minimum vertex cover for $G[V_{1/2}]$.*

Such a partition can be constructed by computing maximum matching of a specially constructed bipartite graph. The algorithm of Hopcroft and Karp is currently the fastest algorithm for maximum matching in bipartite graphs and runs in time $O(|E|\sqrt{|V|})$ (see [13]).

2.2   Algorithm $\mathcal{A}_{\text{SEDS}}$

In order to explain the intuition behind the algorithm, notice that the set S needs to be covered with edges and we want to achieve it by a maximum matching which covers the whole set S. Clearly, we cannot expect that there always exists a perfect matching in $G[S]$. Instead we compute a maximum subset matching with endpoints in $V_1 \cup V_{1/2}$ for which we hope to have good vertex cover properties in $G[V \backslash S]$. The remaining vertices of S will be covered greedily. Finally, we take care of the remaining graph by applying the maximal matching heuristic (MMH).

We now present our main algorithm (see Figure 1).

2.3   Analysis of $\mathcal{A}_{\text{SEDS}}$

We now formulate our main theorem.

Theorem 2. *Given a graph $G = (V, E)$ and $S \subseteq V$, the algorithm $\mathcal{A}_{SEDS}$ has an approximation ratio at most $\min\{2, 3/(1 + 2|S|/|V|)\}$.*

Input: Graph $G = (V, E)$, $S \subseteq V$

Set $M_1 := \emptyset$;
If $|S| > \frac{|V|}{4}$ **Then**
    Compute the NT-Partition $(V_0, V_1, V_{1/2})$ of $G[V \backslash S]$;
    If $|V_0| < 2|V_1|$ **Then**
        Compute $M_1 := MSM(G, S, V \backslash S)$;
    Else
        Compute $M_1 := MSM(G, S, V_1 \cup V_{1/2})$;
    EndIf
EndIf
Cover the remaining vertices of S greedily with edges M_r;
Compute the remaining graph $G' := G[V \backslash V(M_1 \cup M_r)]$;
Construct a maximal matching M_2 in G' by applying the MMH;

Output: $M_1 \cup M_r \cup M_2$

Fig. 1. Algorithm $\mathcal{A}_{\text{SEDS}}$

Proof. Let OPT denote some optimal solution for the MSED problem and $EDS_\mathcal{A}$ the solution produced by algorithm $\mathcal{A}_{\text{SEDS}}$. First, we concentrate on the case $|S| \leq |V|/4$. Then, we show that $\mathcal{A}_{\text{SEDS}}$ computes a solution with approximation ratio $3/(1 + 2|S|/|V|)$ which is better than 2 if $|S| > |V|/4$ holds. We start with

Lemma 1. *If $|S| \leq |V|/4$ holds, then the algorithm $\mathcal{A}_{\text{SEDS}}$ has an approximation ratio at most 2.*

Proof. The algorithm covers the vertices of S greedily with edges, which means that we use at most $|S|$ edges. Since the maximal matching heuristic computes a solution as well for the MEDS problem as for the Minimum Vertex Cover problem (by choosing the endpoints of the constructed matching) with approximation ratio 2, our solution for the graph $G[V \backslash S]$ has at most as many edges as the cardinality of an optimal vertex cover VC_{OPT} of $G[V \backslash S]$. Consequently, the approximation ratio of the algorithm is bounded by

$$\frac{|EDS_\mathcal{A}|}{|OPT|} \leq \frac{|S| + |VC_{OPT}|}{\frac{1}{2}(|S| + |VC_{OPT}|)} = 2.$$

$\square$

In the remaining part of the proof, we will restrict ourselves to instances (G, S) with $|S| > |V|/4$.

For the sake of the analysis, let us now consider a maximum subset matching $M^* := MSM(G^*, S, V(OPT) \cap V')$ of the restricted graph $G^* = (V(OPT), OPT)$, where V' is a subset of $V \backslash S$. We denote by M_R^* the edges contained in OPT to cover the remaining vertices in $S \backslash V(M^*)$, i.e. $M_R^* := \{e \in OPT \mid e \cap (S \backslash V(M^*)) \neq \emptyset\}$. We prove a simple lemma.

Lemma 2. *Let M be a maximum subset matching $MSM(G, S, V')$ and $M_r \subseteq E(G)$ be the edges which are greedily chosen to cover the remaining vertices in $S \backslash V(M)$. Then we have $|M_r| \leq |M_R^*|$.*

Proof. Since OPT is contained in $E(G)$ and by definition of a maximum subset matching, it is clear that $|S \cap V(M^*)| \leq |S \cap V(M)|$ holds. Therefore, we conclude $|S \backslash V(M^*)| \geq |S \backslash V(M)|$ which implies $|M_r| \leq |M_R^*|$.   □

Let us assume that $|S| > |V|/4$ holds, we now show that $\mathcal{A}_{\text{SEDS}}$ has an approximation ratio at most $3/(1 + |S|/|V|)$. We will consider two cases separately.

Case $|V_0| < 2|V_1|$
First of all, the algorithm $\mathcal{A}_{\text{SEDS}}$ computes a maximum subset matching $M_1 := MSM(G, S, V \backslash S)$ of G and then covers the remaining vertices of S greedily with edges M_r.

Let $M^* := MSM(G^*, S, V(OPT) \backslash S)$ be a maximum subset matching of the restricted graph $G^* = (V(OPT), OPT)$ and denote by M_R^* the edges contained in OPT to cover the vertices in $S \backslash V(M^*)$. From Lemma 2 we know that $|M_r| \leq |M_R^*|$ holds.

We analyze the cardinality of $EDS_\mathcal{A}$, the solution produced by $\mathcal{A}_{\text{SEDS}}$, and OPT separately. The maximum subset matching $MSM(G, S, V \backslash S)$ covers in the worst case all the vertices of the remaining graph $G[V \backslash S]$ and $|S| - |M_r|$ vertices of S. Therefore, we can bound the cardinality of $EDS_\mathcal{A}$ as follows:

$$2|EDS_\mathcal{A}| \leq (|S| - |M_r|) + |V \backslash S| + 2|M_r| \leq |V| + |M_r| \leq |V| + |M_R^*|$$

Now we give a lower bound on the optimal solution. Notice that the cardinality of $V(OPT) \backslash S$ is at least $|V_1| + \frac{1}{2}|V_{1/2}|$, since $|V_1| + |V_{1/2}|/2$ is a lower bound on the cardinality of an optimal vertex cover of $G[V \backslash S]$. Therefore, we can assume that a matching in OPT covers the $|V_1| + \frac{1}{2}|V_{1/2}|$ vertices in $G[V \backslash S]$ and $|S| - |M_R^*|$ vertices in S. The remaining vertices in S are covered by $|M_R^*|$ edges. Hence, we get the following:

$$2|OPT| \geq (|S| - |M_R^*|) + |V_1| + \frac{1}{2}|V_{1/2}| + 2|M_R^*| \geq |S| + |V_1| + \frac{1}{2}|V_{1/2}| + |M_R^*|$$

We are ready to analyze the approximation ratio of $\mathcal{A}_{\text{SEDS}}$ by combining the upper and lower bounds. In $(\star)$ we use the property of the case $|V_0| < 2|V_1|$.

$$\frac{2|EDS_\mathcal{A}|}{2|OPT|} \leq \frac{|V| + |M_R^*|}{|S| + |V_1| + \frac{1}{2}|V_{1/2}| + |M_R^*|} \leq \frac{|V|}{|S| + |V_1| + \frac{1}{2}|V_{1/2}|}$$

$$\leq \frac{3}{\frac{3|S| + 3|V_1| + \frac{3}{2}|V_{1/2}|}{|V|}} \underset{(\star)}{\leq} \frac{3}{\frac{|S| + 3|V_1| + |V_{1/2}|}{|S| + 3|V_1| + |V_{1/2}|} + \frac{2|S| + \frac{1}{2}|V_{1/2}|}{|V|}} \leq \frac{3}{1 + 2\frac{|S|}{|V|}}$$

Case $|V_0| \geq 2|V_1|$
Unlike the previous case, the algorithm $\mathcal{A}_{\text{SEDS}}$ computes a maximum subset matching $MSM(G, S, V_1 \cup V_{1/2})$ of G. As before M_r and M_R^* are the sets of edges to cover the remaining vertices of S, where $V(M_R^*) \cap S$ are the vertices left

uncovered by a maximum subset matching $M^* := MSM(G^*, S, (V_1 \cup V_{1/2}) \cap V(OPT))$ of $G^* := (V(OPT), OPT)$. From Lemma 2 we know that $|M_r| \leq |M_R^*|$ holds.

As before, we analyze $EDS_{\mathcal{A}}$ and OPT separately. This time the algorithm $\mathcal{A}_{SEDS}$ computes a maximum subset matching $MSM(G, S, V_1 \cup V_{1/2})$ which contains in the worst case only the vertices in $S \backslash V(M_r)$. Afterwards, the maximal matching heuristic produces a matching which covers $2|V_1| + |V_{1/2}|$ vertices of the remaining graph $G[V \backslash S]$. In this way, we derive the following:

$$2|EDS_{\mathcal{A}}| \leq (|S| - |M_r|) + 2|V_1| + |V_{1/2}| + 2|M_r|$$
$$\leq |S| + 2|V_1| + |V_{1/2}| + |M_r| \leq |S| + 2|V_1| + |V_{1/2}| + |M_R^*|$$

Now we analyze the cardinality of OPT. In contrast to the previous case, the independent set $G[V_0]$ is sufficiently large. Some of the vertices of $V(M_R^*) \cap V_0$ could be used to cover edges between V_0 and V_1. Nevertheless, the number of such edges is bounded by $|V_1|$, since $|V_1| + |V_{1/2}|/2$ is a lower bound on the cardinality of an optimal vertex cover of $G[V \backslash S]$. The crucial fact $|M_R^*| \geq |V_1|$ will be used later on to attain ($\star\star$). We give a lower bound on the cardinality of OPT.

$$2|OPT| \geq (|S| - |M_R^*|) + \frac{1}{2}|V_{1/2}| + 2|M_R^*| = |S| + \frac{1}{2}|V_{1/2}| + |M_R^*|$$

By combining the deduced upper and lower bounds, we analyze the approximation ratio of $\mathcal{A}_{SEDS}$.

$$\frac{2|EDS_{\mathcal{A}}|}{2|OPT|} \leq \frac{|S| + 2|V_1| + |V_{1/2}| + |M_R^*|}{|S| + \frac{1}{2}|V_{1/2}| + |M_R^*|} \underset{(\star\star)}{\leq} \frac{|S| + 2|V_1| + |V_{1/2}| + |V_1|}{|S| + \frac{1}{2}|V_{1/2}| + |V_1|}$$
$$\leq \frac{3}{\frac{3(|S| + |V_1| + \frac{1}{2}|V_{1/2}|)}{|S| + 3|V_1| + |V_{1/2}|}} \leq \frac{3}{1 + \frac{2|S| + \frac{1}{2}|V_{1/2}|}{|V|}} \leq \frac{3}{1 + 2\frac{|S|}{|V|}} \qquad \square$$

3   Dense Instances of the MEDS Problem

In this section, we consider the Minimum Edge Dominating Set problem in dense graphs. Firstly, we start with a observation of fundamental importance to our analysis.

Oberservation 1. *Given a connected graph $G = (V, E)$ and an optimal EDS M of G. There is a vertex $v \in V$ with $N(v) \subseteq V(M)$.*

Proof. If M covers the whole vertex set V, then we have nothing to show. Otherwise the whole neighborhood of a vertex $v \in V \backslash V(M)$ belongs to $V(M)$ to cover the edges incident to v. $\qquad \square$

This observation gives us a simple proof of the analysis of the approximation ratio of the maximal matching heuristic in dense graphs studied by Cardinal et al. (see [4]). Since the cardinality of an optimal EDS of an everywhere ϵ-dense graph $G = (V, E)$ can be lower bounded by $\min_{v \in V}\{|N(v)|\}/2 \geq \epsilon|V|/2$ and the worst case solution of the maximal matching heuristic is a maximum matching, the approximation ratio is bounded by $\min\{2, (|V|/2)/(\epsilon|V|/2)\}$.

Next, we want to derive an equivalent statement for average $\bar{\epsilon}$-dense graphs. We need a lemma which was proven by Karpinski and Zelikovsky.

Lemma 3. *[17] Given an $\bar{\epsilon}$-average dense graph $G = (V, E)$ and let W be the set of $(1 - \sqrt{1 - \bar{\epsilon}})|V|$ vertices with highest degree. Then every vertex of W has degree at least $|W|$.*

As a direct consequence, we get the following

Corollary 1. *Given an $\bar{\epsilon}$-average dense Graph $G = (V, E)$. The cardinality of an optimal EDS M is at least $(1 - \sqrt{1 - \bar{\epsilon}})|V|/2$.*

Proof. If the whole set W of $(1 - \sqrt{1 - \bar{\epsilon}})|V|$ vertices with highest degree belongs to $V(M)$, we have nothing to show. Otherwise the neighborhood of a vertex $v \in W \backslash V(M)$ is a subset of $V(M)$. According to Lemma 3 the degree of this vertex v is at least $(1 - \sqrt{1 - \bar{\epsilon}})|V|$. Therefore, the cardinality of M can be lower bounded by $|N(v)|/2 \geq (1 - \sqrt{1 - \bar{\epsilon}})|V|/2$. $\qquad\square$

Analogously, one can easily deduce similarly to Observation 1 that the maximal matching heuristic computes an EDS in average $\bar{\epsilon}$-dense graphs with approximation ratio at most $\min\{2, (1 - \sqrt{1 - \bar{\epsilon}})^{-1}\}$ as analyzed in [4].

We are ready to state the algorithm for the dense MEDS problem:

Input: Graph $G = (V, E)$

ForAll $v \in V$
    compute $\mathcal{A}_{\mathrm{SEDS}}(G, N(v))$;
EndForAll
Let M_1 be the solution with smallest cardinality among $\{\mathcal{A}_{\mathrm{SEDS}}(G, N(v)) \mid v \in V\}$;
Let W be the set of $(1 - \sqrt{1 - \bar{\epsilon}})|V|$ vertices with highest degree;
Compute $M_2 := \mathcal{A}_{\mathrm{SEDS}}(G, W)$;
ForAll $v \in W$
    compute $\mathcal{A}_{\mathrm{SEDS}}(G, N(v))$;
EndForAll
Let M_3 be the solution with smallest cardinality among $\{\mathcal{A}_{\mathrm{SEDS}}(G, N(v)) \mid v \in W\}$;

Output: The best solution among M_1, M_2 and M_3

Fig. 2. Algorithm $\mathcal{A}_{\mathrm{DEDS}}$

Corollary 2. *The algorithm $\mathcal{A}_{\mathrm{DEDS}}$ has an approximation ratio at most $\min\{2, 3/(1 + 2\epsilon)\}$ for ϵ-everywhere dense graphs and at most $\min\{2, 3/(3 - 2\sqrt{1 - \bar{\epsilon}})\}$ for $\bar{\epsilon}$-average dense graphs. $\mathcal{A}_{\mathrm{DEDS}}$ has a better approximation ratio than 2 if $\epsilon > 1/4$ or $\bar{\epsilon} > 7/16$.*

Proof. Given an ϵ-everywhere dense graph $G = (V, E)$ and an optimal EDS M, $V(M)$ contains always the neighborhood $N(v)$ of a vertex $v \in V$ because of Observation 1. By exhaustive search we find the right vertex v and use the algorithm $\mathcal{A}_{\text{SEDS}}$ for the MSED problem. In case of $\epsilon \leq 1/4$, we know from Theorem 2 that $\mathcal{A}_{\text{SEDS}}$ produces a solution with approximation ratio at most 2. Restricted to ϵ-everywhere dense graphs with $\epsilon > 1/4$, we get a solution with an approximation ratio at most $\frac{3}{1+2|N(v)|/|V|} \leq \frac{3}{1+2\epsilon|V|/|V|}$.

In the case of $\bar{\epsilon}$-average dense graphs, we have to consider two cases. If there is a vertex $v \in W$, which does not belong to $V(M)$, then we use the same argumentation as before. Since the smallest degree of a vertex in W is at least $(1 - \sqrt{1-\bar{\epsilon}})|V|$, the approximation ratio can be bounded as follows:

$$\frac{3}{1 + 2|N(v)|/|V|} \leq \frac{3}{1 + 2(1 - \sqrt{1-\bar{\epsilon}})} = \frac{3}{3 - 2\sqrt{1-\bar{\epsilon}}}$$

Otherwise the whole set W belongs to $V(M)$. Since the cardinality of W is $(1 - \sqrt{1-\bar{\epsilon}})|V|$, the corollary follows from Theorem 2. □

4 Approximation Hardness Results

Assuming the Unique Game Conjecture (see [18]), we provide new lower bounds on efficient approximability for everywhere ϵ-dense (resp. average $\bar{\epsilon}$-dense) instances of the MEDS problem with $1/3 < \epsilon$ (resp. with $5/9 < \bar{\epsilon}$). The starting point of our proof is the hardness result of Khot and Regev [18]. Then we show that the approximation preserving reduction from the Minimum Vertex Cover problem to the dense Vertex Cover problem due to Karpinski and Zelikovsky [16] can be used to derive the claimed inapproximability result for the dense MEDS problem.

We now formulate our inapproximability result.

Theorem 3. *For every $\delta > 0$, it is UGC-hard to approximate the everywhere ϵ-dense MEDS problem for every constant ϵ, $\bar{\epsilon}$ with $\epsilon > \frac{1}{3}$ (resp. average $\bar{\epsilon}$-dense MEDS problem with $\bar{\epsilon} > \frac{5}{9}$) to within $\frac{2}{1+\epsilon} - \delta$ (resp. $\frac{2}{2-\sqrt{1-\bar{\epsilon}}} - \delta$).*

Proof. Khot and Regev ([18]) showed that for every $\delta > 0$ there are instances $G = (V, E)$ of the Vertex Cover problem such that it is UGC-hard to decide whether $|OPT_{VC}| > (1 - \delta)|V|$ or $|OPT_{VC}| \leq (1/2 + \delta)|VC|$. We set $\delta \in (0, \epsilon/(1-\epsilon) - 1/2)$. Given such an instance, we densify it by joining all vertices of a clique of size $\epsilon/(1-\epsilon)|V|$ with all vertices of G. The same reduction was used by Karpinski and Zelikovsky ([16]) to prove that the dense Vertex Cover problem is APX-hard. This new instance G' is ϵ-dense, since every vertex of G' has a vertex degree at least

$$\frac{\epsilon}{1-\epsilon} \cdot n = \frac{\epsilon}{1-\epsilon} \cdot \frac{n'}{1 + \frac{\epsilon}{1-\epsilon}} = \frac{\epsilon}{1-\epsilon} \cdot \frac{n'}{\frac{1-\epsilon}{1-\epsilon} + \frac{\epsilon}{1-\epsilon}} = \epsilon \cdot n'.$$

If the optimal solution of the vertex cover problem $\leq (1/2 + \delta)|V|$, then we can match every vertex in the optimal solution with some vertices in the clique K

which is of size $\epsilon n/(1 - \epsilon) > (1/2 + \delta)n$. Since K is a clique, every remaining vertex in K can be matched by edges in $E(K)$ (We can double the graph G and join it with a twice larger clique K' to obtain a perfect matching in $G'[OPT_{VC} \cup V(K')]$). Therefore, the optimal solution for dense MEDS problem is $\leq n/2(1/2 + \epsilon/(1 - \epsilon) + \delta)$. If the optimal solution of the Vertex Cover problem is larger than $n(1 - \delta)$, we know that the optimal solution of the dense MEDS problem must be at least $n/2(1 + \epsilon/(1 - \epsilon) - \delta)$, since $V(OPT_{EDS})$ is a vertex cover of the graph G'.

Hence, we get the following UGC-hard decision question:

$$\frac{OPT_{EDS}}{n} \geq \frac{1}{2} + \frac{1}{2}\frac{\epsilon}{1 - \epsilon} - \frac{\delta}{2} \quad \text{or} \quad \frac{OPT_{EDS}}{n} \leq \frac{1}{4} + \frac{1}{2}\frac{\epsilon}{1 - \epsilon} + \frac{\delta}{2}$$

This decision question implies directly the following inapproximability factor:

$$\left(\frac{1}{2} + \frac{1}{2}\frac{\epsilon}{1 - \epsilon} - \frac{\delta}{2}\right)\left(\frac{1}{4} + \frac{1}{2}\frac{\epsilon}{1 - \epsilon} + \frac{\delta}{2}\right)^{-1} \leq \frac{1 - \epsilon + \epsilon}{2(1 - \epsilon)} \cdot \frac{4(1 - \epsilon)}{1 - \epsilon + 2\epsilon} - \delta'$$

$$\underset{(\star\star\star)}{\leq} \frac{2}{1 + \epsilon} - \delta'$$

In the case of average $\bar{\epsilon}$-dense instances of the Minimum Edge Dominating Set problem, we set $\epsilon := 1 - \sqrt{1 - \bar{\epsilon}}$ and the claimed inapproximability factor follows from $(\star\star\star)$. It remains to verify that the resulting graph G' is $\bar{\epsilon}$-dense:

$$\sum_{v \in V'} \frac{deg(v)}{|V'|^2} \geq \frac{\epsilon|V'| \overbrace{(|V'|)}^{deg} + n \overbrace{(\epsilon|V'|)}^{deg}}{|V'|^2} = (1 - \epsilon)\epsilon + \epsilon = \epsilon(1 - \epsilon + 1)$$

$$= (1 - \sqrt{1 - \bar{\epsilon}})(1 + \sqrt{1 - \bar{\epsilon}}) = 1 - 1 + \bar{\epsilon} \qquad \square$$

Using the same reduction for the MSED problem with $S = V(K)$, we get the following

Corollary 3. *For every $\delta > 0$ and $3|S| \geq |V|$, it is UGC-hard to approximate the MSED problem within $\frac{2}{1 + |S|/|V|} - \delta$.*

Acknowledgment

We thank Marek Karpinski and Jean Cardinal for a number of interesting discussions and for their support.

References

1. Alon, N., Yuster, R.: Fast Algorithms for Maximum Subset Matching and All-Pairs Shortest Paths in Graphs with a (Not So) Small Vertex Cover. In: Arge, L., Hoffmann, M., Welzl, E. (eds.) ESA 2007. LNCS, vol. 4698, pp. 175–186. Springer, Heidelberg (2007)

2. Bar-Yehuda, R., Kehat, Z.: Approximating the Dense Set-Cover Problem. J. Comput. Syst. Sci. 69, 547–561 (2004)
3. Cardinal, J., Karpinski, M., Schmied, R., Viehmann, C.: Approximating Subdense Instances of Covering Problems, CoRR abs/1011.0078 (2010)
4. Cardinal, J., Labbé, M., Langerman, S., Levy, E., Mélot, H.: A Tight Analysis of the Maximal Matching Heuristic. In: Wang, L. (ed.) COCOON 2005. LNCS, vol. 3595, pp. 701–709. Springer, Heidelberg (2005)
5. Cardinal, J., Langerman, S., Levy, E.: Improved Approximation Bounds for Edge Dominating Set in Dense Graphs. Theor. Comput. Sci. 410, 949–957 (2009)
6. Chlebík, M., Chlebíková, J.: Improvement of Nemhauser-Trotter Theorem and its Applications in Parametrized Complexity. In: Hagerup, T., Katajainen, J. (eds.) SWAT 2004. LNCS, vol. 3111, pp. 174–186. Springer, Heidelberg (2004)
7. Chlebík, M., Chlebíková, J.: Approximation Hardness of Edge Dominating Set Problems. J. Comb. Optim. 11, 279–290 (2006)
8. Clementi, A., Trevisan, L.: Improved Non-Approximability Results for Minimum Vertex Cover with Density Constraints. Theor. Comput. Sci. 225, 113–128 (1999)
9. Eremeev, A.: On some Approximation Algorithms for Dense Vertex Cover Problem. In: Proc. Symp. on Operations Research, pp. 48–52 (1999)
10. Gabow, H., Tarjan, R.: Faster Scaling Algorithms for General Graph-Matching Problems. J. ACM 38, 815–853 (1991)
11. Garey, M., Johnson, D.: Computers and Intractability. W.H. Freeman & Company, New York (1979)
12. Gotthilf, Z., Lewenstein, M., Rainshmidt, E.: A $(2 - c\frac{\log n}{n})$ approximation algorithm for the minimum maximal matching problem. In: Bampis, E., Skutella, M. (eds.) WAOA 2008. LNCS, vol. 5426, pp. 267–278. Springer, Heidelberg (2009)
13. Hopcroft, J., Karp, R.: An $n^{5/2}$ Algorithm for Maximum Matchings in Bipartite Graphs. SIAM J. Comput. 2, 225–231 (1973)
14. Imamura, T., Iwama, K.: Approximating Vertex Cover on Dense Graphs. In: Proc. 16th SODA 2005, pp. 582–589 (2005)
15. Karpinski, M.: Polynomial Time Approximation Schemes for Some Dense Instances of NP-Hard Optimization Problems. Algorithmica 30, 386–397 (2001)
16. Karpinski, M., Zelikovsky, A.: Approximating Dense Cases of Covering Problems, ECCC TR97-004 (1997)
17. Karpinski, M., Zelikovsky, A.: Approximating Dense Cases of Covering Problems. In: Proc. DIMACS Workshop on Network Design: Connectivity and Facilities Location, pp. 169–178 (1997)
18. Khot, S., Regev, O.: Vertex Cover might be Hard to Approximate to within 2-ε. J. Comput. Syst. Sci. 74, 335–349 (2008)
19. Nemhauser, G.L., Trotter, L.E.: Vertex Packings: Structural Properties and Algorithms. Math. Programming 8, 232–248 (1975)
20. Yannakakis, M., Gavril, F.: Edge Dominating Sets in Graphs. SIAM J. Appl. Math. 38, 364–372 (1980)

Near Approximation of Maximum Weight Matching through Efficient Weight Reduction

Andrzej Lingas and Cui Di

Department of Computer Science, Lund University, 22100 Lund, Sweden
Andrzej.Lingas@cs.lth.se, dcdcsunny@gmail.com

Abstract. Let G be an edge-weighted hypergraph on n vertices, m edges of size $\leq s$, where the edges have real weights in an interval $[1, \, W]$. We show that if we can approximate a maximum weight matching in G within factor α in time $T(n, m, W)$ then we can find a matching of weight at least $(\alpha - \epsilon)$ times the maximum weight of a matching in G in time $(\epsilon^{-1})^{O(1)} \times \max_{1 \leq q \leq O(\epsilon \frac{\log \frac{n}{\epsilon}}{\log \epsilon^{-1}})} \max_{m_1 + \ldots m_q = m} \sum_1^q T(\min\{n, sm_j\}, m_j, (\epsilon^{-1})^{O(\epsilon^{-1})})$. We obtain our result by an approximate reduction of the original problem to $O(\epsilon \frac{\log \frac{n}{\epsilon}}{\log \epsilon^{-1}})$ subproblems with edge weights bounded by $(\epsilon^{-1})^{O(\epsilon^{-1})}$. In particular, if we combine our result with the recent $(1 - \epsilon)$-approximation algorithm for maximum weight matching in graphs due to Duan and Pettie whose time complexity has a poly-logarithmic dependence on W then we obtain a $(1 - \epsilon)$-approximation algorithm for maximum weight matching in graphs running in time $(\epsilon^{-1})^{O(1)}(m + n)$.

1 Introduction

A *hypergraph* G consists of a set V of vertices and a set of subsets of V called edges of G. In particular, if all the edges are of cardinality two then G is a graph. A *matching* of G is a set of edges of G without common vertices. If real weights are assigned to the edges of G then a *maximum weight matching* of G is a matching of G whose total weight achieves the maximum.

The problem of finding a maximum weight matching in a hypergraph is a fundamental generalization of that of finding maximum cardinality matching in a graph. The latter is one of the basic difficult combinatorial problems that still admit polynomial-time solutions. For hypergraphs the decision version of the maximum weight matching problem is NP-hard even if the edges are of size $O(1)$ since it is a generalization of the problem of maximum weight independent set for bounded degree graphs [15]. On the other hand, polynomial-time algorithms yielding $(d - 1 + 1/d)$-approximation of maximum weight matching in hypergraphs with edges of size d are known [3].

The fastest known algorithms for maximum weight matching in graphs have substantially super-quadratic time complexity in terms of the number n of vertices of the input graph G [11,12,20]. For these reasons, there is a lot of interest in designing faster approximation algorithms for maximum weight matching [4,5,6,14,18,19].

M. Ogihara and J. Tarui (Eds.): TAMC 2011, LNCS 6648, pp. 48–57, 2011.

Recently, even fast approximation schemes for maximum weight matching in graphs have been presented [1]. The fastest known in the literature is due to Duan and Pettie [7]. It yields a $(1 - \epsilon)$-approximation in time $O(m\epsilon^{-2}\log^3 n)$ for a connected graph on n vertices and m edges with real edge weights. The approximation scheme from [7] is a composition of a $(1-\epsilon)$-approximate reduction of the problem in general edge weighted graphs to that in graphs with small edge weights and an efficient $(1 - \epsilon)$-approximate algorithm for graphs with small edge weights.

1.1   Our Contributions

Let G be an edge-weighted hypergraph on n vertices, m edges of size $\leq s$, where the edges have real weights in an interval $[1, W]$. We show that if we can approximate a maximum weight matching in G within factor α in time $T(n, m, W)$ then we can find a matching of weight at least $\alpha - \epsilon$ times the maximum weight of a matching in G in time $(\epsilon^{-1})^{O(1)} \times$
$\max_{1 \leq q \leq O(\epsilon \frac{\log \frac{n}{\epsilon}}{\log \epsilon^{-1}})} \max_{m_1 + \ldots m_q = m} \sum_1^q T(\min\{n, sm_j\}, m_j, (\epsilon^{-1})^{O(\epsilon^{-1})})$. We obtain our result by an approximate reduction of the original problem to $O(\epsilon \frac{\log \frac{n}{\epsilon}}{\log \epsilon^{-1}})$ subproblems with edge weights bounded by $(\epsilon^{-1})^{O(\epsilon^{-1})}$.

This reduction of maximum weight matching in hypergraphs with arbitrarily large edge weights to that in hypergraphs with small edge weights is incomparable to the aforementioned similar reduction for graphs from [7]. In particular, if we combine our reduction with the aforementioned $(1 - \epsilon)$-approximation algorithm for maximum weight matching in graphs from [7] whose time complexity has a poly-logarithmic dependence on W then we obtain a $(1 - \epsilon)$-approximation algorithm for maximum weight matching in graphs running in time $(\epsilon^{-1})^{O(1)}(m + n)$. In comparison with the approximation scheme from [7], our approximation scheme is more truly linear in $m + n$, as free from the poly-logarithmic in n factor at the cost of larger polynomial dependence on ϵ^{-1}.

As another corollary from our approximate edge-weight reduction for hypergraphs, we obtain also some results on approximating maximum weight independent set in graphs of bounded degree.

1.2   Other Related Results

As the problem of finding maximum weight matching in graphs is a classical problem in combinatorial optimization there is an extensive literature on it. It includes such milestones as an early algorithm of Kuhn [17] just in the bipartite case and an algorithm of Edmond and Karp [8] running in time $O(nm^2)$, where n is the number of vertices and m is the number of edges in the input graph. Hungarian algorithm [17] can be implemented in time $O(mn + n^2 \log n)$ with the help of Fibonacci heaps [9] and this upper bound can be extended to include general graphs [10].

Assuming integer edge weights in $[-W, W]$ and RAM model with $\log(\max\{N, n\})$-bit words, Gabow and Tarjan established $O(\sqrt{n}m\log(nW))$

---

[1] In a preliminary version of this paper presented at SOFSEM Student Forum held in January 2010 (no proceedings), an $O(n^\omega \log n)$-time approximation scheme for maximum weight matching in bipartite graphs has been presented.

and $O(\sqrt{n \log n} m \log(nW))$ time-bounds for maximum weight matching respectively in bipartite and general graphs [11,12].

More recently, Sankowski designed an $O(n^\omega W)$-time algorithm for the weighted matching problem in bipartite graphs with integer weights, where ω stands for the exponent of fast matrix multiplication known to not exceed 2.376 [20]. His result asymptotically improved an earlier upper-time bound for maximum weight matching in bipartite graphs with integer weights of the form $O(\sqrt{n} m W)$ due to Kao [16].

There is also an extensive literature on fast approximation algorithms for maximum weight matching in graphs [4,5,6,14,18,19]. Typically they yield an approximation within a constant factor between $\frac{1}{2}$ and almost $\frac{4}{5}$, running in time of order $m \log^{O(1)} n$. Already the straightforward greedy approach yields $\frac{1}{2}$-approximation in time $O(m \log n)$.

The maximum weight matching problem in hypergraphs is known also as a set packing problem in combinatorial optimization [15]. By duality it is equivalent to maximum weight independent set and hence extremely hard to approximate in polynomial time [13]. The most studied case of maximum weight matching in hypergraphs is that for d-uniform hypergraphs where each edge is of size d. Then a polynomial-time $(d - 1 + 1/d)$-approximation is possible [3]. By duality, one obtains also a polynomial-time $(d - 1 + 1/d)$-approximation of maximum weight independent set in graphs of degree d (cf. [15]).

2   Simple Edge Weight Transformations

In this section, we describe two simple transformations of the edge weights in the input hypergraph G such that an α-approximation of maximum weight matching in the resulting hypergraph yields an $(\alpha - \epsilon)$-approximation of maximum weight matching of G. We assume w.l.o.g. throughout the paper that G has n vertices, m edges, and real edge weights not less than 1. The largest edge weight in G is denoted by W.

Lemma 1. *Suppose that there is an α-approximation algorithm for maximum weight matching in G running in time $T(n, m, W)$. Then, there is an $O(n + m)$-time transformation of G into an isomorphic hypergraph G^* with edge weights in the interval $[1, \frac{n}{\epsilon}]$ such that the aforementioned algorithm run on G^* yields an $(\alpha - \epsilon)$-approximation of maximum weight matching in G in time $T(n, m, \frac{n}{\epsilon})$.*

Proof. We may assume w.l.o.g that $W > \frac{n}{\epsilon}$. Note that the total weight of maximum weight matching in G is at least W. Hence, if we transform G to a hypergraph G' by raising the weight of all edges in G of weight smaller than $\frac{W\epsilon}{n}$ to $\frac{W\epsilon}{n}$ then the following holds:

1. the maximum weight of a matching in G' is not less than that in G;
2. any matching in G' induces a matching in G whose weight is smaller by at most ϵW.

To find an α-approximation of maximum weight matching in G', we can simply rescale the edge weights in G' by multiplying them by $\frac{n}{W\epsilon}$. Let G^* denote the resulting graph. Now it is sufficient to run the assumed algorithm on G^* to obtain an $(\alpha - \epsilon)$-approximation of maximum weight matching in G. Note that the application of the algorithm will take time $T(n, m, \frac{n}{\epsilon})$. $\square$

Lemma 2. *Suppose that there is an $(\alpha - \epsilon)$-approximation algorithm for maximum weight matching in G running in time $T'(n, m, W, \epsilon)$. By rounding down each edge weight to the nearest power of $1 + \epsilon$ and then running the $(\alpha - \epsilon)$-approximation algorithm on the resulting graph, we obtain an $(\alpha - O(\epsilon))$-approximation of maximum weight matching in G in time $T'(n, m, W, \epsilon) + O(n + m)$.*

Proof. Let e be any edge in G. Denote its weight in G by $w(e)$ and its weight in the resulting graph by $w'(e)$. We have $w'(e)(1 + \epsilon) \geq w(e)$. Consequently, we obtain $w'(e) \geq w(e) - \epsilon w'(e) \geq (1 - \epsilon)w(e)$. It follows that a maximum weight matching in the resulting graph has weight at least $1 - \epsilon$ times the weight of a maximum weight matching in G. Thus, if we run the assumed $(\alpha - \epsilon)$-approximation algorithm on the resulting graph then the produced matching with edge weights restored back to their original values will yield an $(\alpha - 2\epsilon)$-approximation. $\square$

3   A Transformation into an $(\alpha - \epsilon)$-Approximation Algorithm

A *sub-hypergraph* of a hypergraph H is any hypergraph that can be obtained from H by deleting some vertices and some edges. Once a vertex is removed all edges containing it are also removed. A class C of hypergraphs such that any subhypergraph of a hypergraph in C also belongs to C is called *hereditary*.

   In this section, we present a transformation of an hypothetical α-approximation algorithm for maximum weight matching in a hereditary family of hypergraphs with edges of size $O(1)$ into a $(\alpha - \epsilon)$-approximation algorithm. The running time of the $(\alpha - \epsilon)$-approximation algorithm is close to that of the α-approximation algorithm in case the largest edge weight is $\epsilon^{-O(\epsilon^{-1})}$.

Theorem 1. *Suppose that there is an algorithm for a maximum weight matching in any hypergraph having edges of size $\leq s$ and belonging to the same hereditary class as G running in time $T(n', m', W') = \Omega(n' + m')$, where n', m' are respectively the number of vertices and edges, and $[1, W']$ is the interval to which all edge weights belong. There is an $(\alpha - \epsilon)$-approximation algorithm for a maximum weight matching in G running in time $(\epsilon^{-1})^{O(1)} \times$ $\max_{1 \leq q \leq O(\epsilon \frac{\log n}{\log \epsilon^{-1}})} \max_{m_1 + \ldots m_q = m} \sum_1^q T(\min\{n, s m_j\}, m_j, (\epsilon^{-1})^{O(\epsilon^{-1})})$.*

Proof. We may assume w.l.o.g that $W = O(n/\epsilon)$ and any edge weight is a nonnegative integer power of $1 + \epsilon$ by Lemmata 1, 2. Order the values of the edge weights in G in the increasing order. Set $k = O(\epsilon^{-1})$ and $l = \lceil \log_{1+\epsilon} \frac{2}{\epsilon} \rceil$. By the form of the edge weights and the setting of l, the following holds.

Remark 1: For any two different edge weights w_1 and w_2, if the number of w_1 is greater than that of w_2 by at least l in the aforementioned ordering then $\frac{\epsilon}{2} w_1 \geq w_2$.

In order to specify our $(\alpha - \epsilon)$-approximation algorithm, we partition the ordered edge weights into consecutive closed basic intervals, each but perhaps for the last, containing exactly l consecutive edge weights, see Fig. 1.

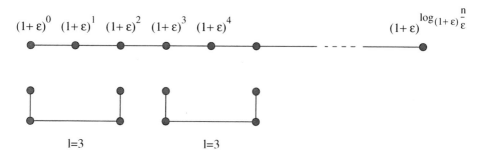

$(1+\varepsilon)^0$ $(1+\varepsilon)^1$ $(1+\varepsilon)^2$ $(1+\varepsilon)^3$ $(1+\varepsilon)^4$ $(1+\varepsilon)^{\log_{(1+\varepsilon)}\frac{n}{\varepsilon}}$

l=3   l=3

Fig. 1. Partitioning of edge weights $(l = 3)$

Next, we group k-tuples of consecutive basic intervals into large intervals composed of $k - 1$ consecutive basic intervals followed by a single basic interval called a gap. This partition corresponds to the situation when the so called shift parameter x is set to 0. For $x \in \{1, .., k-1\}$, the partition into alternating large intervals and gaps is shifted by x basic intervals from the right, so the first large interval from the right is composed solely of $k - 1 - x$ basic intervals, see Fig. 2. The maximal subgraph of G containing solely edges with weights in the large intervals in the partition is denoted by G_x.

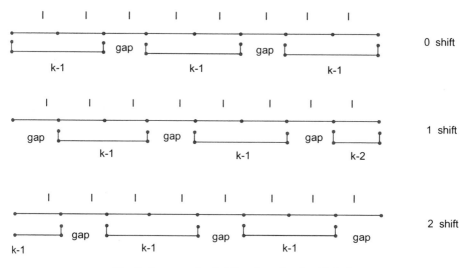

Fig. 2. An example of shift x from 1 to 3 with k=3

For our $(\alpha - \epsilon)$-approximation algorithm for a maximum weight matching in G see Fig. 3. We shall assume the definitions of the subgraphs $G'_x, G_{x,j}, M_x$ given in the algorithm.

Algorithm 1

1. **for** $x \leftarrow 1$ **to** $k - 1$ **do**
2. $\quad M_x \leftarrow \emptyset$;
3. $\quad G'_x \leftarrow G_x$;
4. $\quad$ **for** $j \leftarrow 1$ **to** $O(\log_{1+\epsilon} \frac{n}{\epsilon})$ **do**
5. $\quad$ **begin**
6. $\quad\quad$ Set $G_{x,j}$ to the sub-hypergraph of G'_x induced by the edges whose weights
7. $\quad\quad$ fall in the jth interval from the right;
8. $\quad\quad$ Run the α-approximation algorithm for maximum weight matching $M_{x,j}$ of $G_{x,j}$;
9. $\quad\quad$ $M_x \leftarrow M_x \bigcup M_{x,j}$;
10. $\quad\quad$ Remove all edges incident to $M_{x,j}$ from G'_x;
11. $\quad$ **end**
12. Return the heaviest among the matchings M_x

Fig. 3. The $(\alpha - \epsilon)$-approximation algorithm

Since the union of the gaps over all shifts covers all edge weights, in particular all edge weights occurring in an optimal matching of G, there must be a shift where the gaps cover at most $\frac{1}{k}$ of the weight of the optimal matching of G. Hence, there must be a shift x such that the weight of an optimal matching in G_x is at least $(1 - 1/k)$ of the weight of an optimal matching of G. Thus, it is sufficient to show that M_x closely approximates an α-approximate weight matching of G_x.

Consider a maximum weight matching OM_x of G_x and the α-approximation $M_{x,j}$ of a maximum weight matching of $G_{x,j}$, respectively. Note that $M_{x,j}$ has total weight not smaller than α times the total weight of OM_x restricted to the edges in $G_{x,j}$. On the other hand, each edge e in $M_{x,j}$ can eliminate at most $O(1)$ edges of OM_x from all $G_{x,i}$ for $i > j$. The total weight of the at most $O(1)$ edges is only at most the ϵ fraction of the weight of e by Remark 1. Let EOM_x denote the set of all edges in OM_x eliminated by $M_x = \bigcup_j M_{x,j}$. The following two inequalities follow:

$$weight(M_x) + weight(EOM_x) \geq \alpha \times weight(OM_x)$$

$$\epsilon \times weight(M_x) \geq weight(EOM_x)$$

Consequently, we obtain:

$$weight(M_x) \geq \alpha \times weight(OM_x) - \epsilon \times weight(M_x)$$

$$\geq (\alpha - \epsilon) \times weight(OM_x)$$

Thus, M_x approximates a maximum weight matching of G_x within $(\alpha - \epsilon)$, and consequently the heaviest of the matchings M_x approximates a maximum weight matching of G within $(1 - \epsilon)(1 - 1/k)$. By setting $k = \Omega(\frac{1}{\epsilon})$, we obtain an $(1 - O(\epsilon))$-approximation of the optimum.

It remains to estimate the time complexity of our method. Note that the weight of heaviest edge in $G_{x,j}$ is at most

$$(1 + \epsilon)^{lk} = O(\epsilon^{-1})^{O(\epsilon^{-1})} = (\epsilon^{-1})^{O(\epsilon^{-1})}$$

times larger than that of the lightest one. Let $m_{x,j}$ denote the number of edges in $G_{x,j}$. Next, let $n_{x,j}$ denote the number of vertices in the sub-hypergraph of $G_{x,j}$ induced by the edges of $G_{x,j}$. Note that $n_{x,j} \leq \min\{n, sm_{x,j}\}$ by our assumption on the size of edges in G.

Hence, by rescaling the weights in $G_{x,j}$, we can find $M_{x,j}$ in time $T(\min\{n, sm_{x,j}\}, m_{x,j}, (\epsilon^{-1})^{O(\epsilon^{-1})})$ for $j = 1.., O(\log_{1+\epsilon} \frac{n}{\epsilon}/lk)$ and $x = 0, ..., k - 1$. Note that $\log_{1+\epsilon} \frac{n}{\epsilon} = \frac{\log \frac{n}{\epsilon}}{\log 1+\epsilon} = \Theta(\epsilon^{-1} \log \frac{n}{\epsilon})$ and similarly $lk = \log_{1+\epsilon} \frac{2}{\epsilon}\Theta(\epsilon^{-1}) = \Theta(\frac{\log \frac{2}{\epsilon}}{\log 1+\epsilon}\epsilon^{-1}) = \Theta(\epsilon^{-2} \log \epsilon^{-1})$. It follows that for a given x, the largest value of j, i.e., the number of the subgraphs $G_{x,j}$ is $O(\epsilon \frac{\log \frac{n}{\epsilon}}{\log \epsilon^{-1}})$.

Note that $\sum_j m_{x,j} \leq m$ since each edge of G belongs to at most one hypergraph $G_{x,j}$. Thus, the total time taken by finding all $M_{x,j}$ for $j = 1, ..., O(\epsilon \frac{\log \frac{n}{\epsilon}}{\log \epsilon^{-1}})$, for a fixed x is $\max_{1 \leq q \leq O(\epsilon \frac{\log \frac{n}{\epsilon}}{\log \epsilon^{-1}})} \max_{m_1 + ... m_q = m} \sum_1^q T(\{n, sm_j\}, m_j, (\epsilon^{-1})^{O(\epsilon^{-1})})$. Recall that x ranges over $O(\epsilon^{-1})$ possible values.

By the assumed form of the edge weights in G, we can apply a standard radix sort with $O(\epsilon^{-1} \log \frac{n}{\epsilon})$ buckets to sort the edges of G by their weights in time $O(m + \epsilon^{-1} \log \frac{n}{\epsilon})$. The latter is also $O(\epsilon^{-2} T(n, m, (\epsilon^{-1})^{O(\epsilon^{-1})}))$ by the assumptions on T.

In order to efficiently construct the graphs $G_{x,j}$, the sorted edge list is kept in array and there are double links between an occurrence of an edge in the adjacency lists representing G and its occurrence in the sorted edge list. To determine the edges inducing $G_{x,j}$, we just scan a consecutive fragment of the sorted list from left to right. Given a list of edges of $G_{x,j}$, an adjacency representation of the sub-hypergraph can be constructed in time $O(n + m) = O(T(n, m, (\epsilon^{-1})^{O(\epsilon^{-1})}))$ by using the aforementioned double links.

To remove an edge from G'_x, we locate it on the sorted edge list by using the double links with the adjacency lists and then link its predecessor with its successor on the sorted list. We conclude that the updates of G'_x take time $O(m) = O(T(n, m, (\epsilon^{-1})^{O(\epsilon^{-1})}))$.

Summarizing, our upper time-bound on finding $M_{x,j}$ for all j and x dominates our upper time-bounds for the remaining steps which yields the theorem.   □

4   Applications

There are at least two known exact algorithms for maximum weight matching in bipartite graphs with integer edge weights for which the upper time bounds on their running time in linear fashion depend on the maximum edge weight W [16,20]. Recently, Duan and Pettie have provided substantially more efficient $1 - \epsilon$ approximation algorithm for maximum weight matching in general graphs with integer edge weights, whose running time also depends on W in linear fashion [7]. Furthermore, their final approximation scheme for this problem in fact exhibits poly-logarithmic dependence on W.

Fact 1 (Duan and Pettie, see the proof of Theorem 1 in [7]). *An $(1 - \epsilon)$-approximation of maximum weight matching in a connected graph on m edges and positive integer weights not exceeding W can be found deterministically in time $O(\epsilon^{-2} m \log^3 W)$.*

We can trivially generalize the upper time bound of Fact 1 to include a non-necessarily connected graph by extending it by an additive factor of $O(n)$.

There is one technical difficulty in combining Fact 1 with Theorem 1. In the theorem we assume that there is available an α-approximation algorithm for maximum weight matching for graphs belonging to the same hereditary class as G with arbitrary real edge weights not less than 1 whereas the algorithm of Fact 1 assumes integer weights. In fact, even if the input graph got positive integer weights the preliminary edge weight transformations in the proof of Theorem 1 would result in rational edge weights. There is a simple remedy for this. We may assume w.l.o.g that ϵ is an inverse of a positive integer and through all the steps of our approximation scheme round down the edge weights to the nearest fraction with denominator $O(\epsilon^{-1})$ and then multiply them by the common denominator to get integer weights. This will increase the maximum weight solely by $O(\epsilon^{-1})$ and will preserve close approximability.

Hence, Fact 1 combined in this way with Theorem 1 yield our main application result by straightforward calculations.

Theorem 2. *There is an approximation scheme for a maximum weight matching in a graph on n vertices and m edges running in time $(\epsilon^{-1})^{O(1)}(m+n)$.*

5   Extensions

Note that Theorem 1 includes as a special case the problem of finding a maximum weight independent set in a graph G of maximum degree d which is equivalent to the problem of finding a maximum weight matching in the dual hypergraph with edges corresponding to the vertices of G and *vice versa*.

Several combinatorial algorithms for maximum independent set achieving the approximation ratio of $O(d)$, where d is the maximum or average degree are known in the literature [15]. In the appendix, we demonstrate that by using the method of Theorem 1 they can be simply transformed into good approximation algorithms for maximum weight independent set.

6   Final Remark

In earlier versions of our paper, we presented a simpler formula on the time complexity of our reduction (see Theorem 1) with a single term $T(n, m, ...)$, which resulted in an additional logarithmic factor in the application to maximum weight graph matching.

Acknowledgments

The authors are very grateful to Seth Pettie for his valuable suggestions to apply the second $(1-\epsilon)$-approximation algorithm from [7] instead of the first one and to eliminate an $n \log n$ term in the application in an intermediate version of our paper. They are also very grateful to anonymous referees for valuable comments on the submitted version of the paper as well as the early version of the paper presented solely orally at Student Forum of SOFSEM 2010. (Because of the very short time for preparation of the final version some of the suggestions of the referees had to be postponed to a journal version of the paper). Finally, thanks go to Eva-Marta Lundell for her help with figures.

References

1. Berman, P.: A $d/2$ Approximation for Maximum Weight Independent Set in d-Claw Free Graphs. In: Halldórsson, M.M. (ed.) SWAT 2000. LNCS, vol. 1851, pp. 214–219. Springer, Heidelberg (2000)
2. Cormen, T.H., Leiserson, C.E., Rivest, R.L., Stein, C.: Introduction to Algorithms, 2nd edn. McGraw-Hill Book Company, Boston (2001)
3. Chan, Y.H., Lau, L.C.: On Linear and Semidefinite Programming Relaxations for Hypergraph Matching. Proc
4. Drake, D., Hougardy, S.: A simple approximation algorithm for the weighted matching problem. Info. Proc. Lett. 85, 211–213 (2003)
5. Drake, D., Hougardy, S.: Linear time local improvements for weighted matchings in graphs. In: Jansen, K., Margraf, M., Mastrolli, M., Rolim, J.D.P. (eds.) WEA 2003. LNCS, vol. 2647, pp. 107–119. Springer, Heidelberg (2003)
6. Drake, D., Hougardy, S.: Improved linear time approximation algorithms for weighted matchings. In: Arora, S., Jansen, K., Rolim, J.D.P., Sahai, A. (eds.) RANDOM 2003 and APPROX 2003. LNCS, vol. 2764, pp. 14–23. Springer, Heidelberg (2003)
7. Duan, R., Pettie, S.: Approximating Maximum Weight Matching in Near-linear Time. In: Proc. FOCS (2010)
8. Edmonds, J., Karp, R.M.: Theoretical Improvements in Algorithmic Efficiency for Network Flow Problems. J. ACM 19(2), 248–264 (1972)
9. Fredman, M.L., Tarjan, R.E.: Fibonacci heaps and their uses in improved network optimization algorithms. J. ACM 23(2), 596–615 (1987)
10. Gabow, H.N.: Data structures for weighted matching and nearest common ancestors with linking. In: First Annual ACM-SIAM Symposium on Discrete Algorithms(SODA), pp. 434–443 (1990)
11. Gabow, H.N., Tarjan, R.E.: Faster scaling algorithms for network problems. SIAM J. Comput. 18(5), 1013–1036 (1989)
12. Gabow, H.N., Tarjan, R.E.: Faster scaling algorithms for general graph-matching problems. J. ACM 38(4), 815–853 (1991)
13. Hastad, J.: Clique is Hard to Approximate within $n^{1-\epsilon}$. Acta Math. 182(1), 105–142 (1999)
14. Hanke, Hougardy: $3/4 - \epsilon$ and $4/5 - \epsilon$ approximate MWM algorithms running in $O(mlogn)$ and $O(mlog^2 n)$ time University of Bonn, Research Institute for Discrete Mathematics Report No. 101010
15. Hochbaum, D.S.: Approximating Covering and Packing Problems: Set Cover, Vertex Cover, Independent Set, and Related Problems in Approximation Algorithms for NP-hard Problems. In: Hochbaum, D.S. (ed.) PWS Publishing Company, Boston (1997)
16. Kao, M.-Y., Lam, T.-W., Sung, W.-K., Ting, H.-F.: A Decomposition Theorem for Maximum Weight Bipartite Matchings with Applications to Evolutionary Trees. In: Nešetřil, J. (ed.) ESA 1999. LNCS, vol. 1643, pp. 438–449. Springer, Heidelberg (1999)
17. Kuhn, H.W.: The Hungarian method for the assignment problem. Naval Research Logistics Quarterly 2, 83–97 (1955)
18. Pettie, S., Sanders, P.: A simple linear time 2/3-ϵ approximation for maximum weight matching. Information Processing Letters 91, 271–276 (2004)
19. Preis, R.: Linear time 1/2-approximation algorithm for maximum weighted matching in general graphs. In: Meinel, C., Tison, S. (eds.) STACS 1999. LNCS, vol. 1563, pp. 259–269. Springer, Heidelberg (1999)
20. Sankowski, P.: Weighted bipartite matching in matrix multiplication time. In: Bugliesi, M., Preneel, B., Sassone, V., Wegener, I. (eds.) ICALP 2006. LNCS, vol. 4051, pp. 274–285. Springer, Heidelberg (2006)

Appendix: Approximation Algorithms for Maximum Weight Independent Set in Bounded Degree Graphs

Note that Theorem 1 includes as a special case the problem of finding a maximum weight independent set in a graph G of maximum degree d which is equivalent to the problem of finding a maximum weight matching in the dual hypergraph with edges corresponding to the vertices of G and *vice versa*.

Several combinatorial algorithms for maximum independent set achieving the approximation ratio of $O(d)$, where d is the maximum or average degree are known in the literature [15]. Here, we demonstrate that by using the method of Theorem 1 they can be simply transformed into good approximation algorithms for maximum weight independent set.

Lemma 3. *Suppose that there is an $\alpha(d)$-approximation algorithm for maximum independent set in a graph on n vertices and maximum (or average degree, respectively) degree d running in time $S(n, d)$, where the function S is non-decreasing in both arguments. There is an $\alpha(dW)$-approximation algorithm for maximum weight independent set in a graph on n vertices, maximum (or average degree, respectively) degree d, positive integer vertex weights not exceeding an integer W, running in time $S(nW, dW)$.*

Proof: Let G be the input vertex weighted graph G. We form the auxiliary unweighted graph G^* on the base of G as follows. In G^*, we replace each vertex v in G with the number of its copies equal to the weight of v. We connect each copy of v by an edge with each copy of each neighbor of v. Next, we run the assumed algorithm for maximum unweighted independent set on G^*. Note that any maximal independent set in G^* is in one-to-one correspondence with an independent set in G since whenever a copy of v is in the independent set then all other copies of v can be inserted into it without any conflicts. □

The drawback of Lemma 3 is that the approximation factor and/or the running time of the resulting algorithm for the weighted case can be very large in case the maximum weight W is large. However, we can plug Lemma 3 in the method of Theorem 1 to obtain much more interesting approximation algorithms in the weighted case.

Theorem 3. *Suppose that there is an $\alpha(d)$-approximation algorithm for maximum independent set in a graph on n vertices and maximum degree d running in time $S(n, d)$, where the function S is non-decreasing in both arguments and $S(n, d) = \Omega(nd \log n)$. There is an $(\alpha(d\epsilon^{-1})^{O(\epsilon^{-1})}) - d\epsilon)$-approximation algorithm for maximum weight independent set in a graph on n vertices, with maximum degree d, positive integer vertex weights, running in tim $O(\epsilon \frac{\log(n/\epsilon)}{\log \epsilon^{-1}} S(n(\epsilon^{-1})^{O(\epsilon^{-1})}, d(\epsilon^{-1})^{O(\epsilon^{-1})}))$.*

Proof. sketch. Recall that the problem of maximum (weighted or unweighted) independent set is equivalent to the problem of maximum (weighted or unweighted, respectively) matching in the dual hypergraph. In the dual hypergraph, the edges have size not exceeding the maximum vertex degree in the input graph. We run the method of Theorem 1 on the dual hypergraph using as the black box algorithm the result of the application of Lemma 3 to the assumed algorithm and its adaptation to the maximum matching problem in the dual hypergraph. □

Approximability of
the Subset Sum Reconfiguration Problem

Takehiro Ito[1,*] and Erik D. Demaine[2]

[1] Graduate School of Information Sciences, Tohoku University,
Aoba-yama 6-6-05, Sendai, 980-8579, Japan
`takehiro@ecei.tohoku.ac.jp`
[2] MIT Computer Science and Artificial Intelligence Laboratory,
32 Vassar St., Cambridge, MA 02139, USA
`edemaine@mit.edu`

Abstract. The SUBSET SUM problem is a well-known NP-complete problem in which we wish to find a packing (subset) of items (integers) into a knapsack with capacity so that the sum of the integers in the packing is at most the capacity of the knapsack and at least a given integer threshold. In this paper, we study the problem of reconfiguring one packing into another packing by moving only one item at a time, while at all times maintaining the feasibility of packings. First we show that this decision problem is strongly NP-hard, and is PSPACE-complete if we are given a conflict graph for the set of items in which each vertex corresponds to an item and each edge represents a pair of items that are not allowed to be packed together into the knapsack. We then study an optimization version of the problem: we wish to maximize the minimum sum among all packings in the reconfiguration. We show that this maximization problem admits a polynomial-time approximation scheme (PTAS), while the problem is APX-hard if we are given a conflict graph.

1 Introduction

Reconfiguration problems arise when we wish to find a step-by-step transformation between two feasible solutions of a problem such that all intermediate results are also feasible. Recently, Ito *et al.* [7] proposed a framework of reconfiguration problems, and gave complexity and approximability results for reconfiguration problems derived from several well-known problems, such as INDEPENDENT SET, CLIQUE, MATCHING, etc. In this paper, we study two reconfiguration problems derived from the SUBSET SUM problem.

The SUBSET SUM problem is a well-known NP-complete problem, defined as follows [9]. Suppose that we are given a knapsack with a nonnegative integer capacity c, and a set A of items $a_1, a_2, \ldots, a_n$, each of which has a nonnegative integer size $s(a_i)$, $1 \le i \le n$. We call a subset A' of A a *packing* if the total size of A' does not exceed the capacity c, that is, $\sum_{a \in A'} s(a) \le c$. Given an integer

---

* This work is partially supported by Grant-in-Aid for Scientific Research: 22700001.

M. Ogihara and J. Tarui (Eds.): TAMC 2011, LNCS 6648, pp. 58–69, 2011.

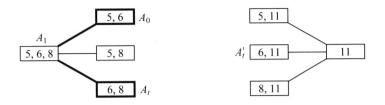

Fig. 1. All packings of total size at least $k = 10$ for $A = \{5, 6, 8, 11\}$ and $c = 20$

threshold k, the SUBSET SUM problem is to find a packing A' whose total size is at least k, that is, $k \leq \sum_{a \in A'} s(a) \leq c$. For a knapsack with capacity $c = 20$ and a set $A = \{5, 6, 8, 11\}$, there are 8 packings of total size at least $k = 10$, as illustrated in Fig.1, where each packing is surrounded by a box. Our definition of SUBSET SUM is known as the decision version of the MAXIMUM SUBSET SUM problem in which we wish to find a packing whose total size is maximum [9][1].

Suppose now that we are given *two* packings A_0 and A_t, both of total size at least k, and we are asked whether we can transform one into the other via packings by moving (namely, either adding or subtracting) a single item to/from the previous one without ever going through a packing of total size less than k. We call this decision problem the SUBSET SUM RECONFIGURATION problem. For two packings $A_0 = \{5, 6\}$ and $A_t = \{6, 8\}$ in Fig.1, the answer is "yes" since they can be transformed into each other via $A_1 = \{5, 6, 8\}$; in Fig.1, two packings (boxes) are joined by a line if and only if one packing can be obtained from the other by moving a single item.

Obviously, we cannot always find such a transformation. For example, there is no transformation between $A_0 = \{5, 6\}$ and $A'_t = \{6, 11\}$ in Fig.1 if we are allowed to use only packings of total size at least $k = 10$. On the other hand, the answer is always "yes" if $k = 0$: we first remove all items of A_0, and obtain the empty packing; and then, add all items of A_t to the knapsack. In turn, we can get a natural optimization problem if we wish to maximize the minimum total size among all packings in a transformation between A_0 and A_t. We call this maximization problem the MAXMIN SUBSET SUM RECONFIGURATION problem. The sequence of packings emphasized by thick lines in Fig.2 is an optimal solution for $A_0 = \{5, 6\}$ and $A'_t = \{6, 11\}$; its objective value is 8.

Reconfiguration problems have been studied extensively in recent literature [2,5,6,7,8], but reconfiguration problems for SUBSET SUM have not been studied yet. One can easily imagine a variety of practical scenarios, where a packing (e.g., representing a feasible display of electronic advertisements on a Web browser) needs to be changed (to show other advertisements) by individual changes (appealing to the user by showing one by one) while maintaining both threshold and capacity of the allowed area on the Web browser (in order to maintain both advertiser and user satisfactions during the transformation). Reconfiguration problems are also interesting in general because they provide a new perspective

---

[1] Note that SUBSET SUM in [4] is slightly different from our definition: SUBSET SUM in [4] is defined as the problem of finding a packing whose total size is exactly k.

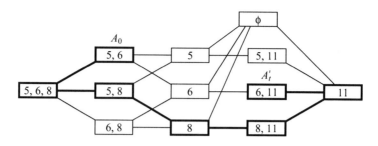

Fig. 2. All packings for $A = \{5, 6, 8, 11\}$ and $c = 20$

and deeper understanding of the solution space and of heuristics that navigate that space.

For the (ordinary) SUBSET SUM and MAXIMUM SUBSET SUM problems, several variants have been studied [9]. In particular, MAXIMUM SUBSET SUM with "conflict graph" [10] is an important variant, because this variant has been studied for several other problems, such as BIN PACKING [3] and SCHEDULING under makespan minimization [1]. In the variant, we are given a *conflict graph* for a set A of items in which each vertex corresponds to an item in A, and each edge represents a pair of items in A that are not allowed to be packed together into the knapsack. It is known that the (ordinary) MAXIMUM SUBSET SUM problem with conflict graph is strongly NP-hard [10].

In this paper, we first show that SUBSET SUM RECONFIGURATION is strongly NP-hard, and is PSPACE-complete for the variant of conflict graph. We then show that MAXMIN SUBSET SUM RECONFIGURATION with conflict graph is APX-hard, and hence there is no polynomial-time approximation scheme (PTAS) for this variant unless P = NP. In contrast, we give a PTAS for the original version of MAXMIN SUBSET SUM RECONFIGURATION. Note that, since this maximization problem is strongly NP-hard, the problem does not admit a fully polynomial-time approximation scheme (FPTAS) unless P = NP; in this sense, a PTAS is the best approximation algorithm we can expect for the problem [11, p. 72].

Our main result of this paper is a PTAS for MAXMIN SUBSET SUM RECONFIGURATION. The strategy of our PTAS is the following: we divide a set A of items into two groups, one is the set of items having "large" sizes, and the other consists of items having "small" sizes; and we deal with the two groups separately. Because such an approximation technique is fairly standard, especially for MAXIMUM SUBSET SUM and BIN PACKING [9,11], one might think that our PTAS could be obtained also straightforwardly by extending several known FPTAS or PTAS [9,11]. However, this is not the case, because the focus of reconfiguration problems is different from the ordinary problems: we seek the *reachability* between two feasible solutions, and hence the placement of items is the central matter. For example, two packings $\{5, 6\}$ and $\{11\}$ in Fig.1 have the same total size 11, and hence we can regard them as an "equivalent" packing in the ordinary SUBSET SUM problem. However, we can*not* identify these two packings in the reconfiguration problems; for example, $\{11\}$ can be transformed into $\{6, 11\}$, but $\{5,6\}$ cannot, when $k = 10$. (See Fig.1.) We thus introduce a "configuration graph" which represents the placements of items and their connectivity. (A formal definition will be given in Section 3, but an example is already shown in Fig.2.) Our main idea is to approximate the configuration graph appropriately.

2   Complexity and Inapproximability

Before showing our results, we introduce some terms and define the problems more formally. In the Introduction, we have defined a packing A_i as a subset of items in a set A such that the total size of A_i is at most the capacity c of a knapsack; the total size of a packing A_i is denoted by $s(A_i)$, that is, $s(A_i) = \sum_{a \in A_i} s(a)$. Note that a packing does not necessarily satisfy a threshold k. We say that two packings A_i and A_j of A are *adjacent* if their symmetric difference is of cardinality 1, that is, $|A_i \bigtriangleup A_j| = |(A_i \setminus A_j) \cup (A_j \setminus A_i)| = 1$; the item a in $A_i \bigtriangleup A_j$ is said to be *moved* between A_i and A_j. A *reconfiguration sequence* between two packings A_0 and A_t is a sequence of packings $A_0, A_1, \ldots, A_t$ such that A_{i-1} and A_i are adjacent for $i = 1, 2, \ldots, t$. For a reconfiguration sequence $\mathcal{P}$, we denote by $f(\mathcal{P})$ the minimum total size among all packings in $\mathcal{P}$, that is, $f(\mathcal{P}) = \min\{s(A_i) : A_i \in \mathcal{P}\}$. Then, for two packings A_0 and A_t, let

$$\mathrm{OPT}(A_0, A_t) = \max\{f(\mathcal{P}) \mid \mathcal{P} \text{ is a reconfiguration sequence between } A_0 \text{ and } A_t\}.$$

Given an integer threshold k and two packings A_0 and A_t with $s(A_0) \geq k$ and $s(A_t) \geq k$, the SUBSET SUM RECONFIGURATION problem is a decision problem to determine whether $\mathrm{OPT}(A_0, A_t) \geq k$. On the other hand, its optimization version is defined as follows: Given two packings A_0 and A_t, the MAXMIN SUBSET SUM RECONFIGURATION problem is to compute $\mathrm{OPT}(A_0, A_t)$. Note that we are asked simply to compute the optimal value $\mathrm{OPT}(A_0, A_t)$, and we need *not* to find an actual reconfiguration sequence.

We first have the following theorem, whose proof is omitted from this extended abstract.

Theorem 1. *Both* SUBSET SUM RECONFIGURATION *and* MAXMIN SUBSET SUM RECONFIGURATION *are strongly NP-hard.*

We then consider the variant with conflict graph. Notice that every feasible packing of A induces an independent set of the conflict graph. Therefore, we have the following theorem.

Theorem 2. SUBSET SUM RECONFIGURATION *with conflict graph is PSPACE-complete.*

Proof. It is easy to see that the problem is in PSPACE. Therefore, we show that SUBSET SUM RECONFIGURATION with conflict graph is PSPACE-hard by giving a polynomial-time reduction from the INDEPENDENT SET RECONFIGURATION problem [7].

Given a graph G of n nodes, an integer threshold k', and two independent sets I_0 and I_t of G, both of cardinality at least k', the INDEPENDENT SET RECONFIG-URATION problem asks whether we can transform I_0 into I_t via independent sets of G, each of which results from the previous one by either adding or subtracting a single node of G, without ever going through an independent set of cardinality less than $k' - 1$. This problem is known to be PSPACE-complete [7].

We now construct the corresponding instance of SUBSET SUM RECONFIGU-RATION with conflict graph. The set A contains n items, and let $s(a) = 1$ for all items a in A. Each item in A corresponds to a node of G, and the conflict graph for A is connected as G. The knapsack is of capacity $c = n$, and let the threshold $k = k' - 1$. Finally, the two packings A_0 and A_t consist of the items which correspond to the nodes in I_0 and I_t, respectively; and hence both A_0 and A_t are of total size at least $k' = k + 1$.

Since every feasible packing of total size at least k induces an independent set in G of cardinality at least $k = k' - 1$, it is obvious that there is a desired transformation between I_0 and I_t if and only if $\mathrm{OPT}(A_0, A_t) \geq k$. □

We finally have the following inapproximability result, whose proof is omitted due to the page limitation.

Theorem 3. MAXMIN SUBSET SUM RECONFIGURATION *with conflict graph is APX-hard, and cannot be approximated within any constant factor unless* $\mathrm{P} = \mathrm{NP}$.

3   PTAS

Since MAXMIN SUBSET SUM RECONFIGURATION with conflict graph is APX-hard, this variant does not admit a PTAS unless $\mathrm{P} = \mathrm{NP}$. However, in this section, we give a PTAS for the original version. Remember that, since we have shown in Theorem 1 that the problem is strongly NP-hard, there is no FPTAS for the problem unless $\mathrm{P} = \mathrm{NP}$; in this sense, a PTAS is the best approximation algorithm we can expect for the problem. We have the following theorem.

Theorem 4. *There is a polynomial-time approximation scheme for* MAXMIN SUBSET SUM RECONFIGURATION.

In the remainder of this section, as a proof of Theorem 4, we give an algorithm which actually finds a reconfiguration sequence $\mathcal{P}$ between two given packings A_0 and A_t such that $f(\mathcal{P}) \geq (1 - \varepsilon')\mathrm{OPT}(A_0, A_t)$ in time polynomial in n (but, exponential in $1/\varepsilon'$) for any fixed constant ε', $0 < \varepsilon' < 1$, where n is the number of items in the set A. Therefore, our approximate objective value $\mathrm{APPRO}(A_0, A_t)$ is $f(\mathcal{P})$, and hence the error is bounded by $\varepsilon'\mathrm{OPT}(A_0, A_t)$, that is,

$$\mathrm{OPT}(A_0, A_t) - \mathrm{APPRO}(A_0, A_t) = \mathrm{OPT}(A_0, A_t) - f(\mathcal{P}) \leq \varepsilon'\mathrm{OPT}(A_0, A_t).$$

As we have mentioned in the Introduction, the placement of items is the central matter in the reconfiguration problem. Therefore, we construct an edge-weighted graph, called a configuration graph, which represents all (feasible) packings to-gether with their adjacency. For a set A of items and a knapsack of capacity c, a *configuration graph* $C = (V, \mathcal{E})$ is defined as follows: each vertex in V corresponds to a packing A_i, and two vertices are joined by an edge e in $\mathcal{E}$ if and only if the corresponding two packings A_i and A_j are adjacent; the *weight* $w(e)$ of e is defined as follows: $w(e) = \min\{s(A_i), s(A_j)\}$. Notice that the weight $w(e)$ of an edge e corresponds to the objective value $f(\mathcal{P}_{i,j})$ for the reconfiguration sequence $\mathcal{P}_{i,j} = \{A_i, A_j\}$ along e. Figure 2 illustrates the configuration graph for a set $A = \{5, 6, 8, 11\}$ and a knapsack of capacity $c = 20$, where each vertex is drawn as a box and each edge as a line. From now on, we may call a packing simply a vertex of a configuration graph if it is clear from the context. Since there is a vertex corresponding to the empty packing, a configuration graph is always connected. Then, MAXMIN SUBSET SUM RECONFIGURATION can be seen as the problem of maximizing the minimum edge-weight in a path between A_0 and A_t in C. It is easy to see that the problem can be solved in time polynomial in $|V| + |\mathcal{E}|$ by the following naive algorithm: delete all edges having the smallest weight from C, and check whether the two vertices A_0 and A_t remain in the same connected component of the resulting graph; if so, let C be the resulting graph and repeat. Note that, however, the size $|V| + |\mathcal{E}|$ of C can be an exponential in n.

We now briefly explain our PTAS together with the organization of this section. For a fixed constant ε', $0 < \varepsilon' < 1$, let

$$\varepsilon = \frac{1}{2}\varepsilon'. \tag{1}$$

(The reason why the coefficient above is $1/2$ will be explained in Section 3.4.) Given a set A of items and a fixed constant ε, $0 < \varepsilon < 1/2$, we divide the items of A into two groups: an item a is called a *large item* if $s(a) \geq \varepsilon c/2$; otherwise the item is called a *small item*. We show in Section 3.1 that the problem can be optimally solved in polynomial time if A contains only large items; in this case, the number of packings (and hence the number of vertices in the configuration graph) can be bounded by a polynomial in n. In Section 3.2 we then explain that small items can be moved greedily with only small error. In Section 3.3 we finally deal with a general instance by combining the techniques above, without losing the reachability and with keeping the small error. Section 3.4 gives the analysis of our algorithm.

3.1   Large Items

In this subsection, we show that MAXMIN SUBSET SUM RECONFIGURATION can be optimally solved in polynomial time if the given set A contains only large items. It suffices to show that we can construct the corresponding configuration graph $C = (V, \mathcal{E})$ in polynomial time for such an instance, and that the size $|V| + |\mathcal{E}|$ of C is a polynomial in n. Formally, we have the following lemma.

Lemma 1. *For a fixed constant $\varepsilon > 0$, suppose that every item in the set A is of size at least $\varepsilon c/2$, where c is the capacity of the knapsack. Then, MAXMIN SUBSET SUM RECONFIGURATION can be optimally solved in polynomial time.*

Proof. Since $s(a) \geq \varepsilon c/2$ for each item $a \in A$, the number of items in any (feasible) packing is bounded by $\lfloor 2/\varepsilon \rfloor$. Let $\gamma = \lfloor 2/\varepsilon \rfloor$, then γ is a fixed constant. We denote by N the number of vertices (packings) in the corresponding configuration graph $\mathcal{C} = (\mathcal{V}, \mathcal{E})$, that is, $N = |\mathcal{V}|$. Since A contains n items and each packing consists of at most γ items, it is easy to see that N can be bounded by $\binom{n+\gamma}{\gamma}$. Therefore, N is a polynomial in n, and hence we can construct $\mathcal{C}$ in time polynomial in n. Since the size $|\mathcal{V}| + |\mathcal{E}|$ of $\mathcal{C}$ is a polynomial in n, we can solve the problem optimally in polynomial time. $\qquad\square$

3.2   Small Items

Suppose in this subsection that the given set A may contain small items. Then, the number of items in a packing cannot be bounded by a constant, and hence the number $N = |\mathcal{V}|$ of vertices in the configuration graph $\mathcal{C} = (\mathcal{V}, \mathcal{E})$ cannot be always bounded by a polynomial in n; more specifically, N can be $O(2^n)$. Therefore, we will later (in Section 3.3) construct an "approximate configuration graph $\mathcal{C}_A$," whose size is bounded by a polynomial in n.

We now explain how to find a reconfiguration sequence greedily when $A_0 \bigtriangleup A_t$ contains only small items for two given packings A_0 and A_t. Let L_ε be the set of large items in A, that is, $L_\varepsilon = \{a \in A \mid s(a) \geq \varepsilon c/2\}$, and let $S_\varepsilon = A \setminus L_\varepsilon$. We have the following lemma.

Lemma 2. *Let A_0 and A_t be an arbitrary pair of packings such that $A_0 \bigtriangleup A_t \subseteq S_\varepsilon$. Then, there exists a reconfiguration sequence $\mathcal{P}_s$ between A_0 and A_t such that*

    (a) *no item in L_ε is moved in $\mathcal{P}_s$; and*
    (b) $f(\mathcal{P}_s) \geq (1 - \varepsilon) \min\{s(A_0), s(A_t)\}.$
Moreover, such a reconfiguration sequence $\mathcal{P}_s$ can be found in linear time.

Proof. We give an $O(n)$-time algorithm which finds a reconfiguration sequence $\mathcal{P}_s$ between A_0 and A_t satisfying (a) and (b), as follows.

Case (i): $s(A_0 \cup A_t) \leq c$.
In this case, we first add all items in $A_t \setminus A_0$ one by one, and obtain the packing $A_0 \cup A_t$; and then, delete all items in $A_0 \setminus A_t$ one by one, and obtain A_t. Note that $A_t \setminus A_0 \subseteq A_0 \bigtriangleup A_t \subseteq S_\varepsilon$ and $A_0 \setminus A_t \subseteq A_0 \bigtriangleup A_t \subseteq S_\varepsilon$, and hence no item in L_ε is moved in this reconfiguration sequence $\mathcal{P}_s$. We clearly have

$$f(\mathcal{P}_s) = \min\{s(A_0), s(A_t)\} > (1 - \varepsilon) \min\{s(A_0), s(A_t)\}.$$

Therefore, $\mathcal{P}_s$ satisfies both (a) and (b). Moreover, $\mathcal{P}_s$ can be found in linear time since we move each item in $A_0 \bigtriangleup A_t$ only once.

Case (ii): $s(A_0 \cup A_t) > c$.
In this case, we first add items in $A_t \setminus A_0$ one by one in arbitrary order as many as possible; let A_j be the current packing. Then, $s(A_j) > (1 - \frac{\varepsilon}{2})c$ because, otherwise, we can add more items to A_j since $s(a) < \varepsilon c/2$ for all items $a \in A_t \setminus A_0$.

We then delete items in $A_0 \setminus A_t$ one by one in arbitrary order until we obtain a packing A'_j such that

$$(1 - \varepsilon)c < s(A'_j) \le \left(1 - \frac{\varepsilon}{2}\right)c. \tag{2}$$

Since $s(a) < \varepsilon c/2$ for all items $a \in A_0 \setminus A_t$, we can always find such a packing A'_j. If $s(A'_j \cup A_t) \le c$, then go to Case (i) above; otherwise, repeat Case (ii). Note that, in this reconfiguration sequence $\mathcal{P}_s$, every addition is executed for an item in $A_t \setminus A_0$ $(\subseteq S_\varepsilon)$ and every deletion is done for an item in $A_0 \setminus A_t$ $(\subseteq S_\varepsilon)$. Thus, $\mathcal{P}_s$ satisfies (a). Furthermore, since each item in $A_0 \triangle A_t$ is moved exactly once, $\mathcal{P}_s$ can be found in linear time. We now show that (b) holds for $\mathcal{P}_s$. By Eq. (2) we have

$$f(\mathcal{P}_s) \ge \min\Big\{(1 - \varepsilon)c, \min\{s(A_0), s(A_t)\}\Big\}.$$

Since $c \ge \min\{s(A_0), s(A_t)\}$, we have $f(\mathcal{P}_s) \ge (1 - \varepsilon) \min\{s(A_0), s(A_t)\}$.   □

3.3   General Instance

We finally deal with a general instance, that is, a set A may contain small items and two packings A_0 and A_t do not necessarily satisfy $A_0 \triangle A_t \subseteq S_\varepsilon$. Our idea is to construct an *approximate configuration graph* $\mathcal{C}_A$, as follows.

Step 1: Configuration graph for L_ε
We first construct a configuration graph $\mathcal{C}_{L_\varepsilon} = (\mathcal{V}_{L_\varepsilon}, \mathcal{E}_{L_\varepsilon})$ for the large item set L_ε of A and the capacity c. Then, as in Lemma 1, $\mathcal{C}_{L_\varepsilon}$ can be constructed in time polynomial in n, and the size $|\mathcal{V}_{L_\varepsilon}| + |\mathcal{E}_{L_\varepsilon}|$ of $\mathcal{C}_{L_\varepsilon}$ can be bounded by a polynomial in n. Figure 3(a) illustrates the configuration graph for L_ε of A, where each box corresponds to a packing consisting of only large items. Note that $\mathcal{C}_{L_\varepsilon}$ contains the vertex corresponding to the empty packing, and hence $\mathcal{C}_{L_\varepsilon}$ is connected.

Step 2: Small items
We then expand $\mathcal{C}_{L_\varepsilon}$ into the approximate configuration graph $\mathcal{C}_A = (\mathcal{V}_A, \mathcal{E}_A)$, as illustrated in Fig.3(b). For each edge in $\mathcal{C}_{L_\varepsilon}$ joining two vertices A_i^L and A_j^L (that consist only of large items), we replace it with an edge e that joins two new vertices $A_{i,x}$ and $A_{j,y}$, called *gate vertices* or *gate packings*, defined as follows. Assume without loss of generality that $A_j^L = A_i^L \cup \{a\}$ for some large item a in L_ε, and hence A_j^L can be obtained by adding one large item a to A_i^L. To extend A_j^L to the gate packing $A_{j,y}$ containing small items, we find a packing $A_j^S \subseteq S_\varepsilon$ of small items for the remaining space $c - s(A_j^L)$ of the knapsack; we employ an FPTAS for the ordinary MAXIMUM SUBSET SUM problem [9] for the fixed constant ε. Then, let $A_{j,y} = A_j^L \cup A_j^S$ and let $A_{i,x} = A_{j,y} \setminus \{a\}$. Note that $A_{i,x} \triangle A_{j,y} = \{a\}$ and hence $A_{i,x}$ and $A_{j,y}$ are adjacent. We call the edge $e = (A_{i,x}, A_{j,y})$ an *external edge*, and the weight $\omega(e)$ is defined as follows: $\omega(e) = \min\{s(A_{i,x}), s(A_{j,y})\} = s(A_{i,x})$. In Fig.3(b), each gate packing

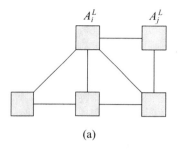
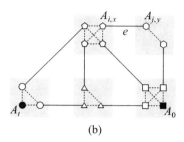

(a)   (b)

Fig. 3. (a) Configuration graph $\mathcal{C}_{L_\varepsilon}$ for the large item set L_ε of A, and (b) approximate configuration graph $\mathcal{C}_A$ for A

is represented by a circle, triangle, square, pentagon, or hexagon, colored with white; all gate packings represented by the same symbol have the same placement of large items; and each external edge is drawn as a (non-dotted) line.

For each vertex A_i^L in $\mathcal{C}_{L_\varepsilon}$, we have thus created the number $d(A_i^L)$ of new gate vertices $A_{i,1}, A_{i,2}, \ldots, A_{i,d(A_i^L)}$, where $d(A_i^L)$ is the degree of A_i^L in $\mathcal{C}_{L_\varepsilon}$. Clearly, $A_{i,x} \cap L_\varepsilon = A_{i,z} \cap L_\varepsilon$ for every pair of vertices. The original vertex A_i^L is deleted, and we connect the $d(A_i^L)$ gate vertices so that they form a clique; for each pair of vertices $A_{i,x}$ and $A_{i,z}$, the edge joining them is called an *internal edge*; in Fig.3(b), each internal edge is drawn as a dotted line. It should be noted that $A_{i,x}$ and $A_{i,z}$ are not necessarily adjacent although they are joined by an internal edge. However, using Lemma 2 we can regard such an internal edge as a reconfiguration sequence $\mathcal{P}_s$ between $A_{i,x}$ and $A_{i,z}$ such that $f(\mathcal{P}_s) \geq (1-\varepsilon)\min\{s(A_{i,x}), s(A_{i,z})\}$. Therefore, the weight $\omega(e)$ of e is defined as follows:

$$\omega(e) = \begin{cases} \min\{s(A_{i,x}), s(A_{i,z})\} & \text{if } A_{i,x} \text{ and } A_{i,z} \text{ are adjacent;} \\ (1-\varepsilon)\min\{s(A_{i,x}), s(A_{i,z})\} & \text{otherwise.} \end{cases} \quad (3)$$

Step 3: A_0 and A_t

The current graph above does not always contain the vertices corresponding to given packings A_0 and A_t. If the graph does not contain A_0, then we add a new vertex A_0 to the graph, and join it with each gate vertex having the same placement $A_0 \cap L_\varepsilon$ of large items by an internal edge. (The case for A_t is similar.) This completes the construction of the approximate configuration graph $\mathcal{C}_A = (\mathcal{V}_A, \mathcal{E}_A)$.

Clearly, a path between the two vertices A_0 and A_t in $\mathcal{C}_A$ corresponds to a reconfiguration sequence between the two packings A_0 and A_t. Since $|\mathcal{V}_A| \leq 2|\mathcal{E}_{L_\varepsilon}| + 2$ and $|\mathcal{E}_{L_\varepsilon}|$ is bounded by a polynomial in n, the size $|\mathcal{V}_A| + |\mathcal{E}_A|$ of $\mathcal{C}_A$ is bounded by a polynomial in n. Therefore, we can find in polynomial time a path between A_0 and A_t whose minimum edge-weight is maximum in $\mathcal{C}_A$; we choose the corresponding reconfiguration sequence $\mathcal{P}$ as our approximate solution.

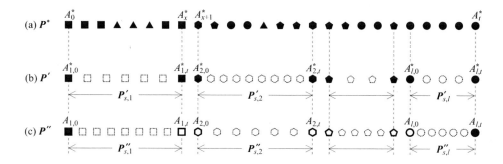

Fig. 4. Reconfiguration sequences $\mathcal{P}^*$, $\mathcal{P}'$ and $\mathcal{P}''$ between A_0 and A_t

3.4   Analysis of the Algorithm

We have shown in Section 3.3 that our algorithm finds a reconfiguration sequence $\mathcal{P}$ between A_0 and A_t in polynomial time. In this subsection, we show that $\mathcal{P}$ satisfies $f(\mathcal{P}) \geq (1 - \varepsilon')\text{OPT}(A_0, A_t)$ for a fixed constant ε', $0 < \varepsilon' < 1$, as required.

Let $\mathcal{P}^* = \{A_0^*, A_1^*, \ldots, A_t^*\}$ be an arbitrary optimal reconfiguration sequence between A_0 and A_t, where $A_0^* = A_0$ and $A_t^* = A_t$. Figure 4(a) illustrates the optimal reconfiguration sequence $\mathcal{P}^*$, where each black symbol corresponds to a packing A_i^* in $\mathcal{P}^*$, and all packings represented by the same symbol have the same placement of large items. Let $A_{\min}^*$ be a packing in $\mathcal{P}^*$ whose total size is minimum, and hence $f(\mathcal{P}^*) = s(A_{\min}^*)$. Then, we have

$$s(A_{\min}^*) = \text{OPT}(A_0, A_t), \tag{4}$$

and

$$s(A_i^*) \geq s(A_{\min}^*) \tag{5}$$

for each packing A_i^*, $0 \leq i \leq t$.

From now on, we transform $\mathcal{P}^*$ into another reconfiguration sequence $\mathcal{P}''$ between A_0 and A_t so that $\mathcal{C}_A$ contains the path corresponding to $\mathcal{P}''$. Remember that our algorithm finds a reconfiguration sequence $\mathcal{P}$ between A_0 and A_t which is optimal in $\mathcal{C}_A$, and hence we have

$$\text{APPRO}(A_0, A_t) = f(\mathcal{P}) \geq f(\mathcal{P}''). \tag{6}$$

We first transform $\mathcal{P}^*$ into a reconfiguration sequence $\mathcal{P}'$ between A_0 and A_t such that the same placement of large items appears consecutively. This can be done by the following algorithm: find the last packing A_x^* in $\mathcal{P}^*$ such that $A_x^* \cap L_\varepsilon = A_0^* \cap L_\varepsilon$; replace the subsequence $\{A_0^*, A_1^*, \ldots, A_x^*\}$ with a reconfiguration sequence $\mathcal{P}_s$ between A_0^* and A_x^* obtained by Lemma 2; set $A_0^* = A_{x+1}^*$ and repeat. We denote by $\mathcal{P}'_{s,i}$ the reconfiguration (sub)sequence obtained by the ith step of the algorithm above. (See Fig.4(b), where the intermediate packings in $\mathcal{P}'_{s,i}$ are represented by dotted symbols.) By Lemma 2(a) all packings in $\mathcal{P}'_{s,i}$ have the same placement of large items, and hence all intermediate packings in $\mathcal{P}'_{s,i}$

are depicted by the same symbol in Fig.4(b). Moreover, from the construction, every two reconfiguration subsequences $\mathcal{P}'_{s,i}$ and $\mathcal{P}'_{s,j}$ with $i \neq j$ have different placements of large items. Let ℓ be the number of iterations of the algorithm, and hence $\mathcal{P}' = \mathcal{P}'_{s,1} \cup \mathcal{P}'_{s,2} \cup \cdots \cup \mathcal{P}'_{s,\ell}$. For each i, $1 \leq i \leq \ell$, let $A^*_{i,0}$ and $A^*_{i,t}$ be the first and last packings in $\mathcal{P}'_{s,i}$, respectively. Then, from the construction, the optimal reconfiguration sequence $\mathcal{P}^*$ contains the packings $A^*_{i,0}$ and $A^*_{i,t}$, $1 \leq i \leq \ell$, and hence they are depicted by black symbols in Fig.4(b). On the other hand, intermediate packings are not necessarily contained in $\mathcal{P}^*$, and hence they are depicted by white dotted symbols. Note that $A^*_{1,0} = A_0$ and $A^*_{\ell,t} = A_t$.

We then transform $\mathcal{P}'$ into the reconfiguration sequence $\mathcal{P}''$ between A_0 and A_t such that $\mathcal{C}_A$ contains the path corresponding to $\mathcal{P}''$. For each i, $1 \leq i \leq \ell-1$, two packings $A^*_{i,t}$ and $A^*_{i+1,0}$ are adjacent; moreover, the item moved between them is a large item $a \in L_\varepsilon$. Therefore, $\mathcal{C}_A$ contains the external edge $e = (A_{i,t}, A_{i+1,0})$ which corresponds to moving the item a from the large item placement $A^*_{i,t} \cap L_\varepsilon$ to $A^*_{i+1,0} \cap L_\varepsilon$. We may regard that the two endpoints (gate packings) $A_{i,t}$ and $A_{i+1,0}$ of e correspond to $A^*_{i,t}$ and $A^*_{i+1,0}$, respectively. Of course, the gate packings $A_{i,t}$ and $A_{i+1,0}$ are not always the same as $A^*_{i,t}$ and $A^*_{i+1,0}$, respectively, and hence they are depicted by (non-dotted) white symbols in Fig.4(c). However, it should be noted that $A^*_{i,t} \cap L_\varepsilon = A_{i,t} \cap L_\varepsilon$ and $A^*_{i+1,0} \cap L_\varepsilon = A_{i+1,0} \cap L_\varepsilon$, and hence $A_{i,t}$ and $A_{i+1,0}$ in Fig.4(c) are depicted by the same symbols as $A^*_{i,t}$ and $A^*_{i+1,0}$, respectively. For the sake of notational convenience, let $A_{1,0} = A^*_{1,0} = A_0$ and $A_{\ell,t} = A^*_{\ell,t} = A_t$. Since $A_{i,0}$ and $A_{i,t}$, $1 \leq i \leq \ell$, have the same large item placement and are contained in $\mathcal{C}_A$, there exists the internal edge $e_{s,i}$ joining them in $\mathcal{C}_A$; let $\mathcal{P}''_{s,i}$ be the reconfiguration subsequence between $A_{i,0}$ and $A_{i,t}$ corresponding to $e_{s,i}$. By Eq. (3) we have

$$f(\mathcal{P}''_{s,i}) = \omega(e_{s,i}) \geq (1 - \varepsilon) \min\{s(A_{i,0}), s(A_{i,t})\} \tag{7}$$

for each i, $1 \leq i \leq \ell$. Let $\mathcal{P}'' = \mathcal{P}''_{s,1} \cup \mathcal{P}''_{s,2} \cup \cdots \cup \mathcal{P}''_{s,\ell}$, then $f(\mathcal{P}'') = \min\{f(\mathcal{P}''_{s,i}) : 1 \leq i \leq \ell\}$. This completes the construction of $\mathcal{P}''$.

We now show the following lemma, whose proof is omitted due to the page limitation.

Lemma 3. $s(A_{i,0}) > (1 - \varepsilon)s(A^*_{i,0})$ and $s(A_{i,t}) > (1 - \varepsilon)s(A^*_{i,t})$ for each i, $1 \leq i \leq \ell$.

Assume that $\mathcal{P}''_{s,k}$ contains the packing whose total size is minimum in $\mathcal{P}''$. Then, by Eq. (7) we have $f(\mathcal{P}'') = f(\mathcal{P}''_{s,k}) \geq (1 - \varepsilon) \min\{s(A_{k,0}), s(A_{k,t})\}$. Therefore, by Lemma 3 and Eqs. (4) and (5) we have

$$\begin{aligned} f(\mathcal{P}'') &> (1 - \varepsilon)^2 \min\{s(A^*_{k,0}), s(A^*_{k,t})\} \geq (1 - \varepsilon)^2 s(A^*_{\min}) \\ &> (1 - 2\varepsilon)s(A^*_{\min}) = (1 - 2\varepsilon)\mathrm{OPT}(A_0, A_t). \end{aligned} \tag{8}$$

By Eqs. (1), (6) and (8) we have

$$\mathrm{APPRO}(A_0, A_t) \geq f(\mathcal{P}'') > (1 - 2\varepsilon)\mathrm{OPT}(A_0, A_t) = (1 - \varepsilon')\mathrm{OPT}(A_0, A_t).$$

This completes the proof of Theorem 4. □

4   Concluding Remark

In the ordinary KNAPSACK problem [4,9], each item is assigned not only a size but also a *profit*, and we wish to find a packing whose total profit is at least a given threshold. Consider the two reconfiguration problems for KNAPSACK, called KNAPSACK RECONFIGURATION and MAXMIN KNAPSACK RECONFIGURATION, which are defined similarly as SUBSET SUM RECONFIGURATION and MAXMIN SUBSET SUM RECONFIGURATION, respectively. Because they are generalizations of our reconfiguration problems for SUBSET SUM, the complexity and inapproximability results in Section 2 hold also for them. However, it remains open to obtain a PTAS for MAXMIN KNAPSACK RECONFIGURATION.

References

1. Bodlaender, H.L., Jansen, K.: On the complexity of scheduling incompatible jobs with unit-times. In: Borzyszkowski, A.M., Sokolowski, S. (eds.) MFCS 1993. LNCS, vol. 711, pp. 291–300. Springer, Heidelberg (1993)
2. Bonsma, P., Cereceda, L.: Finding paths between graph colourings: PSPACE-completeness and superpolynomial distances. Theoretical Computer Science 410, 5215–5226 (2009)
3. Epstein, L., Levin, A.: On bin packing with conflicts. In: Erlebach, T., Kaklamanis, C. (eds.) WAOA 2006. LNCS, vol. 4368, pp. 160–173. Springer, Heidelberg (2007)
4. Garey, M.R., Johnson, D.S.: Computers and Intractability: A Guide to the Theory of NP-Completeness. Freeman, San Francisco (1979)
5. Gopalan, P., Kolaitis, P.G., Maneva, E.N., Papadimitriou, C.H.: The connectivity of Boolean satisfiability: computational and structural dichotomies. SIAM J. Computing 38, 2330–2355 (2009)
6. Hearn, R.A., Demaine, E.D.: PSPACE-completeness of sliding-block puzzles and other problems through the nondeterministic constraint logic model of computation. Theoretical Computer Science 343, 72–96 (2005)
7. Ito, T., Demaine, E.D., Harvey, N.J.A., Papadimitriou, C.H., Sideri, M., Uehara, R., Uno, Y.: On the complexity of reconfiguration problems. Theoretical Computer Science 412, 1054–1065 (2011)
8. Ito, T., Kamiński, M., Demaine, E.D.: Reconfiguration of list edge-colorings in a graph. In: Dehne, F., Gavrilova, M., Sack, J.-R., Tóth, C.D. (eds.) WADS 2009. LNCS, vol. 5664, pp. 375–386. Springer, Heidelberg (2009)
9. Kellerer, H., Pferschy, U., Pisinger, D.: Knapsack Problems. Springer, Heidelberg (2004)
10. Pferschy, U., Schauer, J.: The knapsack problem with conflict graphs. J. of Graph Algorithms and Applications 13, 233–249 (2009)
11. Vazirani, V.V.: Approximation Algorithms. Springer, Berlin (2001)

An Improved Kernel
for Planar Connected Dominating Set[*]

Weizhong Luo[1,2], Jianxin Wang[1], Qilong Feng[1],
Jiong Guo[3], and Jianer Chen[1,4]

[1] School of Information Science and Engineering,
Central South University, Changsha 410083, P.R. China
jxwang@mail.csu.edu.cn
[2] Hunan Financial & Economic University,
Changsha 410205, P.R. China
[3] Universität des Saarlandes,
Campus E 1.4, D-66123 Saarbrücken, Germany
jguo@mmci.uni-saarland.de
[4] Department of Computer Science and Engineering,
Texas A&M University, College Station, Texas 77843-3112, USA
chen@cse.tamu.edu

Abstract. In this paper, we study the Planar Connected Dominating
Set problem, which, given a planar graph $G = (V, E)$ and a non-negative
integer k, asks for a subset $D \subseteq V$ with $|D| \leq k$ such that D forms a
dominating set of G and induces a connected graph. Answering an open
question by S. Saurabh [The 2nd Workshop on Kernelization (WorKer
2010)], we provide a kernelization algorithm for this problem leading to a
problem kernel with $130k$ vertices, significantly improving the previously
best upper bound on the kernel size. To this end, we incorporate a vertex
coloring technique with data reduction rules and introduce for the first
time a distinction of two types of regions into the region decomposition
framework, which allows a refined analysis of the region size and could
be used to reduce the kernel sizes achieved by the region decomposition
technique for a large range of problems.

1 Introduction

The *Dominating Set* problem decides for a given graph $G = (V, E)$ and a non-
negative integer k whether G has a dominating set of at most k vertices. The
problem is NP-complete [6] and has many applications in the fields such as ad-
hoc networks and voting systems. Moreover, from the viewpoint of parameterized
complexity theory, the problem is W[2]-complete, parameterized by k [5]. Therefore, it is unlikely that the problem is fixed-parameter tractable with respect to k.

[*] This work is supported by the National Natural Science Foundation of China under
Grant (61073036, 61070224), the Doctoral Discipline Foundation of Higher Educa-
tion Institution of China under Grant (20090162110056), and the DFG Excellence
Cluster "Multimodal Computing and Interaction (MMCI)".

M. Ogihara and J. Tarui (Eds.): TAMC 2011, LNCS 6648, pp. 70–81, 2011.
© Springer-Verlag Berlin Heidelberg 2011

Equivalently, it is unlikely that the problem admits a kernelization algorithm. A *kernelization algorithm* for a parameterized problem Q runs in polynomial time and, for a given instance (G, k) of Q, produces another instance (G', k') of Q such that (G, k) is a yes-instance if and only if (G', k') is a yes-instance, $k' \leq k$, and the size (i.e., the number of vertices) of G' is bounded by a function of k. The graph G' is called the *problem kernel*. If the size of G' is linearly depending on k, we call G' a linear kernel and the corresponding algorithm is called a linear kernelization algorithm. See [2,9] for more background on kernelization algorithms.

Recently, much attention has been focused on linear kernelization algorithms for problems on planar graphs. In particular, the Planar Dominating Set (PDS) problem has been extensively studied. Alber et al. [1] introduced the concept of *region decomposition* and gave a linear kernel with $335k$ vertices for PDS, which was later improved to $67k$ vertices by Chen et al. [4]. Motivated by [1], Guo and Niedermeier [10] proposed a general framework to obtain linear kernels for planar graph problems. Finally, Bodlaender et al. [3] formally showed the existence of linear kernels for problems satisfying specific properties on bounded-genus graphs. Note that the results from [3] serve mainly as a classification tool and are infeasible for practical applications. Huge constants hide behind the thereby achieved kernel sizes.

Here we study a well-known variant of the Dominating Set problem, namely, the Connected Dominating Set problem, which requires further that the subgraph induced by the dominating set should be connected. Connected Dominating Set is also NP-complete and has applications in various network design settings [6]. The parameterized version of the Connected Dominating Set problem on general graphs is also W[2]-complete [5], parameterized by the solution size k. However, if we restrict the input to planar graphs, we arrive at the main problem of this paper, the Planar Connected Dominating Set (PCDS) problem, where, given a planar graph $G = (V, E)$ and a non-negative integer k, to decide whether there exists a subset $D \subseteq V$ with at most k vertices such that the graph induced by D is connected and every vertex in V is either in D or adjacent to at least one vertex in D.

Kernelization algorithms for PCDS have been studied recently. Lokshtanov et al. [11] showed that the problem has a linear kernel with $3968187 \cdot k$ vertices, based on the method of "reduce or refine". Gu and Imani [7] proposed an improved kernel with $413k$ vertices for the problem. However, a careful examination shows that Gu-Imani's construction in [7] misses some subtle cases of the reduction rules and their revised version of the paper can be found in [8]. Recently, at the 2nd Workshop on Kernelization, Saket Saurabh [12] posed the open question whether there exists a kernel for PCDS with size at most ck for a constant $c \approx 100$. Here, we answer this question by showing a kernel with $130k$ vertices for PCDS, significantly improving the previous results. To this end, we not only extend the reduction rules introduced in [7], but also present new rules that color the vertices of a graph, which enable to reduce some "useless" edges. Furthermore, while analyzing the kernel size based on the region decomposition framework, we distinguish two types of regions which allows to make use of some structural characteristics of the regions, resulting in the improved kernel. We believe that the distinction of region types could be a promising extension to the region decomposition framework and might lead to improved kernels for many planar graph problems.

Due to lack of space, some proofs are deferred to the full paper.

2   Preliminaries

Basic graph notations. Let $G = (V, E)$ be an undirected, simple, connected graph, where $n := |V|$. For two vertices $u, v \in V$, the edge between them is denoted by (u, v). The length of a path is the number of edges in the path. For two vertices $u, v \in V$, let $d(u, v)$ denote the distance between them, that is, the length of the shortest path between u and v. For a vertex w and an edge $e = (u, v)$, let $d(e, w) := \min\{d(v, w), d(u, w)\}$ be the distance between e and w. For a subset $U \subseteq V$, the graph induced by U is denoted by $G[U]$. For two vertices $u, v \in V$, if $u = v$ or $(u, v) \in E$, then v dominates u. Given two sets $V_1, V_2 \subseteq V$, if every vertex of V_2 is dominated by some vertex of V_1, then V_1 dominates V_2. Specially, if $V_2 = \emptyset$, then V_1 still dominates V_2. For a vertex $v \in V$, let $N(v)$ be the set of neighbors of v, that is, $N(v) := \{u \mid (v, u) \in E\}$, and $N[v] := N(v) \cup \{v\}$. If a graph can be drawn in the plane without edge crossings then it is called a *planar* graph. A *plane* graph is a planar graph with a fixed embedding in the plane. Throughout this paper, we assume that we are working with an arbitrary but fixed embedding of G in the plane; whenever this embedding is of relevance, we refer to G as being plane instead of planar.

For a vertex $v \in V$, $N(v)$ is partitioned into the following three subsets:

- $N_1(v) := \{u \mid u \in N(v) \text{ and } N(u) \setminus N[v] \neq \emptyset\}$,
- $N_2(v) := \{u \mid u \in N(v) \setminus N_1(v) \text{ and } N(u) \cap N_1(v) \neq \emptyset\}$, and
- $N_3(v) = N(v) \setminus (N_1(v) \cup N_2(v))$.

For two vertices $v, w \in V$, let $N[v, w] := N[v] \cup N[w]$ and $N(v, w) := N[v, w] \setminus \{v, w\}$. We partition $N(v, w)$ into the following three subsets:

- $N_1(v, w) := \{u \mid u \in N(v, w) \text{ and } N(u) \setminus N[v, w] \neq \emptyset\}$,
- $N_2(v, w) := \{u \mid u \in N(v, w) \setminus N_1(v, w) \text{ and } N(u) \cap N_1(v, w) \neq \emptyset\}$, and
- $N_3(v, w) := N(v, w) \setminus (N_1(v, w) \cup N_2(v, w))$.

Region and region decomposition. In the following, we introduce the region decomposition framework.

Definition 1. *[1] For a plane graph $G = (V, E)$ and two vertices $v, w \in V$, a region $R(v, w)$ between v and w is a closed subset of the plane with the following properties:*

1. The boundary of $R(v, w)$ is formed by two simple paths P_1 and P_2 between v and w, and the length of each path is at most three.
2. All vertices strictly inside (that is, not on the boundary) the region $R(v, w)$ are from $N(v, w)$.

For a region $R = R(v, w)$, v, w are called the endpoints of R. Let $V[R]$ denote the vertices in R, that is, $V[R] = \{u \in V \mid u$ sits strictly inside or on the boundary of $R\}$, and let $V(R) := V[R] \setminus \{v, w\}$. A vertex in $V(R)$ is called *simple*, if it is adjacent to both v and w; otherwise it is called *non-simple*.

Definition 2. *[1] A region $R = R(v, w)$ between two vertices v and w is called simple, if all vertices in $V(R)$ are common neighbors of both v and w, that is, $V(R) \subseteq N(v) \cap N(w)$.*

Definition 3. *A region $R = R(v, w)$ between two vertices v and w is called quasi-simple, if for every non-simple vertex x in R, x satisfies the following conditions:*

1. x sits strictly inside R and
2. x is adjacent to v but not adjacent to w, and is also adjacent to at least one simple vertex in R.

Note that for a quasi-simple region $R = R(v, w)$, it always holds $V(R) \subseteq N(v)$ but $V(R) \setminus N(w)$ might be non-empty. In the following, the boundary of a region R will be denoted by ∂R.

Definition 4. *[1] Let $G = (V, E)$ be a plane graph and $D \subseteq V$. A D-region decomposition of G is a set $\Re$ of regions between pairs of vertices in D such that*

1. for $R(v, w) \in \Re$ no vertex from D is in $V(R(v, w))$ and
2. for two regions $R_1, R_2 \in \Re$, it holds $(R_1 \cap R_2) \subseteq (\partial R_1 \cup \partial R_2)$.

For a D-region decomposition $\Re$, we define $V[\Re] = \bigcup_{R \in \Re} V[R]$. A D-region decomposition $\Re$ is called *maximal* if there is no region R ($R \notin \Re$) such that $\Re' := \Re \cup \{R\}$ is a D-region decomposition with $V[\Re] \subsetneq V[\Re']$.

Definition 5. *[10] A graph problem P on $G = (V, E)$ is said to admit a distance property with constants c_V and c_E if*

1. P asks for a set of vertices or edges satisfying a specified property and
2. for every solution set D with the vertex set $V(D)$, it holds that $\forall u \in V : \exists v \in V(D) : d(u, v) \leq c_V$ and $\forall e \in E : \exists v \in V(D) : d(e, v) \leq c_E$.

3   Data Reduction Rules

We present 7 data reduction rules for Planar Connected Dominating Set. Note that the first 6 rules work also for general graphs. Some of these rules color the vertices in G with two colors, black and white. Assigning the color white to a vertex means that this vertex will be never included into the connected dominating set sought for. Initially, all the vertices are colored black. During the execution of the data reduction rules, if we find a vertex v such that there exists a minimum connected dominating set excluding v, then we color v white. We will show later that by incorporating this vertex coloring we can reduce some "useless" edges and vertices. Finally, the output of our kernelization process is the "uncolored" version of the reduced, colored graph, that is, the same graph without the vertex coloring. Note that we require that the minimum connected dominating set of the input graph containing at least two vertices.

The first rule is a natural consequence of the definition of the coloring.

Rule 1. For two white vertices u and v, if $N(u) \subseteq N(v)$, then remove v from G.

Then, we give a colored version of the first rule for Dominating Set introduced in [1], which deals with the neighborhood of a single vertex.

Rule 2. For a black vertex $v \in V$, if $N_3(v) \neq \emptyset$, then remove $N_2(v) \cup N_3(v)$ from G, and add a new white vertex v' to V and edge (v, v') to E.

The next rule deals with the colors of the neighbors of a single vertex.

Rule 3. For a black vertex $v \in V$, if there exists a black vertex $x \in N_2(v) \cup N_3(v)$, then color x white and remove the edges between x and other white vertices in G.

The next 3 rules consider the "joint neighborhood" $N(v, w)$ of two vertices v and w. Compared to the second rule for Dominating Set introduced in [1], more cases have to be considered.

Rule 4. For two black vertices v, w with $d(v, w) = 1$, if $N_3(v, w)$ is not dominated by a single vertex from $N_3(v, w) \cup N_2(v, w)$, then we distinguish the following 4 cases:

Case 1. $N_3(v, w)$ is dominated by v and also dominated by w: remove $N_3(v, w)$ and $N_2(v, w) \cap N(v) \cap N(w)$ from G; add a new white vertex z to V and edges (v, z) and (w, z) to E.

Case 2. $N_3(v, w)$ is dominated by v but not dominated by w: remove $N_3(v, w)$ and $N_2(v, w) \cap N(v)$ from G; add a new white vertex v' to V and edge (v, v') to E.

Case 3. $N_3(v, w)$ is dominated by w but not dominated by v: remove $N_3(v, w)$ and $N_2(v, w) \cap N(w)$ from G; add a new white vertex w' to V and edge (w, w') to E.

Case 4. $N_3(v, w)$ is not dominated by v and not dominated by w: remove $N_2(v, w) \cup N_3(v, w)$ from G; add two new white vertices v' and w' to V and edges (v, v') and (w, w') to E.

In order to present Rule 5, we need some further notations: For two vertices $u, v \in V$ in a colored graph $G = (V, E)$, let $d_{black}(u, v)$ be the length of the shortest path between u and v such that all vertices on this path with the only exception of u and v are black. We further partition $N_3(v, w)$ for two vertices v and w into the following 3 subsets:

$$Z := N_3(v, w) \cap N(w) \cap N(v),$$
$$X := (N_3(v, w) \cap N(v)) \setminus Z,$$
$$Y := (N_3(v, w) \cap N(w)) \setminus Z.$$

Rule 5. For two black vertices v, w with $d_{black}(v, w) = 2$, if $N_3(v, w)$ cannot be dominated by at most two vertices from $N_2(v, w) \cup N_3(v, w)$, then consider the following 4 cases. Let p be any black vertex adjacent to both v and w.

Case 1. $N_3(v, w)$ can be dominated by at most two vertices from $\{v\} \cup N_2(v, w) \cup N_3(v, w)$ but cannot be dominated by at most two vertices from $\{w\} \cup N_2(v, w) \cup N_3(v, w)$: Set $B := \{x \in Z \mid N(x) \subseteq \{v, w\} \cup Z \cup X\}$. If $X \cup B \setminus \{p\} \neq \emptyset$, then remove $X \cup B \setminus \{p\}$ and add a new white vertex v' to V and edge (v, v') to E.

Case 2. $N_3(v, w)$ cannot be dominated by at most two vertices from $\{v\} \cup N_2(v, w) \cup N_3(v, w)$ but can be dominated by at most two vertices from $\{w\} \cup N_2(v, w) \cup N_3(v, w)$: Set $B := \{x \in Z \mid N(x) \subseteq \{v, w\} \cup Z \cup Y\}$. If $Y \cup B \setminus \{p\} \neq \emptyset$, then remove $Y \cup B \setminus \{p\}$ and add a new white vertex w' to V and edge (w, w') to E.

Case 3. $N_3(v, w)$ cannot be dominated by at most two vertices from $\{v\} \cup N_2(v, w) \cup N_3(v, w)$ and cannot be dominated by at most two vertices from $\{w\} \cup N_2(v, w) \cup N_3(v, w)$: Set $B := \{x \in Z \mid N(x) \subseteq \{v, w\} \cup Z \cup Y \cup X\}$. If $|X \cup Y \cup B \setminus \{p\}| \geq 2$, then remove $X \cup Y \cup B \setminus \{p\}$ and add new white vertices v' and w' to V and edges (v, v') and (w, w') to E.

Case 4. $N_3(v, w)$ can be dominated by at most two vertices from $\{v\} \cup N_2(v, w) \cup N_3(v, w)$ and can be dominated by at most two vertices from $\{w\} \cup N_2(v, w) \cup N_3(v, w)$: Set $B := \{x \in Z \mid N(x) \subseteq \{v, w\} \cup Z\}$. If $B \setminus \{p\} \neq \emptyset$, then remove $B \setminus \{p\}$ and add a new white vertex z and edges (v, z) and (w, z) to E.

The next rule requires a more refined analysis of the structure of $N_3(v, w)$. Further subsets of X and Y are needed:

$$X_1 := \{x \mid x \in X, N(x) \cap N(w) = \emptyset\}, X_2 := X \setminus X_1,$$
$$Y_1 := \{y \mid y \in Y, N(y) \cap N(v) = \emptyset\}, \ Y_2 := Y \setminus Y_1.$$

Further, X_2 and Y_2 are again partitioned into two subsets, respectively:

$$X_2' := \{x \mid x \in X_2, N(x) \cap N_2(v, w) \cap N(w) = \emptyset\}, X_2'' := X_2 \setminus X_2',$$
$$Y_2' := \{y \mid y \in Y_2, N(y) \cap N_2(v, w) \cap N(v) = \emptyset\}, \ Y_2'' := Y_2 \setminus Y_2'.$$

Rule 6. For two black vertices v and w with $d(w, v) \leq 3$ and $d_{black}(w, v) \geq 3$, if $N_3(v, w)$ cannot be dominated by at most three vertices from $N_2(v, w) \cup N_3(v, w)$, then distinguish the following 4 cases.

Case 1. $N_3(v, w)$ can be dominated by at most three vertices from $\{v\} \cup N_2(v, w) \cup N_3(v, w)$ and can also be dominated by at most three vertices from $\{w\} \cup N_2(v, w) \cup N_3(v, w)$: if $Z \neq \emptyset$, then remove Z from G and add a new white vertex z to V and edges (z, v) and (z, w) to E.

Case 2. $N_3(v, w)$ cannot be dominated by at most three vertices from $\{v\} \cup N_2(v, w) \cup N_3(v, w)$ and cannot be dominated by at most three vertices from $\{w\} \cup N_2(v, w) \cup N_3(v, w)$: if $d_{black}(w, v) = 3$, then set $T := \{p, q\}$, where p and q are two black vertices making a path (v, p, q, w); otherwise, $T := \emptyset$. If $|N_3(v, w) \setminus T| \geq 2$, then remove $N_3(v, w) \setminus T$ from G and add two white vertices v' and w' to V and edges (v, v') and (w, w') to E.

Case 3. $N_3(v, w)$ can be dominated by at most three vertices from $\{v\} \cup N_2(v, w) \cup N_3(v, w)$ but cannot be dominated by at most three vertices from $\{w\} \cup N_2(v, w) \cup N_3(v, w)$: let $A_1 = \{x \mid x \in X_2', \nexists y \in N(x) \cap N(w): \{x, y\}$ dominates $Y\}$; if $A_1 \cup X_1 \cup Z \neq \emptyset$, then remove $A_1 \cup X_1 \cup Z$ from G and add a new white vertex v' to V and edge (v, v') to E.

Case 4. $N_3(v, w)$ cannot be dominated by at most three vertices from $\{v\} \cup N_2(v, w) \cup N_3(v, w)$ but can be dominated by at most three vertices from $\{w\} \cup N_2(v, w) \cup N_3(v, w)$: let $A_2 = \{y \mid y \in Y_2', \nexists x \in N(y) \cap N(v): \{y, x\}$ dominates $X\}$; if $A_2 \cup Y_1 \cup Z \neq \emptyset$, then remove $A_2 \cup Y_1 \cup Z$ from G and add a new white vertex w' to V and edge (w, w') to E.

Rule 7. For a quasi-simple region $R = R(v, w)$ between two vertices v and w in G where at least one of v and w is black, say v being black, let (v, y, w, z, v) be the boundary of R. We consider then the following cases:

Case 1. One of y and z is black: If there is a simple black vertex strictly inside R dominating all non-simple vertices in R, then let p denote this vertex and consider the following cases:

Case 1.1. If the non-simple vertices in R can be dominated by a single vertex from $\{y, z\}$, then, for each black vertex x strictly inside R, color x white and remove the edges between x and other white vertices in G.

Case 1.2. If the non-simple vertices in R cannot be dominated by a single vertex from $\{y, z\}$ but can be dominated by $\{y, z\}$, then, for each black vertex x strictly inside R with $x \neq p$, color x white and remove the edges between x and other white vertices in G.

Case 1.3. If the non-simple vertices in R cannot be dominated by $\{y, z\}$, then set $A := \{x \mid x$ is a simple black vertex strictly inside R with $x \neq p\}$. If $A \neq \emptyset$, then remove A from G and add a new white vertex z' to V and edges (z', v) and (z', w) to E.

Case 2. Both y and z are white: If there is a simple black vertex strictly inside R dominating all non-simple vertices in R, then let p be this vertex; otherwise, let p be an arbitrary simple black vertex strictly inside R. For each black vertex x strictly inside R with $x \neq p$, color x white and remove the edges between x and other white vertices in G.

To show the correctness of these rules, it suffices to prove the following two lemmas. Here, for a colored graph G, we call a connected dominating set of G "black", if all its vertices are black. Moreover, we use $\mu(G)$ to denote the size of minimum connected dominating sets of G and $\mu_b(G)$ to denote the size of minimum black connected dominating sets of G.

Lemma 1. *Let G be a planar graph with all vertices being colored white or black and let G' be the graph resulting by one application of one of Rules 1-7 to G. Then, 1. G' is planar and 2. $\mu_b(G) = \mu_b(G')$.*

A graph G is said to be *reduced*, if every vertex in G is colored white or black, and the application of any of Rules 1-7 does not change the color of any vertex in G, nor does it change the structure of G.

Lemma 2. *Let (G, k) be a given instance of Planar Connected Dominating Set and G' denote the reduced graph resulting by first coloring vertices in G with black and exhaustively applying Rules 1-7. Then, $\mu(G) = \mu_b(G') = \mu(G')$.*

The running time of the rules is clearly polynomial.

Theorem 1. *Given an instance (G, k) of the Planar Connected Dominating Set problem, we can construct in $O(n^5)$ time a reduced graph (G', k) with respect to Rules 1-7 such that G has a connected dominating set of size at most k if and only if G' has a connected dominating set that consists of at most k vertices.*

Note that our kernelization algorithm outputs the "uncolored" version of G' from Lemma 2. Due to Lemma 2, this output is equivalent to the input instance. However, while analyzing the kernel size, it is more convenient to consider the colored graph G' and the same bound also holds for the uncolored output.

4   Improved Kernel for Planar Connected Dominating Set

Let $G = (V, E)$ be a planar graph reduced with respect to Rules 1-7. Based on the method of region decomposition [1,10], we show that Planar Connected Dominating Set admits a kernel with at most $130k$ vertices. Firstly, we show that there exits a maximal D-region decomposition $\Re$ of G with at most $O(k)$ regions. Then, we upper-bound the number of vertices contained in the regions of $\Re$. Finally, we upper-bound the number of vertices which do not belong to any of the regions in $\Re$.

Since, the Connected Dominating Set problem admits the distance property with $c_V = 1$ and $c_E = 1$, Lemma 1 from [10] provides an upper bound on the number of regions in a D-region decomposition. However, for Connected Dominating Set, as we will show later, the distance between the two endpoints of a region has a great influence on the upper bound of the number of vertices contained in this region. Hence, we extend the result in [10] and classify the regions in a maximal D-region decomposition into two types, based on the distance between their endpoints: regions with distance one between their corresponding endpoints and regions with distance at least two between their endpoints. By using the connectivity property of the problem solutions, we can bound the number of the regions of the second type.

Lemma 3. *Let $G = (V, E)$ be a plane graph reduced with respect to Rules 1-7, and D be a connected dominating set for G of size at most k. There exists a maximal D-region decomposition $\Re$ of G containing at most $3k - 6$ regions. Furthermore, $\Re$ contains at most $2k - 5$ regions $R(v, w)$ for which it holds $d(v, w) \geq 2$.*

Proof. The size bound on $\Re$ follows easily from Lemma 1 in [10] and the distance constants $c_V = 1$ and $c_E = 1$ for Connected Dominating Set. Let L denote the set of regions $R(v,w) \in \Re$ that satisfy $d(v,w) \geq 2$. In order to prove $|L| \leq 2k - 5$, an auxiliary graph $G' = (D, E_1' \cup E_2')$ is constructed as follows: For any two vertices $v, w \in D$, if $(v,w) \in E$, then add (v,w) to E_1'. Further, if there is a region $R(v,w) \in \Re$ between $v, w \in D$ with $d(v,w) \geq 2$, then add (v,w) to E_2'. Note that $E_1' \cap E_2' = \emptyset$ by definition. Thus, if G' contains multiple edges between a vertex pair, then all these edges must be from E_2'. Moreover, G' must be a planar graph due to the definition of region decompositions. As shown in [1, Lemma 5 and Proposition 1], we have $|E_1' \cup E_2'| \leq 3|D| - 6$. Since $G[D]$ is connected, $|E_1'| \geq |D| - 1$. It follows that $|E_2'| \leq (3|D| - 6) - (|D| - 1) = 2|D| - 5 \leq 2k - 5$. The number of regions $R(v,w)$ in $\Re$ with $d(v,w) \geq 2$ is bounded by $2k - 5$. $\square$

We now investigate the size bound of quasi-simple regions.

Lemma 4. *Let G be a reduced plane graph. Let $R = R(v,w)$ be a quasi-simple region in G and (v, y, w, z, v) be the boundary of R. Assume v is black. Let $L = \{x \mid x$ is a simple vertex in R or a non-simple vertex in R adjacent to $y\}$. Then $|L| \leq 4$.*

Proof. Let L_1 denote the set of black simple vertices strictly inside R, L_2 denote the set of white simple vertices strictly inside R, and L_3 denote the set of non-simple vertices in R adjacent to y. Obviously $L = L_1 \cup L_2 \cup L_3 \cup \{y, z\}$. Since $L_3 \subseteq N_2(v) \cup N_3(v)$, by Rule 3, all vertices in L_3 are white. By Rule 1 and the planarity of G, $|L_3| \leq 1$. By Rule 7, $|L_1| \leq 1$. If $L_1 \neq \emptyset$ and $L_3 \neq \emptyset$, then $L_2 = \emptyset$, since, otherwise, Rule 1 or 7 would be applied, contradicting the fact that G is reduced. If $L_1 \neq \emptyset$ and $L_3 = \emptyset$, then $|L_2| \leq 1$, since, otherwise, Rule 1 or 7 would be applied. If $L_1 = \emptyset$ and $L_3 = \emptyset$, then $|L_2| \leq 2$, since, otherwise, Rule 1 would be applied. In summary, $|L_1 \cup L_2 \cup L_3| \leq 2$. It follows that $|L| \leq 4$. $\square$

Since a simple region is also quasi-simple, Lemma 4 holds also for simple regions. Let $R(v,w)$ be a region of $\Re$. Obviously $V(R) \subseteq N(v,w)$. The following lemma bounds $|N_3(v,w) \cap V(R)|$.

Lemma 5. *Let $G = (V, E)$ be a reduced plane graph and $\Re$ be a maximal D-region decomposition of G.*

- *For any two black vertices $v, w \in V$ with $d(v,w) = 1$, $|N_3(v,w)| \leq 9$.*
- *For every region $R(v,w) \in \Re$ between two black vertices $v, w \in V$ with $2 \leq d(v,w) \leq 3$, $|N_3(v,w) \cap V(R)| \leq 27$.*

Now we are ready to prove the maximal size of regions of a reduced graph. First we consider the regions with two endpoints v and w satisfying $d(v,w) > 1$.

Lemma 6. *Let G be a reduced plane graph with a connected dominating set D excluding all white vertices. Let $\Re$ be a maximal D-region decomposition of G. For a region $R = R(v,w) \in \Re$ with $2 \leq d(v,w) \leq 3$, we have $|V(R)| \leq 41$.*

Proof. Assume the boundary of R is determined by the two paths (v, v_1, w_1, w) and (v, v_2, w_2, w). We will analyze $|N_1(v, w) \cap V(R)|$, $|N_2(v, w) \cap V(R)|$, and $|N_3(v, w) \cap V(R)|$ separately to bound $|V(R)|$.

Due to the definition of regions, $N_1(v, w) \cap V(R) \subseteq \{v_1, v_2, w_1, w_2\}$. If $N_1(v, w) \cap V(R) = \emptyset$, then $N_2(v, w) \cap V(R) \subseteq \{v_1, v_2, w_1, w_2\}$; otherwise, every vertex in $N_2(v, w) \cap V(R)$ is adjacent to at least one vertex in $N_1(v, w) \cap V(R)$. We assume here $N_1(v, w) \cap V(R) = \{v_1, v_2, w_1, w_2\}$. For the case that $N_1(v, w) \cap V(R) \subset \{v_1, v_2, w_1, w_2\}$, better bounds on $|N_2(v, w) \cap V(R)|$ can be achieved. The vertices in $N_2(v, w) \cap V(R)$ can be divided into the following groups:

- $T_1 := \{x \mid x \in N_2(v, w) \cap V(R) \text{ and } x \text{ is adjacent to } v \text{ and } v_1, \text{ but not } w_1\}$,
- $T_2 := \{x \mid x \in N_2(v, w) \cap V(R) \text{ and } x \text{ is adjacent to } v \text{ and } w_1\}$,
- $T_3 := \{x \mid x \in N_2(v, w) \cap V(R) \text{ and } x \text{ is adjacent to } v \text{ and } w_2\}$,
- $T_4 := \{x \mid x \in N_2(v, w) \cap V(R) \text{ and } x \text{ is adjacent to } v \text{ and } v_2\}$,
- $T_5 := \{x \mid x \in N_2(v, w) \cap V(R) \text{ and } x \text{ is adjacent to } w \text{ and } w_2, \text{ but not } v_2\}$,
- $T_6 := \{x \mid x \in N_2(v, w) \cap V(R) \text{ and } x \text{ is adjacent to } w \text{ and } v_2\}$,
- $T_7 := \{x \mid x \in N_2(v, w) \cap V(R) \text{ and } x \text{ is adjacent to } w \text{ and } v_1\}$,
- $T_8 := \{x \mid x \in N_2(v, w) \cap V(R) \text{ and } x \text{ is adjacent to } w \text{ and } w_1\}$.

By planarity of G, one of T_2 and T_7 must be empty and one of T_3 and T_6 must be empty. Without loss of generality, assume that T_3 and T_7 are empty. If one of T_2 and T_6 is empty, then a better bound on $|V(R)|$ can be obtained. Thus, assume that T_2 and T_6 are not empty.

We first bound $|T_1|$ and $|T_2|$. A quasi-simple region $Q_1 = R(v, w_1)$ between v and w_1 is formed by the following vertices: the vertices in $T_1 \cup T_2$, the vertices in R that are adjacent to v and to at least one vertex in T_2 but are not adjacent to w_1, and the vertices $\{v, v_1, w_1\}$. Since v is black, by Lemma 4, $|T_1 \cup T_2 \cup \{v_1\}| \leq 4$. A quasi-simple region $Q_2 = R(w, v_2)$ between w and v_2 is formed by the following vertices: the vertices in $T_5 \cup T_6$, the vertices in R that are adjacent to w and to at least one vertex in T_6 but are not adjacent to v_2, and the vertices $\{w, w_2, v_2\}$. Similarly, $|T_5 \cup T_6 \cup \{w_2\}| \leq 4$.

Next we show that $|T_4| \leq 2$. Assume there are three or more vertices in T_4, and without loss of generality, let a_1, a_2, and a_3 be the three vertices in T_4. Note that v and v_2 together with each of $\{a_1, a_2, a_3\}$ form a triangle and thus a closed area of the embedding of G. Suppose that no vertex lies inside the area formed by (v, a_1, v_2). Then, a_1 is the only vertex in the area formed by (v, a_2, v_2), and a_1 and a_2 are the only two vertices in the area formed by (v, a_3, v_2). Then, a_1 and a_2 must belong to $N_2(v) \cup N_3(v)$. Since v is black, it follows from Rule 3 that a_1 and a_2 must be white and there is no edge between a_1 and a_2, which results in $N(a_1) \subseteq N(a_2)$, contradicting the fact that Rule 1 is not applicable. Therefore, $|T_4| \leq 2$. With the same argument, $|T_8| \leq 2$.

It is easy to get that $(N_2(v, w) \cup N_1(v, w)) \cap V(R)$ consists of vertices in $T_1 \cup T_2 \cup \{v_1\}, T_5 \cup T_6 \cup \{w_2\}, T_4, T_8,$ and v_2, w_1. Therefore, $|(N_2(v, w) \cup N_1(v, w)) \cap V(R)| \leq 4 + 4 + 2 + 2 + 2 = 14$. By Lemma 5, $|N_3(v, w) \cap V(R)| \leq 27$. In conclusion, $|V(R)| = |V(R) \cap (N_1(v, w) \cup N_2(v, w))| + |V(R) \cap N_3(v, w)| \leq 41$. □

Next we show the size bound on regions with adjacent endpoints.

Lemma 7. *Let G be a reduced plane graph with a connected dominating set D excluding all white vertices. Let $\mathfrak{R}$ be a maximal D-region decomposition of G. For a region $R = R(v, w) \in \mathfrak{R}$ with $d(v, w) = 1$, we have $|V(R)| \leq 23$.*

Proof. By a similar argument as in the proof of Lemma 6, $|V(R) \cap (N_1(v, w) \cup N_2(v, w))| \leq 14$. By Lemma 5, $|V(R) \cap N_3(v, w)| \leq 9$. Hence, $|V(R)| \leq 23$.   □

Finally we bound the number of vertices not in $V[\mathfrak{R}]$.

Lemma 8. *Let $G = (V, E)$ be a reduced plane graph with a connected dominating set D excluding all white vertices and let $\mathfrak{R}$ be a maximal D-region decomposition of G. We have $|V \setminus V[\mathfrak{R}]| \leq 8|\mathfrak{R}|$.*

We summarize our findings in the following theorem.

Theorem 2. *Let (G^0, k) be an input instance of Planar Connected Dominating Set. Let $G=(V, E)$ be the reduced graph by applying Rules 1-7 to G^0. If there exists a connected dominating set for G^0 of size at most k, then $|V| \leq 130k$.*

Proof. By Theorem 1, there exists a connected dominating set for G^0 of size at most k if and only if there exists a connected dominating set D for G that consists of at most k black vertices. By Lemma 3, a maximal D-region decomposition $\mathfrak{R}$ of G with at most $3k-6$ regions can be found. Note that for each region $R(v, w) \in \mathfrak{R}$, either $d(v, w) = 1$ or $2 \leq d(v, w) \leq 3$. Lemma 3 upper-bounds the number of regions $R(v, w)$ in $\mathfrak{R}$ with $d(v, w) \geq 2$ by $2k - 5$. Therefore, by Lemmas 6 and 7, $|V(\mathfrak{R})| \leq \sum_{R(v,w) \in \mathfrak{R}, d(v,w) \geq 2} |V(R)| + \sum_{R(v,w) \in \mathfrak{R}, d(v,w)=1} |V(R)| \leq (2k - 5) \cdot 41 + (k - 1) \cdot 23 = 105k - 286$. Therefore, $|V[\mathfrak{R}]| \leq 106k - 286$. By Lemma 8, we have $|V \setminus V[\mathfrak{R}]| \leq 8|\mathfrak{R}| \leq 8 \cdot (3k - 6) = 24k - 48$. Therefore, $|V| = |V[\mathfrak{R}]| + |V \setminus V[\mathfrak{R}]| \leq 130k$.   □

References

1. Alber, J., Fellows, M.R., Niedermeier, R.: Polynomial-time data reduction for dominating set. J. ACM 51, 363–384 (2004)
2. Bodlaender, H.L.: Kernelization: New upper and lower bound techniques. In: Chen, J., Fomin, F.V. (eds.) IWPEC 2009. LNCS, vol. 5917, pp. 17–37. Springer, Heidelberg (2009)
3. Bodlaender, H.L., Fomin, F.V., Lokshtanov, D., Penninkx, E., Saurabh, S., Thilikos, D.M. (Meta)Kernelization. In: Proc. 50th Annual IEEE Symposium on Foundations of Computer Science (FOCS 2009), pp. 629–638 (2009)
4. Chen, J., Fernau, H., Kanj, I.A., Xia, G.: Parametric duality and kernelization: Lower bounds and upper bounds on kernel size. SIAM J. Comput. 37, 1077–1106 (2007)
5. Downey, R.G., Fellows, M.R.: Parameterized Complexity. Springer, New York (1999)
6. Garey, M.R., Johnson, D.S.: Computers and Intractability: A Guide to the Theory of NP-Completeness. W. H. Freeman and Co., San Francisco (1979)

7. Gu, Q.P., Imani, N.: Connectivity is not a limit for kernelization: Planar connected dominating set. In: López-Ortiz, A. (ed.) LATIN 2010. LNCS, vol. 6034, pp. 26–37. Springer, Heidelberg (2010)
8. Gu, Q.P., Imani, N.: Efficient linear kernelization of planar connected dominating set (2010) (submitted to algorithmica)
9. Guo, J., Niedermeier, R.: Invitation to data reduction and problem kernelization. ACM SIGACT News 38(1), 31–45 (2007)
10. Guo, J., Niedermeier, R.: Linear problem kernels for NP-hard problems on planar graphs. In: Arge, L., Cachin, C., Jurdziński, T., Tarlecki, A. (eds.) ICALP 2007. LNCS, vol. 4596, pp. 375–386. Springer, Heidelberg (2007)
11. Lokshtanov, D., Mnich, M., Saurabh, S.: Linear kernel for planar connected dominating set. In: Chen, J., Cooper, S.B. (eds.) TAMC 2009. LNCS, vol. 5532, pp. 281–290. Springer, Heidelberg (2009)
12. Saurabh, S.: Open problem session. In: The 2nd Workshop on Kernelization (WorKer 2010) (2010)

Fast Exact Algorithm for $L(2, 1)$-Labeling of Graphs[*]

Konstanty Junosza-Szaniawski[1], Jan Kratochvíl[2], Mathieu Liedloff[3],
Peter Rossmanith[4], and Paweł Rzążewski[1]

[1] Warsaw University of Technology, Faculty of Mathematics and Information Science,
Pl. Politechniki 1, 00-661 Warszawa, Poland
{k.szaniawski,p.rzazewski}@mini.pw.edu.pl
[2] Department of Applied Mathematics, and Institute for Theoretical Computer
Science, Charles University, Malostranské nám. 25, 118 00 Praha 1, Czech Republic
honza@kam.ms.mff.cuni.cz
[3] Laboratoire d'Informatique Fondamentale d'Orléans, Université d'Orléans,
45067 Orléans Cedex 2, France
mathieu.liedloff@univ-orleans.fr
[4] Department of Computer Science, RWTH Aachen University, Germany
rossmani@cs.rwth-aachen.de

Abstract. An $L(2, 1)$-labeling of a graph is a mapping from its vertex
set into nonnegative integers such that the labels assigned to adjacent
vertices differ by at least 2, and labels assigned to vertices of distance 2
are different. The span of such a labeling is the maximum label used, and
the $L(2, 1)$-span of a graph is the minimum possible span of its $L(2, 1)$-
labelings. We show how to compute the $L(2, 1)$-span of a connected graph
in time $O^*(2.6488^n)$. Previously published exact exponential time algorithms were gradually improving the base of the exponential function
from 4 to the so far best known 3.2361, with 3 seemingly having been
the Holy Grail.

1 Introduction

An $L(2, 1)$-*labeling* of a graph is a mapping from its vertex set into nonnegative
integers such that the labels assigned to adjacent vertices differ by at least 2, and
labels assigned to vertices of distance 2 are different. The *span* of such a labeling is
the maximum label used and the minimum possible span of an $L(2, 1)$-labeling of
a graph G is denoted by $\lambda(G)$. This variant of graph coloring is recently receiving
considerable attention (see [3,7,10,24] for some surveys on the problem and its
generalizations). It is motivated by the Frequency Assignment Problem whose

[*] J. Kratochvíl acknowledges partial support of Czech research project 1M0545. Part
of the research was done when K. Junosza-Szaniawski and P. Rzążewski visited
Charles University supported by Czech project MSM0021620838, and part of the
research was done during Dagstuhl seminar 10441 on *Exact Complexity of NP-hard
Problems*. M. Liedloff acknowledges the support of the French Agence Nationale de
la Recherche (ANR AGAPE ANR-09-BLAN-0159-03).

M. Ogihara and J. Tarui (Eds.): TAMC 2011, LNCS 6648, pp. 82–93, 2011.
© Springer-Verlag Berlin Heidelberg 2011

task is to assign frequencies to transmitters in a broadcasting network while avoiding undesired interference. In the $L(2, 1)$-labeling model the vertices of the input graph correspond to transmitters of the network and the edges indicate which pairs of transmitters are too close to each other so that interference could occur even if the broadcasting channels were just one apart. The second condition follows from a requirement that no transmitter should have two or more close neighbors transmitting on the same frequency.

The concept of distance constrained graph labeling was introduced by Hale [12] and, according to [11], Roberts [21] was the first one who suggested to investigate the $L(2, 1)$ case in particular. In their seminal paper [11], Griggs and Yeh present first complexity results and several inspiring conjectures. Their conjecture that $\lambda(G) \leq \Delta(G)^2$ initiated intensive research and is still not fully resolved. It is known to be true for many special graph classes and quite recently has been proved for graphs of large maximum degree [14]. Yet it is interesting to note that the Petersen and Hoffmann-Singleton graphs are the only two known graphs that satisfy equality in this bound (for maximum degree greater than 2).

From the complexity point of view, Griggs and Yeh showed that determining $\lambda(G)$ is NP-hard and raised the question of computational complexity of determining $\lambda(G)$ for trees. The latter was answered by Chang and Kuo by providing a polynomial time algorithm in [4]. This has been later improved to a linear time algorithm by Hasunuma et al. in [13]. For general graphs, Fiala et al. [6] proved that deciding $\lambda(G) \leq k$ remains NP-complete for every fixed $k \geq 4$, Bodlaender et al. [2] proved NP-completeness for planar inputs for $k = 8$, and Janczewski et al. [16] proved NP-completeness for planar inputs and $k = 4$. The fact that distance constrained labeling is a more difficult task than ordinary coloring is probably most strikingly documented by the NP-completeness of deciding $\lambda(G) \leq k$ for series-parallel graphs [5] (here of course k is part of the input).

Recent trend in algorithmic research is designing exact exponential time algorithms for NP-hard problems while trying to minimize the constant which is the base of the exponential running time function. Kratochvíl et al. [20] gave an $O^*(1.3161^n)$[1] algorithm for $L(2, 1)$-labeling of span 4 (and this algorithm was referenced as one of the examples of the Measure and Conquer branching technique in [9]). A dynamic programming approach can be used to determining the $L(2, 1)$-span (or, in other words, to decide $\lambda(G) \leq k$ even when k is part of the input). The development in this area has been quite interesting. An exact algorithm for the so called Channel Assignment Problem of Kráľ [19] implies an $O^*(4^n)$ algorithm for the $L(2, 1)$-labeling problem. This has been improved by Havet et al. [15] to an $O^*(3.8739^n)$ algorithm by proving and using a bound on the number of 2-packings in a connected graph. That paper concludes with a conjecture on partitioning graphs into stars which would imply a better running time for the $L(2, 1)$-labeling problem when the minimum degree of the input graph is high. This conjecture was later proved by Alon and Wormald [1],

---

[1] Here we use the so called O^* notation: $f(n) = O^*(g(n))$ if $f(n) \leq p(n) \cdot g(n)$ for some polynomial $p(n)$.

however, even for arbitrarily large minimum degree the running time is not better than $O^*(3^n)$. In the meantime, Junosza-Szaniawski and Rzążewski [17,18] modified the algorithm and refined the running time analysis and proved that their algorithm runs in time $O^*(3.2361^n)$. A lower-bound of $\Omega(3.0731^n)$ on the worst-case running-time of their algorithm is also provided. The magic running time of $O^*(3^n)$ still seemed hardly attainable. In this paper we provide a breakthrough in this question by proving the following theorem.

Theorem 1. *The $L(2,1)$-span of a connected graph can be determined in time $O^*(2.6488^n)$.*

Our algorithm is based on a reduction of the number of operations performed in the recursive step of the dynamic programming algorithm, which is in essence similar to Strassen's algorithm for matrix multiplication [23]. This trick itself achieves running time $O^*(3^n)$. Further improvement is obtained by proving in Section 2 an upper bound on the number of pairs of disjoint subset of the vertex set, where one of the sets is a 2-packing. We believe that this bound and the technique which is used for obtaining it are of interest on their own.

2 Auxiliary Combinatorial Results

Throughout the paper we consider finite undirected graphs without multiple edges or loops. The vertex set (edge set) of a graph G is denoted by $V(G)$ ($E(G)$, respectively). The open neighborhood of a vertex u in G is denoted by $N_G(u)$. The set $N_G[u] = N_G(u) \cup \{u\}$ denotes the closed neighborhood of u. The neighborhood of a set X of vertices in G is denoted by $N_G(X) = \bigcup_{v \in X} N_G(v)$ and its closed neighborhood is denoted by $N_G[X] = N_G(X) \cup X$. For a subset $X \subseteq V(G)$ we denote the subgraph of G induced by the vertices in X by $G[X]$. The symbol n is reserved for the number of vertices of the input graph, which will always be denoted by G. The distance $dist_G(x, y)$ between two vertices u and v in a graph G is the length of a shortest path joining v and v.

A subset S of the vertex set of G is called a *2-packing* if the distance of any two distinct vertices of S is at least 3 (i.e., S is an independent set and no two vertices of S have a common neighbor in G). A pair (S, X) of subsets of $V(G)$ is called a *proper pair* if $S \cap X = \emptyset$ and S is a 2-packing in G. The number of proper pairs in G will be denoted by $pp(G)$ and by the definition, we have

$$pp(G) = \sum_{\substack{S \subseteq V(G) \\ S \text{ is a 2-packing}}} 2^{n-|S|}.$$

Finally, we define

$$pp(n) = \max\ pp(H)$$

where the maximum is taken over all connected graphs H with n vertices.

Theorem 2. *The value of $pp(n)$ is upper-bounded by $O(2.6488^n)$.*

Proof. Let $G = (V, E)$ be a connected graph on n vertices such that $pp(G) = pp(n)$. We observe that if S is a 2-packing of G, then for any edge e of G, the set S is also a 2-packing of $G = (V, E \setminus \{e\})$. Thus removing an edge does not decrease the number of proper pairs and we can remove edges from the graph as long as it stays connected. Hence without loss of generality, we assume that G is a tree.

(∗) Suppose in G there are two leaves v_1 and v_2, which have a common neighbor v_3. Notice that every proper pair in G is proper in the graph H obtained from G by removing the edge $v_1 v_3$ and adding the edge $v_1 v_2$ (see Figure 1). Since this operation does not reduce the number of proper pairs, we can assume that there are no two or more leaves with a common neighbor in G.

Fig. 1. Transformation of two leaves with a common neighbor

It is easy to observe that $pp(0) = 1$, $pp(1) = 3$ and $pp(2) = 8$. Assume that $|V(G)| \geq 3$ and let P be a longest path in G. Let v be an end-vertex of the path P, u its neighbor on P, and c a neighbor of u on P other that v (the third vertex on P). By the observation (∗) we can assume that $\deg(u) = 2$.

(A) If $\deg(c) \leq 2$, we can partition all proper pairs (S, X) to two subsets: those in which $v \notin S$ and those in which $v \in S$ (see Figure 2).

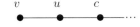

Fig. 2. Case (A) with $\deg(c) \leq 2$

Notice that if $v \notin S$, then v can be in X or outside $S \cup X$. If $v \in S$, then none of the vertices $\{u, c\}$ can belong to S. Each of them can be in X or outside $S \cup X$. Since the graphs $G - v$ and $G - \{v, u, c\}$ are connected, we obtain the following recursion:

$$pp(n) \leq 2\, pp(n-1) + 4\, pp(n-3). \qquad (1)$$

(B) If $\deg(c) > 2$, then all vertices in the set $\{v \in V(G) : dist_G(v, c) = 2\}$ except at most one (the one belonging to the path P) are leaves (since otherwise P is not the longest path) and all neighbors of the vertex c except at most one are of degree 2 (from (∗)). Hence one of the following two cases occurs:

(B0) No neighbor of c is a leaf in G (see Figure 3(a)).
(B1) There exists a vertex $x \in N(c)$ which is a leaf in G (there can be at most one such vertex by the observation (∗)) – (see Figure 3(b)).

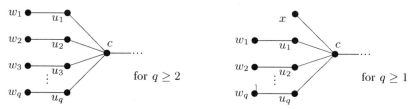

(a) Case (B0) with $\deg(c) > 2$ and no   (b) Case (B1) with $\deg(c) > 2$ and one
neighbor of c is a leaf   neighbor of c is a leaf

Fig. 3. Cases (B0) and (B1)

Let $W = \{w_1, \ldots, w_q\} = \{w \in V(G): w \text{ is a leaf in } G \text{ and } dist_G(w, c) = 2\}$ and $U = N(W)$ in the case (B0) and $U = N(W) \cup \{x\}$ in the case (B1). We can partition the set of proper pairs (S, X) to whose in which $S \cap (W \cup U) = \emptyset$ and the others.

If $S \cap (W \cup U) = \emptyset$, each of the vertices in $W \cup U$ can be in X or outside $S \cup X$.

If $S \cap (W \cup U) = \widehat{S} \neq \emptyset$, $\widehat{S}$ must be a 2-packing in G. Notice that the number of proper pairs $(\widehat{S}, \widehat{X})$ in $G[W \cup U \cup \{c\}]$, such that $\widehat{S} \neq \emptyset$ and $c \notin \widehat{S}$ is equal to:

1. $(3^q - 2^q)2^{q+1} + q \cdot 3^{q-1}2^{q+1} = 3^{q-1}2^{q+1}(3 + q) - 2^{2q+1}$ for $q \geq 2$ in the case (B0).
2. $(3^q - 2^q)2^{q+2} + q \cdot 3^{q-1}2^{q+2} + 3^q2^{q+1} = 3^{q-1}2^{q+1}(9 + 2q) - 2^{2q+2}$ for $q \geq 1$ in the case (B1).

Each of the vertices in $(W \cup U \cup \{c\}) \setminus \widehat{S}$ can be in X or outside $S \cup X$.

Since the graphs $G - (W \cup U)$ and $G - (W \cup U \cup \{c\})$ are connected, we obtain the following recursions:

$$pp(n) \leq 2^{2q} \, pp(n - 2q) + (3^{q-1}2^{q+1}(3 + q) - 2^{2q+1}) \, pp(n - 2q - 1) \quad (2)$$

$$pp(n) \leq 2^{2q+1} \, pp(n - 2q - 1) + (3^{q-1}2^{q+1}(9 + 2q) - 2^{2q+2}) \, pp(n - 2q - 2). \quad (3)$$

We shall prove by induction on n that for $n \geq 0$ the following holds:

$$pp(n) \leq 2 \cdot \tau^n \quad (4)$$

where $\tau = 2.6487..$ is the positive root of the equation $\tau^5 = 16\tau + 88$.

It is easy to observe that the inequality (4) holds for $n \leq 2$. Now assume that the inequality holds for all values smaller than n.

Case (A)
$pp(n) \leq 2 \, pp(n - 1) + 4 \, pp(n - 3) \leq 4\tau^{n-1} + 8\tau^{n-3} = 4(\tau^2 + 2)\tau^{n-3} < 2 \cdot \tau^3 \cdot \tau^{n-3} = 2 \cdot \tau^n$

Case (B0)

$pp(n) \leq 2^{2q}\, pp(n-2q) + (3^{q-1}2^{q+1}(3+q) - 2^{2q+1})\, pp(n-2q-1) \leq 2(2^{2q} \cdot \tau^{n-2q} + (3^{q-1}2^{q+1}(3+q) - 2^{2q+1}) \cdot \tau^{n-2q-1}) = 2 \cdot \tau^n(2^{2q} \cdot \tau^{-2q} + (3^{q-1}2^{q+1}(3+ q) - 2^{2q+1}) \cdot \tau^{-2q-1}) = 2 \cdot \tau^n((\frac{2}{\tau})^{2q} - (\frac{2}{\tau})^{2q+1} + \frac{4(3+q)}{\tau^3}(\frac{6}{\tau^2})^{q-1})$

One can easily verify that the function $h_0(x) = (\frac{2}{\tau})^{2x} - (\frac{2}{\tau})^{2x+1} + \frac{4(3+x)}{\tau^3}(\frac{6}{\tau^2})^{x-1}$ is decreasing for all real $x > 2$ and $h_0(2) = 1$.

Hence $pp(n) \leq 2 \cdot \tau^n((\frac{2}{\tau})^{2q} - (\frac{2}{\tau})^{2q+1} + \frac{4(3+q)}{\tau^3}(\frac{6}{\tau^2})^{q-1}) \leq 2 \cdot \tau^n$.

Case (B1)

$pp(n) \leq 2^{2q+1}\, pp(n-2q-1) + (3^{q-1}2^{q+1}(9+2q) - 2^{2q+2})\, pp(n-2q-2) \leq 2(2^{2q+1}\tau^{n-2q-1} + (3^{q-1}2^{q+1}(9+2q) - 2^{2q+2})\tau^{n-2q-2}) = 2 \cdot \tau^n((\frac{2}{\tau})^{2q+1} - (\frac{2}{\tau})^{2q+2} + \frac{4(9+2q)}{\tau^4}(\frac{6}{\tau^2})^{q-1})$

Since the function $h_1(x) = (\frac{2}{\tau})^{2x+1} - (\frac{2}{\tau})^{2x+2} + \frac{4(9+2x)}{\tau^4}(\frac{6}{\tau^2})^{x-1}$ is decreasing for all real $x > 1$ and $h_1(1) < 1$, we obtain:

$pp(n) \leq 2 \cdot \tau^n((\frac{2}{\tau})^{2q+1} - (\frac{2}{\tau})^{2q+2} + \frac{4(9+2q)}{\tau^4}(\frac{6}{\tau^2})^{q-1}) < 2 \cdot \tau^n$.

We have shown that regardless of the structure of G, the function $2 \cdot \tau^n$ is an upper bound on the number of proper pairs in G. Hence $pp(n) = O(\tau^n) = O(2.6488^n)$. □

One is inclined to conjecture that the worst case is attained in the case of a path P_n on n vertices. A simple calculation shows that $pp(P_n) = \Theta(2.5943..^n)$. The following example shows that intuition fails in this case.

Theorem 3. *The value of $pp(n)$ is bounded from below by $\Omega(2.6117^n)$.*

Proof. We shall prove the theorem by showing a graph with $\Theta(2.6117..^n)$ proper pairs. Let us consider the following graphs:

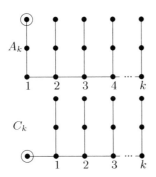
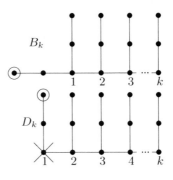

Let a_k, b_k and c_k denote the number of proper pairs in the graphs A_k, B_k and C_k, respectively. Let d_k denote the number of such proper pairs (S, X) in the graph D_k, in which the 2-packing S does not contain the crossed out vertex.

Considering separately the number of proper pairs (S, X), in which S contains and does not contain marked vertices, we obtain the following system of recursions:

$$\begin{cases} a_k = 2b_{k-1} + 4a_{k-1} \\ b_k = 2c_k + 2d_k \\ c_k = 2a_k + 12d_{k-1} \\ d_k = 4d_{k-1} + 12a_{k-1} \end{cases}$$

Solving this system we obtain the result $a_k = \Theta(x^k)$, where $x = 17.8149..$ is the positive solution of the equation $x^3 = 16x^2 + 576$.

Since $k = n/3$, the graph A_k contains $a_k = \Theta(17.8149..^{n/3}) = \Theta(2.6117..^n)$ proper pairs.   □

The main tool in [15] was partitioning a connected input graph into stars of orders at least 2. Our approach is to divide the computation into connected subgraphs of large constant order. The star graph is an example showing that one cannot always find such a partition. However, we can find a *covering* with a small overlap of the connected subgraphs, as shown by the following result.

Theorem 4. *Let G be a connected graph of order n and let $k < n$ be a positive integer. Then there exist connected subgraphs $G_1, G_2, \ldots, G_q$ of G such that (i) every vertex of G belongs to at least one of them, (ii) the order of each of $G_1, G_2, \ldots, G_{q-1}$ is at least k and at most $2k$ (while for G_q we only require $|V(G_q)| \le 2k$), and (iii) the sum of the numbers of vertices of $G_i's$ is at most $n(1 + \frac{1}{k})$.*

Proof. Assume G is rooted in an arbitrary vertex r and consider a DFS-tree T of G. For every vertex v let $T(v)$ be the subtree rooted in v. If $|T(r)| \le 2k$ then add G to the set of desired subgraphs and finish. If there is a vertex v such that $k \le |T(v)| \le 2k$ then add $G[V(T(v))]$ to the set of desired subgraphs and proceed recursively with $G \setminus V(T(v))$. Otherwise there must be a vertex v such that $|T(v)| > 2k$ and for its every child u, $|T(u)| < k$. In such a case find a subset $\{u_1, ..u_i\}$ of children of v such that $k - 1 \le |T(u_1)| + .. + |T(u_i)| \le 2k - 1$. Add $G[\{v\} \cup V(T(u_1)) \cup .. \cup V(T(u_i))]$ to the set of desired subgraphs and proceed with the graph $G \setminus (V(T(u_1)) \cup .. \cup V(T(u_i)))$. This procedure terminates after at most $\frac{n}{k}$ steps and in each of them we have left at most one vertex of the identified connected subgraph in the further processed graph.   □

3   Exact Algorithm for $L(2, 1)$-Labeling

One key ingredient in our algorithm are algebraic manipulations very similar to fast matrix multiplication: If we have $2^k \times 2^k$-matrices A and B we can divide them each into four block matrices of the same size. We can then compute AB very easily by eight matrix multiplications of $2^{k-1} \times 2^{k-1}$-matrices. Doing so recursively leads again to a running time of $O(n^3)$ — just as the naive algorithm itself. It is, however, possible to improve on the running time by using only *seven* matrix multiplications to achieve the same result [23]. It turns out that this technique alone does not work in our case, though. We have to use one other trick: We jump between two representations of partial $L(2, 1)$-labelings in the course of our dynamic programming algorithm. The idea to use different representations of the same data in dynamic programming is not new and was used in a similar way before [22].

We define the partial function $\oplus\colon \{0, \bar{0}, 1, \bar{1}\} \times \{0, 1\} \to \{0, 1, \bar{1}\}$ via this table:

$\oplus$	0	$\bar{0}$	1	$\bar{1}$
0	0	0	1	1
1	$\bar{1}$	–	–	–

The entry "−" signifies that $\oplus$ is not defined on that input.

We generalize $\oplus$ to vectors via

$$a_1 a_2 \ldots a_n \oplus b_1 b_2 \ldots b_n = \begin{cases} (a_1 \oplus b_1) \ldots (a_n \oplus b_n) & \text{if } a_i \oplus b_i \text{ is defined for all} \\ & i \in \{1, \ldots, n\}, \\ \text{undefined} & \text{otherwise,} \end{cases}$$

and to sets of vectors $A \subseteq \{0, \bar{0}, 1, \bar{1}\}^n$, $B \subseteq \{0, 1\}^n$ via

$$A \oplus B = \{\, \mathbf{a} \oplus \mathbf{b} \mid \mathbf{a} \in A,\ \mathbf{b} \in B,\ \mathbf{a} \oplus \mathbf{b} \text{ is defined}\,\}.$$

In a nutshell our algorithm proceeds as follows: Given a graph $G = (V, E)$ of order n, with $V = \{v_1, \ldots, v_n\}$, it computes tables $T_0, T_1, \ldots, T_{2n} \subseteq \{0, \bar{0}, 1, \bar{1}\}^n$. Table T_l contains a vector $\mathbf{a} \in \{0, \bar{0}, 1, \bar{1}\}^n$ if and only if there is a partial labeling $\varphi\colon V \to \{0, \ldots, l\}$ such that :

1. $a_i = 0$ iff v_i is not labeled by φ and there is no neighbor u of v_i with $\varphi(u) = l$,
2. $a_i = \bar{0}$ iff v_i is not labeled by φ and there is a neighbor u of v_i with $\varphi(u) = l$,
3. $a_i = 1$ iff $\varphi(v_i) < l$, and
4. $a_i = \bar{1}$ iff $\varphi(v_i) = l$.

Once we have all tables T_l it is easy to find the smallest l such that T_l contains at least one vector from $\{1, \bar{1}\}^n$ – such vectors correspond to solutions where all vertices are labeled. We then know that such an l is the $L(2,1)$-span of G.

Let $P \subseteq \{0, 1\}^n$ be the encodings of all 2-packings of G. Formally, $\mathbf{p} \in P$ if and only if there is a 2-packing $S \subseteq V$ such that for all i, $1 \le i \le n$, $p_i = 1$ iff $v_i \in S$.

Our strategy is to compute T_{l+1} from $T_l \oplus P$. This is not hard because $T_l \oplus P$ is already almost the same as T_{l+1}: $\mathbf{a} \in T_{l+1}$ iff there is an $\mathbf{a}' \in T_l \oplus P$ such that

1. $a_i = 0$ iff $a_i' = 0$ and there is no $v_j \in N(v_i)$ with $a_j' = \bar{1}$
2. $a_i = \bar{0}$ iff $a_i' = 0$ and there is a $v_j \in N(v_i)$ with $a_j' = \bar{1}$
3. $a_i = 1$ iff $a_i' = 1$, and
4. $a_i = \bar{1}$ iff $a_i' = \bar{1}$.

To compute T_{l+1} from $T_l \oplus P$ is therefore easy: Look at each vector in $T_l \oplus P$ and determine for each 0 whether it remains 0 or has to be changed into $\bar{0}$. What remains is to find a method to compute $T_l \oplus P$ fast.

Towards this end let us fix a constant k (whose size will be specified later). Let $G_1, \ldots, G_q$ be a covering of G by connected subgraphs guaranteed by Theorem 4 and let k' be the order of G_1. Hence, the relation $k \le k' \le 2k$ holds as long as $q > 1$.

We need one more formalism: If $\mathbf{w}$ is a vector and A is a set of vectors, then

$$A_{\mathbf{w}} = \{\, \mathbf{v} \mid \mathbf{w}\mathbf{v} \in A \,\}$$

is the set of all vectors that fall into A after we prefix them with $\mathbf{w}$. Here $\mathbf{w}\mathbf{v}$ denotes the concatenation of vectors $\mathbf{w}$ and $\mathbf{v}$.

If $A \subseteq \{0, \bar{0}, 1, \bar{1}\}^n$ and $B \subseteq \{0,1\}^n$, where $n > k'$, we can compute $A \oplus B$ in the following useful, though perhaps at first sight complicated, manner:

$$A \oplus B = \bigcup_{\substack{\mathbf{u} \in \{0,\bar{0},1,\bar{1}\}^{k'} \\ \mathbf{v} \in \{0,1\}^{k'} \\ \text{s.t. } \mathbf{u} \oplus \mathbf{v} \text{ defined}}} (\mathbf{u} \oplus \mathbf{v})(A_{\mathbf{u}} \oplus B_{\mathbf{v}})$$

$$= \bigcup_{\substack{\mathbf{v} \in \{0,1\}^{k'} \\ \mathbf{w} \in \{0,1,\bar{1}\}^{k'}}} \mathbf{w} \left[\left(\bigcup_{\substack{\mathbf{u} \in \{0,\bar{0},1,\bar{1}\}^{k'} \\ \text{s.t. } \mathbf{u} \oplus \mathbf{v} = \mathbf{w}}} A_{\mathbf{u}} \right) \oplus B_{\mathbf{v}} \right]$$

Let us analyze how long it takes to compute $A \oplus B$ in this manner. We are especially interested in the number of $\oplus$-operations on sets with vectors of length $n - k'$. We can omit such a computation if the first set, i.e.,

$$\bigcup_{\substack{\mathbf{u} \in \{0,\bar{0},1,\bar{1}\}^{k'} \\ \text{s.t. } \mathbf{u} \oplus \mathbf{v} = \mathbf{w}}} A_{\mathbf{u}} \tag{5}$$

is empty. So how many pairs $\mathbf{v}, \mathbf{w}$ are there such that there is at least one $\mathbf{u}$ with $\mathbf{u} \oplus \mathbf{v} = \mathbf{w}$? If we fix $\mathbf{v}$, then obviously $v_i = 1$ implies $w_i = \bar{1}$. So for a fixed $\mathbf{v}$ there are at most $2^{k' - ||\mathbf{v}||}$ many $\mathbf{w}$'s, where $||\mathbf{v}||$ denotes the number of positions i such that $v_i = 1$.

The total number of pairs $\mathbf{v}, \mathbf{w}$ such that $\mathbf{w} = \mathbf{v} \oplus \mathbf{u}$ for some $\mathbf{u}$ and that therefore produce a nonempty contribution in (5) is therefore at most

$$\sum_{\mathbf{v} \in \{0,1\}^{k'}} 2^{k' - ||\mathbf{v}||}.$$

Thus, if we draw the $\mathbf{v}$'s from a set of vectors that represent the 2-packings of a connected graph, then we find at most $pp(k')$ such pairs and, hence, we need to make only such many recursive computations of $\oplus$ on sets of vectors of length $n - k'$.

Since we can cover the input graph by induced subgraphs of orders between k and $2k$, this is indeed possible. Theorem 4 implies that the total length of the vectors is $n' \le n(1 + 1/k)$.

In each recursive computation we have to prepare up to $pp(k')$ many pairs of sets of vectors of length $n' - k'$, where $k \le k' \le 2k$. Then we recursively compute $\oplus$ on these pairs. From the result we get the next table T_{l+1} in linear time. Preparing the recursive calls and combining their results takes only time linear in the sizes of A and B. The size of B is at most $O(n2^{n'})$ bits and the size of A is at most $O(n\,pp(n'))$ bits if we use only our tables T_l for A: The $\bar{1}$'s form a 2-packing and for all other nodes there are only two possibilities, 1 or $0/\bar{0}$.

We arrive at the following recurrence for the running time:

$$t_n \leq O\big(n\,pp(n') + pp(k')t_{n'-k'}\big) \text{ for } k \leq k' \leq 2k$$

It is not hard to see that the solution is $t_n = O^*(pp(n')) = O^*(pp(n(1+1/k)))$. We arrive at our main result by choosing the constant k so big such that $pp(n(1+1/k)) \leq 2.6488^n$ for large n, which is possible because actually $pp(n) = O(\tau^n) = O((2.6488-\epsilon)^n)$ for some $\epsilon > 0$.

4   Pseudocode of the Algorithm

For the sake of completeness, we provide in this section the pseudocode of the algorithm described in Section 3.

Let $G_1, \ldots, G_q$ be a covering of a given graph G by connected subgraphs as ensured by Theorem 4. Let d_i be the order of G_i for $1 \leq i \leq q$. We denote by n' the sum $d_1 + \cdots + d_q$. Let $A \subseteq \{0, \bar{0}, 1, \bar{1}\}^{n'}$ and $B \subseteq \{0, 1\}^{n'}$. We first provide Algorithm MUL which computes $A \oplus B$ using the methods from Section 3.

```
Algorithm MUL(A, B, d₁, ..., d_q):
if q = 1 then return A ⊕ B fi;
k' := d₁;
for each v ∈ {0, 1}^{k'} do
    R := ∅;
    for each w ∈ {0, 1, 1̄}^{k'} do
        A' := ∅;
        for each u ∈ {0, 0̄, 1, 1̄}^{k'} do
            if w = u ⊕ v then A' := A' ∪ A_u fi
        od;
        if A' ≠ ∅ then R := R ∪ MUL(A', B_v, d₂, ..., d_q) fi
    od
od;
return R
```

Obviously, the body of the innermost loop is executed exactly $24^{k'}$ times. All operations can be carried out in constant time except set union and the recursive calls to MUL. A set union $X \cup Y$ takes at most $n(|X|+|Y|)$ steps if we implement sets as simple arrays and remember that we can sort them using radix-sort. Not counting recursive calls the running time is therefore $O(n(|A|+|B|))$ if $d_1 = k' = O(1)$. In the border case that $q = 1$ then a brute force attack to compute $A \oplus B$ can be carried out in $O(n \cdot |A| \cdot |B|) = O(n8^{k'})$, which is constant if $d_1 = k' = O(1)$.

Let $(v_1, \ldots, v_{n'})$ be the vertices of G (with duplicates allowed) such that

$$(v_1, \ldots, v_{n'}) = (u_{11}, \ldots, u_{1d_1}, \ u_{21}, \ldots, u_{2d_2}, \ \ldots\ldots, \ u_{q1}, \ldots, u_{qd_q})$$

and $G_i = (u_{i1}, \ldots, u_{id_i})$. Let k be a constant large enough so that $pp(1 + 1/k) = \tau^{1+1/k} < 2.6488$, where $\tau = 2.6487..$ is the positive root of the equation $\tau^5 = 16\tau + 88$ as provided in the proof of Theorem 2. As the proof of Theorem 4 is constructive and provides a polynomial-time algorithm to compute a cover $G_1, \ldots, G_q$, we can arrange the decomposition of G into $G_1, \ldots, G_q$ in such a way that $k \leq d_i \leq 2k$ for any $1 \leq i < q$. In addition, Theorem 4 ensure that $n' \leq n(1 + 1/k) + O(1)$. While the correctness of the following algorithm does not depend on such an arrangement, it is crucial to the running time, which is closely related to all $pp(G[G_i])$'s. We can only guarantee those to be small if the $G[G_i]$'s are connected. We refer the reader to Section 3 for the running-time analysis.

Finally the following Algorithm T computes tables $T_1, \ldots, T_{2n}$.

Algorithm $T(G, n, v_1, \ldots, v_{n'})$
$T_0 := \{0^{n'}\}$;
$P := \emptyset$;
for each $\mathbf{x} \in \{0, 1\}^{n'}$ **do**
    if $\{\, v_i \mid x_i = 1 \,\}$ is a 2-packing **then** $P := P \cup \{\mathbf{x}\}$ **fi**
od;
for $l = 1, \ldots, 2n$ **do**
    $R := MUL(T_{l-1}, P, d_1, \ldots, d_q)$;
    $T_l := \emptyset$;
    for each $\mathbf{x} \in R$ **do**
        for $i, j = 1, \ldots, n'$ **do**
            if $x_i = 0$, $x_j = \bar{1}$ and $v_i \in N(v_j)$ **then** $x_i := \bar{0}$ **fi**
        od;
        $T_l := T_l \cup \{\mathbf{x}\}$
    od;
od;
return $T_1, \ldots, T_{2n}$

References

1. Alon, N., Wormald, N.: igh degree graphs contain large-star factors. Fete of Combinatorics and Computer Science, Bolyai Soc. Math. Studies (20), 9–21 (2010)
2. Bodlaender, H.L., Kloks, T., Tan, R.B., van Leeuwen, J.: Approximations for lambda-Colorings of Graphs. Computer Journal 47, 193–204 (2004)
3. Calamoneri, T.: The $L(h, k)$-Labelling Problem: A Survey and Annotated Bibliography. Computer Journal 49, 585–608 (2006)
4. Chang, G.J., Kuo, D.: The $L(2, 1)$-labeling problem on graphs. SIAM Journal of Discrete Mathematics 9, 309–316 (1996)

5. Fiala, J., Golovach, P., Kratochvíl, J.: Distance constrained labelings of graphs of bounded treewidth. In: Caires, L., Italiano, G.F., Monteiro, L., Palamidessi, C., Yung, M. (eds.) ICALP 2005. LNCS, vol. 3580, pp. 360–372. Springer, Heidelberg (2005)
6. Fiala, J., Kloks, T., Kratochvíl, J.: Fixed-parameter complexity of λ-labelings. Discrete Applied Mathematics 113, 59–72 (2001)
7. Fiala, J., Kratochvíl, J.: Locally constrained graph homomorphisms - structure, complexity, and applications. Computer Science Review 2, 97–111 (2008)
8. Fomin, F., Grandoni, F., Kratsch, D.: Measure and conquer: Domination – A case study. In: Caires, L., Italiano, G.F., Monteiro, L., Palamidessi, C., Yung, M. (eds.) ICALP 2005. LNCS, vol. 3580, pp. 191–203. Springer, Heidelberg (2005)
9. Fomin, F., Kratsch, D.: Exact Exponential Algorithms. Springer, Heidelberg (2010)
10. Griggs, J.R., Král, D.: Graph labellings with variable weights, a survey. Discrete Applied Mathematics 157, 2646–2658 (2009)
11. Griggs, J.R., Yeh, R.K.: Labelling graphs with a condition at distance 2. SIAM Journal of Discrete Mathematics 5, 586–595 (1992)
12. Hale, W.K.: Frequency assignemnt: Theory and applications. Proc. IEEE 68, 1497–1514 (1980)
13. Hasunuma, T., Ishii, T., Ono, H., Uno, Y.: A Linear Time Algorithm for L(2,1)-Labeling of Trees. In: Fiat, A., Sanders, P. (eds.) ESA 2009. LNCS, vol. 5757, pp. 35–46. Springer, Heidelberg (2009)
14. Havet, F., Reed, B., Sereni, J.-S.: L(2,1)-labellings of graphs. In: Proceedings of SODA 2008, pp. 621–630 (2008)
15. Havet, F., Klazar, M., Kratochvíl, J., Kratsch, D., Liedloff, M.: Exact algorithms for $L(2,1)$-labeling of graphs. Algorithmica 59, 169–194 (2011)
16. Janczewski, R., Kosowski, A., Małafiejski, M.: The complexity of the $L(p,q)$-labeling problem for bipartite planar graphs of small degree. Discrete Mathematics 309, 3270–3279 (2009)
17. Junosza-Szaniawski, K., Rzążewski, P.: On Improved Exact Algorithms for $L(2,1)$-Labeling of Graphs. In: Iliopoulos, C.S., Smyth, W.F. (eds.) IWOCA 2010. LNCS, vol. 6460, pp. 34–37. Springer, Heidelberg (2011)
18. Junosza-Szaniawski K., Rzążewski P.: On the Complexity of Exact Algorithm for L(2,1)-labeling of Graphs. KAM-DIMATIA Preprint Series 2010-992
19. Král, D.: Channel assignment problem with variable weights. SIAM Journal on Discrete Mathematics 20, 690–704 (2006)
20. Kratochvíl, J., Kratsch, D., Liedloff, M.: Exact algorithms for $L(2,1)$-labeling of graphs. In: Kučera, L., Kučera, A. (eds.) MFCS 2007. LNCS, vol. 4708, pp. 513–524. Springer, Heidelberg (2007)
21. Roberts, F.S.: Private communication to J. Griggs
22. van Rooij, J.M.M., Bodlaender, H.L., Rossmanith, P.: Dynamic Programming on Tree Decompositions Using Generalised Fast Subset Convolution. In: Fiat, A., Sanders, P. (eds.) ESA 2009. LNCS, vol. 5757, pp. 566–577. Springer, Heidelberg (2009)
23. Strassen, V.: Gaussian Elimination is not Optimal. Numerische Mathematik 13, 354–356 (1969)
24. Yeh, R.: A survey on labeling graphs with a condition at distance two. Discrete Mathematics 306, 1217–1231 (2006)

An Improved Sufficient Condition for Reconfiguration of List Edge-Colorings in a Tree

Takehiro Ito*, Kazuto Kawamura, and Xiao Zhou

Graduate School of Information Sciences, Tohoku University,
Aoba-yama 6-6-05, Sendai, 980-8579, Japan
{takehiro,kazuto,zhou}@ecei.tohoku.ac.jp

Abstract. We study the problem of reconfiguring one list edge-coloring of a graph into another list edge-coloring by changing only one edge color assignment at a time, while at all times maintaining a list edge-coloring, given a list of allowed colors for each edge. Ito, Kamiński and Demaine gave a sufficient condition so that any list edge-coloring of a tree can be transformed into any other. In this paper, we give a new sufficient condition which improves the known one. Our sufficient condition is best possible in some sense. The proof is constructive, and yields a polynomial-time algorithm that finds a transformation between two given list edge-colorings of a tree with n vertices via $O(n^2)$ recoloring steps. We remark that the upper bound $O(n^2)$ on the number of recoloring steps is tight, because there is an infinite family of instances on paths that satisfy our sufficient condition and whose reconfiguration requires $\Omega(n^2)$ recoloring steps.

1 Introduction

Reconfiguration problems arise when we wish to find a step-by-step transformation between two feasible solutions of a problem such that all intermediate results are also feasible. Recently, Ito *et al.* [5] proposed a framework of reconfiguration problems, and gave complexity and approximability results for reconfiguration problems derived from several well-known problems, such as INDEPENDENT SET, CLIQUE, MATCHING, etc. In this paper, we study the reconfiguration problem for list edge-colorings of a tree, which was introduced by [6].

An (ordinary) *edge-coloring* of a graph G is an assignment of colors from a color set C to each edge of G so that every two adjacent edges receive different colors. In *list edge-coloring*, each edge e of G has a set $L(e)$ of colors, called the *list* of e. Then, an edge-coloring f of G is called an *L-edge-coloring* of G if $f(e) \in L(e)$ for each edge e, where $f(e)$ denotes the color assigned to e by f. Figure 1 illustrates three L-edge-colorings of the same graph with the same list L; the color assigned to each edge is surrounded by a box in the list. Clearly, an edge-coloring is merely an L-edge-coloring for which $L(e) = C$ for every edge e of G, and hence list edge-coloring is a generalization of edge-coloring.

---

* This work is partially supported by Grant-in-Aid for Scientific Research: 22700001.

M. Ogihara and J. Tarui (Eds.): TAMC 2011, LNCS 6648, pp. 94–105, 2011.

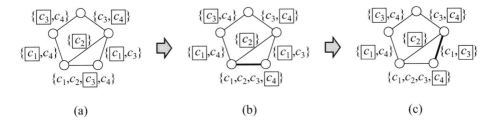

Fig. 1. A sequence of L-edge-colorings of a graph

Suppose now that we are given *two* L-edge-colorings of a graph G (e.g., the leftmost and rightmost ones in Fig.1), and we are asked whether we can transform one into the other via L-edge-colorings of G such that each differs from the previous one in only one edge color assignment. This decision problem is called the LIST EDGE-COLORING RECONFIGURATION problem [6]. For the particular instance of Fig.1, the answer is "yes," as illustrated in Fig.1, where the edge whose color assignment was changed from the previous one is depicted by a thick line. One can imagine a variety of practical scenarios where an edge-coloring (e.g., representing a feasible schedule) needs to be changed (to use a newly found better solution or to satisfy new side constraints) by individual color changes (preventing the need for any coordination) while maintaining feasibility (so that nothing goes wrong during the transformation). Reconfiguration problems are also interesting in general because they provide a new perspective and deeper understanding of the solution space and of heuristics that navigate that space.

Reconfiguration problems have been studied extensively in recent literature [1,3,4,5,6]. It is known that LIST EDGE-COLORING RECONFIGURATION is PSPACE-complete, even for planar graphs of maximum degree 3 and just six colors [6]. On the other hand, Ito *et al.* [6] gave a sufficient condition for which there exists a transformation between any two L-edge-colorings of a tree. Specifically, for a tree T, they proved that any two L-edge-colorings of T can be transformed into each other if $|L(e)| \geq \max\{d(v), d(w)\} + 1$ for each edge $e = vw$ of T, where $d(v)$ and $d(w)$ are the degrees of the endpoints v and w of e, respectively. Their proof yields a polynomial-time algorithm that finds a transformation between two given L-edge-colorings of T via $O(n^2)$ intermediate L-edge-colorings, where n is the number of vertices in T. In addition, they gave an infinite family of instances on paths that satisfy their sufficient condition and whose transformation requires $\Omega(n^2)$ intermediate L-edge-colorings.

As the authors mentioned in [6], their sufficient condition is motivated by the well-known "list coloring conjecture" [7]: it is conjectured that any graph G has an L-edge-coloring if $|L(e)| \geq \chi'(G)$ for each edge e, where $\chi'(G)$ is the chromatic index of G, that is, the minimum number of colors required for an ordinary edge-coloring of G. This conjecture has not been proved yet, but Borodin *et al.* [2] proved that any bipartite graph, and hence any tree, has an L-edge-coloring if $|L(e)| \geq \max\{d(v), d(w)\}$ for each edge $e = vw$. In this sense, there is a gap between the two sufficient conditions [2] and [6]: from the sufficient condition of [6] we cannot say anything about the reconfiguration if a given tree T has an edge $e = vw$ with $|L(e)| = \max\{d(v), d(w)\}$, whereas T has L-edge-colorings.

In this paper, we give a new sufficient condition that improves the known one [6]. In our sufficient condition, a given tree T can have edges $e = vw$ with $|L(e)| = \max\{d(v), d(w)\}$, keeping the reachability of all L-edge-colorings of T. We also show that our sufficient condition is best possible in some sense. Our proof is constructive, and yields a polynomial-time algorithm that finds a transformation between two given L-edge-colorings of T via $O(n^2)$ intermediate L-edge-colorings, where n is the number of vertices in T. We remark that the upper bound $O(n^2)$ on the number of intermediate L-edge-colorings is tight, because the infinite family of instances requiring $\Omega(n^2)$ recoloring steps, given by [6], also satisfies our sufficient condition.

The rest of the paper is organized as follows. In Section 2, we first define some terms, and give our sufficient condition formally in Theorem 2. In the same section, we discuss the tightness of our sufficient condition. The constructive proof of Theorem 2 will be given in Section 3.

2 Improved Sufficient Condition and Its Tightness

In this section, we first introduce some terms and give our main result formally. We then show that our sufficient condition is best possible in some sense.

In Section 1, we have defined an L-edge-coloring of a graph $G = (V, E)$ with a list L. We say that two L-edge-colorings f and f' of G are *adjacent* if $|\{e \in E : f(e) \neq f'(e)\}| = 1$, that is, f' can be obtained from f by changing the color assignment of a single edge e; we say that the edge e is *recolored* between f and f'. A *reconfiguration sequence* between two L-edge-colorings f_0 and f_t of G is a sequence of L-edge-colorings $f_0, f_1, \ldots, f_t$ of G such that f_{i-1} and f_i are adjacent for $i = 1, 2, \ldots, t$. We also say that two L-edge-colorings f and f' are *connected* if there exists a reconfiguration sequence between f and f'. Clearly, any two adjacent L-edge-colorings are connected. For a reconfiguration sequence between two L-edge-colorings, its *length* is defined as the number of L-edge-colorings contained in the reconfiguration sequence. For example, the length of the reconfiguration sequence in Fig.1 is 3.

Let T be a tree with a list L. An edge $e = vw$ is said to be *tight* if $|L(e)| = \max\{d(v), d(w)\}$; *mild* if $|L(e)| = \max\{d(v), d(w)\} + 1$; and *slack* if $|L(e)| \geq \max\{d(v), d(w)\} + 2$. We assume that $|L(e)| \geq \max\{d(v), d(w)\}$ for all edges $e = vw$ in T, and hence each edge in T is tight, mild or slack. For two edges e and e' in T, we denote by $P(e, e')$ the path between e and e' (including themselves) in T. Two tight edges e and e' in T are *neighboring* if $P(e, e')$ contains no tight edge other than e and e'.

We first restate the known sufficient condition, as in the following theorem.

Theorem 1 ([6]). *Every two L-edge-colorings f and f' of a tree T are connected if T contains no tight edge. Moreover, there is a reconfiguration sequence of length $O(n^2)$ between f and f', where n is the number of vertices in T.*

We now give the main result of this paper.

Theorem 2. *Every two L-edge-colorings f and f' of a tree T are connected if either the following condition (a) or (b) holds:*
  (a) T contains at most one tight edge; or
  (b) for every two neighboring tight edges e and e' in T, the path $P(e, e')$ contains at least one slack edge.
Moreover, there is a reconfiguration sequence of length $O(n^2)$ between f and f', where n is the number of vertices in T.

Notice that Condition (a) above is an extension of Theorem 1, and hence our sufficient condition improves the known one. We also remark that our sufficient condition can be checked in linear time: we first delete all slack edges from T, and check whether each subtree contains at most one tight edge. By Condition (b) above, it is obvious that no two tight edges share a vertex.

We will give a proof of Theorem 2 in Section 3. In the remainder of this section, we show that our improved sufficient condition is best possible in the sense that we cannot drop the condition of slack edges: we give an infinite family of instances on paths such that a given path consists only of tight or mild edges and two given L-edge-colorings are *not* connected.

Consider, as an example, the two L-edge-colorings f_0 and f_t in Fig. 2 for the same path P of 5 edges with the same list L. The two end edges e_0 and e_4 of P are tight, while all internal edges are mild. Therefore, P contains no slack edge. Then, we cannot recolor any edge in f_0, and hence f_0 and f_t are not connected.

We can construct such an instance for a path of an arbitrary length. Let $P = v_0, v_1, \ldots, v_{n-1}$ be a path with n vertices, and let $e_i = v_i v_{i+1}, 0 \le i \le n-2$. We first construct an ordinary edge-coloring f of P such that every two edges with distance 2 receive different colors, that is, $f(e_{i-1}) \ne f(e_{i+1})$ for each i, $1 \le i \le n - 3$. Note that f is a proper edge-coloring, and hence every two adjacent edges (i.e. with distance 1) receive different colors. The list L of P is constructed as follows: $L(e_i) = \{f(e_{i-1}), f(e_i), f(e_{i+1})\}$ for each internal edge e_i, $1 \le i \le n - 3$; $L(e_0) = \{f(e_0), f(e_1)\}$ and $L(e_{n-2}) = \{f(e_{n-3}), f(e_{n-2})\}$. Then, all internal edges are mild, and the two end edges e_0 and e_{n-2} are tight. Moreover, f is an L-edge-coloring of P such that any edge in P cannot be recolored. This implies that the condition of slack edges cannot be dropped from

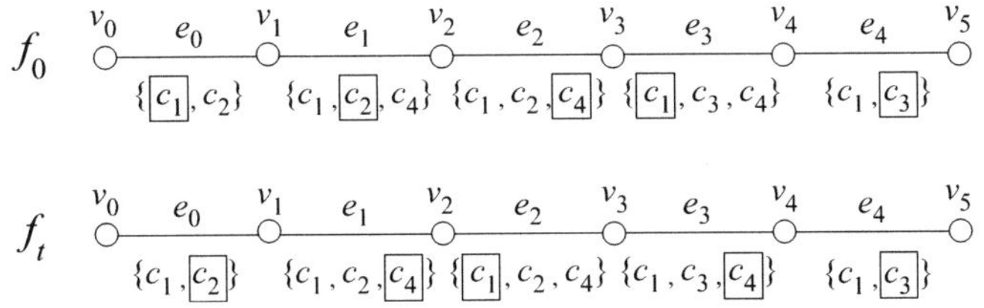

Fig. 2. Two L-edge-colorings f_0 and f_t of a path P which are not connected

a sufficient condition if a tree contains more than one tight edge. (We note that P has at least one L-edge-coloring other than f.)

3   Proof of Theorem 2

In this section, as a proof of Theorem 2, we give a polynomial-time algorithm that finds a reconfiguration sequence of length $O(n^2)$ between two given L-edge-colorings f_0 and f_t of a tree T if our sufficient condition holds.

3.1   Overview and Definitions

We first give an outline of our algorithm. Let T be a tree with a list L for which our sufficient condition holds. We say that an edge e in T is *fixed* if our algorithm decides not to recolor e anymore. Therefore, e must be colored with its target color $f_t(e)$ when it is fixed. The algorithm fixes the edges one by one in a certain order, and terminates when all the edges are fixed. The algorithm consists of the following three steps:

  Step 1: fix each tight edge without recoloring any other tight edges;
  Step 2: modify the tree T so that the degree of each endpoint of all tight edges is not two; and
  Step 3: fix all mild and slack edges.

We will show later in Lemma 1 that every tight edge can be recolored to its target color without recoloring any other tight edges (including non-fixed ones). Therefore, we can fix the tight edges in an arbitrary order in Step 1. After Step 2, every tight edge is fixed, and each endpoint v of all tight edges is of either $d(v) = 1$ or $d(v) \geq 3$; this condition is required in Step 3. In Step 3, we choose an arbitrary vertex r of degree 1, and regard T as a rooted tree whose root is r. We order all the edges $e^1, e^2, \ldots, e^{n-1}$ of T, roughly speaking, by the breadth-first search starting from r, and fix the edges in this order if the edge is not fixed yet. Therefore, e^i is never recolored after the i-th sub-step of Step 3, while non-fixed (mild or slack) edge e^j with $j > i$ may be recolored even if e^j is colored with $f_t(e^j)$. We will show later using Lemma 2 that every non-tight edge of T can be fixed in such a way, and hence we eventually obtain the target L-edge-coloring f_t. Our algorithm recolors each of non-fixed edges at most once for fixing an edge e, and hence e can be fixed by recoloring $O(n)$ edges. Thus, we obtain a reconfiguration sequence of total length $O(n^2)$.

We then give some definitions. For a tree T, we denote by $V(T)$ and $E(T)$ the vertex set and edge set of T, respectively. We sometimes denote by $d(v, T)$ the degree of a vertex v in T. For an L-edge-coloring f of T and a vertex v of T, we say that a color c is *available on v in f* if $c \notin \{f(vx) : vx \in E(T)\}$, that is, c is not assigned to any of the edges incident to v. For an L-edge-coloring f of T, an edge $e = vw$ of T and its endpoint v, we define a subset $C_{av}(f, e, v)$ of $L(e)$, as follows:

$$C_{av}(f, e, v) = L(e) \setminus \{f(vx) : vx \in E(T)\}. \tag{1}$$

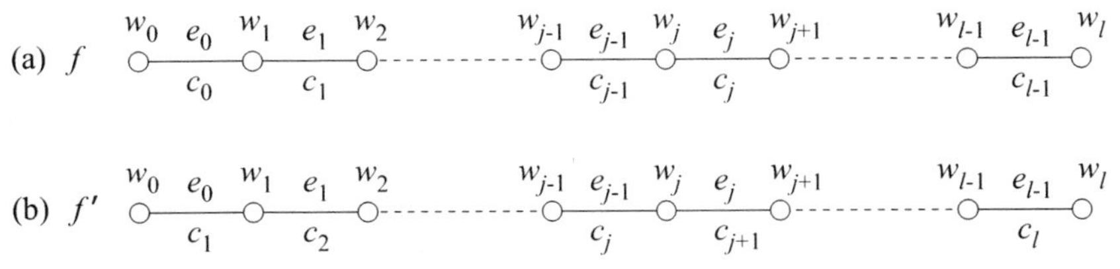

Fig. 3. (a) Recoloring path $P(f, w_0, c_0)$, and (b) L-edge-coloring f' obtained by recoloring the edges in $P(f, w_0, c_0)$

That is, $C_{\mathrm{av}}(f, e, v)$ is the set of all colors in $L(e)$ that are available on v for e. Therefore, $C_{\mathrm{av}}(f, e, v) \cap C_{\mathrm{av}}(f, e, w)$ is the set of all colors in $L(e)$ that are available for $e = vw$ when we wish to recolor e from $f(e)$.

For an L-edge-coloring f of a tree T, a vertex w_0 of T and a color c_0, a path $P(f, w_0, c_0) = w_0, w_1, \ldots, w_l$ in T is called a *recoloring path* (which will make the color c_0 available on w_0) if the following four conditions (i)–(iv) hold (see also Fig. 3(a)):

(i) every edge $e_j = w_j w_{j+1}$, $0 \le j \le l - 1$, in the path is not tight;
(ii) $f(e_0) = c_0$;
(iii) for each edge e_j, $1 \le j \le l - 1$, the color c_j assigned to e_j is available on w_{j-1} for e_{j-1} in f, that is, $f(e_j) = c_j \in C_{\mathrm{av}}(f, e_{j-1}, w_{j-1})$; and
(iv) there exists a color $c_l \in C_{\mathrm{av}}(f, e_{l-1}, w_{l-1}) \cap C_{\mathrm{av}}(f, e_{l-1}, w_l)$.

Along the recoloring path $P(f, w_0, c_0)$, we can obtain a reconfiguration sequence from f to an L-edge-coloring f' of T in which the color c_0 is available on w_0, as follows (see also Fig. 3(b)): at k-th recoloring step, $1 \le k \le l$, we recolor the edge e_{l-k} from the color $c_{l-k} = f(e_{l-k})$ to c_{l-k+1}. By the condition (i) above, we do not recolor any tight edge of T in this reconfiguration sequence. For the sake of convenience, a path $P(f, w_0, c_0) = w_0$ is also called a recoloring path if c_0 is already available on w_0 in f.

3.2 Algorithm

In this subsection, we precisely describe our algorithm and its analysis.

Step 1: Fix all tight edges
In Step 1, we fix all tight edges in a given tree T one by one. Let f be the current L-edge-coloring of T; initially, $f = f_0$. Choose an arbitrary non-fixed tight edge $e = u_0 u_0'$ of T. (See Fig. 4.) Then, there are at most two edges, say $u_0 u_1$ and $u_0' u_1'$, which are colored with $f_t(e)$ and sharing the endpoints u_0 and u_0' with e, respectively. Therefore, we first make the color $f_t(e)$ available on each of u_0 and u_0', and then recolor e to its target color $f_t(e)$. Formally, we have the following lemma.

Lemma 1. *For any L-edge-coloring f of T and an arbitrary tight edge $e = u_0 u_0'$ in T, there exist two recoloring paths $P(f, u_0, f_t(e))$ and $P(f, u_0', f_t(e))$. Moreover, the two recoloring paths can be found in linear time.*

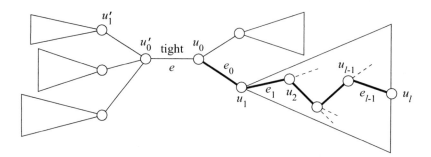

Fig. 4. Tight edge $e = u_0 u_0'$, and a recoloring path $P(f, u_0, f_t(e)) = u_0, u_1, \ldots, u_l$ depicted by thick lines

Proof. We prove only for the endpoint u_0, that is, there exists a recoloring path $P(f, u_0, f_t(e))$. (The proof for the other endpoint u_0' is similar.) Remember that the path $P(f, u_0, f_t(e))$ consisting of a single vertex u_0 is a recoloring path if the color $f_t(e)$ is already available on u_0 in f; in this case, the lemma clearly holds. Therefore, in the remainder of the proof, we may assume that $f_t(e)$ is not available on u_0 in f.

Let $e_0 = u_0 u_1$ be the edge which is colored with $f_t(e)$. (See Fig. 4.) Since e is tight and no two tight edges are adjacent, e_0 is either mild or slack. We give the procedure RECOLORPATH which greedily constructs a recoloring path $P(f, u_0, f_t(e))$. The procedure starts with RECOLORPATH(f, P_1, e) where $P_1 = u_0, u_1$, and always maintains a path $P_k = u_0, u_1, \ldots, u_k$, $k \geq 1$, in T starting from u_0 such that

  (i) every edge $e_j = u_j u_{j+1}$, $0 \leq j \leq k-1$, in the path P_k is not tight;
  (ii) $f(e_0) = f_t(e)$; and
  (iii) for each edge e_j, $1 \leq j \leq k-1$, the color c_j assigned to e_j is available on u_{j-1} for e_{j-1} in f, that is, $f(e_j) = c_j \in C_{av}(f, e_{j-1}, u_{j-1})$.

Therefore, if there is a color c_k in $C_{av}(f, e_{k-1}, u_{k-1}) \cap C_{av}(f, e_{k-1}, u_k)$, then P_k is a recoloring path $P(f, u_0, f_t(e))$, as required. (This corresponds to Lines 3–5 of the procedure.)

Consider the case where $C_{av}(f, e_{k-1}, u_{k-1}) \cap C_{av}(f, e_{k-1}, u_k) = \emptyset$, which corresponds to Lines 7–16 of the procedure. By the condition (i) above, the edge $e_{k-1} = u_{k-1} u_k$ is either mild or slack, and hence we have $|L(e_{k-1})| \geq \max\{d(u_{k-1}), d(u_k)\} + 1$. By Eq. (1) we then have

$$|C_{av}(f, e_{k-1}, u_{k-1})| \geq |L(e_{k-1})| - |\{f(u_{k-1}x) : u_{k-1}x \in E(T)\}|$$
$$\geq \max\{d(u_{k-1}), d(u_k)\} + 1 - d(u_{k-1})$$
$$\geq 1.$$

Therefore, $L(e_{k-1})$ contains at least one color in $C_{av}(f, e_{k-1}, u_{k-1})$, which is available on u_{k-1} for e_{k-1}. Since $C_{av}(f, e_{k-1}, u_{k-1}) \cap C_{av}(f, e_{k-1}, u_k) = \emptyset$, each

---

Procedure 1. RECOLORPATH(f, P_k, e)

1: Let $P_k = u_0, u_1, \ldots, u_k$, $k \geq 1$
2: Let $e_j = u_j u_{j+1}$, $0 \leq j \leq k - 1$, be a mild or slack edge
3: **if** $C_{\mathrm{av}}(f, e_{k-1}, u_{k-1}) \cap C_{\mathrm{av}}(f, e_{k-1}, u_k) \neq \emptyset$ **then**
4:     {The path P_k is a desired recoloring path $P(f, u_0, f_t(e))$}
5:     **return** P_k
6: **else**
7:     **if** $C_{\mathrm{av}}(f, e_{k-1}, u_{k-1})$ contains a color assigned to a non-tight edge $u_k u_{k+1}$ **then**
        {See Case (a) in the proof}
8:         Let $P_{k+1} = u_0, u_1, \ldots, u_k, u_{k+1}$
9:         RECOLORPATH(f, P_{k+1}, e)
10:    **else** {See Case (b) in the proof}
11:        {the color in $C_{\mathrm{av}}(f, e_{k-1}, u_{k-1})$ is assigned to the tight edge $u_k u'$}
12:        Let $e_{s-1} = u_{s-1} u_s$ be the slack edge in P_k closest to $u_k u'$
13:        Let $e'_s = u_s u'_{s+1}$ be a non-tight edge in T such that $u'_{s+1} \neq u_{s+1}$ and $f(e'_s) \in C_{\mathrm{av}}(f, e_{s-1}, u_{s-1})$
14:        Let $P'_{s+1} = u_0, u_1, \ldots, u_s, u'_{s+1}$
15:        RECOLORPATH(f, P'_{s+1}, e)
16:    **end if**
17: **end if**

---

color in $C_{\mathrm{av}}(f, e_{k-1}, u_{k-1})$ is assigned to some edge incident to u_k other than e_{k-1}. Then, there are the following two cases to consider.

Case (a): $C_{\mathrm{av}}(f, e_{k-1}, u_{k-1})$ contains a color assigned to a non-tight edge $u_k u_{k+1}$. In this case, the path $u_0, u_1, \ldots, u_k, u_{k+1}$ satisfies the three conditions (i)–(iii) above. Let $P_{k+1} = u_0, u_1, \ldots, u_k, u_{k+1}$, and we call RECOLORPATH(f, P_{k+1}, e) recursively. This case corresponds to Lines 7–9 of the procedure.

Case (b): the color in $C_{\mathrm{av}}(f, e_{k-1}, u_{k-1})$ is assigned to the tight edge $u_k u'$. Since at most one tight edge is incident to a vertex, $C_{\mathrm{av}}(f, e_{k-1}, u_{k-1})$ contains exactly one color in this case. Remember that u_0 is an endpoint of the tight edge e. Since the path P_k contains no tight edge, the tight edges e and $u_k u'$ are neighboring. Then, by Condition (b) of the sufficient condition, the path $P_k = u_0, u_1, \ldots, u_k$ contains at least one slack edge. Let $e_{s-1} = u_{s-1} u_s$ be the slack edge in P_k closest to $u_k u'$. Since $|L(e_{s-1})| \geq \max\{d(u_{s-1}), d(u_s)\} + 2$, we have

$$|C_{\mathrm{av}}(f, e_{s-1}, u_{s-1})| \geq \max\{d(u_{s-1}), d(u_s)\} + 2 - d(u_{s-1}) \geq 2.$$

Therefore, $C_{\mathrm{av}}(f, e_{s-1}, u_{s-1})$ contains at least two colors, one of which is assigned to the edge $e_s = u_s u_{s+1}$ in P_k. Thus, we have a color c'_s in $C_{\mathrm{av}}(f, e_{s-1}, u_{s-1})$ which is *not* assigned to the edge e_s. Let $e'_s = u_s u'_{s+1}$ be the edge in T such that $f(e'_s) = c'_s$. Note that e'_s is not tight; otherwise, since there must exist a slack edge between e'_s and $u_k u'$, the edge e_{s-1} would not be the closest slack edge to $u_k u'$. Then, the path $P'_{s+1} = u_0, u_1, \ldots, u_s, u'_{s+1}$ satisfies the three conditions (i)–(iii) above. Thus, we recursively call RECOLORPATH(f, P'_{s+1}, e).

To complete the proof, we show that RECOLORPATH(f, P_1, e) will terminate. As in Cases (a) and (b) above, the procedure can always find a next edge (either $u_k u_{k+1}$ or $u_s u'_{s+1}$) for the path P_k, keeping the three conditions (i)–(iii). Thus, it suffices to show that each slack edge is picked up in Line 12 of the procedure at most once. Then, we eventually reach a vertex u_l such that $C_{av}(f, e_{l-1}, u_{l-1}) \cap C_{av}(f, e_{l-1}, u_l) \neq \emptyset$, because any vertex u_l of degree 1 clearly satisfies this condition. (See Fig. 4.)

Suppose for a contradiction that a slack edge e_s is picked up more than once, for two tight edges e' and e''. By the condition (i), both $P(e_s, e')$ and $P(e_s, e'')$ contain no tight edge except for e' and e''. Therefore, e' and e'' are neighboring. But then, there must exist at least one slack edge in $P(e', e'')$. Note that e_s is not in $P(e', e'')$. This contradicts the fact that e_s is the closest slack edge to both e' and e''.   □

By Lemma 1 we can recolor the tight edge e from $f(e)$ to $f_t(e)$ by recoloring $O(n)$ edges in total. Since a recoloring path does not contain any tight edge, e can be fixed without recoloring any other tight edges. We can thus fix the tight edges in an arbitrary order.

Step 2: Modify T so that the degree of each endpoint of all tight edges is not two

Let f be the current L-edge-coloring of T after Step 1. Then, all tight edges are fixed, and hence $f(e) = f_t(e)$ for each tight edge e in T. In Step 2, we modify T so that $d(v) \neq 2$ for each endpoint v of all tight edges in T.

Choose an arbitrary tight edge xy, at least one of whose endpoints is of degree 2, say $d(y) = 2$. Let z be the vertex adjacent to y other than x, as illustrated in Fig. 5(a). We divide T into two subtrees T_x and T_z at y, where T_x and T_z are the subtrees containing the edges xy and yz, respectively. (See Figs. 5(a) and (b).) The list L_x of T_x is defined as the restriction of the list L to T_x, that is, $L_x(e) = L(e)$ for each edge $e \in E(T_x)$. Then, since $d(v, T) \geq d(v, T_x)$ for each vertex $v \in V(T_x)$, our sufficient condition holds for L_x of T_x. On the other hand, the list L_z of T_z is defined as follows:

$$L_z(e) = \begin{cases} L(e) & \text{if } e \neq yz; \\ L(yz) \setminus \{f_t(xy)\} & \text{if } e = yz. \end{cases} \quad (2)$$

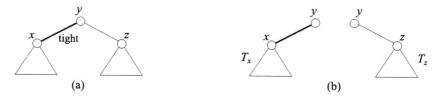

(a)   (b)

Fig. 5. (a) Tight edge xy with $d(y) = 2$, and (b) two subtrees T_x and T_z

We now show that our sufficient condition holds for L_z of T_z. Since no two tight edges are adjacent, yz was not tight in T. Therefore, we have $|L(yz)| \geq \max\{d(y,T), d(z,T)\} + 1$, and hence by Eq. (2) we have

$$|L_z(yz)| \geq \max\{d(y,T), d(z,T)\} \geq \max\{d(y,T_z), d(z,T_z)\}.$$

For any two neighboring tight edges e' and e'' in $E(T_z) \setminus \{yz\}$, the path $P(e', e'')$ in T_z does not contain yz since $d(y, T_z) = 1$. Thus, our sufficient condition clearly holds for L_z if $|L_z(yz)| \geq \max\{d(y,T_z), d(z,T_z)\} + 1$, that is, yz is either mild or slack in T_z. Suppose now that $|L_z(yz)| = \max\{d(y,T_z), d(z,T_z)\}$, and hence yz is tight in T_z. Since no two tight edges are adjacent, yz was mild in T. Then, since xy is tight in T and $d(y, T_z) = 1$, it is easy to see that our sufficient condition holds for T_z.

Since the edge xy is fixed and our sufficient condition holds for both L_x and L_z, by Eq. (2) it suffices to find reconfiguration sequences for T_x and T_z independently. We repeatedly apply Steps 1 and 2 to each subtree if the subtree has either a non-fixed tight edge (which is newly made by Step 2) or a tight edge one of whose endpoints is of degree 2.

Step 3: Fix all mild and slack edges
After Step 2, we may have a forest. However, by the construction, it suffices to find a reconfiguration sequence for each subtree independently. Let T be a subtree in the forest, and let f be the current L-edge-coloring of T. Then, every tight edge e in T is fixed, and either $d(v) = 1$ or $d(v) \geq 3$ for an endpoint v of e.

Choose an arbitrary vertex r of degree 1 in T, and regard T as a rooted tree whose root is r. For an edge $e = uv$ joining a vertex $v \in V(T) \setminus \{r\}$ and its parent u, we denote by T_e the subtree of T which is induced by the edge $e = uv$ and all descendants of v in T. (See Fig.6(a).) It should be noted that T_e does not include the edges incident to u other than e, and hence $d(u, T_e) = 1$. Therefore, T_e consists of a single edge $e = uv$ if v is a leaf of T.

We order all the edges $e^1, e^2, \ldots, e^{n-1}$ of T by the breadth-first search starting from r so that the following two conditions (1) and (2) hold: for each internal vertex u,

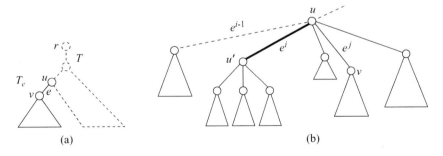

Fig. 6. (a) Subtree T_e in the whole tree T, and (b) i-th sub-step of Step 3

(1) if there exists a tight edge joining u and its child v, then the tight edge uv appears before all the other edges joining u and its children; and

(2) if there exists a mild edge joining u and its child v to which a tight edge vw is incident, then the mild edge uv appears before all the other (non-tight) edges joining u and its children.

Note that, for each internal vertex u, there exists at most one mild edge satisfying Condition (2) above. We fix the edges $e^1, e^2, \ldots, e^{n-1}$ in this order if the edge is not fixed yet.

We now have the following lemma, whose proof is omitted from this extended abstract. (Remember that all tight edges are already fixed.)

Lemma 2. *Let $e = uv$ be an arbitrary non-tight edge in T such that all non-tight edges in T_e are not fixed yet, where u is the parent of $v \in V(T) \setminus \{r\}$. Let c be any color in $C_{av}(f, e, u)$ which is not assigned to a tight edge incident to v. Then, there exists an L-edge-coloring f' of T such that $f'(e) = c$ and f' can be obtained by recoloring each non-tight edge in T_e at most once.*

Remember that $C_{av}(f, e, u) = L(e) \setminus \{f(ux) : ux \in E(T)\}$, and hence c is not assigned to any of the edges incident to u in the whole tree T.

Suppose that we are now in the i-th sub-step of Step 3. Then, the edges $e^1, e^2, \ldots, e^{i-1}$ are all fixed, and we wish to fix $e^i = uu'$. (See Fig. 6(b).) If $f_t(e^i) \in C_{av}(f, e^i, u) \cap C_{av}(f, e^i, u')$, that is, the color $f_t(e^i)$ is already available on both endpoints u and u' for e^i, then we simply recolor e^i to $f_t(e^i)$ and fix it.

We now consider the case $f_t(e^i) \notin C_{av}(f, e^i, u)$ for the endpoint u which is the parent of u'. Then, the color $f_t(e^i)$ is assigned to some edge $e^j = uv$ incident to u. But, in the target L-edge-coloring f_t, the color $f_t(e^i)$ is *not* assigned to any edge incident to u other than $e^i = uu'$. Thus, the edge e^j is not fixed yet, and hence $j > i$ and e^j is either mild or slack. Moreover, all non-tight edges in the subtree T_{e^j} are not fixed. If e^j is mild, then we have

$$|C_{av}(f, e^j, u)| \geq \max\{d(u), d(v)\} + 1 - d(u) \geq 1.$$

Thus, $C_{av}(f, e^j, u)$ contains at least one color c. It is easy to see that there is no tight edge incident to v; otherwise, uv satisfies Condition (2) in the breadth-first search and hence uv must appear before the non-tight edge $e^i = uu'$. (Remember that only one edge satisfies the condition.) Therefore, any color in $C_{av}(f, e^j, u)$ satisfies the assumption of Lemma 2. On the other hand, if e^j is slack, then

$$|C_{av}(f, e^j, u)| \geq \max\{d(u), d(v)\} + 2 - d(u) \geq 2$$

and hence $C_{av}(f, e^j, u)$ contains at least two colors. Therefore, there exists at least one color $c \in C_{av}(f, e^j, u)$ which is not assigned to a tight edge incident to v. (Remember that at most one tight edge is incident to a vertex.) Thus, regardless of mild or slack, we can apply Lemma 2 to the color c and the edge $e^j = uv$, and obtain an L-edge-coloring f' such that $f'(e^j) = c$. Since no edge in $T \setminus T_{e^j}$ is recolored to obtain f' and the color $f_t(e^i)$ was assigned to $e^j = uv$ in f, we thus have $f_t(e^i) \in C_{av}(f', e^i, u)$.

We then consider the case where $f_t(e^i) \notin C_{\text{av}}(f, e^i, u')$. Let f' be the L-edge-coloring of T obtained above; let $f' = f$ if $f_t(e^i) \in C_{\text{av}}(f, e^i, u)$. Then, $f_t(e^i) \in C_{\text{av}}(f', e^i, u)$. Note that, since no edge in $T \setminus T_{e^j}$ is recolored to obtain f', we have $C_{\text{av}}(f', e^i, u') = C_{\text{av}}(f, e^i, u')$. (See also Fig. 6(b).) In addition, $f_t(e^i)$ is not assigned to any fixed (and hence tight) edge incident to u', and all non-tight edges in T_{e^i} are not fixed yet. Therefore, we apply Lemma 2 to the color $f_t(e^i)$ and the edge $e^i = uu'$, and obtain an L-edge-coloring f'' such that $f''(e^i) = f_t(e^i)$. We can now fix the edge e^i, as required.

Note that we recolor each non-tight edge in the subtrees T_{e^i} and T_{e^j} at most once. Thus, the edge e^i can be fixed by recoloring $O(n)$ non-fixed edges in total.

We have shown that our polynomial-time algorithm correctly fixes all the edges in a tree T, and hence it suffices to show that there exists a reconfiguration sequence of length $O(n^2)$ between f_0 and f_t. As in Lemmas 1 and 2, our algorithm fixes each edge by recoloring $O(n)$ edges in total. Thus, all the edges can be fixed via $O(n^2)$ recoloring steps, as required.

4 Conclusion

In this paper, we gave a new sufficient condition that improves the known one [6]. We showed that our sufficient condition is best possible in the sense that we cannot drop the condition of slack edges. Our proof yields a polynomial-time algorithm that finds a reconfiguration sequence of length $O(n^2)$ between two given L-edge-colorings of a tree T, where n is the number of vertices in T.

References

1. Bonsma, P., Cereceda, L.: Finding paths between graph colourings: PSPACE-completeness and superpolynomial distances. Theoretical Computer Science 410, 5215–5226 (2009)
2. Borodin, O.V., Kostochka, A.V., Woodall, D.R.: List edge and list total colourings of multigraphs. J. Combinatorial Theory, B 71, 184–204 (1997)
3. Gopalan, P., Kolaitis, P.G., Maneva, E.N., Papadimitriou, C.H.: The connectivity of Boolean satisfiability: computational and structural dichotomies. SIAM J. Computing 38, 2330–2355 (2009)
4. Hearn, R.A., Demaine, E.D.: PSPACE-completeness of sliding-block puzzles and other problems through the nondeterministic constraint logic model of computation. Theoretical Computer Science 343, 72–96 (2005)
5. Ito, T., Demaine, E.D., Harvey, N.J.A., Papadimitriou, C.H., Sideri, M., Uehara, R., Uno, Y.: On the complexity of reconfiguration problems. Theoretical Computer Science 412, 1054–1065 (2011)
6. Ito, T., Kamiński, M., Demaine, E.D.: Reconfiguration of list edge-colorings in a graph. In: Dehne, F., Gavrilova, M., Sack, J.-R., Tóth, C.D. (eds.) WADS 2009. LNCS, vol. 5664, pp. 375–386. Springer, Heidelberg (2009)
7. Jensen, T.R., Toft, B.: Graph Coloring Problems. Wiley Interscience, New York (1995)

A Compact Encoding of Unordered Binary Trees

Kozue Iwata, Shiro Ishiwata, and Shin-ichi Nakano

Gunma University, Kiryu 376-8515, Japan
nakano@cs.gunma-u.ac.jp

Abstract. A naive encoding of ordered binary trees with n vertices needs $2n$ bits for each tree, and it is asymptotically optimal. In this paper we give a new simple encoding of unordered binary trees. The encoding needs at most $4 + 1.4n$ bits for each tree. Our encoding and decoding algorithms are simple and run in $O(n)$ time.

Keywords: trees, coding, compact representation.

1 Introduction

Studying representation issues on graphs is very natural[7]. In this paper we consider the problem of encoding a given unordered binary tree into a binary string with fewer bits.

A rooted tree is *ordered* if the children of each vertex are ordered. A rooted tree is (ordered)*binary* if each vertex has at most two children, that is, a possible "left child" and a possible "right child". Note that each only child must be designated either as a left child or a right child, so ordered binary trees are not just a subclass of ordered trees. A rooted tree is *unordered binary* if each vertex has at most two children and the children of each vertex are unordered. The expression trees which represent the arithmetic expressions are examples of unordered binary trees. Those trees are among the most fundamental models of computer sciences.

There are a lot of papers for encoding ordered trees and binary trees. For example, see [3,4,1] and referred papers.

The well known naive encoding of ordered trees is as follows. Given an ordered tree T we traverse T starting at the root in a depth first manner. Output an open parenthesis (or "0") when visiting a vertex for the first time, and output a matching closing parenthesis (or "1") when visiting a vertex for the last time. Thus any ordered tree with n vertices has a code with $2n$ bits. One can easily decode the code into the original ordered tree T.

The number of ordered trees with $n \geq 2$ vertices is known as the Catalan number C_{n-1}, and it is defined as follows[6, p.145].

$$C_n = \frac{1}{(n+1)} \frac{(2n)!}{n!n!}$$

Since the Catalan number can be denoted as follows[2, p.495],

$$C_n = \frac{4^n}{(n+1)\sqrt{\pi n}}(1 - \frac{1}{8n} + \frac{1}{128n^2} + \frac{5}{1024n^3} - \frac{21}{32768n^4} + O(n^{-5}))$$

M. Ogihara and J. Tarui (Eds.): TAMC 2011, LNCS 6648, pp. 106–113, 2011.

we need at least $\log C_{n-1} = 2n - o(n)$ bits on average to encode an arbitrary ordered tree, where n is the number of vertices. So the naive encoding with $2n$ bits is asymptotically optimal.

Similarly the number of binary trees with n vertices is C_n. Since $\log C_n = 2n - o(n)$ holds, several known encodings with $2n$ bits are also asymptotically optimal for binary trees.

On the other hand the number of unordered binary trees with n vertices can be computed by a recursive equation and is reported at [8]. The numbers are $1, 1, 2, 3, 6, 11, 23, 46, 98, 207, 451, 983, 2179, \cdots$, for $n = 1, 2, \cdots$, and those are much less than the Catalan numbers. Thus there is a chance to design a code for unordered binary trees with less than $2n$ bits.

In this paper we give a new encoding of unordered binary trees with at most $4 + 1.4n$ bits. Our encoding and decoding algorithms are simple and run in $O(n)$ time.

2   Pair-Single Decomposition

Let T be a given (ordered or unordered) binary tree. We define a partition of the vertices of T into pairs and singles. Each pair consists of two vertices, while each single consists of one vertices.

A *pair* consists of two vertices with the following condition (a) or (b).

(a) two sibling vertices ℓ and r form a pair (ℓ, r).
(b) If (i) two vertices ℓ and r form a pair, (ii) ℓ has exactly one child, say c_ℓ, and (iii) r has exactly one child, say c_r, then the two vertices c_ℓ and c_r form a pair (c_ℓ, c_r).

Each vertex v of T belonging to no pair forms a single (v). Note that each vertex belongs to either a pair or a single, and the partition is unique. See an example in Fig. 1. Each pair and single is surrounded by an ellipse.

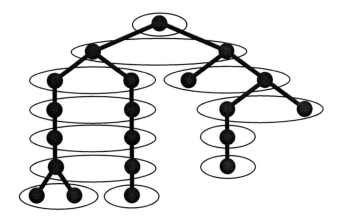

Fig. 1. An example of the pair-single partition

Let T be a given unordered binary tree, and T_o the ordered binary tree derived from T by **Algorithm dfc(v)** below. Then we compute the pair-single decomposition of T_o. When we write a pair (ℓ, r) in T_o we assume ℓ is locating to the left of r.

Algorithm 1. Algorithm dfc(v);

Input   : An unordered binary tree;
Output: An ordered binary tree in which for each pair(ℓ, r) the number
            of children of ℓ is smaller than or equal to the number of
            children of r;
1 **if** v *has no child* **then return** ;
2 **if** v *has exactly one child, say v_c,* **then**
3 $\quad$ dfc(v_c) ;
4 $\quad$ **return** ;
5 **end**
6 // now v has two children ;
7 Set ℓ as the left child of v;
8 Set r as the right child of v;
9 **while** *(ℓ has exactly one child) and (r has exactly one child)* **do**
10 $\quad$ $\ell \leftarrow$ the child of ℓ ;
11 $\quad$ $r \leftarrow$ the child of r ;
12 $\quad$ // each pair in which both ℓ and r have exactly one child is skipped ;
13 **end**
14 **if** ℓ *has more children than r has* **then** exchange ℓ and r;
15 dfc(ℓ);
16 dfc(r);
17 **return** ;

We say a pair (ℓ, r) is $type(i, j)$ if ℓ has exactly i children and r has exactly j children. Now for each pair (ℓ, r) in T_o, the number of children of ℓ is smaller than or equal to the number of children of r, since otherwise the algorithm has swapped the pair. Thus we have the following six types for pairs; $type(0,0)$, $type(0,1)$, $type(0,2)$, $type(1,1)$, $type(1,2)$, $type(2,2)$. Also a single (v) is $type(i)$ if v has exactly i children. We have the following three types for singles; $type(0)$, $type(1)$, $type(2)$.

We need more notations. For $i = 0, 1, 2$, let n_i be the number of vertices in T_o having exactly i children. Also for $i = 0, 1, 2$, let n_i' be the number of singles in T_o wity $type(i)$. For i, j with $0 \leq i \leq j \leq 2$, let p_{ij} be the number of pairs in T_o with $type(i, j)$. Now we have the followings.

$$n = n_0 + n_1 + n_2 \tag{1}$$

$$n_0' = n_0 - 2p_{00} - p_{01} - p_{02} \tag{2}$$

$$n_1' = n_1 - p_{01} - 2p_{11} - p_{12} \tag{3}$$

$$n_2' = n_2 - p_{02} - p_{12} - 2p_{22} \tag{4}$$

$$n_0 = n_2 + 1 \tag{5}$$
$$n_2 = p_{00} + p_{01} + p_{02} + p_{12} + p_{22} \tag{6}$$

3   Encoding

Let T be a given unordered binary tree, and T_o the derived ordered tree from T by **Algorithm dfc**, and $(v_1, v_2, \cdots, v_n)$ the vertices of T_o in preorder. For each single (v_i) define $\ell_A(v_i)$, $\ell_B(v_i)$ and $\ell_C(v_i)$ depending on its type as shown in Table 1. Similarly for each pair (v_i, v), define $\ell_A(v_i)$, $\ell_B(v_i)$ and $\ell_C(v_i)$ depending on its type as shown in Table 2, and define $\ell_A(v) = \ell_B(v) = \ell_C(v) = \varepsilon$ (= empty string). By concatenating those bit strings we define three bit strings $L_A(T), L_B(T)$ and $L_C(T)$ as follows. Later in Section 4 we show one can reconstruct T from any of those three bit strings.

$$L_A(T) = \ell_A(v_1) + \ell_A(v_2) + \ell_A(v_3) + \cdots + \ell_A(v_n)$$
$$L_B(T) = \ell_B(v_1) + \ell_B(v_2) + \ell_B(v_3) + \cdots + \ell_B(v_n)$$
$$L_C(T) = \ell_C(v_1) + \ell_C(v_2) + \ell_C(v_3) + \cdots + \ell_C(v_n)$$

The length of $L_A(T)$ is as follows.

$$
\begin{aligned}
|L_A(T)| &= 2n_0' + 2n_1' + n_2' + 2p_{00} + 3p_{01} + 2p_{02} + 3p_{11} + 3p_{12} + 3p_{22} \\
&= 2n_0 + 2n_1 + n_2 - 2p_{00} - p_{01} - p_{02} - p_{11} + p_{22} \quad \text{by equation (1)–(3)} \\
&= 2 + 2n_1 + 3n_2 - 2p_{00} - p_{01} - p_{02} - p_{11} + p_{22} \quad \text{by equation (4)} \\
&= 2 + 2n_1 + p_{00} + 2p_{01} + 2p_{02} - p_{11} + 3p_{12} + 4p_{22} \quad \text{by equation (5)}
\end{aligned}
$$

Similarly we have the followings.

$$|L_B(T)| = 2 + n_1 + 3p_{00} + 3p_{01} + 3p_{02} + p_{11} + 3p_{12} + 3p_{22}$$
$$|L_C(T)| = 1 + 2n_1 + 4p_{00} + 3p_{01} + 3p_{02} - 2p_{11} + 2p_{12} + p_{22}$$

Table 1. Code for singles

	type (0)	type (1)	type (2)
$l_A(v_i)$	10	11	0
$l_B(v_i)$	10	0	11
$l_C(v_i)$	0	10	11

Table 2. Code for pairs

	type(0,0)	type(0,1)	type(0,2)	type(1,1)	type(1,2)	type(2,2)
$l_A(v_i)$	00	100	01	101	110	111
$l_B(v_i)$	100	00	101	110	01	111
$l_C(v_i)$	100	101	110	00	111	01

Now we show the shortest one among the three bit strings $L_A(T), L_B(T), L_C(T)$ has the length at most $2 + 1.4n$.

Lemma 1. $\min\{|L_A(T)|, |L_B(T)|, |L_C(T)|\} \leq 2 + 1.4n$

Proof. See Appendix A.

We need two bits to denote which bit string among the three to use, then append the chosen bit string. The derived bit string is the code for T. We have the following theorem.

Theorem 1. *One can encode a given unordered binary tree with n vertices into a bitstring of at most $4 + 1.4n$ bits. Encoding runs in $O(n)$ time.*

4 Decoding

Given $L_A(T)$ one can reconstruct the original unordered binary tree T as follows.

Let T_o be the derived ordered tree from T by **Algorithm dfc**, and $(v_1, v_2, \cdots, v_n)$ the vertices of T_o in preorder.

Assume inductively we have reconstructed the subtree T_o' of T_o induced by $(v_1, v_2, \cdots, v_i)$, and we know the pair-single decomposition of T_o' and the type of each pair and single. Assume also we have decoded $L_A(T)$ upto some prefix of $L_A(T)$ corresponding to the codes for $(v_1, v_2, \cdots, v_i)$, and $L_A(T)'$ is the remaining code for $(v_{i+1}, v_{i+2}, \cdots, v_n)$.

For each pair (ℓ, r) of $type(x, y)$ we say ℓ is $type(x)$ and r is $type(y)$. We say ℓ is **the left vertex** of the pair (ℓ, r). Also for each single (v) of $type(x)$ we say v is $type(x)$.

Find the first leaf vertex v of T_o' in preorder which is not $type(0)$. Intuitively this is the vertex from which a new branch to grow to the next vertex v_{i+1}. Now we can know the next vertex v_{i+1} to append belong to a pair or a single as follows. We have the following three cases.

Case 1: v belong to a single.
If v is $type(2)$ then the next vertex v_{i+1} belong to a pair. Otherwise v is $type(1)$ and the next vertex v_{i+1} belong to a single.

Case 2: v belong to a pair, and v is the left vertex of the pair.
If v belong to a pair with $type(1, 1)$ then the next vertex v_{i+1} belong to a pair. Otherwise v is $type(1)$ or $type(2)$. If v is $type(2)$ then the next vertex v_{i+1} belong to a pair. Otherwise the next vertex v_{i+1} belong to a single.

Case 3: v belong to a pair, and v is not the left vertex of the pair.
If v is $type(2)$ then the next vertex v_{i+1} belong to a pair. Otherwise v is $type(1)$ and the next vertex v_{i+1} belong to a single.

Thus we can decide the next vertex v_{i+1} to append belong to a pair or a single. Since the code in Table 1 and 2 are prefix code, we can easily cutoff a suitable code for v_{i+1} from $L_A(T)'$, and append a pair or single to T_o'.

In this way we can reconstruct the original unordered binary tree T from $L_A(T)$. Similarly we can reconstruct T from $L_B(T)$ or $L_C(T)$.

With a suitable data structure the encoding runs in $O(n)$ time.

5   Conclusion

One can encode an unordered binary tree with n vertices into a bitstring of $4 + 1.4n$ bits. Both encoding and decoding run in $O(n)$ time.

References

1. Geary, R.F., Raman, R., Raman, V.: Succinct ordinal trees with level-ancestor queries. ACM Transactions on Algorithms 2, 510–534 (2006)
2. Graham, R.L., Knuth, D.E., Patashnik, O.: Exercise 9.60 in Concrete Mathematics, 2nd edn. Addison-Wesley, Reading (1994)
3. Jacobson, G.: Space-efficient Static Trees and Graphs. In: FOCS, pp.549–554 (1989)
4. Munro, J.I., Raman, V.: Succinct Representation of Balanced Parentheses and Static Trees. SIAM J. Comput. 31, 762–776 (2001)
5. Nakano, S.: Efficient generation of plane trees. Inf. Process. Lett. 84, 167–172 (2002)
6. Rosen, K.H. (ed.): Handbook of Discrete and Combinatorial Mathematics. CRC Press, Boca Raton (2000)
7. Spinrad, J.P.: Efficient Graph Representations. AMS, Providence (2003)
8. Sequence A001190 and A036656 in The On-Line Encyclopedia of Integer Sequences (OEIS), http://oeis.org/

Appendix A

Proof of Lemma 1

We prove a stronger result as follows. Note that the third term in the min increased by one.

$$\min\{|L_A(T)|, |L_B(T)|, 1 + |L_C(T)|\} \leq 2 + 1.4n$$

Remember the followings.

$$|L_A(T)| = 2 + 2n_1 + p_{00} + 2p_{01} + 2p_{02} - p_{11} + 3p_{12} + 4p_{22} \qquad (7)$$
$$|L_B(T)| = 2 + n_1 + 3p_{00} + 3p_{01} + 3p_{02} + p_{11} + 3p_{12} + 3p_{22} \qquad (8)$$
$$1 + |L_C(T)| = 2 + 2n_1 + 4p_{00} + 3p_{01} + 3p_{02} - 2p_{11} + 2p_{12} + p_{22} \qquad (9)$$

Define $L'(T) = \min\{|L_A(T)|, |L_B(T)|, 1 + |L_C(T)|\}$.

Let X be an assignment such that $n_1 = 5n/15$ and $p_{00} = p_{01} = p_{02} = p_{11} = p_{12} = p_{22} = n/15$. We show $L'(T)$ has the maximum value at the assignment X.

Assume otherwise for a contradiction. Now $L'(T)$ has the maximum value at some assignment $Y \neq X$. Denote the assignment Y as follows.

$$n_1 = 5n/15 + \alpha_1 \tag{10}$$
$$p_{00} = n/15 + \alpha_{00} \tag{11}$$
$$p_{01} = n/15 + \alpha_{01} \tag{12}$$
$$p_{02} = n/15 + \alpha_{02} \tag{13}$$
$$p_{11} = n/15 + \alpha_{11} \tag{14}$$
$$p_{12} = n/15 + \alpha_{12} \tag{15}$$
$$p_{22} = n/15 + \alpha_{22} \tag{16}$$
$$\tag{17}$$

By equations (1), (5), (6) we have the following.

$$n = 1 + n_1 + 2(p_{00} + p_{01} + p_{02} + p_{12} + p_{22}) \tag{18}$$

By applying the assignment Y to (18) we have the following.

$$\alpha_1 + 2(\alpha_{00} + \alpha_{01} + \alpha_{02} + \alpha_{12} + \alpha_{22}) = 0$$
$$\alpha_{00} + \alpha_{01} + \alpha_{02} + \alpha_{12} + \alpha_{22} = -\alpha_1/2 \tag{19}$$

Since the assumption means the value of $L'(T)$ at Y is greater than the value of $L'(T)$ at X, we have the following (a),(b) and (c).

(a) The value $|L_A(T)|$ at Y is greater than the value of $|L_A(T)|$ at X.

(b) The value $|L_B(T)|$ at Y is greater than the value of $|L_B(T)|$ at X.

(c) The value $1 + |L_C(T)|$ at Y is greater than the value of $1 + |L_C(T)|$ at X.
Since (b) holds, we have

$$\alpha_1 + 3(\alpha_{00} + \alpha_{01} + \alpha_{02} + \alpha_{12} + \alpha_{22}) + \alpha_{11} > 0 \tag{20}$$

By applying (19) to (20)

$$\alpha_1 + 3(-\alpha_1/2) + \alpha_{11} > 0$$
$$\alpha_{11} > \alpha_1/2 \tag{21}$$

Since (a) and (c) hold, the value $|L_A(T)| + 1 + |L_C(T)|$ at Y is greater than the value $|L_B(T)|$ at X. Now we have the followings.

$$(2\alpha_1 + \alpha_{00} + 2\alpha_{01} + 2\alpha_{02} - \alpha_{11} + 3\alpha_{12} + 4\alpha_{22})$$
$$+(2\alpha_1 + 4\alpha_{00} + 3\alpha_{01} + 3\alpha_{02} - 2\alpha_{11} + 2\alpha_{12} + \alpha_{22}) > 0$$
$$4\alpha_1 + 5(\alpha_{00} + \alpha_{01} + \alpha_{02} + \alpha_{12} + \alpha_{22}) - 3\alpha_{11} > 0 \tag{22}$$

By applying (19) to (22) we have the followings.

$$4\alpha_1 + 5(-\alpha_1/2) - 3\alpha_{11} > 0$$
$$3\alpha_1/2 > 3\alpha_{11}$$
$$\alpha_1/2 > \alpha_{11} \tag{23}$$

Equation (23) contradicts to (21). Thus $L'(T)$ has the maximum value at the assignment X.

The following holds at the assignment X.

$$|L_A(T)| = |L_B(T)| = 1 + |L_C(T)| = 2 + 1.4n$$

We have the following.

$$\min\{|L_A(T)|, |L_B(T)|, 1 + |L_C(T)|\} \leq 2 + 1.4n$$

Low Distortion Metric Embedding into Constant Dimension

András Faragó

The University of Texas at Dallas,
Richardson, TX 75080, USA
farago@utdallas.edu

Abstract. We investigate the possibility of embedding an n-point metric space into a constant dimensional vector space with the maximum norm, such that the embedding is almost isometric, that is, the distortion of distances is kept arbitrarily close to 1. When the source metric is generated by any fixed norm on a finite dimensional vector space, we prove that this embedding is always possible, such that the dimension of the target space remains constant, independent of n. While this possibility has been known in the folklore, we present the first fully detailed proof, which, in addition, is significantly simpler and more transparent, then what was available before. Furthermore, our embedding can be computed in deterministic linear time in n, given oracle access to the norm.

Keywords: computational geometry, low distortion metric embedding.

1 Introduction

It is a frequent situation in many applications/algorithms that data is represented by a large number of points in a metric space. In other words, the data itself forms a finite metric space. In such situations it is very useful if we can simplify the data by mapping the source points into a significantly simpler target metric space, such that the pairwise distances between data points are (approximately) preserved. Such *low distortion metric embeddings* have provided tremendous help in many algorithms and applications.

The task usually comes in two typical flavors. The first one is when the distance metric in the source space is too complicated, which often occurs in various pattern recognition tasks, such as speech recognition, image recognition, fingerprint recognition etc. Here the main goal of the embedding is to replace the complicated metric with a much simpler one, which is typically the ℓ_p metric with $p = 1, 2$, or ∞. It may also happen that the data points themselves are not structured (or we may not be interested in their intrinsic structure, so they can be represented just by abstract labels), then the key information is carried solely by the metric. In that case we are led to the task of embedding a general finite metric space into ℓ_p^d, which is the d-dimensional real vector space $\mathbf{R}^d$, equipped with the ℓ_p norm.

M. Ogihara and J. Tarui (Eds.): TAMC 2011, LNCS 6648, pp. 114–123, 2011.
© Springer-Verlag Berlin Heidelberg 2011

The other flavor is when the source data may reside in a space of simple structure, such as $\mathbf{R}^d$ with the Euclidean norm ℓ_2, but the dimension is excessively high. Even if the metric is simple, the high dimension can dramatically slow down algorithms that often have running times exponential in the dimension (known as the "curse of dimensionality"). In this case the main goal of the embedding is *dimensionality reduction*: the mapping into a much lower dimensional space can provide significant help, even if the metric is not simplified.

Since finding good algorithms for metric embedding with low distortion is far from trivial, it is one of the rare fields where clear practical importance meets the intrinsic mathematical beauty of the question.

2   Previous Results

Embedding problems of finite metric spaces have been the subject of extensive research for a very long time, yielding a large number of results. Below we briefly review some fundamental results about low distortion embeddings of n-point metric spaces into ℓ_p^d with $d \ll n$, as these cases relate most closely to our work.

A classic result is Bourgain's Theorem [5], which says that any n-point finite metric space can be embedded into an Euclidean space, with distortion $O(\log n)$. Linial, London and Rabinovich [15] showed that it also works with target norm ℓ_p for any p. The dimension of the target space was originally exponential in [5], but was later reduced to $O(\log^2 n)$ by Linial, London and Rabinovich [15] and Matoušek [16]. The authors of [15] also proved an $\Omega(\log n)$ lower bound for the distortion. Further improvement was obtained by Abraham, Bartal and Neiman [1], who reduced the target dimension to $O(\log n)$, which is optimal, and also showed that the *average* (but not the worst case!) distortion can be made $O(1)$. A lesson from this chain of results is that when a *general* n-point metric space is embedded into ℓ_p^d, then neither d, nor the distortion can remain bounded as n grows.

Thus, if we hope to achieve the ideal case of almost isometric embedding into bounded dimension, then we must impose some restriction on the source metric space, rather than allowing an arbitrary source metric. A classic result in this direction is the Johnson-Lindenstrauss Lemma [14], which says that for every $\epsilon > 0$, any n points from an ℓ_2 space can be embedded with $1 + \epsilon$ distortion (i.e., almost isometrically) into ℓ_2^d, where $d = O(\log n/\epsilon^2)$. Since it is known (see, e.g., the survey of Indyk and Matoušek [13]) that the *isometric* embedding of the same points is not possible into ℓ_2^d with $d < n - 1$, therefore, it is an appealing and extremely useful fact that allowing a very slight distortion can bring down the target dimension to $O(\log n)$, resulting in exponential dimension reduction. Another useful feature is that the embedding can be computed via a simple random projection. On the other hand, it is not possible to reduce the target dimension to a constant, as there is a known almost matching lower bound of $\Omega(\log n/(\epsilon^2 \log(1/\epsilon)))$ for the dimension, due to Alon [3], see also Matoušek [18].

The lower bound on the dimension shows that we cannot achieve our ideal goal with ℓ_2 target metric, when the source is an arbitrary normed space. How about ℓ_1 as target metric? Then the situation is even worse, as Brinkman and Charikar [6] proves that for every n, the embedding of an n-point ℓ_1 metric into ℓ_1^d with distortion $c > 1$ requires $d = n^{\Omega(1/c^2)}$ in the worst case.

Regarding ℓ_∞ as the target metric, Matoušek [17] showed that any n-point metric can be embedded into ℓ_∞^d with distortion c and of dimension $d = O((c+1)n^{2/(c+1)}\log n)$. Furthermore, an almost matching lower bound can be proved [13], so the target dimension cannot remain bounded for a general source metric with ℓ_∞ target.

One may ask at this point: is there *any* nontrivial embedding result with constant target dimension, independent of the number of source data points? While such results do exist, under special conditions, they are quite rare. A folklore example (see the survey of Indyk [12]) is that ℓ_1^d isometrically embeds into $\ell_\infty^{d'}$ with $d' = 2^d$, regardless of the number of source points. Another example is the theorem of Gupta, Krauthgamer and Lee [10], which says that every doubling tree metric embeds into ℓ_p^d with constant d and constant distortion (which may not be arbitrarily close to 1), for every $p \in [1, \infty]$. Here a *tree metric* is the shortest path metric on the vertices of a tree. A metric is called *doubling*, if its doubling dimension is finite, where the *doubling dimension* is the smallest integer k, such that every ball can be covered by 2^k balls of half the radius.

Another embedding result into constant dimension is Assouad's Theorem [4]. It says that if the source metric has constant doubling dimension, and $0 < \gamma < 1$ is a constant, then there exist constants d, C, such that the "snowflake version" of the source metric embeds into ℓ_2^d with distortion C. Here the *snowflake version* of the source metric is the original metric raised to the power of γ (which remains a metric for $0 < \gamma < 1$). The target dimension and the distortion in Assouad's Theorem depend on the doubling dimension and γ, but not on the number of input points. On the other hand, the embedded metric is a modified version of the source metric, not the original itself. Note that the difference between the original and the snowflake version is not bounded for any $0 < \gamma < 1$. Assouad conjectured that the embedding also applies to the original source metric (corresponding to $\gamma = 1$), but that was disproved by Semmes [20]. As a further step, Gottlieb and Krauthgamer [9] extended Assouad's Theorem by showing that if the source metric is also Euclidean (beyond having constant doubling dimension), then the embedding of the snowflake version into constant dimension can be done with $1 + \epsilon$ distortion with arbitrary $\epsilon > 0$. This again applies to the modified source metric, not the original.

Apparently, the most general result so far on embedding the original n-point source metric (not its snowflake version) into constant dimension is the theorem of Abraham, Bartal and Neiman [2]. It says that any n-point metric space of doubling dimension k can be embedded into ℓ_p^d with $d = O(k/\theta)$ and distortion $O(\log^{1+\theta} n)$, where $\theta \in (0, 1]$. While the target dimension is constant here, the distortion grows to infinity with n.

If, however, the source points are from a fixed finite dimensional normed space, then they can be embedded into ℓ_∞^d with distortion $1 + \epsilon$, and with d constant, that is, independent of the number of points (but, of course, dependent on the source space dimension). This is referred to as a folklore result in the survey of Indyk [12]. Apparently, however, the proof has never been published in full detail, and the claim does not seem easy to prove. The idea, as briefly outlined by Indyk [12], goes as follows. One can first prove that any polyhedral norm, in which the unit ball is a polyhedron with F faces, can be isometrically embedded into ℓ_∞^F. Then one can use a (rather nontrivial) theorem due to Dudley [7], which says that any convex body of diameter 1, in dimension d, can be approximated by a polyhedron that has $O(1/\epsilon)^{d/2}$ faces, such that the Hausdorff distance between the convex body and the approximaitng polyhedron is not more than ϵ. Using this, also applying another result called John's Theorem[1], one can derive the desired embedding.

As mentioned above, the detailed proof of this embedding result, apparently has never been published. The goal of our paper is to fill the gap by providing a new, significantly simpler and more transparent proof of this result. Furthermore, we also address some of the related algorithmic aspects. We show that the embedding is computable by a deterministic linear time algorithm, having oracle access to the norm that induces the source metric.

Part of the mathematical idea is based on an earlier paper of the author, co-authored with T. Linder and G. Lugosi [8], which used the idea for nearest neighbor search, without explicitly targeting metric embeddings. It is interesting to note that while [8] was published in 1993 (preceding many of the above surveyed papers), it went unnoticed for 18 years that some of the ideas could be used for deriving a metric embedding with desirable properties.

3 Preliminaries

We deal with metric spaces that are subsets of finite dimensional normed spaces. As it is well known, norm is a function that assigns a nonnegative real number $\|x\|$ to each vector x in a vector space, satisfying the following properties:

- Homogeneity: $\|cx\| = |c| \, \|x\|$ for any vector x and scalar c.
- Triangle inequality: $\|x + y\| \le \|x\| + \|y\|$ for every x, y.
- Separation of points: if $\|x\| = 0$, then $x = \mathbf{0}$, where $\mathbf{0}$ is the zero vector (Equivalently, $x \ne y$ implies $\|x - y\| > 0$ for every x, y.)

A vector space with a norm is called a *normed space*. If the underlying vector space is finite dimensional, then we call it a *finite dimensional normed space*. A norm induces a distance by

$$d(x, y) = \|x - y\| \tag{1}$$

so any subset of a normed space becomes a metric space with this metric. It is clear that this distance and the norm uniquely determine each other, through

---

[1] For an exposition and proof, see, e.g., Howard [11].

(1) and $\|x\| = d(x, \mathbf{0})$. Note, however, that not every distance on a vector space is induced by some norm; for example, a distance metric does not have to be homogeneous. The restriction we impose is that we only deal with norm-generated distances. Our target norm will be the maximum norm, denoted by ℓ_∞, and defined as $\|v\| = \max_i\{|v_1|, \ldots, |v_d|\}$ for $v = (v_1, \ldots, v_d) \in \mathbf{R}^d$.

One may suspect at this point that perhaps only the algorithmically "easy" distances are generated by a norm, so by restricting ourselves to norm-generated distances, we may exclude all the hard cases. This is, however, not so at all. The reason is that a norm itself can also be hard to compute. Here is an example. Let C be a convex set in a vector space, such that C contains the origin in its interior. It is known [19] that the following construction generates a norm: $\|x\| = \inf\{1/\gamma : \gamma x \in C\}$. Now choosing the convex set C such that it is hard to carry out the above optimization makes the norm hard to compute. For example, we can create a norm that is **NP**-hard to compute, if we choose C as the convex hull of the incidence vectors of cliques in a graph (with a small shifting to include the origin in its interior).

If X and Y are two metric spaces, with metrics d_X, d_Y, respectively, then an *embedding* of X into Y is a mapping $f : X \mapsto Y$. For a real number $D \geq 1$ the embedding has *distortion* (at most) D, if for every $x, y \in X$

$$\frac{1}{D} d_X(x, y) \leq d_Y(f(x), f(y)) \leq D d_X(x, y).$$

Naturally, we are looking for the smallest D that satisfies the requirement. If $D = 1$, the embedding is called an *isometry*. If for every $\epsilon > 0$ there is an embedding with distortion $D \leq 1 + \epsilon$, then we say X can be embedded *almost isometrically* into Y. (The mapping f may depend on ϵ.)

4   Results

Now we prove that an embedding that simultaneously satisfies the following three requirements exists:

(1) The source metric, which is generated by an arbitrary fixed, finite dimensional norm, is embedded into ℓ_∞^d, such that the dimension d is constant, in the sense that it is independent of the number of source points.

(2) The embedding is *almost isometric*, that is, the distortion is bounded by $1 + \epsilon$ for an arbitrarily small fixed $\epsilon > 0$.

(3) The embedding applies to the *original* source metric, not merely a modified version of it (such as the snowflake version).

Theorem 1. *Let $x_1, \ldots, x_n$ be distinct points in an arbitrary finite dimensional normed space S, and let $\epsilon > 0$ be any real number. Then there exists an integer k, independent of n, such that $x_1, \ldots, x_n$ can be embedded into ℓ_∞^k with distortion at most $D = 1 + \epsilon$.*

Proof. Let $\|.\|$ be the norm in S, and let U be the closed unit ball in this space, that is, $U = \{u \in S : \|u\| \leq 1\}$. Fix $\epsilon > 0$, such that $D = 1 + \epsilon$ is the target distortion. Set $\alpha = 1/D$. First we show that one can always find k vectors $z_1, \ldots, z_k \in S$, for some constant k, with

$$\alpha\|x - y\| \leq \max_i |\,\|x - z_i\| - \|y - z_i\|\,| \leq \|x - y\|, \qquad \forall x, y \in U. \qquad (2)$$

The value of k, and the z_i vectors themselves, may depend on ϵ and S, but not on n, the number of input points.

To prove the above statement, let us define a new normed space on $S^2 = S \times S$, by defining the operations and the norm on S^2 componentwise: $c(x, y) = (cx, cy)$, $(x_1, y_1) + (x_2, y_2) = (x_1 + x_2, y_1 + y_2)$, and the norm on S^2 is $\|(x, y)\| = \|x\| + \|y\|$. One can directly check that this way again a finite dimensional normed vector space is obtained. Now for each $z \in S$ define the set $A_z \subset S^2$ by

$$A_z = \left\{ (x, y) \in S^2 : x \neq y, \ \frac{\|y - z\| - \|x - z\|}{\|x - y\|} > \alpha \right\}$$

where $\alpha = 1/D = 1/(1 + \epsilon)$, yielding $0 < \alpha < 1$. Furthermore, let $H \subset S^2$ be the following set:

$$H = \{(x, y) : \|x\| \leq 2, \|y\| \leq 1, \|x - y\| \geq 1/2\}.$$

We are going to use the following properties of these sets:

(i) H is closed and bounded;
(ii) A_z is open;
(iii) $H \subset \bigcup_{z \in S} A_z$.

Properties (i) and (ii) follow directly from the definitions. To see (iii), it is enough to observe that $x \neq y$ implies $(x, y) \in A_x$, since with $z = x$ we have

$$\frac{\|y - z\| - \|x - z\|}{\|x - y\|} = \frac{\|y - x\|}{\|x - y\|} = 1 > \alpha,$$

so any vector $(x, y) \in S^2$ with $x \neq y$ satisfies the definition of some A_z, namely the one with $z = x$. Noting that the requirement $\|x - y\| \geq 1/2$ in the definition of H implies $x \neq y$, we conclude that (iii) must hold.

Thus the family $\{A_z, z \in S\}$ of sets forms an *open cover* of H. As H is closed and bounded, and the space S^2 is also finite dimensional, it follows from the well known Heine-Borel theorem that there exists a *finite* subcover. That is, there must exist finitely many points $z_1, \ldots, z_k$ with $H \subset \bigcup_{i=1}^{k} A_{z_i}$. This means, any $(x, y) \in H$ must satisfy the definition of at least one of these A_{z_i}, that is,

$$\frac{\|y - z_i\| - \|x - z_i\|}{\|x - y\|} > \alpha, \quad \text{for some } i \in \{1, \ldots, k\}. \qquad (3)$$

Note that the above reasoning, including all involved sets, depends only on the space and α, but not on n. Consequently, the value of k, as well as the vectors $z_1, \ldots, z_k$ themselves, can be chosen independently of n.

So far we have shown that (3) holds for any $(x, y) \in H$. Now we prove that (3) remains true for any $x, y \in U$, $x \neq y$. (Recall that U is the closed unit ball in S. Note that, by the definitions, $x, y \in U$ does not imply $(x, y) \in H$.)

If $x, y \in U$ and $\|x - y\| \geq 1/2$, then $(x, y) \in H$, so then we already know (3) holds. To handle the remaining case $0 < \|x - y\| < 1/2$, take two points $x, y \in U$ at distance λ apart, for some $0 < \lambda < 1/2$. That is, $\|x\| \leq 1$, $\|y\| \leq 1$, and $0 < \|x - y\| = \lambda < 1/2$. Set

$$x' = \frac{1}{\lambda}(x - (1 - \lambda)y).$$

Then $x = \lambda x' + (1 - \lambda)y$ holds, that is, x divides the line segment $\overline{x'y}$ such that $\|x - y\| = \lambda \|x' - y\|$. As we have chosen $\|x - y\| = \lambda$, it implies $\|x' - y\| = 1$. This, together with $y \in U$ and with the triangle inequality, gives $2 \geq \|x' - y\| + \|y\| \geq \|x'\|$, yielding $\|x'\| \leq 2$. Collecting these facts, they imply $(x', y) \in H$. But then by (3) there is a z_i with

$$\frac{\|y - z_i\| - \|x' - z_i\|}{\|x' - y\|} > \alpha. \tag{4}$$

Now using $x - z_i = \lambda(x' - z_i) + (1 - \lambda)(y - z_i)$, as well as the fact that the norm axioms ensure that every norm is a convex function, we can write

$$\frac{\|y - z_i\| - \|x - z_i\|}{\|x - y\|} \geq$$

$$\frac{\|y - z_i\| - (\lambda\|x' - z_i\| + (1 - \lambda)\|y - z_i\|)}{\lambda} =$$

$$\|y - z_i\| - \|x' - z_i\| =$$

$$\frac{\|y - z_i\| - \|x' - z_i\|}{\|x' - y\|} > \alpha, \tag{5}$$

where we made use of $\|x' - y\| = 1$. Thus, we have arrived at

$$\frac{\|y - z_i\| - \|x - z_i\|}{\|x - y\|} \geq \alpha \tag{6}$$

also in the case of $0 < \|x - y\| < 1/2$. Hence, we can conclude that for any $x, y \in U$ there must be an $i \in \{1, \ldots, k\}$ with

$$| \|x - z_i\| - \|y - z_i\| | \geq \alpha \|x - y\|. \tag{7}$$

For $\|x - y\| \geq 1/2$ this follows from (3) and the case $0 < \|x - y\| < 1/2$ is covered by the analysis resulting in (6). (The case $\|x - y\| = 0$ is trivial, since then the right-hand side in (7) is 0.) Since the triangle inequality implies $| \|x - z\| - \|y - z\| | \leq \|x - y\|$ for every x, y, z, therefore, combining this with (7), we obtain that (2) must hold for every $x, y \in U$.

Now, let R be the largest norm value that occurs among the input points, i.e., $R = \max_j \|x_j\|$. Then we have $\frac{1}{R}x_r \in U$ for every source point x_r. Taking any two input points x_r, x_s, we can apply (2) to $x = \frac{1}{R}x_r, y = \frac{1}{R}x_s$. Multiplying all sides in (2) by R yields

$$\alpha\|x_r - x_s\| \leq \max_i \big| \, \|x_r - Rz_i\| - \|x_s - Rz_i\| \, \big| \leq \|x_r - x_s\|. \tag{8}$$

We can now define the sought embedding by the following function:

$$f(x) = (\|x - Rz_1\|, \ldots, \|x - Rz_k\|).$$

This maps the input points into ℓ_∞^k, and by (8) it satisfies

$$\frac{1}{D}\|x_r - x_s\| \leq \|f(x_r) - f(x_s)\|_\infty \leq \|x_r - x_s\|.$$

Since k is independent of n, and $D = 1 + \epsilon$ is arbitrarily close to 1, therefore, we indeed obtained an almost isometric embedding into constant dimension, where both the dimension and the distortion are independent of the number of source points.   ♣

Now we show that the embedding of Theorem 1 can be computed by a linear time algorithm. Here we only restrict ourselves to the real number based unit cost algorithmic model, which is the most natural fit for the method. In this model a real number is considered an elementary unit of data, and elementary operations on real numbers are counted as single steps. The algorithm can be transformed, however, into the more conventional bit oriented model (where we operate with bits, so one can only use rational numbers, including the norm values).

Theorem 2. *There exists a linear time algorithm in the real number based unit cost model that computes the embedding derived in Theorem 1.*

Proof. Let us first recall, as pointed out in the proof of Theorem 1, that the vectors $z_1, \ldots, z_k \in S$, used in the embedding, as well as their number k, can be chosen independently of n. They depend only on S and the target distortion $D = 1 + \epsilon$, but not on n. Therefore, for fixed S and D, they can be hardwired in the algorithm[2]. Then computing $f(x_r)$ for an input point x_r only takes computing $R = \max_j \|x_j\|$, and then $\|x_r - Rz_1\|, \ldots, \|x_r - Rz_k\|$, which are the coordinates of $f(x_r)$. All this can clearly be done in linear time, assuming the real number based unit cost model, and oracle access to the norm.   ♣

5   Conclusion

We have presented a new, simpler and more transparent proof for the metric embedding result from an arbitrary fixed finite dimensional normed space into

---

[2] Note that here we do not claim an efficient method to actually *construct* the algorithm, here we only prove that it exists.

ℓ_∞, such that the embedding simultaneously satisfies the following: the target space has constant dimension (independent of the number of source points), the distortion is arbitrarily low, and the embedding applies to the original metric, not a modified version of it. Our embedding can be computed by a linear time algorithm.

Acknowledgement

The author is grateful for the support of NSF Grant CNS-1018760.

References

1. Abraham, I., Bartal, Y., Neiman, O.: Advances in Metric Embedding Theory. In: 38th Annual ACM Symp. on Theory of Computing (STOC 2006), Seattle, WA, USA, pp. 271–286 (May 2006)
2. Abraham, I., Bartal, Y., Neiman, O.: Embedding Metric Spaces in Their Intrinsic Dimension. In: 19th Annual ACM-SIAM Symp. on Discrete Algorithms (SODA 2008), San Francisco, CA, USA, pp. 363–372 (January 2008)
3. Alon, N.: Problems and Results in Extremal Combinatorics – I. Discrete Math. 273, 31–53 (2003)
4. Assouad, P.: Plongements Lipschitziens dans $\mathbf{R}^n$. Bull. Soc. Math. 111, 429–448 (1983)
5. Bourgain, J.: On Lipschitz Embedding of Finite Metric Spaces in Hilbert Space. Israel J. Math. 52, 46–52 (1985)
6. Brinkman, B., Charikar, M.: On the Impossibility of Dimension Reduction in ℓ_1. Journal of the ACM 52, 766–788 (2005)
7. Dudley, R.M.: Metric Entropy of Some Classes of Sets with Differentiable Boundaries. Journal of Approximation Theory 10, 227–236 (1974)
8. Faragó, A., Linder, T., Lugosi, G.: Fast Nearest Neighbor Search in Dissimilarity Spaces. IEEE Transactions on Pattern Analysis and Machine Intelligence 15(9), 957–962 (1993)
9. Gottlieb, L.-A., Krauthgamer, R.: A Nonlinear Approach to Dimension Reduction, arXiv:0907.5477v2 [cs.CG] (April 2010)
10. Gupta, A., Krauthgamer, R., Lee, J.R.: Bounded Geometries, Fractals, and Low-Distortion Embeddings. In: 44th Annual IEEE Symp. on Foundations of Computer Science (FOCS 2003), Cambridge, MA, USA, pp. 534–543 (October 2003)
11. Howard, R.: The John Ellipsoid Theorem, Lecture Note, Dept. of Math., Univ. of South Carolina (November 1997), http://www.math.sc.edu/~howard/Notes/john.pdf
12. Indyk, P.: Algorithmic Applications of Low-Distortion Geometric Embeddings. In: 42nd Annual IEEE Symp. on Foundations of Computer Science (FOCS 2001), Las Vegas, NV, pp. 10–33 (October 2001)
13. Indyk, P., Matoušek, J.: Low Distortion Embeddings of Finite Metric Spaces. In: Goodman, J.E., O'Rourke, J. (eds.) Handbook of Discrete and Combinatorial Geometry, pp. 177–196. Chapman and Hall/CRC, Boca Raton (2004)
14. Johnson, W.B., Lindenstrauss, J.: Extensions of Lipschitz Mappings into a Hilbert Space. Contemp. Math. 26, 189–206 (1984)

15. Linial, N., London, E., Rabinovich, Y.: The Geometry of Graphs and Some of Its Algorithmic Applications. Combinatorica 15, 215–245 (1995)
16. Matoušek, J.: Note on Bi-Lipschitz Embeddings Into Low Dimensional Euclidean Spaces. Comment. Math. Univ. Carolinae 31, 589–600 (1990)
17. Matoušek, J.: On the Distortion Required for Embedding Finite Metric Spaces Into Normed Spaces. Israel J. Math. 93, 333–344 (1996)
18. Matoušek, J.: Lectures on Discrete Geometry. Springer, New York (2002)
19. Rockafellar, R.T.: Convex Analysis. Princeton University Press, Princeton (1970)
20. Semmes, S.: On the Nonexistence Bilipschitz Parametrizations and Geometric Problems about a_∞ Weights. Revista Mathemática Iberoamericana 12, 337–410 (1996)

A Better Upper Bound on Weights of Exact Threshold Functions[*]

Xue Chen[1], Guangda Hu[1], and Xiaoming Sun[2],[**]

[1] Department of Computer Science and Technology, Tsinghua University
[2] Institute for Theoretical Computer Science and Center for Advanced Study, Tsinghua University

Abstract. A Boolean function is called an *exact threshold function* if it decides whether the input vector $x \in \{0,1\}^n$ is on a hyperplane $w^T x = t$ ($w \in \mathbb{Z}^n, t \in \mathbb{Z}$). In this paper we study the upper bound of elements in w required to represent any exact threshold function. Let k be the dimension of the linear subspace spanned by Boolean points on $w^T x = t$. We first give an upper bound $O(n^k)$ for constant k, which matches the lower bound in [2]. Then we prove an upper bound $O(k^{O(k^2)}n^k)$ for general cases, improving the result $\min\{n^{2^k}, n^{n/2+1}\}$ in [2].

1 Introduction

A linear *threshold* function $f(x)$ ($x \in \mathbb{R}^n$) is a Boolean function deciding whether $w^T x \geq t$, where $w = (w_1, w_2, \ldots, w_n) \in \mathbb{R}^n$ is called the *weights* and $t \in \mathbb{R}$ is called the *threshold*. Similarly, a linear *exact threshold* function is a Boolean function deciding whether $w^T x = t$. Threshold functions and exact threshold functions are close related to many areas of studies such as circuit complexity, learning theory, and structural complexity theory etc..

For an (exact) threshold function f, the representation of f (the chosen of the weights w and threshold t) is not unique, for example we can multiply w and t by a factor. Without loss of generality we can assume $w \in \mathbb{Z}^n$ and $t \in \mathbb{Z}$. A natural and very important problem is to find the smallest weights required for representing an (exact) threshold function, i.e. to minimize $\max_{1 \leq i \leq n} |w_i|$. For simplicity, we call this value $\max_{1 \leq i \leq n} |w_i|$ the *key weight* of w.

This problem on threshold function has been studied for a long history. Muroga, Toda and Takasu [5, 6] showed that $(n+1)^{(n+1)/2}/2^n$ is sufficient for the key weight to represent any linear threshold function. For the lower bound part, there were several existence proofs and explicit constructions of linear threshold functions that require the key weight of the order $2^{\Omega(n)}$ [5, 7–9]. For n being

---

[*] This work was supported in part by the National Natural Science Foundation of China Grant 60603005, 61033001, 61061130540, the National Basic Research Program of China Grant 2007CB807900, 2007CB807901, and Tsinghua University Initiative Scientific Research Program 2009THZ02120.
[**] Corresponding author, xiaomings@tsinghua.edu.cn

M. Ogihara and J. Tarui (Eds.): TAMC 2011, LNCS 6648, pp. 124–132, 2011.
© Springer-Verlag Berlin Heidelberg 2011

a power of 2, Håstad [4] gave a threshold function requiring a key weight at least $(1/n)e^{-4n^\beta}n^{n/2}/2^n$, where $\beta = \log(3/2)$. In [1], Alon and Vũ constructed a threshold function which gives a lower bound $n^{n/2}/2^{n(2-o(1))}$ for general n.

For the linear exact threshold functions, the same question was first investigated by Babai et al. [2]. They showed a lower bound $n^{n/2}/2^{n(2-o(1))}$ and an upper bound $n^{\frac{n}{2}+1}$ of the key weight. They also considered a parameterized approach to this problem, which is the basic model in this work. For a linear exact threshold function $f : \{0,1\}^n \to \{0,1\}$ defined by $f(x) = 1$ iff $w^T x = t$, let k be the dimension of the linear subspace spanned by all the vectors in $f^{-1}(1) = \{x \in \{0,1\}^n \mid f(x) = 1\}$. Define the *dimension* of f to be k. For any n and k, $W^E(n,k)$ is used to denote the smallest key weight needed to represent a dimension-k linear exact threshold function with n variables, and a lower bound $(\lfloor \frac{n}{k} \rfloor^k - 1)/2k$ and an upper bound n^{2^k} are given for $W^E(n,k)$ in [2].

In this work, we focus on the upper bound of $W^E(n,k)$. We first show several properties and lemmas in Section 2, which are the basic ideas widely used in our argument. For constant k, we give an explicit construction which provides an upper bound $O(n^k)$ matching the lower bound in [2]. The idea is to reduce the problem into constant size first, and this construction is described in Section 3. And in Section 4, by a more complicated analysis we show our upper bound $O(k^{O(k^2)}n^k)$ for general cases. We conclude the paper with open problems in Section 5.

2   Preliminary

A Boolean function $f : \{0,1\}^n \to \{0,1\}$ is called a linear *exact threshold function* if there exist *weights* $w \in \mathbb{R}^n$ and a *threshold* $t \in \mathbb{R}$ such that $f(x) = 1$ if and only if $w^T x = t$. For any $w = (w_1, w_2, \ldots, w_n)^T \in \mathbb{R}^n$, we use notions $\phi(w)$ to denote $\max_{x \in \{0,1\}^n} |w^T x|$, and $\omega(w)$ to denote the *key weight* $\max_{1 \le i \le n} |w_i|$.

Without loss of generality, we can assume the weights w and the threshold t are all integers. The goal is to find the minimum possible $\omega(w)$ for integer weights w and threshold t representing any given linear exact threshold function. Similarly as in [2], without loss of generality we assume $f(0) = 1$, i.e. the threshold $t = 0$. (This can be seen by rotating the Boolean cube $\{0,1\}^n$ and the hyperplane $w^T x = t$, which does not change the absolute values of weights.) Therefore, the points in $f^{-1}(1)$ form a linear subspace in $\{0,1\}^n$. Suppose we are given points $a_1, a_2, \ldots, a_k \in \{0,1\}^n$ such that $f(x) = 1$ $(x \in \{0,1\}^n)$ if and only if $x \in \text{span}\{a_1, a_2, \ldots, a_k\} \cap \{0,1\}^n$. Without loss of generality, we assume the vectors $a_1, a_2, \ldots, a_k$ are linearly independent in $\mathbb{R}^n$. Hence the *dimension* of f is k.

A vector $w \in \mathbb{Z}^n$ is called a *solution* of $a_1, a_2, \ldots, a_k$ if the linear function $g(x) = w^T x$ $(x \in \{0,1\}^n)$ vanishes exactly on $\text{span}\{a_1, a_2, \ldots, a_k\} \cap \{0,1\}^n$, i.e. $f(x) = 1$ $(x \in \{0,1\}^n)$ if and only if $w^T x = 0$.

We give a basic lemma used in our constructions. It states that a solution w can be constructed from a set of linear equations.

Lemma 1. *Suppose that there exist t vectors $\boldsymbol{w}_1, \boldsymbol{w}_2, \ldots, \boldsymbol{w}_t \in \mathbb{Z}^n$ such that a 0-1 vector $\boldsymbol{x} \in \operatorname{span}\{\boldsymbol{a}_1, \boldsymbol{a}_2, \ldots, \boldsymbol{a}_k\}$ if and only if every $\boldsymbol{w}_i{}^T \boldsymbol{x} = 0$ holds $(1 \le i \le t)$. Then we can construct a solution $\boldsymbol{w} \in \mathbb{Z}^n$ with*

$$\omega(\boldsymbol{w}) \le \sum_{i=1}^{t} \left(\omega(\boldsymbol{w}_i) \prod_{j=1}^{i-1} (\phi(\boldsymbol{w}_j) + 1) \right).$$

Proof. Define the vector $\boldsymbol{w}$ as

$$\boldsymbol{w}_1 + (\phi(\boldsymbol{w}_1) + 1)\, \boldsymbol{w}_2 + \cdots + \left(\prod_{j=1}^{i-1} (\phi(\boldsymbol{w}_j) + 1) \right) \boldsymbol{w}_i + \cdots + \left(\prod_{j=1}^{t-1} (\phi(\boldsymbol{w}_j) + 1) \right) \boldsymbol{w}_t.$$

For every $\boldsymbol{x} \in \{0, 1\}^n$ such that $\boldsymbol{w}^T \boldsymbol{x} = 0$, i.e.

$$\boldsymbol{w}_1{}^T \boldsymbol{x} + (\phi(\boldsymbol{w}_1) + 1)\, \boldsymbol{w}_2{}^T \boldsymbol{x} \tag{1}$$

$$+ \cdots + \left(\prod_{j=1}^{i-1} (\phi(\boldsymbol{w}_j) + 1) \right) \boldsymbol{w}_i{}^T \boldsymbol{x} + \cdots + \left(\prod_{j=1}^{t-1} (\phi(\boldsymbol{w}_j) + 1) \right) \boldsymbol{w}_t{}^T \boldsymbol{x} = 0.$$

one can see that

$$\boldsymbol{w}_1{}^T \boldsymbol{x} \equiv 0 \pmod{\phi(\boldsymbol{w}_1) + 1}.$$

This implies $\boldsymbol{w}_1{}^T \boldsymbol{x} = 0$ by the definition of $\phi(\cdot)$. Then if we divide Eq. (1) by $\phi(\boldsymbol{w}_1) + 1$, and module $\phi(\boldsymbol{w}_2) + 1$, we get $\boldsymbol{w}_2{}^T \boldsymbol{x} = 0$. In this way, we can show that $\boldsymbol{w}_i{}^T \boldsymbol{x} = 0$ for every $i = 1, 2, \ldots, t$. Hence $\boldsymbol{w}$ is a solution. The key weight $\omega(\boldsymbol{w})$ stated in this lemma is straightforward. □

Define a matrix $N \in \{0, 1\}^{k \times n}$ in which the i-th row is $\boldsymbol{a}_i{}^T$ $(1 \le i \le k)$, i.e. $N = [\boldsymbol{a}_1, \ldots, \boldsymbol{a}_k]^T$. Since $\boldsymbol{a}_1, \boldsymbol{a}_2, \ldots, \boldsymbol{a}_k$ are linearly independent, the rank of N must be k, and there exist k linearly independent columns. Without loss of generality, we assume the first k columns of N are linearly independent in $\mathbb{R}^k$.

Let L denote the first k columns of N, and M denote the last $(n - k)$ columns, $N = [L, M]$. We use $\boldsymbol{l}_1, \boldsymbol{l}_2, \ldots, \boldsymbol{l}_k$ and $\boldsymbol{m}_1, \boldsymbol{m}_2, \ldots, \boldsymbol{m}_{n-k}$ to denote the columns of L and M respectively. For a vector $\boldsymbol{x}$, let $\boldsymbol{x}_L$ be the first k elements and $\boldsymbol{x}_M$ be the last $n - k$ elements. We use $x_1, x_2, \ldots, x_k$ to denote the elements of $\boldsymbol{x}_L$ and $x_{k+1}, x_{k+2}, \ldots, x_n$ to denote elements of $\boldsymbol{x}_M$.

For every $\boldsymbol{x}_L \in \{0, 1\}^k$, there is a unique element $\boldsymbol{y} \in \operatorname{span}\{\boldsymbol{a}_1, \boldsymbol{a}_2, \ldots, \boldsymbol{a}_k\}$ with $\boldsymbol{y}_L = \boldsymbol{x}_L$, because L has full rank. Moreover, we can see

$$\boldsymbol{y}^T = \boldsymbol{x}_L{}^T [I_k, L^{-1}M],$$

where I_k is the $k \times k$ identity matrix. Therefore $\boldsymbol{y}_M = (L^{-1}M)^T \boldsymbol{x}_L$ and

$$\boldsymbol{x} \in \operatorname{span}\{\boldsymbol{a}_1, \boldsymbol{a}_2, \ldots, \boldsymbol{a}_k\} \Leftrightarrow \boldsymbol{x} = \boldsymbol{y} \Leftrightarrow \boldsymbol{x}_M = (L^{-1}M)^T \boldsymbol{x}_L.$$

Thus we have the following lemma.

Lemma 2. *A 0-1 vector* $x \in \text{span}\{a_1, a_2, \ldots, a_k\}$ *if and only if*

$$x_M = (L^{-1}M)^T x_L.$$

Based on Lemma 1 and Lemma 2, we give a brute-force construction as the following solvability lemma.

Lemma 3. *For any* $a_1, a_2, \ldots, a_k \in \{0,1\}^n$, *there exits a solution* w *with* $\omega(w) = O(h(k)^{n-k})$, *where* $h(k)$ *is a function only depends on* k.

Proof. The equations set $x_M = (L^{-1}M)^T x_L$ can be split into $(n-k)$ linear equations $x_{k+i} = (L^{-1}m_i)^T x_L$ $(1 \le i \le n-k)$. By Cramer's rule,

$$L^{-1}m_i = \frac{1}{\det(L)} \left(\det(L_1^{(i)}), \det(L_2^{(i)}), \ldots, \det(L_k^{(i)}) \right)^T,$$

where $L_j^{(i)}$ is the matrix obtained by replacing L's j-th column with m_i $(1 \le j \le k)$. Therefore $x_{k+i} = (L^{-1}m_i)^T x_L$ is equivalent to

$$\det(L_1^{(i)})x_1 + \det(L_2^{(i)})x_2 + \cdots + \det(L_k^{(i)})x_k - \det(L)x_{k+i} = 0.$$

By Lemma 1 and Lemma 2, we just need to construct a solution from these equations. In [3], Faddeev and Sominski proved the determinant of a $k \times k$ 0-1 matrix is bounded by $\pm \frac{1}{2^k}(k+1)^{\frac{k+1}{2}}$. Recall that $L, L_1^{(i)}, L_2^{(i)}, \ldots, L_k^{(i)}$ are all $k \times k$ 0-1 matrices, thus we have a solution w with

$$\omega(w) \le \sum_{i=1}^{n-k} \left(\frac{1}{2^k}(k+1)^{\frac{k+1}{2}} \prod_{j=1}^{i-1} \left(\frac{1}{2^k}(k+1)^{\frac{k+1}{2}}(k+1) + 1 \right) \right)$$

$$= O\left(\frac{1}{k+1} \left(\frac{1}{2^k}(k+1)^{\frac{k+3}{2}} + 1 \right)^{n-k} \right).$$

Let $h(k) = \frac{1}{2^k}(k+1)^{\frac{k+3}{2}} + 1$ and this lemma is proved. □

Next we show how to reduce the problem by removing any zero columns in N. Suppose the i_1-th, i_2-th, $\ldots$, and i_t-th columns are $\mathbf{0}$, and w' is a solution for the reduced problem. We use $\tilde{x}$ to denote the sub-vector of x, where the i_1-th, i_2-th, $\ldots$, i_t-th elements are removed. We can construct a solution w for the original problem by two equations $w'^T \tilde{x} = 0$ and $x_{i_1} + x_{i_2} + \cdots + x_{i_t} = 0$ according to Lemma 1. Moreover, if the solution w' itself is constructed by a set of linear equations, we can add $x_{i_1} + x_{i_2} + \cdots + x_{i_t} = 0$ to be the last equation (whose $\omega(\cdot) = 1$). Based on this, a simple use of Lemma 1 gives the following corollary.

Corollary 1. *Suppose* $N = [a_1, \ldots, a_k]^T$ *and* w' *is a solution of the reduced problem (without zero columns). If* w' *is constructed from equation system* $w_1'^T \tilde{x} = 0, w_2'^T \tilde{x} = 0, \ldots, w_t'^T \tilde{x} = 0$, *then there is a solution* w *of* $a_1, a_2, \ldots, a_k$ *with key weight*

$$\omega(w) \le \sum_{i=1}^{t} \left(\omega(w_i') \prod_{j=1}^{i-1} \left(\phi(w_j') + 1 \right) \right) + \prod_{j=1}^{t} \left(\phi(w_j') + 1 \right).$$

3   Constant Dimension Case

In this section we give an upper bound for $W^E(n,k)$ which matches the lower bound in [2] for any constant k. The idea is to remove all duplicated columns in N and solve the reduced problem first.

Suppose there are no zero columns in matrix N. We divide the n columns of N into *groups*, such that every two duplicated columns are in the same group. Suppose there are d groups, we have $d \leq 2^k$. Pick a *leader* column from each group, say they are the r_1-th, r_2-th, ..., and r_d-th columns. We use $s_1, s_2, \ldots, s_d$ to denote the sizes of each group, then $s_1 + \cdots + s_d = n$. For any vector $\boldsymbol{y} \in \mathbb{R}^n$, define the *leader elements* as the r_1-th, r_2-th, ..., and r_d-th elements, and we use $\widetilde{\boldsymbol{y}}$ to denote the sub-vector of all leader elements, i.e. $\widetilde{\boldsymbol{y}} = y_{r_1} y_{r_2} \cdots y_{r_d}$.

For every row $\boldsymbol{a_i}$ in N ($1 \leq i \leq k$), let Q_i be the index set of groups with the i-th row non-zero, i.e. $Q_i = \{j \mid a_i[r_j] = 1, 1 \leq j \leq d\}$ ($a_i[r_j]$ is the r_j-th element of $\boldsymbol{a_i}$, recall that $\boldsymbol{a_i}$ is a 0-1 vector). We first find a solution $\boldsymbol{w'}$ for $\widetilde{\boldsymbol{a}}_1, \widetilde{\boldsymbol{a}}_2, \ldots, \widetilde{\boldsymbol{a}}_k$. The equation $\boldsymbol{w'}^T \widetilde{\boldsymbol{x}} = 0$ guarantees the leader elements $\widetilde{\boldsymbol{x}}$ are in the required linear space. Then we add further linear equations to solve the original problem.

Lemma 4. *A 0-1 vector $\boldsymbol{x}$ satisfies equation system (2) if and only if $\boldsymbol{x} \in \mathrm{span}\{\boldsymbol{a}_1, \boldsymbol{a}_2, \ldots, \boldsymbol{a_k}\}$, where $\boldsymbol{w'}$ is a solution for $\widetilde{\boldsymbol{a}}_1, \widetilde{\boldsymbol{a}}_2, \ldots, \widetilde{\boldsymbol{a}}_k$.*

$$
\begin{cases}
\boldsymbol{w'}^T \widetilde{\boldsymbol{x}} &= 0, \\
\sum_{j \in Q_1} s_j x_{r_j} &= \boldsymbol{a_1}^T \boldsymbol{x}, \\
\sum_{j \in Q_2} s_j x_{r_j} &= \boldsymbol{a_2}^T \boldsymbol{x}, \\
\cdots & \cdots \\
\sum_{j \in Q_k} s_j x_{r_j} &= \boldsymbol{a_k}^T \boldsymbol{x}.
\end{cases} \tag{2}
$$

Proof. For every 0-1 vector $\boldsymbol{x} \in \mathrm{span}\{\boldsymbol{a}_1, \boldsymbol{a}_2, \ldots, \boldsymbol{a_k}\}$, we show that all these equations are satisfied. The first equation is straightforward. For every group, say columns $j_1, j_2, \ldots, j_t$, one can see the j_1-th, j_2-th, ..., and j_t-th elements of $\boldsymbol{x}$ (and also $\boldsymbol{a_i}$, $1 \leq i \leq k$) must be all equivalent. It follows that

$$
\boldsymbol{a_i}^T \boldsymbol{x} = \sum_{j:a_i[j]=1} x_j = \sum_{j \in Q_i} s_j x_{r_j},
$$

where $1 \leq i \leq k$, $a_i[j]$ is the j-th element of $\boldsymbol{a_i}$. Therefore all the equations are satisfied.

On the other hand, we show that if a 0-1 vector $\boldsymbol{x}$ satisfies all these equations, then there must be $\boldsymbol{x} \in \mathrm{span}\{\boldsymbol{a}_1, \boldsymbol{a}_2, \ldots, \boldsymbol{a_k}\}$. By the first equation, we see $\widetilde{\boldsymbol{x}} \in \mathrm{span}\{\widetilde{\boldsymbol{a}}_1, \widetilde{\boldsymbol{a}}_2, \ldots, \widetilde{\boldsymbol{a}}_k\}$. Let $\widetilde{N}$ be the $k \times d$ matrix whose i-th row is $\widetilde{\boldsymbol{a}}_i$ ($1 \leq i \leq k$), i.e. $\widetilde{N} = [\widetilde{\boldsymbol{a}}_1, \ldots, \widetilde{\boldsymbol{a}}_k]^T$, and say $\widetilde{\boldsymbol{x}} = \boldsymbol{\lambda}^T \widetilde{N}$, where $\boldsymbol{\lambda} \in \mathbb{R}^k$. Define $\boldsymbol{x_0} = \boldsymbol{\lambda}^T N$, we claim that $\boldsymbol{x} = \boldsymbol{x_0}$:

By $\boldsymbol{x} \in \{0,1\}^n$ we see $\widetilde{\boldsymbol{x}} \in \{0,1\}^k$, and this implies $\boldsymbol{x_0} \in \{0,1\}^n$ since $\widetilde{\boldsymbol{x}} = \boldsymbol{\lambda}^T \widetilde{N}$, $\boldsymbol{x_0} = \boldsymbol{\lambda}^T N$, and $\widetilde{N}$ is the matrix obtained by removing duplicated columns from N. As a 0-1 vector in $\mathrm{span}\{\boldsymbol{a}_1, \boldsymbol{a}_2, \ldots, \boldsymbol{a_k}\}$, $\boldsymbol{x_0}$ must satisfy all the

$(k + 1)$ equations in (2). The leader elements of $\boldsymbol{x_0}$ are the same as those of $\boldsymbol{x}$, hence the left hand side of the 2nd, 3rd, ..., $(k + 1)$-th equations are the same for $\boldsymbol{x}$ and $\boldsymbol{x_0}$. Therefore $\boldsymbol{a_i}^T \boldsymbol{x} = \boldsymbol{a_i}^T \boldsymbol{x_0}$ for all $i = 1, 2, \ldots, k$, i.e. $N(\boldsymbol{x} - \boldsymbol{x_0}) = \boldsymbol{0}$.

It follows that $\boldsymbol{\lambda}^T N(\boldsymbol{x} - \boldsymbol{x_0}) = 0$, or $\boldsymbol{x_0}^T \boldsymbol{x_0} = \boldsymbol{x_0}^T \boldsymbol{x}$. By a simple observation, that for any $j = 1, 2, \ldots, n$, the j-th element of $\boldsymbol{x_0}$ is 1 implies $x_j = 1$, i.e. $\boldsymbol{x_0} \leq \boldsymbol{x}$, thus $\boldsymbol{x} - \boldsymbol{x_0} \in \{0, 1\}^n$. It must be $\boldsymbol{0}$ due to $N(\boldsymbol{x} - \boldsymbol{x_0}) = \boldsymbol{0}$ and N contains no zero column. Therefore $\boldsymbol{x} = \boldsymbol{x_0}$ and the lemma is proved. $\square$

The reduced problem on $\tilde{\boldsymbol{a}}_1, \tilde{\boldsymbol{a}}_2, \ldots, \tilde{\boldsymbol{a}}_k$ has only constant size (size $d \leq 2^k$). By Lemma 3, there is a solution $\boldsymbol{w}'$ with constant key weight $\omega(\boldsymbol{w}')$. The value $\phi(\boldsymbol{w}') \leq d \cdot \omega(\boldsymbol{w}')$ is also a constant. Let $C = \max\{\omega(\boldsymbol{w}'), \phi(\boldsymbol{w}')\} + 1$. For the 2nd, 3rd, ..., and the $(k+1)$-th equations, the corresponding $\omega(\cdot)$ and $\phi(\cdot)$ values are at most n. By Corollary 1, there is a solution of $\boldsymbol{a_1}, \ldots, \boldsymbol{a_k}$ with key weight

$$\omega(\cdot) \leq C + Cn + Cn(n + 1) + \cdots + Cn(n + 1)^{k-1} + C(n + 1)^k = O(n^k).$$

Theorem 1. *For any linear exact threshold function of constant dimension k, we have an explicit construction of a solution $\boldsymbol{w}$ with $\omega(\boldsymbol{w}) = O(n^k)$, i.e. $W^E(n, k) = O(n^k)$.*

4 General Construction

The previous section actually provides a construction for general cases as well as constant dimension k. However, the key weight $\omega(\boldsymbol{w})$ in that construction is of order $O(k^{O(\exp(k))} n^k)$, which is far from we expected. In this section, we give a more complicated construction works for general k.

We still assume that there are no zero columns in $N = [L, M]$. The idea has some similarities with the previous construction, i.e. to reduce the size of the problem first. As we have shown, for every $\boldsymbol{x_L} \in \{0, 1\}^k$ there is a unique $\boldsymbol{y} \in \text{span}\{\boldsymbol{a_1}, \boldsymbol{a_2}, \ldots, \boldsymbol{a_k}\}$ with $\boldsymbol{y_L} = \boldsymbol{x_L}$, and $\boldsymbol{x} \in \text{span}\{\boldsymbol{a_1}, \boldsymbol{a_2}, \ldots, \boldsymbol{a_k}\}$ if and only if $\boldsymbol{x_M} = \boldsymbol{y_M}$. The vector $\boldsymbol{y_M} = (L^{-1}M)^T \boldsymbol{x_L}$ may not be integers for an arbitrary $\boldsymbol{x_L}$. We say a vector $\boldsymbol{x_L} \in \{0, 1\}^k$ is *forbidden* by $\boldsymbol{m_i}$ if the i-th element of $\boldsymbol{y_M} \notin \mathbb{Z}$, i.e. $(L^{-1}\boldsymbol{m_i})^T \boldsymbol{x_L} \notin \mathbb{Z}$. (recall $\boldsymbol{m_i}$ is the i-th column of M)

We define a reduced matrix $\widetilde{M}$ by the following algorithm. Initially let $\widetilde{M} = M$, then repeat the procedure until no further columns will be deleted.

- If a column $\boldsymbol{m_i}$ forbids nothing, remove it from $\widetilde{M}$;
- If every vector forbidden by $\boldsymbol{m_i}$ is also forbidden by some other column in $\widetilde{M}$, remove $\boldsymbol{m_i}$.

Suppose there are l columns $\widetilde{\boldsymbol{m}}_1, \widetilde{\boldsymbol{m}}_2, \ldots, \widetilde{\boldsymbol{m}}_l$ left in $\widetilde{M}$. One can see that every $\widetilde{\boldsymbol{m}}_i$ $(1 \leq i \leq l)$ must forbid some distinct $\boldsymbol{x_L}$. That is, for every $i = 1, 2, \ldots, l$, there is a vector $\boldsymbol{x}_L^{(i)} \in \{0, 1\}^k$ forbidden only by $\widetilde{\boldsymbol{m}}_i$. Let $\boldsymbol{t_i} = (t_{i1}, t_{i2}, \ldots, t_{il})^T = (L^{-1}\widetilde{M})^T \boldsymbol{x}_L^{(i)}$ (a sub-vector of the corresponding $\boldsymbol{y_M}$ for $\boldsymbol{x}_L^{(i)}$), then we have $t_{ii} \notin \mathbb{Z}$, and $t_{ij} \in \mathbb{Z}$ for all $j \neq i$.

We consider the reduced problem on $[L, \widetilde{M}]$ first. By Lemma 2, the following equation is necessary (but might not be sufficient) for $(\boldsymbol{x_L}, \boldsymbol{x_{\widetilde{M}}})$ being in the linear space spanned by rows of $[L, \widetilde{M}]$.

$$\widetilde{M}(L^{-1}\widetilde{M})^T \boldsymbol{x_L} - \boldsymbol{x_{\widetilde{M}}}) = \boldsymbol{0} \qquad (3)$$

We say a vector $(\boldsymbol{x_L}, \boldsymbol{x_{\widetilde{M}}}) \in \{0,1\}^{k+l}$ is *bad* if it satisfies (3), but not in the linear space spanned by rows of $[L, \widetilde{M}]$, i.e. $(L^{-1}\widetilde{M})^T \boldsymbol{x_L} - \boldsymbol{x_{\widetilde{M}}} \neq \boldsymbol{0}$. The following two Lemmas show how to rule out these bad vectors.

Lemma 5. *For every $\boldsymbol{x_L} \in \{0,1\}^k$, there is at most one 0-1 vector $\boldsymbol{x_{\widetilde{M}}}$ such that $(\boldsymbol{x_L}, \boldsymbol{x_{\widetilde{M}}})$ is bad, i.e. if both $(\boldsymbol{x_L}, \boldsymbol{x_{\widetilde{M}}}')$ and $(\boldsymbol{x_L}, \boldsymbol{x_{\widetilde{M}}}'')$ satisfies Eq. (3), then $\boldsymbol{x_{\widetilde{M}}}' = \boldsymbol{x_{\widetilde{M}}}''$.*

Proof. Suppose $\boldsymbol{x_{\widetilde{M}}}' \neq \boldsymbol{x_{\widetilde{M}}}''$. Pick the i-th column $\widetilde{m_i}$ from $\widetilde{M}$ ($1 \le i \le l$), such that the i-th element of $\boldsymbol{x_{\widetilde{M}}}'$ differs from the i-th element of $\boldsymbol{x_{\widetilde{M}}}''$. Since $t_i = (L^{-1}\widetilde{M})^T \boldsymbol{x_L}^{(i)}$, from Eq. (3) we have

$$\boldsymbol{x_L}^{(i)T} L^{-1}\widetilde{M}\left((L^{-1}\widetilde{M})^T \boldsymbol{x_L} - \boldsymbol{x_{\widetilde{M}}}'\right) = 0, \quad \text{i.e.} \quad t_i^T (L^{-1}\widetilde{M})^T \boldsymbol{x_L} = t_i^T \boldsymbol{x_{\widetilde{M}}}'.$$

$$\boldsymbol{x_L}^{(i)T} L^{-1}\widetilde{M}\left((L^{-1}\widetilde{M})^T \boldsymbol{x_L} - \boldsymbol{x_{\widetilde{M}}}''\right) = 0, \quad \text{i.e.} \quad t_i^T (L^{-1}\widetilde{M})^T \boldsymbol{x_L} = t_i^T \boldsymbol{x_{\widetilde{M}}}''.$$

Therefore $t_i^T \boldsymbol{x_{\widetilde{M}}}' = t_i^T \boldsymbol{x_{\widetilde{M}}}''$. Since only the i-th element of t_i is not an integer, and the i-th elements of $\boldsymbol{x_{\widetilde{M}}}', \boldsymbol{x_{\widetilde{M}}}''$ are different numbers in $\{0,1\}$, we see one of $\{t_i^T \boldsymbol{x_{\widetilde{M}}}', t_i^T \boldsymbol{x_{\widetilde{M}}}''\}$ is an integer while the other is not an integer. This is a contradiction, and the lemma is proved.   □

Together with Eq. (3), the following lemma gives another equation (4) to solve the reduced problem on $[L, \widetilde{M}]$.

Lemma 6. *There exist a matrix $R \in \{0,1\}^{(k+1)\times l}$ such that no bad vector $(\boldsymbol{x_L}, \boldsymbol{x_{\widetilde{M}}})$ satisfies*

$$R((L^{-1}\widetilde{M})^T \boldsymbol{x_L} - \boldsymbol{x_{\widetilde{M}}}) = \boldsymbol{0} \qquad (4)$$

Proof. We consider a random matrix $R \in \{0,1\}^{(k+1)\times l}$ that every element has probability $\frac{1}{2}$ to be 1 and probability $\frac{1}{2}$ to be 0. For a bad vector $(\boldsymbol{x_L}, \boldsymbol{x_{\widetilde{M}}}) \in \{0,1\}^{k+l}$, $(L^{-1}\widetilde{M})^T \boldsymbol{x_L} - \boldsymbol{x_{\widetilde{M}}}$ is a non-zero vector, say the j-th element is not zero ($1 \le j \le l$). For each row r_i^T in R such that $r_i^T((L^{-1}\widetilde{M})^T \boldsymbol{x_L} - \boldsymbol{x_{\widetilde{M}}})$ is 0, we can flip the j-th element of r_i^T to make it non-zero. Thus the probability

$$\Pr_{r_i}\left(r_i^T((L^{-1}\widetilde{M})^T \boldsymbol{x_L} - \boldsymbol{x_{\widetilde{M}}}) = 0\right) \le \frac{1}{2},$$

and it follows that the probability

$$\Pr_{R}\left(R((L^{-1}\widetilde{M})^T \boldsymbol{x_L} - \boldsymbol{x_{\widetilde{M}}}) = \boldsymbol{0}\right) \le \frac{1}{2^{k+1}}.$$

By Lemma 5, there are at most 2^k bad vectors. The existence of matrix R follows directly from a use of union bound.   □

We choose R to be a matrix that no bad vector satisfies (4). One can construct a solution for the reduced problem on $[L, \widetilde{M}]$ from Eq. (3) and (4). The last step is to add Eq. (5) to solve the original problem.

Lemma 7. *For every vector* $x = (x_L, x_M) \in \{0,1\}^n$ *satisfying Eq. (3), (4) and*

$$M((L^{-1}M)^T x_L - x_M) = 0, \tag{5}$$

there must be $(L^{-1}M)^T x_L - x_M = 0$, *i.e.* $x \in \mathrm{span}\{a_1, a_2, \ldots, a_k\}$.

Proof. By x satisfying Eq. (3) and (4) and the above argument, we have $(L^{-1}\widetilde{M})^T x_L - x_{\widetilde{M}} = 0$. Thus x_L is not forbidden by any column in $\widetilde{M}$. Furthermore, by the construction of $\widetilde{M}$, x_L is not forbidden by any column in M, i.e. the elements of $(L^{-1}M)^T x_L$ are all integers.

Similarly as in Section 2 we use $y_M = (y_{k+1}, y_{k+2}, \ldots, y_n)^T$ to denote $(L^{-1}M)^T x_L$, then $y_{k+i} \in \mathbb{Z}$ and therefore $y_{k+i}^2 \geq y_{k+i}$ for $i = 1, 2, \ldots, n-k$. By Eq. (5) we have

$$x_L{}^T L^{-1} M \left((L^{-1}M)^T x_L - x_M \right) = 0,$$

i.e. $y_M{}^T (y_M - x_M) = 0$, hence

$$\sum_{i=1}^{n-k} y_{k+i}^2 = \sum_{i: x_{k+i}=1} y_{k+i},$$

where x_{k+i} is the i-th element of x_M. By $y_{k+i}^2 \geq y_{k+i}$ and $y_{k+i}^2 \geq 0$, we have

- If $x_{k+i} = 0$, then there must be $y_{k+i} = 0$.
- If $x_{k+i} = 1$, then there must be $y_{k+i}^2 = y_{k+i}$, hence $y_{k+i} = 0$ or $y_{k+i} = 1$.

Therefore $x_M \geq y_M$, i.e. $x_M - y_M$ is a 0-1 vector. By Eq. (5) $M(x_M - y_M) = 0$ and M containing no zero columns, $x_M - y_M$ must be the zero vector. Hence $x_M = y_M$ and the lemma is proved. $\qquad\square$

The equation system (3), (4) and (5) can be split into $3k+1$ linear equations. Each equation is a summation of at most l (for Eq. (3) and (4)) or $(n-k)$ (for Eq. (5)) equations from the following $(n-k)$ equations:

$$(L^{-1} m_i)^T x_L = x_{k+i}, \qquad (i = 1, 2, \ldots, n-k).$$

We normalize the coefficients to integers by multiplying $\det(L)$. By a similar argument as in Lemma 3, the values $\omega(\cdot)$ and $\phi(\cdot)$ are bounded by $\frac{l}{2^k}(k+1)^{\frac{k+1}{2}}(k+1) = \frac{l}{2^k}(k+1)^{\frac{k+3}{2}}$ (defined as b_1) for the first $2k+1$ equations, and $\frac{n}{2^k}(k+1)^{\frac{k+3}{2}}$ (defined as b_2) for the last k equations. By Corollary 1, we can construct a solution w for $[L, M]$ with

$$\omega(w) \leq \sum_{i=1}^{2k+1} b_1^i + \sum_{i=2k+2}^{3k+1} \left(b_1^{2k+1} b_2^{i-(2k+1)} \right) + b_1^{2k+1} b_2^k = O\left(b_1^{2k+1} b_2^k \right).$$

We have $l \leq \min\{2^k, n-k\}$ by the definition of $\widetilde{M}$, thus $b_1 \leq \frac{\min\{2^k, n-k\}}{2^k}(k+1)^{\frac{k+3}{2}}$. A simple calculation gives the following theorem.

Theorem 2. *For any linear exact threshold function of dimension k, there is a solution $\boldsymbol{w}$ with $\omega(\boldsymbol{w}) = O\left((\frac{k+1}{4})^{1.5k^2+O(k)} \cdot \min\{2^k, n-k\}^{2k+1} \cdot n^k\right)$.*

Theorem 1 can also be implied directly from this theorem. For general k, Theorem 2 gives an upper bound $O(k^{O(k^2)}n^k) = O(n^{O(k^2)})$.

5   Conclusion

Through out this paper we consider the weights of integer in the representations of linear exact threshold functions. We first show an explicit construction giving a tight upper bound $O(n^k)$ on the weights for constant dimension k. This construction also works for general cases, while the upper bound becomes $O(k^{O(\exp(k))}n^k)$. We then show there exists a solution with upper bound $O(k^{O(k^2)}n^k)$ through a more complicated analysis.

There are still several open problems on this topic. For general cases, our proof go through probabilistic method and the the construction is implicit. It would be interesting to find a deterministic algorithms to construction good solutions. Moreover, we believe that our analysis can be improved to get a better upper bound. And it remains open to close the gap between the upper and lower bounds for large k.

References

1. Alon, N., Văn Vũ, H.: Anti-hadamard matrices, coin weighing, threshold gates, and indecomposable hypergraphs. Journal of Combinatorial Theory, Series A 79(1), 133–160 (1997)
2. Babai, L., Hansen, K.A., Podolskii, V.V., Sun, X.: Weights of exact threshold functions. In: Hliněný, P., Kučera, A. (eds.) MFCS 2010. LNCS, vol. 6281, pp. 66–77. Springer, Heidelberg (2010)
3. Faddeev, D.K., Sominski, I.S.: Problems in Higher Algebra. W. H. Freeman, San Francisco (1965)
4. Håstad, J.: On the size of weights for threshold gates. SIAM Journal on Discrete Mathematics 7(3), 484–492 (1994)
5. Muroga, S.: Threshold Logic and its Applications. John Wiley & Sons, Inc., Chichester (1971)
6. Muroga, S., Toda, I., Takasu, S.: Theory of majority decision elements. Journal of the Franklin Institute 271(5), 376–418 (1961)
7. Parberry, I.: Circuit Complexity and Neural Networks. MIT Press, Cambridge (1994)
8. Smith, D.R.: Bounds on the number of threshold functions. IEEE Transactions on Electronic Computers EC-15(3), 368–369 (1966)
9. Yajima, S., Ibaraki, T.: A lower bound of the number of threshold functions. IEEE Transactions on Electronic Computers EC-14(6), 926–929 (1965)

Submodular Function Minimization under a Submodular Set Covering Constraint

Naoyuki Kamiyama*

Department of Information and System Engineering, Chuo University
kamiyama@ise.chuo-u.ac.jp

Abstract. In this paper, we consider the problem of minimizing a submodular function under a submodular set covering constraint. We propose an approximation algorithm for this problem by extending the algorithm of Iwata and Nagano [FOCS'09] for the set cover problem with a submodular cost function.

1 Introduction

Let N be a finite nonempty set. A real-valued function ρ on 2^N is *submodular* if

$$\rho(X) + \rho(Y) \geq \rho(X \cup Y) + \rho(X \cap Y)$$

for all $X, Y \subseteq N$. Submodular functions play an important role in the field of combinatorial optimization, game theory and so on.

The first polynomial-time algorithm for the submodular function minimization problem was presented by Grötschel, Lovász and Schrijver [1,2] via the ellipsoid method. Combinatorial polynomial-time algorithms for minimizing submodular functions were presented by Iwata, Fleischer and Fujishige [3] and Schrijver [4]. These combinatorial algorithms have been improved in time complexity [5,6,7].

Constrained submodular function minimization problems have also been studied in various contexts. For example, Svitkina and Fleischer [8] considered the submodular function minimization problem with a cardinality lower bound and gave a lower bound for approximability. Iwata and Nagano [9] considered submodular function minimization under a vertex covering constraint, a set covering constraint and an edge covering constraint, and gave approximability and inapproximability. Goel, Karande, Tripathi and Wang [10] also studied the vertex cover problem, the shortest path problem, the perfect matching problem and the minimum spanning tree problem with a monotone submodular cost function.

In this paper, we consider the *submodular function minimization problem under a submodular set covering constraint* (in short SFM-SSCC) which is defined as follows. We are given a non-empty finite set N and a non-negative real-valued

---

* Supported by a Grants-in-Aid from the Ministry of Education, Culture, Sports, Science and Technology of Japan.

submodular function ρ on 2^N and a non-negative real-valued monotone submodular function μ on 2^N such that $\rho(\emptyset) = \mu(\emptyset) = 0$, where μ is called monotone if $\mu(X) \leq \mu(Y)$ for every $X, Y \subseteq N$ such that $X \subseteq Y$. Let $\mathcal{B}$ be the set of $X \subseteq N$ such that $\mu(X) = \mu(N)$. Then, the problem SFM-SSCC asks for finding a minimizer of

$$\min\{\rho(X) \mid X \in \mathcal{B}\}.$$

We assume without loss of generality that $\mu(N) > 0$. Otherwise, we have $\mu(X) = 0$ for all $X \subseteq N$, and thus $\emptyset$ is an optimal solution for the problem SFM-SSCC.

By the result of [11], we can see that the problem SFM-SSCC generalized the set cover problem with a submodular cost function (for the details, see Section 4). The problem SFM-SSCC with a linear cost function was considered in [11,12]. In this paper, we propose an approximation algorithm for the problem SFM-SSCC by extending the algorithm of Iwata and Nagano [9] for the set cover problem with a submodular cost function.

The rest of this paper is organized as follows. In Section 2, we give necessary definitions and basic facts. In Section 3, we propose an approximation algorithm for the problem SFM-SSCC. In Section 4, we consider an application of our result to the set cover problem with a submodular cost function.

2   Preliminaries

We denote by $\mathbb{R}$ and $\mathbb{R}_+$ the sets of real numbers and non-negative real numbers, respectively. Given a subset $X \subseteq N$, we define the characteristic χ_X by $\chi_X(i) = 1$ for $i \in X$ and $\chi_X(i) = 0$ for $i \in N \setminus X$.

2.1   Submodular Functions

Associated with ρ, we consider a submodular polyhedron

$$\mathrm{P}(\rho) = \Big\{ x \in \mathbb{R}^N \mid \sum_{i \in X} x(i) \leq \rho(X) \ (\forall X \subseteq N) \Big\}.$$

Suppose that we are given a non-negative vector $p \in \mathbb{R}_+^N$. Let

$$p_1 > p_2 > \cdots > p_m$$

be the distinct values of p. For each $i \in \{1, \ldots, m\}$, we denote

$$N_i = \{j \in N \mid p(j) \geq p_i\}.$$

Then, we define $\widehat{\rho}(p)$ by

$$\widehat{\rho}(p) = \sum_{i=1}^{m} (p_i - p_{i+1})\rho(N_i),$$

where $p_{m+1} = 0$. A function $\widehat{\rho}$ is called a *Lovász extension* [13]. It is not difficult to see that $\widehat{\rho}(\chi_X) = \rho(X)$ for all $X \subseteq N$. It is known [14] that

$$\widehat{\rho}(p) = \max\{\langle p, z \rangle \mid z \in \mathrm{P}(\rho)\}, \tag{1}$$

where $\langle p, z \rangle$ is an inner product of p and z.

Given a subset $S \subseteq N$, we define a function $\rho_S \colon 2^{N \setminus S} \to \mathbb{R}$ by

$$\rho_S(X) = \rho(X \cup S) - \rho(S)$$

for each $X \subseteq N \setminus S$. The function ρ_S is called a *contraction of ρ on $N \setminus S$*.

2.2 Integer Programming Formulation

Here we formulate the problem SFM-SSCC as an integer programming. Given a subset $T \subseteq N$, it is known [11] that $\mu(T) = \mu(N)$ if and only if

$$\sum_{i \in N \setminus S} \mu_S(\{i\}) \cdot \chi_T(i) \geq \mu_S(N \setminus S)$$

for all $S \subseteq N$. Thus, the problem SFM-SSCC can be formulated as follows.

$$\begin{aligned}
\min \quad & \rho(X) \\
\text{s.t.} \quad & \sum_{i \in N \setminus S} \mu_S(\{i\}) \cdot \chi_X(i) \geq \mu_S(N \setminus S) \quad (\forall S \subseteq N) \\
& X \subseteq N.
\end{aligned}$$

We call this integer programming IP.

3 Algorithm

In this section, we give an approximation algorithm for the problem SFM-SSCC. We use a linear programming relaxation of the problem IP. Recall that $\widehat{\rho}(\chi_X) = \rho(X)$ for all $X \subseteq N$. Thus, the problem LP

$$\begin{aligned}
\min \quad & \widehat{\rho}(x) \\
\text{s.t.} \quad & \sum_{i \in N \setminus S} \mu_S(\{i\}) \cdot x(i) \geq \mu_S(N \setminus S) \quad (\forall S \subseteq N) \\
& x \in \mathbb{R}_+^N
\end{aligned}$$

is a relaxation of the problem IP. Furthermore, it follows from (1) that $\widehat{\rho}(x)$ is equal to the optimal objective value of the problem

$$\begin{aligned}
\min \quad & \sum_{X \subseteq N} \rho(X) \cdot \xi(X) \\
\text{s.t.} \quad & \sum_{X \subseteq N \colon i \in X} \xi(X) = x(i) \quad (\forall i \in N) \\
& \xi \in \mathbb{R}_+^{2^N}
\end{aligned}$$

for each $x \in \mathbb{R}_+^N$. Thus, the problem LP can be written as

$$\min \quad \sum_{X \subseteq N} \rho(X) \cdot \xi(X)$$

$$\text{s.t.} \quad \sum_{i \in N \setminus S} \mu_S(\{i\}) \cdot x(i) \geq \mu_S(N \setminus S) \quad (\forall S \subseteq N)$$

$$\sum_{X \subseteq N:\, i \in X} \xi(X) = x(i) \quad (\forall i \in N)$$

$$x \in \mathbb{R}^N,\ \xi \in \mathbb{R}_+^{2^N}.$$

Here, we neglect the redundant non-negativity constraint of x. Hence, the dual of LP, called DLP, is given as follows.

$$\max \quad \sum_{S \subseteq N} \mu_S(N \setminus S) \cdot y(S)$$

$$\text{s.t.} \quad \sum_{S \subseteq N:\, i \notin S} \mu_S(\{i\}) \cdot y(S) = z(i) \quad (\forall i \in N)$$

$$y \in \mathbb{R}_+^{2^N},\ z \in \mathrm{P}(\rho).$$

Let S^* and opt_D be an optimal solution of the problem SFM-SSCC and the optimal objective value of the problem DLP. Then, we have $\mathrm{opt}_D \leq \rho(S^*)$.

We are now ready to present our algorithm which is called **SFM-SSCC**. The algorithm **SFM-SSCC** can be described as follows. In the algorithm, t is an integer, $y_t \in \mathbb{R}^{2^N}$, $z_t \in \mathbb{R}^N$ and $T_t \subseteq N$.

Step 1: Set $t = 0$, $y_t = \mathbf{0}$, $z_t = \mathbf{0}$ and $T_t = \emptyset$.
Step 2: Repeat the following (2-1) to (2-5) until $\mu(T_t) = \mu(N)$.
(2-1) Compute a vector $a_t \in \mathbb{R}_+^N$ defined by

$$a_t(i) = \begin{cases} \mu_{T_t}(\{i\}), & \text{if } i \notin T_t, \\ 0, & \text{otherwise,} \end{cases}$$

for each $i \in N$.
(2-2) Compute $\lambda_t = \min\{\lambda \in \mathbb{R}_+ \mid z_t + \lambda \cdot a_t \in \mathrm{P}(\rho)\}$.
(2-3) Set $y_{t+1} = y_t + \lambda_t \cdot \chi_{\{T_t\}}$ and $z_{t+1} = z_t + \lambda_t \cdot a_t$.
(2-4) Set T_{t+1} to be the unique maximal set such that

$$\sum_{i \in T_{t+1}} z_{t+1}(i) = \rho(T_{t+1}).$$

(2-5) Update $t = t + 1$.
Step 3: Return T_t.

Notice that it can be proved by using the submodularity of μ that there exists the desired unique maximal set in Step (2-4). For the details, see the proof of Lemma 2.

3.1   Analysis

We first show that the algorithm **SFM-SSCC** is well-defined. We can prove this by the following Statement 2 of Lemma 1.

Lemma 1. *Throughout the algorithm* **SFM-SSCC**,

1. (y_t, z_t) *is feasible to the problem DLP, and*
2. *if* $\mu(T_t) < \mu(N)$, λ_t *exists.*

Proof. We prove the lemma by induction on t. We first show that (y_0, z_0) is feasible to the problem DLP. Since ρ is non-negative, $z_0 = \mathbf{0}$ belongs to $P(\rho)$. Hence, (y_0, z_0) is feasible to the problem DLP.

Next, assuming that (y_t, z_t) is feasible to the problem DLP, we show that λ_t exists. Since z_t belongs to $P(\rho)$ by the induction hypothesis and a_t is non-negative by the monotonicity of μ, it suffices to show that $a_t \neq \mathbf{0}$. Assume that $a_t = \mathbf{0}$. Then,

$$\mu(T_t \cup \{i\}) = \mu(T_t) \tag{2}$$

for all $i \in N \setminus T_t$. By the submodularity of μ,

$$\sum_{i \in N \setminus T_t} \mu(T_t \cup \{i\}) \geq (|N \setminus T_t| - 1)\mu(T_t) + \mu(N). \tag{3}$$

By (2) and (3), $\mu(T_t) \geq \mu(N)$, which contradicts $\mu(T_t) < \mu(N)$.

Finally, assuming that (y_t, z_t) is feasible to the problem DLP and λ_t exists, we show that (y_{t+1}, z_{t+1}) is feasible to the problem DLP. Since it follows from the definition of λ_t that z_{t+1} belongs to $P(\rho)$, we consider the first constraint. Since $y_{t+1}(S) = y_t(S)$ for all $S \in 2^N \setminus \{T_t\}$, it suffice to consider the constraint for $i \in N \setminus T_t$. By the definition of a_t,

$$z_{t+1}(i) - z_t(i) = \lambda_t \cdot \mu_{T_t}(\{i\}).$$

Since $y_{t+1}(S) = y_t(S)$ for all $S \in 2^N \setminus \{T_t\}$,

$$\sum_{S \subseteq N : i \notin S} \mu_S(\{i\}) \cdot y_{t+1}(S) - \sum_{S \subseteq N : i \notin S} \mu_S(\{i\}) \cdot y_t(S)$$

$$= \mu_{T_t}(\{i\}) \cdot (y_{t+1}(T_t) - y_t(T_t)) = \lambda_t \cdot \mu_{T_t}(\{i\}).$$

By these, (y_{t+1}, z_{t+1}) is feasible to the problem DLP. $\qquad\square$

Now we evaluate the time complexity and the approximation ratio of the algorithm **SFM-SSCC**.

Lemma 2. *The algorithm* **SFM-SSCC** *terminates in* $O(|N|^6 \mathrm{EO} + |N|^7)$ *time, where* EO *is the time required for one function evaluation of* ρ *and* μ.

Proof. We first estimate the number of iterations of Step 2. It is not difficult to see that after Step (2-3) there exists $X \subseteq N$ such that

$$\sum_{i \in X} z_{t+1}(i) = \rho(X)$$

and $X \setminus T_t \neq \emptyset$. Otherwise, we can increase λ_t. We define a function $g \colon 2^N \to \mathbb{R}$ by

$$g(S) = \rho(S) - \sum_{i \in S} z_{t+1}(i)$$

for each $S \subseteq N$. It is easy to see that g is a non-negative submodular function since $z_{t+1} \in \mathrm{P}(\rho)$. By $g(T_t) = 0$,

$$0 = g(X) + g(T_t) \geq g(X \cap T_t) + g(X \cup T_t) \geq 0,$$

which implies that $g(X \cup T_t) = 0$, i.e.,

$$\sum_{i \in X \cup T_t} z_{t+1}(i) = \rho(X \cup T_t).$$

By a similar argument, it is easy to see that $X \cup T_t \subseteq T_{t+1}$. Hence, it follows from $X \setminus T_t \neq \emptyset$ that T_{t+1} is larger than T_t. This implies that the number of iterations of Step 2 is at most $|N|$.

Now we consider the time required for one iteration of Step 2. It is clear that Steps (2-1), (2-3) and (2-5) can be done in $O(|N|\mathrm{EO})$ time, $O(|N|)$ time and $O(1)$ time, respectively. Here we consider the time required for Step (2-2). To compute λ_t is equivalent to

$$\min_{S \subseteq N} \frac{\rho(S) - \sum_{i \in S} z_t(i)}{\sum_{i \in S} a_t(i)}.$$

Since a function $\rho - z_t$ is submodular and $a_t \in R_+^N \setminus \{0\}$, it is known [15] that this value can be computed in $O(|N|^5 \mathrm{EO} + |N|^6)$ time. Furthermore, it is known [16, Note 10.11] that in Step (2-4) we can compute the desired unique maximal set in $O(|N|^3 \mathrm{EO})$ time. This completes the proof.   □

In the sequel, we assume that $t = k$ when the algorithm **SFM-SSCC** halts. Let c_k be the objective value for (y_k, z_k), i.e.,

$$c_k = \sum_{t=0}^{k-1} \mu_{T_t}(N \setminus T_t) \cdot \lambda_t.$$

Lemma 3. *If $c_k = 0$, T_k is an optimal solution for the problem SFM-SSCC.*

Proof. By $\mu(N) > 0$, we must have $\lambda_0 = 0$. Thus, T_1 is set to be the maximal set such that $\rho(T_1) = 0$. If $\mu(T_1) = \mu(N)$, the proof is done. From here, we show that if $\mu(T_1) < \mu(N)$, this contradicts the fact that $c_k = 0$. Assume that

$\mu(T_1) < \mu(N)$, i.e., $\mu_{T_1}(N \setminus T_1) > 0$. Since T_1 is set to be the maximal set such that $\rho(T_1) = 0$, $\lambda_1 > 0$. If $\lambda_1 = 0$, it follows from the submodularity of ρ that there exists a set $X \subseteq N$ such that $X \subsetneq T_1$ and $\rho(X) = 0$, which contradicts the maximality of T_1. Hence, $\mu_{T_1}(N \setminus T_1) \cdot \lambda_1 > 0$, which contradicts the fact that $c_k = 0$. $\qquad\square$

Lemma 4. *The approximation ratio of the algorithm* **SFM-SSCC** *is*

$$\max_{S \subseteq N : \mu(S) < \mu(N)} \frac{\sum_{i \in N \setminus S} \mu_S(\{i\})}{\mu_S(N \setminus S)}. \tag{4}$$

Proof. By the definition of c_k,

$$c_k \leq \mathrm{opt}_D \leq \rho(S^*). \tag{5}$$

Since $y_k(S) = 0$ for all $S \in 2^N \setminus \{T_0, \ldots, T_{k-1}\}$,

$$\begin{aligned}
\rho(T_k) = z_k(T_k) &= \sum_{i \in T_k} \sum_{S \subseteq N : i \notin S} \mu_S(\{i\}) \cdot y_k(S) \\
&= \sum_{S \subseteq N} \sum_{i \in T_k \setminus S} \mu_S(\{i\}) \cdot y_k(S) \\
&= \sum_{t \in \{0, \ldots, k-1\}} \sum_{i \in T_k \setminus T_t} \mu_{T_t}(\{i\}) \cdot y_k(T_t).
\end{aligned} \tag{6}$$

Furthermore,

$$c_k = \sum_{S \subseteq N} \mu_S(N \setminus S) \cdot y_k(S) = \sum_{t \in \{0, \ldots, k-1\}} \mu_{T_t}(N \setminus T_t) \cdot y_k(T_t). \tag{7}$$

By (6) and (7),

$$\frac{\rho(T_k)}{c_k} \leq \max_{t \in \{0, \ldots, k-1\}} \frac{\sum_{i \in T_k \setminus T_t} \mu_{T_t}(\{i\})}{\mu_{T_t}(N \setminus T_t)}. \tag{8}$$

By the definition of the algorithm **SFM-SSCC**, $\mu(T_t) < \mu(N)$ for all $t \in \{0, \ldots, k-1\}$. Hence, it follows from (5), (8) and the monotonicity of μ that

$$\begin{aligned}
\frac{\rho(T_k)}{\rho(S^*)} &\leq \max_{t \in \{0, \ldots, k-1\}} \frac{\sum_{i \in T_k \setminus T_t} \mu_{T_t}(\{i\})}{\mu_{T_t}(N \setminus T_t)} \\
&\leq \max_{S \subseteq N : \mu(S) < \mu(N)} \frac{\sum_{i \in N \setminus S} \mu_S(\{i\})}{\mu_S(N \setminus S)}.
\end{aligned}$$

This completes the proof. $\qquad\square$

By Lemmas 1 to 4, we obtain the main result of this paper.

Theorem 1. *The algorithm* **SFM-SSCC** *is a polynomial-time approximation algorithm for the problem SFM-SSCC with a factor* (4).

4   Application

In this section, we consider an application of our result to the set cover problem with submodular cost function which was studied by Iwata and Nagano [9].

4.1   The Submodular Cost Set Cover Problem

In this subsection, we consider the *submodular cost set cover problem*. In this problem, we are given a finite set U, a collection $\mathcal{X} = \{X_1, \ldots, X_n\}$ of subsets of U indexed by $N = \{1, \ldots, n\}$ and a non-negative submodular function ρ on 2^N such that $\rho(\emptyset) = 0$. For each $S \subseteq N$, we denote $X_S = \cup_{i \in S} X_i$. We say that $S \subseteq N$ is a *set cover* if $X_S = U$. We assume without loss of generality that there exists a set cover. Then, this problem asks for finding a set cover $S \subseteq N$ that minimizes $\rho(S)$. Iwata and Nagano [9] showed that this problem can be approximated with a factor of the maximum frequency

$$\eta = \max_{u \in U} |\{i \in N \mid u \in X_i\}|.$$

In this subsection, we prove that our result is an extension of this result.

It is known [11] that the set covering constraint can be represented by the submodular set covering constraint as follows. We define a function μ on 2^N by

$$\mu(S) = \sum_{u \in U} \min\{\sum_{i \in S} \chi_{X_i}(u), 1\}$$

for each $S \subseteq N$. It is not difficult to see that by setting $\mu(\emptyset) = 0$, μ is a non-negative monotone submodular function such that $\mu(\emptyset) = 0$. Hence, what remains is to show that the approximation ratio (4) is no more than η. By the double counting technique,

$$\begin{aligned}
\frac{\sum_{i \in N \setminus S} \mu_S(\{i\})}{\mu_S(N \setminus S)} &= \frac{\sum_{i \in N \setminus S} |X_i \setminus X_S|}{|U| - |X_S|} \\
&= \frac{\sum_{v \in U \setminus X_S} |\{i \in N \setminus S \mid v \in X_i\}|}{|U| - |X_S|} \\
&\leq \max_{v \in U \setminus X_S} |\{i \in N \setminus S \mid v \in X_i\}| \\
&\leq \eta
\end{aligned}$$

for each $S \subseteq N$ such that $\mu(S) < \mu(N)$. This completes the proof.

References

1. Grötschel, M., Lovász, L., Schrijver, A.: The ellipsoid method and its consequences in combinatorial optimization. Combinatorica 1(2), 169–197 (1981)
2. Grötschel, M., Lovász, L., Schrijver, A.: Geometric Algorithms and Combinatorial Optimization. Springer, Heidelberg (1988)

3. Iwata, S., Fleischer, L., Fujishige, S.: A combinatorial strongly polynomial algorithm for minimizing submodular functions. J. ACM 48(4), 761–777 (2001)
4. Schrijver, A.: A combinatorial algorithm minimizing submodular functions in strongly polynomial time. J. Comb. Theory, Ser. B 80(2), 346–355 (2000)
5. Iwata, S.: A faster scaling algorithm for minimizing submodular functions. SIAM J. Comput. 32(4), 833–840 (2003)
6. Iwata, S., Orlin, J.B.: A simple combinatorial algorithm for submodular function minimization. In: SODA, pp. 1230–1237 (2009)
7. Orlin, J.B.: A faster strongly polynomial time algorithm for submodular function minimization. Math. Program. 118(2), 237–251 (2009)
8. Svitkina, Z., Fleischer, L.: Submodular approximation: Sampling-based algorithms and lower bounds. In: FOCS, pp. 697–706 (2008)
9. Iwata, S., Nagano, K.: Submodular function minimization under covering constraints. In: FOCS, pp. 671–680 (2009)
10. Goel, G., Karande, C., Tripathi, P., Wang, L.: Approximability of combinatorial problems with multi-agent submodular cost functions. In: FOCS, pp. 755–764 (2009)
11. Wolsey, L.A.: An analysis of the greedy algorithm for the submodular set covering problem. Combinatorica 2(4), 385–393 (1982)
12. Fujito, T.: On approximation of the submodular set cover problem. Oper. Res. Lett. 25(4), 169–174 (1999)
13. Lovász, L.: Submodular functions and convexity. In: Bachem, A., Grötschel, M., Korte, B. (eds.) Mathematical Programming–The State of the Art, pp. 235–257. Springer, Heidelberg (1983)
14. Edmonds, J.: Submodular functions, matroids, and certain polyhedra. In: Guy, R., Hanani, H., Sauer, N., Schönheim, J. (eds.) Combinatorial Structures and their Applications, pp. 69–87. Gordon and Breach, Reading (1970)
15. Nagano, K.: A faster parametric submodular function minimization algorithm and applications. Technical Report METR 2007-43, The University of Tokyo (2007)
16. Murota, K.: Discrete Convex Analysis. SIAM Monographs on Discrete Mathematics and Applications, vol. 10. Society for Industrial and Applied Mathematics (2003)

Optimal Allocation in Combinatorial Auctions with Quadratic Utility Functions

Akiyoshi Shioura and Shunya Suzuki

Graduate School of Information Sciences, Tohoku University, Sendai 980-8579, Japan
{shioura,shunya}@dais.is.tohoku.ac.jp

Abstract. We discuss the optimal allocation problem in combinatorial auction, where the items are allocated to bidders so that the sum of the bidders' utilities is maximized. In this paper, we consider the case where utility functions are given by quadratic functions; the class of quadratic utility functions has a succinct representation but is sufficiently general. The main aim of this paper is to show the computational complexity of the optimal allocation problem with quadratic utility functions. We consider the cases where utility functions are submodular and supermodular, and show NP-hardness and/or polynomial-time exact/approximation algorithm. These results are given by using the relationship with graph cut problems such as the min/max cut problem and the multiway cut problem.

1 Introduction

Combinatorial auction is an auction such that bidders can place bids on combinations of items, rather than individual items. Combinatorial auctions can be used, for instance, to sell spectrum licenses, pollution permits, land lots, etc., and has emerged as a mechanism to improve economic efficiency when many items are on sale. See [2,3] for comprehensive survey on combinatorial auctions.

In a combinatorial auction, bidders can present bids on bundles of items, and thus may easily express substitutabilities and complementarities among the items on sale. The function that, given a bundle, returns the bidder's value for that bundle is called a utility function. A utility function is associated with each bidder specifying the degree of satisfaction of the bidder for each subset of the items.

Given utility functions of bidders, the auctioneer of a combinatorial auction needs to determine how to allocate items to bidders, which is called the *optimal allocation problem*. One natural objective for the auctioneer is to maximize the economic efficiency of the auction, which is the sum of the utilities of all the bidders. Formally, the optimal allocation problem is defined as follows. Let V be a set of n items, and M a set of m bidders, and assume, for simplicity, that $V = \{1, 2, \ldots, n\}$ and $M = \{1, 2, \ldots, m\}$. Bidder i has a utility function $f_i : 2^V \to \mathbf{R}$ which is *monotone*, i.e., $f_i(X) \geq f_i(Y)$ whenever $X \supseteq Y$. The auctioneer wishes to find a partition $(S_1, S_2, \ldots, S_m)$ of the set V among the m bidders that maximizes the total utility $\sum_{i=1}^{m} f_i(S_i)$.

M. Ogihara and J. Tarui (Eds.): TAMC 2011, LNCS 6648, pp. 142–153, 2011.

Implementation of combinatorial auctions faces several issues to be discussed, including representation of utility functions. A utility function for a bidder requires a value for each subset of items, and therefore requires exponential real values in total. This makes it difficult for bidders to reveal their preference correctly since in practice it is not possible for bidders to submit *correct* values of utilities for a exponential number of subsets of items. This also brings a difficulty to the auctioneer since the input size of utility functions becomes exponential, and the optimal allocation becomes hard to solve in short time.

Thus, we need a restricted class of utility functions which has a succinct representation but sufficiently general[1]. Various such classes of utility functions have been considered in the literature of combinatorial auction (see, e.g., [2], [3, Ch. 9]). Some examples are symmetric functions, (budgeted) additive functions, single-minded functions [23], OR functions, XOR functions, and OR-of-XOR functions [27].

In this paper, we consider one such class of utility functions called *quadratic* functions. In the context of combinatorial auction, the use of quadratic functions is firstly considered independently by Conitzer et al. [6] (as *2-wise dependent* functions) and by Chevaleyre et al. [5] (as *2-additive* functions). A utility function $f : 2^V \rightarrow \mathbf{R}$ is said to be *quadratic* (or *of order 2*) if it is represented as

$$f(X) = \sum_{u,v \in X, u<v} a(u,v) + \sum_{v \in X} b(v) \qquad (X \subseteq V) \qquad (1)$$

by using real values $a(u,v)$ $(u, v \in V, u < v)$ and $b(v)$ $(v \in V)$ (see [9, Sec. 3.6])[2]. While a quadratic utility function is simple and can be represented in a succinct way, it is sufficiently general so that by using the term $a(u,v)$ it can easily express substitutability and complementarity among items. These facts incicate that quadratic utility functions constitute an important class of utility functions.

The main aim of this paper is to reveal the computational complexity of the optimal allocation problem with quadratic utility functions. That is, we consider the case where a utility function $f_i : 2^V \rightarrow \mathbf{R}$ of bidder $i \in M$ is given as

$$f_i(X) = \sum_{u,v \in X, u<v} a_i(u,v) + \sum_{v \in X} b_i(v) \qquad (X \subseteq V) \qquad (2)$$

by using real values $a_i(u,v)$ $(u, v \in V, u < v)$ and $b_i(v)$ $(v \in V)$ The same problem is considered in [5,6], where they only show the NP-hardness of the very general case. In contrast, we classify the optimal allocation problem according to the type of utility functions (substitutes or complements) and the number of bidders (2 or more), analyze the computational complexity of each case, and present exact/approximation algorithms.

Previous Results. We review the computational complexity results of the optimal allocation problem with *general* utility functions. We here consider only the case

---
[1] Representation of utility functions is called a *bidding language*.
[2] A utility function is quadratic if and only if it can be represented by a quadratic polynomial function with $\{0, 1\}$-variables.

Table 1. Summary of Our Results (uf. = utility function)

type of uf. \ # of bidders m	$m = 2$	$m \geq 3$
submodular uf.	NP-hard, 0.879-inapprox. 0.874-approx.	NP-hard, 0.879-inapprox.
gross substitutes uf.	P ($O(n^2 \log n)$ time)	P ($O(mn^2 \log(mn))$ time)
supermodular uf.	P ($O(n^3/\log n)$ time)	NP-hard 0.5-approx. (2/3-approx. for $m = 3$)

where a utility function f is given implicitly by a *value oracle*, which, given a set $S \subseteq V$, returns a function value $f(S)$. Since the value oracle can be easily constructed for quadratic utility functions, all of the results mentioned here are valid for the case of quadratic utility functions.

We firstly consider the case of submodular utility functions. The problem is NP-hard, even if $m = 2$. Moreover, there exists no polynomial-time approximation algorithm with a ratio better than $1 - 1/e$, unless P=NP [18]. Mirrokni et al. [25] also show that an approximation algorithm with a ratio better than $1 - (1 - 1/m)^m$ requires exponentially many calls to the value oracle, implying, without any assumption, that there exists no polynomial-time approximation algorithm with a ratio better than $1 - 1/e$. For the class of gross-substitutes utility functions, which is known to be an important subclass of submodular utility functions [12,16], the optimal allocation problem can be solved in polynomial time [22].

We then consider the case of supermodular utility functions. Compared to the case of submodular utility functions, this case attracts less attention in the literature of combinatorial auction, and much is not known yet for this case. If $m = 2$, then the optimal allocation problem can be easily reduced to the submodular function minimization problem, which can be solved in polynomial time [11]. On the other hand, if $m \geq 3$ then the problem is NP-hard (see, e.g., [6]). While an $O(\frac{\sqrt{\log n}}{n})$-approximation algorithm is given [15], no inapproximability result is known.

Our Results. We analyze the computational complexity of the optimal allocation problem with quadratic utility functions. We consider the two cases where utility functions are *submodular* and *supermodular* (see Section 2 for definitions), and for each case we also consider subcases where the number m of bidders are equal to 2 and more than 2. That is, we consider 4 cases, each of which is denoted as (SUB|m=2), (SUB|m>2), (SUP|m=2), and (SUP|m>2). The results obtained in this paper is summarized in Table 1. These results are shown by using the relationship with graph cut problems such as the min/max cut problem and the multiway (un)cut problem.

For the case of submodular quadratic utility functions, we show the NP-hardness even for the case (SUB|m=2) by using the reduction of the max cut problem in *undirected* graphs. On the other hand, we present the reduction of the case (SUB|m=2) to the max cut problem in *directed* graphs. This reduction yields a 0.874-approximation algorithm for (SUB|m=2), which is better than the approximation ratio $1 - 1/e \simeq 0.632$ for the case of general submodular utility functions. We also consider the special case of gross-substitutes quadratic utility functions as an important subclass of submodular utility functions. It is shown that the problem can be solved efficiently in $O(mn^2 \log(mn))$ time by using the reduction to the minimum quadratic-cost flow problem.

For the case of supermodular quadratic utility functions, we firstly show the polynomial-time solvability of (SUP|m=2) by reducing it to the min cut problem in directed graphs. We then show the NP-hardness of (SUP|m>2) by using the reduction of the multiway (un)cut problem. For this problem, we also present a 0.5-approximation algorithm based on randomized LP rounding, where we use the technique in Langberg et al. [21] for the multiway uncut problem.

The organization of this paper is as follows. Characterizations of submodular/supermodular quadratic utility functions are given in Section 2. In Section 3, we present our results for (SUB|m=2) and (SUB|m>2), while the results for (SUP|m=2), and (SUP|m>2) are given in Section 4. Proofs are given in Appendix due to the page limit.

2 Characterizations of Quadratic Utility Functions

We give characterizations of quadratic utility functions of the form (1) which have submodularity and supermodularity. Throughout this paper we assume $a(v, u) = a(u, v)$ for every $u, v \in V$ with $u < v$.

A utility function $f : 2^V \to \mathbf{R}$ is said to be *submodular* if it satisfies the following condition:

$$f(X \cup \{v\}) - f(X) \ge f(Y \cup \{v\}) - f(Y) \quad (\forall X, Y \in 2^V \text{ with } Y \supset X, \forall v \in V \setminus Y).$$

Intuitively, this condition says that the marginal value of an item decreases as the set of items already acquired increases. A utility function $f : 2^V \to \mathbf{R}$ is said to be *supermodular* if $-f$ is submodular. Submodularity (resp., supermodularity) of utility functions is used to model substitutability (resp., complementarity) of items. A utility function $f : 2^V \to \mathbf{R}$ is said to be *monotone* if it satisfies $f(X) \le f(Y)$ for every $X, Y \in 2^V$ with $X \subseteq Y$.

Theorem 2.1. *Let $f : 2^V \to \mathbf{R}$ be a quadratic utility function of the form (1).*
(i) f is submodular if and only if $a(u, v) \le 0$ ($\forall u, v \in V, u \ne v$).
(ii) Suppose that f is a submodular function. Then, f is monotone if and only if $b(v) + \sum_{u \in V \setminus \{v\}} a(u, v) \ge 0$ ($\forall v \in V$).

Proof. It is well known that f is submodular if and only if the following inequality holds:

$$f(X \cup \{u\}) + f(X \cup \{v\}) \ge f(X \cup \{u, v\}) + f(X) \ (\forall X \subseteq V, \forall u, v \in V \setminus X, u \ne v).$$

This condition is equivalent to the condition $a(u, v) \le 0$ ($\forall u, v \in V, u \ne v$) since

$$\{f(X \cup \{u, v\}) + f(X)\} - \{f(X \cup \{u\}) + f(X \cup \{v\})\} = a(u, v).$$

If f is submodular and monotone, then we have $0 \leq f(V) - f(V \setminus \{v\}) = b(v) + \sum_{u \in V \setminus \{v\}} a(u, v)$ for all $v \in V$. On the other hand, if f is a submodular function satisfying the condition $b(v) + \sum_{u \in V \setminus \{v\}} a(u, v) \geq 0$ $(\forall v \in V)$, then we have

$$f(X) - f(X \setminus \{v\}) = b(v) + \sum_{u \in X \setminus \{v\}} a(u, v) \geq b(v) + \sum_{u \in V \setminus \{v\}} a(u, v) \geq 0$$

for every $X \subseteq V$ and $v \in X$, i.e., f is monotone.   □

Theorem 2.2. *Let $f : 2^V \to \mathbf{R}$ be a quadratic utility function of the form* (1).
(i) *f is supermodular if and only if $a(u, v) \geq 0$ $(\forall u, v \in V, \ u \neq v)$.*
(ii) *Suppose that f is a supermodular function. Then, f is monotone if and only if $b(v) \geq 0$ $(\forall v \in V)$.*

Proof. The statement (i) follows immediately from Theorem 2.1 (i) since f is supermodular if and only if $-f$ is submodular. It is easy to see that f is monotone if $a(u, v) \geq 0$ $(\forall u, v \in V, \ u \neq v)$ and $b(v) \geq 0$ $(\forall v \in V)$. On the other hand, if f is monotone, then we have $0 \leq f(\{v\}) - f(\emptyset) = b(v)$ for all $v \in V$.   □

We also consider an important subclass of submodular utility functions called utility functions with gross substitutes condition [12,16]. The *gross substitutes condition* of a utility function $f : 2^V \to \mathbf{R}$ is described as follows:

$$\forall p, q \in \mathbf{R}^V \text{ with } p \leq q, \ \forall X \in \arg \max_{S \subseteq V} \{f(S) - p(S)\},$$
$$\exists Y \in \arg \max_{S \subseteq V} \{f(S) - q(S)\} \text{ s.t. } X \cap \{v \in V \mid p(v) = q(v)\} \subseteq Y,$$

where p and q are price vectors. Intuitively, the gross substitutes condition says that a bidder still wants to get items that do not change in price after the prices on other items increase.

Theorem 2.3 (cf. [14]). *A quadratic utility function $f : 2^V \to \mathbf{R}$ of the form* (1) *satisfies the gross substitutes condition if and only if the following conditions hold:*

$$a(u, v) \leq 0 \qquad\qquad (\forall u, v \in V, u \neq v),$$
$$a(u, v) \leq \max\{a(u, t), a(v, t)\} \ (\forall u, v, t \in V, u \neq v, u \neq t, v \neq t).$$

In the proof of Theorem 2.3, we use the following characterization of gross-substitutes utility functions.

Theorem 2.4 ([26]; see also [3, Th. 13.5]). *A utility function $f : 2^V \to \mathbf{R}$ satisfies the gross-substitutes condition if and only if f is submodular and satisfies the following condition:*

$$f(X \cup \{u, v\}) + f(X \cup \{t\})$$
$$\leq \max\{f(X \cup \{u, t\}) + f(X \cup \{v\}), f(X \cup \{v, t\}) + f(X \cup \{u\})\} \qquad (3)$$
$$(\forall X \in 2^V, \ \forall u, v, t \in V, u \neq v, u \neq t, v \neq t).$$

Proof (of Theorem 2.3). For every $X \subseteq V$ and distinct $u, v, t \in V \setminus X$, it holds that

$$\{f(X \cup \{u, v\}) - f(X)\} + \{f(X \cup \{t\}) - f(X)\}$$
$$= a(u, v) + \sum_{s \in X} a(s, u) + \sum_{s \in X} a(s, v) + \sum_{s \in X} a(s, t) + \{b(u) + b(v) + b(t)\}.$$

Therefore, the inequality (3) in Theorem 2.4 is equivalent to $a(u, v) \leq \max\{a(u, t), a(v, t)\}$. Hence, the statement of Theorem 2.3 follows from this fact and Theorems 2.1 and 2.4. □

3   Results for Submodular Utility Functions

Hardness. We show the hardness of the problem (SUB|m=2) by the reduction of the max cut problem in undirected graphs. The max cut problem is a famous NP-hard problem; moreover, it is NP-hard to compute a solution with approximation ratio better than 0.879, under the assumption of the unique game conjecture [17].

As an instance of the max cut problem, let us consider an undirected graph $G = (V, E)$ with edge weight $w(u, v) \geq 0$ $((u, v) \in E)$. We define an instance of (SUB|m=2) by regarding V as the item set and by using quadratic utility functions such that

$$a_i(u, v) = \begin{cases} -w(u, v) & ((u, v) \in E), \\ 0 & (\text{otherwise}), \end{cases} \quad (u, v \in V, u < v),$$

$$b_i(v) = (1/2) \sum \{w(u, v) \mid (u, v) \in E, u \in V \setminus \{v\}\} \quad (v \in V)$$

for $i = 1, 2$. The definitions of a_i and b_i imply that the resulting quadratic utility functions f_1 and f_2 are monotone and submodular by Theorem 2.1. Moreover, the objective function value $f_1(V_1) + f_2(V_2)$ of a partition (V_1, V_2) is equal to

$$-\sum_{i=1}^{2} \sum \{w(u, v) \mid (u, v) \in E, \ u, v \in V_i\} + \sum_{(u,v) \in E} w(u, v)$$
$$= \sum \{w(u, v) \mid (u, v) \in E, u \in V_1, v \in V_2\},$$

i.e., the total weight of cut edges in G. Hence, the max cut problem on undirected graphs is reduced to (SUB|m=2), and this reduction preserves the approximation ratio. This fact, together with the result in [17], implies the following:

Theorem 3.1. *The problems* (SUB|m=2) *and* (SUB|m>2) *are NP-hard. Moreover, for both problems it is NP-hard to compute a solution with approximation ratio better than 0.879, under the assumption of the unique game conjecture.*

Approximability. We present an approximability result for the problem (SUB|m=2) by showing the reduction to the max s-t cut problem in directed graphs.

Given an instance of (SUB|m=2), we define a directed graph $G = (V \cup \{s,t\}, E)$ by

$$E = \{(u,v) \mid u,v \in V, u < v\} \cup \{(s,u) \mid u \in V\} \cup \{(v,t) \mid v \in V\}.$$

For each edge $(u,v) \in E$, its weight $w(u,v)$ is defined as follows:

$$\begin{aligned}
w(s,u) &= b_2(u) + \sum\{a_2(u,v) \mid v \in V, \ v > u\} \quad (u \in V),\\
w(v,t) &= b_1(v) + \sum\{a_1(u,v) \mid u \in V, \ u < v\} \quad (v \in V),\\
w(u,v) &= -a_1(u,v) - a_2(u,v) \quad (u,v \in V, u < v).
\end{aligned}$$

Theorem 2.1 (i) implies $w(u,v) \geq 0$, while (ii) implies $w(s,u) \geq 0$ and $w(v,t) \geq 0$. Hence, all of edge weights are nonnegative.

Let (S,T) be a partition of the vertex set $V \cup \{s,t\}$ satisfying $s \in S, t \in T$. Then, the objective function value of (S,T) is equal to

$$\begin{aligned}
&\sum\{w(v,t) \mid v \in S \cap V\} + \sum\{w(s,v) \mid v \in T \cap V\}\\
&\quad + \sum\{w(u,v) \mid u \in S \cap V, v \in T \cap V, u < v\}\\
&= \sum\{b_1(v) \mid v \in S \cap V\} + \sum\{a_1(u,v) \mid u,v \in S \cap V, u < v\}\\
&\quad + \sum\{b_2(v) \mid v \in T \cap V\} + \sum\{a_2(u,v) \mid u,v \in T \cap V, u < v\}\\
&= f_1(S \cap V) + f_2(T \cap V).
\end{aligned}$$

Hence, (SUB|m=2) is reduced to the max s-t cut problem in G, and this reduction preserves the approximation ratio. It is shown by Lewin et al. [24] that a 0.874-approximate solution of the max s-t cut problem can be computed in polynomial time. Therefore, we obtain the following result:

Theorem 3.2. *A 0.874-approximate solution of the problem* (SUB|m=2) *can be computed in polynomial time.*

Exact Polynomial-Time Algorithm for Special Case. We consider a special case where utility functions satisfy the gross substitutes condition, and show that the optimal allocation problem in this case can be reduced to the minimum quadratic-cost flow problem. The reduction is based on the following property of gross-substitutes quadratic utility functions. A set family $\mathcal{F} \subseteq 2^V$ is said to be *laminar* if it satisfies $X \subseteq Y$, $X \supseteq Y$, or $X \cap Y = \emptyset$ holds for every $X, Y \in \mathcal{F}$.

Lemma 3.1 (cf. [14, Cor. 3.4]). *A quadratic utility function $f : 2^V \to \mathbf{R}$ of the form* (1) *satisfies the gross substitutes condition if and only if it is represented as $f(X) = -\sum_{S \in \mathcal{F}} c_S |X \cap S|^2$ by using a laminar family $\mathcal{F} \subseteq 2^V$ and real numbers c_S $(S \in \mathcal{F})$ satisfying $\{v\} \in \mathcal{F}$ $(v \in V)$ and $c_S \geq 0$ $(S \in \mathcal{F}, |S| \geq 2)$.*

This lemma implies that the function value of a gross-substitutes quadratic utility function can be represented as the (quadratic) flow cost on a tree network.

We now explain the reduction to the minimum quadratic-cost flow problem. Suppose that a utility function f_i of bidder i is of the form $f_i(X) = -\sum_{S \in \mathcal{F}_i} c_S^i |X \cap S|^2$, where $\mathcal{F}_i \subseteq 2^V$ is a laminar family and c_S^i $(S \in \mathcal{F}_i)$ are real numbers satisfying the conditions in Lemma 3.1. We construct a graph $\hat{G} = (\hat{V}, \hat{E})$ as follows. Define $\hat{V} = \{r\} \cup V \cup \bigcup_{i=1}^{m} V_i$, where V_i $(i \in M)$ is given as $V_i = \{v_S^i \mid S \in \mathcal{F}_i\}$. Note that $v_{\{u\}}^i \in V_i$ for each $i \in M$ and $u \in V$. We also define $\hat{E} = E_0 \cup \bigcup_{i=1}^{m} E_i$, where

$$E_0 = \{(u, v_{\{u\}}^i) \mid u \in V, \ i \in M\},$$
$$E_i = \{(v_X^i, r) \mid X \in \mathcal{F}, \text{maximal in } \mathcal{F}\}$$
$$\cup \{(v_X^i, v_{\rho(X)}^i) \mid X \in \mathcal{F}_i, \text{not maximal in } \mathcal{F}_i\},$$

and for every non-maximal set $X \in \mathcal{F}_i$, we denote by $\rho(X)$ the unique minimal set $Y \in \mathcal{F}_i$ with $Y \supset X$. Note that edge set E_i for $i \in M$ constitutes a rooted tree with root r. For each edge in E_0, its capacity is given by the interval $[0,1]$, and its cost is 0. For each edge (v_X^i, r) or $(v_X^i, v_{\rho(X)}^i)$ in E_i, its capacity is $[0, +\infty]$, and its cost function is given by $c_X^i \varphi^2$, where φ is the flow value on the edge. We also define supply/demand values of vertices to be 1 for each $u \in V$, $-n$ for r, and 0 for other vertices.

We consider the minimum (quadratic-)cost flow problem on the network $\hat{G}$ under the capacity constraint and the supply/demand constraint. It is not difficult to see that integral feasible flows on the network have one-to-one correspondence to partitions of the set V, and the cost of the flow is equal to the negative of the total utilities for the corresponding partition. Hence, we can obtain an optimal allocation by solving the minimum cost flow problem.

The minimum quadratic-cost flow problem can be solved by iteratively augmenting flows along a shortest path in the so-called "auxiliary network," and the number of iterations is n (see, e.g., [1]). Since the graph $\hat{G}$ has O(mn) vertices and O(mn) edges, the minimum cost flow problem can be solved in O$(mn \log(mn)) \times n =$ O$(mn^2 \log(mn))$ time by using the shortest-path algorithm of Fredman and Tarjan [8] as a subroutine.

Theorem 3.3. *The optimal allocation problem with gross-substitutes quadratic utility functions can be solved in* O$(mn^2 \log(mn))$ *time.*

4   Results for Supermodular Utility Functions

Polynomial-Time Solvable Case. We firstly show that the problem (SUP$|m{=}2$) can be solved in polynomial time.

Lemma 4.1 ([13, Th. 1], [20, Th. 4.1]). *Given an instance of* (SUP$|m{=}2$), *we can construct in* O(n^2) *time an edge-weighted directed graph* $G = (V \cup \{s,t\}, E)$ *such that for every* $X \subseteq V$, *the cut value of* $(X \cup \{s\}, (V \setminus X) \cup \{t\})$ *is equal to* $f_1(X) + f_2(V \setminus X)$.

This lemma shows that (SUP$|m{=}2$) can be reduced to the minimum s-t cut problem in G. Note that the graph G has O(n) vertices and O(n^2) edges. Hence, the minimum s-t cut problem can be solved in O$(n^3 / \log n)$ time by the algorithm of Cheriyan et al. [4].

Theorem 4.1. *The problem* (SUP$|m{=}2$) *can be solved in* O$(n^3 / \log n)$ *time.*

Hardness. To show the NP-hardness of the problem (SUB|m>2), we show that the multiway (un)cut problem on undirected graphs [7,21] can be reduced to (SUB|m>2).

Input of the *multiway (un)cut problem* is an undirected graph $G = (V, E)$ with distinct terminals $s_1, s_2, \ldots, s_k \in V$ ($k \geq 2$) and edge weight $w(u, v) \geq 0$ ($(u, v) \in E$). In the multiway cut problem, we find a partition $(V_1, V_2, \ldots, V_k)$ of V with $s_i \in V_i$ ($i = 1, 2, \ldots, k$) minimizing the total weight of cut edges, i.e., $\sum\{w(u, v) \mid (u, v) \in E, \ u \in V_i, \ , v \in V_j, \ i \neq j\}$, while in the multiway uncut problem, we want to *maximize* the total weight of *uncut* edges. The multiway (un)cut problem is known to be NP-hard, even when $k = 3$ [7].

Given an instance of the multiway (un)cut problem, we define an instance of (SUB|m>2) by regarding V as the item set and by

$$a^i(u, v) = \begin{cases} w(u, v) \ ((u, v) \in E), \\ 0 \qquad\qquad (\text{otherwise}), \end{cases} \quad b^i(v) = \begin{cases} \Gamma \ (v = s_i), \\ 0 \ (\text{otherwise}), \end{cases}$$

where Γ is a sufficiently large positive number. Let $(V_1, V_2, \ldots, V_k)$ be a partition of V which is an optimal solution of this instance. Then, each V_i contains the vertex s_i since $b^i(s_i)$ is a sufficiently large number. Moreover, the objective function value is given as $\sum_{i=1}^{k} \sum \{w(u, v) \mid (u, v) \in E, \ u, v \in V_i\}$, which we want to maximize. Hence, an optimal solution for (SUB|m>2) is an optimal solution for the multiway (un)cut problem, and vice versa.

Theorem 4.2. *The problem* (SUB|m>2) *is NP-hard, even when* $m = 3$.

Approximability. We propose a 0.5-approximation algorithm for the problem (SUB|m>2). Our algorithm is based on a natural linear programming (LP) relaxation:

$$\text{Maximize} \quad \sum_{i=1}^{k} \sum_{u,v \in V, u<v} a^i(u, v) y^i(u, v) + \sum_{i=1}^{k} \sum_{v \in V} b^i(v) x^i(v)$$

$$\text{subject to} \quad \sum_{i=1}^{k} x^i(v) = 1 \quad (v \in V),$$
$$y^i(u, v) = \min\{x^i(u), x^i(v)\} \quad (u, v \in V, u < v),$$
$$x^i(v) \geq 0 \ (v \in V), \quad y^i(u, v) \geq 0 \ (u, v \in V, u < v).$$

The algorithm firstly compute an optimal solution of the LP. Then, the algorithm chooses a bidder $i \in \{1, 2, \ldots, m\}$ and a value $\rho \in [0, 1]$ uniformly at random, and assigns each item $v \in V$ to the bidder i if $x^i(v) \geq \rho$. The algorithm repeats this step until all items are assigned to one of the bidders. Although this algorithm is randomized, it can be derandomized by using the technique in Kleinberg and Tardos [19].

We analyze the performance of the algorithm. For $v \in V$ and $i \in M$, let $X^i(v)$ be a random variable that is 1 if the item v is assigned to the bidder i and 0 otherwise. Similarly, for $u, v \in V$ and $i \in M$, let $Y^i(u, v)$ be a random variable that is 1 if both of the items u and v are assigned to the bidder i and 0 otherwise. We also denote $y(u, v) = \sum_{i=1}^{m} y^i(u, v)$ for $u, v \in V$ with $u < v$.

Lemma 4.2 ([21, Fact 3.1]). *Let $v \in V$ and $i \in M$. Assume that item v is not assigned to any bidder before some iteration. Then, the probability that v is assigned to bidder i in the iteration is $(1/m)x^i(v)$.*

Lemma 4.3 ([21, Claim 3.2]). $\Pr[X^i(v) = 1] = x^i(v)$ *for $v \in V$ and $i \in M$.*

Lemma 4.4. $\Pr[Y^i(u, v) = 1] \geq \frac{y^i(u,v)}{2 - y(u,v)}$ *for $u, v \in V$ with $u < v$ and $i \in M$.*

Proof. Let $\Psi \in [0, 1]$ be the probability that both of u and v are assigned to the bidder i in the *same iteration*. Then, $\Pr[Y^i(u, v) = 1] \geq \Psi$ holds.

The probability that u and v is assigned to an bidder $i \in M$ in some iteration is $(1/m)\min\{x^i(u), x^i(v)\} = (1/m)y^i(u, v)$. Similarly, the probability that at least one of u and v is assigned to any bidder in some iteration is

$$\sum_{i=1}^{m} \frac{1}{m} \cdot \max\{x^i(u), x^i(v)\} = \frac{1}{m}\left[\sum_{i=1}^{m}\{x^i(u) + x^i(v)\} - \sum_{i=1}^{m}\min\{x^i(u), x^i(v)\}\right]$$

$$= \frac{1}{m}\{2 - y(u, v)\}.$$

Hence, the probability that u and v is assigned to a bidder $i \in M$ in the k-th iteration is

$$\left[1 - \frac{1}{m}\{2 - y(u, v)\}\right]^{k-1} \times \frac{1}{m}y^i(u, v).$$

Hence,

$$\Psi = \sum_{k=1}^{\infty}\left[1 - \frac{1}{m}\{2 - y(u, v)\}\right]^{k-1} \times \frac{1}{m}y^i(u, v)$$

$$= \frac{m}{2 - y(u, v)} \cdot \frac{1}{m} \cdot y^i(u, v) = \frac{y^i(u, v)}{2 - y(u, v)}.$$

This implies the claim of the lemma. □

We consider the expected value of the objective function for a solution obtained by the algorithm. By Lemmas 4.3 and 4.4, it holds that

$$\sum_{i=1}^{m}\sum_{u,v\in V, u<v} a^i(u, v)\Pr[Y^i(u, v) = 1] + \sum_{i=1}^{m}\sum_{v\in V} b^i(v)\Pr[X^i(v) = 1]$$

$$\geq \sum_{i=1}^{m}\sum_{u,v\in V, u<v} a^i(u, v) \cdot \frac{y^i(u, v)}{2 - y(u, v)} + \sum_{i=1}^{m}\sum_{v\in V} b^i(v)x^i(v) = 0.5 \cdot \text{OPT}_{\text{LP}},$$

where OPT_{LP} denotes the optimal value of the LP. Since OPT_{LP} is an upper bound of the optimal value of (SUB$|m>2$), we obtain the following result.

Theorem 4.3. *A 0.5-approximate solution of the problem (SUB$|m>2$) can be computed in polynomial time.*

With a more careful analysis we can show that the approximation ratio is $1/(2 - \varepsilon)$ (> 0.5), where $\varepsilon = \min\{y(u,v) \mid (u,v) \in E, \ y(u,v) > 0\}$; this bound is obtained by analyzing the cases $y(u,v) = 0$ and $y(u,v) > 0$ separately.

Our analysis shows that the integrality gap of the LP is at least 0.5. On the other hand, an instance with integrality gap $2/3$ can be easily constructed. An open problem is to close the gap between 0.5 and $2/3$. A possible approach for a better approximation algorithm is to construct a new LP formulation which has a larger value of $\min\{y(u,v) \mid (u,v) \in E, \ y(u,v) > 0\}$.

We then consider alternative approximation algorithms by using the fact that (SUP$|m$=2) can be solved in polynomial time. If $m = 3$, then we can obtain a $2/3$-approximate solution easily by using this fact. We compute an optimal allocation $(V_1^{(12)}, V_2^{(12)}, \emptyset)$ of items to bidders 1 and 2, where bidder 3 is ignored. In the same way, we compute optimal allocations $(V_1^{(13)}, \emptyset, V_3^{(13)})$ for bidders 1 and 3 and $(\emptyset, V_2^{(23)}, V_3^{(23)})$ for bidders 2 and 3. Then, we choose the best allocation among the three, which is a $2/3$-approximate solution of the original problem.

Theorem 4.4. *A $2/3$-approximate solution of the problem* (SUB$|m$>2) *with $m = 3$ can be computed in polynomial time.*

For general case with $m \geq 3$, it is natural to consider the following heuristic based on local search. Given a partition $(V_1, V_2, \ldots, V_m)$ of V and bidders $i, j \in M$, we denote by `realloc`(i, j) an operation which optimally re-allocates items in $V_i \cup V_j$ to bidders i and j. Our heuristic is as follows: start with an arbitrarily chosen initial partition, and repeatedly apply the operation `realloc`(i, j) to arbitrarily chosen two bidders $i, j \in M$ until no improvement is possible by this operation. Although our preliminary computational experiment shows that this heuristic always outputs a near-optimal solution, we can construct a family of instances for which the approximation ratio can be arbitrarily close to 0.

References

1. Ahuja, R.K., Magnanti, T.L., Orlin, J.B.: Network Flows: Theory, Algorithms, and Applications. Prentice-Hall, Englewood Cliffs (1993)
2. Blumrosen, L., Nisan, N.: Algorithmic Game Theory. In: Nisan, N., et al. (eds.) Combinatorial Auction, ch. 11, Cambridge Univ. Press, Cambridge (2007)
3. Cramton, P., Shoham, Y., Steinberg, R.: Combinatorial Auctions. MIT Press, Cambridge (2006)
4. Cheriyan, J., Hagerup, T., Mehlhorn, K.: An o(n)-time maximum-flow algorithm. SIAM J. Comput. 25, 1144–1170 (1996)
5. Chevaleyre, Y., Endriss, U., Estivie, S., Maudet, N.: Multiagent resource allocation in k-additive domains: preference representation and complexity. Annals Oper. Res. 163, 49–62 (2008)
6. Conitzer, V., Sandholm, T., Santi, P.: Combinatorial auctions with k-wise dependent valuations. In: Proc. AAAI 2005, pp. 248–254 (2005)
7. Dahlhaus, E., Johnson, D.S., Papadimitriou, C.H., Seymour, P.D., Yannakakis, M.: The complexity of multiterminal cuts. SIAM J. Comput. 23, 864–894 (1994)

8. Fredman, M.L., Tarjan, R.E.: Fibonacci heaps and their uses in improved network optimization algorithms. J. ACM 34, 596–615 (1987)
9. Fujishige, S.: Submodular Function and Optimization, 2nd edn. Elsevier, Amsterdam (2005)
10. Fujishige, S., Yang, Z.: A note on Kelso and Crawford's gross substitutes condition. Math. Oper. Res. 28, 463–469 (2003)
11. Grötschel, M., Lovász, L., Schrijver, A.: The ellipsoid method and its consequences in combinatorial optimization. Combinatorica 1, 169–197 (1984)
12. Gul, F., Stacchetti, E.: Walrasian equilibrium with gross substitutes. J. Econ. Theory 87, 95–124 (1999)
13. Hammer, P.L.: Some network flow problems solved with pseudo-Boolean programming. Oper. Res. 13, 388–399 (1965)
14. Hirai, H., Murota, K.: M-convex functions and tree metrics. Japan J. Indust. Appl. Math. 21, 391–403 (2004)
15. Holzman, R., Kfir-Dahav, N., Monderer, D., Tennenholtz, M.: Bundling equilibrium in combinatorial auctions. Games Econom. Behav. 47, 104–123 (2004)
16. Kelso, A.S., Crawford, V.P.: Job matching, coalition formation and gross substitutes. Econometrica 50, 1483–1504 (1982)
17. Khot, S., Kindler, G., Mossel, E., O'Donnell, R.: Optimal inapproximability results for max-cut and other 2-variable CSPs? In: Proc. FOCS, pp. 146–154 (2004)
18. Khot, S., Lipton, R.J., Markakis, E., Mehta, A.: Inapproximability results for combinatorial auctions with submodular utility functions. Algorithmica 52, 3–18 (2008)
19. Kleinberg, J., Tardos, É.: Approximation algorithms for classification problems with pairwise relationships: metric labeling and Markov random fields. J. ACM 49, 616–639 (2002)
20. Kolmogorov, V.: What energy functions can be minimized via graph cuts? IEEE Trans. Pattern Anal. Mach. Intell. 26, 147–159 (2004)
21. Langberg, M., Rabani, Y., Swamy, C.: Approximation algorithms for graph homomorphism problems. In: Díaz, J., Jansen, K., Rolim, J.D.P., Zwick, U. (eds.) APPROX 2006 and RANDOM 2006. LNCS, vol. 4110, pp. 176–187. Springer, Heidelberg (2006)
22. Lehmann, B., Lehmann, D., Nisan, N.: Combinatorial auctions with decreasing marginal utilities. Games Econom. Behav. 55, 270–296 (2006)
23. Lehmann, D., O'Callaghan, L., Shoham, Y.: Truth revelation in approximately efficient combinatorial auctions. In: Proc. EC 1999, pp. 96–102 (1999)
24. Lewin, M., Livnat, D., Zwick, U.: Improved rounding techniques for the MAX 2-SAT and MAX DI-CUT problems. In: Cook, W.J., Schulz, A.S. (eds.) IPCO 2002. LNCS, vol. 2337, pp. 67–82. Springer, Heidelberg (2002)
25. Mirrokni, V., Schapira, M., Vondrák, J.: Tight information-theoretic lower bounds for welfare maximization in combinatorial auctions. In: Proc. EC 2008, pp. 70–77 (2008)
26. Reijniese, H., van Gellekom, A., Potters, J.A.M.: Verifying gross substitutability. Econom. Theory 20, 767–776 (2002)
27. Sandholm, T.: Algorithm for optimal winner determination in combinatorial auctions. Artificial Intelligence 135, 1–54 (2002)
28. Vondrák, J.: Optimal approximation for the submodular welfare problem in the value oracle model. In: Proc. STOC 2008, pp. 67–74 (2008)

Energy and Fan-In of Threshold Circuits Computing Mod Functions

Akira Suzuki, Kei Uchizawa*, and Xiao Zhou

Graduate School of Information Sciences, Tohoku University
Aramaki-aza Aoba 6-6-05,
Aoba-ku, Sendai, 980-8579, Japan
{a.suzuki,uchizawa,zhou}@ecei.tohoku.ac.jp

Abstract. In this paper, we consider a threshold circuit C computing the modulus function MOD_m, and investigate a relationship between two complexity measures, fan-in l and energy e of C, where the fan-in l is defined to be the maximum number of inputs of every gate in C, and the energy e to be the maximum number of gates outputting "1" over all inputs to C. We first prove that MOD_m of n variables can be computed by a threshold circuit of fan-in l and energy $e = O(n/l)$, and then provide an almost tight lower bound $e = \Omega((n-m)/l)$. Our results imply that there exists a tradeoff between the fan-in and energy of threshold circuits computing the modulus function.

Keywords: Boolean function, energy complexity, fan-in, MOD function, threshold circuit.

1 Introduction

Neurons in the brain communicate with each other by electrical signals for information processing, and a neuron emitting a signal is said to be "firing." Recent biological study reports the following interesting fact about the energy consumption of neuronal firing: the energy cost of a firing is high while energy supplied to the brain is limited, and hence neural networks must have low firing activity [3,4,5]. Motivated by this fact, Uchizawa, Douglas and Maass propose a new complexity measure, called energy complexity, for a theoretical model of neural network, called a threshold circuit. The *energy* e of a threshold circuit C is defined to be the maximum number of gates outputting "1" in C, where the maximum is taken over all inputs to C [9]. In the previous research, it turns out that the energy complexity has close relationships with other complexity measures such as size (i.e., the number of gates) and depth [8,9,10,11,12]. In particular, it is known that there exists a tradeoff between size and energy of threshold circuit computing the modulus function $\mathrm{MOD}_m : \{0,1\}^n \to \{0,1\}$ [8,12], where $\mathrm{MOD}_m(\boldsymbol{x}) = 0$ if the number of "1"s in an input $\boldsymbol{x} \in \{0,1\}^n$ is a multiple of m and, otherwise, $\mathrm{MOD}_m(\boldsymbol{x}) = 1$. In other words, we know that one of size

* Supported by MEXT Grant-in-Aid for Young Scientists (B) No.21700003.

M. Ogihara and J. Tarui (Eds.): TAMC 2011, LNCS 6648, pp. 154–163, 2011.
© Springer-Verlag Berlin Heidelberg 2011

and energy of a threshold circuit computing the modulus function can be small, while both of them cannot be simultaneously small.

In this paper, we also consider threshold circuits computing the modulus function, and investigate a relationship between the energy complexity and another major complexity measure, called fan-in. The fan-in is defined to be the maximum number of inputs of every gate in the circuit, and is one of intensively studied measures, such as size and depth, in the literature [1,2,6,7]. We show that, similarly to the relationship between size and energy mentioned above, there exists a tradeoff between the fan-in and energy of threshold circuits. More precisely, we first prove that MOD_m of n variables can be computed by a threshold circuit of fan-in l and energy

$$e = O\left(\frac{n}{l}\right). \tag{1}$$

Equation (1) implies that there exists a threshold circuit C of small energy e if every gate in C is allowed to have a large fan-in l (e.g., $l = O(n)$ and $e = O(1)$), and also implies that there exists a threshold circuit C of small fan-in l if C is allowed to use large energy e (e.g., $l = O(1)$ and $e = O(n)$). We then show that the upper bound in Eq. (1) is almost tight by providing a lower bound

$$e = \Omega\left(\frac{n-m}{l}\right). \tag{2}$$

Consider the case where $m = o(n)$, then Eq. (2) implies that a threshold circuit C must have large energy e if every gate in C has a bounded fan-in l (e.g., if $l = O(1)$ then $e = \Omega(n)$), and also implies that a threshold circuit C must have large fan-in l if C has small energy e (e.g., if $e = O(1)$ then $l = \Omega(n)$). Consequently, Eqs. (1) and (2) imply that there exists a tradeoff between the fan-in and energy of threshold circuits computing the modulus function: One of the fan-in and energy can be small, while both of them cannot be simultaneously small.

The rest of this paper is organized as follows. In Section 2, we define some terms on threshold circuits. In Section 3, we first present the upper bound on the energy of a threshold circuit computing the modulus function together with a technical lemma, and then give the lower bound. In Section 4, we prove the technical lemma presented in the last section. In Section 5, we conclude with some remarks.

2   Preliminaries

A *threshold circuit* C is a combinatorial circuit of threshold gates, and is expressed by a directed acyclic graph. Let n be the number of input variables to C. Then each node of in-degree 0 in C corresponds to one of the n input variables $x_1, x_2, \ldots, x_n$, and the other nodes correspond to threshold gates. We define *size* $s(C)$, simply denoted by s, of a threshold circuit C as the number of threshold gates in C. Let $g_1^C, g_2^C, \ldots, g_s^C$ be the gates in C. One may assume without loss of generality that $g_1^C, g_2^C, \ldots, g_s^C$ are topologically ordered with respects to the underlying graph of C. We denote by $l_1, l_2, \ldots, l_s$ the fan-ins of $g_1^C, g_2^C, \ldots, g_s^C$, respectively, and define the maximum number l of $l_1, l_2, \ldots, l_s$ as *fan-in* of C. For g_i^C, $1 \leq i \leq s$, let $w_1, w_2, \ldots, w_{l_i}$ be real numbers as weights and t_i be a real number as a threshold of g_i^C, and let $\boldsymbol{z}_i(\boldsymbol{x}) = (z_1(\boldsymbol{x}), z_2(\boldsymbol{x}), \ldots, z_{l_i}(\boldsymbol{x})) \in \{0,1\}^{l_i}$ be the l_i inputs of g_i^C for an input $\boldsymbol{x} \in \{0,1\}^n$. Then the *output* $g_i^C(\boldsymbol{z}_i(\boldsymbol{x}))$ *of the gate* g_i^C is defined as

$$ g_i^C(\boldsymbol{z}_i(\boldsymbol{x})) = \text{sign} \left(\sum_{j=1}^{l_i} w_j z_j(\boldsymbol{x}) - t_i \right), $$

where $\text{sign}(z) = 1$ if $z \geq 0$ and $\text{sign}(z) = 0$ if $z < 0$. We simply denote $g_i^C(\boldsymbol{z}_i(\boldsymbol{x}))$ by $g_i^C[\boldsymbol{x}]$. Let n' be the number of outputs of C, then C has n' gates with out-degree 0. One may assume without loss of generality that such gates are $g_{s-n'+1}^C, g_{s-n'+2}^C, \ldots, g_s^C$. Then, for every input $\boldsymbol{x} \in \{0,1\}^n$, the *output* $C(\boldsymbol{x})$ of C is denoted by $(u_1, u_2, \ldots, u_{n'})$ where $u_i = g_{s-n'+i}^C[\boldsymbol{x}]$ for each i, $1 \leq i \leq n'$. The gates $g_{s-n'+1}^C, g_{s-n'+2}^C, \ldots, g_s^C$ are called *top gates* of C. Let $f : \{0,1\}^n \to \{0,1\}^{n'}$ be a Boolean function of n inputs and n' outputs. A threshold circuit C *computes* a Boolean function f if $C(\boldsymbol{x}) = f(\boldsymbol{x})$ for every input $\boldsymbol{x} \in \{0,1\}^n$. We define the *energy* $e(C)$ of C as

$$ e(C) = \max_{\boldsymbol{x} \in \{0,1\}^n} \sum_{i=1}^{s(C)} g_i^C[\boldsymbol{x}]. $$

For an input $\boldsymbol{x} = (x_1, x_2, \ldots, x_n) \in \{0,1\}^n$, we define $|\boldsymbol{x}|$ as the hamming weight of the input $\boldsymbol{x}$, that is, $|\boldsymbol{x}| = \sum_{i=1}^{n} x_i$. Then, for positive integers $m \geq 2$ and n, the *modulus function* MOD_m is defined as follows: For every input $\boldsymbol{x} = (x_1, x_2, \ldots, x_n) \in \{0,1\}^n$, $\text{MOD}_m(\boldsymbol{x}) = 0$ if $|\boldsymbol{x}|$ is a multiple of m, and otherwise $\text{MOD}_m(\boldsymbol{x}) = 1$. In the rest of the paper, we consider more general modulus functions MOD_m^r. For positive integers $m \geq 2$, n and r, $0 \leq r \leq m - 1$, MOD_m^r is defined as follows: For every input $\boldsymbol{x} = (x_1, x_2, \ldots, x_n) \in \{0,1\}^n$,

$$ \text{MOD}_m^r(\boldsymbol{x}) = \begin{cases} 0 & \text{if } r \equiv |\boldsymbol{x}| \pmod{m}; \\ 1 & \text{otherwise.} \end{cases} $$

3   Our Results

In this section, we give upper and lower bounds on the energy of threshold circuits computing the modulus function. In Section 3.1, we first provide the upper bound in a form of a theorem, and give a technical lemma. Then, we prove the theorem by the lemma. In Section 3.2, we show a tightness of the upper bound by giving the lower bound.

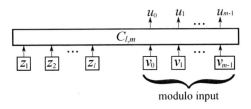

Fig. 1. Overview of the circuit $C_{l,m}$ given in Lemma 1

3.1   Upper Bound

The following theorem is one of our main results, and gives an upper bound on the energy of threshold circuits computing the modulus function.

Theorem 1. *Let $m \geq 2$, n, and l be positive integers. Then, there is a threshold circuit C computing MOD_m^r of n variables such that C has fan-in at most $2l$, energy $e(C) = O\left(n/l\right)$ and size $s(C) = O(mn)$.*

We prove Theorem 1 by construction. For each integer $0 \leq \alpha \leq m-1$, we denote by $\mathbf{1}_\alpha$ be the vector of length m such that only $(\alpha + 1)$st position is one, that is, $\mathbf{1}_0 = (1,0,0,\dots,0)$, $\mathbf{1}_1 = (0,1,0,\dots,0)$, and so on. A threshold circuit with l outputs is called a threshold (l,m)-*circuit* if it receives l input bits $z_1, z_2, \dots, z_l$ and m input bits $v_0, v_1, \dots, v_{m-1}$. (See Fig. 1.) The m inputs is called *modulo input*. The following lemma plays key role in our construction.

Lemma 1. *Let $l \geq 1$ and $m \geq 2$ be two integers. Then, there is a threshold (l,m)-circuit $C_{l,m}$ of fan-in at most $2l$ such that, for each pair of $z = (z_1, z_2, \dots, z_l) \in \{0,1\}^l$ and $\mathbf{1}_\alpha$, $0 \leq \alpha \leq m-1$, the output $C_{l,m}(z, \mathbf{1}_\alpha) = (u_0, u_1, \dots, u_{m-1})$ of $C_{l,m}$ is $\mathbf{1}_\beta$ for some integer β satisfying*

$$\beta \equiv \alpha + |z| \pmod{m}. \tag{3}$$

Moreover, its energy $e(C_{l,m}) = 3$, and its size $s(C_{l,m}) \leq (l+1)(m+1) + m$.

See Fig. 1 for an overview of $C_{l,m}$. We will prove the lemma in Section 4.
     We are now ready to prove Theorem 1.

Proof of Theorem 1. Let $m \geq 2$, n, r and l be positive integers. One may assume that n is a multiple of l, that is, $n = d \cdot l$ for some positive integer d; otherwise, one may increase it so that the condition holds. We construct a desired threshold circuit C computing MOD_m of n variables as follows.
     Let C_i, $1 \leq i \leq d$, be a copy of a (l,m)-circuit $C_{l,m}$ which is given by Lemma 1, and let $(u_{i,0}, u_{i,1}, \dots, u_{i,m-1})$ be the output of C_i. We obtain the desired circuit C by combining $C_1, C_2, \dots, C_d$ as follows (See Fig. 2). We first fix the modulo input of C_1 to $\mathbf{1}_0$, and add the other input l bits of C_1 from $x_1, x_2, \dots, x_l$. Then for each i, $2 \leq i \leq d$, we connect the output $(u_{i-1,0}, u_{i-1,1}, \dots, u_{i-1,m-1})$

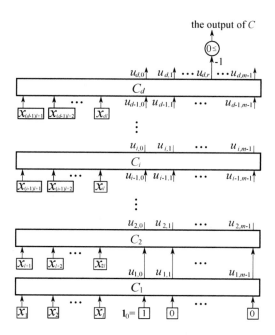

Fig. 2. Overview of the circuit C given in the proof of Theorem 1

of C_{i-1} to C_i as its modulo input, and add the other input l bits of C_i from $x_{(i-1)l+1}, x_{(i-1)l+2}, \ldots, x_{il}$. We complete the construction of C by adding a gate g, as the top gate of C, that has a threshold zero and receives $u_{d,r}$ of C_d with weight -1. See Fig. 2 for an overview of C.

Since we construct C from $C_1, C_2, \ldots, C_d$ which are d copies of $C_{l,m}$ given in Lemma 1, the circuit C clearly has fan-in at most $2l$, energy $e \leq 1 + 3d = 1 + 3n/l = O(n/l)$, and size $s \leq 1 + ((l+1)(m+1) + m) \cdot d = O(nm)$. Thus, it suffices to show that C computes MOD_m of n variables.

Let i be an arbitrary integer such that $2 \leq i \leq d$. Since C_i is a copy of $C_{l,m}$ and C_i receives the output of C_{i-1} as the modulo input, Lemma 1 implies that the output of C_i is $\mathbf{1}_{\beta_i}$ for some β_i such that

$$\beta_i \equiv \beta_{i-1} + \sum_{j=(i-1)l+1}^{il} x_j \pmod{m}. \qquad (4)$$

Furthermore, the circuit C_1 receives $\mathbf{1}_0$ as the modulo input. We thus have

$$\beta_1 \equiv \sum_{j=1}^{l} x_j \pmod{m}. \qquad (5)$$

By Eqs. (4) and (5), we have

$$\beta_d \equiv \beta_{d-1} + \sum_{j=(d-1)l+1}^{dl} x_j \pmod{m}$$

$$\equiv \sum_{j=1}^{l} x_j + \sum_{j=l+1}^{2l} x_p + \cdots + \sum_{j=(d-1)l+1}^{dl} x_p \pmod{m}$$

$$\equiv |\boldsymbol{x}| \pmod{m}.$$

Therefore, $r \equiv |\boldsymbol{x}| \pmod{m}$ if and only if the output $u_{d,r}$ of C_d is one, while the output $u_{d,r}$ of C_d is one if and only if the top gate g outputs zero. Thus, the circuit C computes MOD_m^r of n variables.   □

3.2   Lower Bound

In this section, we show a tightness of our bound given in Theorem 1. The following theorem implies that our upper bound is tight up to a constant factor if $m = o(n)$.

Theorem 2. *Let l, m, n and r be positive integer such that $m \geq 2$, and $r \leq m - 1$. Let C be a threshold circuit computing MOD_m^r of n variables. If C has fan-in at most l, then*

$$e(C) \geq \left\lceil \frac{n - m + 1}{l} \right\rceil = \Omega\left(\frac{n - m}{l} \right). \tag{6}$$

Proof. We prove the theorem by induction on the number n of input variables. Let $m \geq 2$ be an integer. For the inductive base, consider the case where $n \leq m - 1$. In this case, Eq. (6) clearly holds since we have $e(C) \geq 0$ and $\lceil (n - m + 1)/l \rceil = 0$.

For inductive hypothesis, assume that Eq. (6) holds for every threshold circuit C' of fan-in at most l computing $\mathrm{MOD}_m^{r'}$ of n' variables for any n' and r' such that $n' \leq n - 1$ and $0 \leq r' \leq m - 1$.

Let C be a threshold circuit that computes MOD_m^r of n variables and that has fan-in at most l. We say that a threshold gate g computes a non-trivial function if there exist two inputs $\boldsymbol{x}$ and $\boldsymbol{y}$ such that $g[\boldsymbol{x}] = 1$ and $g[\boldsymbol{y}] = 0$. Clearly, the top gate of C computes a non-trivial function, since C computes MOD_m^r of n variables such that $m - 1 < n$. In fact, one may assume without loss of generality that every gate in C computes a non-trivial function. Let $g_1, g_2, \ldots, g_s$ be the gates in C, and assume that $g_1, g_2, \cdots, g_s$ are topologically ordered with respect to the underlying directed acyclic graph of C. Clearly, g_1 only receives a number $k(\leq l)$ of the n input variables, say $x_{i_1}, x_{i_2}, \ldots, x_{i_k}$. Hence, we can fix the output of g_1 to one by determining assignments to the k input variables. Let $a_{i_1}, a_{i_2}, \ldots, a_{i_k} \in \{0, 1\}$ be such assignments to $x_{i_1}, x_{i_2}, \ldots, x_{i_k}$, respectively, that fix the output of g_1 to one, and let C' be the resulting circuit. Clearly, we have

$$e(C) = e(C') + 1. \tag{7}$$

Furthermore, C' computes $\mathrm{MOD}_m^{r'}$ of n' variables where $n' = n - k \leq n - 1$ and

$$r' \equiv r - \sum_{j=1}^{k} a_{i_j} \pmod{m}.$$

Hence, we have by the induction hypothesis

$$e(C') \geq \left\lceil \frac{n' - m + 1}{l} \right\rceil = \left\lceil \frac{n - k - m + 1}{l} \right\rceil. \tag{8}$$

Since $k \leq l$, Eq. (8) implies that

$$e(C') \geq \left\lceil \frac{n - l - m + 1}{l} \right\rceil. \tag{9}$$

By Eqs. (7) and (9), we have

$$e(C) \geq \left\lceil \frac{n - l - m + 1}{l} \right\rceil + 1 = \left\lceil \frac{n - m + 1}{l} \right\rceil.$$

Thus, Eq. (6) holds. $\square$

4   Proof of Lemma 1

In this section, we prove Lemma 1 by constructing the desired (l, m)-threshold circuit $C_{l,m}$ that receives l input bits $z_1, z_2, \ldots, z_l$ together with m input bits $v_0, v_1, \ldots, v_{m-1}$, and outputs m bits $u_0, u_1, \ldots, u_{m-1}$.

First, we recursively make $l + 1$ threshold gates g_i, $0 \leq i \leq l$, as follows.

- A threshold gate g_l has a threshold l and receives the inputs $z_1, z_2, \ldots, z_l$ with weights one. (See Fig. 3(a).)
- For each i, $0 \leq i \leq l-1$, a threshold gate g_i has a threshold i and receives the $2l - i$ inputs from $z_1, z_2, \ldots, z_l$ with weights one and the outputs of $l - i$ gates $g_{i+1}, g_{i+2}, \ldots, g_l$ with weights $-(i + 2), -(i + 3), \ldots, -(l + 1)$, respectively. (See Fig. 3(b).)

For each i, $0 \leq i \leq l - 1$, and each input $z \in \{0, 1\}^l$, we obviously have

$$g_i[z] = \mathrm{sign}\left(|z| - i - \sum_{j=i+1}^{l} (j + 1) \cdot g_j[z] \right). \tag{10}$$

Note that

$$g_l(z) = \mathrm{sign}\left(|z| - l \right) = \begin{cases} 1 & \text{if } |z| = l, \\ 0 & \text{if } |z| \leq l - 1. \end{cases}$$

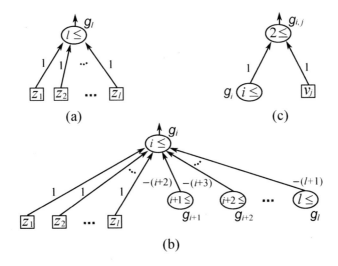

Fig. 3. (a) The gate g_l, (b) the gate g_i, and (c) the gate $g_{i,j}$

For each i, $0 \le i \le l - 1$ and $z \in \{0,1\}^l$, we denote by $a(i, z)$ the value in the sign function of Eq. (10), that is,

$$a(i, z) = |z| - i - \sum_{j=i+1}^{l} (j + 1) \cdot g_j[z]. \tag{11}$$

We now claim that $a(|z|, z) = 0$ and $a(i, z) < 0$ for all $i \ne |z|$, $0 \le i \le l$. By Eq. (11), clearly $a(i, z) < 0$ for all i, $|z| + 1 \le i \le l$, and hence $g_i(z) = 0$ for such i. Therefore,

$$a(|z|, z) = |z| - |z| - \sum_{j=|z|+1}^{l} (j + 1) \cdot g_j[z] = 0,$$

and hence $g_{|z|}[z] = 1$. For each i, $0 \le i \le |z| - 1$, by Eq. (11) we thus have

$$
\begin{aligned}
a(i, z) &= |z| - i - \sum_{j=i+1}^{l} (j + 1) \cdot g_j[z] \\
&= |z| - i - \sum_{j=i+1}^{|z|} (j + 1) \cdot g_j[z] \\
&\le |z| - i - (|z| + 1) \cdot g_{|z|}[z] \\
&= |z| - i - |z| - 1 \\
&= -i - 1 \\
&< 0.
\end{aligned}
$$

We thus complete the proof of the claim. Consequently, for every $z \in \{0,1\}^l$, we have

$$g_i[z] = \begin{cases} 1 & \text{if } i = |z|; \\ 0 & \text{otherwise,} \end{cases} \tag{12}$$

that is, only $g_{|z|}$ of $g_0, g_1, \ldots, g_l$ outputs one.

For every pair of inputs $i \in \{0,1,\ldots,l\}$ and $j \in \{0,1,\ldots,m-1\}$, we then make a threshold gate $g_{i,j}$ computing AND of the output of g_i and v_j, that is, $g_{i,j}$ has a threshold two, and receives the output of g_i and v_j with weights one (See Fig. 3(c)). Clearly, $g_{i,j}$ outputs one if and only if

$$i = |z| \text{ and } v_j = 1. \tag{13}$$

Note that only $g_{|z|}$ of $g_1, g_2, \ldots, g_l$ outputs one, and exactly one of $v_0, v_1, \ldots, v_{m-1}$ equals to one. Hence, for a pair of inputs $z \in \{0,1\}^l$, and 1_α, $0 \le \alpha \le m-1$, only $g_{|z|,\alpha}$ outputs one.

We finally complete the construction of $C_{l,m}$ by adding a threshold gate g'_k for each k, $0 \le k \le m-1$, as a top gate whose output corresponds to u_k, as follows. The gate g'_k computes OR of every output of the gate $g_{i,j}$ such that i, j and k satisfy

$$k \equiv i + j \pmod{m}. \tag{14}$$

More precisely, g'_k receives the output of $g_{i,k-i+mq_i}$ for every i, $0 \le i \le l$, and q_i, $0 \le k - i + mq_i \le m-1$. Clearly

$$\left\lceil \frac{i-k}{m} \right\rceil \le q_i \le \left\lfloor \frac{i-k+m-1}{m} \right\rfloor,$$

and hence $q_i = \lceil (i-k)/m \rceil$. Therefore

$$g'_k[z, 1_j] = \operatorname{sign}\left(-1 + \sum_{i=0}^{l} g_{i,k-i+mq_i}[z, 1_j] \right).$$

Thus, g'_k receives the outputs of the $l+1$ gates. Note that for a pair of $z \in \{0,1\}^l$ and 1_α, $0 \le \alpha \le m-1$, only gate $g_{|z|,\alpha}$ outputs one. Hence, Eq. (14) implies that, for each k, $1 \le k \le m-1$,

$$g'_k[z, 1_\alpha] = \begin{cases} 1 & \text{if } k \equiv |z| + \alpha \pmod{m}; \\ 0 & \text{otherwise,} \end{cases}$$

that is, only g'_β, such that

$$\beta \equiv |z| + \alpha \pmod{m}, \tag{15}$$

outputs one. Since the output of g'_k is u_k for each k, $0 \le k \le m-1$, the output of $C_{l,m}$ is 1_β. Hence, $C_{l,m}$ computes the desired function.

Clearly, $g_0, g_1, \ldots, g_l$ have fan-in at most $2l$, $g_{0,0}, g_{0,1}, \ldots, g_{m-1,l}$ have fan-in two, and $g'_0, g'_1, \ldots, g'_{m-1}$ have fan-in $l+1$. For every pair of z and $\mathbf{1}_\alpha$, the gates outputting ones are $g_{|z|}$, $g_{|z|,\alpha}$ and g'_β satisfying Eq. (15). Hence, the energy e of C is three. Furthermore,

$$s(C) = (l+1) + (l+1)m + m \leq (l+1)(m+1) + m.$$

We thus complete the proof.

5 Conclusions

In this paper, we consider threshold circuits C computing the modulus function MOD_m of n variables, where every gate in C has fan-in at most l. We give an upper bound $e = O(n/l)$ and an almost tight lower bound $e = \Omega((n-m)/l)$ on the energy e of C. A generalization of our results to symmetric functions remains open.

References

1. Franco, L., Cannas, S.A.: Solving arithmetic problems using feed-forward neural networks. Neurocomputing 18, 61–79 (1998)
2. Håstad, J., Goldmann, M.: On the power of small-depth threshold circuits. Computational Complexity 1, 113–129 (1993)
3. Laughlin, S.B., Sejnowski, T.J.: Communication in neuronal networks. Science 301(5641), 1870–1874 (2003)
4. Lennie, P.: The cost of cortical computation. Current Biology 13, 493–497 (2003)
5. Margrie, T.W., Brecht, M., Sakmann, B.: In vivo, low-resistance, whole-cell recordings from neurons in the anaesthetized and awake mammalian brain. Pflugers Arch. 444(4), 491–498 (2002)
6. Shawe-Taylor, J.S., Anthony, M.H.G., Kern, W.: Classes of feedforward neural networks and their circuit complexity. Neural Networks 5, 971–977 (1992)
7. Siu, K.Y., Roychowdhury, V., Kailath, T.: Toward massively parallel design of multipliers. Journal of Parallel and Distributed Computing 24, 86–93 (1995)
8. Suzuki, A., Uchizawa, K., Zhou, X.: Energy-efficient threshold circuits computing mod functions. In: Potanin, A., Viglas, T. (eds.) Proceedings of the 17th Computing: The Australasian Theory Symposium, CRPIT, Perth, Australia, vol. 119, pp. 105–110. ACS (2011)
9. Uchizawa, K., Douglas, R., Maass, W.: On the computational power of threshold circuits with sparse activity. Neural Computation 18(12), 2994–3008 (2006)
10. Uchizawa, K., Nishizeki, T., Takimoto, E.: Energy and depth of threshold circuits. Theoretical Computer Science (to appear)
11. Uchizawa, K., Takimoto, E.: Exponential lower bounds on the size of constant-depth threshold circuits with small energy complexity. Theoretical Computer Science 407(1-3), 474–487 (2008)
12. Uchizawa, K., Takimoto, E., Nishizeki, T.: Size-energy tradeoffs of unate circuits computing symmetric Boolean functions. Theoretical Computer Science 412, 773–782 (2011)

NEXP Does Not Have Non-uniform Quasipolynomial-Size ACC Circuits of $o(\log \log n)$ Depth

Fengming Wang[*]

Department of Computer Science
Rutgers University
New Brunswick, NJ, 08855 USA
fengming@cs.rutgers.edu

Abstract. ACC_m circuits are circuits consisting of unbounded fan-in AND, OR and MOD_m gates and unary NOT gates, where m is a fixed integer. We show that there exists a language in non-deterministic exponential time which can not be computed by any non-uniform family of ACC_m circuits of quasipolynomial size and $o(\log \log n)$ depth, where m is an arbitrarily chosen constant.

1 Introduction

Proving non-uniform circuit lower bounds is a longstanding open problem in complexity theory. The lack of progress in nearly two decades has made it a well-known major challenge in the theoretical computer science community. Recently, Williams [16] proposed a research program which tried to show circuit lower bounds via designing fast satisfiability algorithms for Circuit-SAT problems. A few months ago, Williams [17] succeeded in carrying out the program by proving an ingenious lower bound for NEXP against non-uniform constant-depth ACC circuits of sub-exponential size (For every fixed depth d, he also provided an exponential-size lower bound), thereby solving a notorious long-standing open problem. For more background and history on circuit lower bounds, we refer the readers to [17] which elaborates on this history in detail.

In this paper, we show that Williams' lower bound result can be extended to a broader class of ACC circuits with non-constant depth. A function f is a quasipolynomial if $f = n^{\log^{O(1)} n}$. Formally, our main theorem is stated as:

Theorem 1. NEXP *does not have non-uniform* ACC *circuits of quasipolynomial size and* $o(\log \log n)$ *depth.*

1.1 Related Work

In this section, we survey a few examples of earlier work giving super-polynomial size bounds for circuits of non-constant depth.

---

[*] Supported in part by NSF Grants CCF-0830133 and CCF-0832787.

M. Ogihara and J. Tarui (Eds.): TAMC 2011, LNCS 6648, pp. 164–170, 2011.
© Springer-Verlag Berlin Heidelberg 2011

More than two decades ago, building on his powerful switching lemma, Håstad [10] proved that the parity function can not be computed by families of AC circuits of polynomial size and depth at most $\frac{c \log n}{\log \log n}$ for some positive constant c. This result found many applications in proving lower bounds for the parallel random access machine model (PRAM), which is one of the widely adopted models of parallel computation. For instance, Beame and Håstad [5] exhibited the optimal $\Omega(\frac{\log n}{\log \log n})$ lower bounds on the time for CRCW (Concurrent read and concurrent write) PRAM with polynomially many processors to compute the parity function and related problems.

The classic results of Razborov [13] and Smolensky [14] showed that if p is a prime and q is not a power of p, then the MOD_q function is not computable by any constant-depth and poly-size family of ACC_p circuits. In fact, their technique also works in the regime of non-constant-depth circuit lower bounds. More precisely, one can adapt their polynomial method to show that the same MOD_q function remains hard even for ACC_p circuits of polynomial size and $\Omega(\frac{\log n}{\log \log n})$ depth. Even though the results in this paper are exponentially worse in terms of circuit depth, note that they hold for the more powerful ACC circuit model.

Some other $\Omega(\log \log n)$ depth bounds are known in the setting of *uniform* circuits. Allender and Gore [2] showed that the permanent function is not computable by DLOGTIME-uniform ACC^0 circuits of exponential size. Later Allender [1] proved a smaller (but still super-quasipolynomial) bound for computing the permanent on DLOGTIME-uniform *threshold* circuits. Koiran and Perifel [12] extended this latter result [1], and proved that the permanent function can not be computed by DLOGTIME-uniform threshold or arithmetic circuits of polynomial size and $o(\log \log n)$ depth.

2   Preliminaries

We assume that the readers are familiar with standard notations for complexity classes [3] and circuit complexity classes [15]. General circuits consist of NOT gates and unbounded fan-in AND and OR gates. ACC_m circuits are general circuits equipped with unbounded fan-in MOD_m gates, where m is a fixed integer.

We say a boolean function $g : \Sigma^n \to \{0, 1\}$ is in $\text{ACC}_m(s, d)$ if g can be recognized by some ACC_m circuit of size at most s and depth bounded by d. For any two functions $s(n)$ and $d(n)$, we say a language $L \in \text{ACC}(s(n), d(n))$ if there exists an integer constant m such that for each input length n, its characteristic function L_n is in $\text{ACC}_m(s(n), d(n))$. For any two families of functions $\mathcal{S}$ and $\mathcal{D}$, $\text{ACC}(\mathcal{S}, \mathcal{D}) = \bigcup \text{ACC}(s(n), d(n)) \mid s(n) \in \mathcal{S}, d(n) \in \mathcal{D}$.

SYM^+ circuits have exactly two levels of internal nodes. The top level is a single gate with unbounded fan-in which computes an arbitrary symmetric function and the bottom level contains only AND gates which are connected directly to the input variables. We say a boolean function $g : \Sigma^n \to \{0, 1\}$ is in $\text{SYM}^+(s, t)$ if it can be computed by some SYM^+ circuit of size at most s, where moreover, the fan-in of AND gates is bounded by t. We can define similarly as above the language classes $\text{SYM}^+(s(n), t(n))$ and $\text{SYM}^+(\mathcal{S}, \mathcal{T})$.

For a circuit type $\mathcal{C}$ and a set of associated measures, it will be convenient for us to consider the family of collections of boolean circuits which is denoted as

$$\text{Circuit}_C(s_1(n), s_2(n), .., s_m(n)) = \{G_1, G_2, ..\},$$

where each circuit in G_n has exactly n input variables and its ith measure is bounded by $s_i(n)$ respectively. For general circuits, we only consider the size measure, hence, $\text{Circuit}_{\text{General}}(s(n)) = \{G_1, G_2, ..\}$, where G_n contains all circuits of size at most $s(n)$. We can also give similar definitions for $\text{Circuit}_{\text{ACC}_m}(s(n), d(n))$ with both size and depth measures and for $\text{Circuit}_{\text{SYM}^+}(s(n), t(n))$ where the first measure is the size measure and the second measure is in terms of the bottom fan-in.

For two families $\mathcal{F}_1 = \{G_1, G_2, ..\}$ and $\mathcal{F}_2 = \{G'_1, G'_2, ..\}$, we say $\mathcal{F}_1$ is *transformable to* $\mathcal{F}_2$ if for all sufficiently large $n \in \mathbb{N}$, $\forall C \in G_n$, $\exists C' \in G'_n$ such that $\forall x \in \Sigma^n, C(x) = C'(x)$, namely, C and C' are equivalent. Furthermore, $\mathcal{F}_1$ is *transformable to* $\mathcal{F}_2$ *in time* $t(n)$ if there exists a uniform algorithm which given the standard encoding of C, output C' in time $t(n)$.

For a family $\mathcal{F} = \{G_1, G_2, ..\}$, we say $\mathcal{F}$-SAT is *solvable in time* $t(n)$ if there exists a uniform algorithm $\mathcal{A}$ such that for all sufficiently large $n \in \mathbb{N}$, given an arbitrary C in G_n, $\mathcal{A}$ decides its satisfiability in time $t(n)$.

3   Main Result

3.1   A Fast Satisfiability Algorithm

Transformation between different circuit types is an important building block in our proof. Yao [18], Beigel and Tarui [7] and Allender and Gore [2] studied conversion from $\text{Circuit}_{\text{ACC}_m}(n^{O(1)}, O(1))$ to $\text{Circuit}_{\text{SYM}^+}(n^{\log^{O(1)} n}, \log^{O(1)} n)$. In fact, their strategy works in a more general setting.

Theorem 2 ([7,2]). *For every constant m, there is a constant c such that for any size function $s(n)$ and any depth function $d(n)$, $\text{ACC}_m(s(n), d(n))$ is transformable in time* $2^{O((\log s(n))^{2^{cd(n)}})}$ *to* $\text{SYM}^+(2^{(\log s(n))^{2^{cd(n)}}}, (\log s(n))^{2^{cd(n)}})$.

Corollary 1. *For any constant m, any small constant ϵ, any quasipolynomial $p(n)$ and any depth function $d(n)$ of order $o(\log \log n)$, $\text{ACC}_m(p(n), d(n))$ is transformable to* $\text{SYM}^+(2^{n^\epsilon}, n^\epsilon)$ *in time* $2^{O(n^\epsilon)}$.

In [17], Williams gave an algorithm for solving the satisfiability problem of SYM^+ circuits of size s over n variables in time $O((2^n + s)n^{O(1)})$. Combining it with Corollary 1, the following theorem is immediate.

Theorem 3. *For every constant m, there exists a constant c such that for any quasipolynomial $p(n)$ and any depth function $d(n)$ of order $o(\log \log n)$, the decision problem $\text{Circuit}_{\text{ACC}_m}(p(n), d(n))$-SAT is solvable in time $O(2^n n^c)$.*

The running time in Theorem 3 can be improved using the identical strategy adopted by Williams [16].

Theorem 4. *Let m be a fixed constant. For any positive constant c', any quasipolynomial $p(n)$ and any depth function $d(n)$ of order $o(\log \log n)$, the corresponding decision problem $\text{Circuit}_{\text{ACC}_m}(p(n), d(n))$-SAT is solvable in time $O(\frac{2^n}{n^{c'}})$.*

Proof. Let c be the constant in Theorem 3. Given an $\text{ACC}_m(p(n), d(n))$ circuit over n variables, when the first $(c + c') \log n$ inputs are set to definite values, we simplify it to obtain a circuit over $n - (c + c') \log n$ many variables. Hence, by fixing the first $(c + c') \log n$ input variables to all possible sequences, we get $n^{c+c'}$ many circuits. Create a new circuit by feeding their outputs to a single OR gate. The size of this new circuit is bounded by $p(n)n^{c+c'}$ and its depth is only increased by one. Note that $p(n)n^{c+c'}$ is still a quasipolynomial in $(n - (c + c') \log n)$, and $d(n) + 1$ is in $o(\log \log(n - (c + c') \log n))$ as well. By Theorem 3, its satisfiability can be determined in time $O(2^{n-(c+c') \log n}(n - (c + c') \log n)^c)$ which is $O(\frac{2^n}{n^{c'}})$. This finishes our arguments since the satisfiability problem for the new circuit is equivalent to the one for the original circuit.

Note 1. In [17], there are about n^δ many input variables set in a single copy. However, in order to apply Theorem 3, it is crucial for our work to only fix $O(\log n)$ many input variables, which keeps the size of the final circuit within quasipolynomial (compared to $2^{n^{O(\delta)}}$ in [17]).

3.2   Proof of Main Theorem

In this section, we present our main lower bound result via the framework invented by Williams [17]. The following notions will be useful.

Definition 1. *Let* $x = x_0 x_1 x_2 ... x_{|x|-1}$ *be a binary string, where* $|x|$ *is the size of* x. *We say* x *is succinctly represented by the circuit* C *if* C *has* $\lceil \log(|x| + 1) \rceil$ *many input bits and moreover, for all* $0 \le i \le |x| - 1$, $C(i) = x_i$ *while its output can be arbitrary otherwise. We call such a circuit* C *as a* succinct representations *of* x.

Let ϕ *be a 3-CNF formula with* n *variables and* m *clauses.* ϕ *is succinctly represented by the circuit* C' *if* C' *has* $\lceil \log(m + 1) \rceil$ *many input bits and furthermore, on the input* $0 \le i \le m - 1$, $C'(i)$*'s output is the standard binary encoding of the* ith *clause. Hence,* C' *has roughly* $3(\lceil \log(n + 1) \rceil + 1)$ *output bits, the amount which is needed to encode three literals. We say that* C' *is a* succinct representation *or* compression *of* ϕ.

Theorem 5 (Theorem 1 restated)

$$\text{NEXP} \nsubseteq \text{ACC}(n^{\log^{O(1)} n}, o(\log \log n)).$$

Proof. Suppose $\text{NEXP} \subseteq \text{ACC}(n^{\log^{O(1)} n}, o(\log \log n))$. The first step of our proof is to note that, because of Theorem 4, it is possible to state a slight variant of Lemma 3.1 of [17].

Lemma 1. *There is a universal positive constant* c *with the following property. Assume that* $P \subseteq \text{ACC}(n^{\log^{O(1)} n}, o(\log \log n))$, *then for every* $L \in \text{NTime}[2^n]$, *there is a nondeterministic algorithm* $\mathcal{A}$, *an integer constant* m, *a quasipolynomial* $p'(n)$ *and a depth function* $d'(n)$ *of order* $o(\log \log n)$ *such that*

- $\mathcal{A}$ runs in $O(\frac{2^n}{n^c})$ time,
- for every instance x with $|x| = n$, $\mathcal{A}(x)$ either rejects or prints a circuit $C_x \in G_{n+c\log n}$ where $G_{n+c\log n} \in Circuit_{ACC_m}(p'(n), d'(n))$ such that $x \in L$ if and only if C_x is the compression of a satisfiable 3-CNF formula F_x of size $2^n \cdot n^{O(1)}$, and
- there is at least one computation path $\mathcal{A}(x)$ that outputs C_x.

Hence, Lemma 1 implies that as long as deciding the satisfiability of succinct 3-CNF instances such as C_x can be achieved in nondeterministic time $O(\frac{2^n}{n^{c'}})$ for any c', then $\mathrm{NTime}[2^n] \subseteq \mathrm{NTime}[\frac{2^n}{n^{c'}}]$, in contradiction to the nondeterministic time hierarchy [19]. Therefore, we are done except for showing that the satisfiability of C_x can be tested in this time bound, assuming that $\mathrm{NEXP} \subseteq \mathrm{ACC}(n^{\log^{O(1)} n}, o(\log\log n))$.

The following theorem is a variant of Theorem 5.2 in [17]. It is also implicit in the work of Impagliazzo, Kabanets and Wigderson [11].

Theorem 6 ([11,17]). $\mathrm{NEXP} \subseteq \mathrm{SIZE}(n^{\log^{O(1)} n})$ *implies that for every language L in* NEXP, *there exists a quasipolynomial p such that $\forall x \in L$, there exists a witness w for x with the property that the boolean function whose truth table is given by w can be computed by a general circuit of size at most $p(|x|)$.*

In other words, every instance in L has a succinctly represented witness. In particular, every compressed 3-CNF formula has a succinct satisfying assignment since the Succinct SAT Problem is in NEXP.

Our assumption that $\mathrm{NEXP} \subseteq \mathrm{ACC}(n^{\log^{O(1)} n}, o(\log\log n))$ implies $\mathrm{NEXP} \subseteq \mathrm{SIZE}(n^{\log^{O(1)} n})$, so obviously the conclusion in Theorem 6 holds.

Lemma 2 (Folklore). *If* $\mathrm{P} \subseteq \mathrm{ACC}(n^{\log^{O(1)} n}, o(\log\log n))$, *then there exists a universal constant m' such that for any quasipolynomial $p(n)$, there exists a quasipolynomial $p'(n)$ and a depth function $d(n)$ of order $o(\log\log n)$ such that $Circuit_{General}(p(n))$ is transformable to $Circuit_{ACC_{m'}}(p'(n), d(n))$.*

Proof. The Circuit Value Problem (CVP) is in P, and hence, there exists an integer constant m', a quasipolynomial $q(n)$ and a depth function $d'(n) = o(\log\log n)$ such that CVP is computed by a family of $\mathrm{ACC}_{m'}$ circuits of size at most $q(n)$ and depth bounded by $d'(n)$. Under the standard encoding of circuits, this implies that any general circuit of size at most $p(n)$ has an equivalent $\mathrm{ACC}_{m'}$ circuit of size at most $q(p^2(n))$ and depth bounded by $d'(p^2(n))$. Since $q(p^2(n))$ is still a quasipolynomial in n and $d'(p^2(n)) = o(\log\log n)$, our claim holds.

Theorem 6 tells us that for every x in L, there exists a witness w that is succinctly represented by a circuit of quasipolynomial size. By Lemma 2, this circuit can be assumed to be a quasipolynomial-size ACC'_m circuit C_w of depth $o(\log\log n)$. Thus analogous to the work of Williams [17], our algorithm for deciding the satisfiability of the succinctly represented 3-CNF instance C_x proceeds as the following steps.

1. Guess the circuit C_w of quasipolynomial size and depth $o(\log\log n)$, where w is a witness for C_x being satisfiable.

2. Build a circuit C of the following form: On input i, use C_x to obtain the encoding of the ith clause of the formula F_x. Querying C_w, find the values of the three variables occurring in this clause, according to the witness w.
3. C rejects if and only if these values cause cause the clause to evaluate to 1.

Note that C is unsatisfiable if and only if every clause of F_x is satisfied by w.

Fact 7. *For two fixed integers m and m', there exists a polynomial r such that any* ACC *circuit containing both MOD$_m$ and MOD$_{m'}$ gates of size at most s can be simulated uniformly by an ACC$_l$ circuit of the same depth and size at most $r(s)$ where $l = m \cdot m'$.*

By fact 7, C is a quasi-poly-size ACC$_l$ circuit of depth at most $d(n) + d'(n) + O(1)$ and by Theorem 4, its satisfiability is decidable in time $O(\frac{2^n}{n^{c'}})$ for any c', which concludes our proof for the main theorem.

4   Discussions

We have not fully exploited the strength of the machinery behind Theorem 2. The original form of the transformation provides a large set of parameters which can be tuned smoothly. For instance, one can allow $m(n) = \{l_1, l_2, ...\}$ to be a slowly growing (say, of order $O(\log \log n)$) integer sequence rather than a fixed constant and consider the circuit families ACC$_{m(n)}$ where the nth circuit contains the presence of MOD$_{l_i}$ gates for all $i \leq n$. It is easy for the readers who are familiar with the framework of [18], [7] and [2] to verify that NEXP does not have non-uniform ACC$_{m(n)}$ circuits of quasipolynomial size and non-constant depth. This phenomenon has been observed by several authors, [4], [8] etc, and their further investigations made it explicit that SYM$^+(n^{\log^{O(1)} n}, \log^{O(1)} n)$ actually encompasses a circuit complexity class presumably larger than ACC$(n^{O(1)}, O(1))$, where every ACC$(n^{O(1)}, O(1))$ circuit has an extra symmetric gate at the top. Hence, it is natural to conjecture that NEXP is not contained in this class either. However, the proof of Theorem 4 introduces too many duplicate symmetric gates, which falls beyond the reach of current techniques. Note that Beigel [6] showed that polylog majority gates can be merged into one at the top, but his results would yield the trivial bound 2^{n^c} for some $c > 1$ in our case.

We would like to draw the comparison between this work and [14]. The main technical difficulty which prevents us from obtaining a depth lower bound of order $\Omega(\frac{\log n}{\log \log n})$ is that each application of modulus-amplifying polynomials creates extra AND gates of large fan-in. This in turn causes the snowball effect of the blow-up in the final circuit size. Thus, new ideas are needed in order to improve the current depth lower bound.

In the recent work by Fortnow and Santhanam [9], they defined notions of robust simulations and significant separations, and showed that the theorem of Williams [17] can be strengthened with respect to significant separations. We would like to mention that analogous significant separations can be obtained against the circuit complexity class ACC$(n^{\log^{O(1)} n}, o(\log \log n))$ as well.

Acknowledgement

We thank Eric Allender, Luke Friedman and Ryan Williams for helpful discussions. We are especially grateful to Eric Allender for carefully reading earlier drafts and providing many useful comments.

References

1. Allender, E.: The permanent requires large uniform threshold circuits. Chicago J. Theor. Comput. Sci. (1999)
2. Allender, E., Gore, V.: A uniform circuit lower bound for the permanent. SIAM Journal on Computing 23(5), 1026–1049 (1994)
3. Arora, S., Barak, B.: Computational Complexity, a modern approach. Cambridge University Press, Cambridge (2009)
4. Barrington, D.A.M.: Quasipolynomial size circuit classes. In: Proc. IEEE Conf. on Structure in Complexity Theory, pp. 86–93 (1992)
5. Beame, P., Håstad, J.: Optimal bounds for decision problems on the crcw pram. Journal of the ACM 36(3), 643–670 (1989)
6. Beigel, R.: When do extra majority gates help? polylog() majority gates are equivalent to one. Computational Complexity 4, 314–324 (1994)
7. Beigel, R., Tarui, J.: On ACC. Computational Complexity 4, 350–366 (1994)
8. Beigel, R., Tarui, J., Toda, S.: On probabilistic acc circuits with an exact-threshold output gate. In: Ibaraki, T., Iwama, K., Yamashita, M., Inagaki, Y., Nishizeki, T. (eds.) ISAAC 1992. LNCS, vol. 650, pp. 420–429. Springer, Heidelberg (1992)
9. Fortnow, L., Santhanam, R.: Robust simulations and significant separations, http://arxiv.org/abs/1012.2034 (manuscript)
10. Håstad, J.: Almost optimal lower bounds for small depth circuits. In: Proc. ACM Symp. on Theory of Computing (STOC), pp. 6–20 (1986)
11. Impagliazzo, R., Kabanets, V., Wigderson, A.: In search of an easy witness: exponential time vs. probabilistic polynomial time. Journal of Computer and System Sciences 65(4), 672–694 (2002)
12. Koiran, P., Perifel, S.: A superpolynomial lower bound on the size of uniform non-constant-depth threshold circuits for the permanent. Computational Complexity, 35–40 (2009)
13. Razborov, A.: Lower bounds on the size of bounded-depth networks over a complete basis with logical addition. Mathematical Notes of the Academy of Sciences. of the USSR 41(4), 333–338 (1987)
14. Smolensky, R.: Algebraic methods in the theory of lower bounds for boolean circuit complexity. In: Proc. ACM Symp. on Theory of Computing (STOC), pp. 77–82 (1987)
15. Vollmer, H.: Introduction to Circuit Complexity. Springer, Heidelberg (1999)
16. Williams, R.: Improving exhaustive search implies superpolynomial lower bounds. In: Proc. ACM Symp. on Theory of Computing (STOC), pp. 231–240 (2010)
17. Williams, R.: Non-uniform ACC circuit lower bounds. In: Proc. IEEE Conf. on Computational Complexity (2010), http://www.cs.cmu.edu/~ryanw/acc-lbs.pdf
18. Yao, A.C.-C.: On ACC and threshold circuits. In: Proc. IEEE Symp. on Found. of Comp. Sci (FOCS), pp. 619–627 (1990)
19. Zak, S.: A turing machine hierarchy. Theoretical Computer Science 26, 327–333 (1983)

Quantum Complexity: Some Recent Results, Some Open Problems, Some Thoughts

Richard J. Lipton

College of Computing
Georgia Institute of Technology
Atlanta, GA 30332, USA
rjl@cc.gatech.edu

Abstract. This talk will discuss some recent results on the complexity theory of quantum computing. The long term goal is to understand exactly the relationship between the new quantum complexity classes and the classic ones. It is assumed that you are **not** an expert in: *quantum flavordynamics, quantum geometrodynamics, quantum hydrodynamics, quantum magnetodynamics, quantum triviality, Schrödinger's equation, path integral, Schwinger-Dyson equation, static forces and virtual-particle exchange, and the Ward-Takahashi identity.*

It is assumed that you are interested in hearing some simple ideas that connect basic complexity theory with quantum computation. No previous knowledge is assumed, all will be explained, at least that is the plan of the talk.

M. Ogihara and J. Tarui (Eds.): TAMC 2011, LNCS 6648, p. 171, 2011.
© Springer-Verlag Berlin Heidelberg 2011

Non-adaptive Complex Group Testing with Multiple Positive Sets⋆

Francis Y.L. Chin, Henry C.M. Leung, and S.M. Yiu

Department of Computer Science, The University of Hong Kong,
Pokfulam, Hong Kong

Abstract. Given n items with at most d of them having a particular property (referred as positive items), a single test on a selected subset of them is positive if the subset contains any positive item. The non-adaptive group testing problem is to design how to group the items to minimize the number of tests required to identify all positive items in which all tests are performed in parallel. This problem is well-studied and algorithms exist that match the lower bound with a small gap of $\log d$ asymptotically. An important generalization of the problem is to consider the case that individual positive item cannot make a test positive, but a combination of them (referred as positive subsets) can do. The problem is referred as the *non-adaptive complex group testing*. Assume there are at most d positive subsets whose sizes are at most s, existing algorithms either require $\Omega(\log^s n)$ tests for general n or $O(\binom{s+d}{d} \log n)$ tests for some special values of n . However, the number of items in each test cannot be very small or very large in real situation. The above algorithms cannot be applied because there is no control on the number of items in each test. In this paper, we provide a novel and practical derandomized algorithm to construct the tests, which has two important properties. (1) Our algorithm requires only $O\left((d+s)^{d+s+1}/(d^d s^s) \log n\right)$ tests for all positive integers n which matches the upper bound on the number of tests when all positive subsets are singletons, i.e. $s = 1$. (2) All tests in our algorithm can have the same number of tested items k. Thus, our algorithm can solve the problem with additional constraints on the number of tested items in each test, such as maximum or minimum number of tested items.

Keywords: pooling design, non-adaptive complex group testing, knock-out study, combinatorial group testing.

1 Introduction

In biological studies, there are many situations in which we need to identify a subset of items with a particular property (called positive items) from a large set of items. Instead of testing each item one by one, we can group and test several items together in one experiment. If the outcome is negative, we can conclude

---

⋆ This research was partial supported by HK GRF grant HKU 7117/09E.

M. Ogihara and J. Tarui (Eds.): TAMC 2011, LNCS 6648, pp. 172–183, 2011.

that all items in the group do not have that property using only one experiment. By grouping the items carefully, biologists can save a lot of experiments. For example, during World War II, biologists needed to identify people with syphilitic antigen from a large population using the Wasserman-type blood test [8]. Instead of performing the test on each blood sample, they performed tests on grouped blood samples in order to reduce the total number of tests. In DNA library screening [6,13], biologists need to identify from the DNA library a subset of cloned DNA segments containing a particular substring, called probe. Instead of performing an experiment on each clone-probe pair, biologists group several cloned DNA segments together and perform a single experiment on them. In phenotype knockout studies [17,21,26], biologists need to identify genes causing a particular phenotype from a set of genes. Instead of knocking out genes one by one in each experiment, biologists can knockout several genes at the same time and check whether the phenotype still appears in one experiment.

The *group testing problem* [9,15], which has been studied since World War II on the Wasserman-type blood test mentioned above, is to find the best way of grouping items in each test so as to minimize the total number of tests needed in the worst case. If the tests can be performed sequentially after knowing the results of the previous tests, the problem was solved more than 30 years ago and there exist algorithms [15,18,23] for which the number of tests required is close to the optimal (in term of exact number of tests). However, some experiments are time-consuming, e.g. each phenotype knockout experiment requires several months, and we cannot afford the time to perform tests one after another. Instead, it is desirable to perform all tests in parallel without knowing the results of others. In this case, the *non-adaptive group testing problem*, also called *pooling design* [5,11,12,19], is needed. In this paper, we focus on this non-adaptive version.

Given a set of n items with at most d hidden positive items P, the result of a test on a subset S of items is positive if $P \cap S \neq \emptyset$, otherwise, the result is negative. The *non-adaptive group testing problem* is to design the minimum number of tests t, as a function of n and d, for determining all positive items P from the results of the tests assuming that all tests are performed in parallel and designed without any knowledge of other test results.

The non-adaptive version of the problem seems to be more difficult. Only recently, there were some breakthrough results for solving the problem. Porat and Rothschild [20] solved the problem by constructing an Error Correction Code (ECC). ECC encodes the alphabets in a message into binary strings with the Hamming distance between any pair of strings is at least d. Thus up to $d/2$ errors in each string can be detected and corrected. By picking a suitable alphabet size, they can convert the ECC into $O(d^2 \log n)$ tests for the non-adaptive group testing problem which almost matches the lower bound of $O(d^2 \frac{\log n}{\log d})$ [4]. Indyk et al. [16] provided another solution also with $O(d^2 \log n)$ tests based on concatenated code. They first construct a Reed-Solomon code with suitable parameters, then encode it with another independent random binary code and convert it into

$O(d^2 \log n)$ tests. By decoding the test results in two levels, they can determine the positive items in $O(polylog(n))$ time.

However, there exist many important applications that cannot be modeled by the above group testing problem. First, because of the sensitivity of the experiments, we may not be able to group many items in a test. Similarly, there are cases for which we cannot group too few items. One example is the phenotype knockout experiment. We cannot knock out many genes and leave too few for the test, otherwise the tested individual will die. Therefore, there may be a minimum (or maximum) requirement on the number of tested items in a test. Second, there are many real cases that instead of individual items, a combination of items (forming a *positive subset*) is required to make the test positive. That is, the test will show a positive result only if all items in a positive subset are all present in the test. For example, in DNA hybridization [22], the test result is positive with the presence of some pairs of hybridized DNA strands (positive subset of size 2). In two-hybrid screening [27] for detecting protein-protein interaction, the test result is positive if the test sample contains some pairs of interacting proteins (another example of positive subset of size 2). Similarly in three-hybrid screening [1], the test result is positive if the test sample contains some sets of interacting proteins and RNA (positive subset of size 3). Thus, we need a generalization to model these applications.

Given a set of n items with at most d distinct hidden positive subsets S_i with $|S_i| \leq s$, the result of a test on a subset S of items is positive if there is a positive subset $S_i \subseteq S$, otherwise, the result is negative. The *complex group testing problem* [10,14] is to find the best way of grouping items in each test, so as to minimize the total number of tests needed in the worst case for finding all the hidden positive subsets S_i. In practice, we sometimes require $|S| \geq k$ (or $|S| \leq k$) for some k. In the following, we only consider the case for $|S| \geq k$ as the other case is symmetric (We calculate the optimal k for a given range).

The group testing problem is a special case of the *complex* group testing problem with $s = 1$ and no requirement on $|S|$. However, none of the algorithms [5,11,12,19] for solving the non-adaptive group testing problem can be extended to solve the complex version with $s > 1$ even without any restriction on the size of S. At first glance, one may replace the n items by the $\binom{n}{s}$ combinations of items and apply the above algorithms to design a set of tests. As the $\binom{n}{s}$ combinations of items are not independent, e.g. we cannot test $\{1, 2\}$ without testing $\{1\}$ or $\{2\}$, this reduction does not work. This complex version of the problem seems to be even more difficult and only limited results exist [2,14,24,25]. None of these solutions can handle the requirement on the size of S. And they either require many tests ($\Omega(\log^s n)$ tests) [14], or designed only for specific n values [24,25,7], or there is no guarantee on the running time or the number of tests [2]. For example, Gao et al. [14] represent each item by a distinct polynomial $g_i(x)$ of degree z in a finite field $GF(q)$ with $n \leq q^z$ and $sdz \leq q$. They select sdz distinct elements in $GF(q)$ and perform a test on those items with the same value of $g_i(y)$ for each element y. By choosing the value of q and z carefully, they can solve the problem with $O(s(d \log_q n)^{s+1})$ tests. Stinson et al. [24,25]

construct a set of tests using perfect hash family. They construct a separating hash family with $n = 72^{2^i}$ elements by recursion on integer i. Then they encode this hash family into $O(\binom{s+d}{s} \log n)$ tests. However, their method cannot solve the non-adaptive complex group testing problem (or require much more tests) when $n \neq 72^{2^i}$. Bishop et al. [2] solved the problem with $s = 2$ by assigning each item to a test with probability p. By setting a suitable probability p and number of tests t, they can find all S_i with some false positive subsets, i.e. another round of experiments are needed to identify the false positive subsets.

In this paper, we introduce a deterministic algorithm based on randomization and derandomization to solve the non-adaptive complex group testing problem for all possible number of items n and using no more than $t_0 = O\left(\frac{d+s}{r^s(1-r)^d} \log n\right)$ tests, where $r = \max\{\frac{k}{n-d+1}, \frac{s}{d+s}\}$. When there is no restriction on k i.e. $k = 1$, our algorithm requires $O\left((d+s)^{d+s+1}/(d^d s^s) \log n\right)$ tests which matches the lower bound [4] of $O(d^2 \log n)$ for $s = 1$. When compared with Porat and Rothschild's algorithm [20] and Indyk et al.'s algorithm [16], our algorithm is more flexible because it can handle the cases when $s > 1$ and $k > 1$. Our main contributions can be summarized as follows.

1. Our approach is novel, different from any of the previous work even though the techniques used for this approach are not new. The novelty stems from the following observation. It is known that solving the non-adaptive group testing problem is equivalent to designing a binary $t \times n$ $\bar{d}$-separable matrix [5,11,12,19] with the minimum number of rows. We first extend this concept to a $(\bar{d}, \bar{s})$-separable matrix for the complex version of the problem (see the definition in Section 2), then we show that the probability of a random binary $t \times n$ matrix with $t \geq t_0$ being a $(\bar{d}, \bar{s})$-separable matrix is non-zero, i.e. there always exists such a matrix. We use a greedy approach to fill the matrix row by row and guarantee that every time we fill an entry, there must still exist a solution to fill the rest of the entries to make it $(\bar{d}, \bar{s})$-separable.

2. Our approach can solve more general group testing problem, none of previous approaches can be modified to solve the general problem. In particular, an additional advantage of our solution is that we can guarantee every test has exactly k' items where $k' \geq k$ which can handle the cases when there is a restriction on the size of S.

3. Our approach is practical and gives an optimal design in the sense that the number of tests matches the lower bound of the special case.

The paper is organized as follows. In Section 2, we define what a $(\bar{d}, \bar{s})$-separable matrix is and the relationship between such a matrix and the non-adaptive complex group testing problem. Section 3 shows a sufficient condition for a matrix to be $(\bar{d}, \bar{s})$-separable and proves the the existence of t_0 tests to solve the non-adaptive complex group testing problem. Then, we will describe a derandomized algorithm which constructs no more than t_0 tests in Section 4. Section 5 concludes the paper.

2   Preliminaries

Definition 1. *The Non-adaptive Complex Group Testing (NCGT) Problem: Given n items and d' hidden distinct positive subsets of items, $F = S_1, S_2, \ldots, S_{d'}, d' \leq d, |S_i| \leq s, S_i \not\subseteq S_j$ for all $i \neq j$. The result of a test on a set of items T is positive if and only if there is at least one positive subset $S_i \subseteq T$. The NCGT problem is to design the minimum number of non-adaptive tests for discovering all the positive subsets in F.*

When the size of each positive set $s = 1$, the NCGT problem is equivalent to the classical *non-adaptive group testing problem (pooling design)* [19]. When $s = 2$, it is equivalent to the *non-adaptive group testing for disjoint pairs problem* [2]. Any solution with t tests to the NCGT problem can be represented as a $t \times n$ binary matrix M and item j is included in the i-th test if $M(i, j) = 1$ (column or item and test or row will be used interchangeably if no confusion arises).

For any family $F = S_1, S_2, \ldots, S_{d'}, d' \leq d$ given in the NCGT problem, each S_i corresponds to a subset of at most s columns in M and F corresponds to a collection of at most d subsets of columns. For any F, we first take the and-product of the columns corresponding to each S_i, then take the or-product of all these and-products. The resulting bit vector is denoted as $R(F)$. Note that since the outcome of a test is positive if and only if all items in a positive subset are included in the test, the outcomes of the t tests are the same as $R(F)$ for any family F of positive subsets. If such a matrix represents a solution to the NCGT problem, for any two families $F_1, F_2, R(F_1)$ and $R(F_2)$ must be different otherwise it is no way to distinguish whether F_1 or F_2 is the collection of the positive subsets only based on the outcomes of the tests. This motivates us to define a $(\bar{d}, \bar{s})$-separable matrix as follows.

Definition 2. *A $(\bar{d}, \bar{s})$-separable matrix is a binary matrix, such that for any family F of at most d subsets of columns, each subset S_i has at most s columns, the or-product of the $\leq d$ and-products of $\leq s$ columns corresponding to those subsets in F, denoted as $R(F)$, is distinct.*

It is easy to see that the NCGT problem is equivalent to designing a $t \times n$ $(\bar{d}, \bar{s})$-separable matrix with the minimum number of rows.

3   Existence of $(\bar{d}, \bar{s})$-Separable Matrix

In this section, we show that there always exists a $t \times n$ $(\bar{d}, \bar{s})$-separable matrix with $t \leq t_0$ and all rows have *at least* k '1', i.e. each test has at least k tested items.

$$t_0 = \frac{(d + s) \ln n - d \ln(d + s) - s \ln s + d + 2s}{r^s (1 - r)^d} = O\left(\frac{(d + s) \ln n}{r^s (1 - r)^d}\right)$$

where $r = \max\{\frac{k}{n-d+1}, \frac{s}{d+s}\}$. We first describe a sufficient condition for a matrix with *exactly* k '1' in each row to be a $(\bar{d}, \bar{s})$-separable matrix. (Theorem 1). Based on this sufficient condition, we prove that there always exists such a $t_0 \times n$ $(\bar{d}, \bar{s})$-separable matrix with exactly k '1' in each row (Theorem 2).

Theorem 1. *Given a $t \times n$ binary matrix M, if M has the property that for any $d + s$ distinct columns, there are $\binom{d+s}{s}$ rows such that the induced $\binom{d+s}{s} \times (d+s)$ matrix contains different set of s '1' entries in each row, then M is a $(\bar{d}, \bar{s})$-separable matrix.*

Proof. Let A and B be two distinct families of $\leq d$ subsets of $\leq s$ columns $\{A_i\}$ and $\{B_j\}$ in M respectively. Remove those subsets A_i and B_j with $A_i = B_j$. W.L.O.G. assume subset $A_{min} \in A$ contains the minimum number of columns among the remaining subsets. Since $A_i \not\subseteq A_{min}$ and $S \not\subseteq A_{min}$ for all $A_i \neq A_{min}$ and remaining subsets $S \neq A_{min}$ respectively, $B_j \not\subseteq A_{min}$ for all $B_j \in B$. There always exist at most $s + d$ columns containing all columns in A_{min} and one column from a distinct subset B_j as $|A_{min}| \leq s$ and $|B| \leq d$. Since there are $\binom{d+s}{s}$ rows in M such that the induced $\binom{d+s}{s} \times (d+s)$ matrix contains a row with '1' at these $\leq s$ columns in A_{min} and '0' at the rest $\leq d$ columns. Thus the values of $R(A)$ and $R(B)$ are different (1 and 0 respectively) on that row. □

In particular when $s = 1$, Theorem 1 reduces to the existence of $(d+1) \times (d+1)$ identity matrix for any $d + 1$ columns. It is because given two distinct sets of positive items A and B, we should always find a row such that the $\leq d$ positive items in B is '0' and a positive item in A is '1'. For example, when $n = 9$ and $d = 4$, $A = \{3, 5, 7, 9\}$, $B = \{1, 5, 6, 9\}$, the corresponding $d + 1$ columns can be $\{3$ or $7, 1, 5, 6, 9\}$ and the row must have 1 at positions 3 or 7 and 0 at the others.

By considering a $t \times n$ random matrix where each row is assigned with k '1' randomly, we find that the probability that a $t \times n$ random matrix with exactly k '1' in each row satisfies the sufficient condition of Theorem 1 is non-zero when $t \geq t_0$ (Theorem 2). Thus, there always exists such $t_0 \times n$ $(\bar{d}, \bar{s})$-separable matrix with exactly k '1' in each row; otherwise, the probability should be zero. Similar theorems as Theorem 2 are shown in [3,25]. However, since Bonis proved the theorem by considering a hypergraph while Stinson and Wei proved it by partitioning the matrix into submatrices, these proofs cannot be used to construct the derandomized algorithm.

Lemma 1. *Given a $t \times n$ binary matrix M with exactly k randomly selected '1' in each row, the probability that M being a $(\bar{d}, \bar{s})$-separable matrix is at least*

$$1 - \left(\frac{en}{d+s}\right)^{d+s} \left(\frac{e(d+s)}{s}\right)^s \left(1 - r^s(1-r)^d\right)^t$$

where $r = \frac{k}{(n-d+1)}$ and $e \approx 2.71$ is the Euler's number

Proof. When the k '1 are assigned randomly among n columns, the probability that exactly s '1' are assigned in some particular positions in a subset of $d + s$

columns is $\binom{n-(d+s)}{k-s}/\binom{n}{k}$. Thus the probability that a particular combination of s out of a particular subset of d+s columns are not assigned '1' is $1-\binom{n-(d+s)}{k-s}/\binom{n}{k}$.

Pr(M is $(\bar{d}, \bar{s})$-separable)
$\geq 1 - Pr($There are $d + s$ columns s.t. any induced $\binom{d+s}{s} \times (d + s)$ matrix does not contain all possible combinations of s out of $d + s$ columns)

$$\geq 1 - \binom{n}{d+s}\binom{d+s}{s}\left(1 - \frac{\binom{n-(d+s)}{k-s}}{\binom{n}{k}}\right)^t$$

$$\geq 1 - \left(\frac{en}{d+s}\right)^{d+s}\left(\frac{e(d+s)}{s}\right)^s\left(1 - r^s(1-r)^d\right)^t$$

□

Lemma 2

$$\frac{-1}{\ln(1 - r^s(1-r)^d)} < \frac{1}{r^s(1-r)^d}$$

Theorem 2. There always exists a $t_0 \times n$ $(\bar{d}, \bar{s})$-separable matrix M with exactly k '1's in each row where

$$t_0 = \frac{(d+s)\ln n - d\ln(d+s) - s\ln s + d + 2s}{r^s(1-r)^d} = O\left(\frac{d+s}{r^s(1-r)^d}\log n\right)$$

where $r = \frac{k}{n-d+1}$.

Proof. Consider a random binary $t \times n$ matrix M with exactly k randomly selected '1' in each row. By Lemma 1

$$1 - \left(\frac{en}{d+s}\right)^{d+s}\left(\frac{e(d+s)}{s}\right)^s\left(1 - r^s(1-r)^d\right)^t > 0$$

$\Leftrightarrow -t\ln(1 - r^s(1-r)^d) > (d+s)[\ln n - \ln(d+s) + 1] + s[\ln(d+s) - \ln s + 1]$

$\Leftrightarrow t > \dfrac{(d+s)\ln n - d\ln(d+s) - s\ln s + d + 2s}{-\ln(1 - r^s(1-r)^d)}$

When t satisfies the above inequality, the probability that such a $t \times n$ random binary matrix is a $(\bar{d}, \bar{s})$-separable matrix is larger than 0, i.e. there always exists a $t \times n$ $(\bar{d}, \bar{s})$-separable matrix M. The theorem is proved using Lemma 2.   □

Corollary 1. There always exists a $t_0 \times n$ $(\bar{d}, \bar{s})$-separable matrix M with at least k '1's in each row, where t_0 is defined in Theorem 2 and $r = \max\{\frac{k}{n-d+1}, \frac{s}{d+s}\}$.

Proof. By differentiating the equation in Theorem 2 with respect to r, $r^s(1-r)^d$ has the maximum value and t_0 the minimum value when $r = s/(d+s)$. Thus, when $k/(n-d+1) \leq s/(d+s)$, we can increase the value of k to $s(n-d+1)/(d+s)$ and achieve the minimum $t_0 = (d+s)^{d+s}[(d+s)\ln n - d\ln(d+s) - s\ln s + d + 2s]/(d^d s^s)$. Note that the assumption "at least k 1's" is still satisfied.   □

Note that when solving the classical non-adaptive group testing problem with $s = 1$ and $k = 1$, $t_0 = O(d^2 \log n)$ which matches with the lower bound. When solving the non-adaptive group testing for disjoint pairs problem with $s = 2$ and $k = 1$, $t_0 = O(d^3 \log n)$.

4   Constructing a $(\bar{d}, \bar{s})$-Separable Matrix

Theorem 2 shows that there is a $t_0 \times n$ $(\bar{d}, \bar{s})$-separable matrix with exactly k '1's in each row. In this section, we will introduce a deterministic algorithm for constructing such a $t \times n$ $(\bar{d}, \bar{s})$-separable matrix with $t \le t_0$ by derandomization.

Recall that the sufficient condition for a matrix M being a $(\bar{d}, \bar{s})$-separable matrix is that for any $d + s$ columns, there are $\binom{d+s}{s}$ rows such that the induced $\binom{d+s}{s} \times (d + s)$ matrix represents all possible combinations of s '1' out of $d + s$ columns (Theorem 1). Therefore, if all the $\binom{n}{d+s}$ combinations of columns satisfy this requirement, matrix M is a $(\bar{d}, \bar{s})$-separable matrix. We first show by Lemma 3 that for a random matrix with exactly k entry '1' in each row, the expected number of combinations of $d + s$ columns satisfying the requirement is larger than $\binom{n}{d+s} - 1$. Based on Lemma 4, we can fill in each entry of the matrix with '0' and '1' to each row one by one such that the expected number of groups of $d + s$ columns (out of $\binom{n}{d+s}$) satisfying the requirement does not decrease. Thus, we can construct a $t_0 \times n$ $(\bar{d}, \bar{s})$-separable matrix in a greedy manner.

4.1   The Derandomized Algorithm

Let C be a subset of $d + s$ columns in a $t \times n$ binary matrix M and $M(C)$ be the $t \times (d + s)$ binary matrix by restricting the columns in C.

Lemma 3. *For some $t \le t_0$, the expected number of combinations of columns (out of $\binom{n}{d+s}$) of a random $t \times n$ matrix satisfying the requirement in Theorem 1 is larger than $\binom{n}{d+s} - 1$.*

Proof. For any subset C of $d + s$ columns, the probability that $M(C)$ is a $(\bar{d}, \bar{s})$-separable matrix is at least

$$1 - \binom{d+s}{s}\left(1 - \frac{\binom{n-(d+s)}{k-s}}{\binom{n}{k}}\right)^{t_0}$$

Thus, the expected number of combinations of columns (out of $\binom{n}{d+s}$) satisfying the requirement is at least

$$\binom{n}{d+s}\left[1 - \binom{d+s}{s}\left(1 - \frac{\binom{n-(d+s)}{k-s}}{\binom{n}{k}}\right)^{t_0}\right]$$

$$> \binom{n}{d+s} - 1$$

$\square$

Construct a $t_0 \times n$ matrix M with all entries marked as 'x'';
for $i \leftarrow 1$ **to** t_0 **do**
    $q' \leftarrow 0$;
    for $j \leftarrow 1$ **to** n **do**
        Calculate $E_0(i,j) = \sum_C p(M,C)$ when $M(i,j) = 0$;
        Calculate $E_1(i,j) = \sum_C p(M,C)$ when $M(i,j) = 1$;
        if $E_0(i,j) \geq E_1(i,j)$ *or* $q' \geq k$ **then**
            $M(i,j) \leftarrow 0$;
        else
            $M(i,j) \leftarrow 1$;
            $q' \leftarrow q' + 1$;
        end
        if $\max\{E_0(i,j), E_1(i,j)\} = \binom{n}{d+s}$ **then**
            Assign 0 to all entries marked as 'x';
            Return the first i-th rows of M (an $i \times n$ matrix);
        end
    end
end

Algorithm 1. derandomized algorithm for constructing $(\bar{d}, \bar{s})$-separable matrix

Now, we want to show that we can fill in the matrix in a greedy manner in order to obtain a $(\bar{d}, \bar{s})$-separable matrix with $t \leq t_0$. We order the entries of the matrix from top to bottom and from left to right (i.e., we fill the entry from $M(1,1)$ to $M(t,n)$). Assume that all entries proceeding $M(i,j)$ have been filled. Let $E_0(i,j)$ be the expected number of combinations of columns (out of $\binom{n}{d+s}$) satisfying the requirement in Theorem 1 assuming that we fill the entry $M(i,j)$ with '0'. And $E_1(i,j)$ is defined similarly assuming that we fill the entry $M(i,j)$ with '1'.

For any subset C of $(d+s)$ columns in the matrix M with some entries filled, let $p(M,C)$ be the probability that $M(C)$ contains $\binom{d+s}{s}$ rows such that the induced $\binom{d+s}{s} \times (d+s)$ matrix represents all combinations of s '1' out of $d+s$ positions when each row of M is assigned with exactly k '1' randomly (how to compute $p(M,C)$ will be described in the next subsection). The expected number of subsets C with $M(C)$ satisfying Theorem 1 is $\sum_C p(M,C)$. Thus, $E_0(i,j) = \sum_C p(M,C)$ when all previously assigned entries are fixed and $M(i,j)$ is assigned '0', similarly for $E_1(i,j)$.

Lemma 4. $\max\{E_0(i,j), E_1(i,j)\} \geq \max\{E_0(i',j'), E_1(i',j')\}$ *where* $M(i',j')$ *is the entry just before* $M(i,j)$, *i.e.,* $i' = i$ *and* $j' = j - 1$ *if* $j \leq n$, *otherwise* $i' = i - 1$, $j' = n$ *and* $j = 1$.

Proof. We first let $j = 2, 3, \ldots, n$. Since $E_0(i, j-1)$ and $E_1(i, j-1)$ are calculated based on the assumption that $M(i,j)$ is assigned '0' and '1', $\max\{E_0(i, j-1), E_1(i, j-1)\} = p_0 E_0(i,j) + (1 - p_0)E_1(i,j)$ for some real number $0 \leq p_0 \leq 1$. Thus $\max\{E_0(i,j), E_1(i,j)\} \geq \max\{E_0(i, j-1), E_1(i, j-1)\}$. Similarly, we have $\max\{E_0(i,1), E_1(i,1)\} \geq \max\{E_0(i-1, n), E_1(i-1, n)\}$. $\square$

Based on Lemma 4, we can assign values to $M(i,j)$ according to the larger value of $E_0(i,j), E_1(i,j)$. Algorithm 1 shows the details of the construction. Initially, we mark all unassigned entries by 'x'. Since the value of $\sum_C p(M,C)$ increases monotonically with the assignment of $M(i,j)$ and the initial value of $\sum_C p(M,C)$ with no entry being assigned is larger than $\binom{n}{d+s}-1$, the correctness of Algorithm 1 is guaranteed. The following theorem follows. Note also that when $\max\{E_0(i,j), E_1(i,j)\} = \binom{n}{d+s}$, it means that we can assign anything to the remaining entries, so we assign '0' to these entries.

Theorem 3. *Algorithm 1 outputs a $t \times n$ $(\bar{d}, \bar{s})$-separable matrix with $t \le t_0$.*

4.2 Computing the Probability

In this subsection, we show how to compute $p(M,C)$. Given a $t \times (d+s)$ binary matrix $M(C)$ with all entries in the first $i-1$ rows and the first j entries of the i-th row assigned, the probability $p(M,C)$ that $M(C)$ containing $\binom{d+s}{s}$ rows such that the induced $\binom{d+s}{s} \times (d+s)$ matrix represents all combinations of s '1' out of $d+s$ positions can be calculated by the following arguments.

When the last column of C has been assigned, i.e. we are considering the case when all the entries in the first i rows of $M(C)$ have been assigned, we can identify the set of distinct rows of the $\binom{d+s}{s}$ combinations of s '1' out of $d+s$ positions already existed in the first i rows of $M(C)$. Let R be the set of r out of $\binom{d+s}{s}$ combinations that do not exist in the first i rows of $M(C)$. $p(M,C)$ is equal to the probability $p_{row}(i,r)$ that these r combinations in R appear in the remaining $t-i$ rows of $M(C)$. If all the $\binom{d+s}{s}$ rows have already existed in the first i rows of $M(C)$, then $R = \emptyset$, $r = 0$ and $p(M,C) = 1$. The probability that none of the $t-i$ rows equals to a particular row in R is $(1 - \binom{n-(d+s)}{k-s}/\binom{n}{k})^{t-i}$ and the probability that none of the $t-i$ rows equals to any of the r particular rows in R is $(1 - r\binom{n-(d+s)}{k-s}/\binom{n}{k})^{t-i}$. By inclusion and exclusion principle

$$p_{row}(i,r) = 1 + \sum_{\alpha=1}^{r} (-1)^\alpha \binom{r}{\alpha} \left(1 - \alpha \frac{\binom{n-(d+s)}{k-s}}{\binom{n}{k}}\right)^{t-i}$$

When the last column of C has not been assigned yet, we can calculate $p(M,C) = p_{rc}$ with the following parameters

r = number of rows in R that do not exist in the first $i-1$ rows of $M(C)$
r' = number of rows (out of r) in R can occur in the i-th row by assigning the rest entries properly
w = number of entries in C have not been assigned any value in the i-th row
q = number of unassigned entries in C have to be assigned with '1' such that exactly s '1' appear in the i-th row of C
q' = number of entries in the i-th row have been assigned with '1'

$$p_{rc} = \begin{cases} \left(\dfrac{r'\binom{(n-j)-w}{k-q'-q}}{\binom{n-j}{k-q'}}\right) p_{row}(i, r-1) + \left(1 - \dfrac{r'\binom{(n-j)-w}{k-q'-q}}{\binom{n-j}{k-q'}}\right) p_{row}(i,r) & q+q' \le k \\ 0 & q+q' > k \end{cases}$$

Since the values of $0 \leq w, q \leq d + s$, $0 \leq r' \leq r \leq \binom{d+s}{s} \leq (d + s)^s$, $1 \leq i \leq t$, $1 \leq j < n$ and $1 \leq q' \leq k$, there are $O(t(d + s)^s)$ $p_{row}(i, r)$ and $O(nkt(d + s)^{2+2s})$ p_{rc} needed to be precomputed. All possible values of $\binom{n'}{q'}$ and $(1 - \alpha\binom{n-(d+s)}{k-s}/\binom{n}{k})^{t-i}$ for different parameters can be precomputed in $O(n^2)$ and $O(t(d + s)^s)$ times. Each $p_{row}(i, r)$ can be calculated in $O((d + s)^s)$ time after the above precomputation. Thus, the $O(t(d + s)^s)$ $p_{row}(i, r)$ elements can be calculated in $O(t(d + k)^{2s})$ times. Since each p_{rc} element can be calculated in constant time after the precomputation, the $O(nkt(d + s)^{2+2s})$ possible p_{rc} elements can be calculated in $O(nkt(d+s)^{2+2s})$ times. The total time complexity for pre-calculating all possible $p(M, C)$ is $O(n^2 + nkt(d + s)^{2+2s})$.

5   Conclusions

In this paper, we have introduced a deterministic algorithm for constructing tests with the constraint that at most (or at least) k tested items in each test for the non-adaptive complex group testing problem. The algorithm matches with the lower bound $O(d^2 \log n)$ for the unconstrained classical non-adaptive group testing problem. In the future, more complicated constraints, such as inhibition and errors, should be modeled and considered.

References

1. Bernstein, D.S., Buter, N., Stumpf, C., Wickens, M.: Analyzing mRNA-Protein Complexes Using a Yeast Three-Hybrid System. Methods 26, 123–141 (2002)
2. Bishop, M.A., Macula, A.J., Renz, T.E., Ufimtsev, V.V.: Hypothesis Group Testing for Disjoint Pairs. Jour. Comb. Optim. 15, 7–16 (2008)
3. Bonis, A.: New Combinatorial Structures with Applications to Efficient Group Testing with Inhibiters. Jour. Comb. Optim. 15, 77–94 (2008)
4. Chaudhuri, S., Radhakrishnan, J.: Deterministic Restrictions in Circuit Complexity. In: ACM Symposium on Theory of Computing, pp. 30–36 (1996)
5. Deng, P., Hwang, F.K., Wu, E., MacCallum, D., Wang, F., Znati, T.: Improved Construction for Pooling Design. Jour. Comb. Optim. 15, 123–126 (2008)
6. D'yachkov, A.G., Macula, A.J., Torney, D.C., Vilenkin, P.A.: Two Models of Nonadaptive Group Testing for Designing Screening Experiments. In: Proc. 6th Int. Workshop on Model-Orented Designs and Analysis, pp. 63–75 (2001)
7. D'yachkova, A., Vilenkina, P., Torneyb, D., Macula, A.: Families of Finite Sets in which No Intersection of ℓ Sets Is Covered by the Union of s Others. Jour. Comb. Thy. Series A 99(2), 195–218 (2002)
8. Dorfman, R.: The detection of defective members of large population. The Annals of Mathematical Statistics 14(4), 436–440 (1943)
9. Du, D.Z., Hwang, F.: Combinatorial Group Testing and Its Applications, 2nd edn. World Scientific, Singapore (2000)
10. Du, D.Z., Hwang, F.: Pooling Design and Nonadaptive Group Testing: Important Tools for DNA Sequencing. World Scientific, Singapore (2006)
11. Du, D.Z., Hwang, F.K., Wu, W., Znati, T.: New construction for transversal design. Jour. of Comput. Bio. 13(4), 990–995 (2006)

12. Eppstein, D., Goodrich, M.T., Hirschberg, D.S.: Improved Combinatorial Group Testing Algorithms for Real-World Problem Sizes. SIAM Jour. on Comput. Arch. 36(5), 1360–1375 (2006)
13. Farach, M., Kannan, S., Knill, E., Muthukrishnan, S.: Group Testing Problem with Sequences in Experimental Molecular Biology. In: Proc. Compression and Complexity of Sequences, pp. 357–367 (1997)
14. Gao, H., Hwang, F.K., Thai, M., Wu, W., Znati, T.: Construction of d(H)-disjunct matrix for group testing in hypergraphs. Jour. of Combin. Optim. 12(3), 297–301 (2006)
15. Hwang, F.K.: A Method for Detecting All Defective Members in a Population by Group Testing. Jour. Amer. Statist. Assoc. 67, 605–608 (1972)
16. Indyk, P., Ngo, H.Q., Rudra, A.: Effciently Decodable Non-adaptive Group Testing. In: SODA (2010)
17. Jendreyko, N., Popkov, M., Rader, C., Barbas III, C.F.: Phenotypic Knockout of VEGF-R2 and Tie-2 with an Intradiabody Reduces Tumor Growth and Angiogenesis in vivo. PNAS 102(23), 8293–8298 (2005)
18. Li, C.H.: A Sequential Method for Screening Experimental Variables. Jour. Amer. Statist. Assoc. 57, 455–477 (1962)
19. Ngo, H.Q., Du, D.Z.: A Survey on Combinatorial Group Testing Algorithms with Applications to DNA Library Screening. In: Discrete Mathematical Problems with Medical Applications. DIMACS Series, vol. 55. American Mathematical Society, Providence (2000)
20. Porat, E., Rothschild, A.: Explicit non-adaptive combinatorial group testing schemes. In: Aceto, L., Damgård, I., Goldberg, L.A., Halldórsson, M.M., Ingólfsdóttir, A., Walukiewicz, I. (eds.) ICALP 2008, Part I. LNCS, vol. 5125, pp. 748–759. Springer, Heidelberg (2008)
21. Ruberti, F., Capsoni, S., Comparini, A., Daniel, E.D., Franzot, J., Gonfloni, S., Rossi, G., Berardi, N., Cattaneo, A.: Phenotypic Knockout of Nerve Growth Factor in Adult Transgenic Mice Reveals Severe Deficits in Basal Forebrain Cholinergic Neurons, Cell Death in the Spleen, and Skeletal Muscle Dystrophy. Jour. Neurosci. 20(7), 2589–2601 (2000)
22. Sibley, C.G., Ahlquist, J.E.: The Phylogeny of the Hominoid Primates, as Indicated by DNA-DNA Hybridization. Jour. Mole. Evolu. 20, 2–15 (1984)
23. Sobel, M., Groll, P.A.: Group Testing to Elimate Efficiently All Defectives in a Binormal Sample. Bell System Tech. Jour. 38, 1179–1252 (1959)
24. Stinson, D.R., Trung, T., Wei, R.: Secure Frameproof Codes, Key Distribution Patterns, Group Testing Algorithms and Related Structures. Jour. Stat. Plann. Infer. 86, 595–617 (2000)
25. Stinson, D.R., Wei, R.: Generalized Cover-Free Families. Discrete Mathematics 279, 463–477 (2004)
26. Yang, A.G., Bai, X., Huang, X.F., Yao, C., Chen, S.Y.: Phenotypic Knockout of HIV Type 1 Chemokine Coreceptor CCR-5 by Intrakines as Potential Therapeutic Approach for HIV-1 Infection. Proc. Natl. Acad. Sci. 94, 11567–11572 (1997)
27. Young, K.: Yeast Two-Hybrid: So Many Interactions (in) So Little Time. Biol. Reprod. 58(2), 302–311 (1998)

How to Cut a Graph into Many Pieces

Ruben van der Zwaan, André Berger, and Alexander Grigoriev

Department of Quantitative Economics, Maastricht University,
P.O. Box 616, 6200 MD Maastricht, The Netherlands
{r.vanderzwaan,a.berger,a.grigoriev}@maastrichtuniversity.nl

Abstract. In this paper we consider the problem of finding a graph separator of a given size that decomposes the graph into the maximum number of connected components. We present the picture of the computational complexity and the approximability of this problem for several natural classes of graphs.

We first provide an overview of the hardness of approximation of this problem, which stems mainly from its close relation to the INDEPENDENT SET and to the MAXIMUM CLIQUE problem. Next, we show that the problem is solvable in polynomial time for interval graphs and graphs of bounded treewidth. We also show that MAXINUM COMPONENTS is fixed-parameter tractable on planar graphs with the size of the separator as the parameter. Our main contribution is the derivation of an efficient polynomial-time approximation scheme for the problem on planar graphs.

1 Introduction

In this paper we address the following graph separation problem, called MAXINUM COMPONENTS : Given an n-vertex graph $G = (V,E)$ and an integer $k \leq n$, find a *cut-set* $S \subseteq V$ of size at most k such that the subgraph of G induced by $V \setminus S$ has the maximum number of connected components. Throughout the paper we refer to this problem as MAXINUM COMPONENTS. We remark that the reverse of MAXINUM COMPONENTS : the number of cut-vertices needed to obtain at least ℓ components, is NP-hard to approximate within any factor. This is because any set of vertices separating the graph into at least ℓ components identifies an independent set of size at least ℓ [18].

There are many natural graph-theoretic problems where one has to find a graph separator of a given size satisfying certain properties. This type of problems is very generic and has attracted the attention of many researchers. Moreover, graph separation problems have many applications, e.g., in routing and network reliability. The edge variant of MAXINUM COMPONENTS, k-CUT, is well studied. Surprisingly enough, despite its very basic and quite generic setting, MAXINUM COMPONENTS did not receive much attention in the literature. Marx [27] has investigated the fixed-parameter (in)tractability of MAXINUM COMPONENTS among other related problems. This is in contrast to the amount of research concerning related problems such as BALANCED SEPARATORS [1,15,27], MAXIMUM k-CUT [24,28], MULTIWAY CUT [10,13,14,18,30], MULTICUT [11,12,17,20], VERTEX INTEGRITY [4,5], VALVE LOCATION [7], MINIMUM k-CUT [2,19,21,29], and BOUNDED FRAGMENTATION [22].

M. Ogihara and J. Tarui (Eds.): TAMC 2011, LNCS 6648, pp. 184–194, 2011.
© Springer-Verlag Berlin Heidelberg 2011

Our results. In this paper we present the picture of the computational complexity and the approximability of MAXINUM COMPONENTS for several natural classes of graphs.

In Section 2 we give a summary of complexity results for MAXINUM COMPONENTS, using standard reductions from INDEPENDENT SET and a reduction for split graphs already used by Marx [27]. Also, we show that MAXINUM COMPONENTS can be solved in polynomial time for interval graphs and for graphs of bounded treewidth.

Section 3 contains our main contribution, namely, an efficient polynomial-time approximation scheme (EPTAS) for MAXINUM COMPONENTS on planar graphs. The running time of our algorithm is $O\left(nk^2 \left(\frac{1}{\varepsilon^2} \right)^{O\left(\frac{1}{\varepsilon^2} \right)} \right)$, where k is the upper bound on the size of the cut-set. We also present a fixed-parameter tractable (FPT) algorithm for planar graphs with running time $O\left(nk^{O(k)} \right)$ in the same section. The FPT algorithm is needed for the EPTAS.

Notions and definitions. Let $G(V,E)$ be a simple undirected graph. We use $G[S]$ to denote the subgraph of G induced by $S \subseteq V$. $G \setminus S := G[V \setminus S]$. The neighborhood of $S \subset V$ in $G \setminus S$ is denoted $N(S)$ and $N[S] := N(S) \cup S$.

The set of connected components in G is represented by $\mathscr{C}(G)$, where each element in $\mathscr{C}(G)$, a set of vertices, induces a connected graph. Let $c(G,S)$ denote $|\mathscr{C}(G \setminus S)| - |\mathscr{C}(G)|$ i.e. the number of extra components obtained by removing S from G. Moreover, let *component number* $\kappa(G,k)$ of G be the maximum $c(G,S)$ over all subsets $S \subseteq V$ of size at most k. For the definitions of the tree decomposition and the treewidth we refer to the survey papers by Bodlaender, for example [6].

2   Complexity Results

In this section we summarize several complexity results for MAXINUM COMPONENTS. For brevity, the proofs of Theorems 1, 2, 3 and 4 are postponed to the full version of this paper. The following observation is straightforward.

Observation 1. *The independence number of any graph equals the maximum number of components that can be obtained by removing any number of vertices of the graph.*

The following theorem is now immediate from previous hardness results for INDEPENDENT SET [31, 16].

Theorem 1. *It is NP-hard to approximate* MAXINUM COMPONENTS *within a factor of $n^{1-\varepsilon}$ for every $\varepsilon > 0$. Moreover,* MAXINUM COMPONENTS *is NP-hard on 3-regular planar graphs.*

Using a reduction from MAXIMUM CLIQUE, Marx [27] has shown that MAXINUM COMPONENTS is W[1]-hard with parameter k even on split graphs. We use an almost identical reduction to reduce MAXINUM COMPONENTS to DENSE k-SUBGRAPH in an approximation preserving way. Together with an inapproximability result by Khot [25], this yields the following theorem.

Theorem 2. MAXINUM COMPONENTS *is NP-complete and W[1]-hard when restricted to split graphs. Moreover, it does not admit a PTAS for split graphs unless* $NP \subseteq \cap_{\varepsilon > 0} BPTIME(2^{n^\varepsilon})$.

We complete this section with two positive complexity results. First, we consider MAXINUM COMPONENTS on class of interval graphs.

Theorem 3. *Let* $G = (V, E)$ *be an interval graph and let* k *be an integer. Then,* $\kappa(G, k)$ *can be found in* $O(n^2 k)$ *time.*

Second, we present a dynamic programming algorithm for MAXINUM COMPONENTS on graphs of bounded treewidth. We assume that a tree decomposition is given as part of the input, there are several algorithms for finding the tree decomposition of a graph, for example see Bodlaender [8].

Theorem 4. *Let* $G = (V, E)$ *be a graph with a tree decomposition of width at most* w *and let* k *be an integer. Then,* $\kappa(G, k')$, *for all* $0 \le k' \le k$, *can be found in* $O(nk^2(w+1)^w)$ *time.*

3 An FPT Algorithm and an EPTAS for Planar Graphs

In this section we develop algorithms for the MAXINUM COMPONENTS problem on planar graphs. We assume that a planar embedding is given as a part of the input, otherwise we construct it in linear time using Hopcroft and Tarjan's algorithm [23]. An embedding of a graph G is called outerplanar (or 1-outerplanar) if all vertices of the graph are incident to the outer face. For a positive integer $p \ge 2$, an embedding of G is called p-outerplanar if the removal of all vertices incident to the outer face yields a $(p-1)$-outerplanar graph. Assume without loss of generality, that every connected component in G is incident to the outer face in the embedding. We denote the vertices incident to the outer face by *layer* L_1. Recursively, all vertices incident to the outer face of the graph obtained by removing layers L_1 to L_{j-1} form layer L_j ($j \ge 2$). The set of layers of a plane graph G is denoted as $\mathcal{L} := \{L_1, \ldots, L_z\}$. Given a plane graph G, $\mathcal{L}$ can be found in linear time by a slightly modified BFS-algorithm where the search procedure is done in face-by-face fashion. From now on, we assume that in all algorithms, $\mathcal{L}$ is given as a part of the input.

We will first show that the MAXINUM COMPONENTS on planar graphs is fixed-parameter tractable with respect to parameter k. We will make use of the following lemma.

Lemma 1 (Bodlaender, [6]). *Let* $G = (V, E)$ *be a p-outerplanar graph on n vertices. Then, a tree decomposition of width at most* $3p - 1$ *can be found in* $O(n)$ *time.*

We will first investigate the structure of the optimal solution in case a given family of layers is forbidden in the cut-set. Let $\mathcal{F} := \{L_{i_1}, \ldots, L_{i_r}\} \subset \mathcal{L}$ be a set of *forbidden layers*, i.e., for any $v \in L_{i_t}, 1 \le t \le r$, vertex v cannot be present in the cut-set. Without loss of generality we assume that $1 \le i_1 < \cdots < i_r \le z$. We denote by $\kappa(G, \mathcal{F}, k)$ the maximum of $c(G, S)$ over all $S \subset V$ such that $S \cap \bigcup_{t=1}^{r} L_{i_t} = \emptyset$ and $|S| \le k$. Additionally, we set $i_0 := 1$ and $i_{r+1} := z$. For $0 < t < r$, let $P_t := \bigcup_{l=i_{r-t}}^{i_{r-t+1}} L_l$, and $P_0 := \bigcup_{l=i_r}^{z} L_l$ and $P_r := \bigcup_{l=1}^{i_1} L_l$.

Lemma 2. *Let $G = (V, E)$ be a plane graph, let $k \geq 1$ be an integer and let $\mathcal{F}$ be a set of forbidden layers as denoted above. Then, there exist non-negative integers $k_0, ..., k_r$ such that the following holds:*

- $\kappa(G, \mathcal{F}, k) = \sum_{t=0}^{r} \kappa(G[P_t], \mathcal{F}, k_t),$

- $\sum_{t=0}^{r} k_t = k.$

Proof. Consider an arbitrary cut-set $S \subseteq V \setminus \bigcup_{t=0}^{r} L_{i_t}$.

We use induction in the number r of layers in $\mathcal{F}$ to prove that $c(G, S) = \sum_{t=0}^{r} c(G[P_t], S)$. Since S is an arbitrary cut-set, the claim of the lemma would follow immediately.

For the base case, consider the graph induced by P_r. This graph contains exactly one forbidden layer L_{i_r}. The claim in the base case trivially holds.

Let $P_j' := \bigcup_{t=r-j}^{r} P_t$, $0 \leq j \leq r$. Suppose $c(G[P_j'], S) = \sum_{t=r-j}^{r} c(G[P_t], S)$ holds for some j, $0 \leq j \leq r$, then we claim that

$$c(G[P_{j+1}'], S) = c(G[P_j'], S) + c(G[P_{j+1}], S). \tag{1}$$

Let us remind that by definition $c(G, S)$ is the number of extra components in G after removal of S. Since $G[P_j']$ and $G[P_{j+1}]$ are two intersecting subgraphs, $c(G[P_j' \cup P_{j+1}], S) \leq c(G[P_j'], S) + c(G[P_{j+1}], S)$. We denote the *overcount* of adding up $c(G[P_{j+1} \setminus S])$ and $c(G[P_j' \setminus S])$, the total amount that each component is counted more than once summed over all components, by

$$\Delta := c(G[P_{j+1} \setminus S]) + c(G[P_j' \setminus S]) - c(G[P_j' \cup P_{j+1} \setminus S]). \tag{2}$$

Note that only components that include vertices from L_{i_j} can be counted multiple times. Assume $\Delta = c(G[L_{i_j}])$. Then,

$$c(G[P_{j+1}'], S) = c(G[P_j' \cup P_{j+1}], S) \tag{3}$$

$$= c(G[P_{j+1} \cup P_j' \setminus S]) - c(G[P_{j+1} \cup P_j']) \tag{4}$$

$$= c(G[P_{j+1} \setminus S]) + c(G[P_j' \setminus S]) - \Delta - c(G[P_{j+1} \cup P_j']) \tag{5}$$

$$= c(G[P_{j+1} \setminus S]) + c(G[P_j' \setminus S]) - c(G[L_{i_j}]) - c(G[P_{j+1} \cup P_j']) \tag{6}$$

$$= c(G[P_{j+1} \setminus S]) - c(G[P_{j+1}]) + c(G[P_j' \setminus S]) - c(G[P_j']) \tag{7}$$

$$= c(G[P_{j+1}], S) + c(G[P_j'], S). \tag{8}$$

Here, Equation 3 is true by definition of the set P_{j+1}', Equation 4 follows from the definition of $c(\cdot, \cdot)$, Equation 5 is by Equation 2. By assumption that $\Delta = c(G[L_{i_j}])$, we derive Equation 6. By construction of layers in planar graphs, $c(G[P_j']) = c(G[L_{i_j}])$ and $c(G[P_{j+1}]) = c(G[P_{j+1} \cup P_j'])$, which yields Equation 7. Finally, Equation 8 follows again from the definition of $c(\cdot, \cdot)$.

Thus, to prove Equation 1 it only remains to verify the assumption $\Delta = c(G[L_{i_j}])$. Consider a component $C \in \mathcal{C}(G[P_j' \cup P_{j+1} \setminus S])$ that intersects L_{i_j}. Let $A := G[P_{j+1} \setminus$

$S \cap C$]. Now, we intend to prove that A is connected. For a contradiction, suppose there are at least two components in A, say, A' and A''. By definition, $G[C]$ is connected in $G[P'_j \cup P_{j+1}]$. However, A' and A'' are not connected in $G[P_{j+1}]$. Thus, there must be a path connecting A' and A'' through $G[P'_j \cap C]$ and not $G[P_{j+1} \cap C]$. On the other hand, by construction of layers, every path from A' to A'' through $G[P'_j \cap C]$ must go through L_{i_j}, from which we can shortcut the paths, such that the modified path only goes through $G[P_{j+1} \cap C]$; a contradiction.

Component C can be covered by *one* component from $G[P_{j+1} \cap C \setminus S]$ and several components from $G[P'_j \cap C \setminus S]$. Then, the number of times C is counted too often is precisely the number of components from $G[P'_j \cap C \setminus S]$ that cover C. Further, at most one component from $\mathscr{C}(P'_j \setminus S)$ contains vertices from any component induced by the forbidden layer L_{i_j}. Therefore, $\Delta \leq c(G[L_{i_j}])$. Moreover, from the fact that every component induced by L_{i_j} is contained in some component from $\mathscr{C}(P'_j \setminus S)$, it follows that $\Delta \geq c(G[L_{i_j}])$. Therefore, $\Delta = c(G[L_{i_j}])$.   □

Algorithm 1 is an FPT algorithm for MAXINUM COMPONENTS on planar graphs with respect to parameter k, the pseudo-code is given in Algorithm 1. The main structural property that we use in the algorithm is that among any $k + 1$ consecutive layers there must be one layer without any cut-vertices. Similar ideas are driving all Baker's type polynomial-time approximation schemes [3] for combinatorial problems on planar graphs. By Lemma 2 we can "partition" the graph into smaller pieces, if we know the set of forbidden layers not containing any cut-vertices.

Theorem 5. *Let $G = (V, E)$ be a planar graph and let k be an integer. Then, Algorithm 1 finds $\kappa(G, k)$ in $O\left(nk^{O(k)}\right)$ time algorithm.*

Proof. Given a planar graph $G = (V, E)$ and integer k, let $G[i, j]$ be the graph induced by $\bigcup_{i \leq t \leq j} L_t$. Algorithm 1 consists of two parts. First, we decompose the graph into smaller subgraphs of bounded treewidth. On these subgraphs we run the algorithm from Theorem 4. Using Lemma 2, in the second phase we apply dynamic programming to combine the obtained solutions on the subgraphs, and to return the overall optimal solution. Correctness follows from the following inductive argument. Assume we know $\kappa(G[1, i], k')$ for some i and k', then one of the layers $L_{i+1}, ..., L_{i+k+1}$ does not contain cut-vertices. Therefore, we can extend the solution for $G[1, i]$ to a solutions for $G[1, j]$ where $j \in [i + 1, i + k + 1]$. We can combine these solutions by Lemma 2.

The first part (lines 2 to 7 in Algorithm 1) takes $O(nk^3 k^{O(k)})$ time to calculate $\kappa(G[q, q + d + 1], \mathscr{F}, k')$, as in Algorithm 1. The union of all these small pieces contains at most $2n$ vertices, and the treewidth of each $G[q, q + d + 1]$ is bounded by $3(k + 2) - 1$ and can be found in linear time by Lemma 1. Therefore, by applying the algorithm in Theorem 4 we obtain the optimal solution in each small piece in $O\left(nk^{O(k)}\right)$ time. The second part (lines 9 to 13) combines all parts by dynamic programming. This takes $O(nk^2)$ time. The total running time is $O\left(nk^{O(k)}\right)$.   □

We now move to the presentation of the efficient polynomial-time approximation scheme (EPTAS) for MAXINUM COMPONENTS on planar graphs. Let us remind that an EPTAS (for a maximization problem) is a polynomial-time approximation algorithm

---

Algorithm 1. FPT Algorithm for Planar Graphs

Input: Plane graph $G = (V, E)$, layers $\mathcal{L}$, and $k \geq 0$

1: /* Split the graph into small pieces, and use Theorem 4 on the small pieces */
2: **for** $2 \leq d \leq k+1$ **do**
3:     **for** $0 \leq i < d$ **do**
4:         $\mathcal{F} := \{L_j \in \mathcal{L} \,|\, j \equiv i \bmod d\}$.
5:         For all $0 \leq k' \leq k$, $0 \leq q \leq |\mathcal{L}|$, $q \equiv i \bmod d$, calculate $\kappa(G[q, q+d+1], \mathcal{F}, k')$.
6:     **end for**
7: **end for**
8: /* Use dynamic programming to combine all small pieces. */
9: **for** $k+1 \leq i \leq |\mathcal{L}|$ **do**
10:     **for** $0 \leq k' \leq k$ **do**
11:

$$\kappa(G[1, i], L_i, k') :=$$

$$\max_{j \leq k+1} \max_{k_1 + k_2 = k'} \left\{ \kappa(G[1, i-j], L_{i-j}, k_1) + \kappa(G[i-j, i], L_{i-j} \cup L_i, k_2) \right\}.$$

12:     **end for**
13: **end for**
14: $l := |\mathcal{L}|$
15: **if** $l \leq k$ **then**
16:     **return** $\kappa(G, \emptyset, k)$
17: **else**
18:     $\kappa(G, \emptyset, k) := \max_{0 \leq j \leq k+1} \max_{k_1 + k_2 = k} \left\{ \kappa(G[1, z-j], L_{l-j}, k_1) + \kappa(G[l-j, z], L_{l-j}, k_2) \right\}.$
19: **end if**
20: **return** $\kappa(G, k)$

---

that for any given $\varepsilon > 0$ returns a solution of value at least $(1 - \varepsilon)$ times the optimum in time $poly(n) f(\varepsilon)$, where $poly(n)$ only depends on n and not on ε, and $f(\cdot)$ is some function only depending on ε. For a graph G and a cut-set S, the *degree* of a component $C \in \mathcal{C}(G \setminus S)$ is $|N(C) \cap S|$.

Our approach is as follows. Using a threshold parameter d, we distinguish between components of low and high degree. Then, we analyze the high and low degree components separately. To approximate the maximum number of low degree components we make use of the decomposition Lemma 2. In combination with the Baker's type layer-shifting method [3], we arrive to an EPTAS for the number of low degree components. Secondly, in Lemma 4 we prove that in planar graphs there are not many components of high degree. Thus, by choosing the proper threshold parameter d we ensure that the number of high degree components is at most $\varepsilon \cdot \kappa(G, k)$. Combining both elements, we obtain an EPTAS for MAXINUM COMPONENTS.

In order to proceed further, we need several definitions. For a given cut-set S and an integer $d > 0$, we say a component from $\mathcal{C}(G \setminus S)$ is of *low degree* if it has at most d neighbors in S. Let $c_{\leq d}(G, S)$ be the number of low degree components. Similarly, a component is of *high degree* if it has more than d neighbors in S. We let $c_{>d}(G, S)$ be the number of high degree components. By definition, $c(G, S) := c_{\leq d}(G, S) + c_{>d}(G, S)$. We define $\kappa_{\leq d}(G, k)$ to be the maximum of $c_{\leq d}(S)$ over all $S \subset V$ of cardinality at most k. Likewise, we define $\kappa_{>d}$. Clearly,

$$\kappa(G,k) \le \kappa_{\le d}(G,k) + \kappa_{>d}(G,k). \tag{9}$$

First, we approximate the number of low degree components $\kappa_{\le d}(G,k)$. Following the logic of the layer-shifting method of Baker, for some appropriate number w, we forbid every w-th layer to be in the cut-set S. Then, using Lemma 2 we can combine partial solutions obtained on the subgraphs between the forbidden layers. Choosing an appropriate shift of the forbidden layers, we guarantee that the combined solution contains at least $(1 - \varepsilon)\kappa_{\le d}(G,k)$ components.

Algorithm 2. EPTAS for the maximum number of low degree components in planar graphs

---

Input: Plane graph $G = (V,E)$, layers $\mathscr{L}$, $k \ge 2$, $d \ge 5$, and $\varepsilon > 0$
1: /* Split the graph into small pieces, and use Theorem 4 on the small pieces */
2: $w := d \cdot \left\lceil \frac{1}{\varepsilon} \right\rceil$
3: **for** $0 \le s < w$ **do**
4:      $\mathscr{F} := \{L_j \in \mathscr{L} \mid j \equiv s \bmod w\}$.
5:      For all $0 \le k' \le k$, $0 \le q \le |\mathscr{L}|$, $q \equiv s \bmod w$, calculate $\kappa(G[q, q+w], \mathscr{F}, k')$.
6: **end for**
7: /* Use dynamic programming to combine all small pieces. */
8: **for** $0 \le s < w$ **do**
9:      $\mathscr{F}_s := \{L_j \in \mathscr{L} \mid j \equiv s \bmod w\}$.
10:      **for** $k' \le k$ **do**
11:          Initialize $t_s(0, k') := \kappa(G[0, s], k')$
12:      **end for**
13:      **for** $1 \le i \le \lceil |\mathscr{L}|/w \rceil$ **do**
14:          **for** $k' \le k$ **do**
15:              $t_s(i, k') := \displaystyle\max_{k_1 + k_2 = k'} t_s(i-1, k_1) + \kappa(G[(i-1)w+s, iw+s], \mathscr{F}_s, k_2)$
16:          **end for**
17:      **end for**
18:      $\kappa(G, \mathscr{F}_s, k) := t_s(\lceil |\mathscr{L}|/w \rceil, k)$.
19: **end for**
20: **return** $\displaystyle\max_{0 \le s < w} \kappa(G, \mathscr{F}_s, k)$

Lemma 3. *Let* $G = (V,E)$ *be a planar graph, let* $k \ge 2$ *and* $d \ge 5$ *be integers and let* $0 < \varepsilon < 1$ *be a rational number. Then, Algorithm 2 is an* $O\left(nk^2 \left(\frac{d}{\varepsilon}\right)^{O\left(\frac{d}{\varepsilon}\right)}\right)$ *-time algorithm that approximates* $\kappa_{\le d}(G,k)$ *within a factor of* $(1 - \varepsilon)$.

Proof. Let the distance between two consecutive forbidden layers be $w := d\left\lceil \frac{1}{\varepsilon} \right\rceil$. For any (integer) shift s, $0 \le s < w$, let $V_s := \bigcup\{L_i \in \mathscr{L}(G) \mid i \equiv s \bmod w\}$. Let $S^{OPT} \subset V$ be the optimal cut-set. We say that two distinct components in $G[V \setminus S^{OPT}]$ *break by* V_s if they become connected in $G[V \setminus (S^{OPT} \setminus V_s)]$. The solution that our algorithm returns is therefore at least $c_{\le d}(G, S^{OPT} \setminus V_s)$.

Let b_s be the number of components in $G[V \setminus S^{OPT}]$ that break by V_s. Then, by definition of b_s

$$c_{\leq d}(G, S^{OPT} \setminus V_s) \geq c_{\leq d}(G, S^{OPT}) - b_s. \qquad (10)$$

Observe that a component can only break by V_s if it has a neighboring vertex in $S^{OPT} \cap V_s$. Since we consider only components of degree at most d, there are at most d values $i_1, \ldots, i_d$ such that a component is broken by $V_{i_1}, \ldots, V_{i_d}$. Therefore,

$$\sum_{s=0}^{w-1} b_s \leq d \cdot \kappa_{\leq d}(G, k). \qquad (11)$$

Let $b_{\min} := \min_{s=0,\ldots,w-1} b_s$. From Equation 11 and standard averaging argument

$$b_{\min} \leq \frac{d \cdot \kappa_{\leq d}(G, k)}{w} = \frac{d \cdot \kappa_{\leq d}(G, k)}{d \cdot \lceil \frac{1}{\varepsilon} \rceil} \leq \varepsilon \cdot \kappa_{\leq d}(G, k). \qquad (12)$$

Combining Equations 10 and 12 we derive that Algorithm 2 returns a cut-set splitting the graph in at least $(1 - \varepsilon)\kappa_{\leq d}(G, k)$ components.

It remains to analyze the running time of Algorithm 2. In lines (2-6), we invoke the algorithm from Theorem 4 exactly w times. By Lemma 1, the treewidth of the considered subgraphs is at most $3w - 1 = O(\frac{d}{\varepsilon})$. Therefore, lines (2-6) take $O\left(w \cdot nk^2 \left(\frac{d}{\varepsilon}\right)^{O(\frac{d}{\varepsilon})} \right)$ time. Lines (8-18) take $O(nwk^2)$ time, making the total running time $O\left(nk^2 \left(\frac{d}{\varepsilon}\right)^{O(\frac{d}{\varepsilon})} \right)$.   □

Now, we estimate the number of the high degree components $\kappa_{>d}(G, k)$.

Lemma 4. *Let $G = (V, E)$ be a planar graph, and let $k \geq 2$ and $d \geq 3$ be integers. Then, $\kappa_{>d}(G, k) \leq \frac{2k-4}{d-2}$.*

Proof. Let m be the number of edges in G, and let f be the number of faces in an embedding of G. Let $S \subset V$ be an arbitrary cut-set. Consider a minor $H = (V', E')$ obtained from G by:

1. Contracting all pairs of adjacent vertices that are not in S;
2. Removing all vertices of degree less than d that are not in the cut-set;
3. Removing all edges between cut-set vertices;
4. For each component T of degree more than d, remove arbitrary edges between T and S such that the degree of T becomes d.

In the obtained minor H, each vertex not in S is called a *component vertex*, and each component in $G[V \setminus S]$ of degree at least d has a corresponding component vertex in H. By construction, H is bipartite, all component vertices have degree d. Therefore, $n' = |V'| = |S| + c_{>d}(G, S)$ and $m' = |E'| = d \cdot c_{>d}(G, S)$. Denote by f' the number of faces in a planar embedding of H. As H is bipartite every face is incident to at least four edges, and we can conclude that $f' \leq m'/2$. By Euler's formula,

$$2 = n' - m' + f' \leq k + c_{>d}(G, S) - d \cdot c_{>d}(G, S) + d \cdot c_{>d}(G, S)/2.$$

Thus, $\kappa_{>d}(G, k) \leq \frac{2k-4}{d-2}$.   □

The final ingredient needed for our main result is the following lemma. This lemma provides a simple lower bound on the total number of components $\kappa(G,k)$.

Lemma 5. *Let $G = (V,E)$ be a planar graph and let $k \leq \frac{5}{6}n$. Then, $\lfloor \frac{k}{5} \rfloor \leq \kappa(G,k)$.*

Proof. It is well known that any simple planar graph has a vertex $v \in V$ of degree at most 5. Set an initial cut-set $S = \emptyset$. Find a non-isolated vertex v of the lowest degree in $G[V \setminus S]$. Consider the neighborhood $N(v)$ of v in this graph. If $|S \cup N(v)| \leq k$, then define $S := S \cup N(v)$. Since $k \leq \frac{5}{6}n$, e can make $\lfloor \frac{k}{5} \rfloor$ steps of this procedure, obtaining a cut-set S of at most k vertices while the graph $G[V \setminus S]$ has at least $\lfloor \frac{k}{5} \rfloor$ components.   □

Now we are ready to present our main theorem, an efficient polynomial-time approximation scheme for MAXINUM COMPONENTS.

Theorem 6. *Let $G = (V,E)$ be a planar graph, let $k \geq 2$ be an integer and let $0 < \varepsilon < 1$ be a rational number. Then, there is a $O\left(nk^2 \left(\frac{1}{\varepsilon^2} \right)^{O\left(\frac{1}{\varepsilon^2} \right)} \right)$ time algorithm that approximates $\kappa(G,k)$ within a factor $(1 - \varepsilon)$.*

Proof. Define $\delta := \frac{\varepsilon}{2 - \varepsilon}$. If $k < 55$ we apply the FPT algorithm in Theorem 5 and we find the optimal solution in linear time.

For $55 \leq k$, we observe the following. If $S \subset V$ is a cut-set of cardinality $\lfloor \frac{5}{6}n \rfloor + j \leq k$ where $j > 0$, then $c(G,S) \leq n - \lfloor \frac{5}{6}n \rfloor - j = \lceil \frac{1}{6}n \rceil - j$. By Lemma 5, $\left\lfloor \frac{\lfloor \frac{5}{6}n \rfloor}{5} \right\rfloor = \lfloor \frac{1}{6}n \rfloor \leq \kappa(G, \frac{5}{6}n)$. Hence, we can assume $55 \leq k \leq \frac{5}{6}n$.

Define $d := \lceil \frac{24}{\delta} \rceil$. Consider two solutions to MAXINUM COMPONENTS. The first cut-set is generated by the algorithm in Lemma 5, and the second cut-set is returned by Algorithm 2 with parameters d and δ. From these two cut-sets, we choose the cut-set that splits the graph into the most components. Let the resulting number of components be denoted by ALG. By Lemmas 3 and 5, and by definition of ALG,

$$ALG \geq \max \left\{ (1 - \delta)\kappa_{\leq d}(G,k), \left\lfloor \frac{k}{5} \right\rfloor \right\}. \tag{13}$$

Thus, from Lemma 4 and Equation 13 we obtain

$$\kappa_{>d}(G,k) \leq \frac{2k - 4}{d - 2} \leq \frac{2k}{22}\delta \leq \left\lfloor \frac{k}{5} \right\rfloor \delta \leq \delta ALG. \tag{14}$$

Combining Equations 9, 13 and 14, we derive

$$(1 - \varepsilon)\kappa(G,k) \leq (1 - \varepsilon)\kappa_{\leq d}(G,k) + (1 - \varepsilon)\kappa_{>d}(G,k)$$
$$\leq ALG. \tag{15}$$

The running time for obtaining the solution in Lemma 3 clearly dominates the running time for obtaining the solution in Lemma 5. Therefore, the time requirement in the theorem will be the same as in Lemma 3, making the total running time

$$nk^2 \left(\frac{d}{\delta} \right)^{O\left(\frac{d}{\delta} \right)} = nk^2 \left(\frac{\lceil \frac{24}{\delta} \rceil}{\delta} \right)^{O\left(\frac{\lceil \frac{24}{\delta} \rceil}{\delta} \right)} = nk^2 \left(\frac{1}{\varepsilon^2} \right)^{O\left(\frac{1}{\varepsilon^2} \right)}.$$

□

4   Conclusion

The main result of this paper was an EPTAS for MAXINUM COMPONENTS, a natural graph separation problem which surprisingly did not receive much attention in the literature. Further, we identified two graph classes for which MAXINUM COMPONENTS is polynomially time solvable, namely interval graphs and graphs of bounded treewidth. However, MAXINUM COMPONENTS is W[1]-hard on split graphs and does not allow a PTAS under a reasonable complexity assumption.

References

1. Alon, N., Seymour, P.D., Thomas, R.: A separator theorem for graphs with an excluded minor and its applications. In: Proceedings of the Twenty-Second Annual ACM Symposium on Theory of Computing, pp. 293–299. ACM, New York (1990)
2. Arora, S., Karger, D., Karpinski, M.: Polynomial time approximation schemes for dense instances of np-hard problems. In: Proceedings of the Twenty-Seventh Annual ACM Symposium on Theory of Computing, pp. 284–293 (1995)
3. Baker, B.S.: Approximation algorithms for NP-complete problems on planar graphs. Journal of the ACM 41(1), 153–180 (1994)
4. Barefoot, C., Entringer, R., Swart, H.: Integrity of trees and the diameter of a graphs. Congressus Numerantium (58), 103–114 (1987)
5. Barefoot, C., Entringer, R., Swart, H.: Vulnerability in graphs:a comparative survey. Journal of Combinatorial Mathematics and Combinatorial Computing (1), 12–22 (1987)
6. Bodlaender, H.: A partial k-arboretum of graphs with bounded treewidth. Theoretical Computer Science 209(1-2), 1–45 (1998)
7. Bodlaender, H., Hendriks, A., Grigoriev, A., Grigorieva, N.: The valve location problem in simple network topologies. INFORMS Journal on Computing (to appear)
8. Bodlaender, H.L.: A linear-time algorithm for finding tree-decompositions of small treewidth. SIAM Journal on Computing 25(6), 1305–1317 (1996)
9. Booth, K., Lueker, G.: Testing for the consecutive ones property, interval graphs, and graph planarity using pq-tree algorithms. Journal of Computer and System Sciences 13(3), 335–379 (1976)
10. Chen, J., Liu, Y., Lu, S.: An Improved Parameterized Algorithm for the Minimum Node Multiway Cut Problem. Algorithmica 55(1), 1–13 (2009)
11. Costa, M., Létocart, L., Roupin, F.: Minimal multicut and maximal integer multiflow: A survey. European Journal of Operational Research 162, 55–69 (2005)
12. Călinescu, G., Fernandes, C., Reed, B.: Multicuts in unweighted graphs and digraphs with bounded degree and bounded tree-width. Journal of Algorithms 48(4), 333–359 (2003)
13. Călinescu, G., Karloff, H., Rabani, Y.: An improved approximation algorithm for multiway cut. Journal of Computer and System Sciences 60(3), 564–574 (2000)
14. Dahlhaus, E., Johnson, D.S., Papadimitriou, C.H., Seymour, P.D., Yannakakis, M.: The complexity of multiterminal cuts. SIAM Journal on Computing 23, 864–894 (1994)
15. Feige, U., Mahdian, M.: Finding small balanced separators. In: Kleinberg, J. (ed.) Proceedings of the 38th Annual ACM Symposium on Theory of Computing, STOC, Seattle, WA, USA, May 21-23, pp. 375–384 (2006)
16. Garey, M., Johnson, D.: Computers and Intractability: A Guide to the Theory of NP-Completeness. W. H. Freeman, New York (1979)
17. Garg, N., Vazirani, V., Yannakakis, M.: Primal-dual approximation algorithms for integral flow and multicut in trees. Algorithmica 18, 3–20 (1997)

18. Garg, N., Vazirani, V., Yannakakis, M.: Multiway cuts in node weighted graphs. Journal of Algorithms 50(1), 49–61 (2004)
19. Goldschmidt, O., Hochbaum, D.: Polynomial algorithm for the k-cut problem. In: 29th Annual Symposium on Foundations of Computer Science, pp. 24–26, 444–451 (1988)
20. Guo, J., Hüffner, F., Kenar, E., Niedermeier, R., Uhlmann, J.: Complexity and exact algorithms for vertex multicut in interval and bounded treewidth graphs. European Journal of Operational Research 186(2), 542–555 (2008)
21. Guttmann-Beck, N., Hassin, R.: Approximation algorithms for minimum -cut. Algorithmica 27(2), 198–207 (2000)
22. Hajiaghayi, M.T., Hajiaghayi, M.: A note on the bounded fragmentation property and its applications in network reliability. European Journal of Combinatorics 24(7), 891–896 (2003)
23. Hopcroft, J., Tarjan, R.: Efficient planarity testing. J. ACM 21(4), 549–568 (1974)
24. Kann, V., Khanna, S., Lagergren, J., Panconesi, A.: On the hardness of approximating max k-cut and its dual. Chicago Journal of Theoretical Computer Science - CJTCS-1997-2 (1997)
25. Khot, S.: Ruling out PTAS for graph min-bisection, dense k-subgraph, and bipartite clique. SIAM Journal on Computing 36(4), 1025–1071 (2006)
26. Kloks, T.: Treewidth, Computations and Approximations. LNCS, vol. 842. Springer, Heidelberg (1994)
27. Marx, D.: Parameterized graph separation problems. Theoretical Computer Science 351, 394–406 (2006)
28. Papadimitriou, C., Yannakakis, M.: Optimization, approximation, and complexity classes. Journal of Computer and System Sciences 43(3), 425–440 (1991)
29. Saran, H., Vazirani, V.: Finding k cuts within twice the optimal. SIAM Journal on Computing 24(1), 101–108 (1995)
30. Xiao, M.: Simple and Improved Parameterized Algorithms for Multiterminal Cuts. Theory of Computing Systems 46(7), 723–736 (2010)
31. Zuckerman, D.: Linear degree extractors and the inapproximability of max clique and chromatic number. In: Kleinberg, J.M. (ed.) Proceedings of the Thirty-Eight Annual ACM Symposium on Theory of Computing, pp. 681–690. ACM, New York (2006)

Succinct Dynamic Cardinal Trees with Constant Time Operations for Small Alphabet

Pooya Davoodi[1] and Satti Srinivasa Rao[2]

[1] MADALGO*, Department of Computer Science,
Aarhus University, IT Parken, Åbogade 34,
DK-8200 Aarhus N, Denmark
pdavoodi@cs.au.dk
[2] School of Computer Science and Engineering,
Seoul National University, 599 Gwanakro,
Gwanak-Gu, Seoul 151-744,
Republic of Korea
ssrao@cse.snu.ac.kr

Abstract. A k-ary cardinal tree is a rooted tree in which each node has at most k children, and each edge is labeled with a symbol from the alphabet $\{1, \ldots, k\}$. We present a succinct representation for k-ary cardinal trees of n nodes where $k = O(polylog(n))$. Our data structure requires $2n + n \log k + o(n \log k)$ bits and performs the following operations in $O(1)$ time: parent, child(i) label-child(α), degree, subtree-size, preorder, is-ancestor(x), insert-leaf(α), delete-leaf(α). The update times are amortized. The space is close to the information theoretic lower bound. The operations are performed in the course of traversing the tree. This improves the succinct dynamic k-ary cardinal trees representation of Arroyuelo [1] for small alphabet, by speeding up both the query time of $O(\log \log n)$, and the update time of $O((\log \log n)^2 / \log \log \log n)$ to $O(1)$, solving an open problem in [1].

1 Introduction

In this paper, we present a succinct representation for dynamic k-ary cardinal trees, i.e., rooted trees in which each node has at most k children and each edge is labeled by a symbol from the alphabet $\{1, \ldots, k\}$, for a fixed k. They are also known as tries with degree k. We consider the case where for a k-ary cardinal tree of n nodes, the size of the alphabet is small, in particular $k = (\log n)^{O(1)}$.

A succinct data structure is a representation of an input which uses an amount of space close to the information theoretic lower bound, and supports the required operations efficiently. The information theoretic lower bound for representing a k-ary cardinal tree of n nodes is computed by taking the logarithm of the number of distinct such trees, i.e., $\log \mathcal{C}(n, k) = \log(\binom{kn+1}{n}/(kn + 1)) \approx 2n + n \log k - o(n + \log k)$ bits [1,9]. The required operations are the following:

---

* Center for Massive Data Algorithmics, a Center of the Danish National Research Foundation.

M. Ogihara and J. Tarui (Eds.): TAMC 2011, LNCS 6648, pp. 195–205, 2011.
© Springer-Verlag Berlin Heidelberg 2011

parent: the parent of the current node.

child(i): the i-th child of the current node.

label-child(α): the child of the current node such that the in-between edge is labeled by α.

degree: the number of children of the current node.

subtree-size: the total number of nodes in the subtree rooted at the current node.

is-ancestor(x): true if the current node is an ancestor of a node x; otherwise false.

preorder: the rank of the current node in the preorder traversal of the tree.

insert-leaf(α): insert a new leaf as a child of the current node with an edge labeled by α.

delete-leaf(α): delete the child of the current node with the in-between edge labeled by α (assuming that the child is a leaf).

Our data structure supports the operations in the course of traversing the tree, i.e., a traversal of the tree starts from the root, moves through the tree by performing the navigational operations on the current node, and ends at the root. All the operations have to be performed on the current node of the traversal. This is the same model that was used in [12,15,1]. There are applications where this assumption holds, e.g., constructing Lempel-Ziv indexes which is used for dynamic compressed full-text indexes; and constructing suffix trees, if we supplement the data structure with satellite data. If we do not restrict the operations to be only performed on the current node (if we allow them to be performed on any arbitrary node), Farzan and Munro [6] showed an amortized lower bound of $\Omega(\log n / \log \log n)$ for child, subtree-size, insert-leaf, and delete-leaf.

We use the unit-cost RAM model with word size $w = \Theta(\log n)$ bits.

Previous Work. For static k-ary cardinal trees of n nodes, Benoit et al. [2] gave a representation that requires $2n + n \log k + o(n) + O(\log \log k)$ bits, and supports the navigational operations and queries in $O(1)$ time. The space bound of their structure is $\log \mathcal{C}(n, k) + \Omega(n)$ bits, as k grows. Raman et al. [14] improved the space to $\log \mathcal{C}(n, k) + o(n) + O(\log \log k)$ bits, while supporting all the operations except subtree-size in $O(1)$ time. Recently, a representation with the same space that also supports subtree-size in $O(1)$ time was given in [7].

For the dynamic k-ary cardinal trees, when $k = 2$, i.e., for dynamic binary trees, Munro et al. [12] gave the first representation that uses $2n + o(n)$ bits. This representation, in the course of traversing the tree, supports navigational operations and queries in $O(1)$ time and updates in $O(\log^2 n)$ amortized time. Their structure can also support accessing a b-bit satellite data associated with a node in $O(1)$ time, where $b = \Theta(\log n)$. If $b = O(1)$, they achieve $O(\log n)$ amortized update time, and if no satellite data is associated with the nodes, they obtain $O(\log \log n)$ amortized update time. For $b = O(\log n)$, Raman and Rao [15] improved the update time to amortized $O((\log \log n)^{1+\epsilon})$ while supporting the navigations and queries in $O(1)$ time, in the course of traversing the tree. They also showed how to store the satellite data in $bn + o(n)$ bits. Indeed, the total space of their structure is $2n + bn + o(n)$ bits. More recently, Farzan and Munro [6] proposed the *finger-update* model which is stronger than the traversal pattern that is used in [12,15,1] and this paper as well. In the finger-update model, only the update operations are restricted to be performed on the current node of the traversal (indeed, finger-update is the current node), and all the other operations are allowed to be performed on any node at any time. For $b = O(\log n)$ and for ordinal trees which are the generalized binary trees where there is an order between the children of the nodes, Farzan and Munro [6] presented a data structure that supports all the queries in constant time and updates in constant amortized time. But their structure uses $2n + bn + o(bn)$ bits which is worse than that of [15].

The succinct representation of dynamic k-ary cardinal trees was posed as an open problem in [12]. In 1993, Darragh et al. [5] presented a compact representation of cardinal trees that uses $6n + n\lceil \log k \rceil$ bits of space and achieves $O(1)$ expected time for the operations. Recently, Arroyuelo [1] presented a data structure for this problem that uses $2n + n \log k + o(n \log k)$ bits of space, and supports navigational operations and queries in $O(\log k + \log \log n)$ time and updates in $O((\log k + \log \log n)(1 + (\log k)/(\log(\log k + \log \log n))))$ amortized time. When $k = (\log n)^{O(1)}$, his data structure achieves $O(\log \log n)$ query time and $O((\log \log n)^2 / \log \log \log n)$ amortized update time. Improving this was posed as an open problem by Arroyuelo [1]. We address this problem by presenting a data structure that uses $2n + n \log k + o(n \log k)$ bits and performs the navigation and query operations in $O(1)$ time, and the update operations in $O(1)$ amortized time. Associating satellite data with the nodes is supported by neither our data structure nor the one in [1]. The following theorem states our result.

Theorem 1. *For a k-ary cardinal tree with n nodes, there exists a dynamic data structure of size $2n + n \log k + o(n \log k)$ bits, where $k = (\log n)^{O(1)}$. This data structure supports the operations* parent, child, label-child, degree, subtree-size, preorder, *and* is-ancestor *in $O(1)$ time, and supports* insert-leaf *and* delete-leaf *in $O(1)$ amortized time.*

2 Preliminaries

Dynamic Arrays. A dynamic array [15] is a structure that supports accessing, inserting, and deleting elements in arrays efficiently with a small memory overhead.

Lemma 1. (Dynamic Arrays [15,16]) *There exists a data structure to represent an array of $\ell = w^{O(1)}$ elements, each of size $r = O(w)$ bits, using $\ell r + O(k \log \ell)$ bits, for any parameter $k \leq \ell$. This data structure supports accessing the element of the array in a given index in $O(1)$ time, and inserting/deleting an element in/from a given index in $O(1 + \ell r / kw)$ amortized time. The data structure requires a precomputed table of size $O(2^{\epsilon w})$ bits for any fixed $\epsilon > 0$.*

Searchable Partial Sums. In the searchable partial sums problem for an array A of m numbers from the range $[0, \ldots, k-1]$, we have to maintain A under the following operations:

sum(i): return the value $\sum_{j=1}^{i} A[j]$,

update(i, δ): set $A[i] = A[i] + \delta$, assuming that $A[i] + \delta < k$, and δ is less than a certain fixed number,

search(i): return the smallest j such that sum(j) $\geq i$.

This problem has been considered for different ranges of m and k [13,10]. But we are only interested in solving this problem for small m and k. Raman and Rao [13] gave a data structure that solves the problem for $m = w^{\epsilon}$ and $k \leq w$, for any fixed $0 \leq \epsilon < 1$. Their data structure achieves $O(1)$ time for all the operations and uses $O(mw)$ bits of space. In the following, we show that when both m and k are $O(w^c)$ for a constant $c > 0$, we can obtain a data structure with $O(1)$ time for all the operations that uses $m \log k + o(m \log k)$ bits of space.

Lemma 2. *For any integer $n < 2^w$, there exists a searchable partial sums structure to represent an array of m elements from the range $[0, \ldots, k-1]$, using $m \log k + o(m \log k)$ bits and a precomputed table of size $o(n)$ bits, where m and k are $(\log n)^{O(1)}$. This data structure supports the operations* sum, update, *and* search *in $O(1)$ time.*

Proof. We pack every $w/\log k$ elements of the array into a word. Within each word, every b numbers denote a chunk, where $b = \log^{1/4} n$. Within each chunk, the operations can be supported in $O(1)$ time using a precomputed table of size $o(n)$ bits. The space usage to store all the chunks is $m \log k + o(m \log k)$ bits.

Now, we make a B-tree with branching factor at most b. Each leaf of the B-tree stores a pointer to one of the chunks such that scanning the chunks of the leaves from the left of the B-tree to the right gives the original array. The number of leaves is m/b and the depth of the B-tree is $O(1)$. At each internal node u, we maintain two arrays of length b. The i-th element of the first array maintains the sum of all the elements in the chunks that are descendants of the i-th child of u. The i-th element of the second array maintains the number of all the elements in the chunks that are descendants of the i-th child of u. The operations on these two arrays can be supported in $O(1)$ time, using a precomputed table of size $o(n)$ bits. Since the number of internal nodes is $O(m/b^2)$, the space usage for the B-tree is $O((m/b^2) \cdot (b(\log k + \log m))) = o(m \log k)$ bits.

The operations on the input array, can be performed by traversing the tree top-down and computing the operations at the internal nodes in $O(1)$ time.   □

3   Data Structure and Static Operations

We present a succinct representation for k-ary cardinal trees which uses $2n + n \log k + o(n \log k)$ bits, supports the navigational operations and queries in $O(1)$ time, and the updates in $O(1)$ amortized time. Our structure is similar to the structure of [1]. The input tree is decomposed into disjoint micro trees. Each operation is performed within the micro tree that contains the current node of the traversal, and in the case of the navigational operations, we might traverse to an adjacent micro tree. Each micro tree representation of [1] supports the operations in logarithmic time. We improve the time to $O(1)$ for small alphabet.

Decomposition. We use the greedy decomposition algorithm of [12] to decompose the input tree to micro trees of size in the range $[\log^2 n \ldots k^2 \log^2 n]$. The micro tree containing the root might be smaller than $\log^2 n$. This algorithm performs a postorder traversal of the tree. During the traversal, every at least $\log^2 n$ visited nodes make a micro tree (see [12] for more details). We change the algorithm of [12] a little bit to maintain the following. Let τ be a micro tree. The number of nodes (size) of τ is denoted by $|\tau|$. A frontier node of τ is a node, except the root of τ, that is adjacent to nodes in other micro trees. If the root of τ is adjacent to a frontier node of another micro tree τ', then τ' is the parent micro tree of τ, and τ is a child micro tree of τ'. We duplicate the frontier nodes of τ such that every frontier is also the root of a child micro tree of τ. Therefore, all the children of a frontier node are in the same micro tree, each frontier node is a leaf and is adjacent to only one child micro tree, i.e., the number of frontier nodes of τ denoted by $n_f(\tau)$ equals the number of child micro trees of τ.

Micro Tree Representation. Each micro tree τ is represented with the tuple $(D_\tau, L_\tau, F_\tau, P_\tau, r_\tau, S_\tau)$ defined as follows.

- D_τ: the tree topology of τ, using the DFUDS representation of τ
- L_τ: the edge labels of τ in the DFUDS order
- F_τ: frontiers of τ
- P_τ: pointers to the child micro trees of τ
- r_τ: a pointer to the parent micro tree of τ
- S_τ: the subtree size of all the child micro trees of τ.

Let τ be the micro tree that contains the current node of the traversal. We perform the navigations within τ using D_τ, and for label-child using L_τ. In the case of traversing to a child micro tree of τ, we find the pointer to the child micro tree using F_τ and P_τ. For traversing to the parent micro tree, we use r_τ. To compute subtree-size within τ, we use D_τ. To compute subtree-size of the current node in the whole tree, we use S_τ to compute subtree-size of the root of each child micro tree of τ that is a descendant of the current node. Then we add the subtree size of the current node within τ with all the computed subtrees sizes. The operation is-ancestor can be easily performed using subtree-size.

For each of the six parts except r_τ, we make data structures to perform the corresponding operations on them efficiently. The space usage for D_τ is $2|\tau| + o(|\tau|)$ bits, for L_τ is $|\tau| \log k + o(|\tau| \log k)$ bits, for r_τ is $\log n_f(\tau)$ bits, and for F_τ, P_τ, and S_τ is $o(n_f(\tau))$ bits. Since the micro trees are roughly disjoint, the total space usage is $2n + n \log k + o(n \log k)$ bits. In the following, we describe all the six parts of the micro tree representations.

3.1   Tree Topology of Micro Trees

We make a data structure that maintains a micro tree τ of size at most $k^2 \log^2 n = O(polylog(n))$ nodes using $2|\tau| + o(|\tau|)$ bits, which supports all the required operations within τ (including updates) except label-child in $O(1)$ time. We represent the structure of τ by its DFUDS sequence which is a string of $2 \cdot |\tau|$

parentheses [2]. Benoit et al. [2] showed that the navigation and query operations on a static ordinal tree of size n can be supported in $O(1)$ time using $2n + o(n)$ bits of space by performing rank/select and the balanced parenthesis operations: findclose, findopen, and enclose on the DFUDS sequence of the tree. Let D_τ be the DFUDS sequence of τ. Our data structure supports rank/select and the balanced parenthesis operations as well as update operations all in $O(1)$ time on D_τ. Essentially, our data structure is a dynamic DFUDS sequence of length $O(\log^2 n)$. Note that inserting and deleting of leaves in τ correspond to inserting and deleting of the pair of parentheses "()" in D_τ.

Lemma 3. *There exists a dynamic data structure of size $2m + o(m)$ bits to maintain a sequence of m pairs of balanced parenthesis using precomputed tables of size $o(n)$, where $m = (\log n)^{O(1)}$. This data structure supports the operations: findclose, findopen, and enclose in $O(1)$ time, and supports inserting and deleting of the pair of parentheses "()" in $O(1)$ amortized time.*

Proof. This representation is similar to [4]. We divide the sequence into chunks of size $w\ell$ bits, where $\ell = O(\sqrt{\log n})$. Each chunk is represented by a dynamic array of size $w\ell + O(\sqrt{\log n} \log \ell)$ bits (see Lemma 1), which allows us to access, insert, or delete a parenthesis at a given index in $O(1)$ time (amortized for updates) using a precomputed table of size $o(n)$ bits. Therefore, the total space used for the chunks is $2m + o(m)$ bits.

Now, we make a B-tree with branching factor b, where $b = O(\log^{1/4} n)$. Each leaf of the tree stores a pointer to a sub-chunk of size ℓ such that scanning the sub-chunks of the leaves from the left of the tree to the right gives the original sequence. The number of leaves is $2m/\ell$, and the depth of the tree is $O(1)$. At each internal node u, we maintain an array of length b such that its i-th element stores the number of open parenthesis in the chunks that are descendants of the i-th child of u. Since the array is small (i.e., $O(\log^{\frac{1}{4}} n \cdot \log \log n)$ bits), we can represent it by a searchable partial sums structure using a precomputed table of size $o(n)$ bits. This array is used to perform the operations rank and select in $O(1)$ time by traversing the tree from its root to the appropriate leaf. In addition to this array, similar to [4], we store seven arrays containing different information about the parentheses stored in the subtrees of u. These arrays are used to perform the parenthesis operations. Update operations are also straightforward. See [4] for more details. Since the number of internal nodes is $O(2m/(b\ell))$, the space usage for the B-tree is $O(2m/(b\ell) \cdot b \log m) = o(m)$ bits.   □

The following lemma presents our dynamic DFUDS structure based on the dynamic parenthesis maintenance structure of Lemma 3.

Lemma 4. *There exists a dynamic DFUDS representation of size $2|\tau| + o(|\tau|)$ bits for an ordinal tree τ of $(\log n)^{O(1)}$ nodes using precomputed tables of size $o(n)$ bits. This data structure supports the operations parent, child, degree, subtree-size, is-ancestor, and preorder all in $O(1)$ time, and supports the update operations insert-leaf and delete-leaf in $O(1)$ amortized time.*

Proof. Recall that the DFUDS sequence D_τ contains $2 \cdot |\tau|$ balanced parenthesis. It has been shown that all the operations parent, child, degree, subtree-size, and is-ancestor on τ can be supported using the balanced parenthesis operations and rank/select on D_τ [2]. Also the operation preorder can be supported using the balanced parenthesis operations and rank/select on D_τ [11]. Inserting and deleting leaves correspond to inserting and deleting the pair of parentheses "()".

□

3.2   Edge Labels of Micro Trees

Let L_τ be the sequence containing all the edge labels of τ in the DFUDS ordered. To perform label-child(α) on τ, we find the rank i of α among all the edge labels between the current node and its children, and then we use child(i). To find i, we find the number of α before the current node, and then find the position of the next α using rank/select structure on both D_τ and L_τ. To perform insert-leaf(α), we again need to find i to simply insert the label. But finding i if there is no α among all the edge labels needs more information. For that, we construct a dynamic predecessor structure for all the edge labels below each internal node.

Note that L_τ consists of contiguous sub-sequences s_i, for $i = 1 \cdots I_\tau$, such that s_i represents all the labels below the i-th internal node of τ in preorder, where I_τ is the number of internal nodes in τ. Note that $|s_i| \le k$. We construct the following: (1) a data structure that supports the operations rank, select, insert, and delete on L_τ, (2) a data structure for each s_i, if $|s_i| > \log n / \log \log n$, which supports the operations predecessor, insert, and delete on s_i. In the following, we explain these two structures, and then we combine them.

Dynamic Rank/Select Structure. In the following lemma, we present a data structure which is used to perform label-child in a micro tree.

Lemma 5. *There exists a dynamic representation of size $m \log k + o(m \log k)$ bits for a sequence of m symbols from an alphabet of size k using precomputed tables of size $o(n)$ bits, where m and k are $(\log n)^{O(1)}$. This data structure supports the operations* rank *and* select *in $O(1)$ time, and supports the update operations* insert *and* delete *in $O(1)$ amortized time.*

Proof. There exists a static data structure that supports the operations rank and select in $O(1)$ time for an alphabet of size k, using a multi-ary wavelet tree with $O(1)$ height (Theorem 3.2 of [8]). We dynamize their structure in the following way. We set the branching factor of their wavelet tree to be k', where $k' = O(\sqrt{\log n})$. At each internal node we use a dynamic rank/select structure for an alphabet of size k'. In the following, we explain this data structure. Note that the update operations do not change the structure of the wavelet tree, and thus only the internal node structures should be dynamized.

We pack every ℓ symbols of the sequence into a chunk of size $\ell \log k'$ bits, for $\ell = (w / \log k') \log^{1/4} n$. Each chunk is represented by a dynamic array of size $\ell \log k' + O(\log^{1/4} n \log \ell)$ bits, which allows us to access, insert, or delete a symbol at a given index in $O(1)$ time (amortized for updates) using a precomputed table of size $o(n)$ bits (see Lemma 1). Therefore, the total space used for the chunks is $m \log k' + o(m \log k')$ bits.

Now, we make a B-tree with branching factor at most $\log^{\frac{1}{4}} n$. Each leaf of the B-tree stores a pointer to a sub-chunk of size w bits in one of the chunks such that scanning the sub-chunks of the leaves from the left of the B-tree to the right gives the original sequence. Therefore, each chunk corresponds to $\log^{1/4} n$ leaves. The number of leaves is $m/(\ell \log^{1/4} n)$ and the depth of the B-tree is $O(1)$. At each internal node u, we maintain $k+1$ arrays, each of length $\log^{1/4} n$. One of the arrays is denoted by Size. The i-th element of the array Size maintains the number of symbols in the sub-chunks that are descendants of the i-th child of u. Each of the other k' arrays is for a symbol in the alphabet, and its i-th element maintains the number of the corresponding symbol in the leaves that are descendants of the i-th child of u. We represent each of these arrays by a searchable partial sums structure with $O(1)$ time for the partial sums operations, using a precomputed table of size $o(n)$ bits, since the arrays are small (i.e., $O(\log^{\frac{1}{4}} n \cdot \log \log n)$ bits).

To perform the operation $\mathsf{rank}_\alpha(i)$, we traverse the B-tree top-down starting from the root. Let h be the sub-chunk containing the i-th symbol of the original sequence. At each internal node u, we count the number of α in the sub-chunks that are to the left of h, and are descendants of u. This counting can be performed in $O(1)$ time, using the partial sums structures that are constructed for the array Size and the array corresponding to α. At the leaf level, where we should perform rank in a sub-chunk of size w bits, we read the sub-chunk in $O(1)$ time and perform the rank using word-level computation. The operation $\mathsf{select}_\alpha(i)$ can be performed similarly in $O(1)$ time (the array Size is not required for select).

For the operations insert and delete, we perform them on the appropriate chunks in $O(1)$ amortized time (with the support of the dynamic arrays), and then we update the nodes of the B-tree along the appropriate path in a straightforward manner. Therefore, the total update time is $O(1)$ amortized. □

Dynamic Predecessor. In the following lemma, we present a structure used for $\mathsf{insert}(\alpha)$ to find the rank of α among its siblings.

Lemma 6. *There exists a dynamic predecessor data structure of size $o(m)$ bits for a sorted array of m elements, where $m = (\log n)^{O(1)}$ and each element is from the range $[0 \cdots k-1]$, using a precomputed table of size $o(n)$ bits. This data structure supports the operation* predecessor *in $O(1)$ time, and supports the update operations* insert *and* delete *in $O(1)$ amortized time.*

Proof. For this structure, we use the same packing strategy and dynamic arrays as we used in the proof of Lemma 5. We make a B-tree with branching factor b, where $b = \sqrt{\log n}$. Each leaf maintains b elements from the array, such that concatenating the leaves from left to right, gives the original array. The height of the tree is $O(1)$. At each internal nodes, we maintain b guiding indexes. Every node (including leaves) has $b \log k = o(w)$ bits which can be handled using a precomputed table of size $o(n)$ bits. To perform the operations, we traverse the tree top-down in $O(1)$ time. For the update operations, we also update the internal nodes in a bottom-up traversal. The rebalancing is applied as needed. □

The following lemma combines Lemma 5 and 6, and shows how to perform the operation label-child on τ using the data structures of D_τ and L_τ.

Lemma 7. *For a k-ary cardinal tree τ of at most $k^2 \log^2 n$ nodes where $k = (\log n)^{O(1)}$, there exists a dynamic representation of size $2|\tau| + |\tau| \log k + o(|\tau| \log k)$ bits that supports the operation label-child in $O(1)$ time, and supports the update operations insert-leaf and delete-leaf in $O(1)$ amortized time. The structure uses precomputed tables of size $o(n)$ bits.*

Proof. Similar to [1], we represent the tree τ with D_τ and L_τ. We make a data structure for each of D_τ and L_τ using Lemma 4, 5, and 6 in totally $2|\tau| + |\tau| \log k + o(|\tau| \log k)$ bits. □

3.3   Frontiers of Micro Trees

During performing the operations on a micro tree τ, we need to check whether the current node is a frontier of τ or not, and if it is a frontier, then we may need to traverse to the micro tree rooted at that frontier using a pointer. For the checking, we represent the frontiers of τ with an array F_τ of $n_f(\tau)$ elements. The representation of pointers is explained in Section 3.4. The i-th element of the array F_τ contains the difference between two preorder numbers which belong to the i-th and $i + 1$-st frontiers of τ in the preorder traversal of τ. We make a searchable partial sums structure for F_τ. Since F_τ has $(\log n)^{O(1)}$ elements, each of size $O(\log F_\tau)$ bits, we use the searchable partial sums structure of Lemma 2 that supports the operations sum, update, and search in $O(1)$ time, using $n_f(\tau) \log |\tau| + o(n_f(\tau) \log |\tau|)$ bits. Thus the overall space for all the micro trees is $o(n)$ bits. To check whether the current node is a frontier or not, we use search on F_τ for the preorder number of the current node.

3.4   Pointers to Other Micro Trees

There are two cases where we need to traverse from τ, containing the current node x, to another micro tree: 1) if x is a frontier of τ, then we need to follow a pointer to the child micro tree rooted at x, 2) if x is the root of τ, then we need to follow a pointer to move to the parent micro tree of τ.

For the first case, for each frontier of τ, we store a pointer to another micro tree that is rooted at that frontier. These pointers are represented in the following way. Let τ_i be the micro tree rooted at $F_\tau[i]$, the i-th frontier of τ. We make an array P_τ of $n_f(\tau)$ elements such that $P_\tau[i]$ maintains a pointer to τ_i. Therefore, whenever the current node is $F_\tau[i]$ (that we can check using the representation of Section 3.3), we can traverse to τ_i. The space usage to store P_τ for all the micro trees is $o(n)$ bits. For the second case, since x is a frontier of the parent micro tree τ', we store r_τ such that $F_{\tau'}[r_\tau]$ maintains the preorder number of x.

3.5   Subtree Sizes

We make a data structure that allows us to compute the subtree size of the current node in $O(1)$ time. Let τ be the micro tree containing the current node.

Lemma 4 shows that we can perform subtree-size on the current node within τ in $O(1)$ time. But, to this number, we should add the subtree size of the root of each child micro tree of τ that is a descendant of the current node. For this, we make an array S_τ of length $n_f(\tau)$ such that $S_\tau[i]$ maintains the subtree size of the root of τ_i, where τ_i is the child micro tree rooted at the i-th frontier of τ in the preorder traversal of τ. We represent S_τ by a searchable partial sums structure using $n_f(\tau) \log |\tau| + o(n_f(\tau) \log |\tau|)$ bits (see Lemma 2). The overall space for all the micro trees is $o(n)$ bits. To compute the subtree size, we need to find $\sum_{i=j_\ell}^{j_r} S_\tau[i]$, where τ, τ_{j_ℓ} and τ_{j_r} are the left most and right most child micro trees of τ respectively that are descendants of the current node. To find τ_{j_ℓ}, we do a predecessor search in the array F_τ for the preorder number of the current node. Let e be the left most leaf of τ that is also a descendant of the current node. To find τ_{j_r}, we first find the preorder number of e within τ by adding the preorder number of the current node and its subtree size within τ. Then we do a predecessor search in the array F_τ for the preorder number of e.

4   Update Operations

Operation Insert-Leaf. To perform insert-leaf(α) in a micro tree τ, we update the representation of τ in the following way. We update D_τ by inserting "()" as a leaf into a position i that we find by a predecessor search in s_j of L_τ corresponding to the current node. We update L_τ by inserting α as a new label into position i. The new leaf is not a frontier, but if it is inserted between two frontiers, then it changes the difference between the preorder numbers of them. Therefore, we increment the appropriate element of F_τ. All the above operations are performed in $O(1)$ time.

If $|\tau|$ exceeds the value of $k^2 \log^2 n$, we split τ into micro trees of size in $[2 \log^2 n \cdots 2k \log^2 n]$ using the decomposition algorithm that we used in Section 3. Then we reconstruct the representation of each new micro tree. This can be performed by inserting leaves one by one into the new micro trees. The split and the construction of micro tree representations are both can be performed in $O(|\tau|) = O(k^2 \log^2 n)$ time. Since, this procedure makes micro trees of small enough size (at most $k \log^2 n$), therefore, $k^2 \log^2 n$ number of insert-leaf is required to make any of them full and the insertion time is $O(1)$ amortized.

Operation Delete-Leaf. To perform delete-leaf(α) in a micro tree τ, we update the representation of τ similarly as insert-leaf(α). If $|\tau|$ becomes smaller than $\log^2 n$, then we combine τ with its parent micro tree. This can be performed by inserting the nodes of τ into the parent micro tree, in the preorder traversal of τ using insert-leaf. This procedure takes $O(|\tau|) = O(\log^2 n)$ time. Since the new micro trees that we construct in the split procedure of Section 4 are large enough (at least $2 \log^2 n$ size), the deletion time is $O(1)$ amortized.

Memory Management. We store each micro tree in a separate location of the memory using an Extendible Array [3]. Since the number of micro trees is at most $n/\log^2 n$, and the nominal size of all the micro trees is $s = 2n + n \log k +$

$o(n \log k)$ bits, then the space requirement for the whole collection of micro trees is $s + O(nw/\log^2 n + \sqrt{snw/\log^2 n}) = 2n + n \log k + o(n \log k)$ bits [16].

References

1. Arroyuelo, D.: An improved succinct representation for dynamic k-ary trees. In: Ferragina, P., Landau, G.M. (eds.) CPM 2008. LNCS, vol. 5029, pp. 277–289. Springer, Heidelberg (2008)
2. Benoit, D., Demaine, E.D., Munro, J.I., Raman, R., Raman, V., Rao, S.S.: Representing trees of higher degree. Algorithmica 43(4), 275–292 (2005)
3. Brodnik, A., Carlsson, S., Demaine, E.D., Munro, J.I., Sedgewick, R.: Resizable arrays in optimal time and space. In: Dehne, F., Gupta, A., Sack, J.-R., Tamassia, R. (eds.) WADS 1999. LNCS, vol. 1663, pp. 37–48. Springer, Heidelberg (1999)
4. Chan, H.-L., Hon, W.-K., Lam, T.W., Sadakane, K.: Compressed indexes for dynamic text collections. ACM Transactions on Algorithms 3(2) (2007)
5. Darragh, J.J., Cleary, J.G., Witten, I.H.: Bonsai: a compact representation of trees. Software - Practice and Experience 23(3), 277–291 (1993)
6. Farzan, A., Munro, J.I.: Succinct representation of dynamic trees. Theoretical Computer Science (2010) (in press) (corrected proof)
7. Farzan, A., Raman, R., Rao, S.S.: Universal succinct representations of trees? In: Albers, S., Marchetti-Spaccamela, A., Matias, Y., Nikoletseas, S., Thomas, W. (eds.) ICALP 2009. LNCS, vol. 5555, pp. 451–462. Springer, Heidelberg (2009)
8. Ferragina, P., Manzini, G., Mäkinen, V., Navarro, G.: Compressed representations of sequences and full-text indexes. ACM Transactions on Algorithms 3(2) (2007)
9. Graham, R.L., Knuth, D.E., Patashnik, O.: Concrete Math, 1st edn. Addison-Wesley Longman Publishing Co., Inc., Boston (1988)
10. Hon, W.-K., Sadakane, K., Sung, W.-K.: Succinct data structures for searchable partial sums. In: Ibaraki, T., Katoh, N., Ono, H. (eds.) ISAAC 2003. LNCS, vol. 2906, pp. 505–516. Springer, Heidelberg (2003)
11. Jansson, J., Sadakane, K., Sung, W.-K.: Ultra-succinct representation of ordered trees. In: Proc. 18th Annual ACM-SIAM Symposium on Discrete Algorithms, pp. 575–584. SIAM, Philadelphia (2007)
12. Munro, J.I., Raman, V., Storm, A.J.: Representing dynamic binary trees succinctly. In: Proc. 12th Annual ACM-SIAM Symposium on Discrete Algorithms, pp. 529–536. SIAM, Philadelphia (2001)
13. Raman, R., Raman, V., Rao, S.S.: Succinct dynamic data structures. In: Dehne, F., Sack, J.-R., Tamassia, R. (eds.) WADS 2001. LNCS, vol. 2125, pp. 426–437. Springer, Heidelberg (2001)
14. Raman, R., Raman, V., Satti, S.R.: Succinct indexable dictionaries with applications to encoding -ary trees, prefix sums and multisets. ACM Transactions on Algorithms 3(4) (2007)
15. Raman, R., Rao, S.S.: Succinct dynamic dictionaries and trees. In: Baeten, J.C.M., Lenstra, J.K., Parrow, J., Woeginger, G.J. (eds.) ICALP 2003. LNCS, vol. 2719, pp. 357–368. Springer, Heidelberg (2003)
16. Raman, R., Rao, S.S.: Succinct dynamic dictionaries and trees (2008) (manuscript)

Integer Representations towards Efficient Counting in the Bit Probe Model

Gerth Stølting Brodal[1], Mark Greve[1],
Vineet Pandey[2], and Satti Srinivasa Rao[3]

[1] MADALGO*, Department of Computer Science, Aarhus University,
IT Parken, Åbogade 34, DK-8200 Århus N, Denmark
{gerth,mgreve}@cs.au.dk
[2] Computer Science & Information Systems, BITS Pilani, 333031, India
vineetp13@gmail.com
[3] School of Computer Science and Engineering, Seoul National University,
Republic of Korea
ssrao@cse.snu.ac.kr

Abstract. We consider the problem of representing numbers in close to optimal space and supporting increment, decrement, addition and subtraction operations efficiently. We study the problem in the bit probe model and analyse the number of bits read and written to perform the operations, both in the worst-case and in the average-case. A counter is *space-optimal* if it represents any number in the range $[0, \ldots, 2^n - 1]$ using exactly n bits. We provide a *space-optimal counter* which supports increment and decrement operations by reading at most $n - 1$ bits and writing at most 3 bits in the worst-case. To the best of our knowledge, this is the first such representation which supports these operations by always reading strictly less than n bits. For *redundant counters* where we only need to represent numbers in the range $[0, \ldots, L]$ for some integer $L < 2^n - 1$ using n bits, we define the efficiency of the counter as the ratio between $L + 1$ and 2^n. We present various representations that achieve different trade-offs between the read and write complexities and the efficiency. We also give another representation of integers that uses $n + O(\log n)$ bits to represent integers in the range $[0, \ldots, 2^n - 1]$ that supports efficient addition and subtraction operations, improving the space complexity of an earlier representation by Munro and Rahman [Algorithmica, 2010].

Keywords: Data structure, Gray code, Bit probe model, Binary counter, Integer representation.

1 Introduction

We propose data structures for integer representation which can perform increment, decrement, addition and subtraction with varying trade-offs between the

---

* Center for Massive Data Algorithms, a Center of the Danish National Research Foundation.

M. Ogihara and J. Tarui (Eds.): TAMC 2011, LNCS 6648, pp. 206–217, 2011.

number of bits read or written and the space needed to represent the number. We study the problem in the bit probe model of computation where the complexity measure includes only the bitwise accesses to the data structure and not the resulting computations.

We define a code of dimension n as any cyclic sequence of 2^n distinct binary vectors. For a code of dimension n, we define the operation increment (decrement) as moving the code to its next (previous) code in the cycle. We define a function Val that maps bit sequences to integers, which is used in describing our algorithms. We use B_R and B_W to denote the number of bits read and written respectively. The average number of bits read (written) is computed by summing the number of bits read (written) to perform the operations for each code, and dividing this by the number of different codes. Throughout the paper, $\log n$ denotes $\lceil \log_2 n \rceil$, $\log^{(0)} n = n$ and $\log^{(c)}(n) = \log^{(c-1)}(\log n)$ for $c > 0$.

Previous Work. The Standard Binary Code (SBC) uses n bits to represent an integer in the range $[0, \ldots, 2^n - 1]$ where $b_{n-1} b_{n-2} \ldots b_0$ represents the value $\sum_{i=0}^{n-1} b_i 2^i$. An increment or decrement operation using SBC requires n bits to be read and written in the worst-case but the amortized time per operation is constant. A Gray code is any code in which successive binary vectors in the sequence differ in exactly one component. The Binary Reflected Gray Code (BRGC) [3] requires n bits to be read for each increment operation but only 1 bit to write. Bose et al. [1] have developed a different Gray Code called Recursive Partition Gray Code (RPGC) which requires on an average $O(\log n)$ reads for increment operations. The previous results are summarized in Table 1. For the Gray codes BRGC and RPGC, we define $\mathrm{Val}(X)$ as the number of times one needs to increment the code $0 \ldots 0$ to obtain X. The dimension d of a counter refers to the total number of bits used to represent a number and *space-efficiency* refers to the ratio of number of numbers represented out of all possible bit strings generated (2^d) given the dimension d. Space-efficiency equal to one implies that all possible strings are generated and the counter is *space-optimal*. There could be more than one representation for a given number when efficiency is less than one and such counters are called *redundant counters*.

Our Results. For space-optimal counters, we introduce the notion of an (n, r, w)-*counter* which is a representation of numbers of dimension n where increment and decrement operations can be performed by reading r bits and writing w bits in the worst-case. We obtain a *(4,3,2)-counter* by exhaustive search and use it to construct an $(n, n-1, 3)$-*counter* which performs an increment or decrement operation by reading at most $n - 1$ bits whereas all known results for space-optimal counters read n bits in the worst-case. The codes BRGC and RPGC are examples of $(n, n, 1)$-*counters*. Fredman has conjectured that for Gray codes of dimension n, $B_R = n$ [1,2]. If this conjecture is true, this would imply that if there exists a code with the property that all increments can be made by reading less than n bits, then it would need to write at least 2 bits in the worst-case.

Table 1. Summary of previous results

Space (d)	Space efficiency	Bits read (B_R)		Bits written (B_W)	Inc. &	
		Average-case	Worst-case	Worst-case	Dec.	Ref.
n	1	$2 - 2^{1-n}$	n	n	Y	Binary
		n		1	Y	[3]
		$6 \log n$		1	Y	[1]
		$O(\log^{(2c-1)} n)$		c	N	[1]
$n+1$	$1/2$	$O(1)$	$\log n + 4$	4	Y	[4]
$n + O(t \log n)$	$1 - O(n^{-t})$	$O(\log^{(2c)} n)$	$O(t \log n)$	$2c + 1$	N	[1]

For non-space-optimal counters, the read complexity has been shown to be $\Theta(\log n)$ for a space-efficiency of $1/2$ [1,4]. The best known result so far [1] describes a counter with a space-efficiency of $1 - O(n^{-t})$ to increment a value by reading $O(t \log n)$ bits and writing 3 bits for $t > 0$. Our results shown in Table 2 show that we can reduce the number of bits written to 2 using a representation with space-efficiency $1 - O(2^{-t})$ by reading $\log n + t + 2$ bits where $t \in \mathbb{Z}^+$. By choosing $t = t' \log n$, we can achieve a space-efficiency of $1 - O(n^{-t'})$ by reading $O(t' \log n)$ bits and writing 2 bits. The question that remains open is if redundant counters efficiently allow a representation with 1 write but less than n reads.

For redundant counters with efficiency $1/2$, the best known results were $\log n + 4$ bit reads and 4 bit writes [4]. We reduce the number of bits read and written to $\log n + 3$ and 3. Using the one bit read-write trade-off, we can further reduce the number of bits written to 2 by reading $\log n + 4$ bits.

Table 2. Summary of our results

Space (d)	Space efficiency	Average-case		Worst-case		Inc. &	
		B_R	B_W	B_R	B_W	Dec.	Ref.
4	1	3	1.25	3	2	Y	Th. 1
n		$6 \log(n-4) + O(2^{-n})$	$1 + O(2^{-n})$	$n - 1$	3		Th. 2
$n+1$	$1/2$	$O(\log \log n)$	$1 + O(n^{-1})$	$\log n + 2$	3	N	Th. 3
				$\log n + 3$	2		Th. 4
		$O(\log n)$		$\log n + 3$	3	Y	Th. 6
				$\log n + 4$	2		
n	$1 - \frac{1}{2^{t-1}}$	$O(\log \log n)$	$1 + O(n^{-1})$	$\log n + t + 1$	3	N	Th. 5
				$\log n + t + 2$	2		
		$O(\log n)$		$\log n + t + 2$	3	Y	Cor. 1
				$\log n + t + 3$	2		

2 Space-Optimal Counters with Increment and Decrement

In this section, we describe space-optimal counters which are constructed using a *(4,3,2)-counter* where X denotes the number to be incremented.

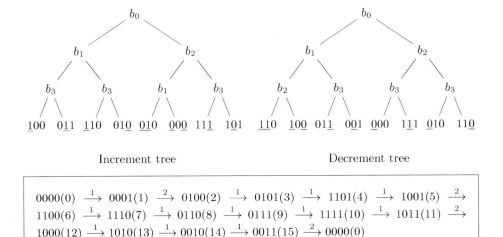

Fig. 1. Sequence generated by the *(4,3,2)-counter* and increment and decrement trees

(4,3,2)-Counter. Fig. 1 shows our *(4,3,2)-counter* obtained through brute force search which represents numbers from $0 \ldots 15$. Assuming the number is of the form $b_3 b_2 b_1 b_0$, the corresponding increment and decrement trees for the *(4,3,2)-counter* are shown in Fig. 1. For any internal node corresponding to bit b_t, the left edge corresponds to $b_t = 0$ and the right edge corresponds to $b_t = 1$. The leaf nodes contain information about the new values for the bits read and the modified bits are shown underlined in the tree and the text.

As an example, for the fifth leaf node from the left in the increment tree, old $b_0 b_2 b_1 = 100$ and new $b_0 b_2 b_1 = 0\underline{1}0$. To increment 9, for example, we take its representation 0111 and go through the path $b_0 b_2 b_3 = 110$ in the increment tree to reach the seventh leaf node; so the new values are $b_0 b_2 b_3 = 11\underline{1}$ and the new number is $\underline{1}111$ which represents 10 (ten). To decrement 9, we go through the path $b_0 b_2 b_3 = 110$ in the decrement tree to reach the seventh leaf node; so the new values are $b_0 b_2 b_3 = 0\underline{1}0$ and the number is $011\underline{0}$ which represents 8.

Theorem 1. *There exists a representation of integers of dimension 4 with efficiency 1 that supports increment and decrement operations with $B_R = 3$ and $B_W = 2$ in the worst-case. On average, an increment/decrement requires $B_R = 3$ and $B_W = 1.25$.*

2.1   Constructing $(n, n-1, 3)$-*Counter* Using *(4,3,2)-Counter*

We can now construct an n-bit space-optimal counter for $n \geq 4$ by dividing the code for a number X into two sections $X_{(4,3,2)}$ and X_G of length 4 and $n-4$ respectively where $X_{(4,3,2)}$ uses the above-mentioned *(4,3,2)-counter* representation and X_G uses the Gray code [3], that is a $(n-4, n-4, 1)$-*counter*. To increment X, we first increment X_G and then check if it represents 0 (which is

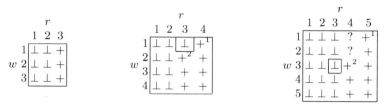

Fig. 2. Exhaustive search results for (n, r, w)-*counter* for $n = 3$, 4 and 5 respectively

possible since we read all bits of X_{G}). If X_{G} is 0, then we increment $X_{(4,3,2)}$. In the worst-case, this requires $n - 4$ reads and 1 write to increment X_{G} and then 3 reads and 2 writes to increment $X_{(4,3,2)}$, providing us with $n - 1$ reads and 3 writes overall.

X_{G} is represented using RPGC where incrementing or decrementing a code of dimension n requires $6 \log n$ average number of reads (although [1, Theorem 2] considers only generating the next code, i.e., increment operation, one can verify that the same analysis holds for the decrement operations as well). The worst-case and hence the average number of writes to increment or decrement a number using RPGC is 1. Since the average number of reads and writes for $X_{(4,3,2)}$ are 3 and 1.25 respectively, and we increment/decrement $X_{(4,3,2)}$ only in one out of every 2^{n-4} codes, the average number of reads and writes are $6 \log(n - 4) + 3/2^{n-4}$ and $1 + 1.25/2^{n-4}$ respectively.

Theorem 2. *There exists a representation of integers of dimension $n \geq 4$ with efficiency 1 that supports increment and decrement operations with $B_R = n - 1$ and $B_W = 3$ in the worst-case. On average, an increment/decrement requires $B_R = 6 \log(n - 4) + O(2^{-n})$ and $B_W = 1 + O(2^{-n})$.*

To the best of our knowledge, this is the first space-optimal counter with B_R strictly less than n.

Exhaustive Search Results. We used exhaustive search to find (n, r, w)-*counters* for small values of n. The results are shown in Fig. 2 for $n = 3$, 4 and 5 respectively. For a combination of n, r and w, a '$\perp$' shows that no counter exists and a '$+$' refers to its existence. A superscript of 1 shows that this is a Gray code while 2 refers to Theorem 2. A '?' shows that the existence of counters remains unknown for the corresponding (n, r, w) value. An enclosed value shows that no counters were found by our brute-force search.

3   Redundant Counters with Increment

To reduce the number of bits read exponentially, counters with space-efficiency less than one have been considered [1,4]. In this section, we discuss redundant counters which show better results and trade-offs for bits read and written and use these in Section 5 to obtain representations that support addition and subtraction efficiently.

Table 3. Transition Table for the increment step where $\ell = \mathrm{Val}(X_L)$ and $p = \log n + \ell$. Underlines show the changed bits and x represents 'don't care' condition.

Previous			New		
ℓ	S	x_p	S	x_p	
$= \ell_{\max}$	0	x	<u>1</u>	x	
$< \ell_{\max}$	0	x	0	x	
$< \ell_{\max}$	1	0	<u>0</u>	<u>1</u>	
$< \ell_{\max}$	1	1	1	<u>0</u>	

3.1   Counters with One Bit Redundancy

To represent numbers from $0 \ldots 2^n - 1$, we select $n + 1$ bits. A number X represented by $x_n x_{n-1} \ldots x_1 x_0$ consists of a *carry bit* $S = x_0$, a lower block X_L of the $\log n$ bits $x_{\log n} \ldots x_1$ and the upper block X_H of the last $n - \log n$ bits. $p = \log n + \ell$ is a location in X_H where ℓ refers to the value represented by X_L. This is used to perform a delayed addition of the carry as explained below. We use Gray codes for representing the numbers in X_L so that increment writes only one bit. The block X_H is represented using SBC. The value of X is given by $(\ell + (\mathrm{Val}(X_H) + 2^\ell \cdot S) \cdot 2^{|X_L|}) \bmod 2^n$.

We determine the number of bits read and written in the worst-case by finding the maximum values of B_R and B_W respectively. The increment step is summarised in Transition Table 3.

Increment: X_L and S are read at every step, therefore B_R is at least $\log n + 1$. $S = 1$ implies that the carry needs to be propagated and we will read one bit from X_H, whereas $S = 0$ implies no carry propagation and we do not need to access X_H. If $\ell > n - \log n$, we reset S to 0. The different cases for increment are described below:

Case 1. $S = 0$ and X_L contains its *largest* value ($100 \ldots 0$ in Gray code): This implies that a new incremental increment of X_H should be initiated. Increment X_L and set the carry bit S to 1. ($B_R = \log n + 1$, $B_W = 2$)

Case 2. $S = 0$ and X_L is any other value: Increment X_L. ($B_R = \log n + 1$, $B_W = 1$)

Case 3. $S = 1$ and $x_p = 1$: Propagation of carry. Change x_p to 0. Increment X_L.($B_R = \log n + 2$, $B_W = 2$)

Case 4. $S = 1$ and $x_p = 0$: Final bit flip in X_H. Change x_p to 1, S to 0 and increment X_L. ($B_R = \log n + 2$, $B_W = 3$).

The average number of reads to increment X_L is $O(\log \log n)$. The bit S is read at every step and it is set to 1 on the average 2 out of every n steps. When $S = 1$, we also need to read $O(\log n)$ bits to find $\mathrm{Val}(X_L)$. Thus the average number of bits read is $O(\log \log n)$. The average number of writes can be shown to be $1 + O(n^{-1})$. Hence we have the following theorem.

Theorem 3. *There exists a representation of integers of dimension $n + 1$ with efficiency $1/2$ that supports increment operations with $B_R = \log n + 2$ and $B_W = 3$. On average, an increment requires $B_R = O(\log \log n)$ and $B_W = 1 + O(n^{-1})$.*

3.2   One Bit Read-Write Trade-Off

We show how to modify the representations of the previous section (Theorem 3) to reduce B_W from 3 to 2 by increasing B_R by 1.

Table 4. Transition Table for the increment step for read-write trade-off where $\ell = \mathrm{Val}(X_L)$, $\ell_{\max} = 2^{|X_L|} - 1$ and $p = \log n + \ell + 1$. Underlines show the changed bits and x represents '*don't care*' condition.

Previous				New	
ℓ	S	x_p	x_{p-1}	S	x_p
$= \ell_{\max}$	0	x	$-$	$\underline{1}$	x
$= 0$	1	1	$-$	1	$\underline{0}$
$= 0$	1	0	$-$	1	$\underline{1}$
> 0	1	x	1	$\underline{0}$	x
> 0	1	0	0	1	$\underline{1}$
> 0	1	1	0	1	$\underline{0}$

The worst-case of B_W for increment is given by *Case* 4 where $B_W = 3$ since S and one bit each in X_H and X_L are modified. As it turns out, we can improve B_W further by delaying the resetting of S by one step if we read another bit. Instead of reading just one bit x_p from X_H when $S = 1$, we can read the pair (x_p, x_{p-1}). If the previously modified bit $x_{p-1} = 1$, then the propagation of carry is complete, else we flip the current bit x_p. The only exception to this case is when $X_L = 0 \ldots 0$ which implies that $p = \log n + 1$ which is the first position in X_H. In this case, only one bit $x_{\log n+1}$ is read and flipped. We modify the increment step as:

Case 3. $S = 1$ and $x_{p-1} = 0$: propagation of carry to continue. $x_{p-1} = 0$ implies that the previous bit was 1 before getting modified. Therefore, flip x_p irrespective of its value and increment X_L. ($B_R = \log n + 3$, $B_W = 2$).

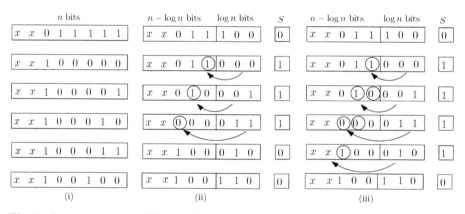

Fig. 3. Increment for a 8-bit number using (i) Standard binary counter (ii) One-bit redundant counter with $B_W = 3$ (iii) with $B_W = 2$. $\log n$ bits are represented using BRGC and x represents '*don't care*' condition.

Case 4. $S = 1$ and $x_{p-1} = 1$: The previous bit was 0 before modification, hence carry has been propagated and x_p is not read. Reset S to 0 and increment X_L. ($B_R = \log n + 2$, $B_W = 2$).

Theorem 4. *There exists a representation of integers of dimension $n + 1$ with efficiency $1/2$ that supports increment operations with $B_R = \log n + 3$ and $B_W = 2$. On average, an increment requires $B_R = O(\log \log n)$ and $B_W = 1 + O(n^{-1})$.*

3.3   Forbidden State Counter with Increment

To increase the space-efficiency of the above proposed representation, we modify the data structure proposed in [1] where a particular value of t bits in a dimension n code is used as a forbidden state. A number $X = x_n \ldots x_1$ consists of X_H ($x_n \ldots x_{\log n+t+1}$), X_F ($x_{\log n+t} \ldots x_{\log n+1}$) and X_L ($x_{\log n} \ldots x_1$) of $n - \log n - t$, t and $\log n$ bits respectively. Similar to the one-bit redundant counter discussed in Section 3.1, X_H and X_L represent the upper and lower blocks in the number while X_F acts as an alternative to the carry bit S. We use ℓ to refer to the value represented by X_L and $F_{\max}$ refers to the value $2^t - 1$.

All the states for which $\text{Val}(X_F) \le F_{\max} - 1$ are considered as normal states for X_F and the state where $\text{Val}(X_F) = F_{\max}$ is used to propagate the carry over X_H (conceptually $X_F = F_{\max}$ corresponds to $S = 1$). This representation will allow us to represent $1 - 1/2^t$ of the 2^n numbers. The block X_H is represented using SBC while X_F and X_L are each individually represented using RPGC. Using X_K to represent $\text{Val}(X_K)$, we obtain $\text{Val}(X) = X_L + (X_F + X_H \cdot F_{\max}) \cdot 2^{|X_L|}$ if $X_F < F_{\max}$ and $\text{Val}(X) = X_L + (X_H + 2^\ell) \cdot 2^{|X_L|} \cdot F_{\max}$ if $X_F = F_{\max}$.

Increment. The increment scheme is similar to the one-bit redundant counter of Section 3.1. We first read X_L and X_F. If $X_F \ne F_{\max}$, we increment X_L. If X_L now becomes 0, we also increment X_F. For the case $X_F = F_{\max}$, X_L is used to point to a position p in X_H. If the bit x_p at position p is equal to 1, it is set to 0 and X_L is incremented to point to the next position in X_H. This corresponds to the increment scheme in the one-bit redundant counter when S is set to 1. If X_L now equals $n - \log n - t$, then we incremement X_F (to set $X_F = 0$ and terminate the propagation of carry). On the other hand, if the value of bit x_p is 0, then we set x_p to 1 and X_F is incremented to the next value (which represents state $X_F = 0$. This corresponds to the carry bit S being set to 0 in Section 3.1.

This scheme gives a representation with $B_R = \log n + t + 1$ and $B_W = 3$. Similar to Section 3.2, we can also obtain a representation with $B_R = \log n + t + 2$ and $B_W = 2$ by reading x_{p-1}. The average number of reads and writes to increment the $\log n$ bits in X_L are $O(\log \log n)$ and 1 respectively. The average number of reads and writes to increment X_F are $O(\log t)$ and 1 respectively. Since X_F is incremented once in every n steps, this adds only $o(1)$ to the average number of reads and writes. Similarly, incrementing X_H also takes $o(1)$ reads and writes on average. In addition, at every step we need to check if $\text{Val}(X_F)$ is equal to either $F_{\max}$ or $F_{\max} - 1$ which requires an average of $O(1)$ reads, and finally the cost of reading X_L to find p on average costs at most $O(\frac{1}{2^t}\frac{1}{n}\log n)$. Thus we have the following theorem.

Theorem 5. *Given two integers n and t such that $t \leq n - \log n$, there exists a representation of an integer of dimension n with efficiency $1 - O(2^{-t})$ that supports increment operations with $B_R = \log n + t + 1$ and $B_W = 3$ or $B_R = \log n + t + 2$ and $B_W = 2$. On average, an increment requires $B_R = O(\log \log n)$ and $B_W = 1 + O(n^{-1})$.*

4   Counters with Increment and Decrement

To support decrement operations interleaved with increment operations, we modify the representation of a number X described in Section 3.1 as follows: a number $X = x_n \ldots x_1 x_0$ consists of an upper block X_H ($x_n \ldots x_{\log n+2}$), a lower block X_L ($x_{\log n+1} \ldots x_1$) and the bit $S = x_0$ which is used as either a carry bit or a borrow bit. We further split the lower block X_L into two parts: an indicator bit I which consists of the bit $x_{\log n+1}$ and a pointer block X_P consisting of the remaining $\log n$ bits. When the indicator bit I is set to 0, S is interpreted as a carry bit, and when the indicator bit is 1, then S interpreted as a borrow bit.

The $\log n$ bits in X_P are used to point to a location in X_H to perform a delayed carry or borrow. We use BRGC for representing X_L so that an increment or decrement writes only one bit. The block X_H is represented using SBC. Since X_L is represented using BRGC, when $\mathrm{Val}(X_L) < 2^{|X_P|}$, the indicator bit I is equal to 0 and I is equal to 1 otherwise. When $I = 1$, incrementing block X_L corresponds to decrementing the block X_P (unless $X_P = 0$) due to the reflexive property of BRGC [3]. We use these observations in our algorithms for increment and decrement.

The main ideas behind the representation and the increment/decrement algorithms are as follows: when the carry bit S is not set, we perform the increment/decrement in the normal way by incrementing/decrementing X_L. When $S = 0$, $\mathrm{Val}(X_L) = 2^{\log n+1} - 1$ and we perform an increment, we set the bit S and reset the block X_L to $0 \ldots 0$. Since I is now set to 0, S will be interpreted as a carry bit untill it is reset again. Similarly, when $S = 0$, $\mathrm{Val}(X_L) = 0$ and we perform a decrement, we set the bit S and decrement X_L to $2^{\log n+1} - 1$. Since I is now set to 1, S will be interpreted as a borrow bit.

To increment X when the carry bit is set, we perform one step of carry propagation in X_H, and then increment X_L. If the propagation finishes in the current step, then we also reset the bit S to 0. To decrement X when the carry bit is set, we first decrement X_L and "undo" one step of carry propagation (i.e., set the bit x_p in X_H to 1). Note that the when performing increments, the carry propagation will finish before we need to change the indicator bit from 0 to 1 (as the length of X_H is less than $2^{\log n}$). The increment and decrement algorithms when the borrow bit are set are similar.

The increment and decrement algorithms are described in the Transition Table 5. Since we read X_L, S and at most one bit in X_H, the read complexity $B_R = \log n + 3$. Since we change at most one bit in each of X_L, X_H and S, the write complexity $B_W = 3$.

The above scheme requires $O(\log n)$ average number of reads as X_L is represented using BRGC and incrementing it requires $O(\log n)$ reads. To get better

Table 5. Transition Table for the increment-decrement counter. For increment, new $p = p + 1$ and for decrement, new $p = p - 1$. x represents '*don't care*' condition and - shows that the value does not exist. $\ell = \mathrm{Val}(X_P)$, $p = \log n + \ell + 1$ and $\ell_{\max} = 2^{|X_P|} - 1$. Underlines show the modified values.

Increment									
Previous					New				Comments
S	ℓ	I	x_p	x_{p-1}	S	I	x_p	x_{p-1}	
0	$= \ell_{\max}$	0	x	x	0	$\underline{1}$	x	x	Increment X_L (sets I)
0	$= \ell_{\max}$	1	x	x	$\underline{1}$	$\underline{0}$	x	x	Increment X_L (resets X_L), Set S
0	$< \ell_{\max}$	x	x	x	0	x	x	x	Only increment X_P
1	$< \ell_{\max}$	0	-	x	$\underline{0}$	0	-	x	(Position p beyond n) Reset S
1	$< \ell_{\max}$	0	0	x	$\underline{0}$	0	$\underline{1}$	x	(Last step of carry propagation) Reset S
1	$< \ell_{\max}$	0	1	x	1	0	$\underline{0}$	x	(Carry propagation)
1	$< \ell_{\max}$	1	x	1	1	1	x	$\underline{0}$	Undo previous borrow
1	$< \ell_{\max}$	1	x	0	-	-	-	-	Does not occur
Decrement									
0	$= 0$	0	x	x	$\underline{1}$	$\underline{1}$	x	x	Decrement X_L, Set S
0	$= 0$	1	x	x	0	$\underline{0}$	x	x	Decrement X_L (Resets I)
0	> 0	x	x	x	0	x	x	x	Only decrement X_P
1	> 0	1	-	x	$\underline{0}$	1	-	x	(Position p beyond n) Reset S
1	> 0	1	0	x	1	1	$\underline{1}$	x	(Borrow Propagation)
1	> 0	1	1	x	$\underline{0}$	1	$\underline{0}$	x	(Last step of borrow propagation) Reset S
1	> 0	0	x	0	$\underline{1}$	0	x	$\underline{1}$	Undo previous carry
1	> 0	0	x	1	-	-	-	-	Does not occur

average-case bounds, we can represent X_P using RPGC. This increases the number of worst-case writes by 1 as now I and X_P are incremented independently. Thus we get a structure with $B_R = O(\log \log n)$ and $B_W = 1 + O(n^{-1})$ on the average but in the worst-case $B_W = 4$.

Theorem 6. *There exists a representation of integers of dimension $n + 1$ with efficiency $1/2$ that supports increment and decrement operations with $B_R = \log n + 3$ and $B_W = 3$. On average, an increment/decrement requires $B_R = O(\log \log n)$ and $B_W = 1 + O(n^{-1})$.*

We can extend the result of Theorem 5 to support decrement operations using an indicator bit as described in Section 4.

Corollary 1. *Given two integers n and t such that $t \leq n - \log n$, there exists a representation of an integer of dimension n with efficiency $1 - O(2^{-t})$ that supports increment and decrement operations with $B_R = \log n + t + 2$ and $B_W = 3$. On average, an increment/decrement requires $B_R = O(\log \log n)$ and $B_W = 1 + O(n^{-1})$.*

5   Addition and Subtraction

In this section, we give a representation for integers which supports addition and subtraction operations efficiently. A number N is said to have a span n if it can take values in the range $[0, \ldots, 2^n - 1]$. Munro and Rahman [4] gave a representation that uses $n + O(\log^2 n)$ bits to represent a number N of span n, and supports adding/subtracting a M of span m to/from N in $O(m + \log n)$ time. We improve the space to $n + O(\log n)$ bits while maintaining the operation time. We describe the data structure and scheme for addition and introduce suitable modifications to support subtaction as well.

We divide the representation of the number into $k = O(\log n)$ *blocks*: $B_1, B_2, \ldots, B_k$ with $b_1, b_2, \ldots, b_k$ bits respectively, where $b_1 = 2$ and for $2 \leq i \leq k$, $b_i = 2^{i-1}$ (if n is not a power of 2, then the last block has size $b_k = n - 2^{\lfloor \log n \rfloor}$ instead of 2^{k-1}). Note that the block sizes satisfy the property that $\sum_{j=1}^{i} b_j = 2^i = b_{i+1}$, for $1 \leq i \leq k - 2$. Each block B_i is maintained using the increment counter of Section 3.1 using $b_i + 1$ bits and a constant number of flag bits as described below. Hence, a number is represented using k blocks of sizes $b_1, b_2, \ldots, b_k$ bits along with $O(k)$ additional bits. The value of the representation is $\text{Val}(B_1) + \sum_{i=2}^{k} \text{Val}(B_i) \cdot 2^{b_i}$. Thus the overall space used is $n + O(k) = n + O(\log n)$ bits.

We now describe the modifications to the increment counter described in Theorem 3. Let X be the counter to be incremented. We introduce two additional bits max and V_H. The bit max indicates whether X represents its maximum value. Assuming $p = \text{Val}(X_L)$ represents a position in X_H, V_H (verifier for block X_H) $= 1$ if all positions in X_H from $0 \ldots p$ are 1. By this definition, when p points to any location beyond X_H and $V_H = 1$ then X_H represents its maximum value. In Section 3.1 we used p to point to a location in X_H only when $S = 1$ but now we use p as a pointer in all steps. When $S = 1$, we perform the delayed increment in X_H and when $S = 0$, we read the bit x_p and use it to set/reset V_H. V_H is set to 0 if $x_p = 0$. If $V_H = 0$, then we set it to 1 if $S = 1$ and $x_p = 0$. This case happens when $X_L = 0 \ldots 0$ for a delayed increment. The bit max is set to 1 when $V_H = 1$ and X_L represents its maximum value. When X represents its maximum value, max $= 1$, $V_H = 1$ and $S = 0$. Incrementing the maximum value of X sets $S = 1$, max $= 0$ and resets X_L to its minimum value. The bit $V_H = 1$ is maintained till S is reset to 0, i.e. throughout the delayed increment process.

We represent every block B_i using the above modified counter. To add M to N, for some $m \leq n$, we first find the largest i such that $\sum_{j=1}^{i-1} b_j < m \leq \sum_{j=1}^{i} b_j$ (i.e., $b_i < m \leq b_{i+1}$). We add M to the number represented by the first i blocks of N in $O(m)$ time. If any of the first i blocks has a carry bit set, then we first perform the necessary work and reset the carry bit in the block, and if necessary propagate the carry to the next block. If there is a carry from B_i to B_{i+1}, we propagate this by modifying the bit max_j of the successive blocks until we find the first block B_j such that max_j is set to 0, and increment B_j, altogether in $O(\log n)$ time. The total running time is $O(m + \log n)$ since incrementing the block and propagation of the carry take $O(\log n)$ time each.

The read and write complexities of the addition algorithm can be shown to be $O(m + \log n)$ and $O(m)$ respectively. Since incrementing a counter of span n has a $\Omega(\log n)$ lower bound for the read complexity, these bounds are optimal.

To support subtraction, we use the increment/decrement counter of Theorem 6 to represent each block, along with additional bits to check for the maximum and minimum values of a number. The details shall be provided in the extended version. M can be subtracted from N in $O(m + \log n)$ time similarly, since the representation of a block supports both increment and decrement operations in $O(\log b)$ time, where b is the length of the block.

Theorem 7. *An integer of span n can be represented by a data structure which uses $n + O(\log n)$ bits such that adding or subtracting an integer of span m can be perfomed by reading $O(m + \log n)$ bits and writing $O(m)$ bits.*

6   Conclusion

We have shown that a number of dimension n can be incremented and decremented by reading strictly less than n bits in the worst-case. For an integer in the range $[0, \ldots, 2^n - 1]$ represented using exactly n bits, our $(n, n-1, 3)$-*counter* reads $n - 1$ bits and writes 3 bits to perform increment/decrement operations. One open problem is to improve the upper bound of $n - 1$ reads for such space-optimal counters. Fredman [2] has shown that performing an increment using BRGC requires n bits to be read in the worst-case but the same is not known for all Gray Codes.

For the case of redundant counters, we have improved the earlier results by implementing increment operations using counters with space-efficiency arbitrarily close to one which write only 2 bits with low read complexity. We have obtained representations which support increment and decrement operations with fewer number of bits read and written in the worst-case and show trade-offs between the number of bits read and written in the worst-case and also between the number of bits read in the average-case and the worst-case. Finally we have also improved the space complexity of integer representations that support addition and subtraction in optimal time.

References

1. Bose, P., Carmi, P., Jansens, D., Maheshwari, A., Morin, P., Smid, M.H.M.: Improved methods for generating quasi-gray codes. In: Kaplan, H. (ed.) SWAT 2010. LNCS, vol. 6139, pp. 224–235. Springer, Heidelberg (2010)
2. Fredman, M.L.: Observations on the complexity of generating quasi-gray codes. SIAM Journal on Computing 7(2), 134–146 (1978)
3. Gray, F.: Pulse code communications. U.S. Patent (2632058) (1953)
4. Ziaur Rahman, M., Ian Munro, J.: Integer representation and counting in the bit probe model. Algorithmica 56(1), 105–127 (2010)

Closed Left-R.E. Sets[*]

Sanjay Jain[1], Frank Stephan[1,2], and Jason Teutsch[3]

[1] Department of Computer Science, National University of Singapore,
Singapore 117543, Republic of Singapore
sanjay@comp.nus.edu.sg
[2] Department of Mathematics, National University of Singapore,
Singapore 119076, Republic of Singapore
fstephan@comp.nus.edu.sg
[3] Institut für Informatik, Universität Heidelberg,
Im Neuenheimer Feld 294, 69120 Heidelberg, Germany
teutsch@math.uni-heidelberg.de

Abstract. A set is called r-closed left-r.e. iff every set r-reducible to it is also a left-r.e. set. It is shown that some but not all left-r.e. cohesive sets are many-one closed left-r.e. sets. Ascending reductions are many-one reductions via an ascending function; left-r.e. cohesive sets are also ascening closed left-r.e. sets. Furthermore, it is shown that there is a weakly 1-generic many-one closed left-r.e. set.

1 Introduction

When studying the limits of computation, one often looks at recursively enumerable (r.e.) and left-r.e. sets. Natural examples of the r.e. sets are Diophantine sets and the word problem of a finitely generated group [8,11,13]. The best-known left-r.e. set is Chaitin's Ω [1,14]. The present work focuses on a special subclass of the left-r.e. sets, namely those which are closed downwards with respect to the many-one or ascending reducibilities. While all r.e. sets exhibit closure under various reducibilities — one-one, many-one, conjunctive, disjunctive, positive truth-table and enumeration [8,11,13] — some left-r.e. sets, such as Chaitin's Ω, fail to do so.

We show that the classes of many-one closed left-r.e. sets and r.e. sets do not coincide: there exist both, cohesive and weakly 1-generic sets, which are many-one closed left-r.e. but not recursively enumerable, see Theorems 4, 15 and Remark 16. We also show that there are cohesive left-r.e. sets which are not many-one closed left-r.e., see Theorem 12.

We introduce the more restrictive notion of ascending reducibility. We show that cohesive and even r-cohesive left-r.e. sets are already ascending closed left-r.e. sets, see Theorem 17.

[*] S. Jain has been supported in part by NUS grants C252-000-087-001 and R252-000-420-112; F. Stephan has been supported in part by NUS grant R252-000-420-112; J. Teutsch has been supported by the Deutsche Forschungsgemeinschaft grant ME 1806/3-1.

M. Ogihara and J. Tarui (Eds.): TAMC 2011, LNCS 6648, pp. 218–229, 2011.

Kolmogorov complexity measures the information content of strings; the applications of this notion range from quantifying the amount of algorithmic randomness [2,7] to establishing lower bounds on the average running time of an algorithm [5]. An important tool to measure the complexity of a set A is the initial segment complexity which maps each n to the Kolmogorov complexity of $A(0)A(1)\dots A(n)$. We show that the initial segment complexity of ascending closed left-r.e. sets has to be sublinear, see Proposition 13. We also show that the initial segment complexity of an ascending closed left-r.e. sets can be $\Omega(n/f(n))$ for any unbounded increasing recursive function f, which is close to optimal, see Theorem 14.

2   Many-One Closed Left-R.E. Sets

Post [9] introduced many-one reducibility by defining that a set B *many-one reduces* to a set A, denoted $A \leq_m B$, if there exists a recursive function f such that $x \in A \iff f(x) \in B$. Below, we formally define a left-r.e. set and many-one closed left-r.e. set.

Definition 1. A set A is *left-r.e.* iff there is a uniformly recursive approximation $A_0, A_1, \dots$ to A such that $A_s \leq_{\text{lex}} A_{s+1}$ for all s. Here $A_s \leq_{\text{lex}} A_{s+1}$ means that either $A_s = A_{s+1}$ or the least element x of the symmetric difference satisfies $x \in A_{s+1}$. If every set many-one reducible to A is left-r.e. then we say that A is a *many-one closed left-r.e.* set.

It is well-known that every set which is many-one reducible to an r.e. set is also itself r.e. [11]; hence every r.e. set is a many-one closed left-r.e. set. Furthermore, a set is recursive iff it is a bounded truth-table (btt) closed left-r.e. set because the complement of any set btt-reduces to the set itself, see [8] for discussion of btt-reductions.

Definition 2 (Friedberg [3], Lachlan [4], Myhill [6] and Robinson [10]). An infinite set A is *cohesive* iff for every r.e. set B either $B \cap A$ or $\overline{B} \cap A$ is finite. An infinite set A is *r-cohesive* iff for every recursive set B either $A \cap B$ or $A \cap \overline{B}$ is finite.

Cohesive sets have been studied widely in recursion theory; they emerged as the culmination of Post's unsuccessful attempts to generate a Turing incomplete r.e. set [13]. The next result gives a cohesive many-one closed left-r.e. set. We remark that Soare [12] already discovered a cohesive left-r.e. set.

The following notational conventions will be useful. Let

$$\varphi_{e,s}(x) = \begin{cases} \varphi_e(x), & \text{if } \varphi_e \text{ halts on input } y \text{ within } s \text{ steps for all } y \leq x; \\ \uparrow, & \text{otherwise.} \end{cases}$$

Note that if φ_e is total, then $\bigcup_s \varphi_{e,s} = \varphi_e$. Otherwise, the domain of $\bigcup_s \varphi_{e,s}$ is some initial segment of $\mathbb{N}$. Let $\varphi_e^{-1}(x) = \min\{y : \varphi_e(y) = x\}$ and $\varphi_{e,s}^{-1}(x) = \min\{y : \varphi_{e,s}(y) = x\}$.

Lemma 3. *Suppose $\varphi_{e_1}, \varphi_{e_2}, \ldots, \varphi_{e_k}$ are total. Furthermore, suppose that the set $S = \text{range}(\varphi_{e_1}) \cap \text{range}(\varphi_{e_2}) \cap \ldots \cap \text{range}(\varphi_{e_k})$ is infinite. Then, for all a, r, there exist $a_1, a_2, \ldots, a_r \in S$ such that $a < a_1 < a_2 < \ldots < a_r$ and, for n, m with $1 \leq n < r$ and $1 \leq m \leq k$ it holds that $\varphi_{e_m}^{-1}(a_n) < \varphi_{e_m}^{-1}(a_{n+1})$.*

Proof. Let a_1 be any member of S which is greater than a. For i with $2 \leq i \leq r$, let $a_i \in S$ be chosen such that $a_i > a_{i-1}$ and for m with $1 \leq m \leq k$, $\varphi_{e_m}^{-1}(a_{i-1}) < \varphi_{e_m}^{-1}(a_i)$. Note that there exist such $a_i \in S$, as S is infinite and only finitely many elements x can have $\varphi_{e_m}^{-1}(x) \leq \varphi_{e_m}^{-1}(a_{i-1})$. □

Theorem 4. *There is a cohesive many-one closed left-r.e. set A.*

Proof. We will use moving markers, $a_0, a_1, \ldots$; let $a_{m,s}$ denote the value of marker a_m as at the beginning of stage s. Let $l_0 = 0$, $l_{d+1} = r_d + 1$, $r_d = l_d + 3^{d+2} + 1$. We let $I_{d,s} = \{a_{m,s} : l_d \leq m \leq r_d\}$. For all m, s, we will have the following property:

(R1): $a_{m,s} < a_{m+1,s}$.

Define the predicate $P_{e,s}(d)$ as

$$P_{e,s}(d) : (\exists a_{m,s}, a_{n,s} \in I_{d,s}) [a_{m,s} < a_{n,s} \text{ and } \varphi_{e,s}^{-1}(a_{m,s}) > \varphi_{e,s}^{-1}(a_{n,s})].$$

For $e \leq d$, let

$$i_{e,s}(d) = \begin{cases} 0, & \text{if } I_{d,s} \not\subseteq \text{range}(\varphi_{e,s}); \\ 1, & \text{if } I_{d,s} \subseteq \text{range}(\varphi_{e,s}) \text{ and } P_{e,s}(d); \\ 2, & \text{if } I_{d,s} \subseteq \text{range}(\varphi_{e,s}) \text{ and not } P_{e,s}(d). \end{cases}$$

For $e \leq d$, let $Q_{e,s}(d) = (i_{0,s}(d), i_{1,s}(d), \ldots, i_{e,s}(d))$. Note that one can consider $Q_{e,s}(d)$ as a number (base 3), with $i_{0,s}(d)$ as being the most significant bit. So one can talk about $Q_{e,s}(d) > Q_{e',s'}(d')$ etc.

We let $a_m = \lim_{s \to \infty} a_{m,s}$, $I_d = \lim_{s \to \infty} I_{d,s}$, $i_e(d) = \lim_{s \to \infty} i_{e,s}(d)$, and $Q_e(d) = \lim_{s \to \infty} Q_{e,s}(d) = (i_0(d), i_1(d), \ldots, i_e(d))$ (we will show later that these limits exist).

Intuitively, the aim of the construction of the moving markers a_m is to maximise the values of $Q_e(e)$ with higher priority given for lower values of e. The required set A will be defined later by choosing one element from each I_e. We define $a_{m,s}$ via the staging construction below. Stage s defines $a_{m,s+1}$.

Initially, let $a_{m,0} = m$.

Stage s: Check if, there exists $e \leq s$, such that, by using $a_{m,s+1} = a_{m,s}$ for $m < l_e$, some values of $a_{m,s+1} \leq s$ for $l_e \leq m \leq r_e$, and any values for $a_{m,s+1}$ for $m > r_e$ such that (R1) is satisfied, we have $Q_{e,s+1}(e) > Q_{e,s}(e)$. If so, then update the values of $a_{m,s+1}$ to the values witnessing above for the least such e. If no such e exists, then $a_{m,s+1} = a_{m,s}$, for all m.

End Stage s

Claim 5. *For all e,*
(a) for all m with $l_e \leq m \leq r_e$, $\lim_{e \to \infty} Q_{e,s}(e)$ and $\lim_{s \to \infty} a_{m,s}$ converge.
(b) $\lim_{s \to \infty} I_{e,s}$ converges.
(c) for all $d \geq e$, $\lim_{s \to \infty} i_{e,s}(d)$ converges.

(a) Follows by induction on e and the fact that $Q_{e,s}(e)$ is bounded. Now (b) and (c) follow by definitions. We let a_m, I_e, $i_e(d)$, and $Q_e(d)$ respectively denote $\lim_{s\to\infty} a_{m,s}$, $\lim_{s\to\infty} I_{e,s}$, $\lim_{s\to\infty} i_{e,s}(d)$, and $\lim_{s\to\infty} Q_{e,s}(d)$.

Claim 6. *For all d and all $e \leq d$, $Q_e(d+1) \leq Q_e(d)$.*

To prove the claim, suppose by way of contradiction that some least d and a corresponding least $e \leq d$ does not satisfy the claim. Let s be large enough such that for all $d' \leq d+1$, $s' > s$, $I_{d',s'} = I_{d',s}$ and $Q_{d',s'}(d') = Q_{d',s}(d')$. Then, in stage s, one could choose $a_{l_d,s+1}, \ldots, a_{r_d,s+1}$ to be $a_{l_{d+1}}, \ldots, a_{r_d+l_{d+1}-l_d}$, which makes $Q_{e,s+1}(d) > Q_{e,s}(d)$, and thus $Q_{d,s+1}(d) > Q_{d,s}(d)$, in contradiction to the choice of s. It follows from Claim 6 that, for all e, for all but finitely many $d \geq e$, $Q_e(d) = Q_e(d+1)$. Thus we get the following:

Claim 7. *For all e, for all but finitely many $d > e$, $i_e(d+1) = i_e(d)$. We let $j_e = \lim_{d\to\infty} i_e(d)$.*

Claim 8. *For all e, $j_e \in \{0, 2\}$.*

To prove the claim, suppose by way of contradiction that $j_e = 1$, for some least e. Choose d large enough such that, for all $e' \leq e$, for all $d' \geq d$, $i_{e'}(d') = j_{e'}$. Consider a large enough stage s such that, for all $d' \leq d$, for all $s' \geq s$, $I_{d',s'} = I_{d',s}$ and $Q_{d',s'}(d') = Q_{d',s}(d')$. Then we could make $Q_{e,s'}(d) > Q_{e,s}(d)$, for large enough $s' > s$ by choosing $a_{l_d,s'}, \ldots, a_{r_d,s'}$ (with $a_{l_d,s'} > a_{l_d}$) appropriately such that for all $e' \leq e$, if $I_d \subseteq \text{range}(\varphi_{e'})$, then $\varphi_{e'}^{-1}(a_{m,s'}) < \varphi_{e'}^{-1}(a_{n,s'})$ for $l_d \leq m < n \leq r_d$. (It is possible to choose such values as, for $e' \leq e$, if $I_d \subseteq \text{range}(\varphi_{e'})$, then $I_{d'} \subseteq \text{range}(\varphi_{e'})$ for all $d' > d$, and then we can use Lemma 3.) But this contradicts the choice of s.

Claim 9. *For all e, for all but finitely many $d \geq e$, $i_e(d) = 0$ implies, for all but finitely many d, $\text{range}(\varphi_e) \cap I_d = \emptyset$.*

To prove the claim, suppose by way of contradiction that e is such that for all but finitely many $d \geq e$, $i_e(d) = 0$, but for infinitely many d, $\text{range}(\varphi_e) \cap I_d \neq \emptyset$. Fix least such e, and let d be such that (i) for all $e' \leq e$, for all $d' \geq d$, $Q_e(d') = Q_e(d)$, and (ii) for all $e' < e$, if $i_{e'}(d) = 0$, then for all $d' \geq d$, $\text{range}(\varphi_{e'}) \cap I_{d'} = \emptyset$. Let s be such that for all $d' \leq d$, for all $s' \geq s$, $I_{d',s'} = I_{d',s}$ and $Q_{d',s'}(d') = Q_{d',s}(d')$. Let $E = \{e' : e' < e, i_{e'}(d) = 2\} \cup \{e\}$. Then, clearly, $\bigcap_{e' \in E} \text{range}(\varphi_{e'})$ is infinite, and thus using Lemma 3, for large enough $s' > s$, we can find, $a_{l_d,s'}, \ldots, a_{r_d,s'}$ such that $i_{e',s'}(d) = 2$ for $e' \in E$, which makes $Q_{d,s'}(d) > Q_{d,s}(d)$, contradicting the choice of s. The claim follows.

Note above that $r_e - l_e \geq Q_{e+1}(e+1)$ for all possible values of $Q_{e+1}(e+1)$, and thus $a_{r_e - Q_{e+1}(e+1)} \in I_e$. Let

$$A = \{a_{r_e - Q_{e+1}(e+1)} : e \in \mathbb{N}\}.$$

Claim 10. *A is cohesive.*

To prove the claim, consider any total φ_e. If for all but finitely many $d > e$, $i_e(d) = 0$, then by Claim 9 range(φ_e) contains elements from only finitely many $I_{e'}$, and thus only finitely many elements of A. On the other hand, if, for all but finitely many $d > e$, $i_e(d) = 2$, then range(φ_e) contains all but finitely many $I_{e'}$, and thus all but finitely many elements of A. The claim follows.

Claim 11. Suppose $B \leq_m A$ as witnessed by φ_e. Then, B is a left-r.e. set.

To prove the claim, first suppose that range(φ_e) $\cap A$ is finite. In this case $B = \{y : \varphi_e(y) \in S\}$ for some finite set S. Thus, B is recursive and a left-r.e. set.

Now suppose that range(φ_e) $\cap A$ is infinite. It follows that, for all but finitely many $d > e$, $i_e(d)$ has value 2 (by Claims 8 and 9). Let d be large enough such that $Q_e(d) = Q_e(d')$, for all $d' \geq d$. Consider a stage s_0 such that for all $d' \leq d$, for all $s \geq s_0$, $I_{d',s} = I_{d',s_0}$ and $Q_{d',s}(d') = Q_{d',s_0}(d')$. Define $s_{k+1} > s_k$ such that, for $d \leq d' \leq d+k+1$, $Q_{e,s_{k+1}}(d') = (j_0, j_1, \ldots, j_e)$. Let

$$\alpha(m, k) = a_{r_m - Q_{m+1,s_k}(m+1)},$$

and define B_k as the characteristic function of $\{y : \varphi_e(y) \in A_{s_k} \cap \bigcup_{r<d+k} I_{r,s_k}\}$ where $A_{s_k} = \{\alpha(m, k) : m < d + k\}$.

The characteristic value of B_k as above converges to characteristic function of B. To show that B is left-r.e., we need to show that $B_k \leq_{\text{lex}} B_{k+1}$. For this consider least d' such that for $m \leq d'$, $I_{m,s_{k+1}} = I_{m,s_k}$ and $Q_{m,s_{k+1}}(m) = Q_{m,s_k}(m)$, but

$$[I_{d'+1,s_{k+1}} \neq I_{d'+1,s_k} \text{ or } Q_{d'+1,s_{k+1}}(d'+1) \neq Q_{d'+1,s_k}(d'+1) \text{ or } d' = d+k+1].$$

Note that $d' \geq d$. If $d' \geq d+k$, then clearly $B_k \leq_{\text{lex}} B_{k+1}$. Otherwise, for $m < d'$, we have that $\alpha(m, k) = \alpha(m, k+1)$. Also, $Q_{d'+1,s_k} < Q_{d'+1,s_k+1}$ and $\alpha(d', k+1) < \alpha(d', k)$, which implies that $\varphi_e^{-1}(\alpha(d', k+1)) < \varphi_e^{-1}(\alpha(d', k))$ (as φ_e^{-1} is monotonic on I_{d',s_k}, due to $Q_{e,s_k}(d') = Q_{e,s_{k+1}}(d') = (j_0, j_1, \ldots, j_e)$, where $j_e = 2$). Thus, $B_k \leq_{\text{lex}} B_{k+1}$. It follows that B is a left-r.e. set. □

Not every left-r.e. set is many-one closed left-r.e.: Besides Ω, a quite easy example can be found by taking an r.e. and nonrecursive set A and considering the set

$$B = \{2x : x \in A\} \cup \{2x + 1 : x \notin A\}.$$

Then the complement of A is many-one reducbible to B but not a left-r.e. set. In contrast to Theorem 4, one can also find cohesive sets with this property.

Theorem 12. *There is a left-r.e. cohesive set A which is not a many-one closed left-r.e. set.*

Proof. In the following let $W_{d,s}$ denote the set of elements of W_d below s which are enumerated within s steps into W_d. Partition $\mathbb{N}$ into intervals I_i of length 2^i: $I_i = \{2^i - 1, 2^i, 2^i + 1, \ldots, 2^{i+1} - 2\}$. Furthermore, assign to every x the e-state given as

$$q_{e,s}(x) = \sum_{d<e} 2^{e-1-d} * W_{d,s}(x).$$

We say that

$q_{e,s}(I_i) = c$ iff $c < 2^e$ is the largest number satisfying $q_{e,s}(x) \geq c$ for at least $2^i - 2^{i-e-1} \cdot (c+1)$ elements of I_i.

Here we let $J_{e,i,s}$ be a witness for the above fact in the way such that $J_{e,i,s} \subseteq I_i$, $|J_{e,i,s}| = 2^i - 2^{i-e-1} \cdot (c+1)$ and $q_{e,s}(x) \geq c$ for all $x \in J_{e,i,s}$. Here we assume that $J_{e,i,s+1} \neq J_{e,i,s}$ implies that $q_{e,s+1}(I_i) > q_{e,s}(I_i)$. It is easy to verify that $\lim_{s\to\infty} q_{e,s}(I_i)$ converges for each e, i and thus, $\lim_{s\to\infty} J_{e,i,s}$ converges for each e, i.

Define $i_{0,s}, i_{1,s}, \ldots$ such that the following properties are satisfied:
  (a) for all e, s: $i_{e,s} < i_{e+1,s}$ and $i_{e,s+1} \geq i_{e,s} > 2e+2$;
  (b) for all e, s, j with $i_{e,s} \leq j \leq s$ it holds that $q_{e,s}(I_{i_{e,s}}) \geq q_{e,s}(I_j)$.
  (c) for all s, for the least e (if any) such that $i_{e,s} \neq i_{e,s+1}$ or $J_{e,i_{e,s},s} \neq J_{e,i_{e,s+1},s+1}$: $q_{e,s+1}(I_{i_{e,s+1}}) > q_{e,s}(I_{i_{e,s}})$.

Note that such $i_{j,s}$ can be recursively defined. It is easy to verify by induction that $i_e = \lim_{s\to\infty} i_{e,s}$ converges. Furthermore, note that $q_{0,s}(I_{i_{0,s}}) = 0$ for all s and $J_{0,i_{0,s},s} = I_{i_{0,s}}$ for all s. Hence, $i_{0,s} = i_{0,0}$ for all s. Now we are ready to define A.

Definition of A_s:
  Let $H_{e,s} = \{x \in J_{e,i_{e,s},s} : q_{e,s}(x) = q_{e,s}(I_{i_{e,s}})\}$ for all e.
  Let $x_{e,s}$ be the $(q_{e+1,s}(I_{i_{e+1,s}}) + 1)$-th element from above of $H_{e,s}$ for all e.
  Let $A_s = \{x_{0,s}, x_{1,s}, \ldots\}$.
End Definition of A_s

Let $A(x) = \lim_{s\to\infty} A_s(x)$. One can verify that $\lim_{s\to\infty} i_{e,s}$, $\lim_{s\to\infty} q_{e,s}(I_{i_{e,s}})$ and $\lim_{s\to\infty} J_{e,i_{e,s},s}$ converge. Thus it is easy to verify that A is well defined. We also let $i_e, J_{e,i_e}, H_e, q_e(x), q_e(I_j)$ denote the limiting values of $i_{e,s}, J_{e,i_{e,s},s}, H_{e,s}, q_{e,s}(x), q_{e,s}(I_j)$, respectively.

Here, it should be noted that $H_{e,s}$ has at least $2^{i_{e,s}-e-1}$ elements. To see this, let $c = q_{e,s}(I_{i_{e,s}})$ and note that $J_{i_{e,s},s}$ has at least $2^i - 2^{i-e-1} \cdot (c+1)$ elements of which less than $2^{i_{e,s}} - 2^{i_{e,s}-e-1} \cdot (c+2)$ many x satisfy $q_{e,s}(x) > c$ while all x satisfy $q_{e,s}(x) \geq c$. So at least $2^{i_{e,s}-e-1}$ elements x of $J_{e,i_{e,s},s}$ satisfy $q_{e,s}(x) = c$ and these are in $H_{e,s}$. As $i_{e,s} \geq 2e+2$, it follows that $|H_{e,s}| \geq 2^{e+1}$ and so there is, for each possible value c' of $q_{e+1,s}(I_{e+1,s}) < 2^{e+1}$, a $(c'+1)$-th largest element of $H_{e,s}$. Thus every $x_{e,s}$ as defined above really exists. For each e, the sequence of the $x_{e,s}$ converges to some value x_e.

To show that $(A_s)_{s\in\mathbb{N}}$ forms a left r.e. approximation, we need to show that $A_s \leq_{\text{lex}} A_{s+1}$. So consider the least e (if any) such that $x_{e,s+1} \neq x_{e,s}$. Note that $i_{e,s+1} = i_{e,s}$ and $J_{e,i_{e,s+1},s+1} = J_{e,i_{e,s},s}$, as otherwise $e > 0$ and $x_{e-1,s+1} \neq x_{e-1,s}$. Hence $H_{e,s+1} \subseteq H_{e,s}$ and, for $s' = s, s+1$, $x_{e,s'}$ is the $(q_{e+1,s'}(I_{i_{e+1,s'}})+1)$-th element of $H_{e,s'}$ from above. As $i_{d,s+1} = i_{d,s}$ and $J_{d,i_{d,s+1},s+1} = J_{d,i_{d,s},s}$ for all $d \leq e$, it follows by rule (c) that $q_{e+1,s+1}(I_{e+1,i_{e+1,s+1},s+1}) \geq q_{e+1,s}(I_{e+1,i_{e+1,s},s})$. Hence $x_{e,s+1} < x_{e,s}$ and that implies that $A_{s+1} >_{\text{lex}} A_s$. So A is a left-r.e. set.

Now we show that A is cohesive. So consider any d, e, k such that $d < e$ and $k \geq 0$. Then, we claim that $q_{d+1}(x_e) \geq q_{d+1}(x_{e+k})$. To see this, suppose $2^k c \leq q_{e+k}(I_{i_{e+k}}) \leq 2^k c + 2^k - 1$. Thus, at least $2^i - 2^{i-e-k-1} \cdot (2^k \cdot c + 2^k)$, many x in $I_{i_{e+k}}$ have $q_{e+k}(x) \geq 2^k c$. Thus, $2^i - 2^{i-e-1}(c+1)$ of x in $I_{i_{e+k}}$ have $q_e(x) \geq c$ and thus $q_e(I_{e+k}) \geq c$. Now, for $x_{e+k} \in H_{e+k}$ and $x_e \in H_e$, $q_{d+1}(x_{e+k}) = \lfloor q_{e+k}(I_{e+k})/2^{k+e-d-1} \rfloor < (c+1)2^k/2^{k+e-d-1}$, and thus $q_{d+1}(x_{e+k}) \leq c/2^{e-d-1}$. On the other hand, $q_{d+1}(x_e) = \lfloor q_e(I_e)/2^{e-d-1} \rfloor \geq \lfloor c/2^{e-d-1} \rfloor$. Thus, $q_{d+1}(x_{e+k}) \leq q_{d+1}(x_e)$.

Thus, as $A = \{x_0, x_1, \ldots\}$, for all d, $q_{d+1}(x_e)$ is same for all but finitely many e. For each d it follows that $W_d(x_e)$ is the same value for almost all e. Thus A is cohesive.

Now consider $B \leq_m A$ via f where, for all i and $x \in I_i$, $f(x) = \max(I_i) + \min(I_i) - x$. Note that $f(x) = f^{-1}(x)$. Thus, f also witnesses $A \leq_m B$. Let $(A_s)_{s \in \mathbb{N}}$ be the left-r.e. approximation of A as given above and $(B_s)_{s \in \mathbb{N}}$ be a left-r.e. approximation of B. Then, the following holds for all e, s:

(∗) If the least $e+1$ elements $x_{0,s}, x_{1,s}, \ldots, x_{e,s}$ of A_s satisfy that $f(x_{0,s})$, $f(x_{1,s}), \ldots, f(x_{e,s})$ are the unique elements of B_s below $\max(\{I_{i_{e,s}}\})$ then $x_0 = x_{0,s}, x_1 = x_{1,s}, \ldots, x_e = x_{e,s}$.

For a proof, assume that the above would be false for some e, s and let d be the least index such that $x_d \neq x_{d,s}$; by the left-r.e.-ness of the approximation, $x_d < x_{d,s}$. Furthermore, by (c), $i_{d,s} = i_d$ as otherwise $d > 0$ and $x_{d-1} \neq x_{d-1,s}$. So $f(x_{d,s}) < f(x_d)$ and $B \cap \{0, 1, \ldots, \max(I_{i_d})\} = \{f(x_0), f(x_1), \ldots, f(x_d)\}$. But $\{f(x_0), f(x_1), \ldots, f(x_d)\} <_{lex} \{f(x_{0,s}), f(x_{1,s}), \ldots, f(x_{d,s})\}$ and hence $B <_{lex} B_s$, a contradiction to $(B_s)_{s \in \mathbb{N}}$ being a left-r.e. approximation of B. So (∗) is true. Now one can determine x_e by searching for the first stage s where $f(x_{0,s})$, $f(x_{1,s}), \ldots, f(x_{e,s})$ are the unique elements of B below $\max(\{I_{i_{e,s}}\})$ and then one knows that $x_e = x_{e,s}$. Thus, we get that A is recursive, in contradiction to A being cohesive.   □

3   Ascending Closed Left-R.E. Sets

An *ascending reduction* is a recursive function f which satisfies $f(x) \leq f(x+1)$ for all x; $B \leq_{asc} A$ iff there is an ascending reduction f with $B(x) = A(f(x))$ for all x. A is called *ascending closed left-r.e.* iff every $B \leq_{asc} A$ is a left-r.e. set.

Let $A[n]$ denote the string $A(0)A(1) \ldots A(n)$. Let $C(x)$ denote the *plain Kolmogorov complexity* for x. That is, $C(x) = \min \{\log(y) : U(y) = x\}$, where U is a fixed universal Turing machine. The function mapping n to $C(A[n])$ is called the initial segment complexity of A and the next result shows that the initial segment complexity of ascending closed left-r.e. sets is sublinear.

Proposition 13. *If A is an ascending closed left-r.e. set then the initial segment complexity $n \mapsto C(A[n])$ is a function of sublinear order.*

Proof. Let c be any constant, and let G_n denote the interval $\{x : x \leq \lceil n/c \rceil\}$. For $d < c$, define B^d by $B^d(x) = A(cx + d)$. Thus $B^d \leq_{asc} A$. Let $(B^d_s)_{s \in \mathbb{N}}$

be left-r.e. approximations of B^d. For each n, let $d_n < c$ be the index for which $(B_s^{d_n} \cap G_n)_{s \in \mathbb{N}}$ converges slowest. Then given d_n and $B^{d_n} \cap G_n$, we can determine $B^d \cap G_n$ for each $d < c$ and therefore $A[n]$ as well. Hence, for some constant b_c and for all n, $C(A[n]) \le n/c + b_c$. This shows that the complexity function $n \mapsto C(A[n])$ has sublinear order. □

Theorem 14. *Let g be a recursive and unbounded non-decreasing function. Then there is an ascending closed left-r.e. set A such that $n \mapsto C(A[n])$ has at least the order $n/g(n)$.*

Proof. Without loss of generality assume $1 \le g(i) \le i$. Partition $\mathbb{N}$ into intervals I_i of length 2^i: $I_i = \{2^i - 1, 2^i, 2^i + 1, \ldots, 2^{i+1} - 2\}$. For each I_i, we will construct a subset $J_i = \lim_{s \to \infty} J_{i,s}$. Let $J_{i,0} = I_i$. At stage s, if there is an $e < \log(g(i))$ (which has not been handled earlier) and an x such that

$$\varphi_e(0)\downarrow \le \varphi_e(1)\downarrow \le \varphi_e(2)\downarrow \le \ldots \le \varphi_e(x)\downarrow \quad \text{and} \quad \varphi_e(x) > \max(I_i).$$

Then, choose one such e and the corresponding x. Determine the two subsets $J_{i,s} \cap \{\varphi_e(y) : y \le x\}$ and $J_{i,s} - \{\varphi_e(y) : y \le x\}$, and let $J_{i,s+1}$ be that one of these two subsets which has the higher cardinality (in case of tie, choose arbitrarily). Note that during the approximation process $J_{i,s}$ gets halved at most $\log(g(i))$ times and therefore the limit J_i has at least $2^i/g(i)$ many elements.

Define A so that the characteristic function of A on the set J_i, in ascending order, is the binary representation of the least number a_i with $C(a_i) \ge 2^i/g(i) - 2$ (where as many leading zeros are added as needed to use up all bits of J_i); A has no elements outside the sets J_i. Note that there is a recursive approximation $a_{i,s}$ to a_i from below.

The set A is left-r.e. as we can have an approximation A_s which takes on each $J_{i,s}$ the characteristic function of the binary representation of $a_{i,s}$ (with sufficiently many leading zeros added in); A_s is 0 on $I_i - J_{i,s}$. If the interval $J_{i,s}$ shrinks to $J_{i,s+1}$, then the bits of $a_{i,s}$ move to the left and some leading zeros are skipped; if $a_{i,s+1} > a_{i,s}$ then the bits are also ascending in lexicographic manner. Hence the resulting approximation is a left-r.e. approximation which runs independently on each interval I_i.

Now suppose $B \le_{\text{asc}} A$ via a recursive non-decreasing function φ_e. If the range of φ_e is finite, then B is clearly recursive. Now suppose that range of φ_e is infinite. Let r be the greatest index satisfying $g(r) \le e$. Let $s_0 = s_1 = s_2 = \ldots = s_r$ be so large that $A_{s_0}(x) = A(x)$ for all $x \le \max(I_r)$. For $k \ge r$, let $s_{k+1} > s_k$ be such that for all $s \ge s_{k+1}$ either $J_{k+1,s} \subseteq \text{range}(\varphi_e)$ or $J_{k+1,s} \cap \text{range}(\varphi_e) = \emptyset$. Note that s_{k+1} can be computed effectively from k.

Define the approximation $(B_k)_{k \in \mathbb{N}}$ of B as

$$B_k(x) = \begin{cases} A_{s_k}(\varphi_e(x)), & \text{if } \varphi_e(x) \le \max(I_k); \\ 0, & \text{if } \varphi_e(x) > \max(I_k). \end{cases}$$

This approximation is a left-r.e. approximation to B as it starts to consider the interval I_k, for $k > r$, only after stage s_k such that for all $s \ge s_k$, $J_{k,s} \subseteq \text{range}(\varphi_e)$

or $J_{k,s} \cap \mathrm{range}(\varphi_e) = \emptyset$. In the first case all the bits of J_{k,s_k} are copied order-preservingly into B_k and the left-r.e. approximation to A on I_k is turned into a left-r.e. approximation to B on the preimage of I_k under φ_e; in the second case all x with $\varphi_e(x) \in I_k$ satisfy $\varphi_e(x) \notin J_{k,s_k}$ and therefore $B_s(x) = 0$ for these x and all stages s. So A is an ascending closed left-r.e. set.

Furthermore, $C(a_i) \geq 2^i/g(i) - 2$. Also, we can compute a_i from the number i, the string $A[\max(I_i)]$ and the number of stages s at which $J_{i,s+1} \neq J_{i,s}$. Hence, for some b and almost all i we have

$$ C(A[\max(I_i)]) \geq C(a_i) - \log i - \log g(i) - b \geq \frac{2^i}{2g(i)}. $$

Taking now n with $\min(I_{i+1}) \leq n \leq \max(I_{i+1})$ and using that g is non-decreasing, we have that $C(A[n]) \geq \frac{n}{8g(n)}$ for almost all n. This proves the postulated bound.   □

Another important type of sets are the 1-generic and weakly 1-generic sets [8]. As one cannot have left-r.e. 1-generic sets, one might ask for which reducibilities r there are r-closed left-r.e. weakly 1-generic sets. The next result shows that one can make such sets for the notion of ascending closed left-r.e. sets.

Recall that a set is *weakly 1-generic* iff for every recursive function f from numbers to strings there exist n and m with $f(n) = A(n+1)A(n+2)\ldots A(n+m)$. The difference between weakly 1-generic and 1-generic is that here one requires the f to be total and independent of the values of A below n.

Theorem 15. *There is an ascending closed left-r.e. weakly 1-generic set A.*

Proof. We will be defining moving markers a_e, b_e and c_e, where $a_e \leq b_e \leq c_e$, $a_0 = 0$ and $a_{e+1} = c_e + 1$. Intuitively, we want to use the part $A(b_e), A(b_e + 1), \ldots, A(c_e)$ to ensure weak 1-genericity (by making $A(b_e)A(b_e + 1)\ldots A(b_e + |\varphi_e(b_e)| - 1) = \varphi_e(b_e)$, if $\varphi_e(b_e)$ is defined). The part $A(a_e), \ldots, A(b_e - 1)$ is used to ensure that A is ascending closed left-r.e.

At the beginning of stage s, the markers have values $a_{e,s}, b_{e,s}$ and $c_{e,s}$ respectively. We will have that $a_e = \lim_{s\to\infty} a_{e,s}$, $b_e = \lim_{s\to\infty} b_{e,s}$, $c_e = \lim_{s\to\infty} c_{e,s}$.

Let $a_{0,s} = 0$ for all s. Let $a_{e+1,s} = c_{e,s} + 1$, for all e, s. Initially $a_{e,0} = b_{e,0} = c_{e,0} = e$, and $A_0 = 0^\infty$. We will also use sets $J_{e',e,s}$, for $e' < e$. These sets are useful for defining A in such a way that, if $\varphi_{e'}$ witnesses an ascending reduction from B to A, then B is left-r.e. Initially, for all e, for $e' < e$, $J_{e',e,0} = \emptyset$. Below, for ease of presentation, we will only describe the changes from stage s to stage $s+1$; all variables which are not explicitly updated will retain the corresponding values from stage s.

Stage s:
1. If there exists an $e \leq s$ such that either Cond $e.1$ or Cond $e.2$ hold, then choose least such e and go to step 2. Otherwise go to stage $s + 1$.
   - Cond $e.1$: There exists $e' < e$ such that, $J_{e',e,s} = \emptyset$ and $\mathrm{range}(\varphi_{e'}) \cap \{x : x > c_{e,s}\}$ contains at least $2e + 2$ elements as can be verified within s steps.
   - Cond $e.2$: $c_{e,s} = b_{e,s}$ and $\varphi_e(b_{e,s})\downarrow$ within s steps.

2. Fix least e such that Cond $e.1$ or Cond $e.2$ holds. If Cond $e.1$ holds, then go to step 3. Otherwise go to step 4.

3. Fix one e' such that Cond $e.1$ holds for e'.
Let $J_{e',e,s+1}$ be $2e+2$ elements from $\text{range}(\varphi_{e'}) \cap \{x : x > c_{e,s}\}$.
Update $b_{e,s+1} = \max(J_{e',e,s+1}) + 1$, $c_{e,s+1} = b_{e,s+1}$.
For $m > e$, let $a_{m,s+1} = b_{m,s+1} = c_{m,s+1} = c_{m-1,s+1} + 1$.
For $m > e$, and $m' < m$, let $J_{m',m,s+1} = \emptyset$.
Let A_{s+1} be obtained from A_s by (i) deleting all elements $\geq b_{e,s}$, and by (ii) inserting, for each $m < e$ such that $J_{m,e,s} \neq \emptyset$, one new element (which was not earlier in A_s) from $J_{m,e,s}$.
Go to stage $s+1$.

4. Suppose $\varphi_e(b_{e,s}) = y$.
Let $c_{e,s+1} = b_{e,s} + |y| + 1$.
For $m > e$, let $a_{m,s+1} = b_{m,s+1} = c_{m,s+1} = c_{m-1,s+1} + 1$.
For $m > e$, and $m' < m$, let $J_{m',m,s+1} = \emptyset$.
Let A_{s+1} be obtained from A_s by (i) deleting all elements $\geq b_{e,s} + |y| + 1$, and by (ii) inserting, for each $m < e$ such that $J_{m,e,s} \neq \emptyset$, one new element (which was not earlier in A_s) from $J_{m,e,s}$, and by setting (iii) $A_{s+1}(b_{e,s}) \ldots A_{s+1}(b_{e,s} + |y| - 1) = y$,
Go to stage $s+1$.

End stage s.

It can be shown by induction on e that $\lim_{s\to\infty} a_{e,s}$, $\lim_{s\to\infty} b_{e,s}$, $\lim_{s\to\infty} c_{e,s}$ indeed exist. For this, for $e' < e$, after $a_{e'}, b_{e'}$ and $c_{e'}$ have reached their final value, a_e does not get modified any further (a_e is set to $c_{e-1} + 1$, in the last stage in which c_{e-1} gets modified). Furthermore, once a_e reaches its final value, b_e can change at most e times due to Cond $e.1$ holding for some $e' < e$ (and thus execution of step 3). Once b_e reaches its final value, c_e gets modified at most once due to success of Cond $e.2$ (and thus execution of step 4). The "$2e+2$" in the algorithm description suffices since each index e has e indices below it, and, after all variables $a_{e'}, b_{e'}, c_{e'}$, with $e' < e$ have stabilised, we encounter Cond $e.1$ at most once for each $e' < e$, and correspondingly Cond $e.2$ once in the beginning, and at most once after each modification of b_e via Cond $e.1$. Also, note that, for $m < e$, $J_{m,e,s} \subseteq \{x : a_{e,s} \leq x < b_{e,s}\}$.

Let $A(x) = \lim_{s\to\infty} A_s(x)$. Now we show that A is weakly 1-generic. Suppose $a_{e'}, b_{e'}, c_{e'}$, for $e' \leq e$, reach their final values before stage s. If $\varphi_e(b_e)$ is defined then Cond $e.2$ succeeds in some stage $s' \geq s$, and step 4 defines $A_{s'+1}(b_e) \ldots A_{s'+1}(b_e + |y| - 1) = y$, where $\varphi_e(b_e) = y$. Furthermore, A never gets modified on inputs $\leq c_{e,s'+1} = c_e$ after stage s'.

Now suppose $B \leq_{\text{asc}} A$ as witnessed by φ_r. If $\text{range}(\varphi_r)$ is finite, then clearly B is recursive. So assume $\text{range}(\varphi_r)$ is infinite. Thus, for each $e > r$, Cond $e.1$ will succeed (eventually) for $e' = r$, after a_e has achieved its final value.

Define s_0 such that a_m, b_m, c_m, (for $m \leq r$) as well as $A[c_r]$ have reached their final values by stage s_0. Let $s_{k+1} > s_k$ such that $J_{r,r+j,s_{k+1}} \neq \emptyset$, for all $j \leq k+1$. Let $B_k = \{x : \varphi_r(x) \in A_{s_k} \text{ and } \varphi_r(x) \leq c_{r+k,s_k}\}$.

Clearly $B(x) = \lim_{k\to\infty} B_k(x)$. Thus, to show that B is left-r.e. it suffices to show that $B_k \leq_{\text{lex}} B_{k+1}$. So consider the least $x \leq c_{r+k,s_k}$, if any, such that in some stage s', $s_k \le s' < s_{k+1}$, Cond e.1 or Cond e.2 succeeds, and $x \ge b_{e,s'}$ (if there is no such x, then we are done). Clearly, $e \ge r$ by hypothesis on s_0. Note that for $j \le k$, $J_{r,r+j,s_k} \ne \emptyset$. Thus, $J_{r,r+j,s'} \ne \emptyset$. Thus, in stage s', $A_{s'+1}(x')$ is set to 1, for some $x' \in J_{r,e,s'}$ such that $A_{s_k}(x') = 0$. Note that $x' < b_{e',s} \le x$. Let y' be least such that $\varphi_r(y') = x'$. Thus,

$$B_k \le_{\text{lex}} A_{s'+1}(\varphi_r(0)) A_{s'+1}(\varphi_r(1)) \ldots A_{s'+1}(\varphi_r(y')) \le_{\text{lex}} B_{k+1}$$

as desired.   □

Remark 16. Note that one can adjust the proof to show that there is a many-one closed left-r.e. and weakly 1-generic set. For this, main change in the construction would be to change Cond e.1 above to:

Cond e.1: There exists $e' < e$ such that $J_{e',e,s} = \emptyset$, and for some z, z', for all $x \le z$, $\varphi_{e'}(x)\downarrow \le z'$ within s steps, and $\{\varphi_{e'}(x) : x \le z\} \cap \{x : c_{e,s} < x \le z'\}$ contains at least $2e + 2$ elements.

Then, setting $J_{e',e,s+1}$ as in step 3, and making b_e to be $> z'$, would achieve the goal, as any element in A which is larger than z' would be able to influence membership in $B = \{x' : \varphi(x') \in A\}$, only for $x > z$. We omit the details.

The next result shows that every r-cohesive set is ascending closed left-r.e. set. Thus, r-cohesive left-r.e. sets form a subclass of ascending closed left-r.e. sets. Recall that every cohesive set is r-cohesive.

Theorem 17. *Every left-r.e. r-cohesive set is an ascending closed left-r.e. set.*

Proof. Suppose A is a left-r.e. r-cohesive set. Suppose $B \le_{\text{asc}} A$. Let $(A_s)_{s \in \mathbb{N}}$ be the left-r.e. approximation of A and f be a non-decreasing recursive function which witnesses that $B \le_{\text{asc}} A$. If range$(f) \cap A$ is finite, then clearly B is recursive. So assume range$(f) \cap A$ is infinite. But then, for some x and for all $y \ge x$, $y \in A$ implies $y \in$ range(f). Fix this x.

Let s_0 be such that for all $s \ge s_0$, for all $y \le x$, $A_s(y) = A(y)$; let $s_{n+1} > s_n$ be such that the least $n + 1$ members of $A_{s_{n+1}} - \{y : y \le x\}$ exist and are in range(f); note that one can effectively find such s_{n+1} from s_n. Let

$$B_n(y) = \begin{cases} A(f(y)), & \text{if } f(y) \le x; \\ 1, & \text{if } f(y) \text{ is among the least } n \text{ members} \\ & \text{of } A_{s_n} \text{ which are greater than } x; \\ 0, & \text{otherwise.} \end{cases}$$

It is easy to verify that B_n is an approximation of B. To see that $(B_n)_{n \in \mathbb{N}}$ form a left-r.e. approximation, we need to show that $B_n \le_{\text{lex}} B_{n+1}$ for all n. So consider any n. If $B_n \not\subseteq B_{n+1}$, then there exists least y such that $y \in B_n - B_{n+1}$. Then $f(y)$ is among the first n members of A_{s_n} which are greater than x and not among the first $n+1$ members of $A_{s_{n+1}}$ which are greater than x. As $(A_s)_{s \in \mathbb{N}}$ is left-r.e. approximation, we have that $A_{s_{n+1}}$ must contain a $f(y')$, $x < f(y') < f(y)$, such that $f(y') \notin A_{s_n}$. But, then $y'' \in B_{n+1} - B_n$, for some $y'' \le y'$. Thus, $(B_n)_{n \in \mathbb{N}}$ is a left-r.e. approximation of B.   □

References

1. Chaitin, G.J.: Incompleteness theorems for random reals. Advances in Applied Mathematics 8(2), 119–146 (1987)
2. Downey, R., Hirschfeldt, D.: Algorithmic Randomness and Complexity. Springer, New York (2010)
3. Friedberg, R.M.: Three theorems on recursive enumeration. I. Decomposition. II. Maximal set. III. Enumeration without duplication. The Journal of Symbolic Logic 23, 309–316 (1958)
4. Lachlan, A.H.: On the lattice of recursively enumerable sets. Transactions of American Mathematical Society 130, 1–37 (1968)
5. Li, M., Vitányi, P.: An introduction to Kolmogorov complexity and its applications, 3rd edn. Springer, New York (2008)
6. Myhill, J.: Solution of a problem of Tarski. The Journal of Symbolic Logic 21(1), 49–51 (1956)
7. Nies, A.: Computability and randomness. Oxford University Press, New York (2009)
8. Odifreddi, P.: Classical recursion theory. Studies in Logic and the Foundations of Mathematics, vol. 125. North-Holland, Amsterdam (1989)
9. Post, E.: Recursively enumerable sets of positive integers and their decision problems. Bulletin of the American Mathematical Society 50, 284–316 (1944)
10. Robinson, R.W.: Simplicity of recursively enumerable sets. The Journal of Symbolic Logic 32, 162–172 (1967)
11. Rogers Jr., H.: Theory of recursive functions and effective computability. MIT Press, Cambridge (1987)
12. Soare, R.I.: Cohesive sets and recursively enumerable Dedekind cuts. Pacific Journal of Mathematics 31(1), 215–231 (1969)
13. Soare, R.I.: Recursively enumerable sets and degrees. Springer, Berlin (1987)
14. Zvonkin, A.K., Levin, L.A.: The complexity of finite objects and the development of the concepts of information and randomness by means of the theory of algorithms. Russian Mathematical Surveys 25(6), 83 (1970)

Π_1^0 Sets and Tilings

Emmanuel Jeandel and Pascal Vanier

Laboratoire d'Informatique Fondamentale de Marseille
{emmanuel.jeandel,pascal.vanier}@lif.univ-mrs.fr

Abstract. In this paper, we prove that given any Π_1^0 subset P of $\{0,1\}^\mathbb{N}$ there is a tileset τ with a countable set of configurations C such that P is recursively homeomorphic to $C \setminus U$ where U is a computable set of configurations. As a consequence, if P is countable, this tileset has the exact same set of Turing degrees.

Introduction

Wang tiles have been introduced by Wang [17] to study fragments of first order logic. Knowing whether a tileset can tile the plane with a given tile at the origin (also known as the origin constrained domino problem) was proved undecidable also by Wang [18]. Knowing whether a tileset can tile the plane in the general case was proved undecidable by Berger [2,3].

Understanding how complex, in the sense of recursion theory, the tilings of a given tileset can be is a question that was first studied by Myers [13] in 1974. Building on the work of Hanf [10], he gave a tileset with no recursive tilings. Durand/Levin/Shen [9] showed, 40 years later, how to build a tileset for which all tilings have high Kolmogorov complexity.

A Π_1^0-set is an effectively closed subset of $\{0,1\}^\mathbb{N}$, or equivalently the set of oracles on which a given Turing machine halts. Π_1^0-sets occur naturally in various areas in computer science and recursive mathematics, see e.g. [6,15] and the upcoming book [7]. It is easy to see that the set of tilings of a given tileset is a Π_1^0-set (up to a recursive coding of $Q^{\mathbb{Z}^2}$ into $\{0,1\}^\mathbb{N}$). This has various consequences. As an example, every non-empty tileset contains a tiling which is not Turing-hard (see Durand/Levin/Shen [9] for a self-contained proof). The main question is how different the sets of tilings are from Π_1^0-sets. In the context of one-dimensional symbolic dynamics, some answers to these questions were given by Cenzer/Dashti/King/Tosca/Wyman [8,4,5].

The main result in this direction was obtained by Simpson [16], building on the work of Hanf and Myers: for every Π_1^0-set S, there exists a tileset whose set of tilings have the same *Medvedev* degree as S. The Medvedev degree roughly relates to the "easiest" Turing degree of S. What we are interested in is a stronger result: *can we find for every Π_1^0-set S a tileset whose set of tilings have the same Turing degrees ?* We prove in this article that this is true if S contains a recursive point. More exactly we build (theorem 2) for every Π_1^0-set S a set of tilings for which the set of Turing degrees is exactly the same as for S, possibly

M. Ogihara and J. Tarui (Eds.): TAMC 2011, LNCS 6648, pp. 230–239, 2011.

with the additional Turing degree of recursive points. In particular, as every *countable* Π_1^0-set contains a recursive point, the question is completely solved for countable sets: the sets of Turing degrees of countable Π_1^0-sets are the same as the sets of Turing degrees of countable sets of tilings. In particular, there exist countable sets of tilings with non-recursive points. This can be thought as a two-dimensional version of theorem 8 in [5].

This paper is organized as follows. After some preliminary definitions, we start with a quick proof of a generalization of Hanf, already implicit in Simpson [16]. We then build a very specific tileset, which forms a grid-like structure while having only countably many tilings. This tileset will then serve as the main ingredient in the theorem in the last section.

1 Preliminaries

1.1 Π_1^0 Sets and Degrees

A Π_1^0 set $P \subseteq \{0,1\}^{\mathbb{N}}$ is a set for which there exists a Turing machine that given $x \in \{0,1\}^{\mathbb{N}}$ as an oracle halts if and only if $x \notin P$. Equivalently, a subset $S \subseteq \{0,1\}^{\mathbb{N}}$ is Π_1^0 if there exists a recursive set L so that $w \in S$ if no prefix of w is in L.

We say that two sets S, S' are *recursively homeomorphic* if there exists a bijective recursive function $f : S \to S'$.

A point x of a set $S \subseteq \{0,1\}^{\mathbb{N}}$ is *isolated* if it has a prefix that no other point of S has. The *Cantor-Bendixson derivative* $D(S)$ of S is the set S without its isolated points. We define inductively $S^{(\lambda)}$ for any ordinal λ:

- $S^{(0)} = S$
- $S^{(\lambda+1)} = D\left(S^{(\lambda)}\right)$
- $S^{(\lambda)} = \bigcap_{\gamma < \lambda} S^{(\gamma)}$ when λ is limit.

The *Cantor-Bendixson rank* of S, noted $CB(S)$, is defined as the first ordinal λ such that $S^{(\lambda)} = S^{(\lambda+1)}$. An element x is of rank λ in S if λ is the least ordinal such that $x \notin S^{(\lambda)}$.

See Cenzer/Remmel [6] for Π_1^0 sets and Kechris [11] for Cantor-Bendixson rank and derivative.

For $x, y \in \{0,1\}^{\mathbb{N}}$ we say that x is *Turing-reducible* to y if y is computable by a Turing machine using x as an oracle and we write $y \leq_T x$. If $x \leq_T y$ and $y \leq_T x$, we say that x and y are *Turing-equivalent* and we write $x \equiv_T y$. The *Turing degree* of $x \in \{0,1\}^{\mathbb{N}}$ is its equivalence class under the relation $\equiv_T$.

1.2 Tilings and SFTs

Wang tiles are unit squares with colored edges which may not be flipped or rotated. A *tileset* T is a finite set of Wang tiles. A *configuration* is a mapping $c : \mathbb{Z}^2 \to T$ assigning a Wang tile to each point of the plane. If all adjacent tiles of a configuration have matching edges, the configuration is called a *tiling*.

The set of all tilings of T is noted $\mathcal{T}(T)$. We say a tileset is *origin constrained* when the tile at position $(0,0)$ is forced, that is to say, we only look at the valid tilings having a given tile t at the origin.

A *Shift of Finite Type (SFT)* $X \subseteq \Sigma^{\mathbb{Z}^2}$ is defined by (Σ, F) where Σ is a finite alphabet and F a finite set of *forbidden patterns*. A *pattern* is a coloring of a finite portion $P \subset \mathbb{Z}^2$ of the plane. A point x is in X if and only if it does not contain any forbidden pattern of F anywhere. In particular, the set of tilings of a Wang tileset is a SFT. Conversely, any SFT is recursively homeomorphic to a Wang tileset. More information on SFTs may be found in Lind and Markus' book [12].

A set of configurations $X \subseteq \Sigma_X^{\mathbb{Z}^2}$ is a *sofic shift* iff there exists a SFT $Y \subseteq \Sigma_Y^{\mathbb{Z}^2}$ and a local map $f : \Sigma_Y \to \Sigma_X$ such that for any point $x \in X$, there exists a point $y \in Y$ such that for all $z \in \mathbb{Z}^2, x(z) = f(y(z))$.

The notion of *Cantor-Bendixson derivative* is defined on configurations in a similar way as with Π_1^0 sets. This notion was introduced for tilings by Ballier/Durand/Jeandel [1]. A configuration c is said to be *isolated* in a set of configurations C if there exists a pattern P such that c is the only configuration of C containing P. The Cantor-Bendixson derivative of C is noted $D(C)$ and consists of all configurations of C except the isolated ones. We define $C^{(\lambda)}$ inductively for any ordinal λ as above.

2 Π_1^0 Sets and Origin Constrained Tilings

A straightforward corollary of Hanf [10] is that Π_1^0 subsets of $\{0,1\}^{\mathbb{N}}$ and origin constrained tilings are recursively isomorphic. This is stated explicitly in Simpson [16].

Theorem 1. *Given any Π_1^0 subset P of $\{0,1\}^{\mathbb{N}}$, there exists a tileset and a tile t such that each origin constrained tiling with this tileset describes an element of P.*

Proof. We take the basic encoding of Turing machines as stated in Robinson [14] for instance. We modify the bottom tiles, ie the tiles containing the initial tape, such that instead of being able to contain only the blank symbol, they can contain only 0s or 1s on the right of the starting head. The Turing machine we encode is the one that given $x \in \{0,1\}^{\mathbb{N}}$ as an input halts if and only if $x \notin P$. Then the constrained tilings, having at the origin the tile with the starting head of the Turing machine, are exactly the runs of the Turing machine on the members of P.   □

Corollary 1. *Any Π_1^0 subset P of $\{0,1\}^{\mathbb{N}}$ is recursively homeomorphic to an origin constrained tileset.*

3   The Tileset

The main problem in the construction of Hanf is that tilings which do not have the given tile at the origin can be very wild : they may correspond to configurations with no computation (no head of the Turing Machine) or computations starting from an arbitrary (not initial) configuration. A way to solve this problem is described in [13] but is unsuitable for our purposes.

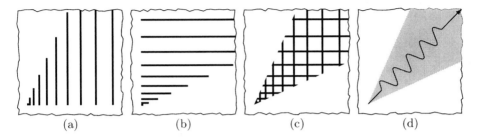

Fig. 1. The tiling in which to encode the Turing machines

Our idea is as follows: We build a tileset which will contain, among others, the *sparse grid* of figure 1c. The main point is that all others tilings of the tileset will have at most one intersection point of two black lines. This means that if we put computation cells of a given Turing machine in the intersection points, every tiling which is not of the form of figure 1c will contain at most one cell of the Turing machine, thus will contain no computation.

To do this construction, we will first draw increasingly big and distant columns as in figure 1a and then superimposing the same construction for rows as in figure 1b, leading to the grid of figure 1c.

It is then fairly straightforward to see how we can encode a Turing machine inside a configuration having the skeleton of figure 1c by looking at it diagonally: time increases going to the north-east and the tape is written on the north west - south east diagonals[1].

Our set of tiles T of figure 2 gives the skeleton of figure 1a when forgetting everything but the black vertical borders. We will prove in this section that it is countable. We set here the vocabulary:

- a vertical line is formed of a vertical succession of tiles containing a vertical black line (tiles 5, 6, 17, 21, 24, 25, 26, 27, 31, 35, 36, 37).
- a horizontal line is formed of a horizontal succession of tiles containing a horizontal black line (tiles 13, 14, 15, 16, 22, 23, 38) or a bottom signal,
- the bottom signal ___ is formed by a connected path of tiles among (30, 31, 27, 14, 7, 36, 38)
- the red signal ▮ is formed by a connected path of tiles containing a red line (tiles among 3 ,7, 10, 12, 14, 19, 22, 32, 33, 38).
- tile 30 is the corner tile
- tiles 30, 32, 33, 34 are the bottom tiles

Lemma 1. *The tileset T admits at most one tiling with two or more vertical lines.*

---

[1] Note that we will have to skip one diagonal out of two in our construction, in order for the tape to increase at the same rate as the time.

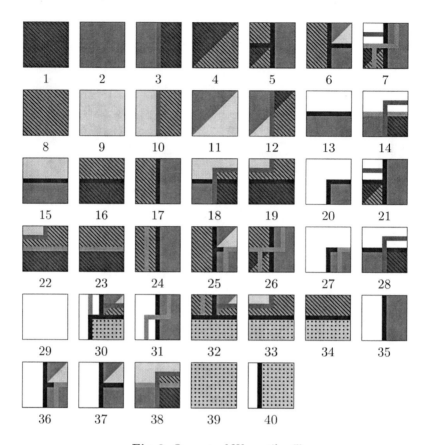

Fig. 2. Our set of Wang tiles T

Proof. The idea of the construction is to force that whenever there are two vertical lines, then the only possible tiling is the one of figure 3. Note that whenever the corner tile appears in a tiling, it is necessarily a shifted version of the tiling on figure 3.

Suppose that we have a tiling in which two vertical lines appear. Suppose they are at distance $k + 1$. Necessarily there must be horizontal lines between them forming squares. Inside these squares there must be a red signal: inside each square, this red signal is vertical, it is shifted to the right each time it crosses a horizontal line. This ensures that there are exactly k squares in this column. Furthermore, the bottom square has necessarily a bottom signal going through its top horizontal line. The bottom signal forces the square of the column before to be of size $k - 1$ and the square of the column after to be of size exactly $k + 1$.

□

Lemma 2. *The tileset T admits a countable number of tilings.*

Proof. Lemma 1 states that there is only one tiling that has more than 2 vertical lines. This means that the other tilings have at most one such line.

- If a tiling has exactly one vertical line, then it can have at most two horizontal lines: one on the left of the vertical one and one on the right. A red signal can then appear on the left or the right of the vertical line arbitrary far from it. There is a countable number of such tilings.
- If a tiling has no vertical line, then it has at most one horizontal line. A red signal can then appear only once. There is a finite number of such tilings.

There is a countable number of tilings that can be obtained with the tileset T. All obtainable tilings are shown in figure 4 and 3.   □

By taking our tileset $T = \{1, \ldots, 40\}$ and mirroring all the tiles along the south west-north east diagonal, we obtain a tileset $T' = \{1', \ldots, 40'\}$ with the exact same properties, except it enforces the squeleton of figure 1b. Remember that whenever the corner tile appeared in a tiling, then necessarily this tiling was α. The same goes for T' and its corner tile. We hence construct a third tileset $\tau = (T \setminus \{30\} \times T' \setminus \{30'\}) \cup \{(30, 30')\}$. The corner tile $(30, 30')$ of τ has the property

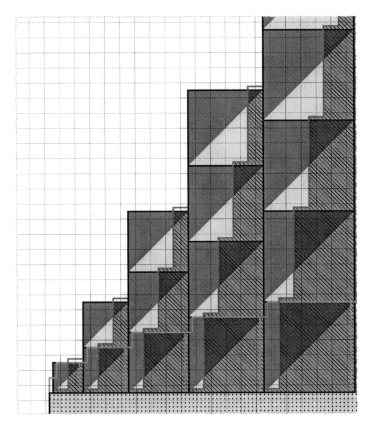

Fig. 3. Tiling α: the unique valid tiling of T in which there are 2 or more vertical lines

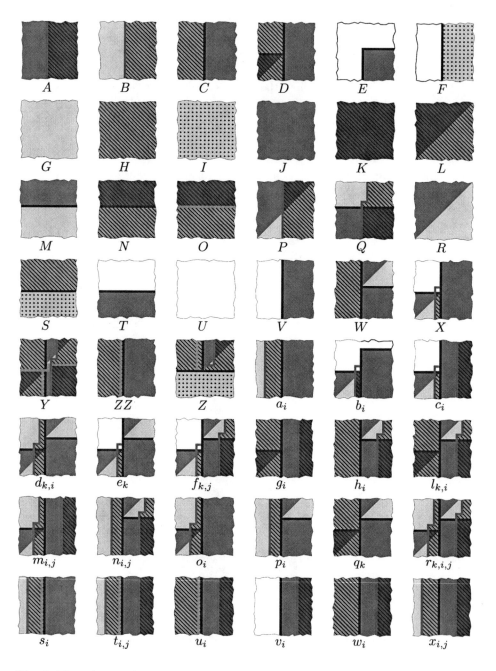

Fig. 4. The other configurations: the $A - ZZ$ configurations are unique (up to shift), and the configurations with subscripts $i, j \in \mathbb{N}, k \in \mathbb{Z}^2$ represent the fact that distances between some of the lines can vary. Note that configuration ZZ cannot have a red signal on its left, because it would force another vertical line.

that whenever it appears, the tiling is the superimposition of the skeletons of figures 1a and 1b with the corner tiles at the same place: there is only one such tiling, call it β.

The skeleton of figure 1c is obtained if we forget about the parts of the lines of the T layer (resp. T') that are superimposed to white tiles, 29' (resp. 29), of T' (resp. T).

As a consequence of lemma 2, τ is countable. And as a consequence of lemma 1, the only tiling by τ in which computation can be embedded is β. The shape of β is the one of figure 1c, the coordinates of the points of the grid are the following (supposing tile $(30, 30')$ is at the center of the grid):

$$\{(f(n), f(m)) \mid f(m)/4 \le f(n) \le 4f(m)\}$$

$$\{(f(n), f(m)) \mid m/2 \le n \le 2m\}$$

where $f(n) = (n+1)(n+2)/2 - 1$.

Lemma 3. *The Cantor-Bendixson rank of $\mathcal{T}(\tau)$ is 12.*

Proof. The Cantor-Bendixson rank of $\mathcal{T}(T) \setminus \{\alpha\}$ is 6, see figure 4, thus the rank of $\mathcal{T}(T) \setminus \{\alpha\} \times \mathcal{T}(T') \setminus \{\alpha'\}$ is 11. Adding the configurations corresponding to the superimposition of α and α', τ is of rank 12.   □

4 Π_1^0 Sets and Tilings

Theorem 2. *For any Π_1^0 subset S of $\{0,1\}^{\mathbb{N}}$ there exists a tileset τ_S such that $S \times \mathbb{Z}^2$ is recursively homeomorphic to $\mathcal{T}(\tau_S) \setminus O$ where O is a computable set of configurations.*

Proof. This proof uses the construction of section 3. Let M be a Turing machine such that M halts with x as an oracle iff $x \notin S$. Take the tileset τ of section 3 and encode in it the Turing machine M having as an oracle x on an unmodifiable second tape. This gives us τ_M, O is the set all tilings except the β ones. To each $(x, p) \in S \times \mathbb{Z}^2$ we associate the β tiling having a corner at position p and having x on its oracle tape. It follows from lemma 2 that O is clearly computable.   □

Corollary 2. *For any countable Π_1^0 subset S of $\{0,1\}^{\mathbb{N}}$, there exists a tileset τ having exactly the same Turing degrees.*

Proof. We know, from Cenzer/Remmel [6], that countable Π_1^0 sets have **0** (computable elements) in their set of Turing degrees, thus the tileset τ_M described in the proof of theorem 2 has exactly the same Turing degrees as S.   □

Theorem 3. *For any countable Π_1^0 subset S of $\{0,1\}^{\mathbb{N}}$ there exists a tileset τ_S such that $CB(\mathcal{T}(\tau_S)) = CB(S) + 11$.*

Proof. Lemma 3 states that $\mathcal{T}(\tau)$ is of Cantor-Bendixson rank 12, 11 without α. In the tileset τ_M of the previous proof, the Cantor-Bendixson rank of the contents of the tape is exactly $CB(S)$, hence $CB(\mathcal{T}(\tau_S)) = CB(S) + 11$.   □

From Ballier/Durand/Jeandel [1] we know that for any tileset X, if $CB(\mathcal{T}(X)) \geq 2$, then X has only recursive points. Thus an optimal construction improves the Cantor-Bendixson rank by at least 2.

Corollary 3. *For any countable Π_1^0 subset S of $\{0,1\}^{\mathbb{N}}$ there exists a sofic subshift X such that $CB(X) = CB(S) + 2$.*

Proof. Take a projection that just keeps the symbols of the Turing machine tape τ_M of the proof of theorem 2 and maps everything else to a blank symbol. Recall the Turing machine tape cells are the intersections of the vertical lines and horizontal lines. This projection leads to 3 possible configurations :

- a completely blank configuration,
- a completely blank configuration with only one symbol somewhere,
- a configuration with a white background and points corresponding to the intersections in the sparse grid of figure 1c.

$\square$

References

1. Ballier, A., Durand, B., Jeandel, E.: Structural aspects of tilings. In: 25th International Symposium on Theoretical Aspects of Computer Science (STACS) (2008)
2. Berger, R.: The Undecidability of the Domino Problem. PhD thesis, Harvard University (1964)
3. Berger, R.: The Undecidability of the Domino Problem. Memoirs of the American Mathematical Society, vol. 66. The American Mathematical Society, Providence (1966)
4. Cenzer, D., Dashti, A., King, J.L.F.: Computable symbolic dynamics. Mathematical Logic Quarterly 54(5), 460–469 (2008)
5. Cenzer, D., Dashti, A., Toska, F., Wyman, S.: Computability of Countable Subshifts. In: Ferreira, F., Löwe, B., Mayordomo, E., Mendes Gomes, L. (eds.) CiE 2010. LNCS, vol. 6158, pp. 88–97. Springer, Heidelberg (2010)
6. Cenzer, D., Remmel, J.B.: Π_1^0 classes in mathematics. In: Handbook of Recursive Mathematics - Volume 2: Recursive Algebra, Analysis and Combinatorics, ch. 13. Studies in Logic and the Foundations of Mathematics, vol. 139, pp. 623–821. Elsevier, Amsterdam (1998)
7. Cenzer, D., Remmel, J.: Effectively Closed Sets. ASL Lecture Notes in Logic (2011) (in preparation)
8. Dashti, A.: Effective Symbolic Dynamics. PhD thesis, University of Florida (2008)
9. Durand, B., Levin, L.A., Shen, A.: Complex tilings. Journal of Symbolic Logic 73(2), 593–613 (2008)
10. Hanf, W.: Non Recursive Tilings of the Plane I. Journal of Symbolic Logic 39(2), 283–285 (1974)
11. Kechris, A.S.: Classical descriptive set theory. Graduate Texts in Mathematics, vol. 156. Springer, New York (1995)
12. Lind, D., Marcus, B.: An introduction to symbolic dynamics and coding. Cambridge University Press, New York (1995)
13. Myers, D.: Non Recursive Tilings of the Plane II. Journal of Symbolic Logic 39(2), 286–294 (1974)

14. Robinson, R.M.: Undecidability and Nonperiodicity for Tilings of the Plane. Inventiones Math. 12 (1971)
15. Simpson, S.: Mass Problems Associated with Effectively Closed Sets (in preparation)
16. Simpson, S.G.: Medvedev Degrees of 2-Dimensional Subshifts of Finite Type. Ergodic Theory and Dynamical Systems (2011)
17. Wang, H.: Proving theorems by Pattern Recognition II. Bell Systems Technical Journal 40, 1–41 (1961)
18. Wang, H.: Dominoes and the $\forall \exists \forall$ case of the decision problem. Mathematical Theory of Automata, 23–55 (1963)

Intuitive Probability Logic

Chunlai Zhou *

Department of Computer Science and Technology,
School of Information, Renmin University of China, Beijing
czhou@ruc.edu.cn

Abstract. In literature, different deductive systems are developed for probability logics. But, for formulas, they provide essentially equivalent definitions of consistency. In this paper, we present a guided maximally consistent extension theorem which says that any probability assignment to formulas in a finite local language satisfying some constraints specified by probability formulas is consistent in probability logics, and hence connects this *intuitive* reasoning with formal reasoning about probabilities. Moreover, we employ this theorem to show two interesting results:

- The satisfiability of a probability formula is equivalent to the solvability of the corresponding system of linear inequalities through a natural translation based on atoms, not on Hintikka sets;
- the Countably Additivity Rule in Goldblatt [6] is necessary for his deductive construction of final coalgebras for functors on **Meas**, the category of measurable spaces.

Keywords: Probability Logic, Belief Types, Modal Logic, Probability Measure.

1 Introduction

Probability logics are motivated by reasoning about knowledge and belief in economics [1], [2], artificial intelligence [4] and formal methods [3]. In this paper, we mainly consider a guided maximally consistent extension theorem which connects *intuitive* reasoning with *formal* reasoning in probability logics.

The building blocks of our syntactic formalism for probability logics are *formulas*. They are constructed from the propositional falsum $\perp$ and propositional letters by the Boolean connectives and a family of belief operators L_r where $r \in \mathcal{Q} \cap [0, 1]$. The characteristic feature of the syntax is this family of operators. The interpretation of $L_r\phi$ is that the agent's belief in the event ϕ is at least r.

Different probability logics have been developed to provide a syntactic definition of *consistency*. In [4], Fagin and Halpern defined a richer langauge for their logic. Their logical language includes not only formulas expressing probabilities but also linear combinations of probability formulas. In order to accommodate this rich syntax with "arithmetic" connectives, they had to formulate an independent system for linear inequalities. In [7], Heifetz and Mongin used a much

---

* This research is partly supported by NSF of China (Grant Number 60905036).

M. Ogihara and J. Tarui (Eds.): TAMC 2011, LNCS 6648, pp. 240–251, 2011.

simpler syntax suggested in [2] by Aumann. Just as ours, its characteristic feature is captured by the belief operators L_r^i for rationals $r \in [0, 1]$ and $i \in I$ where I is a set of agents. In some sense, this syntax is a probabilistic syntax of modal logic. In Zhou [12], we provided a new axiomatization of probability logic by including a new rule capturing the Archimedean property of probability indices. All the above results are only weak completeness. That is to say, a formula is consistent in the system iff it has a model based on a type space.

Deductive systems for probability logics with strong completeness, which would identify satisfied theories as maximally consistent *sets* of formulas, were studied in Goldblatt [6] and Meier [10]. Goldblatt formulated a deductive system for coalgebras over measurable spaces [11] in the same finitary language as ours. But, in order to show the strong completeness, he *postulated* the Lindebaum property. And he showed that the adoption of "Lindenbaum's Lemma" as a postulate rather than a property to be proved is *unavoidable*. The most important contribution of his system is the following Countable Additivity Rule:

$$\Gamma \vdash \phi \text{ implies } \{L_p \psi : \psi \in \bigwedge_\omega \Gamma\} \vdash L_p \phi$$

where Γ may be infinite and $\bigwedge_\omega \Gamma$ is the set of conjunctions of finite subsets of Γ. It is needed in the proof of countable additivity of the measures defined on the canonical models [6], which also depends on the postulate of Lindenbaum property. Meier employed an *infinitary* language including countable conjunction (disjunction) and developed an infinitary probability logic which is strongly complete with respect to the class of type spaces.

From these deductive systems for probability logic, it seems that there is a kind of *inherent* relationship between deductive or *formal* reasoning about probabilities and *intuitive* reasoning about linear inequalities. It is easy to see that probability logics include implicitly or explicitly reasoning about linear inequalities. Specifically, for a probability formula ϕ, its consistency implies through a translation based on atoms the solvability of its corresponding system of linear inequalities. One may wonder whether the converse is true, i.e., the solvability of its corresponding system entails its consistency. In this paper, we will show that the entailment holds. This result means that under a certain translation reasoning about probabilities is equivalent to reasoning about linear inequalities. The crucial step to show this equivalence is a guided maximally consistent extension theorem which says that any probability assignment to formulas in a finite local language satisfying some constraints specified by probability formulas is consistent in probability logics. The reason for the equivalence is that, for a probability formula ϕ, reasoning about its consistency under a certain translation is simply about its probability indices and hence about the solvability of its corresponding system of linear inequalities.

However, the deductive machinery for the consistency of an arbitrary *set* of formulas is much more than reasoning about probability indices. Our first reaction would be to see whether finite satisfiability implies satisfiability, i.e., given an arbitrary set Γ of formulas, if any finite subset is satisfiable, so is Γ. We can find an immediate counterexample $\{\neg M_0 p\} \cup \{L_{\frac{1}{2^n}} p : n \in N\}$, which is finitely satisfiable but not satisfiable. Further one may wonder whether finite satisfiability without such an obvious contradiction *in probability indices* is satisfiable. In this paper, we will show that this is not true by the above guided maximally consistent extension theorem. This negative result also tells us that reasoning about a *set* of probability formulas is more than reasoning about linear inequalities. Moreover, it says that the Countable Additivity Rule in Goldblatt [6] is necessary for the deductive construction of final coalgebras for functors over measurable spaces [11].

The paper is organized as follows. Section 2 introduces the semantics and syntax for probability logic and also presents two probability logics Σ_+ and Σ_s. Σ_s includes the Countable Additivity Rule while Σ_+ does not. Our main result is the guided maximally consistent extension theorem in Section 3. In the last part of this paper, we provide two applications of the theorem. One says that the satisfiability of a probability formula is equivalent to the solvability of the corresponding system of linear inequalities through a certain translation based on atoms not on Hintikka sets. The other tells us that the Countably Additivity Rule in Goldblatt [6] is necessary for his deductive construction of final coalgebras for functors over **Meas**.

2 Semantics and Syntax

The syntax of our logic is very similar to that of modal logic. We start with a fixed infinite set $P := \{p_1, p_2, \cdots\}$ of propositional letters. We also use $p, q, \cdots$ to denote propositional letters. The set of formulas ϕ is built from propositional letters as usual by connectives $\neg$, $\wedge$ and a countably infinite modalities L_r for each $r \in \mathcal{Q} \cap [0, 1]$, where $\mathcal{Q}$ is the set of rational numbers. Equivalently, a formula ϕ is formed by the following syntax:

$$\phi := p \mid \neg\phi \mid \phi_1 \wedge \phi_2 \mid L_r\phi \ (r \in \mathcal{Q} \cap [0, 1])$$

L_r is the primitive modality in our language. But we also use a derived modality M_r which means "at most" in our semantics through the following definition:

$$\text{(DEF M)} \quad M_r\phi := L_{1-r}\neg\phi.$$

Let $\mathcal{L}$ be the formal language consisting of the above components. We use $r, s, \alpha, \beta, \cdots$ (also with subscripts) to denote rationals. Next we describe the semantics of our system. A *probability model* is a tuple

$$M := \langle \Omega, \mathcal{A}, T, v \rangle$$

where

- Ω is a non-empty set, which is called *the universe or the carrier set* of M;
- $\mathcal{A}$ is a σ-field (or σ-algebra) of subsets of Ω;
- T is a measurable mapping from Ω to the space $\Delta(\Omega, \mathcal{A})$ of probability measures on Ω, which is endowed with the σ-field generated by the sets:

$$\{\mu \in \Delta(\Omega, \mathcal{A}) : \mu(E) \geq \alpha\} \text{ for all } E \in \mathcal{A} \text{ and rational } \alpha \in [0, 1],$$

- v is a mapping from P to $\mathcal{A}$, i.e. $v(p) \in \mathcal{A}$.

$\langle \Omega, \mathcal{A}, T \rangle$ is called a *type space*, and T is called a *type function* on the space. A *finite-additive type space* can be defined similarly by simply replacing all (countably-additive) probability measures with finitely additive probability measures.

Definition 1. *For a fixed model M, the satisfaction relation $\models$ between the state w of M and modal formulas ϕ is defined inductively as follows:*

- $M, w \models p$ *iff* $w \in v(p)$ *for propositional letters p;*
- $M, w \models \phi_1 \wedge \phi_2$ *iff* $M, w \models \phi_1$ *and* $M, w \models \phi_2$;
- $M, w \models \neg\phi$ *iff* $M, w \not\models \phi$;
- $M, w \models L_r\phi$ *iff* $T(w)([[\phi]]) \geq r$, *where* $[[\phi]] := \{w \in \Omega : M, w \models \phi\}$.

Note that the associated interpretation $[[\phi]] = \{w \in M : M, w \models \phi\}$ is a measurable set, for all formulas ϕ.

ϕ is *valid* in the probability model M if $M \models \phi$, i.e. for all states $w \in M$, $M, w \models \phi$. ϕ is *valid in a class of probability models $\mathcal{C}$* if, for each $M \in \mathcal{C}$, $M \models \phi$. ϕ is *valid in a class $\mathcal{T}$ of type spaces* if ϕ is valid in all the probability models defined on $\mathcal{T}$.

In this section we will give an axiomatization Σ_+ of probability logic, which is *weakly complete* with respect to the class of type spaces in [12]. Our system is different from that by Heifetz and Mongin in that we don't need the rule (B) there

Probability Logic Σ_+

Axiom Schemata:

- A0: propositional tautologies
- A1: $L_0\phi$
- A2: $L_r\top$
- A3: $L_r(\phi \wedge \psi) \wedge L_t(\phi \wedge \neg\psi) \rightarrow L_{r+t}\phi$, $r + t \leq 1$
- A4: $\neg L_r(\phi \wedge \psi) \wedge \neg L_s(\phi \wedge \neg\psi) \rightarrow \neg L_{r+s}\phi$, $r + s \leq 1$
- A5: $L_r\phi \rightarrow \neg L_s\neg\phi$, $r + s > 1$

Rules:

- (Detachment) ϕ and $\phi \rightarrow \phi'$ infer ϕ'
- (DIS) $\phi \leftrightarrow \psi$ infer $L_r\phi \leftrightarrow L_r\psi$.
- (ARCH): $\gamma \rightarrow L_s\phi$ for all $s < r$ infer $\gamma \rightarrow L_r\phi$.

Probability logic Σ_+ is the smallest set of formulas that contains all propositional tautologies in $\mathcal{L}$ and $(A_1 - A_5)$, and is closed under detachment, DIS, ARCH and uniform substitution. $\vdash_{\Sigma_+} \phi$ denotes the theoremhood of ϕ in Σ_+. We should stress that in this paper $\vdash_{\Sigma_+}$ is used only for theoremhood. In other words, the notations $\phi \in \Sigma_+$ and $\vdash_{\Sigma_+} \phi$ mean the same thing. Also, when the context is clear, we usually omit the subscript Σ_+ in $\vdash_{\Sigma_+}$.

Probability logic Σ_s is the system Σ_+ plus the following additional rules:

- (Assumption Rule) If $\phi \in \Gamma \cup \{A_0, A_1, A_2, A_3, A_4, A_5\}$, then $\Gamma \vdash \phi$,
- (CR) (Cut Rule) If $\Gamma \vdash \psi$ for all $\psi \in \Sigma$ and $\Sigma \vdash \phi$, then $\Gamma \vdash \phi$,
- (DR) Deduction Rule: $\Gamma \cup \{\phi\} \vdash \psi$ implies $\Gamma \vdash \phi \rightarrow \psi$.
- (CAR) (Countable Addivity Rule) $\Gamma \vdash \phi$ implies $\{L_p\psi : \psi \in \bigwedge_\omega \Gamma\} \vdash L_p\phi$ where Γ may be infinite and $\bigwedge_\omega \Gamma$ is the set of conjunctions of finite subsets of Γ.

The rule (CAR) is the characteristic principle for the probability logic Σ_S. The other four rules are simply auxiliary [6].

Definition 2. *A formula ϕ is consistent in Σ_+ (or in Σ_S) if it is a theorem in Σ_+ (or in Σ_S).*

Theorem 1. *For any formula ϕ, it is consistent in Σ_+ iff it is consistent in Σ_s iff it is satisfied in a type space.*

Proof. The interested reader may refer to [6],[13].

3   A Guided Maximally Consistent Extension Theorem

The following theorem says that intuitive reasoning with probabilities can be formalized in probability logics.

3.1   Three-Dimensional Language

In some sense, our following finite language of probabilistic logic is three-dimensional. Enumerate all propositional letters: $p_1, p_2, \cdots$. For a given formula ψ, its accuracy index $q(\phi)$ is the least common multiple of all denominators of the indices that appear in ψ, its depth index $d(\psi)$ is the depth of the formula which refers to the depth of the quantified modality L_r and its width index $w(\psi)$ is the number of propositional letters that occur in ψ. Now we define a finite local language $\mathcal{L}(q, d, w)$. We use index vector (q, d, w) to denote its three dimensions: q is the accuracy index, d the depth index and w the width index. $(q, d, w) \preccurlyeq (q', d', w')$ denotes the combination of the relationships: q is a factor of q', $d \leq d'$ and $w \leq w'$. $(q, d, w) \prec (q', d', w')$ means further that at least one of the three holds: q is a proper factor of q' (q' is divisible by but not equal to q), $d < d'$ and $w < w'$. Define $\mathcal{L}(q, d, w)$ to be the smallest set of formulas that satisfies the following conditions:

1. the indices of any formula is a multiple of $1/q$;
2. the propositional letters that occur are among $p_1, \cdots, p_w$;
3. the depths of formulas are $\leq d$;
4. *tautologically equivalent* formulas are regarded as the same.

So each formula in $\mathcal{L}(q,d,w)$ is an equivalence class. By induction on the depth d, we can show that $\mathcal{L}(q,d,w)$ is finite.

This finite set $\mathcal{L}(q,d,w)$ gives rise to a set $\Omega(q,d,w)$ of maximal consistent sets of formulas in the local language $\mathcal{L}(q,d,w)$, which are called *atoms* of $\Omega(q,d,w)$. Let A_q denote the set of rationals between 0 and 1 which are a multiple of $1/q$. Next we extend each atom Γ to $\mathcal{L}(q',d',w')$ such that $(q,d,w) \prec (q',d',w')$ and define, for any formula ϕ in the language $\mathcal{L}(q,d,w)$,

$$\alpha_\phi^{\Gamma'} := \max\{r \in A_{q'} : L_r\phi \in \Gamma'\} \text{ and } \beta_\phi^{\Gamma'} := \min\{r \in A_{q'} : M_r\phi \in \Gamma'\}.$$

3.2 Guided Maximally Consistent Extension

Fix an index vector (q,d,w). The elements of $\Omega(q,d,w)$ can be enumerated as follows:

$$\Gamma_1(q,d,,w), \Gamma_2(q,d,w), \cdots, \Gamma_{N(q,d,w)}(q,d,w)$$

where $N(q,d,w)$ denotes the number of atoms in $\Omega(q,d,w)$. Let $\gamma_i(q,d,w)$ denote $\bigwedge \Gamma_i(q,d,w)$. We can define a probability space $S(q,d,w)$ on $\Omega(q,d,w)$: $S(q,d,w) := \langle \Omega(q,d,w), 2^{\Omega(q,d,w)}, P(q,d,w) \rangle$ for any *arbitrarily given* probability measure $P(q,d,w)$ on $\Omega(q,d,w)$. Since the parameters q and w are not much relevant in the following reasoning, we denote $P(q,d,w)(\Gamma_i(q,d,w))$ for $(1 \leq i \leq N(q,d,w))$ by $p_i^{(d)}$ for simplicity. So $p_1^{(d)} + \cdots + p_{N(q,d,w)}^{(d)} = 1$. For any formula $\phi \in \Phi(q,d,w)$, we can show by standard normal form theorem in propositional calculus that

$$\phi \leftrightarrow \bigvee_{\phi \in \Gamma_i(q,d,w)} \gamma_i(q,d,w) \text{ is a tautology.}$$

It is easy to see that $P(q,d,w)([\phi]) = \sum_{\phi \in \Gamma_i(q,d,w)} p_i^{(d)}$ where, as usual, $[\phi] = \{\Gamma \in \Omega(q,d,w) : \phi \in \Gamma\}$. Define $l_\phi^{S(q,d,w)}$ to be the largest multiple of $\frac{1}{q}$ which is less than or equal to $P(q,d,w)([\phi])$ and $m_\phi^{S(q,d,w)}$ to be the smallest multiple of $\frac{1}{q}$ which is greater than or equal to $P(q,d,w)([\phi])$. Then such a probability space $S(q,d,w)$ will determine two probability formulas about ϕ:

$$L_{l_\phi^{S(q,d,w)}}\phi \text{ and } M_{m_\phi^{S(q,d,w)}}\phi.$$

Define $\Xi(q,d,w)$ to be the set of all these formulas, i.e.

$$\Xi(q,d,w) := \{L_{l_\phi^{S(q,d,w)}}\phi, M_{m_\phi^{S(q,d,w)}}\phi : \phi \in \Phi(q,d,w)\}.$$

Note that $\Xi(q,d,w) \subseteq \Phi(q,d+1,w)$. Intuitively, the following theorem says, an *arbitrary* probability measure assignment to atoms is consistent with our deduction system Σ_+. Hence it also connects intuitive reasoning with formal reasoning with probabilities.

Theorem 2. *The above defined $\Xi(q, d, w)$ is consistent.*

We relegate the proof of this lemma to Appendix 1.

Remark 1. From Theorem 2, we may summarize our probability logics for consistency as follows:

Probability logic = propositional calculus + reasoning about linear inequalities

Through atoms in finite local languages, probability logic can be reduced to reasoning about linear inequalities (about probability indices) with the help of propositional calculus (See the first application in the next section). So, for consistency of (nested)*formulas*, probability logic is simply the *intuitive* arithmetic of rationals designating probability indices. It is worth noting that the problem of deciding satisfiability is NP-complete [5], no worse than that of propositional logic.

4   Two Applications

In the first application of the above guided maximally consistent extension theorem, we show that the satisfiability of a probability formula is equivalent to the solvability of the corresponding system of linear inequalities through a certain translation based on atoms not on Hintikka sets.

4.1   Satisfiability of a Probability Formula and Solvability of Its Corresponding System of Linear Inequalities

Given a formula ψ, let $\mathcal{L}[\psi] = \mathcal{L}(q(\psi), d(\psi), w(\psi))$ and Ω be the set of maximally consistent sets of formulas in the finite local language $\mathcal{L}[\psi]$. Elements in Ω are called atoms. A *Hintikka set* H over $\mathcal{L}[\psi]$ is a maximal subset of $\mathcal{L}[\psi]$ that satisfies the following conditions:

1. $\perp \notin H$;
2. If $\neg\phi \in \mathcal{L}[\psi]$, then $\neg\phi \in H$ iff $\phi \notin H$;
3. If $\phi_1 \wedge \phi_2 \in \mathcal{L}[\psi]$, then $\phi_1 \wedge \phi_2 \in H$ iff $\phi_1 \in H$ and $\phi_2 \in H$.

Let Ω_H denote the set of Hintikka sets. Note that each atom is a Hintikka set over $\mathcal{L}[\psi]$ but not necessarily the other way around. Here is a counterexample. $L_{1/2}(L_{2/3}p \wedge M_{1/3}p)$ is contained in a Hintikka set but is not contained in any atom because the formula is definitely not satisfiable. This implies that $\Omega \subsetneq \Omega_H$. More importantly, we can easily *use an algorithm* to enumerate all Hintikka sets over $\mathcal{L}[\psi]$. Just as in modal logic, Hintikka sets have advantages over atoms in dealing with computational issues for probability logics.

Let $n = |\Omega|$. Assume that an atom Γ and its maximal consistent extension Γ^+ in the language $\mathcal{L}^+$ (short for $\mathcal{L}(q(\psi), d(\psi) + 1, w(\psi))$) are given. We enumerate all atoms H_i in Ω and associate to each H_i some variable x_i. For any formula ϕ in $\mathcal{L}[\psi]$, if ϕ is propositionally equivalent to the disjunction of conjunctions of formulas in different atoms: $H_1, H_2, \cdots, H_r$, then we associate ϕ with $x_1 + x_2 + \cdots + x_r$. Now we show how to associate each *satisfiable* formula in the language $\mathcal{L}^+$ with a *solvable* system of linear inequalities. Consider a formula χ in $\mathcal{L}^+$ that is satisfiable in a probability model. By propositional calculus, we know that χ is tautologically equivalent to a disjunction of conjunctions of the following form:

$$(*): \bigwedge_k p_{i_k} \wedge \bigwedge_l \neg p_{i'_l} \wedge \bigwedge_m L_{r_m} \gamma_{j_m} \wedge \bigwedge_{m'} \neg L_{s_{m'}} \gamma_{j_{m'}}$$

where p_{i_k} and $p_{i'_l}$ are propositional letters. The first part $\bigwedge_k p_{i_k} \wedge \bigwedge_l \neg p_{i'_l}$ is called *the propositional part* of this conjunction and $\bigwedge_m L_{r_m} \gamma_{j_m} \wedge \bigwedge_{m'} \neg L_{s_{m'}} \gamma_{j_{m'}}$ is called *the probability part*. Since χ is satisfiable, so is at least one disjunct. So, without loss of generality, we simply assume that χ is a conjunction of the above form $(*)$. We associate to χ a system of linear inequalities as follows:

1. For each γ_{j_m}, if it is associated with $x_1 + \cdots + x_{i_{j_m}}$, then the inequality for $L_{r_m} \gamma_{j_m}$ is

$$x_1 + \cdots + x_{i_{j_m}} \geq r_m$$

2. For each $\gamma_{j_{m'}}$, if it is associated with $x_1 + \cdots + x_{i_{j_{m'}}}$, then the inequality for $\neg L_{s_{m'}} \gamma_{j_{m'}}$ is

$$x_1 + \cdots + x_{i_{j_{m'}}} < s_{m'}$$

To sum up, we get an inequality system of the form

$$S_\chi = \begin{cases} \vdots \\ x_1 + \cdots + x_{i_{j_m}} \geq r_m \\ \vdots \\ x_1 + \cdots + x_{i_{j_{m'}}} < s_{m'} \\ \vdots \end{cases}$$

Note that, since χ is satisfiable, S_χ is solvable. Assume that χ is satisfied at some state s of a probability model $M = \langle S, \mathcal{A}, T, v \rangle$. Just as above, for each set Θ of formulas, $[[\Theta]]_M$ denotes the set $\{s' \in S : M, s' \models \gamma$ for all $\gamma \in \Theta\}$. It is easy to check that $x_i = T(s)([[H_i]]_M)(1 \leq i \leq n)$ is a solution to the system S_χ.

In general, we can associate a solvable system of inequalities to a satisfiable set Γ' of formulas in the language $\mathcal{L}^+$. To achieve this, we simply take the conjunction of all formulas in this set and apply the above described procedure to the conjunction. Assume that the associated system of linear inequalities for Γ' is as follows:

$$S_{\Gamma'} = \begin{cases} x_{i_1} + \cdots + x_{i_{\Gamma_i}} \geq r_i \\ \vdots \\ x_{k_1} + \cdots + x_{k_{\Gamma_k}} > r_k \\ -x_{i'_1} - \cdots - x_{i'_{\Gamma_i}} \geq -r'_i \\ \vdots \\ -x_{k'_1} - \cdots - x_{k'_{\Gamma_k}} > -r'_k \end{cases}$$

Implicitly, we assume that all these associated systems of inequalities are probability measures. Actually we can also show the other direction from solvability of $\mathcal{S}_{\Gamma'}$ to satisfiability of Γ' but through atoms in Ω instead. In sum, we have the following equivalence statement:

Theorem 3. *Γ' is satisfiable iff $\mathcal{S}_{\Gamma'}$ is solvable.*

Proof. The left-to-right direction is straightforward from the above translation, while the other direction requires the Hintikka sets here to be atoms and hence needs the Guided Extension Theorem 2.

4.2   Canonical Finitely-Additive Probability Models Are Not a Probability Model

Theorem 4. *Any canonical finitely-additive type space, which consists of all maximally Σ_+-consistent sets of formulas, is not a probability model.*

Proof. The Γ_1 that is constructed in the Appendix according to the Guided Maximally Consistent Extension Theorem is a maximally consistent set of formulas Γ^C whose canonical probability measure is finitely additive but not countably additive.

Theorem 5. *All maximally consistent sets of formulas that are satisfied in probability models form a probability model. Moreover, it is the biggest (or universal) probability model in the sense that any other probability model can be embedded into it.*

Proof. The proof is actually measure theoretical and is analogous to Lemma 4.5 in [8].

Corollary 1. *The set of maximally Σ_+-consistent sets of formulas satisfiable in the class of type spaces is a proper subset of the set of maximally Σ_+-consistent sets.*

Heifetz & Samet showed [9] that the canonical space of all maximally Σ_+-consistent sets of formulas which are *satisfied* in the class of (countably additive) type spaces is the universal type space U in the sense that there is a unique *type morphism* from any type space to this U, which is a map that preserves the structure of the type space. Moss & Viglizzo constructed similar universal type spaces in terms of final coalgebras for polynomial functors on **Meas** from a coalgebraic perspective [11]. In [6], Goldblatt developed deduction systems including the Countable Additivity rule in the introduction to build these final coalgebras from the collections of maximally consistent sets of formulas in these deduction systems. Without this rule, we could apply the same techniques for the above three propositions to find a maximally consistent set of formulas that is not satisfiable in the class of type spaces. This negative result implies that the rule is necessary in his formulation of deduction systems for coalgebras over **Meas**.

References

1. Aumann, R.: Interactive epistemology: Knowledge. Int. J. Game Theory 28, 263–300 (1999)
2. Aumann, R.: Interactive epistemology: Probability. Int. J. Game Theory 28, 301–314 (1999)
3. Doberkat, E.E.: Stochastic Coalgebraic Logic. EATCS Mongraphs in Theoretical Computer Sciences. Springer, Heidelberg (2010)
4. Fagin, R., Halpern, J.: Reasoning about knowledge and probability. J. ACM 41, 340–367 (1994)
5. Fagin, R., Megiddo, N., Halpern, J.: A logic for reasoning about probabilities. Inf. and Comp. 87, 78–128 (1990)
6. Goldblatt, R.: Deduction systems for coalgebras over measurable spaces. J. Logic Comp. 20(5), 1069–1100 (2010)
7. Heifetz, A., Mongin, P.: Probability logic for type spaces. Games Econom. Behav. 35, 31–53 (2001)
8. Heifetz, A., Samet, D.: Topology-free typology of beliefs. J. Econom. Theory 82, 241–324 (1998)
9. Heifetz, A., Samet, D.: Coherent beliefs are not always types. J. Math. Econom. 32, 475–488 (1999)
10. Meier, M.: An infinitary probability logic for type spaces. Israel J. of Math. (to appear)
11. Moss, L., Viglizzo, I.: Final coalgebras for functors on measurable spaces. Inf. and Comp. 204, 610–636 (2006)
12. Zhou, C.: A complete deductive system for probability logic. J. Logic Comp. 19(6), 1427–1454 (2009)
13. Zhou, C.: Probability logics for finitely additive beliefs. J. Logic, Language and Information 19(3), 247–282 (2010)

Appendix 1: Proof of the Guided Extension Lemma

Proof. (Proof of the Guided Extension Theorem) First we simply repeat the construction of a finite canonical space $M^c(q, d, w) = \langle \Omega(q, d, w), 2^{\Omega(q,d,w)}, T^c \rangle$ out of the set $\Omega(q, d, w)$ as in the proof of the completeness of Σ_+ (Theorem 2.21 in [13]). Next we define a new type space by adding one new state w_0 to $\Omega(q, d, w)$ as follows:

- $\Omega'(q, d, w) = \Omega(q, d, w) \cup \{w_0\}$;
- $\mathcal{A}' = 2^{\Omega'(q,d,w)}$
- For any state $w \in \Omega(q, d, w)$ and $A \in 2^{\Omega(q,d,w)}$, $T'^c(w, A) = T^c(w, A)$, and moreover $T'^c(w, \{w_0\}) = 0$.
- For any $w(= \Gamma_i(q, d, w) \in \Omega(q, d, w))$, $T'^c(w_0, w_0) = 0$ and $T'^c(w_0, w) = P(q, d, w)(\Gamma_i(q, d, w))$, where $P(q, d, w)$ is the above given probability measure over $\Omega(q, d, w)$.
- For any propositional letter p, $v'(p) = v(p)$. That is to say, no propositional letter is true at the new state w_0.

Obviously, $M'^c(q, d, w) := \langle \Omega'(q, d, w), \mathcal{A}', T'^c, v' \rangle$ is a type space. It is easy to prove by induction on the complexity of formula $\phi \in \mathcal{L}(q, d, w)$ that, for any atom $\Gamma_i(q, d, w) \in \Omega(q, d, w)$,

$$M^c(q, d, w), \Gamma_i(q, d, w) \models \phi \text{ iff } M'^c(q, d, w), \Gamma_i(q, d, w) \models \phi$$

In other words, $[[\phi]]_{M^c(q,d,w)} = \Omega(q, d, w) \cap [[\phi]]_{M'^c(q,d,w)}$. In particular, this implies

$$[[\gamma_i(q, d, w)]]_{M'^c(q,d,w)} \subseteq \Gamma_i(q, d, w) \cup \{w_0\}, \text{ where } \gamma_i(q, d, w) = \wedge \Gamma_i(q, d, w).$$

Since $T'^c(w_0, w_0) = 0$, it follows directly that $\Xi(q, d, w)$ is satisfiable at w_0 in $M'^c(q, d, w)$. By the Completeness Theorem 1, we know that it is consistent in Σ_+ or in Σ_s.

Appendix 2: Construction of Γ_1

It is easy to see that the following set is finitely consistent:

$$C := \{\neg M_0 p_1\} \cup \{M_{\frac{1}{2^n}} p_1 : n \in \mathcal{N}\}.$$

So there is a maximally finitely consistent set Γ^∞ that contains C as a subset. Note that Γ^∞ is *not* maximally consistent. Fix any index vector (q, d, w), define

$$\Gamma^\infty(q, d, w) := \Gamma^\infty \cap \Phi(q, d, w).$$

In particular, we have a sequence of atoms in the local languages $\mathcal{L}(2^n, 2^n, 2^n)$:

$$\Gamma^\infty(1, 1, 1) \subseteq \Gamma^\infty(2, 2, 2) \subseteq \Gamma^\infty(2^2, 2^2, 2^2) \subseteq \cdots \subseteq \Gamma^\infty(2^n, 2^n, 2^n) \subseteq \cdots$$

such that $\bigcup_n \Gamma^\infty(2^n, 2^n, 2^n) = \Gamma^\infty$. In this section, $M = \langle \Omega, \mathcal{A}, T, v \rangle$ denotes the *canonical* finitely additive probability model based on algebras and, in particular, Ω is the set of all maximally Σ_+-consistent sets of formulas. $[\Gamma^\infty(2^n, 2^n, 2^n)]$ denotes the set $\{\Delta \in \Omega : \Gamma^\infty(2^n, 2^n, 2^n) \subseteq \Delta\}$. For simplicity, let $\mathcal{L}(2^n)$, $\Gamma^\infty(2^n)$ and $\Omega(2^n)$ denote $\mathcal{L}(2^n, 2^n, 2^n)$, $\Gamma^\infty(2^n, 2^n, 2^n)$ and $\Omega(2^n, 2^n, 2^n)$, respectively. Our next step is to apply the above theorem about guided maximal consistent extensions to find a maximal Σ_+-consistent set Γ_1 such that

$$\tfrac{1}{4} = T(\Gamma_1)([\Gamma^\infty(2)]) = T(\Gamma_1)([\Gamma^\infty(2^2)]) = \cdots = T(\Gamma_1)([\Gamma^\infty(2^n)]) = \cdots$$

First we consider the algebra $(\Omega(1), 2^{\Omega(1)})$. Since $\Gamma^\infty(1)$ is an atom in the local language $\mathcal{L}(1)$, $\Gamma^\infty(1) \in \Omega(1)$. Set an arbitrary probability measure $P(1)$ on this algebra satisfying the following conditions:

- $P(1)(\Gamma^\infty(1)) = \tfrac{1}{4}$;
- For all atoms Δ in $\Omega(1)$, $P(1)(\Delta)$ is a multiple of $\frac{1}{2^{m_1}}$ for some positive natural number m_1.

By applying Theorem 2, we know that there exists a maximally consistent set $\Gamma_{2^{m_1}} \in \Omega(2^{m_1})$ such that

- for any formula ϕ in the language $\mathcal{L}(1)$, $\alpha_\phi^{\Gamma_{2^{m_1}}} = \beta_\phi^{\Gamma_{2^{m_1}}} = P(1)([\phi]_1)$; and, in particular,

- $\alpha^{\Gamma_{2^{m_1}}}_{\bigwedge \Gamma^{\infty}(1)} = \beta^{\Gamma_{2^{m_1}}}_{\bigwedge \Gamma^{\infty}(1)} = \frac{1}{4}$, where $[\phi]_1 = \{\Delta \in \Omega(1) : \phi \in \Delta\}$.

Now consider the algebra $(\Omega(2^{m_1}), 2^{\Omega(2^{m_1})})$. Note that $(\Omega(1), 2^{\Omega(1)})$ can be regarded as a quotient algebra of $(\Omega(2^{m_1}), 2^{\Omega(2^{m_1})})$. Set a probability measure $P(2^{m_1})$ on this algebra that satisfies the following conditions:

- $P(2^{m_1})$ agrees with $P(1)$ on $2^{\Omega(1)}$;
- $P(2^{m_1})(\Gamma^{\infty}(2^{m_1})) = \frac{1}{4}$;
- for all atoms Δ in $\Omega(2^{m_1})$, $P(2^{m_1})(\Delta)$ is a multiple of $\frac{1}{2^{m_2}}$ for some $m_2 > m_1$.

By applying Theorem 2, we know that there is a maximal consistent set $\Gamma_{m_2} \in \Omega(2^{m_2})$ such that

- for any formula ϕ in the language $\mathcal{L}(2^{m_1})$, $\alpha^{\Gamma_{2^{m_2}}}_{\phi} = \beta^{\Gamma_{2^{m_2}}}_{\phi} = P(2^{m_1})([\phi]_{m_1})$ where $[\phi]_{m_1} = \{\Delta \in \Omega(2^{m_2}) : \phi \in \Delta\}$;
- $\alpha^{\Gamma_{2^{m_2}}}_{\bigwedge \Gamma^{\infty}(2^{m_1})} = \beta^{\Gamma_{2^{m_2}}}_{\bigwedge \Gamma^{\infty}(2^{m_1})} = \frac{1}{4}$;
- $\Gamma_{2^{m_1}} \subseteq \Gamma_{2^{m_2}}$.

By repeating above process, we get a sequence of maximally consistent sets $\Gamma_{2^{m_i}} \in \Omega(2^{m_i})$:

$$\Gamma_{2^{m_1}} \subseteq \Gamma_{2^{m_2}} \subseteq \cdots$$

that satisfies the following conditions:

- for each $\Gamma^{\infty}(2^{m_i})$, $\alpha^{\Gamma_{2^{m_{i+1}}}}_{\bigwedge \Gamma^{\infty}(2^{m_i})} = \beta^{\Gamma_{2^{m_{i+1}}}}_{\bigwedge \Gamma^{\infty}(2^{m_i})} = \frac{1}{4}$;
- any formula $\phi \in \Phi(2^{m_i})$, $\alpha^{\Gamma_{2^{m_{i+1}}}}_{\phi} = \beta^{\Gamma_{2^{m_{i+1}}}}_{\phi}$

Define $\Gamma_1 := \bigcup_i \Gamma_{2^{m_i}}$. From the above observations, it follows that Γ_1 is a maximally consistent set in $\mathcal{L}$ because it is maximal finitely-consistent and is closed under $(ARCH_t)$ [13]. So $\Gamma_1 \in \Omega$. According to our definition on canonical models,

$$\frac{1}{4} = T(\Gamma_1)([\Gamma^{\infty}(2)]) = T(\Gamma_1)([\Gamma^{\infty}(2^2)]) = \cdots = T(\Gamma_1)([\Gamma^{\infty}(2^n)]) = \cdots$$

Corollary 2. $\bigcap_i [\Gamma^{\infty}(2^{m_i})] = \emptyset$.

Corollary 3. *For such a* Γ_1, $T(\Gamma_1)$ *is finitely additive but not* σ*-additive.*

An Algebraic Characterization of Strictly Piecewise Languages*

Jie Fu, Jeffrey Heinz, and Herbert G. Tanner

University of Delaware
{jiefu,heinz,btanner}@udel.edu

Abstract. This paper provides an algebraic characterization of the Strictly Piecewise class of languages studied by Rogers et al. 2010. These language are a natural subclass of the Piecewise Testable languages (Simon 1975) and are relevant to natural language. The algebraic characterization highlights a similarity between the Strictly Piecewise and Strictly Local languages, and also leads to a procedure which can decide whether a regular language L is Strictly Piecewise in polynomial time in the size of the syntactic monoid for L.

1 Introduction

Rogers et al. [12] study the Strictly Piecewise (SP), which are a proper subclass of the Piecewise Testable (PT) languages of Simon [13]. The Strictly Piecewise languages are interesting for two reasons. First, there are several senses in which the SP class is natural. For example, SP is exactly the class of those languages closed under subsequence [12]. Also, they bear the same relation to Piecewise Testable languages that the Strictly Local (SL) bear to Locally Testable (LT) languages [10, 12]. Second, this class expresses some of the kinds of long-distance dependencies found in natural language [6, 12].

While Rogers et al. provide several characterizations of SP languages, they do not provide an algebraic one. Also, the procedure they give for deciding whether a regular language L belongs to SP is exponential in the size of the smallest deterministic acceptor for L. This paper aims to address these issues. An algebraic characterization for the SP class is provided. This result not only reveals an important similarity between the SP and SL languages, but also leads to a procedure which decides whether L belongs to SP in time quadratic in the size of syntactic monoid for L. However, it remains an open question whether a polynomial time decision procedure exists in the size of the smallest deterministic acceptor.

The rest of this paper is organized as follows. Section 2 reviews foundational concepts and notation. Section 3 defines the Piecewise Testable (PT), Strictly Piecewise (SP), and Stricly Local (SL) classes. Section 4 presents our algebraic characterization of the SP class and Section 5 describes the polynomial-time decision procedure. Finally, Section 6 concludes.

---

* This research is supported by grant #1035577 from the National Science Foundation.

M. Ogihara and J. Tarui (Eds.): TAMC 2011, LNCS 6648, pp. 252–263, 2011.

2   Preliminaries

A *semigroup* is a set with an associative operation. A *monoid* is a semigroup with an identity element (written 1). If S is a semigroup, S^1 denotes the monoid equal to S if $1 \in S$ and to $S \cup \{1\}$ otherwise. A *zero* is an element 0 such that, for every $s \in S$, $s0 = 0s = 0$. The *free* semigroup (monoid) of a set S is the set of all finite sequences of one (zero) or more elements from S.

If x is an element of set S and π a partition of S, the *block* of π containing x is $[x]_\pi$. The partition of S induced by an equivalence relation ρ is S/ρ. A *right (left) congruence* is a partition such that if $[x]_\pi = [y]_\pi$ then $[xz]_\pi = [yz]_\pi$ ($[zx]_\pi = [zy]_\pi$). A *congruence* is both a left and a right congruence.

Following Clifford [2], a *left (right) ideal* of a semigroup S is a non-empty subset T of S such that $ST \subseteq T$ ($TS \subseteq T$). The *left (right) ideal of S generated by T* is $T \cup ST = S^1 T$ ($T \cup TS = TS^1$). The *principal left (right) ideal of S generated by $t \in T$* is $PL(t) = S^1 t$ ($PR(t) = tS^1$).

Let Σ denote a finite set, called the *alphabet*. Sets Σ^+ and Σ^* denote the free semigroup and free monoid of Σ, respectively. We refer to the elements of Σ^+ and Σ^* as *strings* and *words* interchangeably. The unique string of length zero is denoted λ. The set $\Sigma^{\leq k}$ denotes the set of all words of length at most k.

The length of a string u is denoted $|u|$, and $|w|_\sigma$ denotes the number of occurences of σ in w. A string v is a *factor* of w iff there exist strings $x, y \in \Sigma^*$ such that $w = xvy$. A string v is a *prefix (suffix)* of w iff there exist $x \in \Sigma^*$ such that $w = vx$ ($w = xv$). A string v is a *subsequence* of string w iff $v = \sigma_1 \cdots \sigma_n$ and $w \in \Sigma^* \sigma_1 \Sigma^* \cdots \Sigma^* \sigma_n \Sigma^*$, and we write $v \sqsubseteq w$. Languages are subsets of Σ^*. The complement of a language L is $\overline{L} = \{w \in \Sigma^* : w \notin L\}$.

A *semiautomaton* is a tuple $A = \{Q, \Sigma, T\}$, where Q is a non-empty finite set of states and Σ is the alphabet. The transition function is a partial function $T : Q \times \Sigma \to Q$. The domain of the transition function is expanded to $Q \times \Sigma^*$ recursively as follows. For all $q \in Q$, $T(q, \lambda) = q$ and for all $w \in \Sigma^*$ and $\sigma \in \Sigma$, $T(q, wa) = T(T(q, w), a)$. It follows that $T(q, xy) = T(T(q, x), y)$. By definition semi-automata are deterministic.

A *finite-state automaton* (FSA) is a tuple $\mathbf{A} = \{Q, q_0, Q_f, \Sigma, T\}$, where $\{Q, \Sigma, T\}$ is a semi-automaton, $q_0 \in Q$ is the initial state, and $Q_f \subseteq Q$ is a set of final states. The language recognized by $\mathbf{A}$ is $\{w \in \Sigma^* : T(q_0, w) \in Q_f\}$.

A language L is *regular* iff there exists a FSA recognizing it. For every regular language L there is a unique (up to isomorphism) automaton with the fewest number of states recognizing L called the *canonical* FSA for L.

A state q of an automaton is a *sink* state iff, for all $\sigma \in \Sigma$, if $T(q, \sigma)$ is defined then $T(q, \sigma) = q$. One can always make the transition function total by adding a nonfinal sink state and directing all the missing transitions for each state to this sink. When the sink state is added to a canonical acceptor, it is the only state which is both a sink and nonfinal. The resulting automaton is *complete*.

For any automaton $\mathbf{A}$ and state $q \in Q$, let ρ_q be that relation such that, for all elements x and y of Σ^*, $x \rho_q y$ iff $T(q, x) = T(q, y)$. More generally, let

$$f_x = \begin{pmatrix} q_1 & \cdots & q_n \\ T(q_1, x) & \cdots & T(q_n, x) \end{pmatrix} .$$

For all $x, y \in \Sigma^*$, let $x\rho y$ iff $f_x = f_y$. The equivalence relation ρ over Σ^* induces a congruence over Σ^* [15]. The index of ρ is finite because Q is finite.

Let $F_A = \{f_x : x \in \Sigma^*\}$ denote the finite monoid of mappings and $\bar{I}(A) = \Sigma^*/\rho$. Then F_A is isomorphic to $\bar{I}(A)$ under the correspondence of f_x of F_A with $[x]$ of $\bar{I}(A)$, where $[x]$ is the ρ-congruence coset containing x of Σ^*. In this paper, when writing f_x and $[x]$, we choose x to be a shortest-length element in the congruence class without any ambiguity.

For FSA $\mathbf{A}$, where A is the associated semiautomaton of $\mathbf{A}$, F_A is called the *transformation semigroup* and $\bar{I}(A)$ is the *characteristic semigroup* of A. Elements f_x of F_A can also be written in matrix form μ_x, where the rows and columns indicate states in $Q = \{q_1, \ldots, q_n\}$ and $\mu_x[i, j] = 1$ iff $T(q_i, x) = q_j$.

The set of matrices is another semigroup, the *transition semigroup*. The name is derived from the fact that each element in this semigroup is a transition matrix associated to a *walk* x in $\mathbf{A}$. We write $U_A = \{\mu_x : x \in \Sigma^*\}$. Clearly U_A is isomorphic to $\bar{I}(A)$ under the correspondance of μ_x of U_A with $[x]$ of $\bar{I}(A)$.

Definition 1 (Pin 1997). *The* syntactic semigroup *of a regular language L is the transformation semigroup given by its complete canonical semiautomaton.*

In the syntactic semigroup of an automaton A, the *set of generators* of F_A is $Gen(F_A) = \{f_\sigma : \sigma \in \Sigma\}$. The *syntactic monoid* of a regular language L is the syntactic semigroup with identity, $Gen(F_A^1) = \{f_\sigma : \sigma \in \Sigma \cup \lambda\}$.

Pin [11] discusses the equivalence between automata and semigroups. Note that since the transition semigroup U_A of $\mathbf{A}$ is represented as a semigroup of boolean matrices of order $|Q| \times |Q|$, a word w is recognized by $\mathbf{A}$ iff $\mu_x(q_0, q_f) = 1$ for some final state $q_f \in Q_f$. It follows that a finite automaton recognizes a regular language L iff its transition semigroup recognizes L.

A "monoid graph" is a useful method employed by contemporary algebraic theorists to visualize monoids. The nodes of the graph are elements in the monoid, though an initial node labeled "λ" is included by convention. The labels on edges are the elements in the set of generators of the monoid. Given a monoid M, $x \xrightarrow{s} y$ iff $xs = y$, where $x, y \in M$, and $s \in Gen(M)$. The monoid graph of F_A is denoted as $MG(F_A)$. We mark elements x in the monoid graph as final iff $f_x \in F_A$ and there exists a final state q in the canonical acceptor such that $T(q_0, x) = q$ [11]. Examples of monoid graphs are in Figures 1,2, and 3.

Definition 2. *A* unique nonfinal sink state *in an automaton $\mathbf{A}$ is called* zero. *An element f_x is a zero element of the transformation semigroup iff*

$$f_x = \begin{pmatrix} q_1 & \cdots & q_n \\ \mathbf{0} & \cdots & \mathbf{0} \end{pmatrix} .$$

We use the notation $f_x = 0$ for the transformation semigroup, $\mu_x = 0$ for the transition semigroup, and $x = 0$ for the free semigroup Σ^*. The corresponding zero in the characteristic semigroup $\bar{I}(A)$ is denoted $[0]$.

While every complete canonical automaton (except the one recognizing Σ^*) has a unique nonfinal sink state, not every transformation semigroup has a zero.

3   Piecewise Testable and Strictly Piecewise Languages

The concept of a subsequence is central to the notion of piecewise testability.

Definition 3. *The* principle shuffle ideal *of v is the language of all words for which v is a subsequence. We write* $SI(v) = \{w \in \Sigma^* \mid v \sqsubseteq w\}$.

The Piecewise Testable languages is the smallest class of languages including $SI(w)$ for all $w \in \Sigma^*$ and closed under Boolean operations [13]. Similarly, the class of Piecewise k-Testable (PT_k) languages is the smallest class of languages including $SI(w)$ for all $w \in \Sigma^{\leq k}$ and closed under Boolean operations.

A well-known characterization of the PT languages is stated in terms of the sets of subsequences within words. If $P_{\leq k}(w) \overset{\mathrm{def}}{=} \{v : v \sqsubseteq w \text{ and } |v| \leq k\}$ then the following characterization (sometimes taken as the definition of PT [3]) holds.

Theorem 1. *A language L is Piecewise Testable iff there exists k such that, for all words $w_1, w_2 \in \Sigma^*$, if $P_{\leq k}(w_1) = P_{\leq k}(w_2)$ then $w_1 \in L$ iff $w_2 \in L$.*

When k is known, L is said to be Piecewise k-Testable ($L \in PT_k$).

Simon proved one of the first examples of what later became known as Eilenberg's correspondence theorem [11]. One of the relations that Green [4] defines on semigroups is the $\mathcal{J}$ relation, which relates two elements of a semigroup S if they generate the same two-sided principal ideal of S: $a\mathcal{J}b$ iff $S^1aS^1 = S^1bS^1$. A semigroup S is $\mathcal{J}$-*trivial* iff, for all $a, b \in S$, if $a\mathcal{J}b$ then $a = b$. Simon proved the following algebraic characterization of piecewise testable languages.

Theorem 2 (Simon 1975). *A language is Piecewise Testable iff its syntactic monoid is $\mathcal{J}$-trivial.*

As an example, consider the language of all words with exactly one a, $L = \{w : |w|_a = 1\}$. The canonical acceptor for this language is shown in Figure 1.

There are three elements in the monoid $F_{A_1}^1 = \{a, 1, 0\}$, (for simplicity of notation, let x stand for f_x). The $\mathcal{J}$-triviality is established by calculating $F_{A_1}^1 x F_{A_1}^1$, for all $x \in F_{A_1}^1$: $F_{A_1}^1 a F_{A_1}^1 = \{0, a\}$, $F_{A_1}^1 1 F_{A_1}^1 = \{0, a, 1\}$, and $F_{A_1}^1 0 F_{A_1}^1 = \{0\}$. The $\mathcal{J}$-triviality is satisfied, which means this language L is piecewise testable.

Rogers et al. [12] study a proper subclass of the Piecewise Testable languages, the *Strictly Piecewise* class. This paper takes as definition of Strictly Piecewise languages those languages which are closed under subsequence. (Unknown to Rogers et al., languages closed under subsequence were studied forty years earlier by Haines [5] (see also Higman [7]).)

Definition 4. *A language is Piecewise Testable in the Strict Sense ($L \in$ SP) iff, for all $w \in \Sigma^*$, if $w \in L$ and $v \sqsubseteq w$ then $v \in L$.*

Rogers et al. [12] establish the following equivalences (see also [5]).

256 J. Fu, J. Heinz, H.G. Tanner

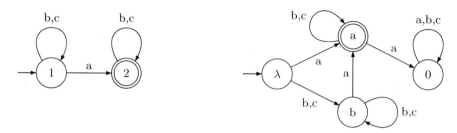

Fig. 1. The canonical automaton and the monoid graph for $L = \{w : |w|_a = 1\}$, which is the language of all words with exactly one a

Theorem 3. *The following are equivalent:*

1. $L \in SP$.
2. $L = \overline{SI(X)}, X \subseteq \Sigma^*$.
3. $L \in \bigcap_{w \in S} \overline{SI(w)}$, *for S finite.*
4. *there exists k such that if $P_{\leq k}(w) \subseteq P_{\leq k}(L)$ then $w \in L$.*

It follows from the third characterization above that any SP language can be characterized by a finite set S. Elements of this set are the *forbidden* subsequences, and the language is all words which do not contain any of these *forbidden* subsequences. The longest word in S is the length k in the 4th characterization above, in which case we say L is Strictly k-Piecewise ($L \in SP_k$)[1].

By forbidding subsequences, SP languages resemble the Stricly Local languages which forbid factors [10]. Any SL language L can be defined as the intersection of the complements of sets defined to be those words which *contain* a forbidden factor. Formally, let the *container* of $w \in \rtimes \Sigma^* \ltimes$ be $C(w) = \{u \in \Sigma^* : w$ is a factor of $\rtimes u \ltimes\}$ then a language $L \in SL$ iff there exists a finite set of forbidden factors $S \subset \rtimes \Sigma^* \ltimes$ such that $L = \bigcap_{w \in S} \overline{C(w)}$[2]. Figure 2 shows the canonical acceptor and the monoid graph for the SL language $L = \overline{\Sigma^* aa \Sigma^*} = \overline{C(aa)}$, i.e. all words except those containing the factor aa.

To illustrate SP languages, consider the language $L = \overline{SI(bb)} \cap \overline{SI(ca)}$, which is the language of all words except those containing either the subsequences bb or ca; i.e., bb and ca are the forbidden subsequences. Thus this SP language can be characterized by the set $\{bb, ca\}$ of forbidden subsequences (or equivalently by the set $\Sigma^{\leq 2}/\{bb, ca\}$ of permissible subsequences [12]). Hence this language belongs to SP_2. Figure 3 shows ths canonical automata and the monoid graph for L. The 0 element is not shown there, but note that all missing edges go to 0.

As with the other piecewise testable languages like the one in Figure 1, it is not difficult to verify that the syntactic monoid of this language is $\mathcal{J}$-trivial.

---

[1] While every SP language is convex [14], it is not the case that all convex languages are SP since, for example, there are nonempty subword-convex languages that do not contain λ but the only SP language not containing λ is the empty one.

[2] The symbols $\rtimes$ and $\ltimes$ invoke left and right word boundaries and are necessary because SL languages make distinctions at word edges [10].

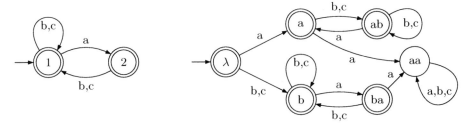

Fig. 2. The canonical acceptor and the monoid for the language $L = \overline{C(aa)}$, which is all words except those containing the factor aa

However, this language, like every other SP language, has two additional properties. Furthermore, no non-SP language has both of these properties.

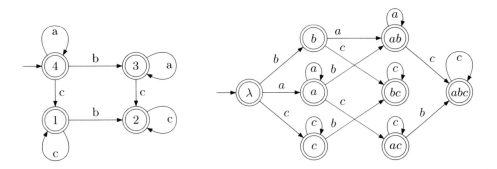

Fig. 3. The canonical automata and the monoid graph of the syntactic monoid of $L = \overline{\mathrm{SI}(bb)} \cap \overline{\mathrm{SI}(ca)}$, i.e. the language where the subsequences bb and ca are forbidden

4 Algebraic Characterization of SP

There are two important concepts that need to be introduced.

Definition 5. *Let L be a regular language recognized by FSA, and consider its characteristic semigroup. Language L is* wholly nonzero *if and only if $\overline{L} = [0]$.*

In other words, a language is wholly nonzero if and only if every word not in the language is in the zero block of the characteristic semigroup. In terms of the transformation semigroup, this means that every word x not in the language is zero; i.e., $f_x = 0$.

Theorem 4. *A language L is wholly nonzero if and only if L is closed under prefix and closed under suffix.*

Proof. Clearly, $[0] \subseteq \overline{L}$. Now suppose L is closed under prefix and suffix, and consider any $x \in \overline{L}$. For contradiction, suppose $f_x \neq 0$. Then in the canonical acceptor $\mathbf{A}$ for L there are states q, q' in $\mathbf{A}$ such that x transforms q to q'. Since $\mathbf{A}$ is canonical, there exist strings w, y such that w transforms q_0 to q and y transforms q' to a final state. Thus $wxy \in L$. Since L is closed under prefix wx belongs to L and since L is closed under suffix, x belongs to L, which contradicts the assumption. Therefore $f_x = 0$, which completes one direction of the proof.

Now suppose $\overline{L} = [0]$ and consider any $w \in L$ and any prefix (suffix) v of w, which means there exists x such that $w = vx$ ($w = xv$). If $v \notin L$ then by assumption $f_v = 0$. It follows that $f_w = f_{vx} = 0 f_x = 0$ ($f_w = f_{xv} = f_x 0 = 0$), which contradicts that $w \in L$.

Observe that $L = \Sigma^*$ and the empty language are wholly nonzero vacuously.

The following two corollaries are almost immediate.

Corollary 1. *The Strictly Piecewise languages are wholly nonzero.*

Proof. The Strictly Piecewise are closed under subsequence by definition and are therefore closed under prefix and suffix.

Corollary 2. *The Strictly Local languages are wholly nonzero.*

Proof. Consider any Strictly Local language L and any $w \in L$. Since $w \in L$, there are no forbidden factors in w and therefore there are none in any prefix or suffix of w. Hence every prefix and suffix of w belongs to L as well.

That both the Strictly Local and Strictly Piecewise are wholly nonzero is a nontrivial property they have in common. To illustrate, recall the SL language $L = \overline{C(aa)}$ (Figure 2). Every string not in this SL language transforms any state in its monoid graph to 0. These are all the strings with the 2-factor aa. Similarly, consider again the language $L = \overline{SI(bb)} \cap \overline{SI(ca)}$ (Figure 3). Every string not in this SP language transforms any state in its monoid graph to 0. These are exactly those strings with either subsequence bb or ca.

The second property is an algebraic characterization of what Rogers et al. describe in automata-theoretic terms as "missing edges propagate down." This means that if some state q in the canonical accepter does not have a transition labeled with symbol σ then no state reachable from q has an outgoing transition labeled with σ. To capture this, we need the following concept relating to zeroes.

Definition 6. *Let M be a monoid. The set of right annihilators of an element $x \in M$, is $RA(x) = \{a \in M : xa = 0\}$.*

In other words, the elements of $RA(x)$ annihilate x from the right. The set of *left annihilators* can be defined similarly, but it does not play a role here.

We now define the following property which captures the notion of "missing edges propagating down."

Definition 7. *A language L is right annihilating iff for any element f_x in the syntactic monoid $F_A(L)$, and for all f_w in the principle right ideal generated by f_x, it is the case that $RA_{F_A(L)}(f_x) \subseteq RA_{F_A(L)}(f_w)$.*

The main theorem of this paper can now be stated and proved.

Theorem 5. *A language L is* SP *iff L is wholly nonzero and right annihilating.*

Proof. By Corollary 1, any SP language is wholly nonzero.

Next consider any $L \in$ SP and any element f_x and any $f_t \in RA_{FA(L)}(f_x)$. It follows that $f_x f_t = 0$; hence, $f_{xt} = 0$. Since $L \in$ SP, there must be some $v \sqsubseteq xt$ such that v is forbidden; i.e $SI(v) \subseteq \overline{L}$. For any f_w in the principal right ideal of f_x, it is the case that there exists f_a such that $f_w = f_x f_a$. Thus $f_w f_t = f_x f_a f_t = f_{xat}$. Since $v \sqsubseteq xt$ it follows that $v \sqsubseteq xat$ and therefore $f_w f_t = f_{xat} = 0$ and so $f_t \in RA_{FA(L)}(f_w)$. The generality of f_w and f_t ensures that $\forall w \in PR(x), RA_{FA(L)}(f_x) \subseteq RA_{FA(L)}(f_w)$.

Now for the other direction. The empty language vacuously satisfies the above conditions and belongs to SP so consider any nonempty regular language L, which is wholly nonzero and right annihilating. We show that L belongs to SP.

By contradiction, suppose L is wholly nonzero and right annihilating, but not in SP. By definition of SP, L is not closed under subsequence. So there is some w and v such that $w \in L$ and $v \sqsubseteq w$ but $v \notin L$. Since $v \sqsubseteq w$, there exists $u_0, u_1, \cdots, u_n$ such that for $v = \sigma_1 \sigma_2 \ldots \sigma_n$, $w = u_0 \sigma_1 u_1 \sigma_2 u_2 \cdots \sigma_n u_n$.

Since $v \notin L$ and since L is wholly nonzero, $v \in [0]$. It will be useful to refer to the suffixes of v as follows: $v_i = \sigma_i \cdots \sigma_n$ for $1 \le i \le n$. For example, $v = v_1 = \sigma_1 \cdots \sigma_n$ and $v_2 = \sigma_2 \cdots \sigma_n$, and $v_n = \sigma_n$.

Now v_2 is a right annihilator of $u_0 \sigma_1$ since $u_0 \sigma_1 v_2 = u_0 v = u_0 0 = 0$. Also, since L is right annihilating, $RA(u_0 \sigma_1)$ is a subset of $RA(u_0 \sigma_1 u_1)$, and so v_2 right annihilates $u_0 \sigma_1 u_1$ as well.

Next consider that v_3 is a right annihilator of $u_0 \sigma_1 u_1 \sigma_2$ since $u_0 \sigma_1 u_1 \sigma_2 v_3 = u_0 \sigma_1 u_1 v_2$ and above we showed that v_2 right annihilates $u_0 \sigma_1 u_1$. Again, since L is right annihilating, $RA(u_0 \sigma_1 u_2 \sigma_2)$ is a subset of $RA(u_0 \sigma_1 u_1 \sigma_2 u_2)$, and so v_3 right annihilates $u_0 \sigma_1 u_1 \sigma_2 u_2$ as well.

Carrying this argument through to its conclusion, we see that $v_n = \sigma_n$ is a right annihilator of $u_0 \sigma_1 u_1 \sigma_2 u_2 \cdots u_{n-1} \sigma_{n-1}$. Therefore σ_n is a right annihilator of $u_0 \sigma_1 u_1 \sigma_2 u_2 \cdots u_{n-2} \sigma_{n-1} u_n$ as well. Hence $u_0 \sigma_1 u_1 \sigma_2 u_2 \cdots u_{n-2} \sigma_{n-1} u_n \sigma_n = 0$.

But this means that $w = u_0 \sigma_1 u_1 \sigma_2 u_2 \cdots u_{n-2} \sigma_{n-1} u_n \sigma_n u_n = 0 u_n = 0$. Since L is wholly nonzero, it follows that $w \notin L$, which contradicts the reduction assumption. Therefore there is no v, w such that $w \in L$, $v \sqsubseteq w$, and $v \notin L$. It follows that regular languages that are wholly nonzero and right annihilating are closed under subsequence and are therefore SP.

We illustrate this property in the context of the decision procedure we present below for deciding whether a regular language is SP.

5   Algorithms for SP Languages

Theorem 3 provides a polynomial-time decision procedure for deciding whether any regular language L is SP, and if it is, it finds the finite set of the shortest forbidden subsequences necessary to define L.

5.1 Deciding SP

The input to the algorithms below is taken to be the monoid graph of the syntactic monoid for a regular language L, with the initial state being the node labeled "λ" and the final states being marked. Since this graph is determinstic, it is possible to obtain the canonical acceptor in time $O(n \log n)$ [9]. Given a minimal DFA $\mathbf{A}$, the syntactic monoid F_A can be obtained through the set of generators $\{f_\sigma\}, \forall \sigma \in \Sigma$. The reader is referred to [1] for the construction method of syntactic monoid F_A.

Theorem 3 provides the basis for the decision procedure, which we call DSP. DSP simply checks whether the syntactic monoid satisfies the wholly nonzero and the right annihilation conditions.

The wholly nonzero condition can be checked in two steps, essentially by checking closure under prefixes and suffixes. To check closure under prefixes, one simply need check whether every state in the canonical accepter $\mathbf{A}$ is final. If they are not, then the syntactic monoid of $\mathbf{A}$ is not wholly nonzero. To check closure under suffixes, both the complete canonical acceptor and the transformation semigroup F_A are examined. Let 0 be the non-final sink state in the complete automaton. If there exists one nonzero element f_x in F_A and one noninitial state q in the canonical acceptor such that $T(q,x) \neq 0$ but $T(q_0, x) = 0$, then the wholly nonzero condition is violated. If no such f_x or q exist, however, we can conclude the language is wholly nonzero.

Whether the right annihilating condition is satisfied can be determined from the Cayley table for F_A. The columns and rows of a Cayley table are labeled with the elements in the syntactic monoid F_A, and the cell is the product$(x \cdot y)$ of the row-th(x) and column-th(y) elements [2]. Then for each $f_x \in F_A$, the principal right ideal generated by x $(PR(x))$ can be found by the union of all distinct elements in the xth row of the table and the right annhilators of x $(RA(x))$ are given by those elements y such that the xth row and y^{th} column is 0.

Then for each $z \in PR(x)$, it is sufficient to check whether $RA(x) \subseteq RA(z)$. If for any $x \in F_A$ and any $z \in PR(x)$, it is the case that $RA(x) \not\subseteq RA(z)$ then the algorithm exits and returns "false". Otherwise it returns "true".

We illustrate these procedures with three examples. Consider first the SP language $L = \overline{\text{SI}(bb)} \cap \overline{\text{SI}(ca)}$ in Figure 3. The elements of its transformation semigroup $F_A(L) = \{f_x, x \in \Sigma^*\}$ are:

$$f_a = \begin{pmatrix} 1\,2\,3\,4 \\ 0\,0\,3\,4 \end{pmatrix} \qquad f_b = \begin{pmatrix} 1\,2\,3\,4 \\ 2\,0\,0\,3 \end{pmatrix} \qquad f_c = \begin{pmatrix} 1\,2\,3\,4 \\ 1\,2\,2\,1 \end{pmatrix}$$

$$f_{ab} = \begin{pmatrix} 1\,2\,3\,4 \\ 0\,0\,0\,3 \end{pmatrix} \qquad f_{bc} = \begin{pmatrix} 1\,2\,3\,4 \\ 2\,0\,0\,2 \end{pmatrix} \qquad f_{ac} = \begin{pmatrix} 1\,2\,3\,4 \\ 0\,0\,2\,1 \end{pmatrix}$$

$$f_{abc} = \begin{pmatrix} 1\,2\,3\,4 \\ 0\,0\,0\,2 \end{pmatrix} \qquad 0 = \begin{pmatrix} 1\,2\,3\,4 \\ 0\,0\,0\,0 \end{pmatrix}.$$

Since F_A is isomorphic to the characteristic semigroup $\bar{I}(A)$, it follows that $\bar{I}(A) = \{[0], [a], [b], [c], [ab], [bc], [ac], [abc]\}$. The transition semigroup $U(A)$ are the set of the adjacency matrices given by each string x in $f_x, f_x \in F_A$.

Table 1. Cayley table for syntactic monoid for $L = \overline{\mathrm{SI}(bb)} \cap \overline{\mathrm{SI}(ca)}$

	λ	a	b	c	ab	bc	ac	abc
λ	λ	a	b	c	ab	bc	ac	abc
a	a	a	ab	ac	ab	abc	ac	abc
b	b	ab	0	bc	0	0	abc	0
c	c	0	bc	c	0	bc	0	0
ab	ab	ab	0	abc	0	0	abc	0
bc	bc	0	0	bc	0	0	0	0
ac	ac	0	abc	ac	0	abc	0	0
abc	abc	0	0	abc	0	0	0	0

The monoid graph for this language is in Figure 3. Recall that although the 0 element is not shown, it is understood that all missing edges go to 0. The Cayley table is given in Table 1. With a little abuse of notation, in the following context, x is used to denote the element f_x in syntactic monoid F_A. The wholly non-zero condition can be checked by examining the syntactic monoid. It is noticed that in this canonical accepter all states are finals and there is no such $f_x \in F_A$ and $q \in Q$ such that $T(q, x) \neq 0$ but $T(q_0, x) = 0$.

The next step is to determine whether the right annihilation condition is satisfied with the help of Cayley table. For example in the Cayley table, the ab-row is all the elements that are in the right ideal generated by ab, $abF_A^1 = \{ab, abc, 0\}$. The elements in those columns corresponding to 0s form the set $RA(ab) = \{b, ab, bc, abc\}$. The right annihilating condition requires that $\forall w \in xF_A$, $RA(x) \subseteq RA(w)$. From the table it is easy to verify that $RA(abc) = \{a, b, ab, bc, ac, abc\}$, which is a superset of $RA(ab)$. Since $RA(0) = F_A$, it likewise follows that $RA(ab) \subseteq RA(0)$. The right annihilation condition for other elements can be verified in the same manner and it can be shown this syntactic monoid is right-annhilating.

Now consider the language $L = \{w : |w|_a = 1\}$ (Figure 1). L is not SP because it does not satisfy the wholly nonzero condition. The element b is not in the language but it is not zero in its syntactic semigroup.

For the language $L = \overline{C(aa)}$ in Figure 2, though it satisfies the wholly nonzero condition, the right annihilating condition is violated. Observe that $aa = 0$ and $ab \in PR(a)$. If L were right annihilating then $RA(a) \subseteq RA(ab)$. However, $aba = a \neq 0$ and thus the right annihilating condition is not met. Therefore, $L = \overline{C(aa)}$ is not SP.

What is the time complexity for DSP? Letting n be the size of the syntactic monoid, the wholly nonzero condition can be checked with time $O(n)$ and right annihilating condition runs in time $O(n^2)$. Thus DSP runs in $O(n^2)$. Holzer studies the size of the syntactic monoid as a natural measure of descriptive complexity for regular languages [8].

5.2   Finding the Shortest Forbidden Subsequences

The following procedure FIND-SSQ takes the syntactic monoid of a SP language as input and finds the finite set of shortest forbidden subsequences which describe the SP language. In order to link the syntactic monoid and the length of forbidden subsequences, the monoid graph is employed to find the set of the shortest paths from the λ node to 0 that covers the graph.

$$\mathcal{P}(F_A) \stackrel{\mathrm{def}}{=} \{x\sigma : f_x \in F_A, \sigma \in \Sigma, x\sigma = 0, \text{ and } \forall f_y = f_x, |x| \le |y|\}$$

FIND-SSQ begins with the syntactic monoid for some $L \in$ SP and $k = 1$.

1. Letting $P_k(S) \stackrel{\mathrm{def}}{=} \{\{P_k(p)\} : p \in S\}$, calculate $P_k(\mathcal{P}(F_A))$, i.e. the set of sets of k-subsequences for each path in $\mathcal{P}(F_A)$.
2. Find all singleton sets in $P_k(\mathcal{P}(F_A))$ and construct the set FS_k, which is the set of hypothesized forbidden subsequences of length k. This set is formed by taking the union of the singleton sets in $P_k(\mathcal{P}(F_A))$. If there is no singleton set found, update k by one and return to step 1.
3. Verify whether each set $P \in P_k(\mathcal{P}(F_A))$ has a nonempty intersection with FS_k. If so then FS_k is a set of forbidden sequences which can define L and $L \in \mathrm{SP_k}$. Otherwise, update k by one and return to step 1.

Theorem 6. FIND-SSQ *terminates at the shortest k for $L \in$ SP.*

Proof. Suppose this k is not the shortest one for the SP language L, and there exists $k' > k$ such that $L \in \mathrm{SP}_{k'}$. This means that there exists at least one path $p' \in \mathcal{P}(F_A)$, with $|p'| > k$, such that $P_k(p') \subseteq P_k(L)$ and $P_{k'}(p') \cap P_{k'}(L) = \emptyset$, for some $k' > k$. The fact that $P_k(p') \subseteq P_k(L)$ implies that $\forall v \in P_k(p')$, $v \in L$, which is guaranteed by the syntactic monoid of L being wholly nonzero.

However, if the algorithm does not terminate at k ensures that there exists at least one element $h \in P_k(p')$ with $h \in FS_k$. Since FS_k is the set of all paths of length k that lead to 0, $h \notin L$. This contradicts the previous statement that $\forall v \in P_k(p')$, $v \in L$. Therefore, no such p' exists and thus the algorithm terminates at the shortest k for the strictly piecewise language L.

We illustrate this algorithm with the automaton in Figure 3, assuming it has already been verified with DSP that it describes an SP language. We refer to the monoid in Figure 3 with F_A. The set of the shortest paths from the λ node to 0 that covers the graph is
$\mathcal{P}(F_A) = \{bb, ca, bab, bcb, bca, abb, aca, cba, cbb, bacb, baca, abcb, abca, acbb, acba\}$.

1. For $k = 1$, all sets in $P_1(\mathcal{P}(F_A))$ are not singleton. Therefore, increase k by 1.
2. For $k = 2$, $P_2(\mathcal{P}(F_A)) = \{\{bb\}, \{ca\}, \{ba, ab, bb\}, \{bc, cb, bb\}, \{bc, ba, ca\}, \{ab, bb\}, \{ac, aa, ca\}, \{cb, ca, ba\}, \{cb, bb\}, \{ba, bc, ac, bb, ab, cb\}, \{ab, bc, ac, cb, bb\}, \{ab, ac, aa, bc, ca, ba\}, \{ab, ac, bb, cb\}, \{ac, ab, ca, ba\}\}$. The singleton sets are $\{bb\}, \{ca\}$ and thus $FS_2 = \{bb, ca\}$. It is easy to verify that for all $P \in P_2(\mathcal{P}(F_A))$, P has a nonempty intersection with FS_2. The algorithm terminates and outputs $\{bb, ca\}$, which are the forbidden subsequences which describe this language.

In sum, this procedure tells us that this language is SP for $k = 2$. Together DSP and FIND-SSQ provide a means to check whether a regular language is SP, and if it is to find the finite set of the shortest forbidden subsequences.

6   Conclusion

Strictly Piecewise languages are wholly nonzero and right annihilating. The wholly nonzero property is shared by the Strictly Local languages and provides a definition for the "Strict" aspect, independent of the relation to the Testable classes. Also, the algebraic characterization for SP provides a polynomial-time decision procedure for a regular language in the size of its syntactic monoid. This paper also leaves open some interesting questions. In particular, we would like to know whether every wholly nonzero, $\mathcal{J}$-trivial language is right annihilating.

References

1. Anderson, J.A.: Automata Theory with Modern Applications. Cambridge University Press (2006)
2. Clifford, A.: The Algebraic Theory of Semigroups. American Mathematical Society, Providence (1967)
3. García, P., Ruiz, J.: Learning k-testable and k-piecewise testable languages from positive data. Grammars 7, 125–140 (2004)
4. Green, J.A.: On the structure of semigroups. The Annals of Mathematics 54(1), pp. 163–172 (1951)
5. Haines, L.H.: On free moniods partially ordered by embedding. Journal of Combinatorial Theory 6, 94–98 (1969)
6. Heinz, J.: Learning long-distance phonotactics. Linguistic Inquiry 41(4), 623–661 (2010)
7. Higman, G.: Ordering by divisibility in abstract algebras. Proceedings of the London Mathematical Society 3(2), 326–336 (1952)
8. Holzer, M., König, B.: Regular languages, sizes of syntactic monoids, graph colouring, state complexity results, and how these topics are related to each other. EATCS Bulletin 83, 139–155 (June 2004)
9. Hopcroft, J.E.: An n log n algorithm for minimizing states in a finite automaton. Tech. rep., Stanford, CA, USA (1971)
10. McNaughton, R., Papert, S.: Counter-Free Automata. MIT Press (1971)
11. Pin, J.É., et A. Salomaa (éd.), G.R.: Syntactic semigroups, vol. 1. Springer Verlag (1997)
12. Rogers, J., Heinz, J., Bailey, G., Edlefsen, M., Visscher, M., Wellcome, D., Wibel, S.: On languages piecewise testable in the strict sense. In: Ebert, C., Jäger, G., Michaelis, J. (eds.) The Mathematics of Language. Lecture Notes in Artifical Intelligence, vol. 6149, pp. 255–265. Springer (2010)
13. Simon, I.: Piecewise testable events. In: Automata Theory and Formal Languages, pp. 214–222 (1975)
14. Thierrin, G.: Convex languages. In: ICALP'72. pp. 481–492 (1972)
15. Watanabe, T., Nakamura, A.: On the transformation semigroups of finite automata. Journal of Computer and System Sciences 26(1), 107–138 (1983)

Deterministic Algorithms for Multi-criteria TSP

Bodo Manthey

University of Twente, Department of Applied Mathematics
P.O. Box 217, 7500 AE Enschede, The Netherlands
b.manthey@utwente.nl

Abstract. We present deterministic approximation algorithms for the multi-criteria traveling salesman problem (TSP). Our algorithms are faster and simpler than the existing randomized algorithms.

First, we devise algorithms for the symmetric and asymmetric multi-criteria Max-TSP that achieve ratios of $1/2k - \varepsilon$ and $1/(4k - 2) - \varepsilon$, respectively, where k is the number of objective functions. For two objective functions, we obtain ratios of $3/8 - \varepsilon$ and $1/4 - \varepsilon$ for the symmetric and asymmetric TSP, respectively. Our algorithms are self-contained and do not use existing approximation schemes as black boxes.

Second, we adapt the generic cycle cover algorithm for Min-TSP. It achieves ratios of $3/2 + \varepsilon$, $\frac{1}{2} + \frac{\gamma^3}{1-3\gamma^2} + \varepsilon$, and $\frac{1}{2} + \frac{\gamma^2}{1-\gamma} + \varepsilon$ for multi-criteria Min-ATSP with distances 1 and 2, Min-ATSP with γ-triangle inequality and Min-STSP with γ-triangle inequality, respectively.

1 Multi-criteria TSP

The traveling salesman problem (TSP) is perhaps the best-studied combinatorial optimization problem. An instance of *Min-TSP* is a complete graph $G = (V, E)$ with edge weights $d : E \to \mathbb{Q}_+$ that satisfy the triangle inequality. The goal is to find a *Hamiltonian cycle* (also called a *tour*) of minimum weight, where the weight of a tour is the sum of its edge weights. (The weight of an arbitrary set of edges is defined analogously.) If G is undirected, we have *Min-STSP* (symmetric TSP). If G is directed, we have *Min-ATSP* (asymmetric TSP). If we restrict the problem to instances that fulfill the γ-triangle inequality for $\gamma \in [1/2, 1)$ (this means $d(u, v) \leq \gamma \cdot (d(u, x) + d(x, v))$ for all distinct $u, v, x \in V$), then we get Min-γ-STSP and Min-γ-ATSP. If we restrict the edge weights to 1 and 2, we get *Min-1/2-STSP* and *Min-1/2-ATSP*. For Max-STSP and Max-ATSP, we have edge weights $w : E \to \mathbb{Q}_+$, and the goal is to find a tour of maximum weight.

All these variants of TSP are NP-hard and APX-hard [5]. Thus, we are in need of approximation algorithms. Min-STSP can be approximated with a ratio of 3/2 [5, Sect. 3.1.3]. Min-ATSP allows for a randomized $O(\log n / \log \log n)$ approximation [4] and for a deterministic $\frac{2}{3} \log_2 n$ approximation [13], where n is the number of vertices. Max-STSP and Max-ATSP can be approximated with ratios of 7/9 [22] and 2/3 [17], respectively. Min-γ-STSP and Min-γ-ATSP can be approximated with constant ratios depending on γ [9–11, 25]. Min-1/2-STSP and Min-1/2-ATSP admit factor 8/7 [6] and 5/4 [7] approximations, respectively.

M. Ogihara and J. Tarui (Eds.): TAMC 2011, LNCS 6648, pp. 264–275, 2011.

In many scenarios, however, there is more than one objective function to optimize. In case of the TSP, we might want to minimize travel time, expenses, number of flight changes, etc., while we want to maximize, e.g., our profit along the route. This gives rise to multi-criteria TSP, where Hamiltonian cycles are sought that optimize several objectives simultaneously. In order to transfer the notion of an optimal solutions to multi-criteria optimization problems, *Pareto curves* have been introduced (cf. Ehrgott [12]). A Pareto curve is a set of all optimal trade-offs between the different objective functions.

In the following, k always denotes the number of objective functions. We assume throughout the paper that $k \geq 2$ is an arbitrary constant. Let $[k] = \{1, 2, \ldots, k\}$. The k-criteria variants of the TSP that we consider are denoted by k-Min-STSP, k-Min-ATSP, k-Min-γ-STSP, k-Min-γ-ATSP, k-Min-1/2-STSP, k-Min-1/2-ATSP as well as k-Max-STSP and k-Max-ATSP.

We define the following terms for Min-TSP only. After that, we briefly point out the differences for Max-TSP. For a k-criteria variant of Min-TSP, we have edge weights $d_1, \ldots, d_k : E \to \mathbb{Q}_+$. For convenience, let $d = (d_1, \ldots, d_k)$. Inequalities of vectors are meant component-wise. A tour H *dominates* another tour $\tilde{H}$ if $d(H) \leq d(\tilde{H})$ and at least one of these k inequalities is strict. This means that H is strictly preferable to $\tilde{H}$. A *Pareto curve* is a set of all solutions that are not dominated by another solution. Since Pareto curves for the TSP cannot be computed efficiently, we have to be satisfied with approximate Pareto curves. A set $\mathcal{P}$ of tours is called an α *approximate Pareto curve* for the instance (G, d) if the following holds: For every tour $\tilde{H}$ of G, there exists a tour $H \in \mathcal{P}$ of G with $d(H) \leq \alpha d(\tilde{H})$. We have $\alpha \geq 1$, and a 1 approximate Pareto curve is a Pareto curve. An algorithm is called an α *approximation algorithm* if it computes an α approximate Pareto curve.

Let us point out the differences for Max-TSP. We have edge weights $w = (w_1, \ldots, w_k)$ (the triangle inequality is not required). Now a tour H dominates $\tilde{H}$ if $w(H) \geq w(\tilde{H})$ and at least one inequality is strict. A set $\mathcal{P}$ of tours is an α approximate Pareto curve if, for every tour $\tilde{H}$, we have an $H \in \mathcal{P}$ with $w(H) \geq \alpha w(\tilde{H})$. Note that $\alpha \leq 1$ for maximization problems.

1.1   Previous Work

Table 1 shows the current approximation ratios for the different variants of multi-criteria TSP. Many of these approximation algorithms can be extended to the case where some objectives should be minimized and others should be maximized [19]. Unfortunately, no deterministic algorithms are known except for k-Min-STSP and 2-Max-STSP. The reason for this is that most approximation algorithms for multi-criteria TSP use cycle covers. A *cycle cover* of a graph is a set of vertex-disjoint cycles such that every vertex is part of exactly one cycle. Hamiltonian cycles are special cases of cycle covers that consist of just one cycle. In contrast to Hamiltonian cycles, cycle covers of optimal weight can be computed in polynomial time. Cycle covers are one of the main tools for designing approximation algorithms for the TSP [7, 8, 13, 17, 22]. However, only a *randomized* fully polynomial-time approximation scheme (FPTAS) for

Table 1. Approximation ratios for multi-criteria TSP. The new deterministic ratio for k-Min-γ-STSP is an improvement for $\gamma \leq 0.58$. The new ratio for k-Min-γ-ATSP is achieved for $\gamma < 1/\sqrt{3}$. The result for k-Min-1/2-STSP is an improvement for $k \geq 3$.

variant	randomized	deterministic	reference	new
2-Min-STSP		2	[15]	
k-Min-STSP		$2+\varepsilon$	[20]	
k-Min-γ-STSP	$\frac{2\gamma^3+2\gamma^2}{3\gamma^2-2\gamma+1}+\varepsilon$, $\frac{1+\gamma}{1+3\gamma-4\gamma^2}+\varepsilon$	$1+\gamma+\varepsilon$, $\frac{2\gamma^2}{2\gamma^2-2\gamma+1}+\varepsilon$	[20]	$\frac{1}{2}+\frac{\gamma^2}{1-\gamma}+\varepsilon$
2-Min-1/2-STSP	4/3	3/2	[1, 20]	
k-Min-1/2-STSP	4/3	$\frac{2k}{k+1}$	[2, 20]	$3/2+\varepsilon$
k-Min-ATSP	$\log n + \varepsilon$		[18]	
k-Min-γ-ATSP	$\frac{1}{1-\gamma}+\varepsilon$		[18]	$\frac{1}{2}+\frac{\gamma^3}{1-3\gamma^2}+\varepsilon$
k-Min-1/2-ATSP	3/2		[20]	$3/2+\varepsilon$
2-Max-STSP	$2/3-\varepsilon$	7/27	[18, 22]	$3/8-\varepsilon$
k-Max-STSP	$2/3-\varepsilon$		[18]	$\frac{1}{2k}-\varepsilon$
2-Max-ATSP	1/2		[14]	$1/4-\varepsilon$
k-Max-ATSP	1/2		[14]	$\frac{1}{4k-2}-\varepsilon$

multi-criteria cycle covers is known [24]. This randomized FPTAS builds on a reduction to a specific unweighted matching problem [23], which is then solved using the RNC algorithm by Mulmuley et al. [21]. Derandomizing this algorithm is assumed to be difficult [3], and these nested reductions make the algorithm quite slow. Hence, it is natural to ask whether there exist deterministic, faster approximation algorithms for multi-criteria TSP.

1.2   New Results

We present deterministic approximation algorithms for several variants of multi-criteria TSP. Our algorithms are considerably simpler and faster than the existing randomized approximation algorithms. Table 1 shows an overview.

First, we devise deterministic and self-contained algorithms for Max-TSP (Sect. 2 and 3). They do not use other algorithms as black boxes except for maximum-weight matching with a single objective function. Furthermore, they do not make any assumption about the representation of the edge weights. The existing algorithms require the (admittedly weak and natural) assumption that the edge weights are encoded in binary. For k-Max-ATSP, we get a ratio of $\frac{1}{4k-2} - \varepsilon$ for any $\varepsilon > 0$ (Sect. 2). For k-Max-STSP, we achieve a ratio of $\frac{1}{2k} - \varepsilon$ (Sect. 3). For the special case of two objective functions, we can improve this to $1/4 - \varepsilon$ for 2-Max-ATSP and $3/8 - \varepsilon$ for 2-Max-STSP. The latter is an improvement over the existing deterministic 7/27 approximation for 2-Max-STSP [18, 22].

Second, we consider the cycle cover algorithm for Min-TSP (Sect. 4). We use a deterministic matching algorithm of Grandoni et al. [16]. The difficulty is that their algorithm does not produce perfect matchings. For k-Min-γ-ATSP, k-Min-γ-STSP, and k-Min-1/2-ATSP, we nevertheless get ratios of $\frac{1}{2} + \frac{\gamma^3}{1-3\gamma^2} + \varepsilon$, $\frac{1}{2} + \frac{\gamma^2}{1-\gamma} + \varepsilon$, and $3/2 + \varepsilon$, respectively. The ratio for k-Min-γ-STSP is an improvement over existing algorithms for $\gamma \leq 0.58$. The result for k-Min-γ-ATSP holds for $\gamma < 1/\sqrt{3}$. The result for k-Min-1/2-ATSP holds of course also for k-Min-1/2-STSP, and it is an improvement for $k \geq 3$.

Due to space limitations, many proofs are omitted from this extended abstract.

2   Max-ATSP

The rough idea behind our algorithm for k-Max-ATSP is as follows: First, we "guess" a few edges that we contract to get a slightly smaller instance. The number of edges that we have to contract depends only on k and ε. Second, we compute k maximum-weight matchings in the smaller instance, each with respect to one of the k objective functions. Third, we compute another matching that uses only edges of the k matchings and that contains as much weight as possible with respect to each objective function. One note is here in order: Usually, cycle covers instead of matchings are used for Max-ATSP. However, although the weight of a cycle cover can be (roughly) twice as large as the weight of a maximum-weight matching, we do not get a better approximation ratio by using cycle covers. The reason is that we lose a factor of roughly $1/2$ if we compute a collection of paths from k initial cycle covers compared to k initial matchings.

The following lemma is a key ingredient of our algorithm. It shows how to get a matching from k different matchings such that a significant fraction of the weight with respect to each matching is preserved. This works as long as no single edge contributes too much weight. The lemma immediately gives a polynomial-time algorithm for this task.

Lemma 1. *Let $G = (V, E)$ be a directed graph, and let $w = (w_1, \ldots, w_k)$ be edge weights. Let $M_1, \ldots, M_k \subseteq E$ be matchings. Let $\eta \in (0, 1)$ be arbitrary such that $w_i(e) \leq \frac{\eta}{2k-2} \cdot w_i(M_i)$ for all $e \in M_i$ and all $i \in [k]$. Then there exists a matching $P \subseteq \bigcup_{i=1}^{k} M_i$ such that $w_i(P) \geq \frac{1-\eta}{2k-1} \cdot w_i(M_i)$ for all $i \in [k]$. Such a matching P can be computed in polynomial time.*

Proof. We construct the matching as follows: We add one heaviest edge $e \in M_1$ with respect to w_1 to P and remove e and all edges adjacent to e from $M_2, \ldots, M_k$. Then we put one heaviest remaining edge from M_2 into P and remove it and all adjacent edges. We proceed with $M_3, \ldots, M_k$ and repeat the process until no edges remain.

Let us analyze $w_i(P)$. In each step, at most two edges of any M_i are removed. Thus, we have removed at most $2i - 2$ edges from M_i until we added the first edge from M_i to P. The weight of these edges is at most $(2i - 2) \cdot \frac{\eta}{2k-2} w_i(M_i) \leq \eta w_i(M_i)$. Now let e be an edge of M_i that we added to P, and let $e_1, \ldots, e_t$ be the $t \leq 2k - 2$ edges that are removed from M_i in the subsequent rounds of the procedure until again an edge of M_i is added. By construction, we have

$w_i(e) \geq w_i(e_j)$ for all $j \in [t]$. Thus, $w_i(e) \geq \frac{1}{2k-1} \cdot (w_i(e) + \sum_{j=1}^{t} w_i(e_j))$. Taking the initial loss of $\eta w_i(M_i)$ into account, we observe that we can put a $\frac{1}{2k-1}$ fraction of $(1 - \eta)w_i(M_i)$ into P for each $i \in [k]$.   □

Now we have to make sure that, for a tour $\tilde{H}$, we can find appropriate matchings $M_1, \ldots, M_k$. For a directed complete graph $G = (V, E)$ and a set $K \subseteq E$ that forms a subset of a tour, we obtain G_{-K} by contracting all edges of K. Contracting an edge (u, v) means that we remove all outgoing edges of u and all incoming edges of v, and then identify u and v. Analogously, for a tour $\tilde{H} \supseteq K$, we obtain a tour $\tilde{H}_{-K}$ by contracting the edges in K.

The following lemma says that, for any tour $\tilde{H}$, there is always a small set K of edges such that, if we contract these edges, the resulting tour $\tilde{H}_{-K}$ consists solely of edges that do not contribute too much to the weight of $\tilde{H}_{-K}$ with respect to any objective function. The proof is identical to the proof of the corresponding lemma for the $(1/2 - \varepsilon)$ approximation for k-Max-ATSP [18, 19]. In the algorithm, we will "guess" good sets K, compute Hamiltonian cycles on G_{-K}, and add the edges of K to get a Hamiltonian cycle of G.

Small set means that $|K| \leq f(k, \varepsilon)$ for some function f that does not depend on the number n of vertices. We can choose $f(k, \varepsilon) \in O(k/ \log(1/(1 - \varepsilon))) = O(k/ \log(1 + \varepsilon)) = O(k/\varepsilon)$ [18, 19] (we have $\log(1 + \varepsilon) = O(1/\varepsilon)$ by Taylor expansion). Moreover, we can choose K such that V_{-K} contains an even number of vertices.

Lemma 2. *Let $G = (V, E)$ be a directed complete graph with edge weights $w = (w_1, \ldots, w_k)$, and let $\varepsilon > 0$. Let $H \subseteq E$ be any tour of G. Then there is a subset $K \subseteq H$ such that $|K| \leq f(k, \varepsilon)$, $|V_{-K}|$ is even, and, for all $i \in [k]$, we have*

1. *$w_i(K) \geq \frac{1}{4} \cdot w_i(H)$ or*
2. *$w_i(e) \leq \varepsilon \cdot w_i(H_{-K})$ for all $e \in H_{-K}$.*

We have to make sure that any edge weighs at most an ε fraction of $w(H)$, provided that $w(e) \leq \varepsilon w(H)$ for all $e \in H$: Let $\beta_i = \max\{w_i(e) \mid e \in H\}$ be the weight of the heaviest edge with respect to w_i. Let $\beta = (\beta_1, \ldots, \beta_k)$. We define new edge weights w^β by setting the weight of edges that are too heavy to 0:

$$w^\beta(e) = \begin{cases} w(e) & \text{if } w(e) \leq \beta \text{ and} \\ 0 & \text{if } w_i(e) > \beta_i \text{ for some } i. \end{cases}$$

Since $w(e) \leq \beta$ for every $e \in H$ by definition, we have $w(H) = w^\beta(H)$. The number of vectors β that result in different weight functions w^β is bounded by n^{2k}: Since the number of edges is less than n^2, there are less than n^2 different edge weights for each objective function. Now we can state and analyze our approximation algorithm for k-Max-ATSP (Algorithm 1).

Theorem 3. *For every $\varepsilon > 0$ and $k \geq 2$, Algorithm 1 is a deterministic approximation algorithm for k-Max-ATSP that achieves an approximation ratio of $\frac{1}{4k-2} - \varepsilon$. Its running-time is $n^{O(k/\varepsilon)}$.*

$\mathcal{P}_{\text{TSP}} \leftarrow \text{MaxATSP-Approx}(G, w, \varepsilon)$
input: directed complete graph $G = (V, E)$, $w : E \to \mathbb{Q}_+^k$, $\varepsilon > 0$
output: $\frac{1}{4k-2} - \varepsilon$ approximate Pareto curve $\mathcal{P}_{\text{TSP}}$ for k-Max-ATSP
1: **for all** $K \subseteq E$ that form a subset of a tour with $|K| \le f(k, \varepsilon)$ and $|V_{-K}|$ even **do**
2:      **for all** $I \subseteq [k]$ and β **do**
3:          compute maximum-weight matchings M_i in G_{-K} w.r.t. w_i^β for $i \in \overline{I} = [k] \setminus I$
4:          compute a matching $P \subseteq \bigcup_{i \in \overline{I}} M_i$ according to Lemma 1
5:          add edges to $K \cup P$ to obtain a Hamiltonian cycle H; add H to $\mathcal{P}_{\text{TSP}}$

Algorithm 1. Approximation algorithm for k-Max-ATSP

Proof. We have to show that, for every tour $\tilde{H}$, there exists a tour $H \in \mathcal{P}_{\text{TSP}}$ with $w(H) \ge (\frac{1}{4k-2} - \varepsilon) \cdot w(\tilde{H})$. By Lemma 2, there exists a subset $K \subseteq \tilde{H}$ of edges and an $I \subseteq [k]$ such that $|K| \le f(k, \varepsilon)$, $|V_{-K}|$ is even, $w_i(K) \ge w_i(\tilde{H})/4$ for all $i \in I$, and $w_i(e) \le \varepsilon w_i(\tilde{H}_{-K})$ for all $e \in \tilde{H}_{-K}$ and $i \in [k] \setminus I$. Let $i \in [k] \setminus I$, and let M_i be a maximum-weight matching in G_{-K} with respect to w_i^β. Then $w_i^\beta(M_i) \ge w_i^\beta(\tilde{H}_{-K})/2$ and $w_i^\beta(e) \le 2\varepsilon w_i(\tilde{H}_{-K})$. Using Lemma 1 with $\eta = (2k-2)2\varepsilon$ we can compute a matching $P \subseteq \bigcup_{i \in [k] \setminus I} M_i$ such that $w_i^\beta(P) \ge \frac{1-\eta}{2k-1} \cdot w_i^\beta(M_i) = \frac{1-(2k-2)2\varepsilon}{2k-1} \cdot w_i^\beta(M_i) \ge (\frac{1}{2k-1} - 2\varepsilon) \cdot w_i^\beta(M_i)$. Now $P \cup K$ is a collection of paths in G. What remains to be done is to estimate the weight of $w(P \cup K)$. For every $i \in I$, we have $w_i(P \cup K) \ge w_i(K) \ge w_i(\tilde{H})/4 \ge (\frac{1}{4k-2} - \varepsilon) \cdot w_i(\tilde{H})$. For every $i \notin I$, we note that $w_i(\tilde{H}) = w_i(K) + w_i(\tilde{H}_{-K})$. This gives us

$$w_i(P \cup K) \ge w_i^\beta(P) + w_i(K) \ge \left(\tfrac{1}{2k-1} - 2\varepsilon\right) \cdot w_i^\beta(M_i) + w_i(K)$$
$$\ge \left(\tfrac{1}{4k-2} - \varepsilon\right) \cdot w_i^\beta(\tilde{H}_{-K}) + w_i(K) \ge \left(\tfrac{1}{4k-2} - \varepsilon\right) w_i(\tilde{H}).$$

The running-time is at most $n^{O(1)+2k+f(k,\varepsilon)} = n^{O(k/\varepsilon)}$. $\qquad\square$

If we have only two objective functions, we can improve the approximation ratio to $1/4 - \varepsilon$. The key ingredient for this is the following lemma, which is the improved counterpart of Lemma 1 for $k = 2$. The lemma can be proved using a cake-cutting argument with one player for each of the two objective functions.

Lemma 4. *Let $G = (V, E)$ be a directed graph with edge weights $w = (w_1, w_2)$ and an even number of vertices. Let $M_1, M_2 \subseteq E$ be two perfect matchings, and let $\eta \in (0, 1/4)$. Suppose that $w_i(e) \le \frac{\eta}{2} \cdot w_i(M_i)$ for all $e \in M_i$ and $i \in \{1, 2\}$. Then there is a matching $P \subseteq M_1 \cup M_2$ with $w_i(P) \ge (\frac{1}{2} - \sqrt{\eta}) w_i(M_i)$ for $i \in \{1, 2\}$. The matching P can be found in polynomial time.*

Proof. Without loss of generality, we assume $M_1 \cap M_2 = \emptyset$. Otherwise, we can simply remove $M_1 \cap M_2$ from both matchings and add it to P. We scale the edge weights such that $w_i(M_i) = 1$ for $i \in \{1, 2\}$. If we ignore the directions of the edges, the graph with edges $M_1 \cup M_2$ is a collection of disjoint cycles. Every cycle has even length and edges from M_1 and M_2 alternate.

Let $c \subseteq M_1 \cup M_2$ be a cycle. We say that c is a light cycle if $w_1(c) \le \sqrt{\eta}$. Otherwise, i.e., if $w_1(c) > \sqrt{\eta}$, we call c a heavy cycle. Note that $M_1 \cup M_2$ has at most $1/\sqrt{\eta}$ heavy cycles.

$\mathcal{P}_{\text{TSP}} \leftarrow$ MaxATSP-Approx-2(G, w, ε)

input: directed complete graph $G = (V, E)$, $w : E \to \mathbb{Q}_+^2$, $\varepsilon > 0$

output: $\frac{1}{4} - \varepsilon$ approximate Pareto curve $\mathcal{P}_{\text{TSP}}$ for k-Max-ATSP

1: **for all** $K \subseteq E$ with $|K| \leq f(2, \varepsilon^2)$ that are a subset of a tour and $|V_{-K}|$ even **do**
2:      **for all** $I \subseteq \{1, 2\}$ and β **do**
3:          compute maximum-weight matchings M_i in G_{-K} w.r.t. w_i^β for $i \in \overline{I}$
4:          compute a matching $P \subseteq \bigcup_{i \in \overline{I}} M_i$ according to Lemma 4
5:          add edges to $K \cup P$ to obtain a Hamiltonian cycle H; add H to $\mathcal{P}_{\text{TSP}}$

Algorithm 2. Improved approximation algorithm for 2-Max-ATSP

We show the lemma by a cake-cutting argument: Player 1 puts cycles (or parts of cycles) into two sets S_1 and S_2, and then Player 2 can choose which set to take. Player i wants to maximize w_i. Player 1 puts light cycles as a whole into S_1 or S_2. Heavy cycles are split into two parts as follows: Player 1 decides to remove one edge of M_1 and one edge of M_2 (these edges are lost also for Player 2). In this way, we get two paths (again disregarding the directions of the edges). Player 1 puts one path into S_1 and the other path into S_2. (It can happen that one of the paths is empty: If we have a cycle of length four, the two edges removed are necessarily adjacent. This, however, does not cause any problem. In particular, cycles of length four are always light cycles.) Finally, Player 2 chooses the set S_i that maximizes w_2. Player 1 has to take S_{3-i}. This yields the matching $P = (S_i \cap M_2) \cup (S_{3-i} \cap M_1)$.

Let us estimate the weight that the players are guaranteed to get. Since we have at most $1/\sqrt{\eta}$ heavy cycles, at most $1/\sqrt{\eta}$ edges from M_2 are removed. The total weight of the edges removed is hence at most $\sqrt{\eta}/2$. Thus, $w_2((S_1 \cup S_2) \cap M_2) \geq w_2(M_2) - \sqrt{\eta}/2 = 1 - \sqrt{\eta}/2$. Hence, Player 2 can always get a weight of at least $\frac{1}{2} \cdot (1 - \sqrt{\eta}/2) \geq \frac{1}{2} - \sqrt{\eta}$.

Let us now focus on Player 1. As for Player 2, we have $w_1((S_1 \cup S_2) \cap M_1) \geq 1 - \sqrt{\eta}/2$. For any heavy weight cycle c, Player 1 can choose to remove edges such that the resulting paths differ by at most $\eta/2$ with respect to w_1. Since light cycles are put as a whole in either S_1 or S_2 and have a weight of at most $\sqrt{\eta}$ with respect to w_1, Player 1 can make sure that $w_1(S_1 \cap M_1)$ and $w_1(S_2 \cap M_1)$ differ by at most $\sqrt{\eta}$. Thus, $w_1(S_i \cap M_1) \geq \frac{1}{2} \cdot \left(1 - \frac{\sqrt{\eta}}{2}\right) - \frac{\sqrt{\eta}}{2} \geq \frac{1}{2} - \sqrt{\eta}$ for both $i \in \{1, 2\}$. Thus, for any choice of Player 2, Player 1 still gets enough weight with respect to w_1. The proof immediately gives a polynomial-time algorithm for computing P. $\qquad\square$

Theorem 5. *For every $\varepsilon > 0$, Algorithm 2 is a deterministic approximation algorithm for 2-Max-ATSP with an approximation ratio of $1/4 - \varepsilon$. Its running-time is $n^{O(1/\varepsilon^2)}$.*

Proof. We have to prove that, for every tour $\tilde{H}$, there is an $H \in \mathcal{P}_{\text{TSP}}$ with $w(H) \geq (\frac{1}{4} - \varepsilon) \cdot w(\tilde{H})$. According to Lemma 2, there is a subset $K \subseteq \tilde{H}$ and an $I \subseteq \{1, 2\}$ such that $|K| \leq f(2, \varepsilon^2)$, $|V_{-K}|$ is even, $w_i(K) \geq w_i(\tilde{H})/4$ for $i \in I$, and $w_i(e) \leq \varepsilon^2 w_i(H_{-K})$ for all $e \in H_{-K}$ and $i \in \{1, 2\} \setminus I = \overline{I}$. Thus, there exists a β such that, first, $w_i^\beta(\tilde{H}_{-K}) = w_i(\tilde{H}_{-K})$ for all $i \in \overline{I}$ and, second, for each $i \in \overline{I}$,

there exists a matching M_i with $w_i^\beta(e) \le 2\varepsilon^2 w_i^\beta(M_i)$ and $w^\beta(M_i) \ge \frac{1}{2} \cdot w^\beta(\tilde{H}_{-K})$.
Using Lemma 4 with $\eta = 4\varepsilon^2$, we can compute a matching $P \subseteq \bigcup_{i \in \bar{I}} M_i$ such
that $w_i^\beta(P) \ge (\frac{1}{2} - 2\varepsilon) w_i^\beta(M_i)$ for each $i \in \bar{I}$. Again, $P \cup K$ is a collection of
paths. For any $i \in I$, we have $w_i(P \cup K) \ge w_i(K) \ge w_i(\tilde{H})/4$, which is sufficient.
For any $i \in \bar{I}$, we have

$$w_i(P \cup K) \ge w_i^\beta(P) + w_i(K) \ge (\tfrac{1}{2} - 2\varepsilon) \cdot w_i^\beta(M_i) + w_i(K)$$
$$\ge (\tfrac{1}{4} - \varepsilon) \cdot w_i^\beta(\tilde{H}_{-K}) + w_i(K) \ge (\tfrac{1}{4} - \varepsilon) \cdot w_i(\tilde{H}).$$

The running-time is bounded by $n^{O(1)+f(2,2\varepsilon^2)} = n^{O(1/\varepsilon^2)}$. □

3   Max-STSP

One key ingredient for our algorithm for k-Max-STSP is the following lemma,
which is the undirected counterpart to Lemma 1. In contrast to k-Max-ATSP,
we now start with k cycle covers rather than k matchings.

Lemma 6. *Let* $G = (V, E)$ *be an undirected graph with edge weights* $w = (w_1, \ldots, w_k)$, *and let* $C_1, \ldots, C_k \subseteq E$ *be cycle covers. Assume that, for some* $\eta > 0$, *we have* $w_i(e) \le \frac{\eta}{2k-1} w_i(C_i)$ *for all* $e \in C_i$ *and all* $i \in [k]$. *Then there exists a collection* $P \subseteq \bigcup_{i=1}^{k} C_i$ *of paths such that* $w_i(P) \ge \frac{1-\eta}{2k} w_i(C_i)$ *for all* i. *Such a collection* P *can be computed in polynomial time.*

As in Sect. 2, we would like to keep a set $K \subseteq E$ of heavy edges. Unfortunately,
it is impossible to contract edges in the same way as in directed graphs [18].
As already done for the randomized algorithms, we circumvent this by setting
the weight along paths of sufficient length to 0 [18, 19]. To do this formally, we
need the following notation: Let $\tilde{H}$ be a Hamiltonian cycle, and let $K \subseteq \tilde{H}$. Let
$L = L(K) = \{v \mid \exists e \in K : v \in e\}$ be the set of vertices that are adjacent to
edges of K. Let $T = T(K) = \{e \in \tilde{H} \mid e \text{ is adjacent to } K \text{ but not in } K\}$. As for
the directed case, let $\beta = (\beta_1, \ldots, \beta_k)$. Now we define

$$w^{-L,\beta}(e) = \begin{cases} w(e) & \text{if } e \cap L = \emptyset \text{ and } w(e) \le \beta \text{ and} \\ 0 & \text{if } e \cap L \ne \emptyset \text{ or there is an } i \text{ with } w_i(e) > \beta_i. \end{cases}$$

This means that under $w^{-K,\beta}$, all edges of K or adjacent to K have weight 0.
Furthermore, all edges that exceed β for some objective are also set to 0.

Now we are prepared to state the undirected counterpart of Lemma 2. As
Lemma 2, its proof is identical to the proof of the corresponding lemma for the
$(\frac{2}{3} - \varepsilon)$ approximation for k-Max-STSP [18, 19]. We can choose the function g
in the lemma such that $g(k, \eta) \in O(\frac{k^3}{\eta \cdot (\log(1-\eta))^2}) = O(k^3/\eta^3)$. We can easily
require that $|V_{-K}|$ is even. The necessary change of the function g is negligible.

$P \leftarrow \text{MaxSTSP-Approx}(G, w, \varepsilon)$

input: undirected complete graph $G = (V, E)$, $w : E \to \mathbb{Q}_+^k$, $\varepsilon > 0$

output: $\frac{1}{2k} - \varepsilon$ approximate Pareto curve $\mathcal{P}_{\text{TSP}}$ for k-Max-STSP

1: **for all** $K \subseteq E$ with $|K| \le g(k, \varepsilon/2)$ that form a subset of a tour **do**
2:     **for all** $I \subseteq [k]$, and β **do**
3:         compute maximum-weight cycle covers C_i in G w.r.t. $w_i^{-K,\beta}$ for $i \in \bar{I}$
4:         compute a collection $P \subseteq \bigcup_{i \in [k] \setminus I} C_i$ of paths according to Lemma 6
5:         remove edges incident to $L(K)$ from P to obtain P'
6:         add edges to $K \cup P'$ to obtain a Hamiltonian cycle H; add H to $\mathcal{P}_{\text{TSP}}$

Algorithm 3. $\frac{1}{2k} - \varepsilon$ approximation for k-Max-STSP

Lemma 7. *Let $G = (V, E)$ be an undirected complete graph with edge weights $w = (w_1, \ldots, w_k)$. Let $\eta > 0$. Let $H \subseteq E$ be any Hamiltonian cycle of G. Then there exists a collection $K \subseteq H$ of paths such that $|K| \le g(k, \eta)$ and $|V_{-K}|$ is even and the following properties hold: Let $L = L(K)$ and $T = T(K)$. For all $i \in [k]$, we have*

1. *$w_i(K) \ge \frac{1}{2} \cdot w_i(H)$ or*
2. *$w_i(e) \le \eta \cdot w_i^{-L}(H)$ for all $e \in H \setminus K$ and $w_i(T) \le \eta \cdot w_i(H)$.*

Now we are prepared to state and analyze our approximation algorithm for k-Max-STSP (Algorithm 3), and we obtain the following theorem.

Theorem 8. *For every $k \ge 2$ and $\varepsilon > 0$, Algorithm 3 is a deterministic approximation algorithm for k-Max-STSP that achieves an approximation ratio of $\frac{1}{2k} - \varepsilon$ and has a running-time of $n^{O(k^3/\varepsilon^3)}$.*

As for 2-Max-ATSP, we can achieve a better approximation ratio of $3/8 - \varepsilon$ for $k = 2$. This improves over the known deterministic $7/27$ approximation [18, 22].

Lemma 9. *Let $G = (V, E)$ be an undirected graph with edge weights $w = (w_1, w_2)$, and let $M_1, M_2 \subseteq E$ be two matchings. Assume that $w_i(e) \le \eta w_i(M_i)$ for $i \in \{1, 2\}$ and all edges $e \in M_i$. Then there exists a collection $P \subseteq M_1 \cup M_2$ of paths such that $w_i(P) \ge (\frac{3}{4} - \eta) \cdot w_i(M_i)$ for $i \in \{1, 2\}$. Such a collection P can be found in polynomial time.*

Theorem 10. *For any $\varepsilon > 0$, Algorithm 4 is a deterministic algorithm for 2-Max-STSP with an approximation ratio of $\frac{3}{8} - \varepsilon$. Its running-time is $n^{O(1/\varepsilon^3)}$.*

4   Cycle Cover Algorithm for Min-TSP

Now we consider multi-criteria Min-TSP. In the following, we need the (natural and weak) assumption that the edge weights are encoded in binary. The main idea is to replace the approximation scheme for cycle covers by the bipartite matching algorithm of Grandoni et al. [16]. Their algorithm does the following: Let $G = (V, E)$ be a bipartite graph, let $\varepsilon > 0$ and k be fixed, let $d = (d_1, \ldots, d_k)$

$\mathcal{P}_{\mathrm{TSP}} \leftarrow \mathrm{MaxSTSP\text{-}Approx\text{-}2}(G, w, \varepsilon)$
input: undirected complete graph $G = (V, E)$, $w : E \to \mathbb{Q}_+^2$, $\varepsilon > 0$
output: $\frac{3}{8} - \varepsilon$ approximate Pareto curve $\mathcal{P}_{\mathrm{TSP}}$ for k-Max-STSP
1: **for all** $K \subseteq E$ with $|K| \le g(2, \varepsilon/2)$ that form a subset of a tour **do**
2:    **for all** $I \subseteq \{1, 2\}$ and β **do**
3:        compute maximum-weight matchings M_i in G w.r.t. $w_i^{-K, \beta}$ for $i \in \overline{I}$
4:        compute a collection $P \subseteq \bigcup_{i \in [k] \setminus I} M_i$ of paths according to Lemma 6
5:        remove edges incident to $L(K)$ from P to obtain P'
6:        add edges to $K \cup P'$ to obtain a Hamiltonian cycle H; add H to $\mathcal{P}_{\mathrm{TSP}}$

Algorithm 4. Improved approximation for 2-Max-STSP

$\mathcal{P}_{\mathrm{TSP}} \leftarrow \mathrm{GenericATSP}(G, d, \varepsilon)$
input: directed complete graph $G = (V, E)$, $d : E \to \mathbb{Q}_+^k$, $\varepsilon > 0$
output: approximate Pareto curve $\mathcal{P}_{\mathrm{TSP}}$ for k-Min-ATSP
1: $w_e \leftarrow 1$ for all $e \in E$
2: compute an $\frac{\varepsilon}{2\beta}$-approximate Pareto curve $\mathcal{C}$ of $\frac{\varepsilon}{2\beta}$-partial cycle covers
3: **for all** $C \in \mathcal{C}$ **do**
4:    break one edge of every cycle of C to obtain a collection P of paths
5:    join the paths with edges to obtain a Hamiltonian cycle H; put H into $\mathcal{P}_{\mathrm{TSP}}$

Algorithm 5. Generic Approximation for k-Min-TSP

be edge lengths, and let w be edge weights. Let $D_1, \ldots, D_k$ be budgets. Let M_{opt} be a matching that maximizes $w(M_{\mathrm{opt}})$ subject to $d_i(M_{\mathrm{opt}}) \le D_i$ for all $i \in [k]$. Then their algorithm outputs a matching M with $w(M) \ge (1 - \varepsilon)w(M_{\mathrm{opt}})$ and $d_i(M) \le (1 + \varepsilon)D_i$ for all $i \in [k]$. We use this algorithm to compute partial cycle covers. An ε-partial cycle cover of a directed graph is a collection of simple cycles and simple paths that contains at least $(1 - \varepsilon) \cdot n$ edges. (A cycle cover in an n vertex graph consists of n edges.) In other words, a partial cycle cover is a subset of a cycle cover. We do this by exploiting that matchings in bipartite graphs stand in one-to-one correspondence to cycle covers in directed graphs. Let $w(e) = 1$ for all edges $e \in E$. Then our goal is simply to maximize the number of edges subject to the budget constraints. We choose all combinations of 0 and $(1 + \varepsilon)^\ell$ (with $\ell \in \{-p, \ldots, p\}$ for some polynomial p) for $D_1, \ldots, D_k$ and run the matching algorithm using these $D_1, \ldots, D_k$ for some small enough ε. This yields $(1 + \varepsilon)$ approximate Pareto curves for ε-partial cycle covers [24].

Let $\beta_d = \max_{i \in [k], e, e' \in E} \frac{d_i(e)}{d_i(e')}$ be the maximum ratio of heaviest to lightest edge with respect to any objective function. We remark that it is crucial that β_d is bounded by a constant in order to our algorithm work satisfactory. The reason is that we have to be content with ε-partial cycle covers. Due to this, we might incur extra costs proportional to $\varepsilon\beta_d$.

Theorem 11. *Fix any $\varepsilon > 0$, $k \ge 2$, and $\beta \ge 1$. If restricted to instances (G, d) with $\beta_d \le \beta$, Algorithm 5 is a $\frac{1+\beta}{2} + \varepsilon$ approximation algorithm for k-Min-ATSP.*

For k-Min-1/2-ATSP, we have $\beta_d \leq 2$. For k-Min-γ-ATSP, we have $\beta_d \leq \frac{2\gamma^3}{1-3\gamma^2}$ for $\gamma < 1/\sqrt{3}$ [11], while for k-Min-γ-STSP, we have $\beta_d \leq \frac{2\gamma^2}{1-\gamma}$ for $\gamma < 1$ [10]. Thus, we get the following derandomized algorithms [18, 20].

Corollary 12. *For every $k \geq 2$ and $\varepsilon > 0$, Algorithm 5 is a deterministic approximation algorithm for multi-criteria Min-TSP. It achieves a ratio of $3/2 + \varepsilon$ for k-Min-1/2-ATSP, a ratio of $\frac{1}{2} + \frac{\gamma^3}{1-3\gamma^2} + \varepsilon$ for k-Min-γ-ATSP for $\gamma < 1/\sqrt{3}$, and a ratio of $\frac{1}{2} + \frac{\gamma^2}{1-\gamma} + \varepsilon$ for k-Min-γ-STSP for $\gamma < 1$.*

5   Open Problems

An obvious question is whether there exists a *deterministic* approximation algorithm for k-Min-ATSP with a non-trivial approximation ratio, which means smaller than $\frac{2}{3} \cdot k \log_2 n$, which is obtained by adding the k weights of each edge to get a single objective function. Furthermore, we would like to know if there are *deterministic* approximation algorithms for k-Max-ATSP and k-Max-STSP that achieve a constant approximation ratio (or at least a ratio of $\omega(1/k)$).

A key step towards improving the deterministic algorithms for multi-criteria Min-TSP would be an approximation scheme for multi-criteria *non-bipartite perfect* matching. Moreover, the algorithms for k-Min-1/2-STSP and k-Min-γ-STSP would yield a better ratio if initialized with undirected cycle covers. However, a derandomization of the randomized FPTAS for general matching [24], which is based on the isolation lemma [21], seems to be difficult [3].

Finally, it is open if there are *deterministic* algorithms for the case where some objectives should be minimized while others should be maximized.

References

1. Angel, E., Bampis, E., Gourvés, L.: Approximating the Pareto curve with local search for the bicriteria TSP(1,2) problem. Theoretical Computer Science 310 (1-3), 135–146 (2004)
2. Angel, E., Bampis, E., Gourvès, L., Monnot, J. (Non)-Approximability for the Multi-criteria TSP(1,2). In: Liśkiewicz, M., Reischuk, R. (eds.) FCT 2005. LNCS, vol. 3623, pp. 329–340. Springer, Heidelberg (2005)
3. Arvind, V., Mukhopadhyay, P.: Derandomizing the Isolation Lemma and Lower Bounds for Circuit Size. In: Goel, A., Jansen, K., Rolim, J.D.P., Rubinfeld, R. (eds.) APPROX and RANDOM 2008. LNCS, vol. 5171, pp. 276–289. Springer, Heidelberg (2008)
4. Asadpour, A., Goemans, M.X., Madry, A., Gharan, S.O., Saberi, A.: An $O(\log n / \log \log n)$-approximation algorithm for the asymmetric traveling salesman problem. In: Proc. 21st Ann. ACM-SIAM Symp. on Discrete Algorithms (SODA), pp. 379–389. SIAM, Philadelphia (2010)
5. Ausiello, G., Crescenzi, P., Gambosi, G., Kann, V., Marchetti-Spaccamela, A., Protasi, M.: Complexity and Approximation. Springer, Heidelberg (1999)

6. Berman, P., Karpinski, M.: 8/7-approximation algorithm for (1, 2)-TSP. In: Proc. 17th Ann. ACM-SIAM Symp. on Discrete Algorithms (SODA), pp. 641–648. SIAM, Philadelphia (2006)
7. Bläser, M.: A 3/4-approximation algorithm for maximum ATSP with weights zero and one. In: Jansen, K., Khanna, S., Rolim, J.D.P., Ron, D. (eds.) RANDOM 2004 and APPROX 2004. LNCS, vol. 3122, pp. 61–71. Springer, Heidelberg (2004)
8. Bläser, M., Manthey, B.: Approximating maximum weight cycle covers in directed graphs with weights zero and one. Algorithmica 42(2), 121–139 (2005)
9. Bläser, M., Manthey, B., Sgall, J.: An improved approximation algorithm for the asymmetric TSP with strengthened triangle inequality. Journal of Discrete Algorithms 4(4), 623–632 (2006)
10. Böckenhauer, H.-J., Hromkovič, J., Klasing, R., Seibert, S., Unger, W.: Approximation algorithms for the TSP with sharpened triangle inequality. Information Processing Letters 75(3), 133–138 (2000)
11. Sunil Chandran, L., Shankar Ram, L.: On the relationship between ATSP and the cycle cover problem. Theoretical Computer Science 370(1-3), 218–228 (2007)
12. Ehrgott, M.: Multicriteria Optimization. Springer, Heidelberg (2005)
13. Feige, U., Singh, M.: Improved Approximation Ratios for Traveling Salesperson Tours and Paths in Directed Graphs. In: Charikar, M., Jansen, K., Reingold, O., Rolim, J.D.P. (eds.) RANDOM 2007 and APPROX 2007. LNCS, vol. 4627, pp. 104–118. Springer, Heidelberg (2007)
14. Glaßer, C., Reitwießner, C., Witek, M.: Balanced combinations of solutions in multi-objective optimization, arXiv:1007.5475v1 [cs.DS] (2010)
15. Glaßer, C., Reitwießner, C., Witek, M.: Improved and derandomized approximations for two-criteria metric traveling salesman. Report 09-076, Rev. 1, Electron. Colloq. on Computational Complexity (ECCC) (2010)
16. Grandoni, F., Ravi, R., Singh, M.: Iterative Rounding for Multi-Objective Optimization Problems. In: Fiat, A., Sanders, P. (eds.) ESA 2009. LNCS, vol. 5757, pp. 95–106. Springer, Heidelberg (2009)
17. Kaplan, H., Lewenstein, M., Shafrir, N., Sviridenko, M.I.: Approximation algorithms for asymmetric TSP by decomposing directed regular multigraphs. Journal of the ACM 52(4), 602–626 (2005)
18. Manthey, B.: On approximating multi-criteria TSP. In: Proc. 26th Int. Symp. on Theoretical Aspects of Computer Science (STACS), pp. 637–648 (2009)
19. Manthey, B.: Multi-criteria TSP: Min and max combined. In: Bampis, E., Jansen, K. (eds.) WAOA 2009. LNCS, vol. 5893, pp. 205–216. Springer, Heidelberg (2010)
20. Manthey, B., Shankar Ram, L.: Approximation algorithms for multi-criteria traveling salesman problems. Algorithmica 53(1), 69–88 (2009)
21. Mulmuley, K., Vazirani, U.V., Vazirani, V.V.: Matching is as easy as matrix inversion. Combinatorica 7(1), 105–113 (1987)
22. Paluch, K., Mucha, M., Mądry, A.: A 7/9 - Approximation Algorithm for the Maximum Traveling Salesman Problem. In: Dinur, I., Jansen, K., Naor, J., Rolim, J. (eds.) APPROX 2009. LNCS, vol. 5687, pp. 298–311. Springer, Heidelberg (2009)
23. Papadimitriou, C.H., Yannakakis, M.: The complexity of restricted spanning tree problems. Journal of the ACM 29(2), 285–309 (1982)
24. Papadimitriou, C.H., Yannakakis, M.: On the approximability of trade-offs and optimal access of web sources. In: Proc. 41st Ann. IEEE Symp. on Foundations of Computer Science (FOCS), pp. 86–92. IEEE, Los Alamitos (2000)
25. Zhang, T., Li, W., Li, J.: An improved approximation algorithm for the ATSP with parameterized triangle inequality. Journal of Algorithms 64(2-3), 74–78 (2009)

Extending Partial Representations
of Interval Graphs

Pavel Klavík*, Jan Kratochvíl*, and Tomáš Vyskočil*

Department of Applied Mathematics and
Institute for Theoretical Computer Science,
Charles University Malostranské náměstí 25,
118 00 Prague, Czech Republic
{klavik,honza,whisky}@kam.mff.cuni.cz

Abstract. We initiate the study of the computational complexity of
the question of extending partial representations of geometric intersection graphs. In this paper we consider classes of interval graphs – given
a collection of real intervals that forms an intersection representation
of an induced subgraph of an input graph, is it possible to add intervals to achieve an intersection representation of the entire graph? We
present an $\mathcal{O}(n^2)$ time algorithm that solves this problem and constructs
a representation if one exists. Our algorithm can also be used to list all
nonisomorphic extensions with $\mathcal{O}(n^2)$ delay.

Although the classes of proper and unit interval graphs coincide, the
partial representation extension problems differ on them. We present
an $\mathcal{O}(mn)$ time decision algorithm for partial representation extension
of proper interval graphs, but for unit interval graphs the complexity
remains open.

Finally we show how our methods can be used for solving the problem
of simultaneous interval representations. We prove that this problem is
fixed-paramater tractable with the size of the common intersection of
the input graphs being the parameter.

1 Introduction

Intersection Graphs. An *intersection representation* of a graph $G = (V, E)$ is
a collection of sets R_v, $v \in V$ such that for any two distinct vertices u and v,
$uv \in E$ if and only if $R_u \cap R_v \neq \emptyset$. If $\mathcal{M}$ is a family of sets, we say that G
is an $\mathcal{M}$-*intersection graph* if it has an intersection representation R such that
$R_v \in \mathcal{M}$ for every $v \in V$. Such a representation is called an $\mathcal{M}$-*representation*
(or, with a slight abuse of notation, a $\mathcal{C}$-*representation* where $\mathcal{C}$ stands for the
class of all $\mathcal{M}$-intersection graphs).

Intersection graphs of geometrical objects, especially of planar ones, are intensively studied for their practical motivations, algorithmic applications, and
interesting theoretical properties. The realm of these classes includes interval

---

* (Supported by Czech research project M10545).

M. Ogihara and J. Tarui (Eds.): TAMC 2011, LNCS 6648, pp. 276–285, 2011.

graphs, circular arc graphs, circle graphs, trapezoid graphs, and many more. Useful overviews can be found in [18,22,10,4]. For many of these classes basic optimization problems, such as finding a maximum clique or independent set, can be solved in polynomial time when a representation is given. It is therefore important to be able to recognize the graphs and construct an intersection representation. Indeed, in many cases these tasks can be achieved in polynomial time.

Many papers and much effort have been devoted to the recognition problem, and generalizations have been considered. For instance, the so called *sandwich* problem refers to the situation when certain edges are mandatory while others are forbidden, cf. [9,11].

Somewhat surprisingly a very natural problem of deciding whether a given representation of a part of a graph can be completed to a representation of the entire one so far has not been addressed. Formally, a *partial $\mathcal{M}$-representation* is an $\mathcal{M}$-representation of an induced subgraph G' of the input graph G. A representation R extends a partial representation R' if it assigns the same sets to the vertices of G'. For an intersection-defined class $\mathcal{C}$ of graphs, we consider the following decision problem.

Problem: REPEXT($\mathcal{C}$) (Partial Representation Extension of $\mathcal{C}$)
   Input: A graph G with a partial $\mathcal{C}$-representation R.
   Output: Does G have a $\mathcal{C}$-representation that extends R?

Extending Partial Solutions. This approach fits well a general paradigm that has been frequently investigated in other areas. It is rather interesting (but not so surprising) that quite often extending a partial solution is provably more difficult than solving a problem from a scratch[1]. Consider for instance graph coloring. By König-Hall Marriage Theorem, it is well known that edges of a cubic bipartite graph can be properly colored by 3 colors. But if some of them are precolored in the input graph, it becomes NP-complete to decide if this precoloring can be extended to a proper 3-edge-coloring of the entire graph [7], even for planar inputs [17].

On the other hand sometimes even the partial solution extension problem remains efficiently solvable. For instance, it has been recently proved by Angelini et al. [2] that planarity of partially embedded graphs can be decided in linear time. (To complete the picture, deciding if a partial straight line drawing of a graph can be extended to a straight line drawing of the entire one is again NP-hard as proven by Patrignani et al. [19].)

Classes of Interval Graphs. *Interval graphs* are intersection graphs of closed intervals of the real line. This class of graphs is denoted by INT. It is probably the most intensively studied class of intersection graphs. It was introduced by Hájos [12] in 1950s, motivated by applications in areas as diverse as biology, sociology, or traffic light scheduling. Aside from many applications, interval graphs have interesting theoretical properties. They can be recognized in polynomial time, in fact several linear time algorithms are known, see [3,6]. They are perfect

---

[1] No wonder architects like to use the metaphor "building on a green meadow".

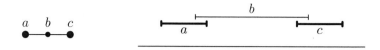

Fig. 1. A partial representation of P_3 extensible only by proper intervals

and hence basic optimization problems such as finding a maximum clique, maximum independent set, or the chromatic number are solvable in polynomial time on them. Thus it is natural to start the investigation of partial representation extension problems with interval graphs. In this paper we prove the following theorem.

Theorem 1. *The* REPEXT(INT) *problem is solvable in time* $\mathcal{O}(n^2)$, *where n is the number of vertices of the input graph.*

Two famous subclasses of interval graphs are *proper interval graphs*, denoted by PROPER INT, and *unit interval graphs*, denoted by UNIT INT. Proper interval graphs have representations in which no interval is a proper subset of another one. Unit interval graphs are those having intersection representations by intervals of a unit length. Roberts [20] proved that these classes are the same. Several linear time algorithms for their recognition are known, see [16,13,5].

Surprisingly, the partial representation extension problem distinguishes these two classes. For example, consider P_3, the path of length 2, with a partial representation of the vertices a and c by intervals placed at distance larger than 1, see Figure 1. As a proper interval representation, it is possible to be extended. On the other hand, it is not possible to extend it as a unit interval representation, the gap between a and c is too big.

In this paper, we show that REPEXT(PROPER INT) can be solved in polynomial time. The complexity of REPEXT(UNIT INT) remains open.

Theorem 2. *The* REPEXT(PROPER INT) *problem can be solved in time* $\mathcal{O}(nm)$, *where n is the number of vertices and m the number of edges of the input graph.*

Both the algorithms are constructive, they not only decide existence of a desired extension but also construct a feasible representation if it exists.

The Structure. The paper is organized as follows. In Section 2, we give an $\mathcal{O}(n^2)$ algorithm for extending interval graphs. In Section 3, we describe an $\mathcal{O}(nm)$ algorithm for REPEXT(PROPER INT). In Section 4, we present an application to a recent open problem described in [14] – using the partial representation extension approach, we construct fixed-parameter tractable algorithms for simultaneous interval graph and proper interval graph recognition. We conclude the paper with open problems and related remarks.

2   Extending Interval Graphs

In this section, we describe an algorithm solving REPEXT(INT) in time $\mathcal{O}(n^2)$. We consider representations having the endpoints of all intervals distinct since it simplifies the description. Every interval graph has such a representation and typically such representations are considered. Moreover, it is possible to modify our algorithm for intervals sharing endpoints.

Recognition. Recognition of interval graphs in linear time was a long-standing open problem, first solved by Booth and Lueker [3] using PQ-trees. Nowadays, there are two main approaches to recognition in linear time. The first one finds a feasible ordering of the maximal cliques which can be done using a data structure called PQ-tree. The second one uses surprising properties of the lexicographic breadth-first search, searches through the graph several times and constructs a representation if the graph is an interval graph [6]. We use the PQ-tree approach to solve REPEXT(INT) in time $\mathcal{O}(n^2)$.

Maximal Cliques. The PQ-tree approach is based on the following characterization of interval graphs, due to Fulkerson and Gross [8]:

Lemma 1 (Fulkerson and Gross). *A graph is an interval graph if and only if there exists an ordering of the maximal cliques such that for every vertex the cliques containing this vertex appear consecutively in this ordering.*

Consider an interval representation of an interval graph. For each maximal clique, consider the intervals representing the vertices of this clique and select a point in their intersection (this intersection is non-empty because intervals of the real line have the Helly property). We call these points *clique-points*. For an illustration, see Figure 2. The ordering of the clique-points from left to right gives the ordering required by Lemma 1. Every vertex appears consecutively since it is represented by an interval. For a clique a, we denote the assigned clique-point by $cp(a)$.

On the other hand, given an ordering of the maximal cliques, we place clique-points in this ordering on the real line. Each vertex is represented by the interval containing exactly the clique-points of the cliques containing this vertex. In this way, we obtain a valid interval representation of the graph.

In the rest of the section, by a clique we mean a maximal clique. The cliques of an interval graph have in total $\mathcal{O}(n + m)$ vertices and can be found in linear time [21]. A feasible ordering of the cliques can be found in linear time using PQ-trees. For readers not familiar with PQ-trees, the description is in [3].

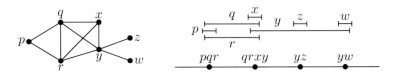

Fig. 2. An interval graph and one of its representations with denoted clique-points

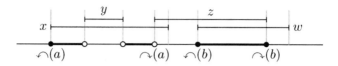

Fig. 3. Clique-points cp(a) and cp(b), having $I(a) = \{x\}$ and $I(b) = \{z, w\}$, can be placed to the bold parts of the real lines

Extending INT. We first sketch the algorithm. We construct a PQ-tree for the input graph, completely ignoring the given partial representation. The partial representation gives another restriction—a partial ordering of the cliques. We search the PQ-tree to find an ordering of the cliques extending this partial ordering. We show that the representation can be extended if and only if such an ordering exists.

To construct a representation, we place clique-points on the real line according to the ordering. We need to be more careful in this step. Since several intervals are pre-drawn, we cannot change their representations. Using the clique-points, we construct a representation in a similar manner as in Figure 2.

Now, we describe everything in detail.

Partial Ordering ◀. For a clique a, let $I(a)$ denote the set of all the pre-drawn intervals that are contained in a. The pre-drawn intervals split the line into several *parts*, traversed by the same intervals. A clique-point cp(a) can be placed only to a part containing exactly the intervals of $I(a)$ and no other pre-drawn intervals.

We denote by $\frown(a)$ (resp. $\frown(a)$) the leftmost (resp. the rightmost) point where the clique-point cp(a) can be placed, formally:

$$\frown(a) = \inf \left\{ x \mid \text{the clique-point cp}(a) \text{ can be placed to } x \right\},$$
$$\frown(a) = \sup \left\{ x \mid \text{the clique-point cp}(a) \text{ can be placed to } x \right\}.$$

For an example, see Figure 3. Notice that it does not mean that the clique-point cp(a) can be placed to all the points between $\frown(a)$ and $\frown(a)$. If a clique-point cannot be placed at all, the given partial representation is not extendible.

For two cliques a and b, we define $a \blacktriangleleft b$ if $\frown(a) \leq \frown(b)$. The obtained relation is a partial ordering. It is quite natural since every correct representation has to place a to the left of b if $a \blacktriangleleft b$. For example, the cliques a and b in Figure 3 satisfy $a \blacktriangleleft b$. This partial ordering can be clearly constructed in time $\mathcal{O}(n^2)$.

The Algorithm

(1) Find maximal cliques and construct a PQ-tree, independently of the partial representation.
(2) Compute $\frown$ and $\frown$ for all the cliques and construct $\blacktriangleleft$.
(3) Search the PQ-tree and find an ordering of the cliques extending $\blacktriangleleft$.
(4) Place the clique-points greedily on the real line, according to the ordering.
(5) Construct a representation using the clique-points.

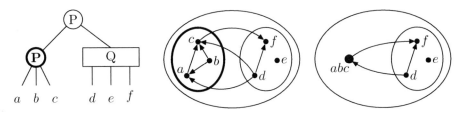

Fig. 4. We show an example how Step 3 works. Consider the highlighted P-node of the PQ-tree on the left. The subgraph induced by a, b and c has a topological sorting $b \rightarrow a \rightarrow c$. We contract these vertices into the vertex abc. Next, we keep the order of Q-node and contract it to the vertex def. Now, we obtain a cycle between abc and def and the algorithm outputs "no".

In the rest of the section, we describe in detail steps three to five and prove the correctness of the algorithm. This proves Theorem 1. All the omitted details will be in a journal version of the paper.

Step 3: Finding an Ordering Extending ◄. A subtree of the PQ-tree contains a set C of cliques. All these cliques appear consecutively in every ordering. Notice that reordering of C influences only this subtree.

This gives the following subroutine. We represent the ordering ◄ by a digraph. We reorder the children from the leaves to the root and modify the digraph by contractions. When we finish reordering of a subtree, the order is fixed and never changed in the future. We process a node when all its subtrees are already processed and represented by single vertices in the digraph. When we finish the reordering, we contract all these vertices, keeping edges in the digraph, see Figure 4.

For a P-node, we check whether the subdigraph induced by vertices corresponding to the children is acyclic. If it is acyclic, we reorder the children according to a topological sorting. Otherwise, there exists a cycle, no feasible ordering exists and the algorithm returns "no". For a Q-node, there are two possible orders. All we need is love, and to check whether one of them is feasible.

This operation can be implemented in linear time depending on the size of the partial ordering ◄ which is $\mathcal{O}(n^2)$.

Step 4: Placing the Clique-Points. The real line has several intervals already pre-drawn by the partial representation. We place clique-points greedily from left to right, according to the ordering. We want to place a clique-point cp(a). Let cp(b) be the last placed clique-point. Consider the infimum over all the points where the clique-point cp(a) can be placed and that are to the right of the clique-point cp(b). We place the clique-point cp(a) to the right of this infimum by an appropriate epsilon, for example the length of the shortest part (see definition of ◄) divided by n. We can easily implement this greedy subroutine in time $\mathcal{O}(n)$.

Lemma 2. *For an ordering $<$ of the cliques compatible with the PQ-tree extending ◄, the greedy subroutine described in Step 4 never fails.*

Due to space limitations, we sketch a proof: For contradiction, let cp(a) be a clique-point which cannot be placed. There exists a clique-point cp(b) such that $b < a$ and cp(a) cannot be placed to the right of cp(b). We question the reason the clique-point cp(b) was not placed more to the left, to allow cp(a) to be placed. We obtain contradiction with ◀ or the constructed PQ-tree. A detailed proof will be in a journal version.

Step 5: Constructing a Representation. We construct a representation of the graph using the clique-points placed in the previous step, similarly to Figure 2. We represent each vertex as an interval containing exactly all the clique-points corresponding to the cliques containing this vertex. Also, we slightly perturbate all the intervals to ensure distinct endpoints.

The intervals placed by the partial representation contain the correct clique-points. Since the ordering of the clique-points is compatible with the PQ-tree, we obtain a correct representation. The last step can be done in time $\mathcal{O}(n)$.

3   Extending Proper Interval Graphs

In this section, we show an algorithm for extending proper interval graphs in time $\mathcal{O}(nm)$, for n vertices and m edges. We modify the algorithm described in Section 2. It is the only algorithm known to us which uses PQ-trees to recognize proper interval graphs. This method is not extremely fast, but we can use the previously described properties. Basically, the algorithm introduces other sets of cliques to appear consecutively. These sets ensure that all edges are realized by single overlapping intervals.

Definitions. We recall that intervals have distinct endpoints. Two intervals of the real line can be in three relative positions, see Figure 5. If they do not intersect, they have *no overlap*. If one is a proper subset of the other one, they *double overlap*. Otherwise, we say they *single overlap*. In proper interval graphs, all the edges are realized by single overlaps.

All the cliques considered in the rest of the section are maximal. For a vertex x, we denote by C_x the set of the cliques containing x. Similarly, $C_{x,\neg y}$ denotes $C_x \setminus C_y$.

Additional Conditions. Let x and y be two intervals that single overlap. Consider an ordering of the maximal cliques given by the representation. Observe that all the cliques of $C_{x,\neg y}$ appear on one side of the intersection of x and y and

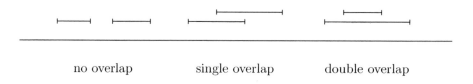

no overlap   single overlap   double overlap

Fig. 5. Types of interval overlapping

similarly $C_{y,\neg x}$ appears on the other side. Together with the consecutivity of C_x and C_y, we know that $C_{x,\neg y}$ and $C_{y,\neg x}$ appear consecutively in the ordering.

We can use this in the opposite direction. To ensure that an edge xy is realized by a single overlap, we add a requirement that each of the sets $C_{x,\neg y}$ and $C_{y,\neg x}$ appears consecutively. For a proper interval graph, we introduce these conditions for every edge of the graph. We introduce $\mathcal{O}(m)$ additional sets, each of size $\mathcal{O}(n)$. Using the PQ-tree algorithm by Booth and Lueker [3], we construct a PQ-tree for all these conditions in time $\mathcal{O}(nm)$.

The Modified Algorithm. We modify the algorithm from Section 2. First, we check whether the given partial representation is correct. In Step 1, we introduce additional sets, as described above. We proceed steps two to four as before.

The clique-points are already fixed at the start of Step 5. We need to place the intervals in a way that all the edges are realized by single overlaps. Unlike interval graphs, proper interval graphs have one specific property. For every representation, the left endpoints appear in the same order as the right endpoints. Denote this ordering by $\lhd$. If two vertices are contained in different cliques, their order in $\lhd$ is fixed. Also, if two intervals are pre-drawn, the order of the corresponding vertices is fixed. These conditions give a partial ordering of the vertices. As $\lhd$, we choose any linear extension of this partial ordering. Using $\lhd$, we place intervals around clique points. The described algorithm proves Theorem 2. Details will be in a journal version.

4   Simultaneous Interval Graphs

In this section, we show an application of the partial representation extension problem. We give an FPT algorithm for a recent open problem by Jampani and Lubiw [14]. Let $G_1, G_2, \ldots, G_k$ be graphs, $G_i = (V_i, E_i)$. Let vertex sets of these graphs intersect as a sunflower, i.e., $V_i \cap V_j = I$ for every $i \neq j$. We would like to find a *simultaneous representation* of all these graphs. This means that we would like to find a representation R_i for every graph G_i such that all these representations assign the same sets to the vertices of I. For an example, see Figure 6. We denote this problem SIM.

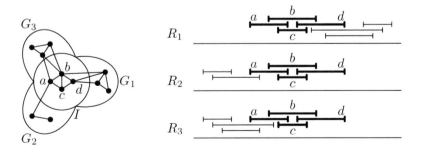

Fig. 6. A simultaneous representation of G_1, G_2 and G_3 with $I = \{a, b, c, d\}$

Jampani and Lubiw [14] show a polynomial time algorithm for SIM(INT) when $k = 2$. They ask whether there exists a polynomial time algorithm for SIM(INT) in general. We show that SIM(INT) and SIM(PROPER INT) are fixed-parameter tracktable for parameter $\ell = |I|$. We describe the algorithm for interval graphs, the algorithm for proper interval graphs is similar.

Proposition 1. SIM(INT) *can be solved in time* $\mathcal{O}(n^2 k \cdot (2\ell)!)$.

Proof. If the partial representation of I were given, we could use the algorithm for REPEXT(INT) described in Section 2 to test whether it is possible to extend it to a simultaneous representation of all the graphs. We just need to test for every graph G_i whether it is possible to extend the partial representation to a representation of entire G_i. This can be done in time $\mathcal{O}(n^2 k)$.

An interval graph with ℓ vertices has $\mathcal{O}((2\ell)!)$ topologically different representations. Therefore, if the algorithm tries all the different representations, the running time is $\mathcal{O}(n^2 k \cdot (2\ell)!)$. □

Similarly, the following holds, the proof is omitted:

Proposition 2. SIM(PROPER INT) *can be solved in time* $\mathcal{O}(mnk \cdot \ell!)$.

5   Conclusions

5.1   Constructing All Representations

Sometimes it is necessary to construct all representations that extend the given partial one. Our algorithms can be used to list all feasible representations with a polynomial delay ($O(n^2)$ both for REPEXT(INT) and REPEXT(PROPER INT)). Details will be discussed in a journal version.

5.2   Allen Algebras

In [1], Allen described thirteen primitive relations between pairs of intervals and initiated the study of computational complexity of the question of deciding if there exists an interval representation satisfying given constraints expressed by subsets of these relations. A full dichotomy for this problem was proved in [15]. Both REPEXT(INT) and REPEXT(PROPER INT) can be formulated in the language of Allen algebras. However, the corresponding constraint satisfaction problems are NP-complete.

5.3   Extending Unit Interval Representations

We conclude by repeating what we consider the main open problem. What is the complexity of REPEXT(UNIT INT)?

References

1. Allen, J.F.: Maintaining knowledge about temporal intervals. Commun. ACM 26(11), 832–843 (1983)
2. Angelini, P., Battista, G.D., Frati, F., Jelínek, V., Kratochvíl, J., Patrignani, M., Rutter, I.: Testing planarity of partially embedded graphs. In: SODA 2010: Proceedings of the Twenty-First Annual ACM-SIAM Symposium on Discrete Algorithms (2010)
3. Booth, K.S., Lueker, G.S.: Testing for the consecutive ones property, interval graphs, and planarity using pq-tree algorithms. Journal of Computational Systems Science 13, 335–379 (1976)
4. Brandstädt, A., Le, V.B., Spinrad, J.P.: Graph classes: a survey. Society for Industrial and Applied Mathematics, Philadelphia (1999)
5. Corneil, D.G.: A simple 3-sweep lbfs algorithm for the recognition of unit interval graphs. Discrete Appl. Math. 138(3), 371–379 (2004)
6. Corneil, D.G., Olariu, S., Stewart, L.: The lbfs structure and recognition of interval graphs. SIAM Journal on Discrete Mathematics 23(4), 1905–1953 (2009)
7. Fiala, J.: NP completeness of the edge precoloring extension problem on bipartite graphs. J. Graph Theory 43(2), 156–160 (2003)
8. Fulkerson, D.R., Gross, O.A.: Incidence matrices and interval graphs. Pac. J. Math. 15, 835–855 (1965)
9. Golumbic, M., Kaplan, H., Shamir, R.: Algorithms and complexity of sandwich problems in graphs (extended abstract). In: van Leeuwen, J. (ed.) WG 1993. LNCS, vol. 790, pp. 57–69. Springer, Heidelberg (1994)
10. Golumbic, M.C.: Algorithmic Graph Theory and Perfect Graphs. North-Holland Publishing Co., Amsterdam (2004)
11. Golumbic, M.C., Shamir, R.: Complexity and algorithms for reasoning about time: a graph-theoretic approach. J. ACM 40(5), 1108–1133 (1993)
12. Hajós, G.: Über eine Art von Graphen. Internationale Mathematische Nachrichten 11, 65 (1957)
13. Hell, P., Huang, J.: Lexicographic orientation and representation algorithms for comparability graphs, proper circular arc graphs, and proper interval graphs. Journal of Graph Theory 20(3), 361–374 (1995)
14. Jampani, K., Lubiw, A.: Simultaneous interval graphs. In: Cheong, O., Chwa, K.-Y., Park, K. (eds.) ISAAC 2010. LNCS, vol. 6506, pp. 206–217. Springer, Heidelberg (2010)
15. Krokhin, A., Jeavons, P., Jonsson, P.: Reasoning about temporal relations: The tractable subalgebras of allen's interval algebra. J. ACM 50(5), 591–640 (2003)
16. Looges, P.J., Olariu, S.: Optimal greedy algorithms for indi erence graphs. Comput. Math. Appl. 25, 15–25 (1993)
17. Marx, D.: NP-completeness of list coloring and precoloring extension on the edges of planar graphs. J. Graph Theory 49(4), 313–324 (2005)
18. McKee, T.A., McMorris, F.R.: Topics in Intersection Graph Theory. SIAM Monographs on Discrete Mathematics and Applications (1999)
19. Patrignani, M.: On extending a partial straight-line drawing. In: Healy, P., Nikolov, N.S. (eds.) GD 2005. LNCS, vol. 3843, pp. 380–385. Springer, Heidelberg (2006)
20. Roberts, F.S.: Indifference graphs. In: Harary, F. (ed.) Proof Techniques in Graph Theory, pp. 139–146. Academic Press, London (1969)
21. Rose, D.J., Tarjan, R.E., Lueker, G.S.: Algorithmic aspects of vertex elimination on graphs. SIAM Journal on Computing 5(2), 266–283 (1976)
22. Spinrad, J.P.: Efficient Graph Representations. Field Institute Monographs (2003)

Using Split Composition to Extend Distance-Hereditary Graphs in a Generative Way
(Extended Abstract)

Serafino Cicerone

Department of Electrical & Information Engineering
University of L'Aquila, L'Aquila, Italy
serafino.cicerone@univaq.it

Abstract. In this paper we introduce a new graph class denoted as $Gen(*; P_3, C_3, C_5)$. It contains all graphs that can be generated via split composition by using paths P_3 and cycles C_3 and C_5 as components. This new graph class extends the well known class of distance-hereditary graphs, which corresponds to $Gen(*; P_3, C_3)$. For the new class we provide efficient algorithms for several basic combinatorial problems: recognition, stretch number, stability number, clique number, domination number, chromatic number, graph isomorphism, and clique width.

Keywords: Distance-hereditary graphs, stretch number, split decomposition, graph algorithms, recognition problem.

1 Introduction

Distance-hereditary graphs have been introduced by Howorka [17], and are defined as those graphs in which every connected induced subgraph is isometric, that is it "inherits" its distance function from the whole graph. Formally, a graph G is a *distance-hereditary graph* if, for each connected induced subgraph G' of G, $d_{G'}(x, y) = d_G(x, y)$ for each $x, y \in G'$. Such graphs have been rediscovered many times (e.g., see [1]) and have different applications. For instance, they can model communication networks [9,12,13] in which node failures may occur: at a given time, if sender and receiver are still connected, any message can be still delivered without increasing the length of the path used to reach the receiver.

Since their introduction, dozens of papers have been devoted to them, and different kind of characterizations have been discovered: metric, forbidden subgraphs, cycle/chord conditions, level/neighborhood conditions, generative, and more. Among such results, the generative properties resulted as the most fruitful for algorithmic applications, since they allowed researchers to solve several combinatorial problems efficiently in the context of distance-hereditary graphs (e.g., see [2,3,9,15,18]). Recently, distance-hereditary graphs have been generalized into the class of k–distance-hereditary graphs:

M. Ogihara and J. Tarui (Eds.): TAMC 2011, LNCS 6648, pp. 286–297, 2011.

Definition 1. *[7] Given a rational number $k \geq 1$, a graph G is a k–distance-hereditary graph if, for each connected induced subgraph G' of G:*

$$d_{G'}(x, y) \leq k \cdot d_G(x, y), \quad \text{for each } x, y \in G'.$$

The class of all the k–distance-hereditary graphs is denoted by DH(k).

Notice the following basic relationships: DH(1) coincides with the class of distance-hereditary graphs; DH$(k_1) \subseteq$ DH(k_2), for each $k_1 \leq k_2$; DH(k) is hereditary for each $k \geq 1$.

The hierarchy formed by classes DH(k) is *fully general*. This means that, for each arbitrary graph G, there exists k' such that $G \in$ DH(k'). In particular, the *stretch number* of G – denoted as $s(G)$ – is the smallest rational number t such that G belongs to DH(t).

Apart from the interesting general results found for the new classes, the original motivation was studying how (if possible) to extend the known algorithmic results from the base class, namely DH(1), to DH(k) for some constant, hopefully large, $k > 1$. This specific task seems to be in contrast with the generality of the DH(k) hierarchy. This consideration leads to redefine the main objective: from the hierarchy DH(k) to some new class $\mathbb{C}$, being $\mathbb{C}$ a superclass of DH(1) and with clear relationships with DH(k).

In this work we introduce a new graph class which can play the role of $\mathbb{C}$. This new class is denoted as $Gen(*; P_3, C_3, C_5)$ and it contains all graphs that can be *generated* by applying the split composition [10], starting from K_2, and using the path P_3, and the cycles C_3 and C_5 as components. We use the notation $Gen(\texttt{opn}; \texttt{list\_of\_components})$ for classes whose elements can be defined in a generative way.

Results: We provide the following results for the graph class $Gen(*; P_3, C_3, C_5)$:

1. Efficient algorithms for the following problems:
   (a) recognition, stretch number (both solved in linear time);
   (b) stability number, clique number, domination number and its variants ($O(n)$ time); chromatic number ($O(n^3)$ time); graph isomorphism (linear time).
2. Clique width is bounded.
3. Relationships with the DH(k) hierarchy.

Some proofs are omitted due to space limitation and they can be found in the full paper [4].

2   Notation and Basic Concepts

In this work we consider finite, simple, loop-less, undirected and unweighted graphs $G = (V, E)$ with vertex set V and edge set E. For sake of simplicity, by graph we mean "connected graphs with at least 3 vertices", although definition and results can be easily extended to any graph. Given $S \subseteq V$, the *induced subgraph* $\langle S \rangle$ of G is the maximal subgraph of G with vertex set S. Given $u \in V$, $N(u)$ denotes the set of *neighbors* of u in G, and $N[u] = N(u) \cup \{u\}$.

We denote by C_k any cycle with k vertices. A *chord* of a cycle is an edge joining two non-consecutive vertices in the cycle. The *chord distance* of a cycle C_k is denoted by $cd(C_k)$, and it is defined as the minimum number of consecutive vertices in C_k such that every chord of C_k is incident to some of such vertices. H_k denotes a *hole*, i.e., the cycle C_k, $k \geq 5$, without chords. We assume $cd(H_k) = 0$. P_k denotes a path with k vertices.

Symbols $p_G(x, y)$ and $P_G(x, y)$ denote a shortest and a longest induced path between x and y, respectively. The lengths of such paths are denoted as $d_G(x, y)$ and $D_G(x, y)$, respectively.

If x and y are two vertices of G such that $d_G(x, y) \geq 2$, then $\{x, y\}$ is a *cycle-pair* if there are two induced paths $p_G(x, y)$ and $P_G(x, y)$ such that $p_G(x, y) \cap P_G(x, y) = \{x, y\}$. In other words, if $\{x, y\}$ is a cycle-pair, then the vertices in $p_G(x, y) \cup P_G(x, y)$ induce a cycle in G; this cycle is denoted by $G[x, y]$.

2.1   Split Composition and Decomposition

Let us recall the concept of *split composition/decomposition* [10]. Let G_1, G_2 be two graphs having disjoint vertex sets $V_1 \cup \{m_1\}$, $V_2 \cup \{m_2\}$ and edge sets E_1, E_2, respectively. The split composition of G_1 and G_2 with respect to the *joining vertices* m_1 and m_2 is the graph G having vertex set $V = V_1 \cup V_2$ and edge set $E = E_1' \cup E_2' \cup \{(x, y) \mid x \in N(m_1), y \in N(m_2)\}$, where $E_i' = \{(x, y) \in E_i \mid x, y \in V_i\}$ for $i = 1, 2$. The composition is denoted as $G = (G_1, m_1) * (G_2, m_2)$, or simply $G = G_1 * G_2$ when we are not interested in the joining vertices.

The *split decomposition* is the inverse operation. Let G_1, G_2 be two graphs having disjoint vertex sets $V_1 \cup \{m_1\}$, $V_2 \cup \{m_2\}$ and edge sets E_1, E_2, respectively. If $G = (G_1, m_1) * (G_2, m_2)$ and $|V_1|, |V_2| \geq 2$, then we say that $\{G_1, G_2\}$ is a *simple decomposition* of G. We call $\{V_1, V_2\}$ the *split* of G associated with the simple decomposition $\{G_1, G_2\}$, being m_1 and m_2 the associated joining vertices. If G has a split we say that G is *split-decomposable*. The *split decomposition* of a graph G is the set $\mathcal{D}(G)$ of graphs obtained by the following recursive procedure:

- if G has a split $\{V_1, V_2\}$, then apply the split decomposition to graphs G_1 and G_2 obtained by the simple decomposition $\{G_1, G_2\}$.
- if G does not have a split then G is called *prime*.

Each element of $\mathcal{D}(G)$ is called *component*. The split decomposition of a graph is not necessarily unique. Cunningham proved the following uniqueness result:

Theorem 1. [10] *Each connected graph has a unique split decomposition into prime graphs, stars, and cliques with a minimum number of components.*

$\mathcal{D}(G)$ denotes the split decomposition of G, while $\mathcal{CD}(G)$ will denote the Cunningham decomposition of G (in which stars and cliques are no further decomposed). All known algorithms to compute the split decomposition compute the Cunningham decomposition. In [11], Dahlhaus given a linear time algorithm for computing the Cunningham decomposition. The following claim follows directly from the definition of split composition:

Property 1. Let $G = (G_1, m_1) * (G_2, m_2)$. If $u \in G_1 \setminus \{m_1\}$ and $v \in G_2 \setminus \{m_2\}$ then

- $D_G(u, v) = D_{G_1}(u, m_1) + D_{G_2}(v, m_2) - 1$;
- $d_G(u, v) = d_{G_1}(u, m_1) + d_{G_2}(v, m_2) - 1$. □

3 Extending Distance-Hereditary Graphs

One of the most popular characterizations of distance-hereditary graphs is based on *one-node extension* operations [1]. Given a graph G, $u \in G$, and $v \notin G$, then:

- $\alpha(G, u; v)$ adds v to G by making v adjacent to u only;
- $\beta(G, u; v)$ adds v to G by making v adjacent to each node in $N(u)$;
- $\gamma(G, u; v)$ adds v to G by making v adjacent to each node in $N[u]$.

Theorem 2. [1] *Every distance-hereditary graph is obtained starting from a single node and by applying a proper sequence of operations α, β, and γ.*

This result can be reformulated in terms of a single operation, i.e. the split composition.

Lemma 1. *Every distance-hereditary graph is obtained starting from K_2 and by applying the split composition using P_3 and C_3 as components.*

Proof. It is easy to verify that the following relationships hold:

- $\alpha(G, u; v) \equiv (G, u) * (P_3, v)$, where v is an external vertex in the path P_3;
- $\beta(G, u; v) \equiv (G, u) * (P_3, v)$, where v is an internal vertex in the path P_3;
- $\gamma(G, u; v) \equiv (G, u) * (C_3, v)$, where v is any vertex in the cycle C_3.

The proof is concluded by observing that these relationships also show that any vertex of the components P_3 and C_3 can be used as joining vertex. □

Previous results that make a connection between distance-hereditary graphs and split composition are [5] and [16]. From the above Lemma it follows that the class of distance-hereditary graphs can be denoted also as $Gen(*; P_3, C_3)$. Now, extending $Gen(*; P_3, C_3)$ via split composition is just matter of selecting a new component to be used along with P_3, C_3. To this aim we recall the following additional characterization of distance-hereditary graphs (see Fig. 1).

Theorem 3. [1] *A graph G is distance-hereditary if and only if it does not contain, as induced subgraph, the following graphs: the hole H_n, $n \geq 5$, the house, the fan, and the domino.*

This result states that only few graphs cannot appear as induced subgraphs in a distance-hereditary graph. Since the smallest ones among such forbidden subgraphs are cycles C_5, then the following definition formalizes the new class:

Definition 2. *Every graph in $Gen(*; P_3, C_3, C_5)$ is obtained starting from K_2 by applying the split composition and using P_3, C_3, and C_5 as components.*

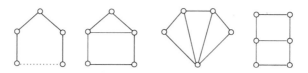

Fig. 1. The hole (chordless cycle) H_n, $n \geq 5$, the house, the fan, and the domino. The dotted line represents a path with one or more edges.

4 Algorithmic Problems

In this section we face the recognition problem and the stretch number computation problem for graph in the new class $Gen(*; P_3, C_3, C_5)$.

4.1 Recognition Problem

The recognition problem for the new graph class can be formulated as follows: given a graph G, decide whether G belongs to $Gen(*; P_3, C_3, C_5)$. This problem can be easily solved by using the split decomposition.

Lemma 2. *A graph G belongs to $Gen(*; P_3, C_3, C_5)$ if and only if each component in $CD(G)$ is a star, a clique, or a cycle C_5.*

Proof. ($\Rightarrow$) Let $G \in Gen(*; P_3, C_3, C_5)$. Then, there exist a sequence $B_1, B_2, \ldots, B_n$ such that:

- $B_i \in \{P_3, C_3, C_5\}$, $1 \leq i \leq n$;
- $G_0 \equiv K_2$;
- $G_i \equiv G_{i-1} * B_i$, $1 \leq i \leq n$;
- $G_n \equiv G$.

We prove the statement by induction on i. The basic case ($i = 1$) holds since $K_2 * B_1 = B_1$, $CD(B_1) = \{B_1\}$, and B_1 is a star (i.e, P_3), a clique (i.e, C_3), or a cycle C_5. Now, let us assume that the statement is true for G_t, $t < n$.

By inductive hypothesis, each component of $CD(G_t)$ is a star, a clique, or a cycle C_5, while B_{t+1}, by definition, is a P_3, a C_3, or a C_5.

Since $G_{t+1} = G_t * B_{t+1}$, then $CD(G_{t+1})$ can be derived from $CD(G_t)$ and B_{t+1} as follows. Let $\bar{C}$ the component in $CD(G_t)$ containing the joining vertex in $G_t * B_{t+1}$. Is is easy to observe that, by construction, $CD(G_{t+1}) = (CD(G_{t+1}) \setminus \bar{C}) \cup CD(\bar{C} * B_{t+1})$. It follows that, to prove the statement it is sufficient to show that each component in $CD(\bar{C} * B_{t+1})$ is a star, a clique, or a cycle C_5. Let us analyze different cases, according to the size of $CD(\bar{C} * B_{t+1})$:

- $|CD(\bar{C} * B_{t+1})| = 1$. Since $\bar{C} * B_{t+1}$ has a split, then, by Theorem 1, it is a clique or a star;
- $|CD(\bar{C} * B_{t+1})| = 2$. In this case it follows that $CD(\bar{C} * B_{t+1}) = \{\bar{C}, B_{t+1}\}$;
- $|CD(\bar{C} * B_{t+1})| \geq 3$. In this case we have $B_{t+1} \equiv C_5$ and C_5 has 2 or more splits. Since such splits divide C_5 into P_3 and C_3, then also in this case each component in $CD(\bar{C} * B_{t+1})$ is a star, a clique, or a cycle C_5.

Fig. 2. *Decomposing a clique K_t, $t \geq 3$, into $C_3 * C_3 * \cdots * C_3$ with $t - 2$ cliques C_3 (black vertices represent joining vertices)*

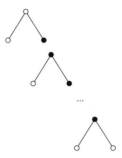

Fig. 3. *Decomposing a star $K_{1,t}$, $t \geq 2$, into $K_{1,2} * K_{1,2} * \cdots * K_{1,2}$ with $t - 1$ stars $K_{1,2}$ (black vertices represent joining vertices)*

($\Leftarrow$) Let us assume that each component of $\mathcal{CD}(G) = \{B_1, B_2, \ldots, B_n\}$ is a star, a clique, or a cycle C_5. Now, perform the following operations on each component B_i, $1 \leq i \leq n$, which is a clique or a star:

- if B_i is a clique K_t, $t \geq 3$, then replace K_t by $C_3 * C_3 * \cdots * C_3$ with $t - 2$ cliques C_3 according to Fig. 2;
- if B_i is a star $K_{1,t}$, $t \geq 2$, then replace $K_{1,t}$ by $K_{1,2} * K_{1,2} * \cdots * K_{1,2}$ with $t - 1$ stars $K_{1,2}$ according to Fig. 3.

From components $B_1, B_2, \ldots, B_n$ and their modifications we get a split decomposition with components in $\{P_3, C_3, C_5\}$. From this split decomposition it is easy to get a generative sequence showing that $G \in Gen(*; P_3, C_3, C_5)$.

Theorem 4. *The recognition problem for the class $Gen(*; P_3, C_3, C_5)$ can be solved in linear time.*

Proof. Let G be a graph in $Gen(*; P_3, C_3, C_5)$. The statement directly follows from Lemma 2 and from the linear time algorithm proposed in [11] for computing the Cunningham decomposition $\mathcal{CD}(G)$.   □

4.2   Computing the Stretch Number

In [7], it has been shown that computing the stretch number for arbitrary graphs is NP-hard. In this section we provide a linear time algorithm to solve this problem in $Gen(*; P_3, C_3, C_5)$. Before presenting it, we give some notation and results. Given u and v in G, then, (1) the stretch number of $\{u, v\}$ is given by $s_G(u, v) = D_G(u, v)/d_G(u, v)$, (2) $s_G(v) = \max\{s_G(v, u) \mid u \in V\}$, and (3) $\bar{v}$ represents any vertex such that $s_G(v) = s_G(v, \bar{v})$. In [7] it has been shown that the *stretch number of* G is given by

$$s(G) = \max_{u,v \in V} s_G(u, v).$$

$\mathcal{S}(G)$ contains all pairs $\{u, v\}$ that give $s(G)$, that is $s_G(u, v) = s(G)$. If $\{x, y\}$ is a cycle-pair that belongs to $\mathcal{S}(G)$, then the cycle $G[x, y]$ is called *inducing-stretch cycle* for G. The following result states which numbers are *admissible* as strecth numbers:

Theorem 5. [8] *Let* $t \geq 1$ *be a rational number.* t *is an admissible stretch number if and only if* $t \geq 2$ *or* $t = 2 - \frac{1}{i}$ *for some integer* $i \geq 1$.

Let G be a graph belonging to $Gen(*; P_3, C_3, C_5)$ such that $G = G_1 * G_2$: following the same approach used by Rao in [19], the next theorem allows us to compute $s(G)$ starting from the stretch numbers of G_1 and G_2. It exploits this technical statement:

Lemma 3. *Let* $G = (V, E)$ *and* $x \in V$. *If* $s_G(x) = 2 - 1/i$, $i \geq 1$ *integers, then there exists a vertex* $y \in V$ *such that* $s_G(x, y) = 2 - 1/i$ *and* $d_G(x, y) = i$.

Theorem 6. *Let* $G_1 = (V_1, E_1)$ *and* $G_2 = (V_2, E_2)$ *be two graphs. Let* $m_1 \in V_1$ *such that* $s_{G_1}(m_1) = 2 - 1/i$, *and let* $m_2 \in V_2$ *such that* $s_{G_2}(m_2) = 2 - 1/j$. *If* $G = (G_1, m_1) * (G_2, m_2)$, *then*

$$s(G) = \max\{\ s(G_1),\ s(G_2),\ s_G(\overline{m_1}, \overline{m_2})\ \},$$

where $s_G(\overline{m_1}, \overline{m_2}) = 2 - \frac{1}{i+j-1}$.

Proof. By assuming that $(a, b) \in \mathcal{S}(G)$, three different cases may occur:

1. both a and b are in V_1;
2. both a and b are in V_2;
3. a belongs to V_1 and b belongs to V_2.

In the first two cases, we get $s(G) = s(G_1)$ and $s(G) = s(G_2)$, respectively. In the remainder, we assume that the third case holds, and we show that $s(G) = s_G(\overline{m_1}, \overline{m_2}) = 2 - \frac{1}{i+j-1}$.

Since $s_{G_1}(m_1) = 2 - 1/i$, by Lemma 3 there exists a vertex $\overline{m_1} \in V_1$ such that $s_{G_1}(\overline{m_1}, m_1) = 2 - 1/i$ and $d_{G_1}(\overline{m_1}, m_1) = i$. Symmetrically, there exists $\overline{m_2} \in V_2$ such that $s_{G_2}(\overline{m_2}, m_2) = 2 - 1/j$ and $d_{G_2}(\overline{m_2}, m_2) = j$. Of course, these relationships imply $D_{G_1}(\overline{m_1}, m_1) = 2i - 1$ and $D_{G_2}(\overline{m_2}, m_2) = 2j - 1$.

According to the definitions of stretch number and split composition, it follows that:

$$\begin{aligned} s_G(\overline{m_1}, \overline{m_2}) &= \frac{D_G(\overline{m_1}, \overline{m_2})}{d_G(\overline{m_1}, \overline{m_2})} \\ &= \frac{D_{G_1}(\overline{m_1}, m_1) + D_{G_2}(\overline{m_2}, m_2) - 1}{d_{G_1}(\overline{m_1}, m_1) + d_{G_2}(\overline{m_2}, m_2) - 1} \\ &= \frac{(2i-1) + (2j-1) - 1}{i+j-1} \\ &= 2 - \frac{1}{i+j-1} \end{aligned}$$

The proof is concluded by showing that the following relationship holds for each $x \in V_1$ and for each $y \in V_2$:

$$2 - \frac{1}{i + j - 1} \geq \frac{D_G(x, y)}{d_G(x, y)} \tag{1}$$

By applying again the notion of split composition, Inequality 1 can be rewritten as:

$$2 - \frac{1}{i + j - 1} \geq \frac{D_{G_1}(x, m_1) + D_{G_2}(y, m_2) - 1}{d_{G_1}(x, m_1) + d_{G_2}(y, m_2) - 1} \tag{2}$$

Now:

- by Theorem 5, since $s_{G_1}(x, m_1) < 2$, we may assume $s_{G_1}(x, m_1) = 2 - 1/k_1$, being $k_1 \geq 1$ an integer. Since $s_{G_1}(\overline{m_1}, m_1) = 2 - 1/i$ and $s_{G_1}(\overline{m_1}, m_1) \geq s_{G_1}(x, m_1)$, it follows that $i \geq k_1$. Moreover, $d_{G_1}(x, m_1) = t_1 k_1$ and $D_{G_1}(x, m_1) = t_1(2k_1 - 1)$, $t_1 \geq 1$. The integer t_1 is due to the fact that from $s_{G_1}(x, m_1) = 2 - 1/k_1$ we cannot derive exact values for the distances $d_{G_1}(x, m_1)$ and $D_{G_1}(x, m_1)$;
- symmetrically, assuming $s_{G_2}(y, m_2) = 2 - 1/k_2$, it follows that $j \geq k_2$. Moreover, $d_{G_2}(y, m_2) = t_2 k_2$ and $D_{G_2}(y, m_2) = t_2(2k_2 - 1)$, $t_2 \geq 1$ integer.

As a consequence, Inequality 2 can be rewritten as:

$$2 - \frac{1}{i + j - 1} \geq \frac{t_1(2k_1 - 1) + t_2(2k_2 - 1) - 1}{t_1 k_1 + t_2 k_2 - 1} \tag{3}$$

By simple algebraic transformations, Inequality 3 can be rewritten as:

$$t_1 i + t_2 j + t_1(j - 1) + t_2(i - 1) \geq t_1 k_1 + t_2 k_2 + (j - 1) + (i - 1) \tag{4}$$

Inequality 4 holds due to the following straightforward observations:

1. $t_1 i \geq t_1 k_1$ (because $t_1 \geq 1$ and $i \geq k_1 \geq 1$);
2. $t_2 j \geq t_2 k_2$ (because $t_2 \geq 1$ and $j \geq k_2 \geq 1$);
3. $t_1(j - 1) \geq (j - 1)$ (because $t_1 \geq 1$);
4. $t_2(i - 1) \geq (i - 1)$ (because $t_2 \geq 1$).

This concludes the proof.   □

Surprisingly, this result shows that that $s(G)$ does not depends on $s(G_1)$ and $s(G_2)$, but only on the maximum stretch number of the joining vertices m_1 and m_2. This leads to an algorithm for computing $s(G)$ based on the following steps:

1. decompose G by split decomposition;
2. compute the stretch number of all components in $\mathcal{CD}(G)$;
3. compute $s_B(u)$ and $s_B(u, v)$ for all joining vertices u and v in B, and for all components $B \in \mathcal{CD}(G)$;
4. rebuild G using the components and, at each rebuilding step, use Theorem 6.

Step 3 is necessary since to use Theorem 6 we need information about the maximum stretch of joining vertices. Notice also that such additional info has to be "updated" as long as the graph G is rebuilt. The next corollary can be used for this purpose. namely to update the value of $s_{G_1}(u)$, $u \in G_1$, after a split composition $G_1 * G_2$.

Corollary 1. *Let $G_1 = (V_1, E_1)$ and $G_2 = (V_2, E_2)$ be two graphs. Let $m_1 \in V_1$ such that $s_{G_1}(m_1) = 2 - 1/i$, and let $m_2 \in V_2$ such that $s_{G_2}(m_2) = 2 - 1/j$. If $G = (G_1, m_1) * (G_2, m_2)$ and $u \in G_1 \setminus \{m_1\}$, then*

$$s_G(u) = \max \left\{ s_{G_1}(u), \frac{D_{G_1}(u, m_1) + D_{G_2}(m_2, \overline{m_2}) - 1}{d_{G_1}(u, m_1) + d_{G_2}(m_2, \overline{m_2}) - 1} \right\}$$

Proof. Omitted.

Notice that, by symmetry, the corollary above can be also used to update the stretch number of a node in G_2. Such a corollary, along with Theorem 6, represents the main tool to give the Algorithm 1, which is able to compute the stretch number of graphs in $Gen(*; P_3, C_3, C_5)$. In this algorithm we use the notation $\mathcal{CD}(G) = \{B_1, B_2, \ldots, B_t\}$ and $\mathcal{JV}(G) = \{x_2, y_2, x_3, y_3, \ldots, x_t, y_t\}$, $t \geq 1$, to represent the result of the Cunningham decomposition, where $\mathcal{JV}(G)$ represents the set of *joining vertices* (i.e., x_i and y_i are the joining vertices of the i-th split, where $x_i \in B_{i-1}$ and $y_i \in B_i$).

Theorem 7. *If $G \in Gen(*; P_3, C_3, C_5)$, then $s(G)$ can be computed in linear time.*

Proof. We prove the statement by analyzing Algorithm 1. Assume that the input graph G has n vertices and m edges. By using the algorithm proposed in [11], step at Line 1 can be performed in linear time. Since $\mathcal{CD}(G)$ has at most $n - 2$ components, and since each component has at most 5 vertices, the whole work performed at cycle 2–7 requires $O(n)$ time.

Consider now cycle at Lines 9–16

- at the end of the t steps of this cycle, step at Line 10 rebuilds the whole graph G, and this requires linear time;
- steps at Lines 11, 13, and 16 requires constant time each. This implies that the complexity of cycle at Line 9, limited to Lines 12–16, is bounded by $|B_1| \times |B_2| \times \cdots |B_t|$. Since each component has at most 5 vertices and $t \leq n - 2$, this complexity results to be $O(n)$.

Hence, the whole work performed at cycle 9–16 requires linear time, and this is the final bound for the algorithm's complexity. Concerning the correctness, it directly follows from the comments at Lines 11, 13, and 16.   □

4.3   Other Problems

By using results in [19], here we show that several basic combinatorial problems can be efficiently solved in the graph class $Gen(*; P3, C3, C5)$. Let $\mathcal{G}_k$, for $k \geq 3$, be the class of graphs for which every prime graph in a split decomposition has at most k vertices.

Algorithm 1. computing the stretch number of $G \in Gen(*; P_3, C_3, C_5)$

Input: a graph $G \in Gen(*; P_3, C_3, C_5)$
Output: $s(G)$

1 compute $\mathcal{CD}(G) = \{B_1, B_2, \ldots, B_t\}$ and $\mathcal{JV}(G) = \{x_2, y_2, x_3, y_3, \ldots, x_t, y_t\}$, $t \geq 1$;

2 **forall** $B \in \mathcal{CD}(G)$ **do**

3   compute $s(B)$;

4   **forall** $u \in B \cap M$ **do**

5   compute and store $d_B(u)$ and $D_B(u)$ (and hence $s_B(u)$) ;

6   **forall** $v \in B \cap M \setminus u$ **do**

7   compute and store $d_B(u, v)$ and $D_B(u, v)$;

8 initialize G_1 as B_1 ;

9 **for** $i = 2$ **to** t **do**

10   compute $G_i = (G_{i-1}, x_i) * (B_i, y_i)$;

11   compute $s(G_i)$;   // by using Theorem 6 and data computed at Line 3

12   **forall** $u \in G_i \cap M$ **do**

13   update $d_{G_i}(u, \overline{u})$ and $D_{G_i}(u, \overline{u})$ (and hence $s_{G_i}(u)$) ;   // by using
                                               // Corollary 1 and data computed at Line 3

14   **forall** $v \in G_i \cap M \setminus u$ **do**

15   **if** ($u \in G_{i-1}$ **and** $v \in B_i$) **or** ($v \in G_{i-1}$ **and** $u \in B_i$) **then**

16   update $d_{G_i}(u, v)$ and $D_{G_i}(u, v)$; // by using Property 1 and
                                            // data computed at Line 3

17 **return** $s(G_t)$

Theorem 8. ([19]) *For any fixed $k \geq 3$:*

- *there is an O(n) algorithm to compute the weighted stability number, the weighted cliques number, the domination number and its variants respectively, and a $O(n^3)$ algorithm to compute the chromatic number of graph in the class $\mathcal{G}_k$, if a split decomposition tree is given with the graph;*
- *the clique-width of graphs in the class $\mathcal{G}_k$ is bounded by $2k + 1$.*

The following results concerning the new class $Gen(*; P_3, C_3, C_5)$ can be easily derived:

Corollary 2. *It follows that:*

- *there is a O(n) algorithm to compute the weighted stability, the weighted cliques number, the domination number and its variants respectively, and a $O(n^3)$ algorithm to compute the chromatic number of graph in the class $Gen(*; P3, C3, C5)$, if a split decomposition tree is given with the graph;*
- *the clique-width of graphs in the class $Gen(*; P3, C3, C5)$ is bounded by 11.*

Theorem 9. *In the class $Gen(*; P3, C3, C5)$, the isomorphism problem can be solved in linear time.*

Sketch of the Proof. By Theorem 1, two graphs G_1 and G_2 belonging to $Gen(*; P3, C3, C5)$ can be transformed into trees T_1 and T_2, in which each node is a prime component in the Cunningham decomposition. It is a well known result that the isomorphism problem for trees can be solved in linear time. Hence, to test the isomorphism between G_1 and G_2 it is sufficient to test the isomorphism between the components. By Theorem 2, each component in T_1 and T_2 is a star, a clique, or a cycle C_5, and hence testing the isomorphism between components is a very easy task.   □

5   Relationships between $Gen(*; P_3, C_3, C_5)$ and DH(k)

We start this section by observing the relationships between the new graph class $Gen(*; P_3, C_3, C_5)$ and the class hierarchy DH(k).

Corollary 3. DH(1) $\subsetneq Gen(*; P_3, C_3, C_5) \subsetneq$ DH(2).

Proof. DH(1) $\subsetneq Gen(*; P_3, C_3, C_5)$ follows from Definition 2. Now, assume that $G \in Gen(*; P_3, C_3, C_5)$. From Lemma 2 and Theorem 6 it follows that $s(G) < 2$, and this implies $Gen(*; P_3, C_3, C_5) \subsetneq$ DH(2).   □

A graph class which can be characterized by the split decomposition enjoys nice algorithmic results (e.g, see [19]). Graph classes that enjoy this property are distance-hereditary [16], circle graphs [14], parity graphs [6], as well graphs in $Gen(*; P_3, C_3, C_5)$ (by definition). In the remainder of the section we investigate the class of all graphs having stretch number less than 2 with respect to the split decomposition. The following result can be considered as a partial answer to this problem.

Theorem 10. *Let G be a graph such that $s(G) < 2$. If $G[x, y]$ is an inducing-stretch cycle of G, then $G[x, y]$ belongs to $Gen(*; P_3, C_3, C_5)$.*

Proof. Omitted.

This result implies that *each* inducing-stretch cycle of G can be decomposed (by Cunningham decomposition) into cliques, stars, and cycles C_5. It would be interesting to complete this investigation by studying how the inducing-stretch cycles of G are joined each other.

References

1. Bandelt, H.J., Mulder, H.M.: Distance-hereditary graphs. J. Comb. Theory, Ser. B 41(2), 182–208 (1986)
2. Brandstädt, A., Dragan, F.F.: A linear-time algorithm for connected r-domination and steiner tree on distance-hereditary graphs. Networks 31(3), 177–182 (1998)
3. Chang, M.S., Hsieh, S.Y., Chen, G.H.: Dynamic programming on distance-hereditary graphs. In: Leong, H.-V., Jain, S., Imai, H. (eds.) ISAAC 1997. LNCS, vol. 1350, pp. 344–353. Springer, Heidelberg (1997)

4. Cicerone, S.: Using split composition to extend distance-hereditary graphs in a generative way. Tech. Rep. TR.12.2011, Department of Electrical and Information Engineering, University of L'Aquila, Italy (2011)
5. Cicerone, S., Di Stefano, G.: Graph classes between parity and distance-hereditary graphs. Discrete Applied Mathematics 95(1-3), 197–216 (1999)
6. Cicerone, S., Di Stefano, G.: On the extension of bipartite to parity graphs. Discrete Applied Mathematics 95(1-3), 181–195 (1999)
7. Cicerone, S., Di Stefano, G.: Graphs with bounded induced distance. Discrete Applied Mathematics 108(1-2), 3–21 (2001)
8. Cicerone, S., Di Stefano, G.: Networks with small stretch number. J. Discrete Algorithms 2(4), 383–405 (2004)
9. Cicerone, S., Di Stefano, G., Flammini, M.: Compact-port routing models and applications to distance-hereditary graphs. J. Parallel Distrib. Comput. 61(10), 1472–1488 (2001)
10. Cunningham, W.H.: Decomposition of directed graphs. SIAM. J. on Algebraic and Discrete Methods 3(2), 214–228 (1982)
11. Dahlhaus, E.: Parallel algorithms for hierarchical clustering and applications to split decomposition and parity graph recognition. J. Algorithms 36(2), 205–240 (2000)
12. Di Stefano, G.: A routing algorithm for networks based on distance-hereditary topologies. In: SIROCCO, pp. 141–151 (1996)
13. Esfahanian, A.H., Oellermann, O.R.: Distance-hereditary graphs and multidestination message-routing in multicomputers. Journal of Combinatorial Mathematics and Combinatorial Computing 13, 213–222 (1993)
14. Gabor, C.P., Supowit, K.J., Hsu, W.L.: Recognizing circle graphs in polynomial time. J. ACM 36(3), 435–473 (1989)
15. Gioan, E., Paul, C.: Dynamic distance hereditary graphs using split decomposition. In: Tokuyama, T. (ed.) ISAAC 2007. LNCS, vol. 4835, pp. 41–51. Springer, Heidelberg (2007)
16. Hammer, P.L., Maffray, F.: Completely separable graphs. Discrete Applied Mathematics 27(1-2), 85–99 (1990)
17. Howorka, E.: Distance-hereditary graphs. The Quarterly Journal of Mathematics 28(4), 417–420 (1977)
18. Rao, M.: Clique-width of graphs defined by one-vertex extensions. Discrete Mathematics 308(24), 6157–6165 (2008)
19. Rao, M.: Solving some NP-complete problems using split decomposition. Discrete Applied Mathematics 156(14), 2768–2780 (2008)

The Complexity and Approximability of Minimum Contamination Problems

Angsheng Li[1,*] and Linqing Tang[1,2,*,**]

[1] State Key Laboratory of Computer Science, Institute of Software, Chinese Academy of Sciences, P.O. Box 8718, Beijing, 100190, P.R. China
[2] Graduate University of Chinese Academy of Sciences, Beijing, China
{angsheng,linqing}@ios.ac.cn

Abstract. In this article, we investigate the complexity and approximability of the Minimum Contamination Problems, which are derived from epidemic spreading areas and have been extensively studied recently. We show that both the Minimum Average Contamination Problem and the Minimum Worst Contamination Problem are NP-hard problems even on restrict cases. For any $\epsilon > 0$, we give $(1 + \epsilon, O(\frac{1+\epsilon}{\epsilon} \log n))$-bicriteria approximation algorithm for the Minimum Average Contamination Problem. Moreover, we show that the Minimum Average Contamination Problem is NP-hard to be approximated within $5/3 - \epsilon$ and the Minimum Worst Contamination Problem is NP-hard to be approximated within $2 - \epsilon$, for any $\epsilon > 0$, giving the first hardness results of approximation of constant ratios to the problems.

1 Introduction

Over the last years, much attention had been paid to algorithmic problems related to information propagation or virus diffusion over large scale networks (such as Internet, social networks, etc). One interesting topic is related to the cascading related problems. The spread of misinformation or flu (resp. virus) in the social network (resp. Internet) may cause panic or epidemic in population (resp. computers), so it is important to look for efficient strategies to limit the effect of the propagation of undesirable instances in large scale networks. One strategy previously considered was trying to find some nodes to get immunization (i.e., remove these nodes from the network), for instance [1,11]. Aspnes et al. [1] considered the inoculation strategies for victims of viruses and defined the Sum-of-Squares Partition Problem, which is to find some fixed number of nodes to get immunization so that after deleting the immunization nodes, the sum of the squares of the size of connected components of the left graph is minimized, they also gave a $(O(1), O((\log n)^{1.5}))$-bicriteria approximation algorithm

---

* The authors are partially supported by Hundred Talent Program of Chinese academy of Sciences under Angsheng Li and the Grand Project "Network Algorithms and Digital Information" of the Institute of Software, Chinese Academy of Sciences.
** Corresponding author.

M. Ogihara and J. Tarui (Eds.): TAMC 2011, LNCS 6648, pp. 298–307, 2011.

for this problem. Kumar et al. [11] gave an $(O(1), O(\log n))$-bicriteria approximation algorithm for the Sum-of-Squares Partition Problem recently. However, in some cases, immunize and separate some node entirely from the network is impractical, one more approachable and general way is to block some links between the vertices (i.e., remove some edges from the network). For example, in order to avoid epidemic, not only the infected people should be isolated from the public, susceptible people also need to restrict their own behaviors (to decrease the opportunity to meet other people) to decrease the chance of being infected. The problem of minimizing the spread of undesirable things by blocking a limited number of links in a network, was first raised by [9]. The goal is to find a links set of fixed size such that the remaining network by blocking those links minimizes the contamination for the undesirable thing (which start from a random vertex of the network and infects all vertices connected with it). Two versions of the Minimum Contamination Problem corresponding to different objects were posed by [9], one is the *Minimum Average Contamination problem* which is to minimize the expectation number of contaminated nodes on the average case and the other is the *Minimum Worst Contamination problem* which is to minimize the expectation of maximum number of contaminated nodes in the worst case (a formal definition will be given later in Section 1). In [9,10], the authors considered these two problems on probabilistic models of networks (precisely the Independent Cascade Model and the Linear Threshold Model) and proposed greedy strategy for efficiently finding good approximation solutions to these problems, they experimentally demonstrated that the greedy methods have better performance than conventional link-removal methods on these two probabilistic models.

In this paper, we consider the Contamination Minimization Problems in deterministic setting. We prove that both the Minimum Average Contamination Problem and the Minimum Worst Contamination Problem are NP-hard even on some restricted cases in the deterministic setting. We show that the greedy strategy, even though works well in some probabilistic setting [9], allows an $\min\{O(K), n\}$ approximation for the two problems and there exists tight examples. We give an $\left(1 + \epsilon, O(\frac{1+\epsilon}{\epsilon} \log n)\right)$-bicriteria approximation algorithm for the Minimum Average Contamination Problem by using linear programming for any $\epsilon > 0$, while the $\left(1 + \epsilon, O(\frac{1+\epsilon}{\epsilon} \log n)\right)$-bicriteria approximation algorithm for the Minimum Worst Contamination Problem was presented by Anshelevich and Kempe [2] (they called the problem Min-Max Component Size Problem). We also show that the LP relaxation admits an $\Omega(n)$ integrality gap if we consider the single criteria approximation. On the hardness side, we show that the Minimum Average Contamination Problem is NP-hard to be approximated within $5/3 - \epsilon$ and that the Minimum Worst Contamination Problem is NP-hard to be approximated within $2 - \epsilon$, for any $\epsilon > 0$.

Definition 1 (The Minimum Contamination Problem). *[9] Given a graph $G = (V, E)$ and a positive number K, the Minimum Contamination Problem is to find a subset $E' \subseteq E$ of size K such that after deleting edges in E' from G, the expected size of contamination (with the virus starting at some single vertex* *which was chosen randomly from the graph) is minimized. Assume $H_1, ..., H_q$ are the connected components of $G \backslash E'$, there are two kinds of object values we considered: (1)$OBJ_A \triangleq \frac{1}{|V|} \sum_{i=1}^{q} |H_i|^2$ which corresponds to the* Minimum Average Contamination Problem *(MACP, for short); (2) $OBJ_W \triangleq \frac{1}{|V|} Max_{i \in [q]} |H_i|$ which corresponds to the* Minimum Worst Contamination Problem *(MWCP, for short).*

We remark that the Minimum Worst Contamination Problem is exactly the Min-Max Component Problem considered by Anshelevich and Kempe [2] and the Minimum Average Contamination Problem is a generalization of the Sum-of-Squares Partition Problem, since we can think vaccinating a node as blocking all edges adjacent to that node. However, the complexity and approximability results of the Sum-of-squares Partition Problem can not be adopt to the Minimum Worst Contamination Problem obviously. One difference between blocking an edge and immunizing a vertex is that blocking one edge can increase the number of connected components of the remaining graph at most one while it is not the case for immunizing one vertex.

Given a graph $G = (V, E)$ and positive numbers K and B, the *decision* version of the Minimum Contamination Problem $\mathcal{I} = (G = (V, E); K, B)$ is to determine whether the Minimum Contamination Problem instance has optimum value at most B. We also consider the weighted version of minimization contamination problems, that with input graph $G = (V, E)$ and a weight vector $\mathbf{w} = (w_v)_{v \in V}$, it only needs to replace $|H_i|$ by $w(H_i)$ in the objective values, where $w(H_i) = \sum_{v \in H_i} w_v$. (Here we think the random contamination occurs on vertices with probability in proportion to their weight).

We will show that all the above problems are NP-hard, so we will focus on the approximation algorithms for these problems.

Definition 2. *An (α, β)-bicriteria approximation algorithm for the* MACP *(resp.* MWCP*) is a poly-time algorithm $\mathcal{A}$ that given a MACP (resp. MWCP) instance $\mathcal{I} = (G = (V, E); K)$, $\mathcal{A}$ find an edges subset of size βK, such that after deleting these edges from G, the Average Contamination (resp. Worst Contamination) is at most $\alpha Opt_A(\mathcal{I})$ (resp. $\alpha Opt_W(\mathcal{I})$), where $Opt_A(\mathcal{I})$ (resp. $Opt_W(\mathcal{I})$) is the optimum value of the* MACP *(resp.* MWCP*) instance $\mathcal{I}$. We also call an $(\alpha, 1)$-bicriteria approximation algorithm an α approximation algorithm.*

2   The Complexity of the Minimum Contamination Problems

In this section, we prove the NP-hardness of the Minimum Contamination Problems by constructing reductions from the 3-Dimensional Matching Problem, which is a well-known NP-hard problem.

Definition 3. *Given 3 disjoint sets A, B and C with $|A| = |B| = |C| = p$, and triplets set $\mathcal{T} \subseteq A \times B \times C$. The* 3-Dimensional Matching Problem *is to ask whether $\mathcal{T}$ contains a perfect matching, which is a subset $S \subseteq \mathcal{T}$ of size p and each elements $g \in A \cup B \cup C$ appears exactly once in coordinates of S.*

The 3-Dimensional Matching Problem is NP-hard even on instances in which each element of $A \cup B \cup C$ appears exactly twice in $\mathcal{T}$ [5], we denote the restrict version of 3-Dimensional Matching problem as 3-DM_2.

We first show that the weighted version of the Minimum Contamination Problems are NP-hard, then we turn the weighted instances into unweighted instances.

Theorem 1. *Both the weighted Minimum Average Contamination Problem and the weighted Minimum Worst Contamination Problem are NP-hard even on bipartite, planar graphs with maximum degree 3.*

Proof (Proof of Theorem 1). We construct polynomial time reductions from the 3-DM_2 to Minimum Contamination problems. Similar reduction appeared in [3] to prove NP-hardness of the Minimum Forest Cover Problem. Given a 3-DM_2 instance $\mathcal{I} = \big((A, B, C); \mathcal{T}\big)$ with $|A| = p$ and $|\mathcal{T}| = 2p$, we construct a weighted graph $G = (V, E)$ as follows:

For each $g \in A \cup B \cup C$, define two vertices g' and g'' of weight 1 respectively, each corresponds to one appearance of g in $\mathcal{T}$ (w.l.o.g., we replace the first appearance of g by g' and the second appearance of g by g''), and a vertex Δ_g with weight 2 if $g \in A$, and with weight $N + 2$ (N is a parameter which will be fixed later) if $g \in B \cup C$. For each $T_l = (a_i, b_j, c_k) \in \mathcal{T}$, define a vertex d_l of weight N. Let $V \triangleq \{g', g'', \Delta_g\}_{g \in A \cup B \cup C} \bigcup \{d_j\}_{T_j \in \mathcal{T}}$.

We define the edges set E as follows: for each $g \in A \cup B \cup C$, add edges (g', Δ_g) and (g'', Δ_g), and for each $T_l \in \mathcal{T}$ and $g' \in T_l$ (resp. $g'' \in T_l$), add an edge (d_l, g') (resp. (d_l, g'')). It is easy to verify that graph $G = (V, E)$ is planar (by Kuratowski's theorem), bipartite and with maximum degree 3.

We define the decision version of MACP instance $\mathcal{I}' = (G = (V, E), K; B_1/|V|)$ and the decision version of MWCP instance $\mathcal{I}'' = (G = (V, E), K; B_2/|V|)$ with $K = 5p$, $B_1 = 4p(N + 3)^2$ and $B_2 = N + 3$.

We claim that the 3-DM_2 instance $\mathcal{I}$ contains a perfect matching if and only if both $\mathcal{I}'$ and $\mathcal{I}''$ are yes instances.

For the "only if" part: if $\mathcal{I}$ has a perfect matching $\mathcal{S} \subseteq \mathcal{T}$, then we define E' as

$$E' = \{(g, \Delta_g)\}_{g \in T_i \in \mathcal{S}} \cup \{(d_l, b_j), (d_l, c_k)\}_{T_l = (a_i, b_j, c_k) \notin \mathcal{S}}$$

since $\mathcal{S}$ is a perfect matching, we have $|\mathcal{S}| = p$ and (since $|\mathcal{T}| = 2p$)

$$|E'| = 3|\mathcal{S}| + 2|\mathcal{T} \backslash \mathcal{S}| = 5p$$

it is easy to verify that each connected component of $G \setminus E'$ is of weight exactly $N + 3$ and there are exactly $4p$ connected components. So we have

$$\sum_i w(H_i)^2 = B_1 = 4p(N + 3)^2$$

and

$$\max_i w(H_i) = B_2 = N + 3$$

For the "if" part: if $\mathcal{I}'$ (resp. $\mathcal{I}''$) is a yes instance, then there exists an edges subset E' of size $5p$, such that let $(H_i)_{i \in [q]}$ be the connected components of $G \backslash E'$, then

$$\sum_i w(H_i)^2 \leq 4p(N+3)^2 \quad (resp. \ \max_i w(H_i) \leq N+3)$$

We say a vertex is a *heavier vertex* if its weight is at least N, then there are $|B| + |C| + |T| = 4p$ different heavier vertices in G. In each connected component of $G \backslash E'$, there can exist at most one heavier vertex if we let N be large enough (say, $N > 24p$) such that

$$(4p+2)N^2 > 4p(N+3)^2 \quad (resp. \ 2N > N+3)$$

since otherwise if there exists some connected component H_j of $G \setminus E'$ has at least 2 heavier vertices, then

$$w(H_j)^2 \geq 4N^2$$

and it implies

$$\sum_i w(H_i)^2 \geq (4p+2)N^2 > 4p(N+3)^2 \quad (resp. \ \max_i w(H_i) \geq 2N > N+3)$$

which contradicts our conditions.

We say that a heavier vertices pair (u, v) is a *neighbor pair*, if there exists a path in G between u and v contains no heavier vertices other than u and v, we call such path *cutoff-needed path*. We know that for each cutoff-needed path, at least one edge should be chosen in E' in order to disconnect the corresponding neighbor pair. By the definition of graph G, there are exactly $5p$ different neighbor pairs (each $g \in B \cup C$ contribute 2 pairs (Δ_g, d_l) and (Δ_g, d_m) while each $g \in A$ contribute 1 pair (d_l, d_m), here assume the two appearances of g is in $T_l, T_m \in T$).

On the other hand, since $|E'| = 5p$ and cutoff-needed paths are edge disjoint, which means that exactly one edge should be chosen to block on each cutoff-needed path and the left graph should have exactly $4p$ connected components (each heavier vertex in one connected component and each connected component should have exactly one heavier node). Since

$$\sum_i w(H_i) = w(G) = 4p(N+3)$$

we know that

$$\sum_i w(H_i)^2 \leq 4p(N+3)^2 \quad (resp. \ \max_i w(H_i) \leq N+3)$$

is true if and only if $w(H_i) = N+3$ for each $i \in [4p]$.

We say that a connected component of the left graph $G \setminus E'$ is a *proper component* if it is a star graph with vertex set $\{d_l, a_i, b_j, c_k\}$ and center d_l. We claim that for any $g \in B \cup C$, exactly one of vertex g' and g'' is in some proper component.

Consider a component H_i contains g': if the heavier node in H_i is d_l for the T_l contains g', then H_i must be a proper component, since it is the only possible way for H_i to have weight exactly $N + 3$; otherwise the heavier node in H_i is Δ_g, in this case, g'' should be in a component H_j with heavier node d_m for the T_m contains g'', similar argument to that in the first case implies that H_j is a proper component.

So there are exactly p different proper components, we claim that these proper components correspond to a perfect 3-dimension matching of $\mathcal{I}$. Since each proper component corresponds to a $T_l \in \mathcal{T}$, by the above argument we only need to verify that for each $a \in A$, exactly one of a' and a'' is in some proper component of the left graph. For any $a \in A$, if a' is in some proper component H_i, then a'' and Δ_a should be in the same component H_j and since $w(a) + w(\Delta_a) = 3$, we know that $H_j \setminus \{a'', \Delta_a\} = \{d_l\}$ for the T_l contains a'' is not a proper component.

In all, if the MACP instance $\mathcal{I}'$ (resp. the MWCP instance $\mathcal{I}''$) is a yes instance, then there exists a perfect matching for the $3\text{-}DM_2$ instance $\mathcal{I}$.

By a simple gadget reduction, we are able to prove

Theorem 2. *Both the unweighted Minimum Average Contamination Problem and the unweighted Minimum Worst Contamination Problem are NP-hard even on bipartite graphs.*

We are also able to show

Theorem 3. *The Minimum Worst Contamination Problem is NP-hard even on power law graphs, while there exists a polynomial time algorithm for the weighted Minimum Worst Contamination Problem on trees.*

3   Approximation Algorithms for the Minimum Contamination Problems

Since the problems we concern are NP-hard, we consider the approximation algorithms of these problems. Kimura et al. [9,10] made use of the bond percolation methods to estimate the decrease of each edges' deletion and experimentally showed that the greedy strategy (choose the edges which decrease the objective most up now) works well in the Independent Cascade Models and the Linear Threshold Models. We show that the greedy strategy may works poorly.

Theorem 4. *The greedy strategy strategy is a $\min\{O(K), n\}$ approximation algorithm for both the Minimum Average Contamination Problem and the Minimum Worst Contamination Problem. There exists instances on which the greedy algorithm has performance $\min\{\Omega(K), n\}$ for both the Minimum Average Contamination Problem and the Minimum Worst Contamination Problem.*

3.1   Bicriteria Approximation Algorithms for the Minimum Contamination Problems

Given a Minimum Average Contamination Problem instance $\mathcal{I} = (G = (V, E), K)$ with $V = \{1, ..., n\}$, we define a 0-1 variable y_{ij} for each vertices pair (i, j) with $y_{ij} = 1$ means that i and j are in the same connected component of the remaining graph after the edge blocking, we also define a 0-1 variable x_e for each edge $e \in E$ with $x_e = 1$ means edge e is chosen to be blocked in the optimum solution. Denote $H_1, ..., H_q$ as the connected components of the remaining graph, then $\sum_{k=1}^{q} |H_k|^2 = \sum_{i,j} y_{ij}$. So we are able to present the natural linear relaxation for $\mathcal{I}$ as follows:

$$\min \quad \frac{1}{|V|} \sum_{i,j \in V} y_{ij}$$

$$st. \quad y_{ij} + \sum_{e \in P} x_e \geq 1 \quad \forall P \in \mathcal{P}_{ij}$$

$$- \sum_{e \in E} x_e \geq -K$$

$$x_e \geq 0, y_{ij} \geq 0 \quad \forall e \in E, \forall i, j \in V$$

Even though there may be exponentially many constraints, we can find a separation oracle (the shortest path) in polynomial time, so the above linear programming can be solved in polynomial time.

Theorem 5. *There exists an* $\left(1 + \epsilon, O(\frac{1+\epsilon}{\epsilon} \log n)\right)$*-bicriteria approximation algorithm for the Minimum Average Contamination Problem, for any* $\epsilon > 0$.

Proof. We use the linear relaxation above to get an optimum fractional solution $\left(x_e^*\right)_{e \in E}$ and $\left(y_{ij}^*\right)_{i,j \in V}$. The rounding technique we used here was introduced for the first time by [11] to get approximation algorithms for the Sum-of-Squares Partition Problem. For any fixed $\epsilon > 0$ we proceed as follows: for each $y_{ij}^* \geq \frac{1}{1+\epsilon}$ define $\bar{y}_{ij} \triangleq 1$, otherwise define $\bar{y}_{ij} \triangleq 0$, for each $e \in E$ define $\bar{x}_e = \min\{\frac{(1+\epsilon)}{\epsilon} x_e^*, 1\}$. Then we have

$$\sum_{i,j \in V} \bar{y}_{ij} \leq (1 + \epsilon) \sum_{i,j \in V} y_{ij}^* \quad and \quad \sum_{e \in E} \bar{x}_e \leq \frac{(1 + \epsilon)K}{\epsilon}$$

moreover, for each $i, j \in V$

$$\bar{y}_{ij} + \sum_{e \in P} \bar{x}_e \geq 1 \quad \forall P \in \mathcal{P}_{ij}$$

Now we construct a Minimum Multi-cut Problem instance $\mathcal{I}'$ on graph G as follows: for each i, j with $\bar{y}_{ij} = 0$, there is a terminal pair (i, j) needs to be separated. The goal of the problem is to find an edges subset of minimum size whose deletion separates all the terminal pairs. Consider the following Linear relaxation of $\mathcal{I}'$:

$$\min \quad \sum_{e \in E} x_e$$

$$st. \quad \sum_{e \in P} x_e \geq 1 \quad \forall P \in \mathcal{P}_{ij} \ with \ \overline{y}_{ij} = 0$$

$$x_e \geq 0 \quad \forall e \in E$$

Denote $OPT_{Lin}(\mathcal{I}')$ as the optimum value of the above relaxation for $\mathcal{I}'$. It is easy to check that $(\overline{x}_e)_{e \in E}$ is a feasible solution of the relaxation. Make use of the region grown techniques approximation algorithm [8] for the Minimum Multicut problem on $\mathcal{I}'$, we find an edges subset E' of size at most

$$O(\log n)OPT_{Lin}(\mathcal{I}') \leq O(\log n) \sum_{e \in E} \overline{x}_e \leq O(\frac{1+\epsilon}{\epsilon} \log n)K$$

which separates all (i, j) with $\overline{y}_{ij} = 0$.

It is easy to verify that if we take E' as the solution of the Minimum Average Contamination Problem instance $\mathcal{I}$, then it satisfies:

$$|E'| \leq O(\frac{1+\epsilon}{\epsilon} \log n)K$$

$$Sol = \sum_{i,j \in V} \overline{y}_{ij} \leq (1+\epsilon) \sum_{i,j \in V} \overline{y}_{ij}^* \leq (1+\epsilon)Opt(\mathcal{I})$$

$Opt(\mathcal{I})$ is the optimum value of MACP instance $\mathcal{I}$ which satisfies $Opt(\mathcal{I}) \geq \sum_{i,j \in V} \overline{y}_{ij}^*$.

Even we use the linear relaxation to get an $(1+\epsilon, O(\frac{1+\epsilon}{\epsilon} \log n))$-bicriteria approximation algorithm for the MACP with good performance, we are able to show that in the single criteria sense the linear relaxation we used has integrality gap $\Omega(n)$ even for MACP instances on star graphs.

4 Approximation Hardness of the Minimum Contamination Problems

In this section, we prove the hardness of approximation results for the Minimum Contamination Problems. To our best knowledge, they are the first nontrivial results about hardness of approximating these problems.

Definition 4 (The Multiway Cut Problem). *Given a graph $G = (V, E)$ and a set of terminals set $S = \{s_1, ..., s_k\} \subseteq V$, a Multiway Cut is a set of edges whose removal disconnects the terminals from each other. The problem is to find the Multiway Cut set of minimum size.*

The problem of finding minimum size multiway cut is APX-hard for any fixed $k \geq 3$ [6].

Theorem 6. *For any $\epsilon > 0$, it's NP-hard to approximate the Minimum Average Contamination Problem and the Minimum Worst Contamination Problem within $5/3 - \epsilon$ and $2 - \epsilon$ respectively.*

Proof (Proof of Theorem 6). We construct gap-preserving reductions from the Multiway Cut Problem with $k = 3$ to the Minimum Contamination Problems. Given a Multiway Cut instance $\mathcal{I} = (G = (V, E); \{s_1, s_2, s_3\}; K)$, the problem is to decide whether there exists a multiway cut of size at most K. We construct the Minimum Contamination Problem instances as follows: Let $N > (\max\{|E|, |V|\})^2$ be a parameter which will be fixed later, for each s_i, construct a clique $D_i = (V_i, E_i)$ with $|V_i| = N$. Define graph $G' = (V', E')$ as: $V' \triangleq \cup_{i=1}^3 V_i \bigcup V$ and $E' = \cup_{i=1}^3 E_i \bigcup E \bigcup \cup_{i=1}^3 W_i$ where $W_i \triangleq \{(v_{ij}, w) | v_{ij} \in V_i, w$ satisfies $(w, s_i) \in E\}$.

We define the MACP instance as $\mathcal{I}_A = (G' = (V', E'); K)$ and the MWCP Problem instance as $\mathcal{I}_W = (G' = (V', E'); K)$.

If there exists a multiway cut of size K, then we can delete the K corresponding edges from E', then every D_i are separated from other $D_j ((i \neq j))$ in the remaining graph $G' \setminus E'$, so the maximum size of the connected component of the remaining graph is at most $N + n$, so in the first case

$$Opt(\mathcal{I}_A) \leq \frac{3(N + n - 2)^2}{|V'|} \quad \text{and} \quad Opt(\mathcal{I}_W) \leq \frac{(N + n)}{|V'|}$$

On the other side, if the Minimum Multiway Cut of G is larger than K, we claim there must be two D_i's are connected in the remaining graph after any deletion of K edges from G'. Since for each i, $|D_i| = N$ and s_i connects to all nodes in D_i, any deletion of at most $K < |E| < N$ edges can not separate s_i from D_i and can not partite D_i for all $i \in [3]$. So we may assume the K deleted edges all from E, since any minimum multiway cut of G is larger than K, after deleting K edges from G, there must be some s_i and s_j are still connected, which means D_i and D_j are connected in G', in this case the size of the maximum connected component is at least $2N + 2$, i.e., in the second case

$$Opt(\mathcal{I}_A) \geq \frac{(2N + 2)^2 + (N + 1)^2 + n - 3}{|V'|} \quad \text{and} \quad Opt(\mathcal{I}_W) \geq \frac{2N + 2}{|V'|}$$

The NP-hardness of the Multiway Cut Problem with $k = 3$ implies that it's NP-hard to decide whether the Minimum Average Contamination Problem instance $\mathcal{I}_A = (G' = (V', E'); K)$ is in the first case or in the second case, which means that it NP-hard to approximate the Minimum Average Contamination Problem within

$$\frac{(2N + 2)^2 + (N + 1)2 + n - 3}{3(N + n/3)^2} \geq 5/3 - \epsilon$$

For any $\epsilon > 0$, we can choose large enough N such that the last inequality is true. Similarly we can prove the $2 - \epsilon$ hardness of approximation result for the Minimum Worst Contamination Problem.

5   Conclusion

In the paper, we study the complexity and approximability of the Minimum Contamination Problems in deterministic setting, which has various applications in different areas, such as Epidemic and Panic control in social networks, virus spreading control in networks, etc. We show that the Minimum Average Contamination Problem admits $(1 + \epsilon, O(\frac{1+\epsilon}{\epsilon} \log n))$-approximation algorithms, and is NP-hard to be approximated within $(5/3 - \epsilon, \beta)$ while the Minimum Worst Contamination Problem is NP-hard to be approximated within $(2 - \epsilon, \beta)$, here $\beta > 1$ is the hardness of approximating the Multiway Cut Problem with $k = 3$, similar techniques directly $5/3 - \epsilon$ hardness of approximation results for the Sum-of-Squares Partition Problem. There are still gaps which need to be closed for these problems. An improved single criteria approximation algorithm would also be interesting. The complexity of the Minimum Average Contamination Problem on trees and power law graphs is still not fixed yet.

References

1. Aspnes, J., Chang, K., Yampolskiy, A.: Inoculation Strategies for Victims of Viruses and the Sum-of-Squares Partition Problem. In: Proceedings of the Sixteenth Annual ACM-SIAM Symposium on Discrete Algorithms, pp. 43–52 (2005)
2. Anshelevich, E., D. Kempe. Prophylactic Vaccination of Networks (to appear)
3. Bazgan, C., Couëtoux, B., Tuza, Z.: Covering a Graph with a Constrained Forest (Extended Abstract). In: Dong, Y., Du, D.-Z., Ibarra, O. (eds.) ISAAC 2009. LNCS, vol. 5878, pp. 892–901. Springer, Heidelberg (2009)
4. Cheeger, J.: A Lower Bound for the Smallest Eigenvalues of the Laplacian. In: Problem in Analysis (Papers dedicated to Salomon Bochner,1969), pp. 195–199. Princeton Univ. Press, Princeton (1970)
5. Chlebik, M., Chlebikova, J.: Approximation hardness for small occurrence instances of NP-hard problems. In: Petreschi, R., Persiano, G., Silvestri, R. (eds.) CIAC 2003. LNCS, vol. 2653, pp. 152–164. Springer, Heidelberg (2003)
6. Dahlhaus, E., Johnson, D., Papadimitriou, C., Seymour, P., Yannakakis, M.: The complexity of multiterminal cuts. SIAM J. Comp. 23, 864–894 (1994)
7. Ferrante, A., Pandurangan, G., Park, K.: On the hardness of optimization in power-law graphs. Theoretical Computer Science 393(1-3), 220–230 (2008)
8. Garg, N., Vazirani, V.V., Yannakakis, M.: Approximate max-flow min-(multi)cut theorems and their applications. SIAM J. Comput. 25(2), 235–251 (1996)
9. Kimura, M., Saito, K., Motoda, H.: Minimizing the spread of contamination by blocking links in a network. In: Proceedings of the 23rd AAAI Conference on Artificial Intelligence, pp. 1175–1180 (2008)
10. Kimura, M., Saito, K., Motoda, H.: Blocking links to minimize contamination spread in a social network. ACM Transactions on Knowledge Discovery from Data (TKDD 3(2), 1–23 (2009)
11. Kumar, V., Rajaraman, R., Sun, Z., Sundaram, R.: Existence Theorems and Approximation Algorithms for Generalized Network Security Games. In: The 30th International Conference on Distributed Computing Systems (2010) (to appear)

On the Low-Dimensional Steiner Minimum Tree Problem in Hamming Metric

Ernst Althaus[1], Joschka Kupilas[2], and Rouven Naujoks[3]

[1] Johannes-Gutenberg-Universität Mainz,
ernst.althaus@uni-mainz.de
[2] Max-Planck-Institut für Informatik*
joschka.kupilas@uni-mainz.de
[3] Max-Planck-Institut für Informatik
naujoks@mpi-inf.mpg.de

Abstract. It is known that the d-dimensional Steiner Minimum Tree Problem in Hamming metric is NP-complete if d is considered to be a part of the input. On the other hand, it was an open question whether the problem is also NP-complete in fixed dimensions. In this paper we answer this question by showing that the problem is NP-complete for any dimension strictly greater than 2. We also show that the Steiner ratio is $2 - \frac{2}{d}$ for $d \geq 2$. Using this result, we tailor the analysis of the so-called k-LCA approximation algorithm and show improved approximation guarantees for the special cases $d = 3$ and $d = 4$.

1 Introduction and Related Work

In this paper we discuss the d-dimensional Steiner minimum tree problem in Hamming metric. This problem with applications in many fields of science like computational biology or computational linguistics reads as follows: given a set of d-dimensional strings T (called the *terminals*) over some finite alphabet Σ, one seeks – possibly using other strings in Σ^d (called the *Steiner nodes*) – a tree of minimum cost, spanning the terminal set. As implied by the name, the distance between two strings is given by their Hamming distance. The complexity of the problem naturally depends on which sizes are considered to be part of the input. If d and $|\Sigma|$ are sufficiently small, the problem is in P, in particular, if d and $|\Sigma|$ are considered to be constants. In [4] Foulds and Graham have shown that for $|\Sigma| = 2$ and for d being part of the input, the problem becomes NP-complete.

As discussed for example in [1] and [6], algorithms trying to find a solution by exhaustive search can be heavily accelerated if one is able to determine good lower bounds for a problem instance. In [5] the authors show, how such a computation can be reduced to the computation of low dimensional subproblems and thus, the question whether subproblems of low (i.e. fixed) dimension can be solved in polynomial time is of great practical importance. It is known and easy

---

* This work was done while the second author was at the Johannes-Gutenberg-Universität Mainz.

M. Ogihara and J. Tarui (Eds.): TAMC 2011, LNCS 6648, pp. 308–319, 2011.
© Springer-Verlag Berlin Heidelberg 2011

to see that the one-dimensional problem just corresponds to the minimum spanning tree of the input sequences, i.e. the number of distinct input strings minus one. Furthermore, in [1] the authors have shown that also in two dimensions, a minimum spanning tree is a Steiner minimum tree. Since then, it was an open question whether the problem is in P for any fixed dimension d. In the first part of this paper, we will close this discussion by showing that the problem becomes NP-complete in dimensions strictly greater than two. Later, we show that the Steiner ratio is $2 - \frac{2}{d}$ for $d \geq 2$, allowing us to tailor the analysis of the approximation algorithm of Robins & Zelikovsky (see [7]) to the Steiner minimum tree problem in Hamming metric.

2 Preliminaries

We assume that we are given a alphabet Σ and a positive integer d. We denote by s_i the i-th character of a string $s \in \Sigma^d$. For a set $S \subseteq \Sigma^d$ and an $i \in \{1, \ldots, d\}$ we call $\{s_i \mid s \in S\}$ the i-th column of S.

Definition 1 (Hamming distance). *Given two strings* $s, s' \in \Sigma^d$, *the value*

$$\|s, s'\| := |\{ i \in \{1, \ldots, d\} \mid s_i \neq s'_i \}|$$

is called the Hamming distance *of the strings* s *and* s'.

Definition 2 (difference index). *For* $u, v \in \Sigma^d$ *with* $\|u, v\| = 1$ *we call the index* $i \in \{1, \ldots, d\}$ *such that* $u_i \neq v_i$, *the* difference index *of* u *and* v.

Definition 3 (Steiner (minimum) tree in Hamming metric). *Given a set* $T \subseteq \Sigma^d$ *of terminals, a Steiner tree in Hamming metric* $\mathcal{T}$ *over* T *is a tree, spanning* $T \cup S$ *for a subset* $S \subseteq \Sigma^d$, *called the set of* Steiner nodes. *The cost of* $\mathcal{T}$ *(denoted by* $\mathrm{cost}(\mathcal{T})$*) is defined as the sum of its edge costs, where the cost of an edge* (u, v) *is defined as* $\|u, v\|$. *A Steiner minimum tree for* T *is a Steiner Tree with minimal cost. We write* $\mathrm{SMT}(T)$ *for such a tree and denote by* SMTH-d *the problem of computing it.*

Similar to a Steiner minimum tree, we write $\mathrm{MST}(T)$ for a minimum spanning tree over T.

Definition 4 (full component). *A* full component *of a Steiner tree* $\mathcal{T}$ *is a maximal subtree of* $\mathcal{T}$ *in which all leafs are terminals and no inner node is a terminal. We call a full component* trivial *if it consists only of two terminals and an edge connecting them.*

Definition 5 (Steiner ratio). *For a* $T \subseteq \Sigma^d$ *we define* $\mathcal{R}(T) := \frac{\mathrm{cost}(\mathrm{MST}(T))}{\mathrm{cost}(\mathrm{SMT}(T))}$ *The* Steiner ratio *is then defined as* $\sup_{T \subseteq \Sigma^d} (\mathcal{R}(T))$.

3   The Complexity in Low Dimensions

Theorem 1. *SMTH-3 is NP-hard.*

Proof. Clearly, the decision version of SMTH-3 is in NP as the cost of a tree can be verified in polynomial time. Thus, what is left to show is that SMTH-3 is NP-hard. We show NP-hardness by reducing 3SAT to SMTH-3, i.e. for a given boolean 3SAT formula F we show how to construct a terminal set T such that cost(SMT(T)) falls below some threshold if and only if F is satisfiable. Let $V = \{V_1, V_2, \dots, V_n\}$ be the variables in F and let $C = \{C_1, C_2, \dots, C_m\}$ be the clauses in F with $C_i = (L_{i_1} \vee L_{i_2} \vee L_{i_3})$ where $L_{i_j} \in \{V_1, \neg V_1, V_2, \neg V_2, \dots, V_n, \neg V_n\}$ are literals. We are looking for a truth assignment $\pi : V \to \{true, false\}$ satisfying F. Using the alphabet $\Sigma = \{V_1, \dots, V_n, V_1^T, \dots, V_n^T, V_1^F, \dots, V_n^F, C_1, \dots, C_m, 0\}$, we construct T as follows:

- For each variable V_i, we create the terminals $V_i V_i^T 0$ and $V_i V_i^F 0$. We call such a pair a *variable block*.
- For each clause $C_i = (L_{i_1} \vee L_{i_2} \vee L_{i_3})$, we create the terminals $C_i X_1 C_i$, $C_i X_2 C_i$, $C_i X_3 C_i$, where $X_k = V_{i_k}^T$ if $L_{i_k} = V_{i_k}$, and $X_k = V_{i_k}^F$ if $L_{i_k} = \neg V_{i_k}$. We call these three terminals a *clause block*. Furthermore, we create for each clause C_i the terminal $0C_i C_i$.
- We create the terminal 000.

Note that T contains exactly $2n + 4m + 1$ terminals (see Fig. 1). We claim that F is satisfiable if and only if cost(SMT(T)) $\le 5m + 3n$.

We first prove that cost(SMT(T)) $\le 5m + 3n$ if F is satisfiable. Given a satisfying truth assignment $\pi : V \to \{true, false\}$ for F, we construct a Steiner tree with cost $5m + 3n$ from π using only edges of length 1. For each variable V_i, we create the Steiner node $0V_i^T 0$ if $\pi(V_i) = true$ or $0V_i^F 0$ if $\pi(V_i) = false$. Since π is a satisfying truth assignment, every clause C_i must contain at least one literal L_k that evaluates to true under π. For one such L_k per clause we create the Steiner node $0V_j^T C_i$ if $L_k = V_j$ or $0V_j^F C_i$ if $L_k = \neg V_j$. Thus, in total we have $n + m$ Steiner nodes that we connect to the terminals as follows: For each variable V_i, we connect $V_i V_i^T 0$ and $V_i V_i^F 0$ via an edge. By construction, there is a Steiner node $0V_i^X 0$ with $X \in \{T, F\}$ which we connect to the corresponding terminal $V_i V_i^X 0$. For each clause, we connect the three terminals $C_i V_j^X C_i$ to each other via two edges and connect the Steiner node $0V_j^X C_i$ (for some $X \in \{T, F\}$) to the corresponding terminal in the corresponding clause block and via another edge to the terminal $0C_i C_i$. Furthermore, we connect the Steiner nodes $0V_j^X C_i$ to the corresponding Steiner nodes $0V_j^X 0$ and the Steiner nodes $0V_j^X 0$ to the terminal 000. Since each added edge has a length of 1, this yields a tree of cost $5m + 3n$.

Let us now prove the other direction, i.e. F is satisfiable if cost(SMT(T)) $\le 5m + 3n$. So, let cost(SMT(T)) $\le 5m + 3n$. We assume without loss of generality that SMT(T) consists only of edges of length 1, which is a safe assumption since by splitting longer edges and by adding further Steiner nodes each SMT can be transformed into another SMT with this property. We will show that there exists a SMT whose Steiner nodes are of certain types, which will then lead to a truth assignment satisfying F.

First, we will show that there exists a Steiner Minimum Tree whose (non-trivial) full components all contain at least one terminal with the first coordinate being 0. In order to see this, notice that there is a spanning tree S over the terminals consisting of edges of length 2 such that at least one endpoint of each edge has 0 as the first coordinate (connect 000 to each $V_i V_i^X 0$ and to each $0 C_j C_j$, and each $0 C_j C_j$ to the three clauses of the corresponding clause block). Now assume that there is a SMT with a non-trivial full component (of t terminals) containing no terminal that has 0 as its first coordinate. Since all these terminals are then pairwise different in the first coordinate (except pairs in the clause blocks or in the variable blocks, which are wlog. directly connected and therefore form trivial full components), the cost of this full component in the first coordinate is at least $t - 1$. Likewise, all terminals are pairwise different in the last two coordinates, so the cost for the last two coordinates is $t - 1$ too, yielding an overall cost of at least $2t - 2$ for a full component of t terminals that does not contain any terminals that have 0 as their first coordinate. Hence we can remove that full component and reconnect the terminals using edges of the spanning tree S.

We now assume that every non-trivial full component of the SMT contains at least one terminal with 0 as its first coordinate. Let F be such a full component. Since all letters except 0 appear at most once as the first coordinate of terminals of F, we can change the first coordinate of all Steiner nodes in F to 0 without increasing its overall cost. This may however create edges of length 2 between Steiner nodes and terminals (and double Steiner nodes, which we delete). Insert a new Steiner node with first coordinate 0 into those edges. Now we have a Steiner Minimum Tree with all Steiner nodes of the form $0XX$ and edges of length 1. Since we supposed $cost(SMT(T)) \leq 5m + 3n$, there are $\leq m + n$ Steiner nodes.

Let S_i be a Steiner node connected to $V_i V_i^X 0$, where $X \in \{T, F\}$. There must be such a Steiner node, since no terminal has distance 1 to $V_i V_i^X 0$. Since its first coordinate is 0, it has the form $0V_i^X 0$. Likewise, let T_j be a Steiner node connected to the C_j-clause block, it must have the form $0V_k^Z C_j$. The Steiner nodes S_i and T_j are pairwise different and together $m + n$ nodes, so there are no further Steiner nodes, and there is exactly one Steiner node for each clause and variable block. Now examine the path from some terminal $C_j Y C_j$ to $V_1 V_1^T 0$ in the Steinertree. This path will possibly first visit other terminals in the clause block, let $C_j V_k^X C_j$ be the last of them. Then it will visit the Steiner node T_j, possibly other Steiner nodes T_l and eventually for the first time a Steiner node S_i. All the T_x on the path have the same letter in the second coordinate, since they differ in the third coordinate and the path consists of edges of length 1. For the same reason, also S_i and $C_j V_k^X C_j$ have the same second coordinate, so $i = k$. Since this holds for every clause block, the Steiner nodes S_i give us a fullfilling assignment for the boolean formula: Let $\pi(V_i) = True$ if $S_i = 0V_i^T 0$ and $\pi(V_i) = False$ if $S_i = 0V_i^F 0$. Then the clause j is fulfilled by this assignment, since the terminal $C_j V_k^X C_j$, with $X = T$ if $S_i = 0V_i^T 0$ and $X = F$ if $S_i = 0V_i^F 0$,

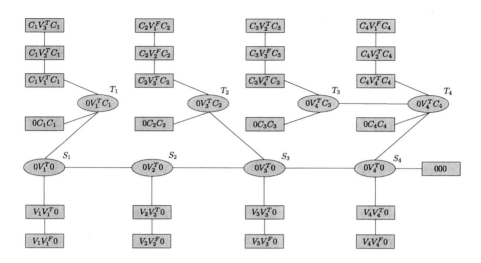

Fig. 1. The terminal set and a SMT for the 3SAT formula $(V_1 \vee V_2 \vee V_3) \wedge (\neg V_1 \vee \neg V_2 \vee V_3) \wedge (V_2 \vee \neg V_3 \vee V_4) \wedge (\neg V_1 \vee V_2 \vee V_4)$ with the truth assignment $\pi(V_1) = true$, $\pi(V_2) = true$, $\pi(V_3) = true$, $\pi(V_4) = true$

is part of the clauseblock j, meaning that V_k or $\neg V_k$ resp., was a literal in the clause j, and is therefore fulfilled.

4   The Steiner Ratio

In this section, we will examine the Steiner ratio $\mathcal{R}$, i.e. the approximation guarantee yielded by a minimum spanning tree. This ratio plays an important role in the analysis of state of the art approximation algorithms for the Steiner tree problem. First, we will present a sequence of lemmas allowing us to restrict the discussion to certain instances for which we will analyse in the following theorems to establish the Steiner ratio of $2 - \frac{2}{d}$ for $d \geq 2$. As we know that for $d \leq 2$ a minimum spanning tree is also a Steiner minimum tree (see [1]), we restrict the following the discussion to the case that $d \geq 3$. Furthermore, we assume that all edges have length 1, so $\text{cost}(\text{SMT}(T)) = |T| + |S| - 1$, where $|S|$ are the Steiner nodes. We say that a Steiner minimum tree $\mathcal{T}$ over some terminal set T satisfies

- property Π_1 if $\mathcal{T}$ consists of one full component
- property Π_2 if T has no elements with distance 1
- property Π_3 if for each Steiner node s in $\mathcal{T}$ there is a terminal $t \in T$ such that $\|s, t\| = 1$
- property Π_4 if there exists a spanning tree over T with edge weights 2

Note that the properties Π_2 and Π_4 intentionally only depend on the terminals that are spanned by $\mathcal{T}$. The main idea in this section is to show in the Lemmas

1 to 4 that for each terminal set T there is a terminal set T' with $\mathcal{R}(T') \geq \mathcal{R}(T)$ such that there is a Steiner minimum tree over T' satisfying the properties Π_1, Π_2, Π_3 and Π_4. Exploiting the existence of such trees, we will show in Theorem 2 an upper bound on the Steiner ratio for such terminal sets T' and show its tightness in Theorem 3.

Lemma 1. *For any terminal set T, there is a terminal set T' with $\mathcal{R}(T') \geq \mathcal{R}(T)$ such that an $\mathrm{SMT}(T')$ satisfies Π_1.*

Proof. Let $C_i, 1 \leq i \leq k$ be the full components of an $\mathrm{SMT}(T)$ and let T_i be the terminals in the components C_i. Since $\mathrm{cost}(\mathrm{SMT}(T)) = \sum_{i=1}^{k} \mathrm{cost}(\mathrm{SMT}(T_i))$ and $\mathrm{cost}(\mathrm{MST}(T)) \leq \sum_{i=1}^{k} \mathrm{cost}(\mathrm{MST}(T_i))$,

$$\mathcal{R}(T) \leq \frac{\sum_{i=1}^{k} \mathrm{cost}(\mathrm{MST}(T_i))}{\sum_{i=1}^{k} \mathrm{cost}(\mathrm{SMT}(T_i))}$$

which means that there is at least one i with $\mathcal{R}(T) \leq \mathcal{R}(T_i)$. Thus, the set of terminals $T' := T_i$ establishes the claim, since there is an $\mathrm{SMT}(T_i)$ consisting of one full component.

Lemma 2. *For any terminal set T for which there is an $\mathrm{SMT}(T)$ satisfying Π_1, there is a terminal set T' with $\mathcal{R}(T') \geq \mathcal{R}(T)$ such that an $\mathrm{SMT}(T')$ satisfies Π_1 and Π_2.*

Proof. If T also satisfies Π_2 we are done. Thus, let us assume otherwise and let t_1, t_2 be two terminals in T with distance 1. Then there is an $\mathrm{SMT}(T)$ with an edge $\{t_1, t_2\}$, yielding a full component C_1 with the terminal set $T_1 := \{t_1, t_2\}$. Clearly, $\mathrm{cost}(\mathrm{MST}(T_1)) = \mathrm{cost}(\mathrm{SMT}(T_1)) = 1$ and thus, there must be a full component C_i other than C_1 spanning a terminal set T_i with $\mathcal{R}(T) \leq \mathcal{R}(T_i)$. We can now assume wlog. that C_i satisfies property Π_2 since otherwise we could simply repeat this decomposition into smaller full components. Setting $T' := T_i$ yields the claim.

Lemma 3. *For any terminal set T for which there is an $\mathrm{SMT}(T)$ satisfying Π_1 and Π_2, there is a terminal set T' with $\mathcal{R}(T') \geq \mathcal{R}(T)$ such that an $\mathrm{SMT}(T')$ satisfies Π_1, Π_2 and Π_3.*

Proof. If such an $\mathrm{SMT}(T)$ also satisfies Π_3 we are done. Thus, we assume that there is a Steiner node s, such that $\forall\, t \in T \colon \|s, t\| \geq 2$. Furthermore, we will assume that s is the only Steiner node with this property. This can be done, since otherwise the following argument could be applied to each such Steiner node separately.

We add a new terminal t_0 with $\|s, t_0\| = 1$ and $\|t, t_0\| \geq 2$ for all $t \in T$. Note that such a terminal must exist, since we can obtain t_0 by changing a character of s an element that is not contained in Σ. Since $\forall\, t \in T \colon \|s, t\| \geq 2$, we also have that $\forall\, t \in T \colon \|t_0, t\| \geq 2$. Then, $\mathrm{MST}(T \cup \{t_0\})$ has a cost at least larger by 2 than an MST(T) and SMT($T \cup \{t_0\}$) is larger by exactly 1 than SMT(T). Thus,

$$\mathcal{R}(T \cup \{t_0\}) = \frac{\text{cost}(\text{MST}(T)) + 2 + \eta}{\text{cost}(\text{SMT}(T)) + 1} > \mathcal{R}(T).$$

for some $\eta \in \mathbb{N}_0$. Here the strict inequality holds, because $\mathcal{R}(T) < 2 - \frac{2}{|T|} < 2$. Setting $T' := T \cup \{t_0\}$ establishes the claim, which can be seen as follows: extending the SMT(T) by adding the edge (s, t_0) with cost 1 yields a Steiner tree $\mathcal{T}$ spanning $T \cup \{t_0\}$. Note that $\mathcal{T}$ must also be a Steiner minimum tree, since t_0 contains a character that did not appear in any of the terminals in T. Thus, $\text{cost}(\text{SMT}(T \cup \{t_0\})) = \text{cost}(\text{SMT}(T)) + 1$. Furthermore, $\mathcal{T}$ consists of one full component, the terminals spanned by $\mathcal{T}$ have pairwise distances of at least 2 and for s, there is a terminal with distance 1, namely the added terminal t_0.

Lemma 4. *For any terminal set T for which there is an SMT(T) satisfying Π_1, Π_2 and Π_3, there is a terminal set T' with $\mathcal{R}(T') \geq \mathcal{R}(T)$ such that an SMT(T') satisfies Π_1, Π_2, Π_3 and Π_4.*

Proof. Assume there is no such spanning tree. Then T can be decomposed into subsets $T_1, T_2, \ldots, T_q$, so that the terminals in T_i are connect-able with a spanning tree of edge length 2, but there is no connection of length 2 between T_i and T_j with $i \neq j$.

We claim that there are $i, j \in \{1, \ldots, q\}$ with $i \neq j$ such that, there are at most two Steiner nodes s_i, s_j on some path between a terminal $t_i \in T_i$ and a terminal $t_j \in T_j$. For this to be seen, let $(t_j, s_1, s_2, \ldots, s_l, t_i)$ be a path from t_j to t_i with Steiner nodes s_i. We stated that there must be a terminal $t_k \in T_k$ with $\|s_2, t_k\| = 1$. Now we can distinguish two cases. If T_k is unequal to T_j, the claim holds for the sets T_j and T_k with the path (t_j, s_1, s_2, t_k). If T_k is equal to T_j, then the path $(t_k, s_2, \ldots, s_l, t_i)$ is shorter than the path $(t_j, s_1, s_2, \ldots, s_l, t_i)$. By repeating this argument inductively, either the first case holds at some node or we obtain a path with exactly two Steiner nodes, connecting two distinct sets T_i and T_j, showing the claim.

Thus, consider the path (t_i, s_i, s_j, t_j) connecting T_i and T_j and denote by a the difference index of (s_i, s_j). Now we claim that there exists neither for s_i nor for s_j a terminal that differs only in the a-th element from s_i or s_j respectively: Assume without loss of generality that there is a t'_j that differs only in the a-th element from s_j. Then $\|t_j, t'_j\| \leq 2$ and thus, $t'_j \in T_j$. But since the difference index of both (t'_j, s_j) and (s_j, s_i) is a, $\|t'_j, s_i\| = 1$ and therefore $\|t'_j, t_i\| = 2$, which conflicts with the assumption that the distance between T_i and T_j is ≥ 3. So, there is no terminal that differs from either s_i or s_j in the a-th element.

Now we can add a terminal t_0 that only differs from s_i in the a-th element (we choose a character that is not used by any terminal in T) and obtain $T' = T \cup \{t_0\}$. t_0 has distances of at least 2 to all existing terminals, thus this increases the MST by 2, but the SMT only by 1. Thus,

$$\mathcal{R}(T') = \frac{\text{cost}(\text{MST}(T)) + 2}{\text{cost}(\text{SMT}(T)) + 1} > \mathcal{R}(T)$$

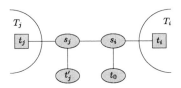

Fig. 2. Proof illustration for Lemma 4

Analogously to the arguments used in the proof of Lemma 3, there is now a Steiner minimum tree $\mathcal{T}$ spanning $T \cup \{t_0\}$ satisfying the properties Π_1, Π_2 and Π_3. Repeatedly applying this argument by adding a finite number (at most $q-1$) of terminals, establishes also property Π_4 and thereby the claim.

Theorem 2. *For any set T of d-dimensional strings, $\mathcal{R}(T) \leq 2 - \frac{2}{d}$.*

Proof. Let T be an arbitrary terminal set. Then the above lemmas show that there is a terminal set T' with $\mathcal{R}(T') \geq \mathcal{R}(T)$ such that there is an SMT(T') satisfying the properties Π_1, Π_2, Π_3 and Π_4. Let $\mathcal{T}' = (T' \cup S, E)$ be an SMT over T' with Steiner nodes S satisfying Π_1, Π_2, Π_3 and Π_4. Then by Lemma 4, cost(MST(T')) $= 2|T'| - 2$. Let us assume that the claim does not hold, i.e. $\mathcal{R}(T') > 2 - \frac{2}{d}$ Then

$$\mathcal{R}(T') = \frac{2|T'| - 2}{|T'| + |S| - 1} > \frac{2d - 2}{d} \Leftrightarrow |S| < \frac{1}{d-1}|T'| - \frac{1}{d-1} \tag{1}$$

Since all edges in $\mathcal{T}'$ have length 1, an edge connects two nodes that differ in exactly 1 character. So, the edges of the SMT can be partitioned according to their difference index $i \in \{1, 2, \ldots, d\}$. $\mathcal{T}'$ consists of $|T'| + |S| - 1$ edges, so there is at least one $i \in \{1, 2, \ldots, d\}$, such that the $\mathcal{T}'$ has $d_i \geq \frac{1}{d}(|T'| + |S| - 1)$ edges of difference index i. Consider the forest consisting of $S \cup T'$ and the d_i edges of type i. In general, for a forest we have $k = m - n$, where k is the number of isolated components, m is the number of vertices and n is the number of edges. In our case, this yields:

$$k = |T'| + |S| - d_i \leq |T'| + |S| - \frac{1}{d}(|T'| + |S| - 1)$$

$$= \frac{d-1}{d}|T'| + \frac{d-1}{d}|S| + \frac{1}{d} \overset{(I)}{<} \frac{d-1}{d}|T'| + \frac{1}{d}|T'| - \frac{1}{d} + \frac{1}{d} < |T'|,$$

where the strict inequality I follows from inequality 1. But if the number of isolated components is strictly less than $|T'|$, then at least one isolated component contains at least 2 terminals. These terminals are connected only by edges of the same difference index, and so, they have distance 1, which is a contradiction to our assumption.

In [3], the author has shown that there is a terminal set T of d-dimensional strings of cardinality d such that the Steiner ratio for this problem instance is

$2 - \frac{2}{d}$. In the following theorem, we will show, that there are even arbitrary large problem instances, for which the Steiner ratio is achieved.

Theorem 3. *The bound $\frac{\text{cost(MST)}}{\text{cost(SMT)}} \leq 2 - \frac{2}{d}$ is tight: There exists an arbitrarily large input instance $T \subseteq \Sigma^d$ with $\mathcal{R}(T) \geq 2 - \frac{2}{d}$.*

Proof. We start by constructing a rooted tree $\mathcal{T}$ with the following properties: a) the root has degree d, b) all other inner nodes have degree $d + 1$ and c) all leaves have the same depth m. Then we assign strings from Σ^d to the tree nodes as follows. We assign to the root node an arbitrary string in Σ^d. For all nodes v_i of depth $n \in \{0, \ldots, m - 1\}$ starting from $n = 0$, we assign strings to the adjacent nodes $v_i^1, v_i^2, \ldots, v_i^d$ of v_i of depth $n + 1$ in the following way: if $(A_1, A_2, \ldots, A_d)$ is the string corresponding to v_i, we assign the string $(A_1, A_2, \ldots, A_{j-1}, X, A_{j+1}, \ldots, A_d)$ to v_i^j for a so far not used element $X \in \Sigma$. We identify the set of strings assigned to the leaves of this tree with the input instance T. Since $\mathcal{T}$ is a tree spanning over T,

$$\text{cost(SMT}(T)) \leq \text{cost}(\mathcal{T}) = \sum_{i=1}^{m} d^i = \frac{d^{m+1} - d}{d - 1}$$

Note that the pairwise distances of elements in T are at least 2 which can easily be seen by induction over m and thus, $\text{cost(MST}(T)) \geq 2(|T| - 1) = 2(d^m - 1)$. But then $\mathcal{R}(T) = \frac{\text{cost(MST}(T))}{\text{cost(SMT}(T))} \geq 2(d^m - 1)\frac{d-1}{(d^{m+1}-d)} = 2 - \frac{2}{d}$. Since we can choose m arbitraily large, the finishes the proof.

From Theorem 2 and Theorem 3 follows immediately:

Theorem 4. *The Steiner ratio of Steiner minimum trees in Hamming metric in dimension $d \geq 2$ is $2 - \frac{2}{d}$*

4.1 Applications

In [2], the authors give an polynomial time algorithm for the Steiner tree problem with an expected approximation ratio of $\ln(4) \approx 1.39$. However, in this section we will use the Steiner ratio to derive better approximation guarantees for the algorithm by Robins and Zelikovsky [7] in the case of Hamming Metric. Let us first recall some important definitions.

Definition 6 (k-restricted Steiner Trees). *A k-restricted Steiner tree $\mathcal{T}$ is a Steiner tree in which every full component contains at most k terminals. A shortest k-restricted Steiner tree for a terminal set T is denoted by $\text{OPT}_k(T)$.*

Definition 7 (Loss). *A minimum cost connection of the Steiner nodes of a full component K of $\mathcal{T}$ to its terminals is denoted by $LOSS(K)$ and $LOSS(\mathcal{T})$ is defined as the sum over the $LOSS(K)$ of the full components K of $\mathcal{T}$. Finally, $\text{LOSS}_k(T)$ is then defined as the loss of $\text{OPT}_k(T)$.*

In the following we will omit the input set T for the sake of readability, i.e. we write for example LOSS_k for $\text{LOSS}_k(T)$.

In [7], the authors propose the so-called k-LCA approximation algorithm for the Steiner minimum tree problem in graphs which constructs a tree with a cost of at most

$$\text{LOSS}_k \cdot \ln \left(1 + \frac{\text{cost(MST)} - \text{cost(OPT}_k)}{\text{LOSS}_k} \right) + \text{cost(OPT}_k)$$

As a consequence one can show:

Corollary 1. *The k-LCA algorithm has an approximation ratio of*

$$1 + \alpha \cdot \ln \left(\frac{\alpha + \mathcal{R} - 1}{\alpha} \right)$$

for the Steiner ratio $\mathcal{R}$ and for an α such that $\alpha \geq \frac{\text{LOSS}(T)}{\text{cost}(T)}$ for all Steiner trees T.

Proof. The proof follows the proof of Theorem 2 in [7]. As mentioned before, the authors show that the tree constructed by the k-LCA algorithm has a cost of at most

$$\text{LOSS}_k \cdot \ln \left(1 + \frac{\text{cost(MST)} - \text{cost(OPT}_k)}{\text{LOSS}_k} \right) + \text{cost(OPT}_k) \qquad (2)$$

Since $\alpha \geq \frac{\text{LOSS}(T)}{\text{cost}(T)}$ for any Steiner tree T, in particular $\text{LOSS}_k \leq \alpha \cdot \text{cost(OPT}_k)$. The partial derivative $\text{LOSS}_k \cdot \ln \left(1 + \frac{\text{cost(MST)} - \text{cost(OPT}_k)}{\text{LOSS}_k} \right)'_{\text{LOSS}_k}$ is always positive and thus, the bound attains its maximum for $\text{LOSS}_k = \alpha \cdot \text{cost(OPT}_k)$. Together with the fact that $\text{cost(MST)} \leq \mathcal{R} \cdot \text{cost(SMT)} \leq \mathcal{R} \cdot \text{cost(OPT}_k)$, this yields a cost of

$$\text{cost(OPT}_k) \left(1 + \alpha \cdot \ln \left(1 + \frac{(\mathcal{R} - 1) \, \text{cost(OPT}_k)}{\alpha \cdot \text{cost(OPT}_k)} \right) \right)$$

which shows the claim for $k \to \infty$, as $\text{cost(OPT}_k)$ goes to $\text{cost(SMT)})$ for $k \to \infty$.

Together with the Steiner ratio that we have derived in the previous section for the SMTH-d problem, we have:

Theorem 5. *For SMTH-d, there is a polynomial time $(1 + \frac{1}{2} \ln(3 - \frac{4}{d}))$-approximation algorithm.*

Proof. In [8] the authors show that for any Steiner tree, its loss is not larger than $\frac{1}{2}$ times its cost and thus, $\alpha = \frac{1}{2}$. Furthermore, we have shown in Theorem 4 that the Steiner ratio is $\mathcal{R} = 2 - \frac{2}{d}$. Then by Corollary 1, the k-LCA algorithm gives the desired approximation guarantee.

For the case $d = 3$, the algorithm achieves an approximation ratio of ≈ 1.255, and ≈ 1.347 for $d = 4$.

We will now prove that in the three dimensional case, the analysis of the k-LCA algorithm can be improved to a ratio of ≈ 1.231. For this, we use a Lemma from [1], which allows us to restrict the following discussion to input sequences I, such that there is always a minimum spanning tree over I with edge lengths strictly less than 3:

Lemma 5. *For $T \subseteq \Sigma^3$, let $C_1, \ldots C_k$ be the connected components of the graph $G = (T, E)$ with $E = \{ (u, v) \in T \times T \mid \|u, v\| \le 2 \}$. Furthermore, let c_i be an arbitrary representative of the component C_i. Then $\bigcup_{i=1}^{k} \mathrm{SMT}(C_k) \cup \mathrm{MST}(\{c_1, c_2, \ldots, c_k\})$ is a Steiner minimum tree over T.*

Theorem 6. *For SMTH-3, there is a polynomial time $(1 + \frac{1}{3}\ln(2))$-approximation algorithm.*

Proof. If we only use edges of length 1, we know that $\mathrm{LOSS}_k = |S_k| = \mathrm{cost}(\mathrm{OPT}_k) - |T| + 1$, where S_k are the Steiner nodes of OPT_k. Since we can assume without loss of generality that $\mathrm{cost}(\mathrm{MST}) \le 2|T| - 2$, we obtain for the cost A of the approximate solution:

$$A \le \mathrm{LOSS}_k \cdot \ln\left(1 + \frac{\mathrm{cost}(\mathrm{MST}) - \mathrm{cost}(\mathrm{OPT}_k)}{\mathrm{LOSS}_k}\right) + \mathrm{cost}(\mathrm{OPT}_k)$$

$$\le |S_k| \cdot \ln\left(1 + \frac{2|T| - 2 - |T| - |S_k| + 1}{|S_k|}\right) + \mathrm{cost}(\mathrm{OPT}_k)$$

$$< |S_k| \cdot \ln\left(\frac{|T|}{|S_k|}\right) + \mathrm{cost}(\mathrm{OPT}_k)$$

$$= \mathrm{cost}(\mathrm{OPT}_k) \cdot \left(\frac{|S_k|}{\mathrm{cost}(\mathrm{OPT}_k)} \cdot \ln\left(\frac{|T|}{|S_k|}\right) + 1\right)$$

$$= \mathrm{cost}(\mathrm{OPT}_k) \cdot \left(\frac{|S_k|}{|S_k| + |T| - 1} \cdot \ln\left(\frac{|T|}{|S_k|}\right) + 1\right)$$

$$\xrightarrow{|T| \to \infty} \mathrm{cost}(\mathrm{OPT}_k) \cdot \left(\frac{|S_k|}{|S_k| + |T|} \cdot \ln\left(\frac{|T|}{|S_k|}\right) + 1\right)$$

The single maximum of $\frac{|S_k|}{|S_k| + |T|} \cdot \ln\left(\frac{|T|}{|S_k|}\right) + 1$ for fixed $|T|$ is at 1.279 for $\frac{|S_k|}{|T|} = 0.279$. But we know that $|S| \ge \frac{|T|}{2} - \frac{1}{2}$, which is also true for full components, so $|S_k| \ge \frac{|T|}{2} - \frac{1}{2}$ and thus, $\frac{|S_k|}{|T|} \ge \frac{1}{2}$ for $|T| \to \infty$. So the maximal value of $\frac{|S_k|}{|S_k| + |T|} \cdot \ln(\frac{|T|}{|S_k|}) + 1$ is at $\frac{|S_k|}{|T|} = \frac{1}{2}$, which yields the claim.

We conclude this section with a negativity result. As Corollar 1 shows, the approximation guarantee of the k-LCA algorithm depends on two parameters, namely the Steiner ratio and the upper bound α on the ratio $\frac{\mathrm{LOSS}(\mathcal{T})}{\mathrm{cost}(\mathcal{T})}$ for Steiner trees $\mathcal{T}$ over T. In Theorem 5 we make use of the Steiner ratio derived in this

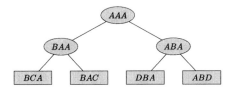

Fig. 3. A terminal set T with $\frac{\text{LOSS}(T)}{\text{cost}(T)} = \frac{3}{6} = \frac{1}{2}$

section and use for α the value $\frac{1}{2}$ since it was shown in [8] that the loss of a Steiner tree in graphs is never larger than $\frac{1}{2}$ times its cost. A natural question is now whether this bound is tight for Steiner minimum trees in Hamming metric. Unfortunately the answer is yes as one can construct a problem instance T for which the ratio is exactly $\frac{1}{2}$. For an example, see Figure 3 (which can easily be extended to larger instances): recall that $\text{LOSS}(T)$ is defined as the cost of a tree connecting the Steiner nodes to to terminals. Such a tree must consist of at least $|S|$ many edges where S is the set of Steiner nodes. Because each edge in T has cost 1, $\text{LOSS}(T) \geq |S| = |T| - 1$. As $\text{cost}(T) = 2|T| - 2$, $\frac{\text{LOSS}(T)}{\text{cost}(T)} \geq \frac{1}{2}$.

5   Conclusion

In this paper we have shown that the d-dimensional Steiner minimum tree in Hamming metric problem becomes NP-complete in dimensions strictly larger than 2. Additionally, we have shown a Steiner ratio of $2 - \frac{2}{d}$ and how this result can be used to tailor the analysis of existing approximation algorithms.

References

1. Althaus, E., Naujoks, R.: Computing Steiner minimum trees in Hamming metric. In: SODA 2006: Proceedings of the Seventeenth Annual ACM-SIAM Symposium on Discrete Algorithm, pp. 172–181. ACM, New York (2006)
2. Byrka, J., Grandoni, F., Rothvoß, T., Sanità, L.: An improved lp-based approximation for steiner tree. In: STOC, pp. 583–592 (2010)
3. Cieslik, D.: The Steiner ratio of several discrete metric spaces. Discrete Mathematics 260(1-3), 189–196 (2003)
4. Foulds, L.R., Graham, R.L.: The Steiner problem in phylogeny is NP-complete. Advances in Applied Mathematics 3, 43–49 (1982)
5. Holland, B.R., Huber, K.T., Penny, D., Moulton, V.: The minmax squeeze: Guaranteeing a minimal tree for population data. Molecular Biology and Evolution 22(2) (2005)
6. Naujoks, R.: NP-hard Networking Problems Exact and Approximate Algorithms. PhD thesis, Saarland University (2008)
7. Robins, G., Zelikovsky, A.: Improved Steiner tree approximation in graphs. In: Proceedings of the Eleventh Annual ACM-SIAM Symposium on Discrete Algorithms, pp. 770–779 (2000)
8. Robins, G., Zelikovsky, A.: Tighter bounds for graph Steiner tree approximation. SIAM Journal on Discrete Mathematics 19, 122–134 (2005)

Lower Bounds for Testing Computability by Small Width OBDDs

Joshua Brody⋆, Kevin Matulef⋆, and Chenggang Wu⋆

IIIS, Tsinghua University
{joshua.e.brody,matulef,wuchenggang0316}@gmail.com

Abstract. We consider the problem of testing whether a function $f : \{0,1\}^n \to \{0,1\}$ is computable by a read-once, width-2 *ordered binary decision diagram* (OBDD), also known as a *branching program*. This problem has two variants: one where the variables must occur in a fixed, known order, and one where the variables are allowed to occur in an arbitrary order. We show that for both variants, any nonadaptive testing algorithm must make $\Omega(n)$ queries, and thus any adaptive testing algorithm must make $\Omega(\log n)$ queries.

We also consider the more general problem of testing computability by width-w OBDDs where the variables occur in a fixed order. We show that for any constant $w \geq 4$, $\Omega(n)$ queries are required, resolving a conjecture of Goldreich [15].

We prove all of our lower bounds using a new technique of Blais, Brody, and Matulef [6], giving simple reductions from known hard problems in communication complexity to the testing problems at hand. Our result for width-2 OBDDs provides the first example of the power of this technique for proving strong nonadaptive bounds.

1 Introduction

In this work we consider the problem of testing whether a function $f : \{0,1\}^n \to \{0,1\}$ is computable by a very limited type of computational device, a read-once *Ordered Binary Decision Diagram* (OBDD), also known as a *Branching Program*. We formalize this question in the language of *property testing*. The goal of a property tester is to distinguish objects which have a property from those that are "far" from having the property, with limited access to the object. Here, the object is a boolean function $f : \{0,1\}^n \to \{0,1\}$, and we would like a randomized algorithm that accepts f with high probability if it is computable by a read-once OBDD, and rejects f with high probability if it disagrees with any OBDD on an ϵ fraction of inputs. Property testing algorithms are *adaptive* if queries are chosen based on answers to previous queries; otherwise, testers are *non-adaptive*. The complexity of the algorithm is the number of times it queries f, which should hopefully be a small function of n and ϵ.

⋆ Supported in part by the National Basic Research Program of China Grant 2007CB807900, 2007CB807901, and the National Natural Science Foundation of China Grant 61033001, 61061130540, 61073174.

M. Ogihara and J. Tarui (Eds.): TAMC 2011, LNCS 6648, pp. 320–331, 2011.

Property testing, and in particular testing of boolean functions, has a long history. Over the last two decades researchers have studied algorithms for testing many different properties of functions, such as the property of being linear [7], being monotone [16,13], being a dictator, a monomial [22] or a k-junta [12,4,5], or being expressible in various different "concise" forms such as an s-sparse polynomial, a size-s decision tree, etc. [9] (see, e.g., the survey of [24]).

The class of OBDDs has been studied in many areas of theoretical computer science, particularly in computational learning theory (e.g. [11,23,3,14,8]), where there is a well-known connection to testing. In particular, Goldreich et. al. [17] observed that any proper learning algorithm for a class of functions $\mathcal{C}$ can be used to test whether f is in $\mathcal{C}$ versus far from $\mathcal{C}$. Thus, the complexity of a proper learning algorithm serves as an upper bound on the number of queries required to test. However, this bound is often weak, since for many interesting classes, the query complexity of testing is much smaller than the complexity of learning (for instance, the class of linear functions is testable with $O(1/\epsilon)$ queries [7], independent of n, even though any learning algorithm must see at least $\Omega(n)$ values of f). It is natural to ask whether this is also the case for OBDDs.

1.1   Results for Width-2 and Width-3 OBDDs

The problem of testing whether a function f is computable by an OBDD was first studied by Ron and Tsur in [25]. They point out that although previous testing results have looked at the problem of testing whether f has a simple *form*, it seems reasonable instead to fix a simple *model of computation*, and test whether f is computable within the model. They focused on the model of *width-2 read-once OBDDs*, because this class has a simple structure, yet still generalizes some previously well-studied classes, such as linear functions and monomials.

Ron and Tsur identified two variants of this problem. In the first, the tester wishes to determine whether f is computable by an OBDD where the variables must appear in a fixed, known order (say, $x_1 \ldots x_n$). In the second, studied in [26], the tester wishes to determine whether f is computable by an OBDD where the variables can appear in any arbitrary order. Note that the complexity of the former problem is not a priori related to the complexity of the latter, since a function can be far from an OBDD with variables in order $x_1 \ldots x_n$, but still be equal to an OBDD when the variables are rearranged[1].

For the first variant of the width-2 OBDD testing problem, where the variables must occur in a fixed order, Ron and Tsur gave an adaptive upper bound of $\tilde{O}(\log n) \cdot \text{poly}(1/\epsilon)$ queries [25] (here the $\tilde{O}$ notation hides factors of $\log \log n$). This is an exponential improvement over the $\Omega(n)$ queries required for learning the same class[2]. Their upper bound raises the obvious question of whether the

---

[1] In general, if $\mathcal{C}'$ is a subset of $\mathcal{C}$, the complexity of testing membership in $\mathcal{C}'$ may be quite different than the complexity of testing membership in $\mathcal{C}$. However, they are often the same when the classes are conceptually related.

[2] The learning bound is easy to show, since the class of width-2 read-once OBDDs contains, for instance, all linear functions- a set of 2^n functions which are all far from each other.

$\tilde{O}(\log n)$ dependence on n is necessary, and indeed, whether any dependence on n is necessary at all. Our first result provides an answer to this question.

Theorem 1. *Any nonadaptive testing algorithm requires $\Omega(n)$ queries (and thus any adaptive testing algorithm requires $\Omega(\log n)$ queries[3]) to test*

   i. width-2 OBDDs with variables in a fixed order.
  ii. width-3 OBDDs with variables in a fixed order.

For width-2 OBDDs, Theorem 1 is essentially tight in terms of n. As mentioned, for adaptive algorithms Ron and Tsur gave an upper bound of $\tilde{O}(\log n)$, and for nonadaptive algorithms, a simple Occam learning algorithm yields an upper bound of $O(n)$[4].

A few words are in order regarding the history of Theorem 1. In [25], Ron and Tsur gave a lower bound of $\Omega(\log n)$ for testing fixed-order width-2 OBDDs with one-sided error. However, this proof was recently found to contain a flaw, which we discuss in the full version of this paper. In subsequent work [28], the same authors gave a different proof which improved the bound to $\Omega(n)$ for nonadaptive testers, even with two-sided error, thus achieving the same bound as in Theorem 1. The "NO" instances (functions that are far from width-2 OBDDs) we use in our proof are partially inspired by [28]. However, our proof technique uses a different, more modular approach, as we shall discuss below.

For the problem of testing width-2 OBDDs when the variable order is unknown, the lower bound of Ron and Tsur does not apply. Tackling this case was posed as an open question in [28]. Our technique is able to handle this case as well, as shown in our second main result.

Theorem 2. *Any two-sided error, nonadaptive algorithm for testing computability by width-2 OBDDs with variables in arbitrary order requires $\Omega(n)$ queries (and thus, any adaptive algorithm requires $\Omega(\log n)$ queries).*

Remark: An earlier version of this paper claimed that the lower bound in Theorem 2 was essentially optimal, since it matched an upper bound given by Ron and Tsur in [26]. However, after reading a draft of this paper, Ron and Tsur discovered a flaw in their upper bound in [26], and showed that in the arbitrary-order case, Theorem 2 can be strengthened to $\Omega(n)$, even for adaptive algorithms [27]. Their improvement is nearly optimal, since in this case one can test using a simple Occam learning algorithm which makes $O(n \log n)$ queries.

---

[3] It is well known, and simple to show, that any adaptive testing algorithm which makes q queries can be transformed into a nonadaptive algorithm which makes 2^q queries. Hence a nonadaptive bound of $\Omega(g(n))$ implies an adaptive bound of $\Omega(\log g(n))$. As our result shows, this exponential gap is sometimes necessary.

[4] An Occam learning algorithm for a class $\mathcal{C}$ is simply a nonadaptive algorithm that draws a set of random examples, and searches for a member of $\mathcal{C}$ consistent with f evaluated on those examples. It is well known that such an algorithm will produce an ϵ-accurate hypothesis with constant probability after seeing $O(\log |\mathcal{C}|/\epsilon)$ examples.

1.2   Results for Width-w OBDDs, for Constant $w \geq 4$

We also consider the problem of testing computability by width-w read-once OBDDs, for constant $w \geq 4$, where the variables must appear in a fixed order. This problem was previously studied by Goldreich in the $w = 4$ case [15]. He showed that unlike the width-2 case, where testing can be done with exponentially fewer queries than learning, testing width-4 OBDDs requires $\Omega(\sqrt{n})$ queries. He conjectured that the true bound is $\Omega(n)$. In this case, the complexity of testing would be essentially the same as the complexity of learning, since for any constant w, a simple Occam learning algorithm implies an $O(n)$ upper bound. Our final result confirms Goldreich's conjecture.

Theorem 3. *For any constant $w \geq 4$, any adaptive algorithm for testing computability by width-w OBDDs with variables in fixed order requires $\Omega(n)$ queries.*

1.3   Techniques

All of our lower bounds are proven using a new technique developed by the first two authors, along with Blais [6]. The result in [6] shows to reduce communication problems to testing problems, and thus leverage known lower bounds in communication complexity to prove lower bounds in testing.

Traditionally, property testing lower bounds are proven using Yao's Minimax Lemma. One starts by designing two families of objects- one family of YES instances (objects which have the property), and another of NO instances (objects which are far). In the next step, one defines a distribution over each family, and then one must show that no deterministic algorithm can distinguish with high probability whether a function is drawn from the YES or NO family. This step often involves nontrivial technical analysis. The philosophy put forth in [6] is that one can often eliminate the work required in this step, by creating YES and NO instances which correspond to the YES and NO instances of a communication problem. If strong lower bounds for the communication problem are already known, no new technical analysis is required.

Roughly speaking, the technique given in [6] translates *one-way* communication lower bounds to *nonadaptive* testing bounds, and *two-way* communication lower bounds to *adaptive* testing bounds. This was observed in [6], though no examples were given there of nonadaptive bounds that did not also apply to the adaptive case. For testing width-2 OBDDs in the fixed-order case, the nonadaptive complexity is much higher than the adaptive complexity. Thus, to prove Theorem 1, we must reduce from a communication problem that is hard when one-way communication is allowed, but easier when two-way communication is allowed. Our solution is to use an asymmetric communication problem for which this is known to be the case. Namely, we use the AUGMENTED-INDEX problem, a variant of the well-known INDEX problem. These problems have long been useful for proving streaming lower bounds [19,21,10], and now find a use in property testing as well. (We note that our proof of Theorem 2 has a similar flavor, and features a reduction from the INDEX problem. However, as we remarked earlier,

Ron and Tsur have subsequently improved this to an adaptive lower bound of $\Omega(n)$, via a reduction from the symmetric SET-DISJOINTNESS problem.)

2   Preliminaries

For two boolean functions $f, g : \{0,1\}^n \rightarrow \{0,1\}$, the *distance* between f and g, denoted $d(f, g)$ is equal to $\Pr_x[f(x) \neq g(x)]$.

We will work in the standard property testing model. Let $\mathcal{P}$ denote a property (a subset) of boolean functions. Then a *tester* for $\mathcal{P}$ is a randomized algorithm which when given query access to a function $f : \{0,1\}^n \rightarrow \{0,1\}$ and a distance parameter ϵ, outputs "accept" with probability at least $2/3$ if $f \in \mathcal{P}$, and "reject" with probability at least $2/3$ if $d(f, g) > \epsilon$ for all $g \in \mathcal{P}$. A *one-sided* tester is a tester which outputs "accept" with probability 1 if $f \in \mathcal{P}$. A *nonadaptive* tester is a tester which chooses its entire query set at the start of the algorithm.

We use $Q(\mathcal{P})$ to denote the minimum query complexity of a (possibly adaptive) tester for $\mathcal{P}$, and $Q^{\mathrm{na}}(\mathcal{P})$ to denote the minimum query complexity of a nonadaptive tester for $\mathcal{P}$.

A two-party *communication problem* is defined by a function $C : A \times B \rightarrow \{0,1\}$. Alice has an input $a \in A$ and Bob has an input $b \in B$, and they would like to compute the value of $C(a, b)$. We work in the *public randomness* model, where Alice and Bob generate messages based on random bits they both see.

We use $R(C)$ to denote the minimum number of bits they must communicate for them both to compute $C(a, b)$ with probability at least $2/3$ on any input pair (a, b). We use $R^{\rightarrow}(C)$ to denote the *one-way* complexity- that is, the number of bits Alice must communicate for Bob to compute $C(a, b)$ with high probability (without sending any message to Alice).

For more details on communication complexity, consult the standard work of Kushilevitz and Nisan [20].

2.1   Reducing Communication Problems to Testing Problems

In this section, we give a sketch of the lower bound approach in [6]. We refer the reader to that paper for a more formal treatment.

Lemma 1. *Let C be a communication problem and $\mathcal{P}$ a property. Given functions $f : \{0,1\}^n \rightarrow \{0,1\}$ and $g : \{0,1\}^{n+1} \rightarrow \{0,1\}$, define the function $h = h(f, g) : \{0,1\}^n \rightarrow \{0,1\}$ as*

$$h(x) := g(f(x), x)$$

Suppose there is a way for Alice and Bob to create functions f_a and g_b based on their inputs such that (i) $h(f_a, g_b) \in \mathcal{P}$ if $C(a, b) = 1$, and (ii) $h(f_a, g_b)$ is $\Omega(1)$-far from $\mathcal{P}$ if $C(a, b) = 0$. Then,

1. *$R^{\rightarrow}(C) \leq Q^{\mathrm{na}}(\mathcal{P})$, and*
2. *$R(C) \leq 2Q(\mathcal{P})$.*

Remark: In most reductions, h is defined as $h(x) := f_a(x) \oplus g_b(x)$ for some g_b independent of $f(x)$; however, this need not always be the case. In Section 4, we crucially rely on the asymmetry of the general construction.

Proof. (sketch) For the latter case, suppose there is an adaptive testing algorithm A for $\mathcal{P}$. Alice and Bob can cooperatively run this algorithm on h, using shared randomness to decide which values of h to query. Every time A queries $h(x)$ on some x, Alice sends $f(x)$ to Bob, who then computes $h(x)$ and returns it to Alice. In this way, both players maintain the list of query results $\{h(x)\}$ and can therefore generate successive values to query. If A is indeed a testing algorithm for $\mathcal{P}$, then at the end of the protocol they will know the value of $C(a,b)$ with high probability; the number of bits exchanged is twice the number of queries made by A.

The former case is similar. The only difference is that since queries are generated nonadaptively, Alice can send Bob $\{f(x)\}$ in a single message. Thus, the communication cost equals the query complexity of the testing algorithm.

2.2   Communication Complexity Problems

To prove our lower bounds, we give reductions from some standard communication problems. First, the SET-DISJOINTNESS problem:

Set-Disjointness. Alice and Bob are given n-bit inputs a and b and must compute

$$\text{DISJ}(a,b) := \bigvee_{i=1}^{n} a_i \wedge b_i$$

It is well-known that the two-way communication complexity, $R(\text{DISJ})$, is $\Omega(n)$, even with the promise that $a_i \wedge b_i = 1$ for at most one i [18].

For our width-2 OBDD bounds, where the nonadaptive complexity is larger than the adaptive complexity, it is necessary for us to reduce from communication problems where the one-way complexity (where Alice is allowed to send to Bob, but not vice-versa) is larger than the two-way complexity. We use the following two problems:

Index. Alice has an n-bit input a, and Bob has a $\log n$-bit index $i \in [n]$ (throughout, we will use the shorthand $[n]$ to denote the set $\{1, \ldots, n\}$). Bob's goal is to compute

$$\text{INDEX}(a,i) = a_i$$

It is well-known that the one-way complexity, $R^{\rightarrow}(\text{INDEX})$, is $\Omega(n)$ [1].

Augmented-Index. AUGMENTED-INDEX is nearly identical to INDEX, but Bob is also allowed to see bits $a_1 \ldots a_{i-1}$ without incurring any communication cost. Even with this additional "free" information, the one-way communication complexity, $R^{\rightarrow}(\text{AUGMENTED-INDEX})$, remains $\Omega(n)$ [2].

2.3 Branching Program Basics

A *binary decision diagram* (BDD), or *branching program*, is a layered directed acyclic graph with a distinguished source vertex and two sink vertices, labeled 0 and 1. Each internal vertex is labeled with an input variable and has outgoing edges to vertices in the next layer of the BDD. These edges are labeled with possible outcomes for the variable. In an *ordered binary decision diagram* (OBDD), all vertices in a layer are labeled with the same input variable. The *width* of an OBDD is the maximum number of vertices in any layer. An OBDD is *read-once* if vertices in different layers are labeled by different variables. In this work, we are only concerned with read-once OBDDs, and so we will often drop the "read-once" modifier.

We will use some basic definitions and claims developed by Ron and Tsur. The following can be found in [25]. The proofs of the facts are left as exercises.

Definition 1. *A function* $f : \{0,1\}^n \to \{0,1\}$ *is a* linear *function of* $x_1 \ldots x_n$ *if it can be written in the form* $f(x) = b_0 + \sum_{i=1}^{n} b_i x_i$ *where* $b_0 \ldots b_n \in \{0,1\}^5$.

Fact 4. *A function* $f : \{0,1\}^n \to \{0,1\}$ *is computable by a width-2 OBDD with variables in* fixed *order* $x_1 \ldots x_n$ *if and only if it can be written as*

$$f(x) = f_n(x_n, f_{n-1}(x_{n-1}, \ldots f_2(x_2, f_1(x_1))))$$

where f_1 *is a boolean function on one bit and* $f_2, \ldots, f_n$ *are boolean functions on two bits.* f *is computable by a width-2 OBDD with variables in* arbitrary *order if and only if it can be written in the same form after some permutation* $\pi \in S_n$ *is applied to the variables.*

We will slightly abuse notation, and use f_i both to refer to a function on 2 variables (the i'th variable and f_{i-1}), as well as a function on the first i variables.

Definition 2. *Consider a function* $f : \{0,1\}^n \to \{0,1\}$ *expressed in the form given in Fact 4. We say that a level i is* relevant *if the function f_i depends on the value of x_i. Relevant levels can be either* linear *or* blocking. *A relevant level is a* linear *level if f_i is a linear function of x_i and f_{i-1}. Otherwise, it is* blocking.

Fact 5. *If i is a blocking level, then f_i is either the* AND *or* OR *of $a \in \{x_i, \overline{x_i}\}$ and $b \in \{f_{i-1}, \overline{f_{i-1}}\}$. Thus, if i is blocking level, there exists a setting $t \in \{0,1\}$ for x_i such that f_i is constant, regardless of the value of f_{i-1}.*

Definition 3. *Let* $f : \{0,1\}^n \to \{0,1\}$. *We define the* influence *of variable i in f as* $\mathrm{Inf}_f(i) := \Pr_x[f(x) \neq f(x^{\oplus i})]$, *where $x^{\oplus i}$ denotes x with the i'th bit flipped.*

Lemma 2. *Let M be a width-2 OBDD with variables in order $x_1 \ldots x_n$. Let $j < i$ and suppose level i is a blocking level. Then* $\mathrm{Inf}_{f_i}(j) \leq \frac{1}{2}\mathrm{Inf}_{f_{i-1}}(j)$.

Fact 6. *Let $f, g : \{0,1\}^n \to \{0,1\}$ and suppose there is a variable x_i such that* $|\mathrm{Inf}_f(i) - \mathrm{Inf}_g(i)| = \tau$. *Then* $d(f,g) \geq \tau/2$.

---

[5] In this definition and throughout the paper, addition is taken to be over $GF(2)$.

3   A Lower Bound for Testing Fixed-Order Width-2 and Width-3 OBDDs

In this section we prove Theorem 1. The proof will be via a reduction from AUGMENTED-INDEX. The main idea is that Alice will use her input to form a linear function, and Bob will use his index (plus extra knowledge) to form a linear function plus an AND of two consecutive variables. They then take the XOR of their two functions. If $a_i = 1$, then in the resulting function the AND will appear before the linear part, so it will be computable by a width-2 OBDD. However if $a_i = 0$, then in the resulting function the AND will appear after a variable in the linear part, so it will be far from any width-2 or width-3 OBDD (where the variables must appear in a fixed order). To show that our "no" instances are far, we make use of the following lemma.

Lemma 3. *Let $h : \{0,1\}^n \to \{0,1\}$ be a function of the form $h(x) = x_i + (x_{i+1} \wedge x_{i+2}) + \sum_{k \in S} x_k$ for some $i \in [n-2]$ and $S \subseteq \{i+3, \ldots, n\}$. Then,*

    i. h is $1/4$-far from any width-2 OBDD, with variables in fixed order $x_1 \cdots x_n$.
    ii. h is $1/8$-far from any width-3 OBDD, with variables in fixed order $x_1 \cdots x_n$.

In the case of width-2 OBDDs, Lemma 3 essentially appears as Claim 6 in [25]. We leave the proof of Lemma 3 to the full version of this paper. We are now ready to prove Theorem 1.

Theorem (Restated). *Any nonadaptive testing algorithm requires $\Omega(n)$ queries (and thus any adaptive algorithm requires $\Omega(\log n)$ queries) to test*

    i. width-2 OBDDs with variables in a fixed order.
    ii. width-3 OBDDs with variables in a fixed order.

Proof. Let $n' = \lfloor (n-3)/4 \rfloor$. We will show a reduction from AUGMENTED-INDEX on n' variables. Since the one-way communication complexity of AUGMENTED-INDEX is $\Omega(n')$, and n' is linear in n, this will imply an $\Omega(n)$ testing bound.

First, Alice uses her input $a \in \{0,1\}^{n'}$ to form the function f, and Bob uses his input $i \in [n']$ (plus knowledge of $a_1 \ldots a_{i-1}$) to form the function g as follows

$$f(x) := \sum_{k=1}^{n'} x_{4k+3a_k} \qquad \text{and} \qquad g(x) := \left(\sum_{k=1}^{i-1} x_{4k+3a_k} \right) + (x_{4i+1} \wedge x_{4i+2})$$

Bob can then solve AUGMENTED-INDEX by running a testing algorithm on the joint function $h(x) = f(x) + g(x)$ and having Alice send him the value of $f(x)$ whenever he needs to query h. It is easy to see that if $a_i = 1$, then $h(x) = (x_{4i+1} \wedge x_{4i+2}) + x_{4i+3} + \sum_{k=i+1}^{n'} x_{4k+3a_k}$, so h is a width-2 OBDD. On the other hand, if $a_i = 0$ then $h(x) = x_{4i} + (x_{4i+1} \wedge x_{4i+2}) + (\sum_{k=i+1}^{n'} x_{4k+3a_k})$, so by Lemma 3, h is far from any width-2 or width-3 OBDD.

4   A Lower Bound for Testing Arbitrary-Order Width-2 OBDDs

In this section we prove Theorem 2. The proof will be via a reduction from INDEX. The main idea is that Alice will use her inputs to form a width-2 OBDD on pairs of variables corresponding to the indices where $a_k = 1$. Then Bob will use his index i to append two more variables to the end of the OBDD. If $a_i = 0$, then no variable will be used more than once, so the resulting function will remain a width-2 OBDD. If $a_i = 1$, then two variables will be used twice, which will cause the resulting function to be far from any (read-once) width-2 OBDD.

Theorem (Restated). *Any two-sided error, nonadaptive algorithm for testing computability by width-2 OBDDs with variables in arbitrary order requires $\Omega(n)$ queries (and thus, any adaptive algorithm requires $\Omega(\log n)$ queries).*

Proof. Let $n' = \lfloor (n-1)/2 \rfloor$. We will show a reduction from INDEX on n' variables. Since the one-way communication complexity of INDEX is $\Omega(n')$, and n' is linear in n, this will imply an $\Omega(n)$ testing bound.

First, Alice uses her input $a \in \{0,1\}^{n'}$ to form the function

$$f(x) := x_1 + \sum_{k=1}^{n'} a_k(x_{2k} + x_{2k+1})$$

Then Bob uses his index $i \in [n']$ to form the combined function

$$h(x) := (f(x) \wedge x_{2i}) + x_{2i+1}$$

We claim that Bob can solve INDEX by running a testing algorithm on h. To see this, first note that f is just a linear function on some subset of variables. If $a_i = 0$, then the variables x_{2i} and x_{2i+1} do not appear in f, so the resulting h is clearly a read-once width-2 OBDD.

If $a_i = 1$, then the variables x_{2i} and x_{2i+1} do appear in f. In this case, we can write $f(x) = f'(x) + x_{2i} + x_{2i+1}$, where $f'(x)$ is a linear function on some non-empty subset of variables not involving x_{2i} or x_{2i+1} (note that the variable x_1 is included in f just to guarantee that f' is always a linear function on at least one variable). We can thus express the resulting h as $h(x) = ((f'(x) + x_{2i} + x_{2i+1}) \wedge x_{2i}) + x_{2i+1}$, which simplifies to the following

$$h(x) = \begin{cases} x_{2i+1} & \text{if } x_{2i} = 0 \\ f'(x) + 1 & \text{if } x_{2i} = 1 \end{cases}$$

We claim that h is $1/8$-far from a read-once, width-2 OBDD. To see this, first note that any variable x_k relevant to h has $\mathrm{Inf}_h(k) = 1/2$. This is easily checked, since (i) x_{2i+1} is influential if and only if $x_{2i} = 0$, (ii) the variables relevant to f' are influential if and only if $x_{2i} = 1$, and (iii) x_{2i} is influential if and only if $x_{2i+1} \neq f'(x) + 1$, which happens exactly half the time (as x_{2i+1} and $f'(x)$ are linear functions on disjoint subsets of variables).

Assume for contradiction that h is 1/8-close to some width-2 OBDD ℓ. Without loss of generality, we can assume that ℓ's irrelevant variables come at the beginning. This means that the last level of ℓ must be a blocking level. Otherwise, if it is a linear level, the variable at that level will have influence 1 in ℓ, but only influence 1/2 or 0 in h, so by Fact 6, h and ℓ will be 1/4-far.

Since the last level of ℓ is blocking, there must exist a setting $t \in \{0,1\}$ for the variable at the last level which makes the function ℓ constant. However, for any fixed setting of any single variable in h, the resulting function is at least 1/4-far from constant. This is easily checked, since (i) fixing the value of x_{2i} induces either x_{2i+1} or $f'(x) + 1$, both of which are 1/2-far from constant, (ii) fixing the value of x_{2i+1} still allows h to be balanced when $x_{2i} = 1$, so in this case h is 1/4-far from constant, and (iii) fixing the value of a variable which appears in f' still allows h to be balanced when $x_{2i} = 0$, so in this case h is also 1/4-far from constant. Since h and ℓ disagree on at least 1/4 of the settings where ℓ's last variable is set to t, this implies they must be 1/8-far.

5 A Lower Bound for Testing Fixed-Order Width-w OBDDs, for Constant $w \geq 4$.

In this section we prove Theorem 3, via a reduction from SET-DISJOINTNESS. To perform the reduction, Alice and Bob will use the elements in their sets to produce a set of AND clauses. They then run a testing algorithm on the XOR of these clauses. Crucially, they will construct the clauses in such a way that when their sets intersect, they produce at least $k := \lfloor \log_2 w \rfloor$ clauses with *interleaving* variables. This large number of interleaving variables forces the hard instances to be far from computable by (fixed-order) width-w OBDDs.

In order to show that our hard instances are far, we make use of the following technical lemma, which we prove in the full version of this paper. The lemma is a generalization of the $w = 4$ case proved in Theorem 4.2 of Goldreich [15].

Lemma 4. *Let* $k := \lfloor \log_2 w \rfloor$ *and* $n' := \lfloor n/2k \rfloor - 1$. *For boolean variables* $x_1, \ldots, x_{2k}$, *define predicates* $\sigma_1, \sigma_2, \sigma_3, \sigma_4$ *as*

$$\sigma_1(x_1, \ldots, x_{2k}) := 0, \quad \sigma_2(x_1, \ldots, x_{2k}) := (x_1 \wedge x_{k+1}) + \ldots + (x_{k-1} \wedge x_{2k-1})$$

$$\sigma_3(x_1, \ldots, x_{2k}) := x_k \wedge x_{2k}, \quad \sigma_4(x_1, \ldots, x_{2k}) := (x_1 \wedge x_{k+1}) + \ldots + (x_k \wedge x_{2k})$$

and for $v_1, \ldots, v_{n'} \in \{1, 2, 3, 4\}$, *define a boolean function* $h = h_{v_1, \ldots, v_{n'}}$ *as*

$$h(x) := x_1 + \sum_{i=1}^{n'} \sigma_{v_i}(x_{2ki+1}, \ldots, x_{2ki+2k})$$

i. If $v_i \in \{1, 2, 3\}$ *for all* i, *then* h *is computable by a width-w OBDD.*
ii. If $v_j = 4$ *for a unique* j, *then* h *is* $\frac{1}{2w^2}$*-far from any width-w OBDD.*

Theorem (Restated). *For any constant* $w \geq 4$, *any adaptive algorithm for testing computability by width-w OBDDs with variables in fixed order requires* $\Omega(n)$ *queries.*

Proof. Let $k := \lfloor \log w \rfloor$ and $n' := \lfloor n/2k \rfloor - 1$. We will reduce from SET-DISJOINTNESS (with the promise that Alice and Bob's sets intersect in at most one place) on n' variables. Since n' is $\Theta(n)$, this will imply an $\Omega(n)$ lower bound.

Let $S, T \subseteq [n']$ denote Alice's and Bob's inputs respectively. Then Alice constructs the function f and Bob constructs the function g as follows

$$f(x) := \sum_{i \in S} (x_{2ki+1} \wedge x_{2ki+k+1}) + \ldots + (x_{2ki+k-1} \wedge x_{2ki+2k-1})$$

$$g(x) := \sum_{i \in T} (x_{2ki+k} \wedge x_{2ki+2k})$$

We will show that they can solve SET-DISJOINTNESS by running a testing algorithm for width-w OBDDs on $h(x) := x_1 + f(x) + g(x)$. Note that each set of $2k$ consecutive variables $(x_{2ki+1}, x_{2ki+2}, \ldots, x_{2ki+2k})$ depends on a unique coordinate from $[n']$, and that h contains $\sigma_4(x_{2ki+1}, \ldots, x_{2ki+2k})$ if and only if $i \in S \cap T$. The theorem thus follows from Lemma 4.

Acknowledgments

We would like to thank Dana Ron and Gilad Tsur for helpful discussions, and for sharing their manuscript [28] and their improvement to our Theorem 2.

References

1. Ablayev, F.: Lower bounds for one-way probabilistic communication complexity and their application to space complexity. Theoretical Computer Science 175(2), 139–159 (1996)
2. Bar-Yossef, Z., Jayram, T.S., Krauthgamer, R., Kumar, R.: The sketching complexity of pattern matching. In: Jansen, K., Khanna, S., Rolim, J.D.P., Ron, D. (eds.) RANDOM 2004 and APPROX 2004. LNCS, vol. 3122, pp. 261–272. Springer, Heidelberg (2004)
3. Bergadano, F., Bshouty, N.H., Tamon, C., Varricchio, S.: On learning branching programs and small depth circuits. In: Ben-David, S. (ed.) EuroCOLT 1997. LNCS, vol. 1208, pp. 150–161. Springer, Heidelberg (1997)
4. Blais, E.: Improved bounds for testing juntas. In: Proc. 12th International Workshop on Randomization and Approximation Techniques in Computer Science, pp. 317–330 (2008)
5. Blais, E.: Testing juntas nearly optimally. In: Proc. 41st Annual ACM Symposium on the Theory of Computing, pp. 151–158 (2009)
6. Blais, E., Brody, J., Matulef, K.: Property testing lower bounds via communication complexity (2011), http://web.mit.edu/matulef/www/papers/PTviaCC.pdf
7. Blum, M., Luby, M., Rubinfeld, R.: Self-testing/correcting with applications to numerical problems. J. Comput. Syst. Sci. 47, 549–595 (1993); Earlier version in STOC 1990
8. Bshouty, N.H., Tamon, C., Wilson, D.K.: On learning width two branching programs. Inf. Process. Lett. 65, 217–222 (1998)

9. Diakonikolas, I., Lee, H., Matulef, K., Onak, K., Rubinfeld, R., Servedio, R., Wan, A.: Testing for concise representations. In: Proc. 48th Annual IEEE Symposium on Foundations of Computer Science, pp. 549–558 (2007)

10. Do Ba, K., Indyk, P., Price, E., Woodruff, D.P.: Lower bounds for sparse recovery. In: Proc. 21st Annual ACM-SIAM Symposium on Discrete Algorithms (2010)

11. Ergün, F., Kumar, R., Rubenfeld, R.: On learning boundedwidth branching programs. In: Proc. 8th International Conference on Learning Theory, pp. 361–368 (1995)

12. Fischer, E., Kindler, G., Ron, D., Safra, S., Samorodnitsky, A.: Testing juntas. J. Comput. Syst. Sci. 68, 753–787 (2004)

13. Fischer, E., Lehman, E., Newman, I., Raskhodnikova, S., Rubinfeld, R., Samorodnitsky, A.: Monotonicity testing over general poset domains. In: Proc. 34th Annual ACM Symposium on the Theory of Computing, pp. 474–483 (2002)

14. Gavald'a, R., Guijarro, D.: Learning ordered binary decision diagrams. In: Jantke, K.P., Shinohara, T., Zeugmann, T. (eds.) ALT 1995. LNCS, vol. 997, Springer, Heidelberg (1995)

15. Goldreich, O.: On testing computability by small width OBDDs. In: Proc. 14th International Workshop on Randomization and Approximation Techniques in Computer Science, pp. 574–587 (2010)

16. Goldreich, O., Goldwasser, S., Lehman, E., Ron, D., Samorodnitsky, A.: Testing monotonicity. Combinatorica 20(3), 301–337 (2000)

17. Goldreich, O., Goldwasser, S., Ron, D.: Property testing and its connection to learning and approximation. J. ACM 45(4), 653–750 (1998)

18. Håstad, J., Wigderson, A.: The randomized communication complexity of set disjointness. Theory of Computing, 211–219 (2007)

19. Kane, D.M., Nelson, J., Woodruff, D.P.: On the exact space complexity of sketching and streaming small norms. In: Proc. 21st Annual ACM-SIAM Symposium on Discrete Algorithms (2010)

20. Kushilevitz, E., Nisan, N.: Communication Complexity. Cambridge University Press, Cambridge (1997)

21. Magniez, F., Mathieu, C., Nayak, A.: Recognizing well-parenthesized expressions in the streaming model. In: Proceedings of the 42nd ACM Symposium on Theory of Computing, STOC 2010 (2010)

22. Parnas, M., Ron, D., Samorodnitsky, A.: Testing basic boolean formulae. SIAM J. Disc. Math. 16(1), 20–46 (2002)

23. Raghavan, V., Wilkins, D.: Learning branching programs with queries. In: Proc. 6th Annual Workshop on Computational Learning Theory (1993)

24. Ron, D.: Algorithmic and analysis techniques in property testing. Foundations and Trends in Theoretical Computer Science 5(2), 73–205 (2009)

25. Ron, D., Tsur, G.: Testing computability by width two oBDDs. In: Dinur, I., Jansen, K., Naor, J., Rolim, J. (eds.) APPROX 2009. LNCS, vol. 5687, pp. 686–699. Springer, Heidelberg (2009)

26. Ron, D., Tsur, G.: Testing computability by width-2 oBDDs where the variable order is unknown. In: Calamoneri, T., Diaz, J. (eds.) CIAC 2010. LNCS, vol. 6078, pp. 131–142. Springer, Heidelberg (2010)

27. Ron, D., Tsur, G.: Personal communication (2011)

28. Ron, D., Tsur, G.: Testing computability by width-two obdds (Journal version) (2011) (manuscript)

On the Amount of Nonconstructivity in Learning Recursive Functions

Rūsiņš Freivalds[1,*] and Thomas Zeugmann[2,**]

[1] Institute of Mathematics and Computer Science, University of Latvia
Raiņa Bulvāris 29, Riga, LV-1459, Latvia
`Rusins.Freivalds@mii.lu.lv`
[2] Division of Computer Science, Hokkaido University
N-14, W-9, Sapporo 060-0814, Japan
`thomas@ist.hokudai.ac.jp`

Abstract. Nonconstructive proofs are a powerful mechanism in mathematics. Furthermore, nonconstructive computations by various types of machines and automata have been considered by e.g., Karp and Lipton [17] and Freivalds [11]. They allow to regard more complicated algorithms from the viewpoint of much more primitive computational devices. The amount of nonconstructivity is a quantitative characterization of the distance between types of computational devices with respect to solving a specific problem.

In the present paper, the amount of nonconstructivity in learning of recursive functions is studied. Different learning types are compared with respect to the amount of nonconstructivity needed to learn the whole class of general recursive functions. Upper and lower bounds for the amount of nonconstructivity needed are proved.

Keywords: inductive inference, recursive functions, nonconstructivity.

1 Introduction

Nonconstructive methods of proof in mathematics have a rather long and dramatic history. The debate was especially passionate when mathematicians tried to overcome the crisis concerning the foundations of mathematics.

The situation changed slightly in the forties of the last century, when nonconstructive methods found their way even to discrete mathematics. In particular, Paul Erdős used nonconstructive proofs masterly, beginning with the paper [7].

Another influential paper in this regard was Bārzdiņš [2], who introduced the notion of advice in the setting of Kolmogorov complexity of recursively enumerable sets. Karp and Lipton [17] introduced the notion of a Turing machine that

---

* This research was performed while this author was visiting the Division of Computer Science at Hokkaido University. The research of the first author was supported by Grant No. 09.1570 from the Latvian Council of Science and by Project 2009/0216/1DP/1.1.2.1.2/09/IPIA/VIA/004 from the European Social Fund.
** Supported by MEXT Grant-in-Aid for Scientific Research on Priority Areas under Grant No. 21013001.

M. Ogihara and J. Tarui (Eds.): TAMC 2011, LNCS 6648, pp. 332–343, 2011.
© Springer-Verlag Berlin Heidelberg 2011

takes advice to understand under what circumstances nonuniform upper bounds can be used to obtain uniform upper bounds. Damm and Holzer [6] adapted the notion of advice for finite automata.

A further step was taken by Freivalds [11, 12], who introduced a qualitative approach to measure the amount of nonconstructivity (or advice) in a proof. Analyzing three examples of nonconstructive proofs led him to a notion of nonconstructive computation which can be easily used for many types of automata and machines and which essentially coincides with Karp and Lipton's [17] notion when applied to Turing machines.

As outlined by Freivalds [11, 12], there are several results in the theory of inductive inference of recursive functions which suggest that the notion of nonconstructivity may be worth a deeper study in this setting, too.

In the present paper we prove several upper and lower bounds for the amount of nonconstructivity in learning classes of recursive functions. When learning recursive functions growing initial segments $(f(0), \ldots, f(n))$ are fed to the learning algorithm, henceforth called *strategy*. For each initial segment the strategy has then to compute a hypothesis i_n which is a natural number. These hypotheses are interpreted with respect to a suitably chosen hypothesis space ψ which is a numbering. The interpretation of the hypothesis i_n is that the strategy conjectures program i_n in the numbering ψ to compute the target function f. One requires the sequence $(i_n)_{n \in \mathbb{N}}$ of all computed hypotheses to converge to a program i correctly computing the target function f, i.e., $\psi_i = f$. A strategy learns a class of recursive functions provided it can learn every function from it. The model just explained is basically *learning in the limit* as introduced by Gold [14]. Many variations of this model have been studied (cf., e.g., [4, 10, 15, 25], and the references therein).

For many of these variations it was shown that the class $\mathcal{R}$ of all recursive functions is *not* learnable. Several attempts have been undertaken to classify the difficulty of learning the class $\mathcal{R}$. Adleman and Blum [1] showed the degree of unsolvability of the problem to learn the class $\mathcal{R}$ to be strictly less than the degree of the halting problem. A further approach was to characterize the difficulty of learning classes of recursive functions by using oracles (cf., e.g., [5, 19]).

We introduce a *new measure*, i.e., the amount of nonconstructivity needed to learn the class $\mathcal{R}$. That is, the strategy receives as a second input a bitstring of finite length which we call *help-word*. If the help-word is correct, the strategy learns in the desired sense. Since there are infinitely many functions to learn, a parameterization is necessary, i.e., we allow for every n a possibly different help-word and we require the strategy to learn every recursive function contained in $\{\psi_0, \ldots, \psi_n\}$ with respect to the numbering ψ (cf. Definition 4). The difficulty of the learning problem is then measured by the length of the help-words needed, i.e., in terms of the growth rate of a function d bounding this length.

As in previous approaches, the help-word does *not* just provide an answer to the learning problem. There is still much work to be done by the strategy. The usefulness of this approach is nicely reflected by our results which show that the function d may vary from arbitrarily slow growing (for learning in the limit) to $n + 1$ (for minimal identification).

2   Preliminaries

Unspecified notations follow Rogers [22]. In addition to or in contrast with [22] we use the following. Let $\mathbb{N} = \{0, 1, 2, \ldots\}$ be the set of all natural numbers, and let $\mathbb{N}^+ = \mathbb{N} \setminus \{0\}$. We use $\mathbb{N}^*$ for the set of all finite sequences of natural numbers. By $|S|$ and $\wp(S)$ we denote the cardinality and power set of a set S, respectively. Let $\emptyset$, $\in$, $\subset$, $\subseteq$, $\supset$, $\supseteq$, and $\#$ denote the empty set, element of, proper subset, subset, proper superset, superset, and incomparability of sets, respectively.

By $\mathfrak{T}$ we denote the set of all total functions of one variable over $\mathbb{N}$. The set of all partial recursive and recursive functions of one respectively two variables over $\mathbb{N}$ is denoted by $\mathcal{P}$, $\mathcal{R}$, $\mathcal{P}^2$, $\mathcal{R}^2$, respectively. Let $f \in \mathcal{P}$, then we use $\mathrm{dom}(f)$ for the *domain* of the function f, i.e., $\mathrm{dom}(f) = \{x \mid x \in \mathbb{N}, \ f(x) \text{ is defined}\}$. By $\mathrm{Val}(f)$ we denote the *range* of f, i.e., $\mathrm{Val}(f) = \{f(x) \mid x \in \mathrm{dom}(f)\}$.

A function $f \in \mathcal{P}$ is said to be *strictly monotonic* provided for all $x, y \in \mathbb{N}$ with $x < y$ we have, if both $f(x)$ and $f(y)$ are defined then $f(x) < f(y)$. By $\mathcal{R}_{mon}$ we denote the set of all strictly monotonic recursive functions.

Any function $\psi \in \mathcal{P}^2$ is called a *numbering*. Let $\psi \in \mathcal{P}^2$, then we write ψ_i instead of $\lambda x.\psi(i, x)$ and set $\mathcal{P}_\psi = \{\psi_i \mid i \in \mathbb{N}\}$ as well as $\mathcal{R}_\psi = \mathcal{P}_\psi \cap \mathcal{R}$. Consequently, if $f \in \mathcal{P}_\psi$, then there is a number i such that $f = \psi_i$. If $f \in \mathcal{P}$ and $i \in \mathbb{N}$ are such that $\psi_i = f$, then i is called a ψ–*program* for f. Let ψ be any numbering, and $i, x \in \mathbb{N}$; if $\psi_i(x)$ is defined (abbr. $\psi_i(x)\downarrow$) then we also say that $\psi_i(x)$ *converges*. Otherwise, $\psi_i(x)$ is said to *diverge* (abbr. $\psi_i(x)\uparrow$). Let $\psi \in \mathcal{P}^2$ and $f \in \mathcal{P}$; then we use $\min_\psi f$ to denote the least number i such that $\psi_i = f$.

A numbering $\varphi \in \mathcal{P}^2$ is called a *Gödel numbering* (cf. Rogers [22]) iff $\mathcal{P}_\varphi = \mathcal{P}$, and for any numbering $\psi \in \mathcal{P}^2$, there is a *compiler* $c \in \mathcal{R}$ such that $\psi_i = \varphi_{c(i)}$ for all $i \in \mathbb{N}$. We use *Göd* to denote the set of all Gödel numberings.

By $\mathcal{NUM} = \{\mathcal{U} \mid (\exists \psi \in \mathcal{R}^2)\, [\mathcal{U} \subseteq \mathcal{P}_\psi]\}$ we denote the family of all subsets of all recursively enumerable classes of recursive functions. Let $\mathcal{NUM}! = \{\mathcal{U} \mid (\exists \psi \in \mathcal{R}^2)\, [\mathcal{U} = \mathcal{P}_\psi]\}$ denote the family of all recursively enumerable classes of recursive functions. The elements of $\mathcal{NUM}!$ are referred to as *indexed families*.

We call (φ, Φ) a measure of computational complexity (cf. [20]) if $\varphi \in$ *Göd* and $\Phi \in \mathcal{P}^2$ satisfies Blum's [3] axioms. That is, (1) $\mathrm{dom}(\varphi_i) = \mathrm{dom}(\Phi_i)$ for all $i \in \mathbb{N}$ and (2) the predicate "$\Phi_i(x) = y$" is uniformly recursive for all $i, x, y \in \mathbb{N}$.

Let $\langle \ldots \rangle$ be any recursive encoding of $\mathbb{N}^*$ onto $\mathbb{N}$. We write f^n instead of $\langle (f(0), \ldots, f(n)) \rangle$, for all $n \in \mathbb{N}$, $f \in \mathcal{R}$. A sequence $(j_n)_{n \in \mathbb{N}}$ of natural numbers is said to *converge* to the number j if $j_n = j$ for all but finitely many $n \in \mathbb{N}$. Moreover, $(j_n)_{n \in \mathbb{N}}$ is said to *finitely converge* to the number j if it converges in the limit to j and for all $n \in \mathbb{N}$, $j_n = j_{n+1}$ implies $j_k = j$ for all $k \geq n$.

Definition 1 (Gold [13, 14]). *Let $\mathcal{U} \subseteq \mathcal{R}$ and let $\psi \in \mathcal{P}^2$. The class $\mathcal{U}$ is said to be learnable in the limit with respect to ψ if there is a strategy $S \in \mathcal{P}$ such that for each function $f \in \mathcal{U}$,*

(1) *for all $n \in \mathbb{N}$, $S(f^n)$ is defined,*
(2) *there is a $j \in \mathbb{N}$ with $\psi_j = f$ and the sequence $(S(f^n))_{n \in \mathbb{N}}$ converges to j.*

If $\mathcal{U}$ is learnable in the limit w.r.t. ψ by S, we write $\mathcal{U} \in \mathcal{LIM}_\psi(S)$. Let $\mathcal{LIM}_\psi = \{\mathcal{U} \mid \mathcal{U}$ is learnable in the limit w.r.t. $\psi\}$, and let $\mathcal{LIM} = \bigcup_{\psi \in \mathcal{P}^2} \mathcal{LIM}_\psi$.

In the following modification of Definition 1 we require the strategy to converge to $\min_\psi f$ instead of converging to any program for the target function f.

Definition 2 (Freivalds [9], Kinber [18]). *Let $\mathcal{U} \subseteq \mathcal{R}$ and let $\psi \in \mathcal{P}^2$. The class $\mathcal{U}$ is said to be ψ-minimal learnable in the limit with respect to ψ if there is a strategy $S \in \mathcal{P}$ such that for each function $f \in \mathcal{U}$,*

(1) *for all $n \in \mathbb{N}$, $S(f^n)$ is defined,*
(2) *the sequence $(S(f^n))_{n \in \mathbb{N}}$ converges to $\min_\psi f$.*

If $\mathcal{U}$ is ψ-minimal learnable in the limit w.r.t. ψ by a strategy S, we write $\mathcal{U} \in \mathcal{MIN}_\psi(S)$. Furthermore, let $\mathcal{MIN}_\psi = \{\mathcal{U} \mid \mathcal{U}$ is ψ-minimal learnable in the limit w.r.t. $\psi\}$, and let $\mathcal{MIN} = \bigcup_{\psi \in \mathcal{P}^2} \mathcal{MIN}_\psi$.

In general it is not decidable whether or not a strategy has already converged when successively fed some graph of a function. With the next definition we consider a special case where it has to be decidable whether or not a strategy has learned its input function. That is, we replace the requirement that the sequence of all created hypotheses "has to *converge*" by "has to *converge finitely.*"

Definition 3 (Gold [14], Trakhtenbrot and Barzdin [23]). *Let $\mathcal{U} \subseteq \mathcal{R}$ and let $\psi \in \mathcal{P}^2$. The class $\mathcal{U}$ is said to be finitely learnable with respect to ψ if there is a strategy $S \in \mathcal{P}$ such that for any function $f \in \mathcal{U}$,*

(1) *for all $n \in \mathbb{N}$, $S(f^n)$ is defined,*
(2) *there is a $j \in \mathbb{N}$ such that $\psi_j = f$ and the sequence $(S(f^n))_{n \in \mathbb{N}}$ finitely converges to j.*

If $\mathcal{U}$ is finitely learnable w.r.t. ψ by a strategy S, we write $\mathcal{U} \in \mathcal{FIN}_\psi(S)$. The learning types $\mathcal{FIN}_\psi$ and $\mathcal{FIN}$ are defined analogously to the above.

Of course, we can also combine ψ-minimal learnability and finite identification resulting in the learning types $\mathcal{MIN}\text{-}\mathcal{FIN}_\psi$ and $\mathcal{MIN}\text{-}\mathcal{FIN}$.

The strategies used for nonconstructive inductive inference take as input not only the encoded graph of a function $f \in \mathcal{R}$ but also a help-word w. The help-words are assumed to be encoded in binary. So, for such strategies we write $S(f^n, w)$ to denote the program output by S. Then, for all the inference types defined above, we say that S nonconstructively identifies f with the help-word w provided the sequence $(S(f^n, w))_{n \in \mathbb{N}}$ (finitely) converges to a number j such that $\varphi_j = f$ (for $\mathcal{LIM}$ and $\mathcal{FIN}$) and $j = \min_\psi f$ (for $\mathcal{MIN}$), respectively.

Definition 4. *Let $\psi \in \mathcal{P}^2$, let $\mathcal{U} \subseteq \mathcal{R}$, and $d \in \mathcal{R}$. A strategy $S \in \mathcal{P}^2$ infers $\mathcal{U}$ with nonconstructivity $d(n)$ in the limit with respect to ψ, if for each $n \in \mathbb{N}$ there is a help-word of length at most $d(n)$ such that for every $f \in \mathcal{U} \cap \{\psi_0, \psi_1, \ldots, \psi_n\}$ the sequence $(S(f^n, w))_{n \in \mathbb{N}}$ converges to a program i satisfying $\psi_i = f$.*

Nonconstructive finite and minimal inference is defined in analogue to the above.

Looking at Definition 4 as well as at the definition of nonconstructive finite and minimal inference, it should be noted that the strategy may need to know either an appropriate upper bound for n or even the precise value of n in order to exploit the fact that the target function is from $f \in \mathcal{U} \cap \{\psi_0, \psi_1, \ldots, \psi_n\}$.

To simplify notation in several theorems and proofs given below, we make the convention that logarithmic function is to the base 2 and that it is replaced by its integer valued counterpart $\lfloor \log n \rfloor + 1$.

3   Results

Already Gold [13] showed that $\mathcal{R} \notin \mathcal{LIM}$. So, we start our investigations by asking for the amount of nonconstructivity needed to identify the set $\mathcal{R}$ of all recursive functions in the limit with respect to any Gödel numbering φ.

Using an idea from Freivald and Wiehagen [8], we prove that the needed amount of nonconstructivity is surprisingly small. To show this result, for every function $f \in \mathcal{R}_{mon}$ we define its *inverse* f_{inv} as follows: $f_{inv}(n) = \mu y[f(y) \geq n]$ for all $n \in \mathbb{N}$. Recall that $\mathtt{Val}(f)$ is recursive for all $f \in \mathcal{R}_{mon}$. Thus, for all $f \in \mathcal{R}_{mon}$ we can conclude that $f_{inv}(n) \in \mathcal{R}$.

Theorem 1. *Let $\varphi \in G\ddot{o}d$ be arbitrarily fixed, and let $d \in \mathcal{R}_{mon}$ be any function. Then there is a strategy $S \in \mathcal{P}^2$ such that the class $\mathcal{R}$ can be identified with nonconstructivity $\log d_{inv}(n)$ in the limit with respect to φ.*

Proof. Let $\varphi \in G\ddot{o}d$ be arbitrarily fixed. Without loss of generality, we can also assume any complexity function $\Phi \in \mathcal{P}^2$ such that (φ, Φ) is a complexity measure.

The key idea of the proof is that, in order to learn any function from $\mathcal{R}$, it suffices to have an upper bound for $\min_{\varphi} f$. So, assuming any help-word w of length precisely $\log d_{inv}(n)$, the strategy S uses the length of the help-word w to create a bitstring that contains only 1s and has the same length as the help word. This bitstring is interpreted in the usual way as a natural number k. By construction, we then have $k \geq d_{inv}(n)$. Furthermore, since $d \in \mathcal{R}_{mon}$, we directly obtain that $d(k) \geq d(d_{inv}(n)) \geq n$. Consequently, the strategy S uses k to compute

$$u_* =_{df} d(k) \, ,$$

and by construction, we have $u_* \geq n$.

Assume any function $f \in \mathcal{R} \cap \{\psi_0, \psi_1, \ldots, \psi_n\}$, and let f^m and w be the input to the strategy S. Then, S initializes the index set I_{init} to be $I_{init} = \{0, \ldots, u_*\}$ and checks whether or not $\Phi_i(x) \leq m$ for every $i \in I_{init}$ and $0 \leq x \leq m$. For all i and x that passed this test successfully, S checks whether or not $\varphi_i(x) = f(x)$. If this is not the case, i is removed from I_{init}. Let I_m be the resulting index set.

The strategy uses the amalgamation technique (cf. [4, 24]). That is, let amal be a recursive function mapping any finite set I of φ-programs to a φ-program such that for any $x \in \mathbb{N}$, $\varphi_{\mathrm{amal}(I)}(x)$ is defined by running $\varphi_i(x)$ for every $i \in I$ in parallel and taking the first value obtained, if any.

So, the output of $S(f^m, w)$ is $\text{amal}(I_m)$.

We have to show that the sequence $(\text{amal}(I_m))_{m \in \mathbb{N}}$ converges to a φ-program for f. By construction we know that I_{init} contains at least one φ-program for f. This program and any other φ-program computing a subfunction of f can never be removed from I_{init}. But if a φ-program j from I_{init} does not compute a subfunction of f, then there must be an x such that $\varphi_j(x) \downarrow \neq f(x)$. So, as soon as $m \geq \max\{x, \Phi_j(x)\}$, the program j is removed from I_{init}. Since I_{init} is finite, there must be an m_* such that I_{m_*} contains only φ-programs for f or a subfunction of f. We conclude that $\text{amal}(I_{m_*})$ is a φ-programs for f. Furthermore, $I_\ell = I_{m_*}$ for all $\ell \geq m_*$, and thus the strategy S learns f in the limit. $\square$

So there is no smallest amount of nonconstructivity needed to learn $\mathcal{R}$ in the limit. But the amount of nonconstructivity cannot be zero, since then we would have $\mathcal{R} \in \mathcal{LIM}$. One can define a total function $t \in \mathfrak{T}$ such that $t(n) \geq d(n)$ for all $d \in \mathcal{R}_{mon}$ and all but finitely many n. Consequently, $\log t_{inv}$ is then a lower bound for the amount of nonconstructivity needed to learn $\mathcal{R}$ in the limit when using the algorithm from the proof of Theorem 1.

We continue by asking what amount of nonconstructivity is needed to obtain φ-minimal identification in the limit of the class $\mathcal{R}$. Now, the situation is intuitively more complex, since $\mathcal{LIM}_\varphi \setminus \mathcal{MIN}_\varphi \neq \emptyset$ for every $\varphi \in G\ddot{o}d$. Interestingly, there are even Gödel numberings φ such that $\mathcal{MIN}_\varphi$ contains only classes of finite cardinality (cf. Freivalds [10]). On the other hand, the sufficient amount of nonconstructivity given in Theorem 2 does *not* depend on the Gödel numbering. Theorem 2 below is not the best possible and we shall improve it below, but it shows an easy way to achieve φ-minimal learning of $\mathcal{R}$ in the limit.

Theorem 2. *Let $\varphi \in G\ddot{o}d$ be arbitrarily fixed. Then there is a strategy $S \in \mathcal{P}^2$ such that the class $\mathcal{R}$ can be φ-minimal identified with nonconstructivity $n + 1$ in the limit with respect to φ.*

Proof. Let $\varphi \in G\ddot{o}d$ be arbitrarily fixed, and let $n \in \mathbb{N}$. The help-word w is a bitstring b of length $n + 1$ defined as follows. If $\varphi_i \in \mathcal{R}$, then the ith entry of b is 1, and 0 otherwise. So, the length of w allows the strategy to compute n.

Assume any function $f \in \mathcal{R} \cap \{\psi_0, \psi_1, \ldots, \psi_n\}$, and let f^m and w be the input to S. Then S only considers those functions φ_i, $0 \leq i \leq n$, for which the ith entry in the help-word is 1. Since all these remaining functions are total, the strategy searches for the least index j among these functions for which $\varphi_j^m = f^m$. That is, it essentially uses the identification by enumeration principle (cf. Gold [14]). $\square$

The proof of Theorem 2 was easy which may be an indication that a smaller amount of nonconstructivity may suffice. So far we could not show a lower bound for the amount of nonconstructivity needed to achieve φ-minimal inference in the limit of $\mathcal{R}$. As Theorem 4 shows, we can achieve a much better result when using nonconstructivity $n + 1$, again an indication that we used a too great amount of nonconstructivity in Theorem 2. And indeed, we can do exponentially better.

Theorem 3. *Let $\varphi \in G\ddot{o}d$ be arbitrarily fixed. Then there is a strategy $S \in \mathcal{P}^2$ such that the class $\mathcal{R}$ can be φ-minimal identified with nonconstructivity $2 \cdot \log n$ in the limit with respect to φ.*

Proof. The key observation for the proof is that it suffices to know the *number* of recursive functions in the set $\{\varphi_0, \ldots, \varphi_n\}$. To use this information appropriately, the first half of the help-word w is the binary encoding of n and the second half of w provides the number, say k, of recursive functions in the set $\{\varphi_0, \ldots, \varphi_n\}$. This number is written in binary but leading zeros are added to ensure that both parts of w have the same length. Thus $2 \cdot \log n$ many bits suffice to represent w.

Assume any function $f \in \mathcal{R} \cap \{\psi_0, \psi_1, \ldots, \psi_n\}$, and let f^m and w be the input to the strategy S. Then S, by dovetailing its computations, first tries to compute $\varphi_i(0), \ldots, \varphi_i(m)$ for all $0 \leq i \leq n$ until it finds the first k programs $i_1, \ldots, i_k$ such that $\varphi_i(0), \ldots, \varphi_i(m)$ turn out to be defined for every $i \in \{i_1, \ldots, i_k\}$. Once S has found these programs $i_1, \ldots, i_k$, it outputs the least program $i \in \{i_1, \ldots, i_k\}$ for which it verifies $\varphi_i^m = f^m$ provided there is such a program, and m otherwise.

By construction, there are $n + 1 - k$ many programs $j \in \{0, \ldots, n\}$ such that $\varphi_j \in \mathcal{P} \setminus \mathcal{R}$. For each of these programs j there is a least y_j such that $\varphi_j(y_j)\uparrow$. Let $y_{\max}$ be the maximum of all these y_j. Hence, as soon as $m \geq y_{\max}$, the strategy S must find precisely the programs $i_1, \ldots, i_k$ such that $\varphi_i \in \mathcal{R}$ for all $i \in \{i_1, \ldots, i_k\}$. By assumption, the target function f possesses a program i with $0 \leq i \leq n$, and so for all $m \geq y_{\max}$, the strategy must output $\min_\varphi f$. □

Next we provide the theorem already mentioned above which shows that with nonconstructivity $n + 1$ a much stronger result is possible.

Theorem 4. *Let $\varphi \in G\ddot{o}d$ be arbitrarily fixed. Then there is a strategy $S \in \mathcal{P}^2$ such that the class $\mathcal{R}$ can be φ-minimal finitely identified with nonconstructivity $n + 1$ with respect to φ.*

Proof. Let $\varphi \in G\ddot{o}d$, and let $n \in \mathbb{N}$. The help-word w is a bitstring b of length $n + 1$ defined as follows. If $\varphi_i \in \mathcal{R}$ and $\varphi_i \neq \varphi_j$ for all $0 \leq j < i$, then the ith entry of b is 1, and 0 otherwise. So, the length of the help-word directly allows the strategy to compute n. Note that now the help-word allows for implicitly having a one-to-one enumeration for the functions $f \in \mathcal{R} \cap \{\varphi_0, \ldots, \varphi_n\}$.

Assume any function $f \in \mathcal{R} \cap \{\varphi_0, \varphi_1, \ldots, \varphi_n\}$, and let f^m and w be the input to S. Then S only considers those functions φ_i, $0 \leq i \leq n$, for which the ith entry in the help-word is 1. For all these i, the strategy computes φ_i^m and checks whether or not they are pairwise different. As long as this is not the case, the strategy outputs m. If all these φ_i^m are pairwise different, then the strategy outputs the i for which it could verify $f^m = \varphi_i^m$.

By construction, it is obvious that S finitely converges to $\min_\varphi f$. □

So far we could not prove the amount of nonconstructivity given in Theorem 4 to be the best possible. We thus look at the case, where we have to learn an indexed family $\mathcal{U}$ of recursive functions. Note that for every indexed family $\mathcal{U}$ and any of its numberings $\psi \in \mathcal{R}^2$ we have $\mathcal{U} \in \mathcal{MIN}_\psi$ (cf. Gold [14]). In contrast, $\mathcal{NUM} \# \mathcal{FIN}$ (see e.g., [25] and the references therein). So, it is only natural to ask for the amount of nonconstructivity needed to finitely learn ψ-minimal programs. The answer is provided by our theorems below.

Theorem 5. *Let $\mathcal{U}$ be any indexed family, and let $\psi \in \mathcal{R}^2$ be any numbering for $\mathcal{U}$. Then there is a strategy $S \in \mathcal{P}^2$ such that the class $\mathcal{U}$ can be ψ-minimal finitely identified with nonconstructivity $2 \cdot \log n$ with respect to ψ.*

Proof. The key observation is that it suffices to know the number k of distinct functions in $\{\psi_0, \ldots, \psi_n\}$. The help-word w is divided in two halves, where the first half is the binary encoding of n and the second half encodes k in binary (again including leading zeros). So $2 \cdot \log n$ many bits suffice for representing w.

On input f^m and w the strategy S computes, by dovetailing its computations, $\psi_i(x)$ for all $i \in \{0, \ldots, n\}$ and $x = 0, 1, 2, \ldots$ until it has verified that there are exactly k different functions. Let $i_1, \ldots, i_k$ be the least indices of these k different functions. Next, it checks whether or not there is precisely one $i \in \{i_1, \ldots, i_k\}$ such that $f^m = \psi_i^m$. If this is the case, S outputs this i. Otherwise, it outputs m.

By construction, it is obvious that S finitely converges to $\min_\psi f$.   □

Next we show that the amount of nonconstructivity given in Theorem 5 cannot be substantially reduced.

Theorem 6. *There is an indexed family $\mathcal{U}$ and a numbering $\psi \in \mathcal{R}^2$ for it such that no strategy $S \in \mathcal{P}^2$ can ψ-minimal finitely identify the class $\mathcal{U}$ with nonconstructivity $c \cdot \log n$ with respect to ψ, where $c \in (0, 1)$ is any constant.*

Proof. We construct the indexed family $\mathcal{U}$ by defining the numbering $\psi \in \mathcal{R}^2$ for it. For this purpose, we use the following pairing function $c \colon \mathbb{N} \times \mathbb{N} \to \mathbb{N}$, where $c(x, y) = 2^x(2y + 1) - 1$. Note that this pairing function is a bijection. It may be traced back to Pepis [21] and Kalmár [16]. Furthermore, we interpret every function in $\mathcal{P}^2$ as a strategy and obtain thus an effective enumeration $S_0, S_1, S_2, \ldots$ of all possible strategies. Below, for $\ell \in \mathbb{N}$, we use the shortcut $i^{\ell+1}$ to denote the encoding f^ℓ of the initial segment of the function f for which $f(z) = i$ for all $i = 0, \ldots, \ell$.

For every $i \in \mathbb{N}$ we define two functions ψ_{2i} and ψ_{2i+1} as follows. Let x and y be the uniquely determined numbers such that $i = c(x, y)$. Now, we successively define for $k = 1, 2, 3, \ldots$ the functions values $\psi_{2i}(k - 1) = \psi_{2i+1}(k - 1) = i$ and input i^k and y to the strategy S_x until we find the smallest k such that the following Conditions (A) and (B) are satisfied.

(A) There is an $\ell < k$ such that each of the values $S_x(i, y), \ldots, S_x(i^{\ell+1}, y)$ turns out to be computable in at most k steps.

(B) $S_x(i, y) \neq S_x(i^2, y) \neq \cdots \neq S_x(i^\ell, y) = S_x(i^{\ell+1}, y)$.

If Conditions (A) and (B) never turn out to be satisfied then the function values $\psi_{2i}(k)$ and $\psi_{2i+1}(k)$ are defined for all $k \in \mathbb{N}$, and thus $\psi_{2i}, \psi_{2i+1} \in \mathcal{R}$.

On the other hand, if Conditions (A) and (B) turn out to be satisfied then Condition (B) implies that the sequence $S_x(i, y), \ldots, S_x(i^{\ell+1}, y)$ tends to converge finitely. That is, it either converges finitely or it cannot converge finitely at all. Now, we continue to define the functions ψ_{2i} and ψ_{2i+1} as follows.

(C) If $S_x(i^\ell, y) = 2i$, then we define $\psi_{2i}(z) = i + k + z$ for all $z \geq k$.
   Furthermore, we set $\psi_{2i+1}(z) = i$ for all $z \geq k$.
(D) If $S_x(i^\ell, y) = 2i + 1$, then we define $\psi_{2i+1}(z) = i + k + z$ for all $z \geq k$.
   Furthermore, we set $\psi_{2i}(z) = i$ for all $z \geq k$.
(E) If $S_x(i^\ell, y) \notin \{2i, 2i + 1\}$, then we define $\psi_{2i}(z) = \psi_{2i+1}(z) = z$ for all $z \geq k$.

So we have $\psi_{2i}, \psi_{2i+1} \in \mathcal{R}$, and thus, $\psi \in \mathcal{R}^2$. Finally, we set $\mathcal{U} = \mathcal{R}_\psi$.

We show that there is no strategy $S \in \mathcal{P}^2$ that ψ-minimal finitely infers $\mathcal{U}$ with nonconstructivity $c \cdot \log n$ with respect to ψ, where $c \in (0, 1)$ is any constant.

Suppose the converse, i.e., there is a strategy $S \in \mathcal{P}^2$ that ψ-minimal finitely infers $\mathcal{U}$ with nonconstructivity $c \cdot \log n$ with respect to ψ. Then there must be a $v \in \mathbb{N}$ such that $S = S_v$ in our enumeration $S_0, S_1, S_2, \ldots$ of all possible strategies. Let d be the function from Definition 4. Furthermore, for every $n \in \mathbb{N}$ and every $f \in \{\psi_0, \ldots, \psi_n\}$ there has to be a help-word w of length at most $d(n)$ and depending only on n such that the sequence $(S_v(f^m, w))_{m \in \mathbb{N}}$ finitely converges to the minimal ψ-program of f.

By assumption, there is a $c \in (0, 1)$ such that $d(n) \leq c \cdot \log n$. We fix this c and conclude that for n large enough we have

$$d(n) > 1 \quad \text{and} \quad \frac{1 - c}{2} > \frac{2 + (v + 1)}{\log n} . \tag{1}$$

Now, we obtain successively

$$1 > \frac{c + 1}{2}$$

$$1 > \frac{2c + 1 - c}{2}$$

$$1 > \frac{c \cdot \log n}{\log n} + \frac{1 - c}{2}$$

$$1 > \frac{d(n)}{\log n} + \frac{2 + (v + 1)}{\log n} , \quad \text{since } c \cdot \log n \geq d(n) \text{ and by (1)}$$

$$\log n > d(n) + 2 + (v + 1)$$

$$\log n - \log 2^{v+1} > d(n) + 2$$

$$\log \frac{n}{2^{v+1}} > d(n) + 2$$

$$\frac{n}{2^{v+1}} > 2^{d(n)+2}$$

$$\frac{n + 2}{2^{v+1}} > 2 \cdot 2^{d(n)} + 1$$

$$\frac{n + 2}{2^{v+1}} > 2w + 1 , \quad \text{since } 2^{d(n)} \geq w$$

$$\frac{n}{2} > 2^v (2w + 1) - 1$$

$$\frac{n}{2} > c(v, w) .$$

Now, let $i = c(v, w)$ and consider the functions ψ_{2i} and ψ_{2i+1}. By our choice of n, these functions must be among the functions $\{\psi_0, \ldots, \psi_n\}$.

Let $\ell \in \mathbb{N}^+$ be the least number such that S_v on two successive inputs outputs the same hypothesis, i.e., $S_v(i, w) \neq \cdots \neq S_v(i^\ell, w) = S_v(i^{\ell+1}, w)$. Such an ℓ has to exist, since otherwise S_v can neither finitely identify ψ_{2i} nor ψ_{2i+1}.

If $S_v(i^\ell, w) \notin \{2i, 2i + 1\}$ we are already done, since ψ_{2i} and ψ_{2i+1} are the only functions from $\mathcal{U}$ having an initial segment where all values are equal to i. Finally, if $S_v(i^\ell, w) \in \{2i, 2i+1\}$ then by construction (cf. Condition (C) and (D), respectively), we have $\psi_{2i}(z) = \psi_{2i+1}(z)$ for all $z = 0, \ldots, \ell$ but $\psi_{2i} \neq \psi_{2i+1}$. So the strategy S_v fails to finitely learn either function ψ_{2i} or ψ_{2i+1}. □

As the proof of Theorem 6 shows, the failure to ψ-minimal finitely identify the indexed family $\mathcal{U}$ with respect to the numbering ψ with nonconstructivity $c \cdot \log n$, for $c \in (0, 1)$, is caused by the requirement to finitely identify the functions from $\mathcal{U}$. Thus, we directly obtain the following corollary.

Corollary 1. *There is an indexed family $\mathcal{U}$ and a numbering $\psi \in \mathcal{R}^2$ for it such that no strategy $S \in \mathcal{P}^2$ can finitely identify the class $\mathcal{U}$ with nonconstructivity $c \cdot \log n$ with respect to ψ, where $c \in (0, 1)$ is any constant.*

4   Conclusions and Open Problems

We have presented a model for the inductive inference of recursive functions that incorporates a certain amount of nonconstructivity. In our model, the amount of nonconstructivity needed to solve the learning problems considered has been used as a quantitative characterization of their difficulty.

We studied the problem of learning the whole class $\mathcal{R}$ under various postulates. These postulates range from learning in the limit to finite and minimal identification. As far as learning in the limit is concerned, the amount of nonconstructivity needed to learn $\mathcal{R}$ can be very small and there is no smallest amount that can be described in a computable way (cf. Theorem 1).

This result is nicely contrasted by the fact that we needed nonconstructivity $2 \cdot \log n$ to φ-minimal identify the class $\mathcal{R}$ in the limit and nonconstructivity $n+1$ to φ-minimal finitely identify $\mathcal{R}$ (cf. Theorems 3 and 4, respectively). That is, each additional postulate exponentially increased the amount of nonconstructivity needed. It remains, however, open whether or not these results can be improved.

Furthermore, we investigated the amount of nonconstructivity needed to φ-minimal finitely identify any indexed family of recursive functions. In this setting we obtained an upper bound of $2 \cdot \log n$ for the amount of nonconstructivity needed and showed that this amount *cannot* be substantially improved (cf. Theorems 5 and 6).

Unfortunately, so far we could only show the lower bound presented in Theorem 6. Proving lower bounds for the other cases studied in this paper remains open and should be addressed in the future.

Acknowledgments. We would like to thank the anonymous referees for their careful reading and their valuable comments.

References

[1] Adleman, L.M., Blum, M.: Inductive inference and unsolvability. The Journal of Symbolic Logic 56(3), 891–900 (1991)

[2] Bārzdiņš, J.M.: Complexity of programs to determine whether natural numbers not greater than n belong to a recursively enumerable set. Soviet Mathematics Doklady 9, 1251–1254 (1968)

[3] Blum, M.: A machine independent theory of the complexity of recursive functions. Journal of the ACM 14(2), 322–336 (1967)

[4] Case, J., Smith, C.: Comparison of identification criteria for machine inductive inference. Theoretical Computer Science 25(2), 193–220 (1983)

[5] Cholak, P., Downey, R., Fortnow, L., Gasarch, W., Kinber, E., Kummer, M., Kurtz, S., Slaman, T.A.: Degrees of inferability. In: Proc. 5th Annual ACM Workshop on Computational Learning Theory, pp. 180–192. ACM Press, New York (1992)

[6] Damm, C., Holzer, M.: Automata that take advice. In: Hájek, P., Wiedermann, J. (eds.) MFCS 1995. LNCS, vol. 969, pp. 149–158. Springer, Heidelberg (1995)

[7] Erdős, P.: Some remarks on the theory of graphs. Bulletin of the American Mathematical Society 53(4), 292–294 (1947)

[8] Freivald, R.V., Wiehagen, R.: Inductive inference with additional information. Elektronische Informationsverarbeitung und Kybernetik 15(4), 179–185 (1979)

[9] Freivald, R.: Minimal gödel numbers and their identification in the limit. In: Bečvář, J. (ed.) MFCS 1975. LNCS, vol. 32, pp. 219–225. Springer, Heidelberg (1975)

[10] Freivalds, R.: Inductive inference of recursive functions: Qualitative theory. In: Bārzdiņš, J., Bjørner, D. (eds.) Baltic Computer Science. LNCS, vol. 502, pp. 77–110. Springer, Heidelberg (1991)

[11] Freivalds, R.: Amount of nonconstructivity in finite automata. In: Maneth, S. (ed.) CIAA 2009. LNCS, vol. 5642, pp. 227–236. Springer, Heidelberg (2009)

[12] Freivalds, R.: Amount of nonconstructivity in deterministic finite automata. Theoretical Computer Science 411(38-39), 3436–3443 (2010)

[13] Gold, E.M.: Limiting recursion. The Journal of Symbolic Logic 30, 28–48 (1965)

[14] Gold, E.M.: Language identification in the limit. Inform. Control 10(5), 447–474 (1967)

[15] Jain, S., Osherson, D., Royer, J.S., Sharma, A.: Systems that Learn: An Introduction to Learning Theory, 2nd edn. MIT Press, Cambridge (1999)

[16] Kalmár, L.: On the reduction of the decision problem. First Paper. Ackermann prefix, a single binary predicate. The Journal of Symbolic Logic 4(1), 1–9 (1939)

[17] Karp, R.M., Lipton, R.J.: Turing machines that take advice. L' Enseignement Mathématique 28, 191–209 (1982)

[18] Кинбер, Е.Б.: О предельном синтезе почти минимальных геделевских номеров. In: Bārzdiņš, J. (ed.) Теория Алгоритмов и Программ, vol. I, pp. 212–223. Latvian State University (1974)

[19] Kummer, M., Stephan, F.: On the structure of the degrees of inferability. Journal of Computer and System Sciences 52(2), 214–238 (1996)

[20] Landweber, L.H., Robertson, E.L.: Recursive properties of abstract complexity classes. Journal of the ACM 19(2), 296–308 (1972)

[21] Pepis, J.: Ein Verfahren der mathematischen Logik. The Journal of Symbolic Logic 3(2), 61–76 (1938)

[22] Rogers Jr., H.: Theory of Recursive Functions and Effective Computability. McGraw-Hill, New York (1967); reprinted, MIT Press (1987)

[23] Trakhtenbrot, B.A., Barzdin, Y.M.: Finite Automata, Behavior and Synthesis. North Holland, Amsterdam (1973)

[24] Wiehagen, R.: Zur Theorie der Algorithmischen Erkennung. Dissertation B, Humboldt-Universität zu Berlin (1978)

[25] Zeugmann, T., Zilles, S.: Learning recursive functions: A survey. Theoretical Computer Science 397(1-3), 4–56 (2008)

A Bad Instance for k-Means++*

Tobias Brunsch and Heiko Röglin

Department of Computer Science, University of Bonn, Germany
brunsch@cs.uni-bonn.de, heiko@roeglin.org

Abstract. k-means++ is a seeding technique for the k-means method with an expected approximation ratio of $O(\log k)$, where k denotes the number of clusters. Examples are known on which the expected approximation ratio of k-means++ is $\Omega(\log k)$, showing that the upper bound is asymptotically tight. However, it remained open whether k-means++ yields an $O(1)$-approximation with probability $1/\text{poly}(k)$ or even with constant probability. We settle this question and present instances on which k-means++ achieves an approximation ratio of $(2/3 - \varepsilon) \cdot \log k$ only with exponentially small probability.

1 Introduction

In the *k-means problem* we are given a set of *data points* $X \subseteq \mathbb{R}^d$ and the objective is to group these points into k mutually disjoint *clusters* $C_1, \ldots, C_k \subseteq X$. Each of those clusters should contain only 'similar points' that are close together in terms of Euclidean distance. In order to evaluate the quality of a clustering, we assign a *cluster center* $c_i \in \mathbb{R}^d$ to each cluster C_i and consider the potential $\Phi = \sum_{i=1}^{k} \sum_{x \in C_i} \|x - c_i\|^2$. The goal of the k-means problem is to find clusters and cluster centers that minimize this potential.

Aloise et al. showed that the k-means problem is $\mathcal{NP}$-hard, even for $k = d = 2$ [2]. To deal with this problem in practice, several heuristics have been developed over the past decades. Probably the "most popular" one [5] is Lloyd's algorithm [7], usually called the *k-means method* or simply *k-means*. Starting with k arbitrary cluster centers, each data point is assigned to its nearest center. In the next step each center is recomputed as the center of mass of the points assigned to it. This procedure is repeated until the centers remain unchanged.

Though Vattani showed that the running time of k-means can be exponential in the number of input points [8], speed is one of the most important reasons for its popularity in practice. This unsatisfying gap between theory and practice was narrowed by Arthur, Manthey, and Röglin who showed that the running time of k-means is polynomially bounded in the model of smoothed analysis [3].

Another problem is that k-means may yield poor results if the initial centers are badly chosen. The approximation ratio can be arbitrarily large, even for small input sets and $k = 2$. To improve the quality of the solutions found by the k-means method, Arthur and Vassilvitskii proposed the following seeding technique called k-means++, which has an expected approximation ratio of $O(\log k)$ [4].

* A part of this work was done at Maastricht University and was supported by a Veni grant from the Netherlands Organisation for Scientific Research.

M. Ogihara and J. Tarui (Eds.): TAMC 2011, LNCS 6648, pp. 344–352, 2011.

1. Choose center c_1 uniformly at random from the input set X.
2. For $i = 2$ to k do:
   Let $D_i^2(x)$ be the square of the distance between point x and the nearest already chosen center $c_1, \ldots, c_{i-1}$. Choose the next center c_i randomly from X, where every $x \in X$ has a probability of $\frac{D_i^2(x)}{\sum_{y \in X} D_i^2(y)}$ of being chosen.

Previous work. In [4] instances are given on which k-means++ yields in expectation an $\Omega(\log k)$-approximation, showing that the bound of $O(\log k)$ for the approximation ratio of k-means++ is asymptotically tight. However, the expected approximation ratio of a heuristic is not the only useful quality criterion if the variance is large. If, for example, an $O(1)$-approximation is obtained with probability $1/\mathrm{poly}(k)$, then after a polynomial number of restarts an $O(1)$-approximation is reached with high probability even if the expected approximation ratio is $\Omega(\log k)$. Interestingly, k-means++ achieves a constant factor approximation for the instance given in [4] with constant probability, and no instance was known on which k-means++ does not yield an $O(1)$-approximation with constant probability.

For this reason Aggarwal, Deshpande, and Kannan called the lower bound of $\Omega(\log k)$ "misleading" [1]. In the same paper they showed that sampling $O(k)$ instead of k centers with the k-means++ seeding technique and selecting k good points among them yields an $O(1)$-approximation with constant probability. Unfortunately the selection step is done with LP-based algorithms, which makes this approach less simple and efficient than k-means++ in practice.

Therefore, both Arthur and Vassilvitskii [4] and Aggarwal, Deshpande, and Kannan [1] raise the question whether k-means++ yields an $O(1)$-approximation with constant probability. Aggarwal et al. call this a "tempting conjecture" which "would be nice to settle". So far the only known result in this direction is due to Arthur and Vassilvitskii who mention that the probability to achieve an $O(1)$-approximation is at least $c \cdot 2^{-k}$ for some constant $c > 0$ [4].

Our contribution We modify the instances given in [4] and show that it is very unlikely that k-means++ achieves an approximation ratio of $(2/3 - \varepsilon) \cdot \log k$ on this modified example.

Theorem 1. *Let* $r \colon \mathbb{N} \to \mathbb{R}^+$ *be a real function.*

1. *If* $r(k) = \delta^* \cdot \ln(k)$ *for a fixed real* $\delta^* \in (0, 2/3)$, *then there is a class of instances on which* **k-means++** *achieves an* $r(k)$-*approximation with probability at most* $\exp(-k^{1 - 3/2 \cdot \delta^* - o(1)})$.
2. *If* $r = o(\log k)$, *then there is a class of instances on which* **k-means++** *achieves an* $r(k)$-*approximation with probability at most* $\exp(-k^{1 - o(1)})$.

2   Construction and Analysis of a Bad Instance

2.1   Construction

Throughout the paper "log" denotes the natural logarithm. Let $r = r(k) > 0$ be a function where $r(k) = \delta^* \cdot \log k$ for a fixed real $\delta^* \in (0, 2/3)$ or $r = o(\log k)$.

Without loss of generality let $r(k) \to \infty$ in the latter case. Additionally, let $\delta = \delta(k) := r(k)/\log k$ be the ratio of $r(k)$ and $\log k$. Based on the function r, we introduce a parameter $\Delta = \Delta(k)$. In Section 2.3 we describe the details of this choice. In this section we present the instances used for proving Theorem 1, which are a slight modification of the instances given in [4].

We first choose k centers $c_1, \ldots, c_k$, each with squared distance $\Delta^2 - (k-1)/k$ to each other. For each point c_i we construct a regular $(k-1)$-simplex with center c_i and with side length 1. We denote the vertices of this simplex by $x_1^{(i)}, \ldots, x_k^{(i)}$, and we assume that the simplices for different points c_i and $c_{i'}$ are constructed in orthogonal dimensions. Then we get

$$\|x_j^{(i)} - c_i\|^2 = \frac{k-1}{2k}, \tag{1}$$

and for $x_j^{(i)} \neq x_{j'}^{(i')}$ we get

$$\|x_j^{(i)} - x_{j'}^{(i')}\|^2 = \begin{cases} 1 & : \quad i = i', \\ \Delta^2 & : \quad i \neq i', \end{cases} \tag{2}$$

due to the fact that for $i \neq i'$ the squared distance between $x_j^{(i)}$ and $x_{j'}^{(i')}$ is

$$\|x_j^{(i)} - x_{j'}^{(i')}\|^2 = \|x_j^{(i)} - c_i\|^2 + \|c_i - c_{i'}\|^2 + \|x_{j'}^{(i')} - c_{i'}\|^2 = \Delta^2$$

because of orthogonality and Equation (1). Let $C_i = \{x_1^{(i)}, \ldots, x_k^{(i)}\}$ for $i = 1, \ldots, k$. As input set for our k-means problem we consider the union $X = \bigcup_{i=1}^{k} C_i$ of these sets.

In the remainder we show that, with a good choice of Δ, X is a bad instance for k-means++. Note that the only relevant difference to the example given in [4] is the choice of Δ. While in [4] it was sufficient to choose Δ large enough, we have to tune Δ much more carefully to prove Theorem 1.

2.2   Reduction to a Markov Chain

We consider the k-clustering $C^* = (C_1, \ldots, C_k)$ induced by the centers $c_1, \ldots, c_k$. Note that for small Δ this might be a non-optimal solution, but its potential is an upper bound for the optimal potential. Due to Equation (1), the potential Φ^* of C^* is

$$\Phi^* = \sum_{i=1}^{k} \sum_{x \in C_i} \|x - c_i\|^2 = k^2 \cdot \frac{k-1}{2k} \leq \frac{k^2}{2}$$

as for any point of C_i the nearest center is c_i. Now let C' be a clustering with distinct centers $c'_1, \ldots, c'_t$, $1 \leq t \leq k$, chosen from X. For each center c'_i let l_i be the index of the set C_{l_i} that c'_i belongs to. Let $s := |\{l_1, \ldots, l_t\}|$ denote the number of *covered* sets C_i and let $X_u := X \setminus \bigcup_{i=1}^{t} C_{l_i}$ denote the set of the points of *uncovered* sets. Furthermore, let Φ denote the potential of X induced by the centers $c'_1, \ldots, c'_t$ and let $\Phi(X_u)$ be the part of Φ contributed by the uncovered sets. Applying Equations (1) and (2) we get $\Phi(X_u) = (k - s) \cdot k \cdot \Delta^2$ and $\Phi = (s \cdot k - t) \cdot 1^2 + \Phi(X_u) \geq (s - 1) \cdot k + \Phi(X_u)$.

The inequality $\frac{\Phi}{\Phi^*} \leq r$ is necessary for C' being an r-approximation. This implies

$$r \geq \frac{\Phi}{\Phi^*} \geq \frac{\Phi(X_u)}{\Phi^*} \geq \frac{2(k - s) \cdot \Delta^2}{k},$$

i.e. at least $s^* := \left\lceil k \cdot \left(1 - \frac{r}{2\Delta^2}\right)\right\rceil$ of the k sets C_i have to be *covered* to get an r-approximation.

Let us assume that we are in step 2 of k-means++ (see introduction) and let s denote the number of covered sets C_i. The probability of covering an uncovered set in this step is

$$\frac{\Phi(X_u)}{\Phi} \leq \frac{\Phi(X_u)}{(s - 1) \cdot k + \Phi(X_u)} = \frac{1}{1 + \frac{s - 1}{(k-s) \cdot \Delta^2}} =: p_s. \tag{3}$$

Hence, we can upper-bound the probability that k-means++ yields an r-approximation by the probability of reaching vertex v_{s^*} within k steps in the following Markov chain, starting from vertex v_0.

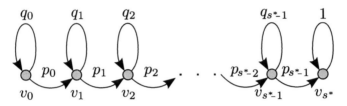

Here, p_s are the probabilities defined in Inequality (3), $p_0 = 1$ and $q_s = 1 - p_s$.

2.3   How to Choose Δ?

Arthur and Vassilvitskii [4] have shown that choosing Δ large enough results in instances on which k-means++ has an expected approximation ratio of $\Omega(\log k)$. This does not suffice for proving Theorem 1 because if we choose Δ too large, the probability that we do not cover every cluster becomes small. Hence, if we choose Δ too large, we have a good probability of covering every cluster and thus of obtaining a constant-factor approximation. On the other hand, if we choose Δ too small, already a single covered cluster might suffice to obtain a constant-factor approximation.

We first define a function $\varepsilon \colon \mathbb{N} \to (0, 1)$ as follows:

$$\varepsilon = \varepsilon(k) := \begin{cases} 1/3 & : & r = o(\log k), \\ \frac{2}{3} \cdot \frac{\log r}{r} & : & r = \delta^* \cdot \log k. \end{cases}$$

Now we set $\tilde{\Delta} = \tilde{\Delta}(k) := \sqrt{r} \cdot e^{r \cdot (1+\varepsilon)/4} = \exp(\Theta(r))$ and $\Delta := \lceil \tilde{\Delta} \rceil$. In the analysis of the Markov chain in the following section, we will assume that k is chosen sufficiently large such that the following inequalities hold:

$$\Delta^2 > r, \tag{4}$$

$$\frac{\Delta^2}{k} \leq \frac{r}{2}, \tag{5}$$

$$\left(\frac{r+2}{2}\right)^\Delta \geq \left(\frac{2\Delta^2}{r}\right)^2, \tag{6}$$

$$\Delta^6 \leq \frac{19}{18} r^3 \cdot e^{3r(1+\varepsilon)/2}, \tag{7}$$

$$k - 1 \geq \left(1 - \frac{\varepsilon}{9}\right) \cdot k, \tag{8}$$

$$r + 2 \leq \left(1 + \frac{\varepsilon}{3}\right) \cdot r, \tag{9}$$

$$\frac{r}{2\Delta^2} + \frac{\varepsilon}{3} \cdot \left(1 + \frac{\varepsilon}{3}\right) \cdot \left(\frac{r}{2\Delta^2}\right)^2 \leq \left(\frac{\varepsilon}{3}\right)^2, \tag{10}$$

$$\log r \leq \frac{3}{2}\varepsilon \cdot r. \tag{11}$$

In the appendix we show that, for our choice of ε, Inequalities (4) to (11) are satisfied for every sufficiently large k. We have not made any attempt to simplify these inequalities as they appear in exactly this form in the analysis in the next section.

2.4   Analysis of the Markov Chain

Now we concentrate on bounding the probability to reach vertex v_{s^*} in the Markov chain above. For this we introduce geometrically distributed random variables $X_0, \ldots, X_{s^*-1}$. Variable X_s describes the number of trials that are required to move from vertex v_s to vertex v_{s+1}. We would like to show that the expected value of $X := \sum_{s=0}^{s^*-1} X_s$ is much greater than k and then conclude that it is unlikely to reach v_{s^*} within k steps. Unfortunately, Hoeffding's Inequality [6] which is often used for drawing such a conclusion requires random variables with bounded domain. So we make a technical detour by introducing additional random variables $Y_s := \min\{X_s, \Delta\}$, $s = 0, \ldots, s^* - 1$, and $Y := \sum_{s=0}^{s^*-1} Y_s$. We will see that the differences caused by truncating the variables X_s are negligible for our purpose.

The expected value of X_s is $1/p_s$, the expected value of Y_s is $(1 - q_s^\Delta)/p_s$ (see Appendix A). If we express p_s as $p_s = \frac{1}{1 + \frac{1}{z_s}}$ for $z_s = \frac{(k-s) \cdot \Delta^2}{s-1}$, then

$$1 - \frac{\mathbf{E}[Y_s]}{\mathbf{E}[X_s]} = q_s^\Delta = (1 - p_s)^\Delta = \left(1 - \frac{1}{1 + \frac{1}{z_s}}\right)^\Delta = \left(\frac{1}{z_s + 1}\right)^\Delta.$$

As z_s is decreasing with s and $s \le s^* - 1 \le k \cdot \left(1 - \frac{r}{2\Delta^2}\right)$, we can bound z_s for $s \ge 1$ by

$$z_s \ge \frac{\left(k - k \cdot \left(1 - \frac{r}{2\Delta^2}\right)\right) \cdot \Delta^2}{k \cdot \left(1 - \frac{r}{2\Delta^2}\right) - 1} = \frac{\frac{r}{2}}{1 - \frac{r}{2\Delta^2} - \frac{1}{k}} \ge \frac{r}{2} .$$

The non-negativity of the second last denominator follows from Inequalities (4) and (5). By applying Inequality (6), we get

$$\frac{\mathbf{E}[Y_s]}{\mathbf{E}[X_s]} = 1 - \left(\frac{1}{z_s + 1}\right)^{\Delta} \ge 1 - \left(\frac{1}{\frac{r}{2} + 1}\right)^{\Delta} \ge 1 - \left(\frac{r}{2\Delta^2}\right)^2 . \tag{12}$$

Due to Inequality (12) a lower bound for $\mathbf{E}[X]$ implies a lower bound for $\mathbf{E}[Y]$. The former one can be bounded as follows.

$$\mathbf{E}[X] = \sum_{s=0}^{s^*-1} \mathbf{E}[X_s] = \sum_{s=0}^{s^*-1} \frac{1}{p_s} = 1 + \sum_{s=1}^{s^*-1} \left(1 + \frac{s-1}{(k-s) \cdot \Delta^2}\right)$$

$$= s^* + \sum_{i=k-s^*+1}^{k-1} \frac{k-i-1}{i \cdot \Delta^2} = s^* - \frac{s^*-1}{\Delta^2} + \frac{k-1}{\Delta^2} \cdot \sum_{i=k-s^*+1}^{k-1} \frac{1}{i}$$

$$\ge s^* \cdot \left(1 - \frac{1}{\Delta^2}\right) + \frac{k-1}{\Delta^2} \cdot \log\left(\frac{k}{k-s^*+1}\right) .$$

Using $s^* \ge k \cdot \left(1 - \frac{r}{2\Delta^2}\right)$, we can lower bound this by

$$\mathbf{E}[X] \ge k \cdot \left(1 - \frac{r}{2\Delta^2}\right) \cdot \left(1 - \frac{1}{\Delta^2}\right) + \frac{k-1}{\Delta^2} \cdot \log\left(\frac{k}{k - k \cdot \left(1 - \frac{r}{2\Delta^2}\right) + 1}\right)$$

$$\ge k \cdot \left[\left(1 - \frac{r+2}{2\Delta^2}\right) + \frac{k-1}{k\Delta^2} \cdot \log\left(\frac{\Delta^2}{\frac{r}{2} + \frac{\Delta^2}{k}}\right)\right] .$$

Inequalities (5), (8), (9) and the choice of Δ yield

$$\frac{\mathbf{E}[X]}{k} \ge 1 - \frac{\left(1 + \frac{\varepsilon}{3}\right) \cdot r}{2\Delta^2} + \frac{1 - \frac{\varepsilon}{9}}{\Delta^2} \cdot \log\left(\frac{\Delta^2}{r}\right)$$

$$\ge 1 - \frac{\left(1 + \frac{\varepsilon}{3}\right) \cdot r}{2\Delta^2} + \frac{1 - \frac{\varepsilon}{9}}{\Delta^2} \cdot r \cdot \frac{1 + \varepsilon}{2}$$

$$\ge 1 + \frac{r}{2\Delta^2} \cdot \frac{\varepsilon}{3} \cdot \left(1 + \frac{\varepsilon}{3}\right) ,$$

where the last inequality holds because $\varepsilon \in (0, 1)$. Applying Inequality (12), we can show that even the expected value of Y is significantly larger than k.

$$\frac{\mathbf{E}[Y]}{k} \ge \left(1 - \left(\frac{r}{2\Delta^2}\right)^2\right) \cdot \frac{\mathbf{E}[X]}{k} \ge \left(1 - \left(\frac{r}{2\Delta^2}\right)^2\right) \cdot \left(1 + \frac{r}{2\Delta^2} \cdot \frac{\varepsilon}{3} \cdot \left(1 + \frac{\varepsilon}{3}\right)\right)$$

$$= 1 + \frac{r}{2\Delta^2} \cdot \left(\frac{\varepsilon}{3} \cdot \left(1 + \frac{\varepsilon}{3}\right) - \frac{r}{2\Delta^2} - \frac{\varepsilon}{3} \cdot \left(1 + \frac{\varepsilon}{3}\right) \cdot \left(\frac{r}{2\Delta^2}\right)^2\right)$$

$$= 1 + \frac{r}{2\Delta^2} \cdot \left(\frac{\varepsilon}{3} + \left(\frac{\varepsilon}{3}\right)^2 - \left(\frac{r}{2\Delta^2} + \frac{\varepsilon}{3} \cdot \left(1 + \frac{\varepsilon}{3}\right) \cdot \left(\frac{r}{2\Delta^2}\right)^2\right)\right) .$$

Hence, we get $\mathbf{E}[Y] \geq k \cdot (1 + \frac{r}{2\Delta^2} \cdot \frac{\varepsilon}{3}) = k + k \cdot f$ for $f = f(k) = \frac{\varepsilon r}{6\Delta^2}$ because of Inequality (10). Using Hoeffding's Inequality [6], we can now bound the probability to reach vertex v_{s^*} within k steps in the Markov chain above.

$$\mathbf{Pr}[X \leq k] \leq \mathbf{Pr}[Y \leq k] \leq \mathbf{Pr}[\mathbf{E}[Y] - Y \geq k \cdot f] \leq \exp\left(-\frac{2 \cdot (k \cdot f)^2}{s^* \cdot \Delta^2}\right)$$

$$\leq \exp\left(-\frac{2k^2 f^2}{k \cdot \Delta^2}\right) = \exp\left(-k \cdot \frac{2f^2}{\Delta^2}\right).$$

Because of Inequalities (7) and (11) we can bound the fraction $2f^2/\Delta^2$ by

$$\frac{2f^2}{\Delta^2} = \frac{\varepsilon^2 r^2}{18\Delta^6} \geq \frac{\varepsilon^2 r^2}{19r^3 \cdot e^{3r(1+\varepsilon)/2}} = \frac{\varepsilon^2}{19} \cdot \frac{1}{e^{3r(1+\varepsilon)/2+\log r}} \geq \frac{\varepsilon^2}{19} \cdot \frac{1}{e^{(3/2+3\varepsilon)\cdot r}}$$

$$= \frac{\varepsilon^2}{19} \cdot \frac{1}{e^{(3/2+3\varepsilon)\cdot\delta\cdot\log k}} = \frac{\varepsilon^2}{19} \cdot k^{-(3/2+3\varepsilon)\cdot\delta}.$$

If $r = o(\log k)$, then $\delta \in o(1)$ and $\mathbf{Pr}[X \leq k] \leq \exp\left(-k^{1-o(1)}\right)$. If $r = \delta^* \cdot \log k$ for some fixed real $\delta^* \in (0, 2/3)$, then we get

$$\mathbf{Pr}[X \leq k] \leq \exp\left(-k^{-o(1)} \cdot k^{1-\left(\frac{3}{2}+\frac{2\log r}{r}\right)\cdot\delta^*}\right)$$

$$= \exp\left(-k^{1-\frac{3}{2}\delta^*-o(1)} \cdot k^{-\frac{2\log r}{\log k}}\right)$$

$$= \exp\left(-k^{1-\frac{3}{2}\delta^*-o(1)}\right).$$

This concludes the proof of Theorem 1.

3 Conclusion

We proved that, in general, k-means++ yields an $o(\log k)$-approximation only with negligible probability. The proof of this result is based on instances with fairly high dimension. Since we constructed the simplices in orthogonal dimensions, our instances have dimension $\Theta(k^2)$.

It remains open how k-means++ behaves on instances in small dimensions. One intriguing question is whether there exists an upper bound for the expected approximation ratio of k-means++ that depends only on the dimension of the instance. Currently we cannot exclude the possibility that the expected approximation ratio of k-means++ is $O(\log d)$ where d is the dimension of the instance.

References

1. Aggarwal, A., Deshpande, A., Kannan, R.: Adaptive sampling for k-means clustering. In: Dinur, I., Jansen, K., Naor, J., Rolim, J. (eds.) APPROX 2009. LNCS, vol. 5687, pp. 15–28. Springer, Heidelberg (2009)

2. Aloise, D., Deshpande, A., Hansen, P., Popat, P.: NP-hardness of Euclidean sum-of-squares clustering. Machine Learning 75(2), 245–248 (2009)
3. Arthur, D., Manthey, B., Röglin, H.: k-means has polynomial smoothed complexity. In: Proc. of the 50th Annual IEEE Symposium on Foundations of Computer Science (FOCS), pp. 405–414 (2009)
4. Arthur, D., Vassilvitskii, S.: k-means++: The advantages of careful seeding. In: Proc. of the 18th Annual ACM-SIAM Symposium on Discrete Algorithms (SODA), pp. 1027–1035. SIAM, Philadelphia (2007)
5. Berkhin, P.: Survey of Clustering Data Mining Techniques. Technical report, Accrue Software (2002)
6. Hoeffding, W.: Probability inequalities for sums of bounded random variables. Journal of the American Statistical Association 58(301), 13–30 (1963)
7. Lloyd, S.P.: Least squares quantization in PCM. IEEE Transactions on Information Theory 28(2), 129–136 (1982)
8. Vattani, A.: k-means requires exponentially many iterations even in the plane. In: Proc. of the 25th ACM Symposium on Computational Geometry (SoCG), pp. 324–332. ACM, New York (2009)

A   The Expected Value of Truncated Geometrically Distributed Random Variables

Let X be a geometrically distributed random variable with parameter p, let $q := 1 - p$ and let M be a non-negative integer. The expected value of the truncated random variable $Y := \min\{X, M\}$ is

$$\mathbf{E}[Y] = \sum_{i=1}^{\infty} \min\{i, M\} \cdot p \cdot q^{i-1} = \sum_{i=1}^{\infty} i \cdot p \cdot q^{i-1} - \sum_{i=M+1}^{\infty} (i - M) \cdot p \cdot q^{i-1}$$

$$= \mathbf{E}[X] - q^M \cdot \sum_{i=1}^{\infty} i \cdot p \cdot q^{i-1} = \left(1 - q^M\right) \cdot \mathbf{E}[X] = \frac{1 - q^M}{p}.$$

B   Inequalities (4) to (11)

Throughout this section any inequality $f(k) \leq g(k)$ is a short hand for $f(k) \leq g(k)$ for sufficiently large k. First note that regardless of the choice of r the inequalities $r \leq \frac{2}{3}\log k$ and $\frac{2}{3}(\log r)/r \leq \varepsilon \leq 1$ hold. The latter one immediately implies Inequality (11).

- Inequality (4) follows from $\Delta^2 \geq \tilde{\Delta}^2 \geq r \cdot \exp\left(\frac{r}{2}\right)$ which is greater than r because $r > 0$.
- As $\tilde{\Delta} \leq \sqrt{r} \cdot \exp(r/2) \leq \sqrt{r} \cdot \exp\left((\log k)/3\right) = \sqrt{r} \cdot \sqrt[3]{k} \leq \sqrt{r} \cdot \sqrt{k/2} - 1$, we get Inequality (5): $\Delta^2 \leq (\tilde{\Delta} + 1)^2 \leq r \cdot k/2$.
- Due to the fact $\Delta \to \infty$ we get $2^\Delta \geq \Delta^4$. The inequalities $(r+2)/2 \geq 2$ and $2\Delta^2/r \leq \Delta^2$ then imply the correctness of Inequality (6).

- Inequality (7) is a consequence of $\tilde{\Delta} \to \infty$. This yields $\tilde{\Delta} + 1 \leq \sqrt[6]{19/18} \cdot \tilde{\Delta}$ and hence $\Delta^6 \leq (\tilde{\Delta} + 1)^6 \leq \frac{19}{18} \tilde{\Delta}^6 = \frac{19}{18} r^3 \cdot \exp(3r \cdot (1 + \varepsilon)/2)$.

- Inequalities (8) and (9) hold if $\varepsilon \geq 9/k$ and $\varepsilon \geq 6/r$. This is true since $\varepsilon = \Omega\left((\log r)/r\right)$, $k = \exp(\Omega\left(r\right))$ and $1/r = O(1/r)$.

- Let us consider Inequality (10). As $\Delta^2 > r$ (see Inequality (4)), $r/(2\Delta^2) + \varepsilon/3 \cdot (1 + \varepsilon/3) \cdot (r/(2\Delta^2))^2 \leq r/(2\Delta^2) + 4/9 \cdot r/(2\Delta^2) \leq r/\Delta^2$. The correctness follows from $\Delta^2 \geq r \cdot \exp(r/2)$, i.e. $r/\Delta^2 \leq 1/\exp(r/2)$, whereas $(\varepsilon/3)^2 = \Omega\left(1/r^2\right)$.

Catching a Fast Robber on Interval Graphs*

Tomáš Gavenčiak

Department of Applied Mathematics,
Charles University, Prague
gavento@kam.mff.cuni.cz

Abstract. We analyse the Cops and ∞-fast Robber game on the class of interval graphs and show it to be polynomially decidable on such graphs. This solves an open problem posed in paper "Pursuing a fast robber on a graph" by Fomin et al. [4] The game is known to be already NP-hard on chordal graphs and split-graphs.

The game is played by two players, one controlling k cops, the other a robber. The players alternate in turns, all the cops move at once to distance at most one, the robber moves along any cop-free path. Cops win by capturing the robber, the robber by avoiding capture.

The analysis relies on the properties of an interval representation of the graph. We show that while the game-state graph is generally exponential, every cops' move can be decomposed into simple moves of three types, and the states are reduced to those defined by certain cuts of the interval representation. This gives a restricted game equivalent to the original one together with a winning strategy computable in polynomial time.

Keywords: cop and robber game, pursuit game, combinatorial game, interval graph, interval graph representation.

1 Introduction

The recent development in the area of combinatorial "Cop and Robber" games (also called pursuit-evasion games) includes results on games with various speeds of the players and their computational complexity.

The Cops and s-fast Robber game is a generalisation of the original Cops and Robber game introduced by Nowakowski and Winkler [7] and by Quilliot [8] allowing the Robber to make s steps instead of 1.

In their paper "Pursuing a fast robber on a graph" [4], Fomin et al. propose the complexity of the ∞-fast Robber game on interval graphs as an open question which we answer. The proof is constructive and shows how to decompose seemingly complicated moves of the cops into simple basic moves.

---
* This work was partially supported by Charles University grant GAUK 64110. I would like to thank Jan Kratochvíl, Andrzej Proskurowski and Peter Golovach for insightful remarks.

M. Ogihara and J. Tarui (Eds.): TAMC 2011, LNCS 6648, pp. 353–364, 2011.
© Springer-Verlag Berlin Heidelberg 2011

Founding the boundary graph classes and restrictions for which the games are NP-hard or polynomially decidable shows which are the aspects of the problems that make it easy or hard, providing further insight into the problem.

In this paper we draw the line between interval graphs and split and chordal graphs. Our result is not only an example of a game reduction technique, but also indicates that the hardness of the problem on chordal graphs lies in their unbounded asteroidal number (and not in i.e. unbounded clique size or diameter).

See Section 7 for further discussion.

The **Cops and s-fast Robber game** is defined as follows: The game is played by k cops controlled by one player and one robber controlled by the second player on a given simple undirected graph G. The cops and the robber are positioned on vertices of G at all times, more cops may share a vertex. Both players have a complete information about G and the game state.

First, the cops choose starting vertices, then the robber chooses a starting vertex. One turn then consists of each of the cops moving to distance at most 1, and then by the robber moving along a cop-free path of length at most s. The robber may never move through a vertex occupied by a cop. Note that this restricts the robber to one component of the cop-free subgraph of G.

Should a cop be at or move to the robber's vertex, the cops immediately win. The robber wins by avoiding the capture indefinitely.

This game is equivalent to the original Cop and Robber game for $s = 1$.

Some of the complexity aspects of Cops and s-fast Robber games are examined by Fomin et al. [5,4]. They show that for all s, these games are NP-hard (and even W[2]-hard in the version parametrised by k) to decide even on chordal graphs while still being polynomially decidable on interval graphs.

The Cops and ∞-fast Robber game is the limit of the sequence of Cop and s-fast Robber game, but is an interesting game on its own. The hardness results of Fomin et al. easily extend to this game but the polynomiality proofs fail.

One of the main ideas of this paper is that any cops' winning strategy on an interval graph *essentially* consists only of *sweeping* and *splitting* moves on the interval representation of the graph. We use a formal version of this statement to show that the game can be decided in polynomial time.

Theorem 1. *There is a polynomial-time algorithm deciding a Cop and ∞-fast Robber game with k cops on a given graph G.*

2   Preliminaries

In this paper, we use "the" standard graph- and game-theoretic notation. For introduction to these areas, we recommend the books Modern Graph Theory [2] and Lessons in Play: An Introduction to Combinatorial Game Theory [1].

We use $N(v)$ and $N[V]$ to denote the closed neighbourhood (including v, resp. V) of a vertex or a set of vertices, respectively. We briefly mention some of the less-known graph classes and their properties:

A graph G is *chordal* (also called *bridged*) if there are no induced cycles of length at least 4. A graph is a *split graph* if its vertices can be partitioned into two sets I and K, I an independent set and K a complete subgraph. Every split graph is also chordal.

A graph is an *interval graph* if it can be realised as an intersection graph of a family of intervals on a real line. For a family of intervals $\mathcal{I}$, the associated intersection graph $G(\mathcal{I})$ has one vertex for each of the intervals and an edge between the vertices corresponding to intervals I_1 and I_2 if and only if $I_1 \cap I_2 \neq \emptyset$. Every interval graph is chordal.

Throughout the article we assume that every interval graph comes with a fixed interval representation consisting of *open* intervals with integral endpoints $1, 2, \ldots 2|V_G|$ and such that no two intervals share an endpoint. Note that such an interval representation can be reconstructed from G in linear time, as shown by Korte and Mhring [6].

We identify the intervals of the chosen representation with the vertices, interchanging these frequently.

In the rest of the paper, we fix G to be a connected interval graph with a given interval representation, V be the set of vertices of G and E be the set of its edges. We also fix k to be the given number of cops.

Note that in a disconnected graph, the cops have to decide on a distribution among the components and the robber then chooses a component to play in. Later in game only the cops in that component matter. Assuming we can compute the minimum number of cops necessary for each separate connected component in polynomial time, we can just sum up these numbers to get the minimum number of cops that can capture the robber in the entire graph.

We slightly modify the game: all the cops must start at the leftmost interval of G. This is equivalent to the game defined above, since the cops can first move to any desired position (ignoring the robber) and then pretend that the game started from that position.

Formally we call the game state before cops' move $\mathcal{C}(C, r)$, where C is a multiset of vertices occupied by the cops and r is the vertex the robber moved to. The game state before robber's move is $\mathcal{R}(C, A)$ with C as above and A is the set of all vertices the robber may move to (the connected component of $G - C$ containing the robber).

This slightly reduces the complexity of the examined states. Note that before any robber's turn, two states with robber in the same component of $G - C$ offer the same moves to the robber. Also note that A is always connected including the first round.

The initial state of the game is $\mathcal{R}(C, V - C)$, where C is the multiset containing k times the leftmost vertex of G. We also add a special state $\mathcal{WIN}$, which denotes the capture of the robber.

3   Specifics of the Game

In this section, we introduce the definitions necessary for the game strategy reduction. These allow a more precise look at the game.

The main idea of this paper is to *simulate* arbitrarily complex cops' winning strategy by a simplified, *restricted* strategy using only three kinds of actions

– *sweep, split* and *endgame*. The definitions of the actions are provided below together with a formal definition of a restricted strategy.

Informally, an action is a predefined part of cops' strategy starting at a distinguished game state and ending at another distinguished game state. A restricted cops' strategy may only decide the next action in such distinguished game state.

Before defining the restricted actions, let us introduce some useful concepts and make few observations.

Proposition 1. *At any point in the game, the cops restrict robber's movement to an interval, the robber may move to arbitrary vertex of this interval except those occupied by the cops.*

The union of vertices (as intervals) reachable by the robber is obviously a single interval (this interval does not generally correspond to a vertex). We call this interval the *playground*. Any cop-free vertex inside or incident with the playground is adjacent to a vertex accessible to the robber.

The left and right endpoints of the playground are called the left and right *barriers*. We denote the playground with barriers L and R, $L < R$ by (L, R). Note that L and R are not contained in the playground (as all the intervals are open) and all intervals incident with these barriers are be occupied by the cops (otherwise the playground would be bigger).

The vertices incident to the barrier are called the *barrier's support*, denoted $V(L)$. Note that the support of either barrier may be empty. The vertices contained *inside* a playground (L, R) are called the *playground support*, denoted $V(L, R)$. Support of a playground is always disjoint from the supports of the barriers.

From the cops occupying a barrier support, choose and fix one cop per vertex. Let us call these cops the cops *holding* the barrier. The choice and fixing of the cops is is mostly symbolic since the cops are indistinguishable, but is useful in some proofs. Note that a cop may hold both barriers at once, but as we will see, that this may happen only just before capturing the robber.

A playground (L, R) is *feasible*, if $|V(L) \cup V(R)| \le k$, that is if the cops are able to hold both barriers at once. For every *nontrivial* feasible playground (L, R) (nonempty and not containing all the vertices), we fix a *canonical* game state

$$\S(L, R) = \mathcal{R}(V(L) \cup V(R), V(L, R))$$

with all extra cops not explicitly holding any vertices positioned at the leftmost vertex outside $V(L, R)$. A game state corresponding to an empty playground would be $\mathcal{WIN}$, as the cops must have just caught the robber.

There is a minor technicality – there is no game state corresponding to the *full* playground containing all the vertices, as there is no way to place the cops and get the playground to be the entire graph. We fix the state canonical for $(-\infty, \infty)$ to be the state $\mathcal{R}(\{v_0\}, V - v_0)$ with v_0 the left-most vertex. This makes the playground $(-\infty, \infty)$ always feasible.

Proposition 2. *If (L, R) is a feasible playground then the playground corresponding to its canonical game state is the same, or a subset in case of (L, R)*

full playground. As the full playground occurs only in the first move, we usually omit that case.

The cops in position C *threaten* to take a vertex set B if the cops can occupy every vertex of B after one move. This is equivalent to an existence of a matching of C (as a multiset) with all vertices of B.

In our strategy, B is usually a barrier and there may be additional explicit conditions – usually, some other vertices must be held both before and after the move. When considering a set of cops threatening B, we fix a matching between the threatening cops and the vertices of B for the moment.

Now we are ready to define cops' restricted strategy. While a general cops' strategy is mapping from *every* valid state of the game to a move valid in that state, a *restricted cop's strategy* is a mapping from the robber states canonical to some feasible playground to actions valid in that state. Therefore, the restricted strategy can influence the game only after an action has finished (which it always does either in a state canonical to some playground or in $\mathcal{WIN}$).

4   Essential Cops' Moves

Formally, a *restricted cops' strategy* is a cops' strategy that, in a canonical game state, may only choose one valid action or pass. The chosen action is then played as a sub-strategy. After the action finishes, the strategy may choose another action.

An *action* is a sequence of moves starting at a canonical state and ending in one of defined canonical states or in $\mathcal{WIN}$. An action is *valid* if it can be played out by k cops.

The actions are ordered by preference. If an *endgame* move is valid from some playground, then we do not consider the possibility of other moves. Also, if a *sweep* from (L, R) to (L', R) is valid, we consider no *splits* to (L', R) or (L, R') for any R', and symmetrically for *sweep* to (L, R'). This saves us few cases in the technical parts of the proofs.

Endgame from §(L, R). From the state §(L, R), the cops assume a position C holding both L and R. C dominates $V(L, R)$. In the next move, the cops capture the robber. Such a multiset C with $|C| \leq k$ is a *witness* of the validity of the action.

Sweep from §(L, R) **to** §(L', R) (or to §(L, R')). From the state §(L, R), the cops assume a position C holding L and R. In C they threaten barrier L' (or R') while holding R (or L). If $L < L'$ (or $R' < R$) then C also dominates $V(L, L')$ (or

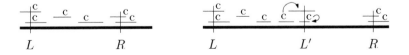

Fig. 1. Illustrations of *endgame* and *sweep* actions

Fig. 2. Split action with $R' < L'$ and $R' \geq L'$

$V(R', R))$. Then the cops in one move stop holding L (or R) and either capture the robber or take and start holding L' (or R') and move to the state canonical for the goal state.

The action is valid if such set C with $|C| \leq k$ exists and an *endgame* from (L, R) is not possible. After the action, the robber is either in the new playground or captured, as (L, L') (or (R', R)) was dominated.

Split from $\S(L, R)$ to either $\S(L', R)$ or to $\S(L, R')$. From the state $\S(L, R)$, the cops assume a position C holding both L and R. In C they threaten barrier L' while holding R *and* threaten R' while holding L. If $R' < L'$ then C also dominates $V(R', L')$.

After moving to C, the cops let the robber move and decide according to robber's position r. If r is adjacent to a cop, they capture the robber. If $r \in V(L, R')$, the cops stop holding R and take and start holding R'. Otherwise, the cops stop holding L and take and start holding L'. If the cops did not capture the robber, they then assume the state canonical for either $\S(L, R')$ or $\S(L', R)$.

The action is valid only if such witness set C with $|C| \leq k$ exists and neither *endgame* from $\S(L, R)$ nor *sweep* from $\S(L, R)$ to one of $\S(L, R')$, $\S(L', R)$ is possible. This last condition is added only for simplicity and could be dropped.

The existence of a witness of any of the action is decidable in polynomial time. The proof of the lemma below is somewhat technical and we present it to Section 6. The algorithms are mostly straightforward modifications of the well-known greedy minimum dominating set algorithms for interval graphs.

Lemma 1. *There are polynomial-time algorithms deciding the validity of actions endgame, sweep and split.*

This allows us to decide the existence of a cops' restricted strategy.

Theorem 2. *There is a polynomial time algorithm that, given an interval graph G and k, decides the existence of a winning restricted strategy for k cops.*
The length of a winning strategy (measured in game turns) is $O(|V_G|^3)$.

Proof. We construct a game-state digraph D representing the game with cops using restricted strategy.

The *cop-states* of D are all feasible playgrounds and $\mathcal{WIN}$. The empty playground is the initial cop-state, $\mathcal{WIN}$ is the cop-win state.

The *robber-states* of D are quadruples (L, L', R', R) of barriers such that *split* from (L, R) to either (L, R') or (L', R) is a valid action.

For every valid *endgame* from $\S(L, R)$, there is a move from (L, R) to $\mathcal{WIN}$. For every valid *sweep* from $\S(L, R)$ to $\S(L', R')$, there is a cop-move from (L, R) to

(L', R'). For every valid *split* from $\S(L, R)$ to either $\S(L, R')$ or $\S(L', R)$, there is a cop-move from (L, R) to (L, R, L', R') and two robber-moves from (L, R, L', R') to (L, R') and (L', R).

We decide the game given by D using a general state-marking algorithm, giving us either a winning strategy for the cops or a non-losing strategy for the robber.

If we get a cops' winning strategy in D, it is also a restricted cops' strategy for the general game, where the cops play the original action of the move dictated by the strategy.

On the other hand, the moves of any restricted cops' strategy are present in the game-state digraph. Both cops' restricted strategy and the game encoded by D ignore the position of the robber in the playground except for the final move in *split* (that corresponds to the robber choosing one of the playgrounds). The game of D does not allow the robber to get captured prematurely, but that possibility is ruled out by assuming the robber to play optimally.

The game digraph D has $O(n^4)$ states and moves. The feasibility of every state and the validity of every move can be decided in polynomial time according to Lemma 1. The general combinatorial game-decision algorithm then runs in time polynomial in the size of D.

To see the maximum length of a strategy, note that an optimal cops' strategy visits every state at most once, there are $O(|V_G|^2)$ cop-states and every robber-state is followed by a cop-state. It is easy to see that playing out any single action takes $O(n)$ moves.   □

5   Simulating General Cops' Strategy

In this section we prove the following theorem.

Theorem 3. *For an interval graph G and an integer k, k cops have a winning strategy for the Cops and ∞-fast Robber game if and only if k cops have a restricted winning strategy.*

With this, we may immediately prove Theorem 1.

Proof of Theorem 1. According to Theorem 2, we can decide the existence of a winning restricted cops' strategy in polynomial time. By Theorem 3, such strategy exists if and only if a general winning cops' strategy exists.   □

Proof of Theorem 3. The "if" part is straightforward, as the cops can play out the actions of a restricted winning strategy. The action properties ensure that the actions are possible and that the robber is inside the respective playground or captured.

For the other direction, let $\mathcal{S}$ be an optimal (in the terms of length) cop's winning strategy for k cops. We may assume that the strategy wins from any state, as it may move the cops to the state canonical to the full playground and then play from there. Note that if the cops play according to $\mathcal{S}$, the game will never revisit a game state.

Let S be the subgraph of the game state digraph representing the strategy. The vertices are all the cop- and robber-states of the game. From each cop-vertex, there is exactly one cop-move as dictated by S. From each robber-vertex, all the robber-moves are present. Note that S is acyclic.

Fix any total ordering o of the states of S extending the partial order given by the moves.

The restricted cops' winning strategy is obtained using the following lemma:

Lemma 2. *Given any robber state $\mathcal{R}(C, A)$ of S, let $\mathcal{R}(C_i, A_i)$ be the robber states of S reachable from $\mathcal{R}(C, A)$ in two moves (a robber move and a cop move). Let $P = \S(L, R)$ be a playground corresponding to $\mathcal{R}(C, A)$ and $P_i = \S(L_i, R_i)$ be the playgrounds corresponding to $\mathcal{R}(C_i, A_i)$.*

There is a restricted cops' strategy that starts in the cop-state canonical to $\S(L, R)$ and moves a cop-state canonical to one of $\S(L_i, R_i)$ in $O(n)$ actions, or wins.

The desired restricted strategy $\mathcal{T}$ is as follows:

For a cop-state canonical to $\S(L, R)$, find the latest (w.r.t. o) robber-state $\mathcal{R}(C, B)$ with playground $\S(L, R)$. Let $\mathcal{C}(C, r_i)$ be all the possible following cop-states and let $\mathcal{R}(C_i, R_i)$ and $\S(L_i, R_i)$ be the resulting states of the cop-moves dictated by S and the corresponding playgrounds.

The restricted strategy $\mathcal{T}$ should play the actions given by Lemma 2. This leaves the game in a state canonical to one of $\S(L_i, R_i)$ or won for the cops. Note that all $\S(L_i, R_i)$ are different from $\S(L, R)$, because the state $\mathcal{R}(C, B)$ is latest such state.

Now we have that the latest occurrence of a robber-state with playground $\S(L_i, R_i)$ is (w.t.r. o) larger than that of $\S(L, R)$. Therefore, by playing $\mathcal{T}$, the latest state (w.r.t. o) with the same playground as the current one only increases.

This proves that $\mathcal{T}$ is acyclic and therefore winning for the cops.   □

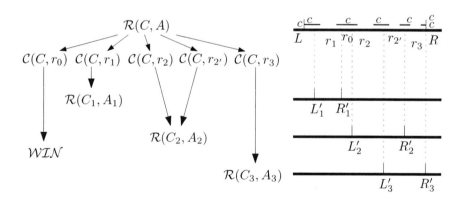

Fig. 3. An illustration of one robber state $\mathcal{R}(C, R)$ of S with some of the robber moves r_i followed by the cop moves dictated by S, and the corresponding playgrounds

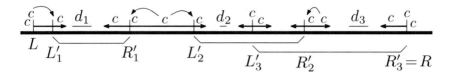

Fig. 4. An illustration of the playgrounds of $\mathcal{P}$. Note that the playgrounds may overlap. This also illustrates how the d_j separate the groups of cops of C.

Proof of Lemma 2. Let D be the non-dominated vertices of P, that is, $D = P \setminus N[C]$. If $D = \emptyset$, then C is a witness for *endgame* and the strategy may immediately win.

Otherwise, let $\mathcal{P}$ be an inclusion-minimal sub-collection of playgrounds covering D, indexed by $j \in \{1\ldots\}$ from left to right. Denote these playgrounds $P'_j = \S(L'_j, R'_j)$ and the corresponding states $\mathcal{R}(C'_j, B'_j)$. For every P'_j, there is some vertex d_j of D covered only by P'_j, otherwise $\mathcal{P}$ is not minimal.

The cops of C threaten every individual barrier L'_j and R'_j. As every P'_j contains an non-dominated vertex, the cops threatening L'_j and R'_j are disjoint for every j.

The strategy construction is algorithmic and reduces $\mathcal{P}$ and modifies C on the way. However, multiset C is modified only the left of d_1 and to the right of $d_{|\mathcal{P}|}$. After the first two actions, the playground is always $\S(L'_1, R'_{|\mathcal{P}|})$.

The strategy starts with a *sweep* to $\S(L'_1, R)$ followed by a *sweep* from $\S(L'_1, R)$ to $\S(L'_1, R'_{|\mathcal{P}|})$. The first *sweep* is valid as the cops of C to the left of d_1 together with $V(R)$ witness. The validity of the second *sweep* is witnessed by the cops of C to the right of $d_{|\mathcal{P}|}$ together with $V(L'_1)$. This changes the position

If there is only one playground $\S(L'_1, R'_1)$ in $\mathcal{P}$, we have that the current playground is $\S(L'_1, R'_1)$ and we are done.

If there are multiple playgrounds in $\mathcal{P}$, let the strategy play a *split* to either $\S(L'_1, R'_1)$ or $\S(L'_2, R'_{|\mathcal{P}|})$. The *split* is witnessed by C, as the original C threatened both $V(R'_1)$ and $V(L'_2)$, dominating $V(R'_1, L'_2)$ if $R'_1 < L'_2$. Note that the cops of C between d_1 and d_2 are unmodified.

This either moves the game to state canonical to $\S(L'_1, R'_1)$, or reduces $\mathcal{P}$. Multiset C is modified by taking $V(L'_2)$, but (as this was possible in the original C) this may be done only using the cops to the left of d_2 and right of d_1. In every step, we renumber the elements of $\mathcal{P}$ from 1 up.

Repeating this step, the strategy either wins or moves to the state canonical to some $\S(L'_j, R'_j)$. □

6   Complexity of Finding Witness

In this section we prove the complexity of the three decision problems in Lemma 5, splitting it into the following three lemmas. The statement of Lemma 5 then directly follows.

The algorithms are not given explicitly, but follow straightforwardly by following the order of assumptions in the proofs of the lemmas.

Lemma 3. *A smallest witness for endgame from §(L, R) can be computed in polynomial time.*

Proof. Both $V(L)$ and $V(R)$ have to be occupied and $D := V(L, R) - N[V(L) \cup V(R)]$ has to be dominated (note that all the vertices of G can be used for the domination). The size of a smallest witness is then $|V(L) \cup V(R)| + \mathrm{dom}_G(D)$.

The size of $\mathrm{dom}_G(D)$ is computed by a greedy sweep algorithm for a minimum dominating set in interval graphs. See a paper by Brandstädt [3] for details. □

Lemma 4. *A smallest witness for sweeps from §(L, R) to §(L', R) and from §(L, R) to §(L, R') can be computed in polynomial time.*

Proof. We prove only the direction from §(L, R) to §(L', R) as the other is symmetrical.

Let C be a smallest witness and M a matching of every cop of C with the threatened vertex such that C holds R (at every vertex of $V(R)$ there is a loop in the matching). Note that C occupies L.

We can ensure that M uses the maximum number of cops from $V(L)$ just by changing M. Let the vertices threatened by cops from $V(L)$ be $L'_C \subseteq V(L')$. We fix this part of M. M now threatens $L'' := V(L') - L'_C$ without cops at $V(L)$ or $V(R)$ (and optionally dominates $V(L, L')$).

If $L' < L$, then C has to contain $|L''|$ additional cops threatening L''. These cops can be positioned at the threatened vertices and altogether, $|V(L) \cup V(R)| + |L''|$ cops are required.

If $L' > L$, let $D := V(L, L') - N[V(L) \cup V(R)]$ and let $c_1, c_2, \ldots$ be the cops dominating D ordered left to right. We can ensure that every cop c_i is on the rightmost-ending vertex such that D is still dominated, since moving c_i to such vertex v preserves both domination of D and (possible) threatening of some vertex $l \in L''$. For the latter note that v lies either left of L or $v \in V(L)$. Note that such c_i are exactly those returned by a left-to-right sweep greedy algorithm computing $\mathrm{dom}_G(D)$.

Now we may assume that M is maximal between L'' and c_i's. Therefore, C must contain at least one extra cop for every $v \in L''$ not matched to some c_i. This gives the total size of C. □

Lemma 5. *A smallest witness for split from §(L, R) to either §(L', R) or to §(L, R') can be found in polynomial time.*

Proof. Let C be a smallest witness and M_L (resp. M_R) be matchings of cops of C with $V(L') \cup V(R)$ (resp. $V(R') \cup V(L)$) such that C holds R in M_L (resp. C holds L in M_R) as in the proof of Lemma 4.

We can ensure that both M_L and M_R use the maximum number of cops from $V(L)$ and $V(R)$ just by changing M. Let the vertices of $V(L')$ threatened by cops from $V(L)$ be L'_C and symmetrically, let R'_C be the vertices of $V(R')$ threatened by cops from $V(R)$. We fix these parts of M_L and M_R. Let $L'' := V(L') - L'_C$ and

$R'' := V(R') - R'_C$. C now threatens both L'' (in M_L) and R'' (in M_R) without the cops at $V(L)$ and $V(R)$ (resp. those occupying/holding these barriers) and optionally dominates $V(R', L')$.

Case 0. If there is an interval (vertex) v of G containing both R' and L', then the cops of C threatening L'' and R'' can be assumed to be positioned at v. This also makes $V(R', L')$ dominated. In this case we can take $C = V(L) \cup V(R) \cup (\{v\} \times \max(|L''|, |R''|))$.

In the remaining cases we assume that there is no such vertex v.

Case $L' < R'$. For every $l \in L''$, we may assume that the cop c threatening l in M_L is positioned on the neighbour v of l with the rightmost right end. If cop c also threatens some vertex $r \in R''$ then we still have $v \in N[r]$. Note that if v would lie to the right of R' then vertex l would satisfy conditions of Case 0. Let T_L be the cops threatening L'' in M_L.

We can make M_R maximal between R'' and T_L. Now C has to contain one extra cop for every vertex of R'' not threatened by some T_L. Let T_R be set of these vertices. The cops threatening T_R in M_R can be positioned on the vertices they threaten.

Now we can take $C := V(L) \cup V(R) \cup T_L \cup T_R$.

Case $L' \geq R'$. Let $D := V(R', L') - N[V(L) \cup V(R)]$ (the vertices left to be dominated). Order the vertices of $D \cup L''$ left-to-right by the right ends as $d_1, d_2, \ldots$. We show that we may assume that the cops of C dominating D and threatening R'' are in the positions generated by the following algorithm:

Start with $C' := \emptyset$, $M' := \emptyset$ and take the vertices d_i in turn. If $d_i \in R''$ and not threatened in M', place a cop into C on the right-most neighbour of d_i and add the matching into M'. Otherwise if $d_i \notin N[C']$, add the right-most neighbour (wrt. right endpoint) v of d_i to C'. If there is $r \in R''$ non-threatened neighbour of v, add the left-most such r (wrt. right endpoints) to M' to threaten d_i. In other cases, proceed with next d_{i+1}.

Take the left-most cop c_i of C threatening R'' or dominating D positioned differently than the corresponding c'_i in C'. If c_i is left of c'_i (wrt. right endpoint), we may move c_i right to c'_i. This preserves domination and threatening R'' and (optional) threatening L'' (as c'_i can not be to the right of L'). If c_i is right of c'_i (wrt. right endpoint), then the vertex d_j that caused c'_i to be in C' is either not dominated (if the reason was domination of d_j) or not threatened (if the reason was threatening some $r \in R''$).

Again, we may assume that M is maximal between C' and L''. Then C has to contain extra cops T_L to threaten L'', these cops may be assumed to be on the threatened vertices. Finally, we can take $C := V(L) \cup V(R) \cup C' \cup T_L$. $\square$

7 Conclusion

We have shown an algorithm deciding the Cop and ∞-fast Robber game on interval graphs, therefore answering an open question of Fomin et al. posed in their paper "Pursuing a fast robber on a graph" [4].

Since the game is already NP-hard for general chordal graphs and even split graphs, it would be interesting to consider the complexity of the game on chordal graphs with bounded asteroidal number (or the number of leaves of the underlying tree for the standard intersection representation of chordal graphs) and the class of circular-arc graphs.

The notion of playgrounds of the reduced game can be extended to such graphs and seem to have some common properties, but the analysis does not extend in a straightforward way.

We propose the complexity of the game on such graphs as an open question. Even an algorithm exponential in the asteroidal number would be of interest.

References

1. Albert, M.H., Nowakowski, R.J., Wolfe, D.: Lessons in Play: An Introduction to Combinatorial Game Theory. AK Peters, USA (2007)
2. Bollobás, B.: Modern graph theory. Graduate Texts in Mathematics, vol. 184. Springer, New York (1998)
3. Brandstädt, A.: The computational complexity of feedback vertex set, hamiltonian circuit, dominating set, steiner tree, and bandwidth on special perfect graphs. Elektronische Informationsverarbeitung und Kybernetik 23(8/9), 471–477 (1987)
4. Fomin, F.V., Golovach, P., Kratochvil, J., Nisse, N., Suchan, K.: Pursuing a fast robber on a graph. Theoretical Computer Science (2009) (submitted)
5. Fomin, F.V., Golovach, P.A., Kratochvíl, J.: On tractability of cops and robbers game. In: IFIP TCS, pp. 171–185 (2008)
6. Korte, N., Möhring, R.H.: An incremental linear-time algorithm for recognizing interval graphs. SIAM J. Comput. 18(1), 68–81 (1989)
7. Nowakowski, R., Winkler, P.: Vertex to vertex pursuit in a graph. Discrete Math. 43(2), 235–239 (1983)
8. Quilliot, A.: A short note about pursuit games played on a graph with a given genus. Journal of Combinatorial Theory, Series B 38(1), 89–92 (1985)

Some Tractable Win-Lose Games

Samir Datta and Nagarajan Krishnamurthy

Chennai Mathematical Institute, India
{sdatta,naga}@cmi.ac.in

Abstract. Finding a Nash equilibrium in a bimatrix game is PPAD-hard (Chen and Deng, 2006 [5], Chen, Deng and Teng, 2009 [6]). The problem, even when restricted to win-lose bimatrix games, remains PPAD-hard (Abbott, Kane and Valiant, 2005 [1]). However, there do exist polynomial time tractable classes of win-lose bimatrix games - such as, very sparse games (Codenotti, Leoncini and Resta, 2006 [8]) and planar games (Addario-Berry, Olver and Vetta, 2007 [2]).

We extend the results in the latter work to $K_{3,3}$ minor-free games and a subclass of K_5 minor-free games. Both these classes strictly contain planar games. Further, we sharpen the upper bound to unambiguous logspace UL, a small complexity class contained well within polynomial time P. Apart from these classes of games, our results also extend to a class of games that contain both $K_{3,3}$ and K_5 as minors, thereby covering a large and non-trivial class of win-lose bimatrix games. For this class, we prove an upper bound of nondeterministic logspace NL, again a small complexity class in P. Our techniques are primarily graph theoretic and use structural characterizations of the considered minor-closed families.

Keywords: $K_{3,3}$-minor-free, K_5-minor-free, Win-lose bimatrix game, Nash equilibrium.

1 Introduction

In his seminal paper [15] in 1951, Nash proved that every finite non-cooperative game always has at least one equilibrium point. In recent years, there has been a flurry of activity in the algorithmic and complexity theoretic community addressing various questions regarding the computation of Nash equilibria. On the one hand, related hardness results have been proved. Daskalakis, Goldberg and Papadimitriou [9] showed that finding a Nash equilibrium is PPAD complete for an n-player game ($n \geq 4$). (Ref. [16] for a definition of PPAD). Chen and Deng [4] extended this result to the 3-player case and subsequently to the 2-player case too [5] (also ref. [6]). Abbott, Kane and Valiant [1] showed that restricting the payoffs of the players to 0 or 1 does not make the game easier. In other words, bimatrix $(0,1)$ or win-lose games are as hard as the general-sum case. Chen, Teng and Valiant [7] showed that even finding approximate Nash equilibria (correct upto a logarithmic number of bits) for win-lose bimatrix games is PPAD hard.

On the other hand, search is on for classes of games where efficient (polynomial time, say) algorithms exist. Codenotti, Leoncini and Resta [8] proposed

M. Ogihara and J. Tarui (Eds.): TAMC 2011, LNCS 6648, pp. 365–376, 2011.

an algorithm to efficiently compute Nash equilibria in sparse win-lose bimatrix games. They showed that win-lose bimatrix games that have at most two winning positions per pure strategy can be solved in linear time. Addario-Berry, Olver and Vetta [2] showed that planar win-lose bimatrix games are polynomial time tractable. Planar win-lose bimatrix games are games where the bipartite graph obtained by considering the payoff matrices as adjacency matrices is planar. See Section 3 for formal definitions. They proved that:

Theorem 1. *(Theorem 3.6 in [2]) There is a polynomial time algorithm for finding a Nash equilibrium in a two-player planar win-lose game.*

We sharpen the above result by observing that polynomial time can be replaced by deterministic logarithmic space L and more importantly, we extend the result from planar games to the following classes of win-lose bimatrix games where we prove solvability in unambiguous logarithmic space UL for Classes 1 and 2, and in nondeterministic logarithmic space NL for Class 3. (Recall: UL ⊆ NL).

1. $K_{3,3}$ minor-free games,
2. K_5-minor free games where the triconnected components are planar or V_8,
3. Games whose triconnected components are K_5, V_8 or planar.

For definitions of these classes, see Section 3. Notice that all these classes strictly include planar graphs since from Kuratowski's Theorem [14] (also ref. [19] by Wagner), it follows that planar graphs are exactly the class of graphs which exclude both $K_{3,3}$ and K_5 as minors. Class 3 above strictly contains Classes 1 and 2 (Ref. [3], [19]), and also contains games that are neither $K_{3,3}$-minor-free nor K_5-minor-free. Hence, our results cover a large and non-trivial superclass of planar win-lose bimatrix games. Our results are motivated by a remark in [2] which indicates that planarity is a necessary condition for their method to work and exhibits an oriented $K_{3,3}$ as an example where their method does not apply.

Our proofs for all these classes use the essential ingredients of the proof in [2], in particular, showing that an undominated induced cycle exists for each of these classes. Such a cycle corresponds to a Nash equilibrium. (We outline these results from [2], in Section 3). We also show that such a cycle can be found in UL for Classes 1 and 2, and in NL for Class 3 (and hence, in polynomial time for all three classes). Our proofs, further, build on a triconnected decomposition of planar graphs and $K_{3,3}$-minor-free graphs (respectively, K_5-minor-free graphs).

We discuss the proof outline in the following section. For details of these proofs, refer Section 4. Section 3 contains background and preliminaries, and Section 5 concludes with some open problems.

2   Outline of Proof

We outline the proof for the $K_{3,3}$ minor-free case. The proof for the subclass of K_5 minor-free games is similar. The proof for games whose triconnected components are K_5, V_8 or planar follows from the proofs for the other two classes.

Given a $K_{3,3}$ minor-free game we consider its underlying graph. The essential idea carried over from [2] is to identify an *undominated induced cycle* in this graph. Such a cycle corresponds to a Nash equilibrium. (Ref. Section 3 for definitions). Since we are looking for cycles of a particular kind we need to focus on any biconnected component of the underlying undirected graph (since even undirected cycles cannot span biconnected components). We further decompose the underlying biconnected graph into triconnected components using a method from [11]. We show that if we can produce undominated induced cycles in each triconnected component (which inherits orientation to edges from the original graph), then there is a way to "stitch" them together to obtain at least one undominated induced cycle in the original graph. For non-planar components (that is, K_5 and V_8 (Ref. [3], [19])), we prove the following. We show that every strongly-connected bipartite subdivision of K_5 and V_8 has an undominated induced cycle, and we show how to find one.

The problem thus reduces to finding undominated induced cycles in the triconnected components of a $K_{3,3}$ minor-free graph which by a lemma of Asano [3] are exactly triconnected planar graphs or K_5's. For the former, we know from [2] how to find such cycles and for the latter, we explicitly show how to find such cycles. Notice that in the process of finding triconnected components we have lost bipartiteness and possibly altered the notion of domination. So we have to carefully deal with subdivisons of triconnected graphs, instead which preserve bipartiteness, domination and most properties of 3-connectivity.

3   Background and Preliminaries

3.1   Win-Lose Bimatrix Games

Definition 1. *A win-lose (or $(0,1)$-) bimatrix game is specified by two $R \times C$ matrices M_R and M_C with entries from $\{0,1\}$ where the payoff for the row (column) player playing action $r_i \in R$ (respectively, $c_j \in C$) is $M_R(i,j)$ (respectively, $M_C(i,j)$) if the column (respectively, row) player plays action $c_j \in C$ (respectively, $r_i \in R$) in response.*

Pure strategies of the row (column) player are just rows (respectively, columns) that index the payoff matrices of the players. A mixed strategy of the row (column) player is a probability distribution over the set of rows (respectively, columns). The aim of each player is to choose a strategy that maximizes his/ her expected payoff.

Definition 2. *A best response of the row (column) player is a mixed strategy x (respectively, y) of the row (respectively, column) player that maximizes his (her) expected payoff given a mixed strategy y (respectively x) of the other player.*

A Nash equilibrium is a pair of strategies that are mutual best responses.

Definition 3. *Given a win-lose bimatrix game specified with the matrices M_R, M_C, its associated (or underlying) graph is the bipartite directed graph,*

$G_{R,C} = (V, E)$ with bipartitions R, C (that is, $V = R \cup C$) and the following adjacency matrix:

$$\begin{pmatrix} 0 & M_R \\ M_C^T & 0 \end{pmatrix}$$

Definition 4. An undominated induced cycle in $G_{R,C}$ is a cycle $\mathcal{C}$ such that there are no vertices $\{u, v, w\}$, where v and w ($v \neq w$) are on $\mathcal{C}$, such that $(u, v) \in \mathcal{C}$ and $(u, w) \in \mathcal{C}$. (An induced cycle $\mathcal{C}$ is dominated by a vertex v (not on $\mathcal{C}$), if there are two or more edges from v to vertices on $\mathcal{C}$).

We use the following results from [2].

Claim 1. (From Sections 2.1 and 2.2 in [2]) It suffices to look at the case when $G_{R,C}$ is strongly connected and is free of digons (cycles of length 2).

Lemma 1. (Corollary 2.3 in [2]) Let $S \subset V$ and let the subgraph restricted to S, $G_{R,C}[S]$, be an induced cycle. S corresponds to a (uniform) Nash equilibrium if the cycle $G_{R,C}[S]$ is undominated.

Lemma 2. (Theorem 3.4 in [2]) Any non-trivial, strongly connected, bipartite, planar graph contains an undominated facial cycle.

Lemma 3. (Theorem 3.1 in [2]) Any non-trivial, strongly connected, bipartite, planar graph has an undominated induced cycle.

3.2   Minor-Free Graphs

Definition 5. A subdivision of a graph G is a graph obtained by subdividing its edges by adding more vertices.

Definition 6. Given a graph G, a minor of G is a subgraph H that can be obtained from G by a finite sequence of edge-removal and edge-contraction operations. If G does not contain $K_{3,3}$ (K_5) as a minor, it is called $K_{3,3}$-minor-free (K_5-minor-free).

We are interested in $K_{3,3}$-minor-free graphs and K_5-minor-free graphs due to Kuratowski's (and Wagner's) theorems.

Theorem 2. (Kuratowski, 1930 [14], Wagner, 1937 [19]) A finite graph is planar if and only if it contains neither a $K_{3,3}$-minor nor a K_5-minor.

Definition 7. A k-connected graph is one which remains connected on removing less than k vertices. A 2-connected graph is called biconnected. A graph is called triconnected if it is either 3-connected or a cycle.

Definition 8. Consider two graphs G_1 and G_2 each containing cliques of equal size. The clique-sum of G_1 and G_2 is a graph formed from their disjoint union by identifying pairs of vertices in the corresponding equal-sized cliques to form a shared clique, and possibly deleting some of the clique edges. A k-clique-sum is a clique-sum in which both cliques have at most k-vertices.

We use the following results regarding $K_{3,3}$-minor-free and K_5-minor-free graphs.

Lemma 4. *(Asano, 1985 [3]. Also ref. Vazirani, 1989 [18]) Let G be a $K_{3,3}$-minor-free graph. The triconnected components of G are either planar or K_5.*

Lemma 5. *(Wagner, 1937 [19]. Also ref. Khuller, 1988 [13]) Any K_5-minor-free graph can be obtained by repeatedly taking 3-clique-sums of planar graphs and V_8. (V_8 is the 4-rung Mobius ladder, on 8 vertices).*

The class of K_5-minor free graphs where we prove existence of undominated induced cycles, namely K_5-minor free graphs whose triconnected components are either V_8 or planar, can also be obtained by taking 3-clique-sums of planar graphs and V_8, but for every set of shared-clique-vertices, we require the clique-sum to be between two graphs. Given 3 or more cliques with a common set of vertices, repeatedly taking clique-sums may lead to graphs with no undominated induced cycles. E.g., $K_{3,3}$ can be obtained by a 3-clique-sum operation on three K_4's. We can orient these edges to obtain the counter example shown in [2].

3.3   Triconnected Decomposition of a Graph

Given a graph, we can find its cut-vertices and biconnected components in L (e.g. see [10]). If we start with a digraph instead, there is virtually no change in the procedure since two biconnected components do not share any edge. In 1973, Hopcroft and Tarjan [12] proposed a technique to decompose a graph into its triconnected components. Finding triconnected components can also be accomplished in L - e.g. see [11]. This procedure outputs a triconnected-component-separating-pair tree which has triconnected components and separating pairs[1] as nodes with an edge between a triconnected component and a separating pair if and only if the separating pair belongs to the component.

4   Finding a Nash Equilibrium

From Claim 1, it suffices to look at bipartite digraphs that are strongly connected and free of digons. It is easy to see that this initial pre-processing can be done in L. If the graph $G_{R,C}$ has a digon, a pure equilibrium can easily be found in L. Checking if $G_{R,C}$ is strongly connected can be done in L as well. If $G_{R,C}$ is not strongly-connected, we can find, in L, one strongly connected component, say S, that is a source (hence "undominated") in the strongly-connected-component-dag of $G_{R,C}$. (The strongly-connected-component-dag, H, of $G_{R,C}$ is such that strongly connected components of $G_{R,C}$ are vertices of H. For $S_1, S_2 \in H$, (S_1, S_2) is an edge if $\exists v_1 \in S_1, v_2 \in S_2$ such that (v_1, v_2) is an edge in $G_{R,C}$).

Now, we find the triconnected components of $G_{R,C}$ and as discussed in Section 3, this can be done in L. It is important to notice that while computing a

---

[1] Pairs of vertices removing which disconnects the graph. Only triconnected separating pairs are present in the triconnected-component-separating-pair tree - Ref. [11].

triconnected component, we add a spurious edge between the vertices of a separating pair if it is not already present. These spurious edges pose three kinds of problems when we consider bipartite digraphs: (1) How do we orient these edges? (2) How do we ensure bipartiteness in the resulting triconnected components? (3) How do we make sure that no new domination is introduced?

For handling (1), note that if there is a directed path from u to v, (where $\{u, v\}$ is a separating pair) in some component C_0 formed after removing $\{u, v\}$ from the graph, we can orient the edge as (u, v) in every other component C formed by removing the same separating pair. Then, if we find a cycle in a triconnected component, it will correspond to a cycle in the original graph. However, to do this orientation, we must check for reachability in the corresponding directed graphs, which is in NL. For $K_{3,3}$-minor-free and K_5-minor-free graphs, we know from [17] that this is in UL. Problem (2) is easy to solve: we just need to subdivide the spurious edge appropriately i.e. if both its endpoints are in the same partition then subdivide the edge by introducing a single vertex, else subdivide it by introducing two vertices. Observe that, to solve (2), if both endpoints are in different partitions, we do not need to subdivide the edges at all. But this may introduce new dominations. Subdividing spurious edges even if the endpoints are in different partitions takes care of (3). (As existing edges are neither subdivided nor removed, existing domination relationships are preserved). Hence:

Lemma 6. *Given a digraph G whose underlying undirected graph is biconnected, we can obtain a triconnected-component-separating-pair tree T with edges in triconnected components directed in such a way that every directed cycle present in a triconnected component corresponds to some directed cycle in G. This procedure is in NL, and in UL if G is $K_{3,3}$-minor-free or K_5-minor-free.*

Now, we prove the following lemma.

Lemma 7. *Let G be a strongly connected orientation of a subdivision of a triconnected planar graph which is bipartite. Then at least one of the faces of G is bounded by an undominated, induced (directed) cycle.*

Proof. Using Lemma 2 we know there exists an undominated, facial (directed) cycle. Since in a triconnected graph every facial cycle is induced, the same follows for any subdivision thereof.

We prove the following lemmas for non-planar triconnected components, in subsequent sections.

Lemma 8. *Every strongly-connected bipartite subdivision of K_5 has an undominated induced cycle.*

Lemma 9. *Every strongly-connected bipartite subdivision of V_8 has an undominated induced cycle.*

Finally, we need to stitch together various cycles across triconnected components.

Lemma 10. *Given the triconnected-component-separating-pair tree T of (the underlying undirected graph of) a strongly connected bipartite graph G, it is possible to find an undominated induced cycle in the original graph G in L.*

Thus, we obtain our main result:

Theorem 3. *There is an NL procedure for finding a Nash equilibrium in a two-player win-lose game which belongs to one of the following classes:*

1. *$K_{3,3}$ minor-free games,*
2. *K_5-minor free games where the triconnected components are planar or V_8,*
3. *Games whose triconnected components are K_5, V_8 or planar.*

For Classes 1 and 2, a Nash equilibrium can, in fact, be computed in UL.

Proof. We are done by invoking the above Lemmas. Apart from Lemmas 6, 7, 10 and 1, we invoke 8 and 4 for the proof for Class 1, Lemmas 9 and 5 for Class 2 and for Class 3, Lemmas 8, 9 and the observation that undominated induced cycles in K_5 and V_8 can be "stitched" together in L too.

Remark 1. Class 3 in Theorem 3 includes games that are neither $K_{3,3}$-minor-free nor K_5-minor-free, and contains Classes 1 and 2 (and planar games) too.

4.1   Stitching Cycles Together

Before we complete the proof of Lemma 10, we prove the following lemma:

Lemma 11. *Given a graph G with two triconnected components C_1 and C_2 which share a separating pair $S = \{u, v\}$ and suppose $O_i \in C_i$ (for $i \in \{1, 2\}$) are two undominated induced cycles both passing through the (undirected) edge $\{u, v\}$ but in opposite directions, then the cycles can be "stitched" together to obtain an undominated induced cycle in G.*

Proof. It is easy to see that dropping the oriented copies of S, we get a directed cycle O consisting of all other directed edges in O_1 and O_2. O is an undominated induced cycle as there are no edges between vertices in C_1 and C_2, and O_i is already undominated and induced in C_i (for $i \in \{1, 2\}$).

Now, we prove Lemma 10.

Proof. (of Lemma 10) We first modify the triconnected-component-separating-pair tree T as obtained from Lemma 6, as follows.

Consider a separating pair $S = \{u, v\}$ and the corresponding triconnected components, say, $C_1, C_2, \ldots, C_i$. If component C_j ($1 \le j \le i$) has a vertex w that dominates S (that is, G has edges from w to both u and v), then we remove edges (in T) between S and C_k, $\forall k \ne j$. If more than one component contains such vertices that dominate S, we pick one component arbitrarily. If none of them contain vertices that dominate S, then we leave the edges between S and C_k ($1 \le k \le i$) untouched. Repeat the above for all separating pairs.

Now, we pick a tree, say T_1 from the resulting forest. It suffices to find an undominated induced cycle in T_1 because of the following. If C_j $(1 \leq j \leq i)$ has a vertex w that dominates S, then undominated cycle(s) in C_j do not pass through u and v. Hence, no vertex outside T_1 can dominate any cycle within T_1.

Notice that we can stitch together only two cycles at a separating pair. Hence, we further modify T_1 as follows. We arbitrarily pick one undominated induced cycle per component in T_1. We keep the edge (C, S) in T_1 if and only if the picked cycle in triconnected component C passes through the separating pair S. If an S node has just one edge incident on it, remove it. If it has two or more bidirected edges incident on it, keep exactly two of them and remove the rest. Note that vertices from components we have "disconnected" from S cannot dominate (in G) cycles in the components we have retained. This is because each "disconnected" component itself has an undominated cycle passing though S.

The trees in the resulting forest are just paths. Pick one of these paths, say P. If P has just one component C, we are done as any undominated induced cycle in C is an undominated induced cycle in G as well. On the other hand, if P has two or more components, we repeatedly invoke Lemma 11 to stitch together undominated induced cycles across all these components, thereby obtaining an undominated induced cycle in G.

Since the steps of the above procedure can be performed by a L-transducer, we have completed the proof of the lemma.

4.2   Undominated Induced Cycle in K_5

We prove Lemma 8. The key idea is that removing an edge of K_5 makes it planar.

Proof. (of Lemma 8) Let G be a strongly connected bipartite (oriented) subdivision of K_5. Using Lemma 12 below, there exists an edge (or a subdivided edge) $e = (w_1, w_2)$ such that $G \setminus e$ is strongly connected. As $G \setminus e$ is also planar, by the result of Addario-Berry, Olver and Vetta [2] (Lemma 3 above), it has an undominated induced cycle, say U. Notice that U continues to be induced in G. We also prove in Lemma 12 below that adding back $e = (w_1, w_2)$ introduces no new domination(s) in G, because either w_1 and w_2 belong to the same color class (R or C) or w_2 is the only vertex in its color class. It follows that U is undominated in G as well.

Lemma 12. *Let $G = (V, E)$ be an orientation of K_5 that is strongly connected and such that each $w \in V$ belongs to one of the color classes R or C. There exists an edge $e \in E$, $e = (w_1, w_2)$ such that $G' = G \setminus e$ is strongly connected and such that one of the following holds:*

(i) w_1 and w_2 belong to the same color class.

(ii) One of the color classes (say, C) has exactly one vertex and e is an incoming edge to that vertex. That is, $C = \{w_2\}$.

Proof. Let $V = \{1, 2, 3, 4, 5\}$. It suffices to look at the number of vertices belonging to each class:

(I) All vertices belong to the same color class, say R. It suffices to prove that $\exists e = (w_1, w_2) \in E$ such that $G' = G \setminus e$ is strongly connected. Condition (i) of the theorem holds no matter what e we choose.

(II) $|R| = 4, |C| = 1$. Here, apart from proving that $G \setminus e$ is strongly connected, we need to prove that either $w_1, w_2 \in R$ or $C = \{w_2\}$.

(III) $|R| = 3, |C| = 2$. We need to prove that $G \setminus e$ is strongly connected and $w_1, w_2 \in R$ or $w_1, w_2 \in C$.

If there does not exist any edge whose removal ensures that the rest of the graph is strongly connected, then $\forall e = (u, v) \in E$, either u is a source in $G \setminus e$ or v is a sink. Without loss of generality, let v be a sink. (The case when u is a source is symmetric and a similar argument works). Consider the subgraph $G \setminus u$. As shown in Fig. 1, only two non-isomorphic cases arise. (We denote the remaining vertices by a, b, c).

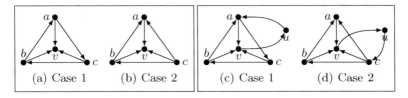

(a) Case 1   (b) Case 2   (c) Case 1   (d) Case 2

Fig. 1. (a) and (b) show Cases 1 and 2 in the proof for K_5-subdivision. (c) and (d): After adding the vertex u and some edges to Cases 1 and 2 respectively.

Consider case 1. As G is strongly connected, $(v, u) \in E$ and at least one of $(u, a), (u, b), (u, c) \in E$. Without loss of generality, let $(u, a) \in E$. This is shown in Fig. 1(c). Now, in case (I), independent of the orientation of the edges between u, b and u, c, the graphs $G \setminus (a, v)$, $G \setminus (b, v)$ and $G \setminus (c, v)$ are strongly connected. In case (II), if $C = \{v\}$ or $C = \{u\}$, again choose e to be either (a, v), (b, v) or (c, v). If $C = \{a\}$, drop (b, v) or (c, v) so that condition (i) is satisfied. Similarly for $C = \{b\}$ or $C = \{c\}$. In case (III), at least one of a, b, c is in R. Say $a \in R$. If $v \in R$, then choose $e = (a, v)$. If $v \in C$ and one of $a, b, c \in C$ (say $b \in C$), choose $e = (b, v)$. If $C = \{u, v\}$, choose $e = (b, a)$.

Consider case 2. As G is strongly connected, $(v, u) \in E$ and $(u, c) \in E$. Ref. Fig. 1(d). Choose $e = (c, a)$ or (b, a) or (c, v) or (b, v). $G \setminus e$ remains strongly connected. The proof for case (I) follows rightaway. In cases (II) and (III) too, we are done as one of these pairs of vertices must belong to the same color class.   □

4.3   Undominated Induced Cycle in V_8

Proof. (of Lemma 9) Let $G = (V, E)$ be a counter example and let $G' = (V', E')$ be its underlying V_8. (G' is obtained from G by ignoring subdividing vertices). As G is strongly connected, so is G'. However, though G is bipartite, G' is not. (It is easy to see that G must have at least 3 sub-dividing vertices). Also, G' has no undominated induced cycle (otherwise G has one too). The following *Claim* is easy to prove.

Claim. The following hold for cycles in G' (and in G):

1. All cycles in G' (and hence in G) are of length > 3.
2. 4-cycles in G' correspond to undominated induced cycles in G.
3. 5-cycles in G' are induced but may be dominated (in both G' and G).
4. If G' has a 6-cycle, it has a 4-cycle too.
5. If G' has a 7-cycle, it either has a 4-cycle or a 5-cycle.
6. If G' has an 8-cycle, it has a 5-cycle too.

As G is a counter-example, because of the above *Claim*, G' contains no 4-cycle.
    $\Rightarrow G'$ contains a 5-cycle.
Without loss of generality, let this cycle be (a, b, c, d, e, a). This cycle may either be dominated by f or by h. This leads to two cases as shown in Fig. 2.

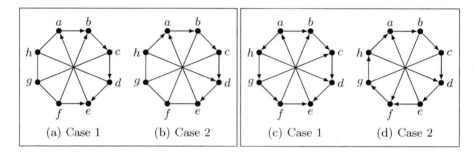

(a) Case 1   (b) Case 2   (c) Case 1   (d) Case 2

Fig. 2. Orienting edges in G', given that (a, b, c, d, e, a) is dominated. (a) Case 1: Cycle dominated by f. (b) Case 2: Cycle dominated by h. (c) Case 1 forces this orientation, having an undominated induced cycle (a, h, d, e, a). (d) Case 2 forces this orientation, having an undominated induced cycle whether (c, g) or (g, c) is an edge.

Case 1: f dominates (a, b, c, d, e, a). That is, (f, b) and $(f, e) \in E'$.
    $\Rightarrow (g, f) \in E'$ (otherwise f is a source in G' but G' is strongly connected).
    $\Rightarrow (g, c) \in E'$ (else (c, g, f, b, c) is a 4-cycle, hence undominated, induced).
    $\Rightarrow (h, g) \in E'$ (owing to strong-connectivity, again)
    $\Rightarrow (h, d) \in E'$ (otherwise we have a 4-cycle, again)
    $\Rightarrow (a, h) \in E'$ (due to strong connectivity)
    $\Rightarrow (a, h, d, e, a)$ is a 4-cycle (an undominated induced cycle), a contradiction to the fact that G is a counter example. So f cannot dominate (a, b, c, d, e, a).

Case 2: h dominates (a, b, c, d, e, a). That is, (h, a) and $(h, d) \in E'$.
    $\Rightarrow (g, h) \in E'$ (since G' is strongly connected).
Now, $(f, g) \in E'$ (because, if $(c, g) \in E'$, f has to dominate the resulting 5-cycle, and if $(g, c) \in E'$, (f, g) ensures strong-connectivity). Similarly, $(e, f) \in E'$.
    $\Rightarrow (b, f) \in E'$ (otherwise (f, b, c, d, e, f) is an undominated induced cycle).
Now, if $(g, c) \in E'$, (h, d, e, f, g, h) is an undominated induced cycle. Otherwise, (c, g, h, a, b, c) is an undominated induced cycle.
Hence, h cannot dominate (a, b, c, d, e, a) either.
    $\Rightarrow (a, b, c, d, e, a)$ itself is an undominated induced cycle.   □

5   Conclusion and Future Work

We have shown that finding a Nash equilibrium is in UL for win-lose bimatrix games that are $K_{3,3}$ minor-free or K_5-minor free whose triconnected components are planar or V_8. Further, we have shown that finding a Nash equilibrium is in NL for win-lose bimatrix games whose triconnected components are K_5, V_8 or planar. As win-lose bimatrix games are PPAD hard in general, finding or narrowing down on what actually causes the hardness is an interesting open problem.

Acknowledgement

The authors wish to thank T. Parthasarathy, K. V. Subrahmanyam, G. Ravindran, Prajakta Nimbhorkar and Tanmoy Chakraborty for useful discussions. The first author would also like to thank the organizers and participants of the Aarhus-Chennai Computational Complexity Workshop 2010, for their feedback on a talk he had given, on the topic.

References

[1] Abbott, T.G., Kane, D.M., Valiant, P.: On the complexity of two-player win-lose games. In: Proc. Symp. Found. of Comp. Sc (FOCS), pp. 113–122. IEEE Computer Society, Los Alamitos (2005)

[2] Addario-Berry, L., Olver, N., Vetta, A.: A polynomial time algorithm for finding Nash equilibria in planar win-lose games. J. Graph Algo. Appl. 11(1), 309–319 (2007)

[3] Asano, T.: An approach to the subgraph homeomorphism problem. Theoretical Comp. Sc. 38, 249–267 (1985)

[4] Chen, X., Deng, X.: 3-NASH is PPAD-complete. Electronic Colloquium on Computational Complexity (ECCC) (134) (2005)

[5] Chen, X., Deng, X.: Settling the complexity of two-player Nash-equilibrium. In: Proc. Symp. Found. of Comp. Sc. (FOCS), pp. 261–272 (2006)

[6] Chen, X., Deng, X., Teng, S.-H.: Settling the complexity of computing two-player Nash equilibria. J. ACM 56(3) (2009)

[7] Chen, X., Teng, S.-H., Valiant, P.: The approximation complexity of win-lose games. In: Proc. Symp. Discrete Algo (SODA), pp. 159–168. SIAM, Philadelphia (2007)

[8] Codenotti, B., Leoncini, M., Resta, G.: Efficient computation of Nash equilibria for very sparse win-lose bimatrix games. In: Azar, Y., Erlebach, T. (eds.) ESA 2006. LNCS, vol. 4168, pp. 232–243. Springer, Heidelberg (2006)

[9] Daskalakis, C., Goldberg, P.W., Papadimitriou, C.H.: The complexity of computing a Nash equilibrium. SIAM J. Comput. 39(1), 195–259 (2009)

[10] Datta, S., Limaye, N., Nimbhorkar, P., Thierauf, T., Wagner, F.: Planar graph isomorphism is in log-space. In: Proc. Conf. Comput. Compl (CCC), pp. 203–214 (2009)

[11] Datta, S., Nimbhorkar, P., Thierauf, T., Wagner, F.: Graph isomorphism for $K_{3,3}$-free and K_5-free graphs is in log-space. In: Kannan, R., Kumar, N. (eds.) IARCS Annual Conf. on Found. of Software Tech. and Theor. Comp. Sc (FSTTCS 2009). LIPIcs, vol. 4, pp. 145–156. Schloss Dagstuhl - Leibniz-Zentrum fuer Informatik (2009)

[12] Hopcroft, J.E., Tarjan, R.E.: Dividing a graph into triconnected components. SIAM J. Comput. 2(3), 135–158 (1973)

[13] Khuller, S.: Parallel algorithms for K_5-minor free graphs. Technical Report TR88-909, Cornell Univ., Comp. Sc. Dept. (1988)

[14] Kuratowski, K.: Sur le probleme des courbes gauches en topologie. Fund. Math. 15, 271–283 (1930)

[15] Nash, J.F.: Non-cooperative games. Ann. Math. 54(2), 286–295 (1951)

[16] Papadimitriou, C.H.: On the complexity of the parity argument and other inefficient proofs of existence. J. Comput. Syst. Sci. 48(3), 498–532 (1994)

[17] Thierauf, T., Wagner, F.: Reachability in $K_{3,3}$-free graphs and K_5-free graphs is in unambiguous log-space. In: Kutyłowski, M., Charatonik, W., Gębala, M. (eds.) FCT 2009. LNCS, vol. 5699, pp. 323–334. Springer, Heidelberg (2009)

[18] Vazirani, V.V.: NC algorithms for computing the number of perfect matchings in $K_{3,3}$-free graphs and related problems. Info. and Comput. 80(2), 152–164 (1989)

[19] Wagner, K.: Über eine Eigenschaft der ebenen Komplexe. Mathematische Annalen 114, 570–590 (1937)

A Note on Obfuscation for Cryptographic Functionalities of Secret-Operation Then Public-Encryption

Ning Ding and Dawu Gu

Department of Computer Science and Engineering
Shanghai Jiao Tong University, China
{dingning,dwgu}@sjtu.edu.cn

Abstract. Obfuscating programs has been a fascinating area of theoretical cryptography in recent years. Hohenberger et al. in TCC'07 and Hada in EUROCRYPT'10 showed that re-encryption and encrypted signature are obfuscateable and their constructions are dedicated and the security proofs are complicated. Whereas, obfuscation for other complicated cryptographic functionalities still remains unknown.

In a recent breakthrough on fully homomorphic encryption in STOC'09, Gentry noted that one algorithm in his construction, called Recrypt, is a re-encryption program. Along Gentry's sight, we observe that with fully homomorphic encryption, we can obfuscate a category of functionalities, including re-encryption and encrypted signature and even signature-then-encryption, which can be characterized as first secret operation and then public encryption. We formally demonstrate the obfuscation for this category of functionalities, in which the construction and security proof are general and simple. We then show the applicability of this obfuscation that it is secure in the contexts of the three functionalities.

Keywords: Cryptography, Obfuscation, Fully Homomorphic Encryption.

1 Introduction

Question. In recent years, theoretical cryptography community has focused on a fascinating research line of obfuscating programs. Loosely speaking, obfuscating a program P is to construct a new program which can preserve the functionality of P, but its code is fully "unintelligent". Any adversary can only use the functionality of P and cannot learn anything more than this, i.e. cannot reverse-engineering nor understand it. In other words, an obfuscated program should not reveal anything useful beyond executing it.

Barak et al. [2] formalized the definition of obfuscation through a simulation-based definition called the virtual black-box property. Following [2], many works focused on how to obfuscate different cryptographic functionalities. Among them, there are some negative results, e.g. [2,8]. [2] showed there doesn't exist any general obfuscation method for all programs. [8] showed many natural cryptographic

M. Ogihara and J. Tarui (Eds.): TAMC 2011, LNCS 6648, pp. 377–389, 2011.
© Springer-Verlag Berlin Heidelberg 2011

functionalities cannot be obfuscated. On the other hand, there exist some positive results, e.g. [4,6,9,10,11]. Among these positive results, [9,11] demonstrated how to securely obfuscate two complicated functionalities in cryptography, i.e. re-encryption (RE) and encrypted signature (ES), while others focused on a very basic and simple primitive, i.e. (multiple-bit) point functions, traditionally used in some password based identification systems. In summary, so far only a few primitives are obfuscateable and it is actually very hard to construct an obfuscation for any other primitive.

Recently, Gentry [7] proposed the first fully homomorphic encryption scheme and noted that one algorithm, called Recrypt, is actually a re-encryption program. Briefly, let FHES = (KeyGen, Enc, Dec, Evaluate) be a fully homomorphic encryption scheme. Let $(pk_1, sk_1), (pk_2, sk_2) \leftarrow$ KeyGen(1^n). Assume Recrypt is now given c_m, an encryption of m under pk_1, and c_{sk_1}, an encryption of sk_1 under pk_2, and its goal is to output an encryption of m under pk_2. It first encrypts c_m using pk_2 and runs Evaluate on input pk_2, Dec, c_{sk_1} and the new encryption of c_m and finally outputs what Evaluate outputs. By the functionality of Evaluate, we have the output is an encryption of m under pk_2, and thus the re-encryption functionality is achieved. (But we still need to investigate the security of this method in the context of RE formally.)

Let us take a closer look at RE and ES. First consider the following naive program P implementing RE: P has sk_1, pk_2 hardwired; Given a ciphertext encrypted under pk_1 P first decrypts it using sk_1 and then encrypts the decryption using pk_2. One can see that it is insecure to let anyone other than the sk_1-holder obtain P since it exposes sk_1 (while executing P by any third party securely is the motivation of obfuscating P). Notice that P's computation can be divided to two steps: first decrypt the ciphertext using secret key sk_1 (secret operation) and then encrypt the decryption using pk_2 (public encryption). We also consider the naive program P' implementing ES, which first signs the message using a signing key (secret operation) and then encrypts the signature using a public encryption key (public encryption). Even for Sign-then-Encrypt (StE), it can also be executed by a naive program identically to P' except that in the second step it encrypts both the signature and the message.

To securely implement these functionalities, one can first construct the corresponding naive programs and then obfuscate them. Actually, this is the route adopted by [11] for RE and [9] for ES. Notice that these naive programs share a common character. That is, each of them first performs a secret operation and then performs a public encryption operation. For convenience of statement we say a functionality which can be executed by a such naive program is of this character. Actually, there are many cryptographic functionalities which are of this character. Since Gentry's Recrypt shows a different and conceptually simpler way to implement RE, we have naturally the following question: *Can we securely obfuscate those cryptographic functionalities of this character using fully homomorphic encryption?*

Our Result. We show that with fully homomorphic encryption, all the functionalities which possess the required character are obfuscateable. The security of the obfuscation holds even adversaries possess some dependent information on obfuscated programs (e.g. (sk_1, pk_1, pk_2)). Further, we investigate the security of the obfuscation in the contexts of RE, ES and StE. We enhance the security in [11] for RE and [1] for StE by letting adversaries obtain the obfuscated programs while in the original literature, adversaries can only access the programs in the oracle manner. Our investigation shows that when instantiated in these contexts, the obfuscation can achieve the security proposed by [9] for ES and the enhanced security from [11] for RE and the enhanced security from [1] for StE except the EU-CMA insider security w.r.t. obfuscator. As explained in [9,11], obfuscation for ES, RE and StE etc. mainly aims to letting a third party, rather than sender and receiver, execute the programs for the sender, usually the sk_1-holder. Our investigation means that these delegation tasks can indeed be executed by any third party who doesn't know the secret of the receiver.

Our Technique. Let GF be a general functionality of the required character, and P denote a naive program implementing GF. P on input x first performs a secret (or non-secret) operation, denoted C, and then encrypts the output of $C(x)$ with a public key pk_e. With fully homomorphic encryption, we can also output an encryption of $C(x)$ by running Evaluate on pk_e, a universal circuit U conforming to $U(C, x) = C(x)$, two encryptions of C and x encrypted under pk_e. Thus, we can construct a program Q which has the encryption of C hardwired and on input x first encrypts x and then outputs Evaluate's output on all the required arguments. We then show Q implementing GF is indeed an obfuscated program. This is essentially the entire construction and it is quite simple. We then consider the applicability of this construction by showing that our obfuscation is secure in the contexts of ES, StE and RE.

Organizations. In the rest of this paper, Section 2 presents the preliminaries required in the paper and Section 3 presents our result.

2   Preliminaries

Due to space limitations we assume familiar with the notions of IND-CPA encryption schemes and EU-CMA (existential unforgeability under chosen message attack) signature schemes etc. and refer readers to [7] for the notion of fully homomorphic encryption as well as the construction of IND-CPA security. We present the definition of obfuscation as follows.

Obfuscation. We use the following definition which strengthens the one in [9] by requiring the ACVBP holds even D can obtain some auxiliary input which may depend on the obfuscation, as shown in [8].

Definition 1. *(Average-case secure obfuscation w.r.t. dependent input). A PPT machine Obf is a circuit obfuscator for a class of probabilistic circuits $\mathcal{C} = \{\mathcal{C}_n\}_{n \in \mathbb{N}}$ w.r.t. dependent input z (may depend on C), if for every probabilistic circuit $C \in \mathcal{C}_n$, the following holds:*

Functionality: $\Pr[C' \leftarrow Obf(C) : \forall x; \mathsf{StaDiff}(C(x); C'(x)) = 0] = 1$, *where* $\mathsf{StaDiff}$ *denotes the statistical difference.*

Average-case virtual black-box property (ACVBP): There exists a PPT oracle machine S (simulator) such that, for every non-uniform PPT oracle machine D (distinguisher), every polynomial $p(\cdot)$, all sufficiently large n, dependent input z,

$$\left| \Pr \begin{bmatrix} C \leftarrow \mathcal{C}_n; \\ C' \leftarrow Obf(C); \\ b \leftarrow D^C(C', z) \end{bmatrix} : b = 1 \right] - \Pr \begin{bmatrix} C \leftarrow \mathcal{C}_n; \\ C'' \leftarrow S^C(1^n); \\ b \leftarrow D^C(C'', z) \end{bmatrix} : b = 1 \right] \right| < \frac{1}{p(n)}$$

3   Our Result

In Section 3.1 we propose our obfuscation for the generate functionality GF. In Sections 3.2, 3.3, 3.4, we show the applicability of the obfuscation to the category of functionalities by illustrating its applications to ES, StE and RE.

3.1   Obfuscation for the General Functionality

In this subsection we first present GF, which is the abstract of the category of functionalities we are interested in and then provide an obfuscated program implementing GF.

Let $C = \{C_n\}_{n\in\mathbb{N}}$ be an arbitrary ensemble of circuits and sk be a secret that was generated by some key generator Gen, i.e. $(pk, sk) \leftarrow \mathsf{Gen}(1^n)$ (in the public key setting, Gen may be the key generator KeyGen of a public-key encryption scheme PKE or the key generator SKG of a signature scheme DS, where in the latter case we usually use notation vk to substitute pk). Let $C_{n,sk}$ denote C_n having sk hardwired (if there is no requirement that C_n should have a secret hardwired let sk be empty). Let $(pk_e, sk_e) \leftarrow \mathsf{KeyGen}(1^n)$ and Enc be the encryption algorithm of PKE. The general functionality GF_{sk,pk_e} is as follows.

Functionality. GF_{sk,pk_e} (and $\mathcal{C}_{sk,pk_e}$): input $x \in \{0,1\}^n \cup \{\mathsf{RetrieveKey}\}$.
  1. If $x = \mathsf{RetrieveKey}$, output pk_e;
  2. Let $y \leftarrow C_{n,sk}(x)$;
  3. Let $z \leftarrow \mathsf{Enc}(pk_e; y)$ and output z.

It can be seen that GF_{sk,pk_e} indeed consists of the two desired operations: first a secret operation (in line 2) and then a public encryption (in line 3) (line 1 is proposed for constructing obfuscation). To obtain an obfuscation for GF_{sk,pk_e}, one usually needs to first construct a naive program $\mathcal{C}_{sk,pk_e}$ implementing GF_{sk,pk_e} (e.g. simply execute each line of GF_{sk,pk_e}), and then obfuscate $\mathcal{C}_{sk,pk_e}$.

Since the goal of GF_{sk,pk_e} is to output an encryption of y, we can use Evaluate to generate an encryption of y too if we instantiate PKE with a FHES. That is, we adopt an alternative program implementing GF_{sk,pk_e}. Let ObC_{sk,pk_e} denote this program which works as follows: given the encryption of $C_{n,sk}$, it computes an encryption of x and executes Evaluate on input the two encryptions, which output is an encryption of $C_{n,sk}(x) = y$ under pk_e.

Let us present this more precisely. First, it is reasonable to assume the size of $C_{n,sk}$ can be computed given index n for the publicly known C_n. Thus let $U =$

$\{U_n\}_{n\in\mathbb{N}}$ be an ensemble of universal circuits satisfying that U_n on input $C_{n,sk}$ and $x \in \{0,1\}^n$ emulates $C_{n,sk}$'s computation on input x to output $C_{n,sk}(x)$. Similarly, the depth of U_n can be estimated in advance. Thus due to [7] there exists a fully homomorphic encryption scheme FHES = (KeyGen, Enc, Dec, Evaluate) in which Evaluate can handle any circuit which depth is bounded by U_n's depth. Thus we first construct the naive program $\mathcal{C}_{sk,pk_e}$ where PKE is instantiated with FHES. Then construct obfuscator Obf and ObC_{sk,pk_e} as follows.

Obfuscator. Obf: input: the description of $\mathcal{C}_{sk,pk_e}$.

1. Obtain pk_e and $C_{n,sk}$ from $\mathcal{C}_{sk,pk_e}$'s description.
2. Compute $c_{C_{n,sk}} \leftarrow \mathsf{Enc}(pk_e; C_{n,sk})$;
3. Generate the program ObC_{sk,pk_e}: input: $x \in \{0,1\}^n \cup \{\mathsf{RetrieveKey}\}$.
   (a) If $x = \mathsf{RetrieveKey}$, output pk_e;
   (b) Let $c_x \leftarrow \mathsf{Enc}(pk_e; x)$;
   (c) Compute $z \leftarrow \mathsf{Evaluate}(pk_e; U_n, c_{C_{n,sk}}, c_x)$ and output z.

Note that there is a slight difference between the outputs of $\mathcal{C}_{sk,pk_e}$ and ObC_{sk,pk_e}. That is, $\mathcal{C}_{sk,pk_e}(x)$ is a fresh encryption of y output by Enc while $ObC_{sk,pk_e}(x)$ is an encryption of y output by Evaluate. Due to the current constructions of FHES in e.g. [7], the two encryptions are not identically distributed. However, this difference is actually indifferent since GF only needs to output a normal encryption of y under pk_e. Thus it is very reasonable to say that ObC_{sk,pk_e} essentially preserves the functionality of $\mathcal{C}_{sk,pk_e}$ (Gentry [7] thought Recrypt is a re-encryption program due to the same consideration). So we say Obf satisfies a *relaxed functionality* property, or just the functionality property.

Proposition 1. *For any polynomial-time algorithm f, ObC_{sk,pk_e} is an obfuscation of $\mathcal{C}_{sk,pk_e}$ w.r.t. (the relaxed functionality property and) dependent input $f(sk, pk, pk_e)$.*

Proof. We show the relaxed functionality property and ACVBP can be satisfied.

1. Relaxed Functionality. Since both $ObC_{sk,pk_e}(x)$ and $\mathcal{C}_{sk,pk_e}(x)$ are normal encryptions of $y = C_{n,sk}(x)$ under pk_e, this property obviously holds.

2. ACVBP. To show this, we need to construct a simulator S and show that for any non-uniform PPT D, letting $z' \leftarrow f(sk, pk, pk_e)$, $|\Pr[D^{\mathcal{C}_{sk,pk_e}}(ObC_{sk,pk_e}, z') = 1] - \Pr[D^{\mathcal{C}_{sk,pk_e}}(S^{ObC_{sk,pk_e}}(1^n), z') = 1]|$ is negligible, where $S^{ObC_{sk,pk_e}}(1^n)$ denotes a fake program being input to D. In the following we first present S's description and then establish this indistinguishability.

Simulator. S: input n and oracle access to ObC_{sk,pk_e}.

1. Query the oracle ObC_{sk,pk_e} with RetrieveKey to obtain pk_e.
2. Sample $(pk', sk') \leftarrow \mathsf{Gen}(1^n)$ and generate $C_{n,sk'}$.
3. Compute $c_{C_{n,sk'}} \leftarrow \mathsf{Enc}(pk_e; C_{n,sk'})$.

4. Generate the program ObC_{sk',pk_e}, which description is as follows:
   input: $x \in \{0,1\}^n \cup \{\mathsf{RetrieveKey}\}$
   (a) If $x = \mathsf{RetrieveKey}$, output pk_e;
   (b) Let $c_x \leftarrow \mathsf{Enc}(pk_e; x)$;
   (c) Compute $z \leftarrow \mathsf{Evaluate}(pk_e; U_n, c_{C_{n,sk'}}, c_x)$ and output z.

Now we turn to the indistinguishability. Suppose D on oracle access to $\mathcal{C}_{sk,pk_e}$ and input $z' \leftarrow f(sk, pk, pk_e)$ can distinguish ObC_{sk,pk_e} from ObC_{sk',pk_e} with non-negligible probability. Then we can construct a PPT A' which can distinguish $c_{C_{n,sk}}$ from $c_{C_{n,sk'}}$ still with non-negligible probability. A' works in the following way. Given pk_e from the challenger, A' samples $(pk, sk), (pk', sk') \leftarrow$ $\mathsf{Gen}(1^n)$ and generates $C_{n,sk}$ and $C_{n,sk'}$. A' sends $C_{n,sk}$ and $C_{n,sk'}$ to the challenger which randomly encrypts one of them and responds with the ciphertext, which is either $c_{C_{n,sk}}$ or $c_{C_{n,sk'}}$. Then A' adopts Obf's strategy in line 3 to generate a program which is either ObC_{sk,pk_e} or ObC_{sk',pk_e}. A' sends the program as well as $z' \leftarrow f(sk, pk, pk_e)$ to D as input and emulates D's computation. During this emulation, A' adopts $\mathcal{C}_{sk,pk_e}$'s strategy to answer all D's oracle queries. Output what D outputs finally. By A''s description, we can see that the probability that A' distinguishes the two encryptions is equal to the probability that D distinguishes the two programs, which contradicts the security of IND-CPA of FHES. Thus, the ACVBP holds. The proposition follows.   □

We remark that Proposition 1 still holds even if D is given access to oracle $C_{n,sk}$ since we only need to let A further adopt $C_{n,sk}$'s strategy to answer all D's $C_{n,sk}$-queries ($C_{n,sk}$ will be instantiated with the signing algorithm in the context of ES), and that the consideration on $f(sk, pk, pk_e)$ is necessary, since in the later applications D (adversary) has some auxiliary inputs besides the obfuscation which are, for instance, (sk, pk, pk_e) or (pk, pk_e) (the corresponding f can be chosen as the one which simply outputs all the inputs or the latter two inputs). In the following subsections, we will illustrate the applicability of the result in this subsection by showing ES, StE and RE can be implemented using this result simply and securely.

3.2   Application to Encrypted Signature

The ES functionality for Alice and Bob generates a signature on a given message under Alice's signing key and then encrypts the signature under Bob's public encryption key. Hada [9] proposed the first obfuscation for ES. We will use our result to obfuscate this functionality too and the advantage of ours is the simpleness in the construction and security proof. Let $\mathsf{DS} = (\mathsf{SKG}, \mathsf{S}, \mathsf{V})$ be an EU-CMA signature scheme and PKE be as before. Let $(vk, sk) \leftarrow \mathsf{SKG}(1^n)$ and $(pk_e, sk_e) \leftarrow \mathsf{KeyGen}(1^n)$. The ES functionality can be shown as follows.

ES Functionality. ES_{sk,pk_e} (and $\mathcal{C}_{sk,pk_e}$): input $m \in \{0,1\}^n \cup \{\mathsf{RetrieveKey}\}$.
   1. If $m = \mathsf{RetrieveKey}$, output pk_e;

2. Compute $\sigma \leftarrow \mathsf{S}(sk; m)$;
3. Compute $z \leftarrow \mathsf{Enc}(pk_e; \sigma)$ and output z.

It can be seen that ES_{sk,pk_c} is an instantiation of GF_{sk,pk_c}, in which the circuit $C_{n,sk}$ is now instantiated with $\mathsf{S}_{n,sk}$, where $\mathsf{S}_{n,sk}$ denotes the circuit representation of S w.r.t. the security parameter n and having sk hardwired. Now we instantiate PKE with a FHES to construct the naive program $\mathcal{C}_{sk,pk_c}$ implementing ES_{sk,pk_c}. Then run $\mathsf{Obf}(\mathcal{C}_{sk,pk_c})$ (in which we use $c_{\mathsf{S}_{n,sk}}$ to denote the encryption of $\mathsf{S}_{n,sk}$ under pk_e) to generate the obfuscated program implementing ES_{sk,pk_c}, denoted $\mathcal{ES}_{sk,pk_c}$, which is shown as follows.

Obfuscation for ES_{sk,pk_c}: $\mathcal{ES}_{sk,pk_c}$. input: $m \in \{0,1\}^n \cup \{\mathsf{RetrieveKey}\}$
  1. If $m = \mathsf{RetrieveKey}$, output pk_e;
  2. Let $c_m \leftarrow \mathsf{Enc}(pk_e; m)$;
  3. Compute $z \leftarrow \mathsf{Evaluate}(pk_e; U_n, c_{\mathsf{S}_{n,sk}}, c_m)$ and output z.

By Proposition 1, $\mathcal{ES}_{sk,pk_c}$ is an obfuscation for ES_{sk,pk_c} w.r.t. $f(sk, vk, pk_e)$ for any f. Now we turn to investigate if it can achieve the security in the context of ES. We first review the security requirement presented by [9], as Definition 2 shows, and then show $\mathcal{ES}_{sk,pk_c}$ can achieve the required security.

Definition 2. *(EU w.r.t. ES Obfuscator [9]) Let DS be a signature scheme and PKE be a public-key encryption scheme. Also, let Obf be the obfuscator for ES. The DS scheme is existentially unforgeable w.r.t. Obf if for every non-uniform PPT machine A (adversary), every polynomial $p(\cdot)$, all sufficiently large n, the following condition holds where A doesn't query the oracle with m:*

$$\Pr \left[\begin{array}{l} (vk, sk) \leftarrow SKG(1^n); (pk_e, sk_e) \leftarrow KeyGen(1^n); \\ C' \leftarrow Obf(\mathcal{C}_{sk,pk_c}); (m, \sigma) \leftarrow A^{\mathsf{S}_{n,sk}}(vk, pk_e, C'); \\ V(vk; m, \sigma) = Accept. \end{array} \right] < \frac{1}{p(n)}$$

Proposition 2. *The DS scheme is EU w.r.t. our Obf (where PKE is instantiated with FHES).*

Proof. Suppose there exists an A which can violate the proposition. Then we can construct a B which can break the EU-CMA security of DS. B on input vk samples $(pk_e, sk_e) \leftarrow \mathsf{KeyGen}(1^n)$, adopts the simulator S's strategy on pk_e to generate a fake program, denoted C'', feeds (vk, pk_e, C'') to A (now f is chosen as the one that receives only two inputs (vk, pk_e) and just outputs them) and performs the experiment. During the experiment, for each A's oracle query m', B queries its signing oracle with m' which then responds with a signature for m'. On receiving the signature, B transits it to A as the response. Finally output what A outputs. As shown in the remarks following Proposition 1, A cannot distinguish C' from C'' given access to oracle $\mathsf{S}_{n,sk}$, so A on input C'' can still forge a signature with non-negligible probability and B breaks the EU-CMA security of DS. This is impossible. $\qquad\square$

We lastly remark that we could add the IND-CPA requirement in Definition 2, i.e., when A outputs two messages (m_0, m_1) and m_b for random bit b is encrypted A cannot guess correctly b. Actually, this IND-CPA security can be proved using a proof similar to that of Proposition 3, thus we adopt the original statement in [9] to present the security definition (while [1] explicitly required the encryption security, as Definition 3 shows). Further, we discuss why Definition 2 doesn't allow A to access the decryption-oracle $\mathsf{Dec}(sk_e; \cdot)$ to investigate the IND-CCA security. It can be seen that if A is given access to the decryption oracle, then A can easily forge a signature by first sending a m to the obfuscation C' and then querying the decryption oracle with C''s output. Then the response of the decryption oracle is a signature for m. Thus [9] didn't provide A with the decryption oracle and we only need to investigate the IND-CPA security.

3.3   Application to Sign-Then-Encrypt

The StE functionality differs from ES in that StE should encrypt both the signature and message instead of the signature only in ES. An et al. [1] showed that the sequential composition of "sign then encrypt" can achieve some security. Hada [9] informally stated an obfuscation for StE based on his obfuscation for ES. We will propose an obfuscation for StE and discuss the security that our obfuscation can achieve. Let DS and PKE be as before. Let $(vk, sk) \leftarrow \mathsf{SKG}(1^n)$ and $(pk_e, sk_e) \leftarrow \mathsf{KeyGen}(1^n)$. The StE functionality can be shown as follows.

StE Functionality. StE_{sk,pk_e} (and $\mathcal{C}_{sk,pk_e}$): input $m \in \{0,1\}^n$.
  1. If $m = \mathsf{RetrieveKey}$, output pk_e;
  2. Compute $\sigma \leftarrow \mathsf{S}(sk; m)$;
  3. Compute $z \leftarrow \mathsf{Enc}(pk_e; \sigma, m)$ and output z.

Also, StE_{sk,pk_e} is an instantiation of GF_{sk,pk_e}, in which the circuit $C_{n,sk}$ is now instantiated with $\mathsf{S}'_{n,sk}$, where $\mathsf{S}'_{n,sk}$ is identical to $\mathsf{S}_{n,sk}$ except that it outputs the signature as well as the message. Also, we instantiate PKE with a FHES to construct the naive program $\mathcal{C}_{sk,pk_e}$ implementing StE_{sk,pk_e}. Then run $\mathsf{Obf}(\mathcal{C}_{sk,pk_e})$ (in which we use $c_{S'_{n,sk}}$ to denote the encryption of $\mathsf{S}'_{n,sk}$) to generate the obfuscated program implementing StE_{sk,pk_e}, denoted $\mathcal{StE}_{sk,pk_e}$, which is shown as follows.

Obfuscation for StE$_{sk,pk_e}$: $\mathcal{StE}_{sk,pk_e}$: input $m \in \{0,1\}^n \cup \{\mathsf{RetrieveKey}\}$.
  1. If $m = \mathsf{RetrieveKey}$, output pk_e;
  2. Let $c_m \leftarrow \mathsf{Enc}(pk_e; m)$;
  3. Compute $z \leftarrow \mathsf{Evaluate}(pk_e; U_n, c_{S'_{n,sk}}, c_m)$ and output z.

By Proposition 1, $\mathcal{StE}_{sk,pk_e}$ is an obfuscation for StE_{sk,pk_e} w.r.t. $f(sk, vk, pk_e)$. Now we turn to investigate the security of $\mathcal{StE}_{sk,pk_e}$ in the context of StE. For convenience of stating the security in [1], we use the phrase "a sign-then-encrypt scheme" to denote the signcrypt scheme that is sequential composition of a signature scheme (e.g. DS) and a public key encryption scheme (e.g. FHES).

Using our obfuscation, the sign-then-encrypt scheme works as follows. Initially, Alice obtains $(vk_1, sk_1) \leftarrow \mathsf{SKG}(1^n)$ and $(pk_{e1}, sk_{e1}) \leftarrow \mathsf{KeyGen}(1^n)$, and Bob obtains $(vk_2, sk_2) \leftarrow \mathsf{SKG}(1^n)$ and $(pk_{e2}, sk_{e2}) \leftarrow \mathsf{KeyGen}(1^n)$. To signcrypt a message m, Alice runs $z \leftarrow \mathcal{StE}_{sk_1, pk_{e2}}(m)$, which is the signcryption, and sends z to Bob. On receiving z, Bob runs the decrypt-then-verify program (functionality) $\mathrm{DtV}_{sk_{e2}, vk_1}$, which first decrypts z using sk_{e2} and verifies the validness of signature using vk_1 and then outputs m if the verification succeeds. (Abusing notation a little) we still use $\mathcal{StE}$ to denote our sign-then-encrypt scheme.

Now we can introduce the security in [1]. [1] proposed two security: outsider security and insider security, where the outsider security is proposed w.r.t. any adversary not belonging to the two parties and the insider security is proposed w.r.t. one adversarial party. We first present the outsider security as follows.

Definition 3. *(IND-CPA and EU-CMA outsider security w.r.t. Obf, enhanced from [1]) Let* **PKE** *and* **DS** *be as before and* **StE** *be the sign-then-encrypt scheme of sequential composition of* **PKE** *and* **DS**. *Let Obf be the obfuscator for* **StE**. *Let the random variable* **IND**(StE, A, k) *where* $A = (A_1, A_2)$, *b is a uniformly random bit, denote the result of the following probabilistic experiment where A doesn't query the oracle with m:*

> $\mathsf{IND}(\mathsf{StE}, A, n)$
> $(vk_1, sk_1) \leftarrow \mathsf{SKG}(1^n), (pk_{e2}, sk_{e2}) \leftarrow \mathsf{KeyGen}(1^n);$
> $C' \leftarrow Obf(\mathcal{C}_{sk_1, pk_{e2}}), (m_0, m_1, z) \leftarrow A_1^{\mathcal{C}_{sk_1, pk_{e2}}}(vk_1, pk_{e2}, C');$
> $y \leftarrow \mathsf{Enc}(pk_{e2}; m_b); (b', m, \sigma) \leftarrow A_2(y, z);$
> *Output* b'.

We say **StE** *is of the IND-CPA and EU-CMA outsider security w.r.t. Obf if for any non-uniform A both* $\Pr[b' = b] - 1/2$ *and the probability* (m, σ) *is a valid pair of message and signature are negligible.*

We give two remarks on Definition 3. First, the outsider (and insider) security can be defined w.r.t. the strongest encryption security IND-CCA2 as shown in [1]. But our scheme cannot achieve this security due to the IND-CPA security of **FHES**, so we only present the outsider (and insider) security w.r.t. the IND-CPA security. Second, Definition 3 enhances the original definition in [1] by letting A further obtain C''s description while the original definition in [1] only allows A to access $\mathcal{C}_{sk_1, pk_{e2}}$-oracle (similarly for Definition 4 and Definition 5).

We now turn to the insider security. [1] showed that from any sign-then-encrypt scheme **StE**, we can induce a signature scheme and an encryption scheme. The induced signature scheme **DS'** is identical to **StE** except that receiver's sk_{e2} now becomes a part of sender's verifying key of **DS'**. To sign a message m, Alice runs $z \leftarrow \mathrm{StE}_{sk_1, pk_{e2}}(m)$, which is the signature, and publishes z. On receiving z, anyone can verify z by running $m \leftarrow \mathrm{DtV}_{sk_{e2}, vk_1}(z)$ (notice that sk_{e2} is publicly known) and accepts if $m \neq \perp$. The induced encryption scheme **PKE'** is also identical to **StE** except that sender's signing key sk_1 now becomes a part of receiver's public key of **PKE'**. To encrypt a message m for Bob, anyone can run $z \leftarrow \mathrm{StE}_{sk_1, pk_{e2}}(m)$, which is the ciphertext and sends z to Bob (notice that sk_1

is publicly known). On receiving z, Bob runs $m \leftarrow \mathrm{DtV}_{sk_{e2},vk_1}(z)$ to obtain m. So far we can present the notion of the insider security as follows.

Definition 4. *(IND-CPA and EU-CMA insider security w.r.t. Obf, enhanced from [1]) We say a sign-then-encrypt scheme StE is IND-CPA and EU-CMA insider secure w.r.t. Obf if the induced DS' is EU-CMA and PKE' is IND-CPA where any adversary A can further obtain the description of $C' \leftarrow Obf(\mathcal{C}_{sk_1,pk_{e2}})$.*

Proposition 3. *The scheme $St\mathcal{E}$ is of IND-CPA and EU-CMA outsider security and IND-CPA insider security w.r.t. our Obf.*

Proof. Outsider security. Suppose there exists an $A = (A_1, A_2)$ which can break the security. First consider the case that $\Pr[b' = b] - 1/2$ is non-negligible. Then we construct a B which can break the IND-CPA security of FHES. B on input pk_{e2} samples $(vk_1, sk_1) \leftarrow \mathsf{SKG}(1^n)$, generates $C' \leftarrow Obf(\mathcal{C}_{sk_1,pk_{e2}})$, feeds (vk_1, pk_{e2}, C') to A_1 (f is now chosen as the one outputting the latter two inputs) and performs the experiment. During the experiment, B adopts $\mathcal{C}_{sk_1,pk_{e2}}$'s strategy to answer A_1's oracle queries. When A_1 proposes two messages (m_0, m_1) and z, B transits (m_0, m_1) to the challenger which encrypts a random one and responds with the encryption. Then B sends the encryption and z to A_2. Finally B outputs the b' in A_2's output. By the assumption on A, B breaks the IND-CPA security of FHES.

Second consider the case that (m, σ) is a valid pair of message and signature with non-negligible probability. Then we can construct a B which can break the EU-CMA security of DS. B on input vk_1 samples $(pk_{e2}, sk_{e2}) \leftarrow \mathsf{KeyGen}(1^n)$, adopts the simulator S's strategy on pk_{e2} to generate C'', feeds (vk_1, pk_{e2}, C'') to A_1 and performs the experiment identically as above except that for each A_1's oracle query m', B queries its signing oracle with m' which then responds with a signature for m' and on receiving the signature B encrypts it and m' using pk_{e2} and sends the encryption to A_1 as the response. Finally output the (m, σ) in A_2's output. By the security of obfuscation, A on input C'' can still forge a signature with non-negligible probability and thus B breaks the EU-CMA security of DS.

IND-CPA insider security. Suppose A can break the IND-CPA security of PKE'. We can construct a B which can break the IND-CPA security of FHES. Given pk_{e2}, B samples $(vk_1, sk_1) \leftarrow \mathsf{SKG}(1^n)$ and generates $C' \leftarrow Obf(\mathcal{C}_{sk_1,pk_{e2}})$, feeds $(sk_1, vk_1, pk_{e2}, C')$ to A_1 (f is now chosen as the one just outputting all inputs) and performs the experiment identically as above except that when A_1 proposes two messages (m_0, m_1), B signs both of them using sk_1 and transits the two pairs of message and signature to the challenger which randomly encrypts one of them and responds with the encryption (notice that the encryption is identical to $\mathsf{StE}_{sk_1,pk_{e2}}(m_b)$ for the random bit b) and on receiving the encryption, B sends it and z to A_2. Lastly, output A_2's guess. By the assumption on A, B breaks the IND-CPA security of FHES. This is impossible. □

We lastly remark that $St\mathcal{E}$ is not EU-CMA insider secure w.r.t. our Obf, since an adversary holding sk_{e2} against DS' can decrypt $c_{S'_{n,sk_1}}$ in the obfuscation and extract sk_1 and thus can easily forge a signature (note that DS' is EU-CMA and $St\mathcal{E}$ is EU-CMA insider secure w.r.t. the original definitions in [1]).

3.4   Application to Re-encryption

The RE functionality for Alice and Bob takes a ciphertext for a message m under Alice's public key, and transforms it into a ciphertext for the same message m under Bob's public key. Hohenberger et al. [11] proposed the first obfuscation for RE. Gentry [7] gave a remark on his fully homomorphic encryption that the algorithm Recrypt is actually a realization of RE. We will present Gentry's method in our term formally and show it can meet the security requirement in [11]. Let PKE be as before and Alice owns $(pk_1, sk_1) \leftarrow$ KeyGen(1^n) and Bob owns $(pk_2, sk_2) \leftarrow$ KeyGen(1^n). The RE functionality can be shown as follows.

RE Functionality. RE_{sk_1, pk_2} (and $\mathcal{C}_{sk_1, pk_2}$): input $c \in \{0,1\}^{l(n)} \cup \{\mathsf{RetrieveKey}\}$.
1. If $c = \mathsf{RetrieveKey}$, output pk_2;
2. Let $m \leftarrow \mathsf{Dec}(sk_1; c)$;
3. Let $z \leftarrow \mathsf{Enc}(pk_2; m)$ and output z.

It can be seen that RE_{sk_1, pk_2} is also an instantiation of GF_{sk, pk_e}, in which (sk, pk_e) is now replaced by (sk_1, pk_2) and $C_{n, sk}$ is now instantiated with Dec_{n, sk_1}, which denotes the circuit representation of $\mathsf{Dec}(sk_1; \cdot)$ w.r.t. security parameter n. Also, we instantiate PKE with a FHES to construct the naive program $\mathcal{C}_{sk_1, pk_2}$ implementing RE_{sk_1, pk_2}. Then run $\mathsf{Obf}(\mathcal{C}_{sk_1, pk_2})$ (in which we use $c_{Dec_{n, sk_1}}$ to denote the encryption of Dec_{n, sk_1} under pk_2) to generate the obfuscated program implementing RE_{sk_1, pk_2}, denoted $\mathcal{RE}_{sk_1, pk_2}$, which is shown as follows.

Obfuscation for RE_{sk_1, pk_2} $\mathcal{RE}_{sk_1, pk_2}$: input $c \in \{0,1\}^{l(n)} \cup \{\mathsf{RetrieveKey}\}$.
1. If $c = \mathsf{RetrieveKey}$, output pk_2;
2. Let $c_c \leftarrow \mathsf{Enc}(pk_2; c)$;
3. Compute $z \leftarrow \mathsf{Evaluate}(pk_2; U_n, c_{Dec_{n, sk_1}}, c_c)$ and output z.

By Proposition 1, $\mathcal{RE}_{sk_1, pk_2}$ is an obfuscation for RE_{sk_1, pk_2} w.r.t. $f(sk_1, pk_1, pk_2)$ for any f. Now we first enhance the security requirement in [11] as Definition 5 shows and then show our Obf satisfies the requirement as Proposition 4 shows.

Definition 5. *(IND-CPA w.r.t. obfuscator Obf, enhanced from [11].) Let PKE be a public-key encryption scheme and Obf be an obfuscator for RE. Let the random variable $\mathsf{IND}_b(PKE, A, n)$ where $b, b' \in \{0,1\}$, $A = (A_1, A_2)$, denote the result of the following probabilistic experiment:*

$$\mathsf{IND}_b(PKE, A, n)$$
$$(pk_1, sk_1) \leftarrow KeyGen(1^n), (pk_2, sk_2) \leftarrow KeyGen(1^n);$$
$$C' \leftarrow Obf(\mathcal{C}_{sk_1, pk_2}), (m_0, m_1, i, z) \leftarrow A_1^{\mathcal{C}_{sk_1, pk_2}}(pk_1, pk_2, C');$$
$$y \leftarrow Enc(pk_i; m_b); b' \leftarrow A_2(y, z);$$
$$Output\ b'.$$

We say that scheme PKE is IND-CPA w.r.t. Obf if for any non-uniform PPT A, $|\Pr[\mathsf{IND}_0(PKE, A, n)] - \Pr[\mathsf{IND}_1(PKE, A, n)]|$ is negligible.

Proposition 4. *The FHES is IND-CPA w.r.t. our Obf.*

Proof. The proof is similar to the part for IND-CPA outsider security of the previous proof. Suppose $A = (A_1, A_2)$ is an adversary which can violate the proposition. First consider the case that $i = 2$ occurs with non-negligible probability. The desired B on given a public key from the challenger, views it as pk_2, samples $(pk_1, sk_1) \leftarrow \mathsf{KeyGen}(1^n)$, generates $C' \leftarrow Obf(\mathcal{C}_{sk_1, pk_2})$, feeds (pk_1, pk_2, C') to A_1 and proceeds as shown previously. Thus if A_2's outputs b's in the two experiments differ with non-negligible probability, B can distinguish the encryptions of (m_0, m_1) with non-negligible probability. Second in the case that $i = 1$ occurs with non-negligible probability, B views its public key as pk_1 and samples $(pk_2, sk_2), (pk', sk') \leftarrow \mathsf{KeyGen}(1^n)$, generates $C'' \leftarrow Obf(\mathcal{C}_{sk', pk_2})$, feeds (pk_1, pk_2, C'') to A_1, adopts $\mathcal{C}_{sk', pk_2}$'s strategy to answer A_1's oracle queries and encrypts m_b using pk_1. It can be seen that A_1 cannot distinguish C' from C'' either and thus A_2's outputs are indistinguishable in the two experiments. $\square$

Acknowledgments

This work is supported by China Postdoctoral Science Foundation funded project (20100480595) and Shanghai Postdoctoral Scientific Program (11R21414500).

References

1. An, J.H., Dodis, Y., Rabin, T.: On the security of joint signature and encryption. In: Knudsen, L.R. (ed.) EUROCRYPT 2002. LNCS, vol. 2332, pp. 83–107. Springer, Heidelberg (2002)
2. Barak, B., Goldreich, O., Impagliazzo, R., Rudich, S., Sahai, A., Vadhan, S., Yang, K.: On the (im)possibility of obfuscating programs. In: Kilian, J. (ed.) CRYPTO 2001. LNCS, vol. 2139, pp. 1–18. Springer, Heidelberg (2001)
3. Canetti, R.: Towards realizing random oracles: hash functions that hide all partial information. In: Kaliski Jr., B.S. (ed.) CRYPTO 1997. LNCS, vol. 1294, pp. 455–469. Springer, Heidelberg (1997)
4. Canetti, R., Dakdouk, R.R.: Obfuscating point functions with multibit output. In: Smart, N.P. (ed.) EUROCRYPT 2008. LNCS, vol. 4965, pp. 489–508. Springer, Heidelberg (2008)
5. Canetti, R., Varia, M.: Non-malleable obfuscation. In: Reingold, O. (ed.) TCC 2009. LNCS, vol. 5444, pp. 73–90. Springer, Heidelberg (2009)
6. Canetti, R., Rothblum, G.N., Varia, M.: Obfuscating hyperplane membership. In: Micciancio, D. (ed.) TCC 2010. LNCS, vol. 5978, pp. 72–89. Springer, Heidelberg (2010)
7. Gentry, C.: Fully homomorphic encryption using ideal lattices. In: Proc. STOC 2009, pp. 169–178. ACM, New York (2009)
8. Goldwasser, S., Kalai, Y.T.: On the impossibility of obfuscation with auxiliary input. In: Proc. FOCS 2005, pp. 553–562. IEEE, Los Alamitos (2005)
9. Hada, S.: Secure Obfusction for encrypted signature. In: Gilbert, H. (ed.) EUROCRYPT 2010. LNCS, vol. 6110, pp. 92–112. Springer, Heidelberg (2010)

10. Hofheinz, D., Malone-Lee, J., Stam, M.: Obfuscation for cryptographic purposes. Journal of Cryptology 23(1), 121–168 (2010)
11. Hohenberger, S., Rothblum, G.N., Shelat, A., Vaikuntanathan, V.: Securely Obfuscating Re-encryption. In: Vadhan, S.P. (ed.) TCC 2007. LNCS, vol. 4392, pp. 233–252. Springer, Heidelberg (2007)
12. Wee, H.: On obfuscating point functions. In: Proceedings of the 37th STOC, pp. 523–532. ACM, New York (2005)

Grey-Box Steganography*

Maciej Liśkiewicz, Rüdiger Reischuk, and Ulrich Wölfel

Institut für Theoretische Informatik, Universität zu Lübeck,
Ratzeburger Allee 160, 23538 Lübeck, Germany
{liskiewi,reischuk,woelfel}@tcs.uni-luebeck.de

Abstract. We propose a new model of steganography which combines partial knowledge about the type of covertext channel with machine learning techniques to learn the covertext distribution. Stegotexts are constructed by either modifying covertexts or creating new ones, based on the learned hypothesis. We illustrate our concept with channels that can be described by monomials. A generic construction is given showing that besides the learning complexity, the efficiency of secure grey-box steganography depends on the complexity of membership tests and suitable modification procedures. For the concept class monomials we present an efficient algorithm for changing a covertext into a stegotext.

1   Introduction

The aim of steganography is to hide secret messages in unsuspicious covertexts such that the mere existence of this message is concealed. The basic scenario assumes two communicating parties Alice (sender) and Bob (receiver) plus an adversary Eve, also called a "warden" due to Simmons' [22] scenario of secret communication among prisoners. Eve wants to find out whether Alice and Bob exchange hidden messages among their covertext communication.

A "useful" stegosystem should not only be *secure* (against Eve finding out about the presence of hidden communication), but also *reliable* (i.e. with high probability, encoded messages can be correctly decoded), *computationally efficient* (i.e. the time, space and oracle query complexities should be polynomial in the length of the hidden message) and *rate efficient* (i.e. the transmission rate should be close to the covertext entropy).

In the past few years significant advances have been achieved in developing a theoretical foundation of steganography [4,6,7,12,2,13,14,16,18]. Using notions from cryptography such as *indistinguishability* and adapting them to a steganographic scenario, Hopper et al. have constructed stegosystems that are provably secure against passive and active attacks [12,2]. Their constructions are based on the assumption that Alice and Bob know nothing about the covertext channel. They are only given access to a *black-box* oracle that samples according to the channel distribution. By repeatedly sampling based on a history of previously sampled covertexts these schemes try to find samples that already "contain"

---

* Supported by DFG research grant RE 675/5-1.

M. Ogihara and J. Tarui (Eds.): TAMC 2011, LNCS 6648, pp. 390–402, 2011.

the message bits to be embedded, hence this method has been named "rejection sampling". While Hopper et al. only embed one bit per covertext document, Le and Kurosawa [16] increase this rate by means of a coding scheme similar to arithmetic coding that they call "$\mathcal{P}$-Codes".

However, all black-box stegosystems suffer from several drawbacks. Lysyanskaya and Meyerovich first pointed out that sampling based on the full history might be too difficult and analysed under which conditions stegosystems that sample with restricted length histories become insecure [18]. Furthermore, Hundt et al. have shown that the construction of such a history-based sampling oracle, a core component of all black-box stegosystems, can lead to an intractable problem for practically relevant covertext channels [14]. Moreover, the scheme in [12] embeds only one bit per document, so each convertext consists of a large number of documents. In order to achieve a reasonable transmission rate, i.e. the average number of hiddentext bits per bit sent, one either has to choose documents of small size or embed more than one bit per document.

Dedić et al. have analysed a generalisation of the scheme in [12] to embed an arbitrary number of bits per document [7] They have shown that for a reliable and secure black-box stegosystem the number of sample documents drawn from the covertext channel grows exponentially in the number of bits embedded per document. Note that this exponential bound also holds for the construction by Le and Kurosawa [16] which uses black-box sampling, too.

In *white-box* steganography, on the other hand, the stegoencoder is assumed to have full knowledge about the covertext channel. The availability of a cumulative distribution function for the covertext channel enables Le and Kurosawa [16] to modify their encoding procedure for black-box sampling and turn it into a white-box stegosystem. Although this makes their construction much more efficient, it seems unlikely that in practice the cumulative distribution is known.

In our study we want to overcome the exponential sampling complexity of the black-box approach without assuming too much knowledge about the covertext channel, as in white-box steganography. The model that we propose here will be called *grey-box* steganography, as the encoder has *partial knowledge* of the covertext channel, making it lie between the black- and white-box scenarios. We will investigate whether efficient and secure grey-box steganography is possible and extract the different properties required for this purpose. Equipped with partial knowledge, the encoder still has to gather more information about the covertext channel to select as stegotexts only those documents that appear in the covertext channel. We will model this situation as an algorithmic learning problem (for an introduction to learning theory see [1]). A priori, Alice knows that the covertext channel belongs to some class of channels, but does not know which covertext documents lie in the support of the actual channel. This is where algorithmic learning comes into play: Alice considers covertext samples and computes a hypothesis that describes the support of the channel. Based on this hypothesis, she actively tries to construct suitable stegotexts that encode her hidden message instead of passively waiting for the sampling oracle to give her a covertext with the desired properties (i.e. using *rejection sampling*).

This construction can be done by modifying an existing covertext or creating a new one. In both cases the distribution of output stegotexts should look like "normal" samples from the oracle. We give a proof of concept with channels that can be described by monomials and concentrate on learning the support of such a channel. To avoid further complications of the learning process due to highly unbalanced distributions, a uniform distribution on the support will be assumed. A generic construction is given showing that apart from the learning complexity, the efficiency of grey-box steganography depends on the complexity of the membership test, and suitable covertext modification procedures. For the concept class monomials we present an efficient algorithm for changing a covertext into a stegotext. Obviously, membership tests for such concepts can be done fast. An additional feature of our construction is that only the sender needs access to the sampling oracle (to learn the concept class), while the receiver only decodes, as in [12,7] and unlike [16], where both sender and receiver require the sampling oracle (black-box) or the cumulative distribution function (white-box).

2   Basic Notation and Definitions

Let Σ be a finite alphabet and $\sigma := \log |\Sigma|$. As usual, Σ^ℓ denotes the set of strings of length ℓ over Σ, and $\Sigma^\star$ the set of strings of finite length over Σ. We denote the length of a string u by $|u|$ and the concatenation of two strings u_1 and u_2 by $u_1 || u_2$, or by $u_1 u_2$ if this does not lead to ambiguities.

Symbols $u \in \Sigma$ will be called *documents* and a finite concatenation of documents a *communication sequence* or *covertext*. Typically, the document models a piece of data (e.g. a digital image or fragment of the image) while the communication sequence $c \in \Sigma^\star$ models the complete message sent to the receiver in a single communication exchange.

If $\mathcal{P}$ is a probability distribution with finite support denoted by $\text{supp}(\mathcal{P})$, we define the *min-entropy* of $\mathcal{P}$ as $H_\infty(\mathcal{P}) = \min_{x \in \text{supp}(\mathcal{P})} - \log \Pr_{\mathcal{P}}[x]$. This notion provides a measure of the minimal amount of randomness present in $\mathcal{P}$.

Definition 1 (Channel). *A channel $\mathcal{C}$ is a function that takes a history $\mathcal{H} \in \Sigma^\star$ as input and produces a probability distribution $\mathcal{C}_{\mathcal{H}}$ on Σ. A history $\mathcal{H} = c_1 c_2 \ldots c_m$ is legal if each subsequent symbol is obtainable given the previous ones, i.e., $\Pr_{\mathcal{C}_{c_1 c_2 \ldots c_{i-1}}}[c_i] > 0$ for all $i \leq m$. The min-entropy of $\mathcal{C}$ is the value $\min_{\mathcal{H}} H_\infty(\mathcal{C}_{\mathcal{H}})$ where the minimum is taken over all legal histories $\mathcal{H}$.*

This gives a very general definition of covertext distributions which allows dependencies between individual documents that are present in typical real-world communications. In order to embed additional information into covertexts, one has to assume that the covertext channel distribution has a sufficiently large min-entropy.

To get information about the covertext distribution *sampling oracles* can be used. $EX_{\mathcal{C}}(\mathcal{H})$ denotes an oracle that generates documents according to a channel $\mathcal{C}$ with history $\mathcal{H}$, i.e. each call of $EX_{\mathcal{C}}(\mathcal{H})$ returns a document c with probability $\Pr_{\mathcal{C}_{\mathcal{H}}}[c]$ and the responses are independent of each other.

A steganographic information transmission is thought of as taking a finite sequence $C_1, C_2, \ldots \in \Sigma^*$ of covertexts and based on them to construct a stegotext $S \in \Sigma^*$ such that the sequence additionally encodes an independent message M. This encoding is done by Alice who then sends the stegotext to the receiver Bob over a public channel. Let b denote the message encoding rate, i.e. a single stegodocument can encode up to b bits of M. Longer messages M have to be split into blocks of b bits each and for each block a separate stegodocument is generated. Their concatenation yields the stegotext.

Definition 2 (Stegosystem). *In the following, let $n = \ell \cdot b$ denote the length of the messages to be embedded, thus ℓ stegodocuments each hiding b bits are needed. A stegosystem S for the message space $\{0,1\}^n$ is a triple of probabilistic algorithms $[SK, SE, SD]$ with the following functionality:*

- *SK is the key generation procedure that on input 1^n outputs a key K of length κ, where κ is a security parameter that depends on n;*
- *SE is the encoding algorithm that takes as input a key $K \in \{0,1\}^\kappa$, a message $M \in \{0,1\}^n$ (called hiddentext), a channel history $\mathcal{H}$, and accesses the sampling oracle $EX_\mathcal{C}()$ of a given covertext channel $\mathcal{C}$ and returns a stegotext $S \in \Sigma^\ell$;*
- *SD is the decoding algorithm that takes K, S, and $\mathcal{H}$, and having access to the sampling oracle $EX_\mathcal{C}()$ returns a message M'.*

S is called a black-box stegosystem *if SE and SD have no a priori knowledge about the distribution of the covertext channel and can obtain information about it only by querying the sampling oracle.*

The application of SK is shared by Alice and Bob beforehand and its result is kept secret from an adversary. All further actions of Alice are specified by SE, those of Bob by SD. For all stegosystems discussed in this paper SK generates keys with a uniform distribution, thus when specifying a stegosystem we skip the description of SK.

The time complexities of the algorithms SK, SE, SD are measured with respect to n, κ, and the document size (specified formally by $\sigma = \log |\Sigma|$), where an oracle query is charged as one unit step. A stegosystem is *computationally efficient* if its time complexities are polynomially bounded. By convention, the running time of an algorithm includes the so called *description size* of that algorithm with respect to some standard encoding.

Ideally, one would expect that the encoder always succeeds in encoding the original message M and that the decoder always succeeds in extracting M from the stegotext. Since this may not always be possible, we define the unreliability of a stegosystem.

Definition 3 (Unreliability). *The unreliability $\mathtt{UnRel}_{\mathcal{C},S}$ of S with respect to $\mathcal{C}$ is given by $\max_{M \in \{0,1\}^n, \mathcal{H}} \Pr_{K \leftarrow SK(1^n)}[SD(K, SE(K, M, \mathcal{H}), \mathcal{H}) \neq M]$.*

Next, let us measure the security of a stegosystem. How likely is it that an adversary, the warden W, can discover that the covertext channel is used for transmitting additional information? If we put no algorithmic restrictions on W

(i.e. information-theoretic security) it is necessary that (1) the stegotext S lies in the support of the covertext channel, otherwise W could test S for membership in $\text{supp}(\mathcal{C})$, and (2) the probability of producing a stegotext S equals the probability of drawing S according to $\mathcal{C}$. Cachin has proposed the following information-theoretic model of steganographic security [6].

Definition 4 (Information-theoretic Security). *Let $\mathcal{C}$ be a covertext channel with distribution $P_{\mathcal{C}}$ and let $P_{S,\mathcal{C}}$ be the output distribution of the steganographic embedding function SE having access to the sampling oracle $EX_{\mathcal{C}}()$. The stegosystem $[SK, SE, SD]$ is called perfectly secure for the channel $\mathcal{C}$ (against passive adversaries) if the relative entropy satisfies $D(P_{\mathcal{C}}||P_{S,\mathcal{C}}) = 0$.*

To simplify the analysis, for the systems given later we will assume that the distribution on the support is uniform. Thus, we concentrate on the problem how the encoder can learn the support of the channel and then uniformly generate stegotexts. The constructions given below can be extended to a wider class of distributions using statistical learning techniques [15].

For a security analysis in the complexity-theoretic sense, W is assumed to be polynomially time-bounded. Thus, Alice has to make sure that an adversary cannot detect deviations from the two conditions above in polynomial time. However, the adversary may actively perform a *chosen hiddentext attack* [12,7]. Let $SE(K, M, \mathcal{H})$ with access to $EX_{\mathcal{C}}(\mathcal{H})$ be denoted by $SE^{\mathcal{C}}(K, M, \mathcal{H})$. In contrast, we define an oracle OC that for given message $M \in \{0,1\}^n$ and channel history $\mathcal{H}$ returns a truly random covertext $c_1 c_2 \ldots c_\ell$ of length $\ell = |SE^{\mathcal{C}}(K, M, \mathcal{H})|$ from the covertext channel $\mathcal{C}$ with the history $\mathcal{H}$, i.e. each c_i is drawn according to the probability distribution $\mathcal{C}_{\mathcal{H}||c_1 c_2 \ldots c_{i-1}}$.

Definition 5 (Warden, Chosen Hiddentext Attack (CHA)). *A probabilistic algorithm W is a (t, q, λ)-warden for the stegosystem $\mathcal{S} = [SK, SE, SD]$ if*

- *W runs in time t and accesses a reference oracle $EX_{\mathcal{C}}()$ that he can query for samples from the covertext channel $\mathcal{C}$ with a history $\mathcal{H}$;*
- *W can make a number of q queries of total length λ bits on a challenge oracle CH which is either $SE^{\mathcal{C}}(K, M, \mathcal{H})$ or $OC(M, \mathcal{H})$, where M and $\mathcal{H}$ can be chosen by W;*
- *the task of W is to determine the use of the stegosystem $\mathcal{S}$ with the help of the challenge oracle: $W^{\mathcal{C},CH} = 1$ means that W decides on "stegotext", resp. $W^{\mathcal{C},CH} = 0$ means that W decides on "covertext".*

We define the *advantage* of W over random guessing for a given covertext channel $\mathcal{C}$ as

$$\text{Adv}_{\mathcal{C},\mathcal{S}}^{\text{cha}}(W) \; := \; \left| \Pr_{K \leftarrow SK(1^n)}[W^{\mathcal{C},SE^{\mathcal{C}}(K,\cdot,\cdot)} = 1] - \Pr[W^{\mathcal{C},OC(\cdot,\cdot)} = 1] \right| \; .$$

Note that in order to maximize the advantage, W may depend on the channel $\mathcal{C}$. In the most favourable case, W may possess a complete specification of $\mathcal{C}$, so that he even does not need to query the reference oracle. The amount of such information about $\mathcal{C}$ is part of the description size of W. This knowledge may put the adversary in a much better situation than the encoder.

Definition 6 (Steganographic Security against CHA). *The insecurity of a stegosystem $\mathcal{S}$ with respect to a covertext channel $\mathcal{C}$ and complexity bounds t, q, λ is defined by*

$$\text{InSec}^{cha}_{\mathcal{C},\mathcal{S}}(t, q, \lambda) := \max_{W}\{\text{Adv}^{cha}_{\mathcal{C},\mathcal{S}}(W)\} \ ,$$

where the maximum is taken over all adversaries W working in time at most t and making at most q queries of total length λ bits to the challenge oracle CH.

Note that we do not explicitly mention the description size of the adversary, but assume this to be included in the running time t (W has to read this information at least once).

Below we recall some notions from cryptography required for the specification of the encoding function SE. Let $F : \{0,1\}^k \times \{0,1\}^l \to \{0,1\}^L$ be a function. Here $\{0,1\}^k$ is considered as the key space of F. For each key $K \in \{0,1\}^k$ we define the subfunction $F_K : \{0,1\}^l \to \{0,1\}^L$ by $F_K(x) = F(K, x)$. Thus, F specifies a family of functions, and is called a family of permutations if $l = L$ and for each key K the subfunction F_K is a permutation on $\{0,1\}^l$. For such an F we define the advantage of a probabilistic distinguisher D having access to a challenging oracle as

$$\text{PRP-Adv}_F(D) = \left| \Pr_{K \in_R \{0,1\}^k}[D^{F_K(\cdot)} = 1] - \Pr_{P \in_R PERM(l)}[D^{P(\cdot)} = 1] \right|,$$

where $PERM(l)$ denotes the family of all permutations on $\{0,1\}^l$. The insecurity of a pseudorandom family of permutations F is given by $\text{PRP-InSec}_F(t, q) = \max_D\{\text{PRP-Adv}_F(D)\}$, where the maximum is taken over all probabilistic distinguishers D running in at most t steps and making at most q oracle queries. F is called a (t, q, ϵ)-pseudorandom family if $\text{PRP-InSec}_F(t, q) \leq \epsilon$. Let the length l grow polynomially with respect to k. A sequence $\{F_k\}_{k \in \mathbb{N}}$ of families $F_k : \{0,1\}^k \times \{0,1\}^l \to \{0,1\}^l$ is called pseudorandom if for all polynomially bounded distinguishers D, $\text{PRP-Adv}_F(D)$ is negligible in k (for more formal definition of pseudorandom permutations see e.g. [5]).

3   A Grey-Box Model for Steganography

Previous steganographic models have considered computationally restricted adversaries W that possess *full knowledge* of the covertext channel. Dedić et al. [7] consider this "a meaningful strengthening of the adversary". We think that such strengthening is not appropriate to model Alice and Eve's basic knowledge about a covertext channel. In practice, encoders and wardens get an idea about typical covertexts by observing samples. They do not and likely will never possess any short advice that fully describes the channels they are looking at. Furthermore, there may be different families of channels (e.g. images, texts, audio-signals) and Alice may preselect one specific family from which the actual channel is then drawn without any outside influence. This more realistic setting strengthens the encoder and may be a chance to overcome the negative results for the black-box scenario. We do not know any steganographic system used in practice that is based on rejection-sampling, instead stegotexts typically are derived by slight modifications of given covertexts.

In the grey-box model Alice has some *partial knowledge* about the covertext channel. Therefore, we use the notion of concept classes from machine learning and define a *channel family* $\mathcal{F}$ as a set of covertext channels that share some common characteristics, such as e.g. all pseudo-random sequences, digital photographs from a certain camera, or all English literary texts. In the context of pseudo-random sequences, a single channel $\mathcal{C}_i$ contains strings output by a specific pseudo-random number generator with a fixed seed and the channel family $F_{PRS} = \{\mathcal{C}_1, \mathcal{C}_2, \ldots\}$ contains channels with different seeds.

Note that both the encoder and the warden know the concept class, the family of channels. For the actual channel $\mathcal{C}$, one member is selected at random, which is unknown to the encoder. Depending on the modelled strength of the warden, W may also lack knowledge about $\mathcal{C}$ or he may have additional information about $\mathcal{C}$. Here, we do not investigate this question further and allow the adversary to have full knowledge. The decoder, on the other hand, is not involved in the learning process, he does not need any information about the concept class.

As before, the encoding SE may access the sampling oracle $EX_{\mathcal{C}}()$, but now we clearly differentiate between accesses to the oracle for learning purposes to construct a hypothesis for the covertext channel, and accesses to get a covertext that – using the hypothesis – can be modified into a stegotext.

Depending on the concept class, Alice may be able to derive a good hypothesis – an exact or very close description of the channel – or not. Even if the concept class is not known to be efficiently learnable it makes sense to consider a situation where a precise description of the channel is given to Alice for free. Still, even in this favourable case it is not clear how Alice can construct stegotexts. She must be able to efficiently modify covertexts and test the modifications for membership in the support of the channel.

Definition 7. *The* insecurity *and* unreliability *of a stegosystem S with respect to the channel family $\mathcal{F}$ are defined by*

$$\mathrm{InSec}^{\mathrm{cha}}_{\mathcal{F},S}(t,q,\lambda) \; := \; \max_{\mathcal{C} \in \mathcal{F}} \mathrm{InSec}^{\mathrm{cha}}_{\mathcal{C},S}(t,q,\lambda) \quad and \quad \mathrm{UnRel}_{\mathcal{F},S} \; := \; \max_{\mathcal{C} \in \mathcal{F}} \mathrm{UnRel}_{\mathcal{C},S} \, .$$

We think this definition, which specifies the insecurity of stegosystems with respect to *families* of channels instead of *all* channels, corresponds better to real life intuition of insecurity than the commonly used definition. In fact, in real life steganalysis, our grey-box steganography model is already implicitly used to analyse the insecurity of particular stegosystems with respect to specific channel families. For example, it is easy to see that the steganographic algorithm F5 for JPEG images [23] is insecure with respect to the common insecurity definition, because $\mathrm{InSec}^{\mathrm{cha}}_{\mathcal{C},F5}$ is huge for almost all channels $\mathcal{C}$ deviating significantly from images compressed by JPEG. But this observation seems to be useless for a stegoanalyst, for whom a much more appropriate approach to analyse the insecurity of F5 would be to use our definition and restrict the channels to the family of JPEG-compressed images, like it was done e.g. in [10].

4   The Monomial Covertext Channels

In the rest of the paper we will present an example of a stegosystem showing that the issues discussed above are relevant and the grey-box model makes sense. In our study we consider a family of channels that can be described by monomials.

Consider a concept class over the document space $\Sigma = \{0,1\}^\sigma$ consisting of channels $\mathcal{C}$ where for each history $\mathcal{H}$, $\mathcal{C}_\mathcal{H}$ is a uniformly distributed subset of Σ that can be defined by a monomial. We denote such a channel family by MONOM.

A monomial over $\{0,1\}^\sigma$ will be represented by a vector $\mathbf{H} = (\mathbf{h}_1, \ldots, \mathbf{h}_\sigma) \in \{0,1,\times\}^\sigma$ and defines the subset of all 0-1-vectors, for which the i-th component is 0 if $\mathbf{h}_i = 0$, and 1 if $\mathbf{h}_i = 1$. The other components are called free variables. So, e.g. the monomial represented as "0×0×1" describes the set of strings $\{00001, 00011, 01001, 01011\}$. We denote the subset defined by a monomial $\mathbf{H}$ by $\boldsymbol{H}$.

One of the novel ingredients of our grey-box-stegosystems is a procedure called **Monomial-modify**, which for a given monomial $\mathbf{H}$ and a cover-document $c \in \boldsymbol{H}$, modifies c to a stego-document $s \in \boldsymbol{H}$ that encodes a b-bit message M in a way that preserves the uniform probability distribution over $\boldsymbol{H}$ to guarantee indistinguishability of stegotexts. This nontrivial task is described below.

Let, for short, $\sigma_b := \lfloor \sigma/b \rfloor$ and define for a permutation π of $\{1,2,\ldots,\sigma\}$ and $1 \le j \le b$ the subset $I_\pi(j)$ as follows: $I_\pi(j) := \{\pi(\sigma_b \cdot (j-1) + 1), \pi(\sigma_b \cdot (j-1) + 2), \ldots, \pi(\sigma_b \cdot j)\}$. These subsets partition a document $c = a_1 \ldots a_\sigma$ into b subsequences of length σ_b, where the j-th set contains all elements a_i with index i in $I_\pi(j)$. Let $FV_\pi(j)$ denote those indices in $I_\pi(j)$ that belong to free variables. Each subsequence embeds one bit of the message M as the parity of all its elements. If the parity does not match we want to flip at least one of these bits. If a free variable is chosen for this purpose it is guaranteed that the modified string still belongs to $\boldsymbol{H}$.

Procedure Monomial-modify$(M, c, \mathbf{H}, K)$
Input: hiddentext $M = m_1, m_2, \ldots, m_b \in \{0,1\}^b$; covertext document
$\quad\quad c = a_1 a_2 \ldots a_\sigma \in \{0,1\}^\sigma$; hypothesis monomial
$\quad\quad \mathbf{H} = \mathbf{h}_1 \mathbf{h}_2 \ldots \mathbf{h}_\sigma \in \{0,1,\times\}^\sigma$; private key K;
let π be the permutation specified by key K;
for $j := 1, \ldots, b$ **do**
$\quad$ **if** $[m_j \ne \bigoplus_{i \in I_\pi(j)} a_i$ **and** $FV_\pi(j) \ne \emptyset]$ **then** $a_{\nu_j} = 1 - a_{\nu_j}$, where
$\quad \nu_j := \min FV_\pi(j)$
end
Output: $s = a_1 a_2 \ldots a_\sigma$

The following procedure is used to decode a stegotext document.

Procedure Document-decode(s, K)
Input: stegotext document $s = a_1 a_2 \ldots a_\sigma \in \{0,1\}^\sigma$; private key K;
let π be the permutation specified by key K;
For $j := 1, \ldots, b$ **do** $m_j := \bigoplus_{i \in I_\pi(j)} a_i$;
Output: $m_1 m_2 \ldots m_b$

The crucial property of the procedure **Monomial-modify** says that if $\mathbf{H}$ and $\mathbf{C}$ are monomials (corresponding to a hypothesis, respectively to a concept) such that $\mathbf{H} \subseteq \mathbf{C}$ and if c is chosen randomly in $\mathbf{C}$, then **Monomial-modify** preserves the uniform probability distribution over $\mathbf{C}$. This is described formally by the following claim.

Lemma 1. *Let* $\mathbf{H}$ *and* $\mathbf{C}$ *be given monomials such that* $\mathbf{H} \subseteq \mathbf{C}$ *and let* K *be an arbitrary private key. Then for every* $s \in \mathbf{C}$ *it holds*

$$\Pr[\textbf{Monomial-modify}(M, c, \mathbf{H}, K) = s] \;=\; 1/|\mathbf{C}| \;,$$

where the probability is taken over random choices of $c \in \mathbf{C}$ *and* $M \in \{0,1\}^b$. *Moreover, for every* M, *every* $\mathbf{H}$ *with* φ *free variables, and* $c \in \mathbf{C}$ *the probability* $\Pr[\textbf{Document-decode}(\textbf{Monomial-modify}(M, c, \mathbf{H}, K), K) \neq M] \leq b \cdot e^{-\varphi/b+1}$ *over random choices of* K. *The time complexity of both procedures is linear in* σ.

5   Stegosystem for Monomial Channels

Now, we are ready to construct a stegosystem for monomial channels. The system, denoted by $\mathcal{S}$, is based on the following encoding and decoding procedures using families of permutations $F : \{0,1\}^k \times \{0,1\}^n \to \{0,1\}^n$. For a stegosystem $\mathcal{S}$ that is perfectly secure in the information-theoretic setting we choose $k = n$ and the function $F_K(x) = x \oplus K$. In the complexity-theoretic case a family F of efficiently computable pseudorandom permutations is used in order to prevent chosen hiddentext attacks. The following procedure is used by Alice to encode the message M.

Procedure Encode(M, K)
Input: hiddentext $M = m_1 \ldots m_n \in \{0,1\}^n$; private key $K = K_0, \ldots, K_{2\ell}$;
let $\mathcal{H}$ be a current history;
choose $T_0 \in_R \{0,1\}^n$ and let $T_1 := F_{K_0}(T_0 \oplus M)$;
parse $T_0 T_1$ into $t_1 t_2 \ldots t_{2\ell}$, where $|t_i| = b$;
for $i := 1, \ldots, 2\ell$ **do**
 $c_i := EX_{\mathcal{C}}(\mathcal{H})$;
 access $EX_{\mathcal{C}}(\mathcal{H})$ to learn a hypothesis $\mathbf{H}_i$ for $\mathcal{C}_{\mathcal{H}}$;
 $s_i := \textbf{Monomial-modify}(t_i, c_i, \mathbf{H}_i, K_i)$;
 let $\mathcal{H} := \mathcal{H} || s_i$;
end
Output: $s_1 \ldots s_{2\ell}$

The procedure below is used by Bob to decode a stegotext s.

Procedure Decode(s, K)
Input: stegotext $s = s_1 \ldots s_{2\ell} \in \{0,1\}^{2n}$; private key $K = K_0, \ldots, K_{2\ell}$;
for $i := 1, \ldots, 2\ell$ **do**
 $t_i := \textbf{Document-decode}(s_i, K_i)$;
end
$M := F_{K_0}^{-1}(t_{\ell+1} \ldots t_{2\ell}) \oplus t_1 \ldots t_\ell$;
Output: $M = m_1 \ldots m_\ell$

Using the definition of perfect security according to Definition 4 and the security against chosen hiddentext attack given in Definition 6 we can now apply the new framework and state the following practical result.

Theorem 1. *Let the min-entropy of every channel C in MONOM be at least h. Let b denote the rate of the stegoencoding and n the length of the secret message to be embedded. Assume Alice has no a priori knowledge of C, but both Alice and the warden have access to a sampling oracle $EX_C()$. Then the stegosystem S is computationally efficient and achieves the following reliability and security:*

$$\mathtt{UnRel}_{\mathtt{MONOM},S} \ \leq \ 2n \cdot e^{-h/b+1} + 2^{-n} \qquad and$$

– with encrypting function $F_K(x) = x \oplus K$ perfect security, and
– with a family F of pseudorandom permutations

$$\mathtt{InSec}^{\mathtt{cha}}_{\mathtt{MONOM},S}(t, q, \lambda) \ \leq \ 2 \cdot \mathtt{PRP\text{-}InSec}_F(p(t), \lambda/n) + \xi(\lambda, n)$$

where p is a fixed polynomial and the function $\xi(\lambda, n) := \left(\frac{\lambda^2}{n^2} - \frac{\lambda}{n} \right) 2^{-n}$ is a function related to the insecurity of the family F of pseudorandom permutations used in S.

Note that this stegosystem is secure in both cases even if the adversary has complete knowledge of the channel.

A parity-based approach to steganography has previously been suggested by Anderson and Petitcolas [3]. They argue that the more bits are used for calculating the parity, the less likely one can distinguish the stegotext from an unmodified covertext. In our case, Alice produces stegotexts that are always consistent with her hypothesis and thus cannot be distinguished from covertexts by construction (modulo the error Alice makes when learning). Alice could also use a pseudo-random function f_K with key K instead of the parity, in which case she would eventually have to try changing different free variables before obtaining the desired value to be embedded, thus increasing the time complexity of her embedding algorithm.

Monomial concept classes may look too simple to describe covertexts in practice. However, in this setting we do not have to restrict the variables, in learning theory also called attributes, to properties of the physical medium. If one can efficiently implement a modification of a simple attribute, these attributes may also represent semantic properties of a document. For example, pictures may be classified according to their content – whether they were taken in summer or winter, contain objects like lakes, mountains, etc. Thus, in a simple way one can construct a secure system that may be called *semantic steganography*.

Recall the properties that were needed to achieve efficient and secure steganography for the concept class of monomials: 1.) monomials are efficiently learnable from positive examples, 2.) for each monomial **H** with enough entropy there is an efficient embedding function for the hiddentext on the support of **H**, and 3.) one can efficiently compute a uniformly selected stegotext (in this case the procedure **Monomial-modify**). This generic construction can be applied to other concept classes fulfilling these properties.

For the class of monomials one actually does not need the modification procedure Monomial-modify to generate a stegotext from a given covertext. In this case, the hypothesis space even allows a direct generation of stegotexts by selecting for all, but one free variable in each group values at random.

6   Conclusions and Future Work

This paper introduces a new approach to modeling and analysing steganography. Previous models (e.g. [12], [7] or [16] either treat the covertext channel as a black-box – resulting in a sampling complexity exponential in the number of bits per covertext document – or assume a priori full knowledge about the covertext distribution, which seems unrealistic. We overcome this situation by allowing the encoder to *modify* covertexts, as done in most practical stegosystems. Our grey-box model is more realistic in the sense that we assume the encoder to have some partial knowledge about the channel.

Furthermore, a finer-grained distinction between the different ingredients for securely hiding information into covertexts provides new insights and helps in constructing stegosystems. We have shown that for efficiently learnable covertexts secure and efficient steganography is possible by presenting a construction for monomial channels, which are efficiently PAC-learnable. So far, our construction is restricted to monomial channels with the uniform distribution. For general distributions, note that the actual distribution on the support of the channels has to be learned in addition to the support in order to achieve information theoretic security. For arbitrary distributions this cannot be done efficiently. However, in the complexity theoretic setting we think that our construction can at least be generalized to the case where each free variables x_i independently of the others takes the value 1 with some arbitrary probability p_i, so called product distributions.

Even for channels that are hard to learn in the PAC-sense, assuming that by some other means the encoder can get hypotheses about the channel, one can design efficient stegosystems if the modification problem has an efficient solution.

Steganographic techniques like LSB-flipping for digital images can easily be expressed by this approach. They can be viewed as variants of **Monomial-modify**, with all but the last bits of each pixel being fixed and the least significant bit being a free variable. The support of the covertext channel for a given image I thus consists of all images that only differ in their least significant bits. However, digital images taken by modern cameras do not tend to generate truly random LSBs. Thus, representing the hypothesis as a monomial may be inappropriate for camera channels and the monomial stegosystem insecure. An important future task will be the implementation of grey-box steganography with practically relevant covertext channels.

In the grey-box setting there may still be a huge advantage for the adversary if he has complete knowledge of the covertext channel. As a next step one should investigate more carefully the case that the knowledge of the adversary is limited similar to the situation of the stegoencoder.

References

1. Angluin, D.: Computational Learning Theory: Survey and Selected Bibliography. In: Proc. STOC 1992, pp. 351–369. ACM, New York (1992)
2. von Ahn, L., Hopper, N.J.: Public-key steganography. In: Cachin, C., Camenisch, J.L. (eds.) EUROCRYPT 2004. LNCS, vol. 3027, pp. 323–341. Springer, Heidelberg (2004)
3. Anderson, R.J., Petitcolas, F.A.P.: On the limits of steganography. IEEE Journal of Selected Areas in Communications 16(4), 474–481 (1998)
4. Backes, M., Cachin, C.: Public-Key Steganography with Active Attacks. In: Kilian, J. (ed.) TCC 2005. LNCS, vol. 3378, pp. 210–226. Springer, Heidelberg (2005)
5. Bellare, M., Desai, A., Jokipii, E., Rogaway, F.: A Concrete Security Treatment of Symmetric Encryption. In: Proc. FOCS 1997, pp. 394–403. IEEE, Los Alamitos (1997), full paper available under
http://www-cse.ucsd.edu/~adesai/papers/pubs.html#BDJR97
6. Cachin, C.: An information-theoretic model for steganography. Information and Computation 192(1), 41–56 (2004)
7. Dedić, N., Itkis, G., Reyzin, L., Russell, S.: Upper and lower bounds on black-box steganography. Journal of Cryptology 22(3), 365–394 (2009)
8. Denis, F.: PAC Learning from Positive Statistical Queries. In: Richter, M.M., Smith, C.H., Wiehagen, R., Zeugmann, T. (eds.) ALT 1998. LNCS (LNAI), vol. 1501, pp. 112–126. Springer, Heidelberg (1998)
9. Ehrenfeucht, A., Haussler, D.: Learning decision trees from random examples. Information and Computation 82(3), 231–246 (1989)
10. Fridrich, J.J., Goljan, M., Hogea, D.: Steganalysis of JPEG Images: Breaking the F5 Algorithm. In: Petitcolas, F.A.P. (ed.) IH 2002. LNCS, vol. 2578, pp. 310–323. Springer, Heidelberg (2003)
11. Haussler, D.: Bias, version spaces and Valiant's learning framework. In: Proc. of the 4th International Workshop on Machine Learning, pp. 324–336. University of California, Irvine (1987)
12. Hopper, N.J., Langford, J., von Ahn, L.: Provably secure steganography. In: Yung, M. (ed.) CRYPTO 2002. LNCS, vol. 2442, pp. 77–92. Springer, Heidelberg (2002)
13. Hopper, N.J.: On Steganographic Chosen Covertext Security. In: Caires, L., Italiano, G.F., Monteiro, L., Palamidessi, C., Yung, M. (eds.) ICALP 2005. LNCS, vol. 3580, pp. 311–323. Springer, Heidelberg (2005)
14. Hundt, C., Liśkiewicz, M., Wölfel, U.: Provably secure steganography and the complexity of sampling. In: Asano, T. (ed.) ISAAC 2006. LNCS, vol. 4288, pp. 754–763. Springer, Heidelberg (2006)
15. Kearns, M.: Efficient Noise-Tolerant Learning from Statistical Queries. In: Proc. STOC 1993, pp. 392–401. ACM, New York (1993)
16. Le, T.V., Kurosawa, K.: Bandwidth optimal steganography secure against adaptive chosen stegotext attacks. In: Camenisch, J.L., Collberg, C.S., Johnson, N.F., Sallee, P. (eds.) IH 2006. LNCS, vol. 4437, pp. 297–313. Springer, Heidelberg (2007)
17. Letouzey, F., Denis, F., Gilleron, R.: Learning From Positive and Unlabeled Examples. In: Arimura, H., Sharma, A.K., Jain, S. (eds.) ALT 2000. LNCS (LNAI), vol. 1968, pp. 71–85. Springer, Heidelberg (2000)
18. Lysyanskaya, A., Meyerovich, M.: Provably secure steganography with imperfect sampling. In: Yung, M., Dodis, Y., Kiayias, A., Malkin, T. (eds.) PKC 2006. LNCS, vol. 3958, pp. 123–139. Springer, Heidelberg (2006)

19. Quinlan, J.R.: Induction of decision trees. Machine Learning 1(1), 81–106 (1986)
20. Quinlan, J.R.: C4.5: Programs for Machine Learning. Morgan Kaufmann, San Mateo (1993)
21. Rogaway, P.: Nonce-based symmetric encryption. In: Roy, B., Meier, W. (eds.) FSE 2004. LNCS, vol. 3017, pp. 348–359. Springer, Heidelberg (2004)
22. Simmons, G.J.: The prisoners' problem and the subliminal channel. In: Crypto 1983, pp. 51–67. Plenum Press, New York (1984)
23. Westfeld, A.: F5-A Steganographic Algorithm. In: Moskowitz, I.S. (ed.) IH 2001. LNCS, vol. 2137, pp. 289–302. Springer, Heidelberg (2001)

Tight Bounds on Communication Complexity of Symmetric XOR Functions in One-Way and SMP Models[*]

Ming Lam Leung, Yang Li, and Shengyu Zhang

Department of Computer Science Engineering, The Institute of Theoretical Computer
Science and Communications, The Chinese University of Hong Kong
{mlleung,syzhang}@cse.cuhk.edu.hk, danielliy@gmail.com

Abstract. We study the communication complexity of symmetric XOR functions, namely functions $f : \{0,1\}^n \times \{0,1\}^n \to \{0,1\}$ that can be formulated as $f(x,y) = D(|x \oplus y|)$ for some predicate $D : \{0, 1, ..., n\} \to \{0,1\}$, where $|x \oplus y|$ is the Hamming weight of the bitwise XOR of x and y. We give a public-coin randomized protocol in the Simultaneous Message Passing (SMP) model, with the communication cost matching the known lower bound for the *quantum* and *two-way* model up to a logarithm factor. As a corollary, this closes a quadratic gap between the previous quantum lower bound and the randomized upper bound in the one-way model. This answers an open question raised in Shi and Zhang [SZ09], and disqualifies the problem from being a candidate to separate randomized and quantum one-way communication complexities.

1 Introduction

Communication complexity quantifies the minimum amount of communication needed for two (or sometimes more) parties to jointly compute some function f. Since introduced by Yao [Yao79], it has attracted significant attention in the last three decades, not only for its elegant mathematical structure but also for its numerous applications in other computational models [KN97, LS09].

The two parties involved in the computation, usually called Alice and Bob, can communicate in different manners, and here we consider the three well-studied models, namely the two-way model, the one-way model and the simultaneous message passing (SMP) model. In the two-way model, Alice and Bob are allowed to communicate interactively in both directions, while in the one-way model, Alice can send message to Bob and Bob does not give feedback to Alice. An even weaker communication model is the SMP model, where Alice and Bob are prohibited to exchange information directly, but instead they each send a message to a third party Referee, who then announces a result. A randomized protocol is called private-coin if Alice and Bob each flip their own and private random coins. If they share the same random coins, then the protocol is called

---

[*] This work was supported by Hong Kong General Research Fund No. 419309 and No. 418710.

M. Ogihara and J. Tarui (Eds.): TAMC 2011, LNCS 6648, pp. 403–408, 2011.

public-coin. The private-coin model differs from the public-coin model by at most an additive factor of $O(\log n)$ in the two-way and one-way models [New91].

We use $R^{priv}(f)$ to denote the communication complexity of a best private-coin randomized protocol that computes f with error at most $1/3$ in the two-way protocol. Similarly, we use the $R^{\|,priv}(f)$ to denote the communication complexity in the private-coin SMP model, and $R^{1,priv}(f)$ for the private-coin one-way model. Changing the superscript "priv" to "pub" gives the notation for the communication complexities in the public-coin models. If we allow Alice and Bob to use quantum protocols, then $Q(f), Q^1(f), Q^{\|}(f)$ represent the quantum communication complexity in two-way model, one-way model and SMP model, separately. In the quantum case the communication complexity is evaluated in terms of the number of qubits in the communication. If Alice and Bob share prior entanglement, then we use a star in the superscript to denote the communication complexity.

Arguably the most fundamental issue in communication complexity is to determine the largest gap between the quantum and classical complexities. In particular, there is no super-constant separation between quantum and classical complexities in the one-way model; actually, it could well be the truth that they are the same up to a constant factor for all total Boolean functions.

One way to understand the question is to study special classes of functions. An important class of Boolean functions is that of XOR functions, namely those in the form of $f(x \oplus y)$ where $x \oplus y$ is the bitwise XOR of x and y. Some well-studied functions such as the Equality function and the Hamming Distance function are special cases of XOR functions. XOR functions belong to a larger class of "composed functions"; see [LZ10] for some recent studies.

While the general XOR function seems hard to study, recently Shi and Zhang [SZ09] considered symmetric XOR functions, i.e. $f(x \oplus y) = D(|x \oplus y|)$ for some $D : \{0, 1, ..., n\} \to \{0, 1\}$. Define r_0 and r_1 to be the minimum integers such that $r_0, r_1 \leq n/2$ and $D(k) = D(k + 2)$ for all $k \in [r_0, n - r_1)$ and set $r = \max\{r_0, r_1\}$. Shi and Zhang proved that the quantum lower bound for symmetric XOR functions in the two-way model is $\Omega(r)$, and on the other hand, they also gave a randomized protocol in communication of $\tilde{O}(r)$ in the two-way model and $\tilde{O}(r^2)$ in the one-way model. Pinning down the quantum and randomized communication complexities of symmetric XOR functions in the one-way model was raised as an open problem.

In this work, we close the quadratic gap by proving a randomized upper bound of $\tilde{O}(r)$, which holds even for the SMP model. Namely,

Theorem 1. *For any symmetric XOR function f,*

$$R^{\|,pub}(f) = O(r\log^3 r / \log\log r) \qquad (1)$$

Combining this upper bound with Shi and Zhang's quantum lower bound in the two-way model, we have the following.

Corollary 1. *The (public-coin) randomized and quantum communication complexities of symmetric XOR functions are $\tilde{\Theta}(r)$ in the two-way, the one-way, and the SMP models.*

A good question for further exploration is the communication complexity of XOR functions in the *private-coin* SMP model.

2   Preliminaries

In this part we review some known results on the randomized and quantum communication complexities of the Hamming Distance function and the Equality function.

Let $\mathsf{Ham}_n^{(d)}$ be the boolean function such that $\mathsf{Ham}_n^{(d)}(x,y) = 1$ if and only if the two n-bit strings x and y have Hamming distance at most d. Yao [Yao03] showed a randomized upper bound of $O(d^2)$ in the public-coin SMP model, later improved by Gavinsky, Kempe and de Wolf [GKdW04] to $O(d \log n)$ and further by Huang, Shi, Zhang and Zhu [HSZZ06] to $O(d \log d)$. Let $\mathcal{HD}_{d,\epsilon}$ denote the $O(d \log d \log(1/\epsilon))$-cost randomized protocol by repeating the [HSZZ06] protocol for $O(\log(1/\epsilon))$ times so that the error probability is below ϵ.

The parity function $\mathsf{Parity}(x)$ is defined as $\mathsf{Parity}(x) = 1$ if and only if $|x|$ is odd.

A function $f : \{0,1\}^n \times \{0,1\}^n \rightarrow \{0,1\}$ is a symmetric XOR function if $f(x,y) = S(x \oplus y)$ for some symmetric function S. That is, $f(x,y) = D(|x \oplus y|)$ where $D : \{0,1,\ldots,n\} \rightarrow \{0,1\}$. Let $\tilde{D}(k) = D(n-k)$ and $\tilde{S}(x,y) = \tilde{D}(|x \oplus y|)$. Define r_0 and r_1 to be the minimum integers such that $r_0, r_1 \leq n/2$ and $D(k) = D(k+2)$ for all $k \in [r_0, n-r_1)$; set $r = \max\{r_0, r_1\}$. By definition, $D(k)$ only depends on the parity of k when $k \in [r_0, n-r_1]$. Suppose $D(k) = T(\mathsf{Parity}(k))$ for $k \in [r_0, n-r_1]$ (for some function T).

All the logarithms in this paper are to base 2.

3   A Public-Coin Protocol in the SMP Model

This section gives the protocol in Theorem 1. We will first give a subprocedure $\mathcal{P}_k$ which computes the function in the special case of $|x \oplus y| \leq k$. It is then used as a building block for the general protocol $\mathcal{P}$.

In the protocols we will use random partitions. A *random k-partition* of $[n]$ is a random function p mapping $[n]$ to $[k]$, *i.e.* mapping each element in $[n]$ to $[k]$ uniformly at random and independently. We call the set $\{i \in [n] : p(i) = j\}$ the *block $B(j)$*. A simple fact about the random partition is the following.

Lemma 1. *For any string $z \in \{0,1\}^n$ with at most k 1's, a random k-partition has*

$$\Pr[\text{All } k \text{ blocks have less than } c \text{ 1's}] \geq 1 - O(1/k^2). \tag{2}$$

where $c = 4 \log k / \log \log k$.

Proof. Consider the complement event. There are k possible blocks to violate the condition, $\binom{k}{c}$ choices for the c 1's (out of k 1's) put in the "bad" block, and for each of these 1's, the probability of it mapped to the block is $1/k$. Thus the union bound gives

$$\textbf{Pr}[\text{There exists a block with } c \text{ 1's}] \leq k \cdot \binom{k}{c} \cdot \frac{1}{k^c} \qquad (3)$$

$$\leq \left(\frac{ek}{c}\right)^c \cdot \frac{1}{k^{c-1}} = \left(\frac{e}{c}\right)^c \cdot k \qquad (4)$$

It is easily verified that the chosen c makes this bound $O(1/k^2)$.

Now the protocol P_k is as in **Box** $\mathcal{P}_k$. Recall that $\mathcal{HD}_{d,\epsilon}$ is the $O(d \log d \log(1/\epsilon))$-communication randomized protocol for the Hamming Distance problem $\text{Ham}_n^{(d)}$, with error probability below ϵ.

Box $\mathcal{P}_k$:

A public-coin randomized protocol $\mathcal{P}_k$ for functions $f(x,y) = D(|x \oplus y|)$, with promise $|x \oplus y| \leq k$, in the SMP model

Input: $x \in \{0,1\}^n$ to Alice and $y \in \{0,1\}^n$ to Bob, with promise $|x \oplus y| \leq k$
Output: One bit $\bar{f}$ by Referee satisfying $\bar{f} = f(x,y)$ with probability at least 0.9.

Protocol:
Alice and Bob:

1. Use public coins to generate a common random k-partition $[n] = \biguplus_{i=1}^k B(i)$.
2. **for** $i = 1$ **to** k
     for $j = 0$ **to** $c = 4 \log k / \log \log k$
        run (Alice and Bob's part of) the protocol $\mathcal{HD}_{j,\epsilon}$ on input $(x_{B(i)}, y_{B(i)})$ with $\epsilon = 1/(10k \log c)$, sending a pair of messages $(m_{a,i,j}(x_{B(i)}), m_{b,i,j}(y_{B(i)}))$.

Referee:

1. **for** $i = 1$ **to** k
    (a) On receiving $\{(m_{a,i,j}(x_{B(i)}), m_{b,i,j}(y_{B(i)})) : j = 1, \ldots, c\}$, run (Referee's part of) the protocol $\mathcal{HD}_{j,\epsilon}$ which outputs h_{ij}.
    (b) Use binary search in $(h_{i1}, \ldots, h_{ic})$ to find the Hamming distance h_i of $(x_{B(i)}, y_{B(i)})$.
2. Output $D(\sum_{i=1}^k h_i)$.

Lemma 2. *If $|x \oplus y| \leq k$, then Referee outputs $D(|x \oplus y|)$ with probability at least 0.9. The cost of protocol $\mathcal{P}_k$ is $O(k \log^3 k / \log \log k)$.*

Proof. First, by Lemma 1, each block contains at most c different indices i s.t. $x_i \neq y_i$. Namely, the Hamming distance of $x_{B(i)}$ and $y_{B(i)}$ is at most c. Thus running the protocols $\mathcal{HD}_{j,\epsilon}$ for $j = 0, ..., c$ would give information to find the Hamming distance h_i of $(x_{B(i)}, y_{B(i)})$. In each block $B(i)$, h_i is correctly computed as long as each of the $\lceil \log c \rceil$ values h_{ij} on the (correct) path of the binary search is correct. Thus a union bound gives the overall error probability upper bounded by $k(\log c)\epsilon = 1/10$. The cost of the protocol is $O(k \cdot c \cdot c \log c \log(1/\epsilon)) = O(k \log^3 k / \log \log k)$.

With the protocol $\mathcal{P}_k$ in hand, we now construct the general protocol as in **Box $\mathcal{P}$.**

Box $\mathcal{P}$:
A public-coin randomized protocol $\mathcal{P}$ for functions $f(x, y) = S(x \oplus y)$ in the SMP model

Input: $x \in \{0, 1\}^n$ to Alice and $y \in \{0, 1\}^n$ to Bob
Output: One bit b which equals to $f(x, y)$ with probability at least $2/3$.

Protocol:

1. Run the protocol $\mathcal{HD}_{r_0, 1/10}$ on (x, y) and the protocol $\mathcal{HD}_{r_1, 1/10}$ on $(\bar{x}, y)$.
2. Run the protocol $\mathcal{P}_{r_0}$ for function S on (x, y) and the protocol $\mathcal{P}_{r_1}$ for function $\tilde{S}$ on $(\bar{x}, y)$.
3. Alice: send $\mathsf{Parity}(x)$
4. Bob: send $\mathsf{Parity}(y)$.
5. Referee:
   (a) If $\mathcal{HD}_{r_0, 1/10}$ on (x, y) outputs 1, then output $\mathcal{P}_{r_0}$ on (x, y) and halt.
   (b) If $\mathcal{HD}_{r_1, 1/10}$ on $(\bar{x}, y)$ outputs 1, then output $\mathcal{P}_{r_1}$ on $(\bar{x}, y)$ and halt.
   (c) Output $T(\mathsf{Parity}(x) \oplus \mathsf{Parity}(y))$.

Theorem 2. *The protocol $\mathcal{P}$ outputs the correct value with probability at least $2/3$, and the complexity cost is $O(r \log^3 r / \log \log r)$.*

Proof. Correctness: If $|x \oplus y| \leq r_0$, then with probability at least 0.9, the protocol $\mathcal{HD}_{r_0, 1/10}(x, y)$ outputs 1, thus Referee outputs $\mathcal{P}_{r_0}$ on (x, y), which equals to $f(x, y)$ with probability at least 0.9 by the correctness of the protocol $\mathcal{P}_{r_0}$. Thus the overall success probability is at least $0.81 > 2/3$.

If $|x \oplus y| \geq n - r_1$, then $|\bar{x} \oplus y| \leq r_1$ and with probability at least 0.9, the protocol $\mathcal{HD}_{r_1, 1/10}(\bar{x}, y)$ outputs 1, thus Referee outputs $\mathcal{P}_{r_1}(\tilde{S}, \bar{x}, y)$, which equals to

$$\tilde{S}(\bar{x} \oplus y) = \tilde{D}(n - |x \oplus y|) = D(|x \oplus y|) \tag{5}$$

with probability at least 0.9 by the correctness of the protocol $\mathcal{P}_{r_1}$. Thus the overall success probability is at least $0.81 > 2/3$.

If $r_0 < |x \oplus y| < n - r_1$, then the protocol proceeds to the very last step with probability at least $1 - 0.1 - 0.1 = 0.8$. And once this happens, then Referee outputs the correct value with certainty, since $f(x,y) = T(\mathsf{Parity}(x \oplus y)) = T(\mathsf{Parity}(x) \oplus \mathsf{Parity}(y))$.

Complexity: The cost is twice of the cost of the protocol P_r, plus twice of the cost of the protocol $\mathcal{HD}_{r,1/10}$, plus 2, which in total is $O(r \log^3 r / \log \log r)$.

References

[GKdW04]   Gavinsky, D., Kempe, J., de Wolf, R.: Quantum communication cannot simulate a public coin, arXiv:quant-ph/0411051 (2004)

[HSZZ06]   Huang, W., Shi, Y., Zhang, S., Zhu, Y.: The communication complexity of the hamming distance problem. Information Processing Letters 99(4), 149–153 (2006)

[KN97]   Kushilevitz, E., Nisan, N.: Communication Complexity. Cambridge University Press, Cambridge (1997)

[LS09]   Lee, T., Shraibman, A.: Lower bounds on communication complexity. Foundations and Trends in Theoretical Computer Science 3(4), 263–398 (2009)

[LZ10]   Lee, T., Zhang, S.: Composition theorems in communication complexity. In: Abramsky, S., Gavoille, C., Kirchner, C., Meyer auf der Heide, F., Spirakis, P.G. (eds.) ICALP 2010. LNCS, vol. 6198, pp. 475–489. Springer, Heidelberg (2010)

[New91]   Newman, I.: Private vs. common random bits in communication complexity. Information Processing Letters 39(2), 67–71 (1991)

[SZ09]   Shi, Y., Zhang, Z.: Communication complexities of XOR functions. Quantum Information and Computation 9(3&4), 255–263 (2009)

[Yao79]   Yao, A.: Some complexity questions related to distributive computing. In: Proceedings of the Eleventh Annual ACM Symposium on Theory of Computing (STOC), pp. 209–213 (1979)

[Yao03]   Yao, A.: On the power of quantum fingerprinting. In: Proceedings of the Thirty-Fifth Annual ACM Symposium on Theory of Computing, pp. 77–81 (2003)

The Hardness of Median in the Synchronized Bit Communication Model

Karolina Sołtys

Faculty of Mathematics, Informatics and Mechanics
University of Warsaw
ksoltys@students.mimuw.edu.pl

Abstract. The synchronized bit communication model, defined recently by Impagliazzo and Williams in [1], is a communication model which allows the participants to share a common clock. The main open problem posed in this paper was the following: does the synchronized bit model allow a logarithmic speed-up for all functions over the standard deterministic model of communication? We resolve this question in the negative by showing that the Median function, whose communication complexity is $O(\log n)$, does not admit polytime synchronized bit protocol with communication complexity $O\left(\log^{1-\varepsilon} n\right)$ for any $\varepsilon > 0$. Our results follow by a new round-communication trade-off for the Median function in the standard model, which easily translates to its hardness in the synchronized bit model.

Keywords: communication complexity, median, synchronized bit model, round complexity.

1 Introduction

Communication complexity, introduced by Yao in 1979 [6], is an important concept in complexity theory which tries to determine the amount of communication needed to compute a function whose input has been distributed among two or more participants. A natural question, especially in the context of the protocols used in distributed computing, is whether *synchronous* protocols, in which the participants can use a common clock, are more powerful than *asynchronous* ones, in which the players do not have this ability (note that the standard protocols studied in communication complexity are asynchronous). The synchronization allows the participants to convey some information by *not* sending a message at a given moment of time, which naturally leads us to examine the time-communication trade-off of the protocol.

In their recent paper [1], Impagliazzo and Williams formalized the notion of synchronous protocols and partially solved several interesting questions related to them by introducing two models of communication, called the synchronized bit model and the synchronized connection model, and studying their complexity. We can briefly summarize the synchronized bit complexity model as an extension of the standard deterministic model of communication, where a player

M. Ogihara and J. Tarui (Eds.): TAMC 2011, LNCS 6648, pp. 409–415, 2011.
© Springer-Verlag Berlin Heidelberg 2011

in one step can send 0, 1 or a blank, and where blanks do not count towards the communication complexity of the protocol. It is interesting to consider the polytime bit complexity of a problem Π: the minimum complexity of any synchronized bit protocol for the problem Π using a polynomial number of steps. This function is denoted by $PB(\Pi)$.

The authors prove in [1] the following bounds for the polytime bit complexity:

$$\Omega\left(\frac{D(\Pi)}{\log n}\right) \leq PB(\Pi) \leq O\left(\frac{D(\Pi)}{\log\log n}\right) \tag{1}$$

and conclude their paper with the following questions:

Question 1. Can the upper bound on the polytime bit complexity in (1) be improved?

Question 2. What is the complexity of Median in this model?

We solve Question 2 and give a partial answer to Question 1 by proving the following result:

Theorem 1. *The Median function does not admit a polytime synchronized bit protocol with communication complexity* $O\left(\log^{1-\varepsilon} n\right)$ *for any* $\varepsilon > 0$.

Since the deterministic complexity of the Median function is $O(\log n)$, we get the following lower bound for the synchronized bit complexity of this problem:

$$\omega\left(\frac{D(\text{MEDIAN})}{\log^{\varepsilon} n}\right) \leq PB(\text{MEDIAN})$$

for each $\varepsilon > 0$, which of course forbids any significant improvement to the upper bound in (1) in general.

Theorem 1 can be easily translated, using methods established in [1], in terms of the round-communication trade-off in the standard deterministic model.

Theorem 2. *The Median function does not admit a deterministic protocol using* $O\left(\log^{1-\varepsilon} n\right)$ *rounds and a logarithmic amount of communication at each round for any* $\varepsilon > 0$.

Our result provides a new round-communication trade-off for the Median function. The study of the round-communication trade-offs for various functions is an important area of communication complexity with significant applications to streaming algorithms (lower bounds for the rounds-communication trade-off in the deterministic communication complexity model imply the same bounds for the number of passes-memory trade-off in the streaming model; the opposite implication usually does not hold). With Median being a central problem in the streaming model, the fact that our approach can be used to prove some lower bounds for it in this model (albeit slightly weaker than those already known, [3]) in a completely different manner, can potentially be quite fruitful.

To facilitate our proofs we define a natural problem, Strategy, which is easily seen as complete for the class of communication problems solvable in $O(\log n)$

rounds and $O(\log n)$ communication. Our reduction from Strategy to Median allows us to show that Median is also a complete problem for that class.

The structure of the paper is as follows. In section 2 we briefly describe the relation between synchronized bit complexity and round complexity, and define the problem central to our paper, Strategy. In the following section we prove a round-communication trade-off for the Strategy problem by showing a reduction from the $k(\cdot)$-Pointer-Jumping problem, whose round-communication trade-off has been extensively studied ([4], [5]). The reduction uses an intermediate problem – $k(\cdot)$-Level-Strategy. Although both reductions are quite straightforward, they do not preserve the size of the instance, which leads us to a system of asymptotic inequalities on the lower bound for Strategy. We then show a reduction from the Strategy problem to Median.

2   Preliminaries

We assume that the reader is familiar with the basic notions of communication complexity [2].

We will use the straightforward two-way translation between the synchronized bit protocols running in time $O(t)$ and using $O(b)$ bits of communication and deterministic protocols using $O(b)$ rounds and $O(\log t)$ communication at each round, which is explained in detail in [1]. We can therefore approach the problem in terms of round complexity: the existence of polytime synchronized bit protocol using $O(b)$ bits of communication is equivalent to the existence of a deterministic protocol using $O(b)$ rounds and $O(\log n)$ communication at each round.

Definition 1. *We define the problem Strategy of size n as follows. Let T be a full binary tree with n vertices. A function f assigns to each vertex of the tree a number from the set $\{0, 1\}$. Alice knows the values of f in the vertices in the odd layers of the tree (the vertices with odd depth), and Bob knows the values of f in the vertices in the even layers. We define the* leaf reached by f *to be the leaf which is an endpoint of the path starting at the root and going always downwards – to the left son of v if $f(v) = 0$ and to the right son otherwise. The players' goal is to determine the index of the leaf they reach.*

Remark 1. The Strategy problem is complete for the class of communication problems solvable in $O(\log n)$ rounds and $O(\log n)$ communication.

Proof. We will reduce an arbitrary problem Π of this class to the Strategy problem. The problem Π has a deterministic protocol using $O(\log n)$ rounds and $O(\log n)$ communication. Without loss of generality we can assume that the protocol makes the players alternate in sending messages containing just one bit. In each Alice's (and analogously Bob's) vertex of the protocol tree and corresponding to a communication history consistent with her input, the message she sends depends only on her input and the communication history (in all the other vertices of Alice we fix her message in an arbitrary way). We can now transform the protocol tree into a Strategy tree by setting the value of the function f in each vertex of the tree to the message sent in that vertex, and assigning to the leaves the outputs of the protocol for the given communication history.   □

Remark 2. The Strategy problem can be solved in $O\left(\frac{\log n}{\log\log n}\right)$ rounds and $O(\log n)$ communication at each round.

This upper bound may be easily obtained by using the reductions described in [1] to change the model to the synchronized bit model, use the upper bound proved therein for this model, and then translate the model back to the standard deterministic model.

Definition 2. *We define the problem $k(\cdot)$-Level-Strategy of size n as follows. We have an n-ary tree T of height $k(n)$, with leaves indexed from 1 to $n^{k(n)}$. There is a function $f : T \to [n]$ for each vertex $v \in T$, Alice knows this function for the vertices in the odd layers and Bob knows it for the vertices in the even layers. As in Strategy, they want to determine the index of the leaf they descend to starting from the root and following the function f ($f(v) = l$ means that if they arrive to the vertex v they descend to the l-th son of v). Note that n is a parameter, and not the input size; both Alice and Bob have input of size $O\left(n^k(n)\log n\right)$.*

Definition 3. *The problem $k(\cdot)$-Pointer-Jumping is defined as follows. Alice and Bob each hold a list of n pointers, each pointing to a pointer in the list of the other. An initial pointer v_0 is marked. They want to determine the $k(n)$-th pointer they reach after following the pointers starting from v_0.*

In [4] it was proved that if we allow just $k(n) - 1$ rounds then $k(n)$-Pointer-Jumping requires $\Omega(n)$ communication.

3   Round Complexity of Strategy

We will prove the following theorem by showing a sequence of reductions from Pointer Jumping to Strategy:

Theorem 3. *The Strategy function does not admit a deterministic protocol using $O\left(\log^{1-\varepsilon} n\right)$ rounds and a logarithmic amount of communication at each round for any $\varepsilon > 0$.*

Let $r(n)$ be some function such that $r(\Theta(n)) = \Theta(r(n))$ (we will fix it later). We will prove the following easy lemmas:

Lemma 1. *If the Strategy problem of size m can be solved in $O(r(m))$ rounds using $O(r(m)\log m)$ communication, then, for each $k(\cdot)$, $k(\cdot)$-Level-Strategy can be solved in $O\left(r\left(n^{k(n)}\right)\right)$ rounds using $O\left(r\left(n^{k(n)}\right)\log n^{k(n)}\right)$ communication.*

Proof. For each $k(\cdot)$, there is a simple reduction from $k(\cdot)$-Level-Strategy of size n (i. e. with $n^{k(n)}$ leaves) to Strategy of size $O\left(n^{k(n)}\right)$. We will create a Strategy tree S by replacing each Alice's vertex v of the tree T in $k(\cdot)$-Level-Strategy with a binary tree of height $\lceil \log n \rceil$, with n leftmost leaves corresponding to the sons of v in T. On every vertex of even depth in the subtree we fix Bob's input to be 0, and we fix Alice's input in the subtree so that the correct leaf is reached. Note that if n is not a power of two, then the Strategy tree S will be slightly larger than T, with additional vertices and not corresponding to vertices of T, but that is inconsequential, because our construction assures that these vertices are never reached by the protocol. It is easy to see that this is a proper reduction.   □

Lemma 2. *If $k(\cdot)$-Level-Strategy of size n can be solved in $O\left(r\left(n^{k(n)}\right)\right)$ rounds using $O\left(r\left(n^{k(n)}\right) \log n^{k(n)}\right)$ communication, then $k(\cdot)$-Pointer-Jumping of size n can also be solved in $O\left(r\left(n^{k(n)}\right)\right)$ rounds using $O\left(r\left(n^{k(n)}\right) \log n^{k(n)}\right)$ communication.*

Proof. Given an instance G of $k(\cdot)$-Pointer-Jumping of size n we will create an instance of $k(\cdot)$-Level-Strategy of size n (with $n^{k(n)}$ leaves), which will be, informally speaking, a tree of possible paths of length $k(n)$ in the graph G. In every odd layer (resp. even layer) every vertex that is an i-th son will point its j-th son if and only if $g_A(i) = j$, where g_A is the Alice's (resp. Bob's) input in the Pointer-Jumping instance. It is easy to see that in this reduction every player can locally compute their input, and that the vertex reached by their functions in $k(\cdot)$-Level-Strategy is an i-th son if and only if the output for the $k(\cdot)$-Pointer-Jumping is i.   □

By combining the two lemmas we obtain the following

Corollary 1. *If the Strategy problem of size m can be solved in $O(r(m))$ rounds using $O(r(m) \log m)$ communication, then, for each $k(\cdot)$, $k(\cdot)$-Pointer-Jumping of size n can be solved in $O\left(r\left(n^{k(n)}\right)\right)$ rounds using $O\left(r\left(n^{k(n)}\right) \log n^{k(n)}\right)$ communication.*

We can now prove the main theorem of this section.

Proof (of Theorem 3). In [4] it was shown that if we allow no more than $k(n) - 1$ rounds then $k(\cdot)$-Pointer-Jumping of size n requires $\Omega(n)$ communication. We thus know that for every $k(\cdot)$ the protocol for Strategy must yield, after using the reductions described, a protocol for $k(\cdot)$-Pointer-Jumping using either a greater number of rounds:

$$r\left(n^{k(n)}\right) \geq k(n)$$

or a greater amount of communication:

$$r\left(n^{k(n)}\right) \log n^{k(n)} \geq \Omega(n).$$

A function $r(n)$ which for any $k(n)$ violates both of these inequalities is thus a viable lower bound for Strategy, that is no $O(r(n))$-round, $O(r(n) \log n)$-communication protocol for Strategy may exist.

It is easy to check that for all $\varepsilon > 0$ the function $r(n) = \log^{1-\varepsilon} n$ fails to satisfy both of these inequalities when we set $k(n) = \sqrt{\frac{n}{\log n}}$, which proves Theorem 3.   □

Note here that if we were to prove a lower bound tightly matching the upper bound for Strategy, that is $O\left(\frac{\log n}{\log\log n}\right)$ (proved in Remark 2), which we believe may be the case, we would need to use a different method, because close examination of the inequalities obtained by setting $r(n) = \frac{\log n}{\log\log n}$ reveals that, regardless of the function $k(n)$, at least one of the inequalities must be satisfied.

4 Reduction from Strategy to Median

Proposition 1. *If the Median problem of size n can be solved in $O(r(n))$ rounds using $O(c(n))$ communication Strategy of size n can also be solved in $O(r(n))$ rounds using $O(c(n))$ communication.*

Proof. We will show here a reduction from Strategy(T, f) to Median(S, A, B), where S is a set of natural numbers and A, B are subsets of this set held by Alice and Bob respectively. The reduction will work inductively on the height k of the tree T of the Strategy problem.

It is easy to reduce Strategy on trees of height 1 to Median over $S = \{0, 1\}$: we give the empty subset to Bob and to Alice either the subset $\{0\}$ or $\{1\}$ depending on the value of f in the root. The two possible values of Median will correspond to the two possible leaves reached by f.

Let us denote by l_i the size of the set S produced by the reduction for the trees of size i, and by w_i the the number of elements given to Alice and Bob ($w_i = |A \cup B|$); we will construct the reduction inductively so that l_i and w_i are well defined.

We will now show the induction step for the trees of height k. Let T_l be the tree of height $k - 1$ rooted at the left son of the root, ant T_r be the tree rooted at the right son. We denote by r the root of the tree T and by A_l the subset of S_{k-1} given to Alice by the reduction from Strategy$(T_l, f|_{T_l})$ (we define A_r, B_l, B_r analogously). The reduction will create the sets:

$$S = S_k = \{1, ..., l_k\}, \text{ where } l_k = 2w_{k-1} + 2l_{k-1}$$

$$A = (B_l + w_{k-1}) \cup (B_r + (w_{k-1} + l_{k-1})) \cup$$
$$\cup \begin{cases} \{1, ..., w_{k-1}\} & \text{if } f(r) = 0 \\ \{l_k - w_{k-1} + 1, ..., l_k\} & \text{otherwise} \end{cases}$$

$$B = (A_l + w_{k-1}) \cup (A_r + w_{k-1} + l_{k-1})$$

where $C + d = \{c + d \mid c \in C\}$.

It is easy to check that this is a proper reduction. The basic idea is that we give Alice some amount of small numbers if she turns left in r, and the same amount of big numbers if she turns to the right, so that the problem reduces either to finding the median in the subproblem of Median corresponding to T_l or to finding the median in the subproblem corresponding to T_r, with the roles of the players reversed (because it is now Bob who holds the roots of T_l and T_r).

Easy calculations of the recurrence relations for w_k and l_k show that these functions are exponential in k, so the reduction from Strategy on the tree of size n produces an instance of Median of size $O(n)$.   □

Combining this reduction with Theorem 3 yields the proof of our main result, Theorem 1.

It is worth noticing that this reduction, together with Remark 1, proves as well that Median is complete for the class of communication problems solvable in $O(\log n)$ rounds and $O(\log n)$ communication.

5   Conclusions

Our results still hold in the randomized case.

It is possible that a stronger lower bound for Median in the synchronized bit complexity model, tightly matching the upper bound of $O\left(\frac{\log n}{\log \log n}\right)$, may be proven. It would also be interesting to extend the synchronized bit model to the multiparty case and to study the complexity of the model in this setting.

In [1] the authors define also another synchronized model: the connection complexity model, which is based on the assumption that in each timestep every party decides whether to try to establish a connection. Some information is exchanged if and only if a connection has been established, and only successful connections count toward the communication cost of the protocol. The model turns out to be surprisingly powerful, enabling the participants to solve the Disjointness problem in polynomial time and only one bit of communication. We believe that also for this model the possible multiparty extension seems worth further examination.

Acknowledgements. I would like to thank Prof. Iordanis Kerenidis for encouraging me to write this paper and for fruitful discussions.

References

1. Impagliazzo, R., Williams, R.: Communication complexity with synchronized clocks. In: Proceedings of the 2010 IEEE 25th Annual Conference on Computational Complexity, CCC 2010, pp. 259–269. IEEE Computer Society, Washington, DC, USA (2010)
2. Kushilevitz, E., Nisan, N.: Communication complexity (1997)
3. Munro, J.I., Paterson, M.S.: Selection and sorting with limited storage. Technical report (1978)
4. Nisan, N., Wigderson, A.: Rounds in communication complexity revisited. SIAM J. Comput. 22, 211–219 (1993)
5. Ponzio, S.J., Radhakrishnan, J., Venkatesh, S.: The communication complexity of pointer chasing. J. Comput. Syst. Sci. 62, 323–355 (2001)
6. Yao, A.C.-C.: Some complexity questions related to distributive computing (preliminary report). In: Proceedings of the Eleventh Annual ACM Symposium on Theory of Computing, STOC 1979, pp. 209–213. ACM, New York (1979)

Lower Bounds for the Smoothed Number of Pareto Optimal Solutions*

Tobias Brunsch and Heiko Röglin

Department of Computer Science, University of Bonn, Germany
brunsch@cs.uni-bonn.de, heiko@roeglin.org

Abstract. In 2009, Röglin and Teng showed that the smoothed number of Pareto optimal solutions of linear multi-criteria optimization problems is polynomially bounded in the number n of variables and the maximum density ϕ of the semi-random input model for any fixed number of objective functions. Their bound is, however, not very practical because the exponents grow exponentially in the number $d+1$ of objective functions. In a recent breakthrough, Moitra and O'Donnell improved this bound significantly to $O\big(n^{2d}\phi^{d(d+1)/2}\big)$.

An "intriguing problem", which Moitra and O'Donnell formulate in their paper, is how much further this bound can be improved. The previous lower bounds do not exclude the possibility of a polynomial upper bound whose degree does not depend on d. In this paper we resolve this question by constructing a class of instances with $\Omega\big((n\phi)^{(d-\log(d))\cdot(1-\Theta(1/\phi))}\big)$ Pareto optimal solutions in expectation. For the bi-criteria case we present a higher lower bound of $\Omega(n^2\phi^{1-\Theta(1/\phi)})$, which almost matches the known upper bound of $O(n^2\phi)$.

1 Introduction

In multi-criteria optimization problems we are given several objectives and aim at finding a solution that is simultaneously optimal in all of them. In most cases the objectives are conflicting and no such solution exists. The most popular way to deal with this problem is to just concentrate on the relevant solutions. If a solution is *dominated* by another solution, i.e., it is worse than the other solution in at least one objective and not better in the others, then this solution does not have to be considered for our optimization problem. All solutions that are not dominated by any other solution are called *Pareto optimal* and form the so-called *Pareto set*. For a general introduction to multi-criteria optimization problems, we refer the reader to the book of Matthias Ehrgott [Ehr05].

Smoothed Analysis. For many multi-criteria optimization problems the worst-case size of the Pareto set is exponential. However, worst-case analysis is often too pessimistic, whereas average-case analysis assumes a certain distribution

---

* A part of this work was done at Maastricht University and was supported by a Veni grant from the Netherlands Organisation for Scientific Research.

M. Ogihara and J. Tarui (Eds.): TAMC 2011, LNCS 6648, pp. 416–427, 2011.
© Springer-Verlag Berlin Heidelberg 2011

on the input universe. Usually it is hard if not impossible to find a distribution resembling practical instances. *Smoothed analysis*, introduced by Spielman and Teng [ST04] to explain the efficiency of the simplex algorithm in practice despite its exponential worst-case running time, is a combination of both approaches and has been successfully applied to a variety of fields like machine learning, numerical analysis, discrete mathematics, and combinatorial optimization in the past decade (see [ST09] for a survey). Like in a worst-case analysis the model of smoothed analysis still considers adverserial instances. In contrast to the worst-case model, however, these instances are subsequently slightly perturbed at random, for example by Gaussian noise. This assumption is made to model that often the input an algorithm gets is subject to imprecise measurements, rounding errors, or numerical imprecision. In a more general model of smoothed analysis, introduced by Beier and Vöcking [BV04], the adversary is even allowed to specify the probability distribution of the random noise. The influence he can exert is described by a parameter ϕ denoting the maximum density of the noise.

Optimization Problems and Smoothed Input Model. Beier and Vöcking [BV04] have initiated the study of binary bi-criteria optimization problems. In their model, which has been extended to multi-criteria problems by Röglin and Teng [RT09], one considers optimization problems that can be specified in the following form. There are an arbitrary set $\mathcal{S} \subseteq \{0,1\}^n$ of *solutions* and $d+1$ objective functions $w^j : \mathcal{S} \to \mathbb{R}, j = 0, \dots, d$, given. While w^0 can be an arbitrary function, which is to be minimized, the functions $w^1, \dots, w^d$, which are to be maximized, are linear of the form $w^j(s) = w_1^j s_1 + \dots + w_n^j s_n$ for $s = (s_1, \dots, s_n) \in \mathcal{S}$. Formally, the problem can be described as follows:

minimize $w^0(s)$,   and   **maximize** $w^j(s)$ for all $j = 1, \dots, d$
subject to s in the feasible region $\mathcal{S}$.

As there are no restrictions on the set $\mathcal{S}$ of solutions, this model is quite general and can encode many well-studied problems like, e.g., the multi-criteria knapsack, shortest path, or spanning tree problem. Let us remark that the choice which objective functions are to be maximized and minimized is arbitrary and just chosen for ease of presentation. All results also hold for other combinations of objective functions.

In the framework of smoothed analysis the coefficients $w_1^j, \dots, w_n^j$ of the linear functions w^j are drawn according to (adversarial) probability density functions $f_{i,j} : [-1,1] \to \mathbb{R}$ that are bounded by the maximum density parameter ϕ, i.e., $f_{i,j} \leq \phi$ for $i = 1, \dots, n$ and $j = 1, \dots, d$. The adversary could, for example, choose for each coefficient an interval of length $1/\phi$ from which it is chosen uniformly at random. Hence, the parameter ϕ determines how powerful the adversary is. For large ϕ he can specify the coefficients very precisely, and for $\phi \to \infty$ the smoothed analysis becomes a worst-case analysis. The coefficients are restricted to the interval $[-1,1]$ because otherwise, the adversary could diminish the effect of the perturbation by choosing large coefficients.

Previous Work. Beier and Vöcking [BV04] showed that for $d = 1$ the expected size of the Pareto set of the optimization problem above is $O(n^4\phi)$ regardless of how the set $\mathcal{S}$, the objective function w^0 and the densities $f_{i,j}$ are chosen. Later, Beier, Röglin, and Vöcking [BRV07] improved this bound to $O(n^2\phi)$ by analyzing the so-called *loser gap*. Röglin and Teng [RT09] generalized the notion of this gap to higher dimensions, i.e., $d \geq 2$, and gave the first polynomial bound in n and ϕ for the smoothed number of Pareto optimal solutions. Furthermore, they were able to bound higher moments. The degree of the polynomial, however, was $d^{\Theta(d)}$. Recently, Moitra and O'Donnell [MO10] showed a bound of $O(n^{2d}\phi^{d(d+1)/2})$, which is the first polynomial bound for the expected size of the Pareto set with degree polynomial in d. An "intriguing problem" with which Moitra and O'Donnell conclude their paper is whether their upper bound could be significantly improved, for example to $f(d, \phi)n^2$. Moitra and O'Donnell suspect that for constant ϕ there should be a lower bound of $\Omega\left(n^d\right)$. In this paper we resolve this question almost completely.

Our Contribution. For the bi-criteria case, i.e., $d = 1$, we prove a lower bound of $\Omega\left(\min\left\{n^2\phi^{1-\Theta(1/\phi)}, 2^{\Theta(n)}\right\}\right)$. This is the first bound with dependence on n and ϕ and it nearly matches the upper bound $O(\min\left\{n^2\phi, 2^n\right\})$. For $d \geq 2$ we prove a lower bound of $\Omega\left(\min\left\{(n\phi)^{(d-\log(d))\cdot(1-\Theta(1/\phi))}, 2^{\Theta(n)}\right\}\right)$. Note that throughout the paper "log" denotes the binary logarithm. This is the first bound for the general multi-criteria case. Still, there is a significant gap between this lower bound and the upper bound of $O(\min\left\{n^{2d}\phi^{d(d+1)/2}, 2^n\right\})$, but the exponent of n is nearly $d - \log(d)$. Hence our lower bound is close to the lower bound of $\Omega\left(n^d\right)$ conjectured by Moitra and O'Donnell.

Restricted Knapsack Problem. To prove the lower bounds stated above we consider a variant of the knapsack problem where we have n objects $a_1, \ldots, a_n$, each with a weight w_i and a profit vector $p_i \in [0, 1]^d$ for a positive integer d. By a vector $s \in \{0, 1\}^n$ we describe which objects to put into the knapsack. In contrast to the unrestricted variant not all combinations of objects are allowed. Instead, all valid combinations are described by a set $\mathcal{S} \subseteq \{0, 1\}$. We want to simultaneously minimize the total weight and maximize all total profits of a solution s. Thus, the *restricted knapsack problem*, denoted by $K_{\mathcal{S}}(\{a_1, \ldots, a_n\})$, can be written as

$$\textbf{minimize } \sum_{i=1}^n w_i \cdot s_i, \quad \text{and} \quad \textbf{maximize } \sum_{i=1}^n (p_i)_j \cdot s_i \text{ for all } j = 1, \ldots, d$$
$$\textbf{subject to } s \text{ in the feasible region } \mathcal{S}.$$

For $\mathcal{S} = \{0, 1\}^n$ we just write $K(\{a_1, \ldots, a_n\})$ instead of $K_{\mathcal{S}}(\{a_1, \ldots, a_n\})$.

Note that the instances of the restricted knapsack problem that we use to prove the lower bounds are not necessarily interesting on its own because they have a somewhat artificial structure. However, they are interesting as they show that the known upper bounds in the general model cannot be significantly improved.

2   The Bi-criteria Case

In this section we present a lower bound for the expected number of Pareto optimal solutions in bi-criteria optimization problems that shows that the upper bound of Beier, Röglin, and Vöcking [BRV07] cannot be significantly improved.

Theorem 1. *There is a class of instances for the restricted bi-criteria knapsack problem for which the expected number of Pareto-optimal solutions is lower bounded by*

$$\Omega\left(\min\left\{n^2\phi^{1-\Theta(1/\phi)}, 2^{\Theta(n)}\right\}\right),$$

where n is the number of objects and ϕ is the maximum density of the profits' probability distributions.

Note that the exponents of n and ϕ in this bound are asymptotically the same as the exponents in the upper bound $O(\min\{n^2\phi, 2^n\})$ proved by Beier, Röglin, and Vöcking [BRV07].

For our construction we use the following bound from Beier and Vöcking.

Theorem 2 ([BV04]). *Let $a_1, \ldots, a_n$ be objects with weights $2^1, \ldots, 2^n$ and profits $p_1, \ldots, p_n$ that are independently and uniformly distributed in $[0, 1]$. Then, the expected number of Pareto optimal solutions of $K(\{a_1, \ldots, a_n\})$ is $\Omega(n^2)$.*

Note that scaling all profits does not change the Pareto set and hence Theorem 2 remains true if the profits are chosen uniformly from $[0, a]$ for an arbitrary $a > 0$. We will exploit this observation later in our construction.

The idea how to create a large Pareto set is what we call the copy step. Let us consider an additional object b with weight 2^{n+1} and fixed profit q. In Figure 1 all solutions are represented by a weight-profit pair in the weight-profit space. The set of solutions using object b can be considered as the set of solutions that do not use object b, but shifted by $(2^{n+1}, q)$. If the profit q is chosen sufficiently large, i.e., larger than the sum of the profits of the objects $a_1, \ldots, a_n$, then there is no domination between solutions from different copies and hence the Pareto optimal solutions of $K(\{a_1, \ldots, a_n, b\})$ are just the copies of the Pareto optimal solutions of $K(\{a_1, \ldots, a_n\})$. Lemma 3 formalizes this observation.

Lemma 3. *Let $a_1, \ldots, a_n$ be objects with weights $2^1, \ldots, 2^n$ and non-negative profits $p_1, \ldots, p_n$ and let b be an object with weight 2^{n+1} and profit $q > \sum_{i=1}^n p_i$. Furthermore, let $\mathcal{P}$ denote the Pareto set of $K(\{a_1, \ldots, a_n\})$ and let $\mathcal{P}'$ denote the Pareto set of $K(\{a_1, \ldots, a_n, b\})$. Then, $\mathcal{P}'$ is the disjoint union of $\mathcal{P}'_0 := \{(s, 0) : s \in \mathcal{P}\}$ and $\mathcal{P}'_1 := \{(s, 1) : s \in \mathcal{P}\}$ and thus $|\mathcal{P}'| = 2 \cdot |\mathcal{P}|$.*

Now we use the copy idea to construct a large Pareto set. Let $a_1, \ldots, a_{n_p}$ be objects with weights $2^1, \ldots, 2^{n_p}$ and with profits $p_1, \ldots, p_{n_p} \in P := [0, 1/\phi]$ where $\phi > 1$, and let $b_1, \ldots, b_{n_q}$ be objects with weights $2^{n_p+1}, \ldots, 2^{n_p+n_q}$ and with profits $q_i \in Q_i := (m_i - \lceil m_i \rceil / \phi, m_i]$, where $m_i = (n_p + 1)/(\phi - 1) \cdot ((2\phi - 1)/(\phi - 1))^{i-1}$. The choice of the intervals Q_i is due to the fact that we have to ensure $q_i > \sum_{j=1}^{n_p} p_j + \sum_{j=1}^{i-1} q_j$ to apply Lemma 3 successively for the objects

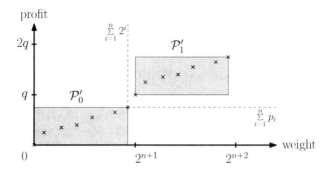

Fig. 1. The copy step. The Pareto set $\mathcal{P}'$ consist of two copies of the Pareto set $\mathcal{P}$.

$b_1, \ldots, b_{n_q}$. We will prove this inequality in Lemma 4. More interesting is the fact that the size of an interval Q_i is $\lceil m_i \rceil / \phi$ which might be larger than $1/\phi$. To explain this consider the case $m_i > 1$ for some index i. For this index the interval Q_i is not a subset of $[-1, 1]$ as required for our model. Instead of avoiding such large values m_i by choosing n_q small enough, we will split Q_i into $\lceil m_i \rceil$ intervals of equal size which must be at least $1/\phi$. This so-called split step will be explained later.

Lemma 4. *Let* $p_1, \ldots, p_{n_p} \in P$ *and let* $q_i \in Q_i$. *Then,* $q_i > \sum_{j=1}^{n_p} p_j + \sum_{j=1}^{i-1} q_j$ *for all* $i = 1, \ldots, n_q$.

Note that with Lemma 4 we implicitly show that the lower boundaries of the intervals Q_i are non-negative.

Proof. Using the definition of m_i, we get

$$q_i > m_i - \frac{\lceil m_i \rceil}{\phi} \geq m_i - \frac{m_i + 1}{\phi} = \frac{\phi - 1}{\phi} \cdot m_i - \frac{1}{\phi}$$

$$= \frac{n_p + 1}{\phi} \cdot \left(\frac{2\phi - 1}{\phi - 1}\right)^{i-1} - \frac{1}{\phi}.$$

On the other hand we have

$$\sum_{j=1}^{n_p} p_j + \sum_{j=1}^{i-1} q_j \leq \sum_{j=1}^{n_p} \frac{1}{\phi} + \sum_{j=1}^{i-1} m_j = \frac{n_p}{\phi} + \sum_{j=1}^{i-1} \frac{n_p + 1}{\phi - 1} \cdot \left(\frac{2\phi - 1}{\phi - 1}\right)^{j-1}$$

$$= \frac{n_p}{\phi} + \frac{n_p + 1}{\phi - 1} \cdot \frac{\left(\frac{2\phi-1}{\phi-1}\right)^{i-1} - 1}{\frac{2\phi-1}{\phi-1} - 1}$$

$$= \frac{n_p}{\phi} + \frac{n_p + 1}{\phi} \cdot \left(\left(\frac{2\phi - 1}{\phi - 1}\right)^{i-1} - 1\right)$$

$$= \frac{n_p + 1}{\phi} \cdot \left(\frac{2\phi - 1}{\phi - 1}\right)^{i-1} - \frac{1}{\phi}. \qquad \square$$

Combining Theorem 2, Lemma 3 and Lemma 4, we immediately get a lower bound for the knapsack problem using the objects $a_1, \ldots, a_{n_p}$ and $b_1, \ldots, b_{n_q}$ with profits chosen from P and Q_i, respectively.

Corollary 5. *Let $a_1, \ldots, a_{n_p}$ and $b_1, \ldots, b_{n_q}$ be as above, but the profits p_i are chosen uniformly from P and the profits q_i are arbitrarily chosen from Q_i. Then, the expected number of Pareto optimal solutions of $K(\{a_1, \ldots, a_{n_p}, b_1, \ldots, b_{n_q}\})$ is $\Omega\left(n_p^2 \cdot 2^{n_q}\right)$.*

Proof. Because of Lemma 4, we can apply Lemma 3 for each realization of the profits $p_1, \ldots, p_{n_p}$ and $q_1, \ldots, q_{n_q}$. This implies that the expected number of Pareto optimal solutions is 2^{n_q} times the expected size of the Pareto set of $K(\{a_1, \ldots, a_{n_p}\})$ which is $\Omega\left(n_p^2\right)$ according to Theorem 2.   □

The profits of the objects b_i grow exponentially and leave the interval $[0, 1]$. As mentioned earlier, we resolve this problem by splitting each object b_i into $k_i :=$ $\lceil m_i \rceil$ objects $b_i^{(1)}, \ldots, b_i^{(k_i)}$ with the same total weight and the same total profit, i.e., each with weight $2^{n_p+i}/k_i$ and profit $q_i^{(l)} \in Q_i/k_i := (m_i/k_i - 1/\phi, m_i/k_i]$. As the intervals Q_i are subsets of $\mathbb{R}_+$, the intervals Q_i/k_i are subsets of $[0, 1]$. It remains to ensure that for any fixed index i all objects $b_i^{(l)}$ are treated as a group. This can be done by restricting the set $\mathcal{S}$ of solutions. Let $\mathcal{S}_i = \{(0, \ldots, 0), (1, \ldots, 1)\} \subseteq \{0, 1\}^{k_i}$. Then, the set $\mathcal{S}$ of solutions is defined as $\mathcal{S} := \{0, 1\}^{n_p} \times \prod_{i=1}^{n_q} \mathcal{S}_i$. By choosing the set of solutions that way, the objects $b_i^{(1)}, \ldots, b_i^{(k_i)}$ can be viewed as substitute for object b_i. Thus, a direct consequence of Corollary 5 is the following.

Corollary 6. *Let $\mathcal{S}$, $a_1, \ldots, a_{n_p}$ and $b_i^{(l)}$ be as above, let the profits $p_1, \ldots, p_{n_p}$ be chosen uniformly from P and let the profits $q_i^{(1)}, \ldots, q_i^{(k_i)}$ be chosen uniformly from Q_i/k_i. Then, the expected number of Pareto optimal solutions of $K_{\mathcal{S}}(\{a_1, \ldots, a_{n_p}\} \cup \{b_i^{(l)} : i = 1, \ldots, n_q, \, l = 1, \ldots, k_i\})$ is $\Omega\left(n_p^2 \cdot 2^{n_q}\right)$.*

The remainder contains just some technical details. First, we give an upper bound for the number of objects $b_i^{(l)}$.

Lemma 7. *The number of objects $b_i^{(l)}$ is upper bounded by $n_q + \frac{n_p+1}{\phi} \cdot \left(\frac{2\phi-1}{\phi-1}\right)^{n_q}$.*

Proof. The number of objects $b_i^{(l)}$ is $\sum_{i=1}^{n_q} k_i = \sum_{i=1}^{n_q} \lceil m_i \rceil \le n_q + \sum_{i=1}^{n_q} m_i$, and

$$
\begin{aligned}
\sum_{i=1}^{n_q} m_i &= \frac{n_p+1}{\phi-1} \cdot \sum_{i=1}^{n_q} \left(\frac{2\phi-1}{\phi-1}\right)^{i-1} \le \frac{n_p+1}{\phi-1} \cdot \frac{\left(\frac{2\phi-1}{\phi-1}\right)^{n_q}}{\frac{2\phi-1}{\phi-1} - 1} \\
&= \frac{n_p+1}{\phi} \cdot \left(\frac{2\phi-1}{\phi-1}\right)^{n_q}.
\end{aligned}
$$

□

Now we are able to prove Theorem 1.

Proof (Theorem 1). Without loss of generality let $n \geq 4$ and $\phi \geq \frac{3+\sqrt{5}}{2} \approx$ 2.62. For the moment let us assume $\phi \leq \left(\frac{2\phi-1}{\phi-1}\right)^{\frac{n-1}{3}}$. This is the interesting case leading to the first term in the minimum in Theorem 1. We set $\hat{n}_q :=$ $\frac{\log(\phi)}{\log((2\phi-1)/(\phi-1))} \in [1, \frac{n-1}{3}]$ and $\hat{n}_p := \frac{n-1-\hat{n}_q}{2} \geq \frac{n-1}{3} \geq 1$. All inequalities hold because of the bounds on n and ϕ. We obtain the numbers n_p and n_q by rounding, i.e., $n_p := \lfloor \hat{n}_p \rfloor \geq 1$ and $n_q := \lfloor \hat{n}_q \rfloor \geq 1$. Now we consider objects $a_1, \ldots, a_{n_p}$ with weights 2^i and profits chosen uniformly from P, and objects $b_i^{(l)}$, $i = 1, \ldots, n_q$, $l = 1, \ldots, k_i$, with weights $2^{n_p+i}/k_i$ and profits chosen uniformly from Q_i/k_i. Observe that P and all Q_i/k_i have length $1/\phi$ and thus the densities of all profits are bounded by ϕ. Let N be the number of all these objects. By Lemma 7, this number is bounded by

$$N \leq n_p + n_q + \frac{n_p+1}{\phi} \cdot \left(\frac{2\phi-1}{\phi-1}\right)^{n_q} \leq \hat{n}_p + \hat{n}_q + \frac{\hat{n}_p+1}{\phi} \cdot \left(\frac{2\phi-1}{\phi-1}\right)^{\hat{n}_q}$$

$$= \hat{n}_p + \hat{n}_q + \frac{\hat{n}_p+1}{\phi} \cdot \phi = 2\hat{n}_p + \hat{n}_q + 1 = n.$$

Hence, the number N of objects we actually use is at most n, as required. As set of solutions we consider $\mathcal{S} := \{0,1\}^{n_p} \times \prod_{i=1}^{n_q} \mathcal{S}_i$. Due to Corollary 6, the expected size of the Pareto set of $K_{\mathcal{S}}(\{a_1, \ldots, a_{n_p}\} \cup \{b_i^{(l)} : i = 1, \ldots, n_q, l = 1, \ldots, k_i\})$ is

$$\Omega\left(n_p^2 \cdot 2^{n_q}\right) = \Omega\left(\hat{n}_p^2 \cdot 2^{\hat{n}_q}\right) = \Omega\left(\hat{n}_p^2 \cdot 2^{\frac{\log(\phi)}{\log\left(\frac{2\phi-1}{\phi-1}\right)}}\right) = \Omega\left(n^2 \cdot \phi^{\frac{1}{\log\left(\frac{2\phi-1}{\phi-1}\right)}}\right)$$

$$= \Omega\left(n^2 \cdot \phi^{1-\Theta(1/\phi)}\right),$$

where the last step holds because

$$\frac{1}{\log\left(2 + \frac{c_1}{\phi-c_2}\right)} = 1 - \frac{\log\left(1 + \frac{c_1}{2\phi-2c_2}\right)}{\log\left(2 + \frac{c_1}{\phi-c_2}\right)} = 1 - \frac{\Theta\left(\frac{c_1}{2\phi-2c_2}\right)}{\Theta(1)} = 1 - \Theta\left(\frac{1}{\phi}\right)$$

for any constants $c_1, c_2 > 0$. We formulated this calculation slightly more general than necessary as we will use it again in the multi-criteria case.

For $\phi > \left(\frac{2\phi-1}{\phi-1}\right)^{\frac{n-1}{3}}$ we construct the same instance as above, but for maximum density $\phi' > 1$ where $\phi' = \left(\frac{2\phi'-1}{\phi'-1}\right)^{\frac{n-1}{3}}$. Since $n \geq 4$, ϕ' exists, is unique and $\phi' \in \left[\frac{3+\sqrt{5}}{2}, \phi\right)$. This yields $\hat{n}'_p = \hat{n}'_q = \frac{n-1}{3}$ and, as above, the expected size of the Pareto set is $\Omega\left((\hat{n}'_p)^2 \cdot 2^{\hat{n}'_q}\right) = \Omega\left(n^2 \cdot 2^{\Theta(n)}\right) = \Omega\left(2^{\Theta(n)}\right)$. $\qquad\square$

3   The Multi-criteria Case

In this section we present a lower bound for the expected number of Pareto optimal solutions in multi-criteria optimization problems. We concentrate our attention to $d \geq 2$ as we discussed the case $d = 1$ in the previous section.

Theorem 8. *For any fixed integer $d \geq 2$ there is a class of instances for the restricted $(d + 1)$-dimensional knapsack problem for which the expected number of Pareto-optimal solutions is lower bounded by*

$$\Omega \left(\min \left\{ (n\phi)^{(d-\log(d)) \cdot (1-\Theta(1/\phi))}, 2^{\Theta(n)} \right\} \right),$$

where n is the number of objects and ϕ is the maximum density of the profit's probability distributions.

Unfortunately, Theorem 8 does not generalize Theorem 1. This is due to the fact that, though we know an explicit formula for the expected number of Pareto optimal solutions if all profits are uniformly chosen from $[0, 1]$, we were not able to find a simple non-trivial lower bound for it. Hence, in the general multi-criteria case, we concentrate on analyzing the copy and split steps.

In the bi-criteria case we used an additional object b to copy the Pareto set (see Figure 1). For that we had to ensure that every solution using this object has higher weight than all solutions without b. The same had to hold for the profit. Since all profits are in $[0, 1]$, the profit of every solution must be in $[0, n]$. As the Pareto set of the first $n_p \leq n/2$ objects has profits in $[0, n/(2\phi)]$, we could fit $n_q = \Theta(\log(\phi))$ copies of this initial Pareto set into the interval $[0, n]$.

In the multi-criteria case, every solution has a profit in $[0, n]^d$. In our construction, the initial Pareto set consists only of a single solution, but we benefit from the fact that the number of mutually non-dominating copies of the initial Pareto set that we can fit into the hypercube $[0, n]^d$ grows quickly with d.

Let us consider the case that we have some Pareto set $\mathcal{P}$ whose profits lie in some hypercube $[0, a]^d$. We will create $\binom{d}{d_h}$ copies of this Pareto set; one for every vector $x \in \{0, 1\}^d$ with exactly $d_h = \lceil d/2 \rceil$ ones. Let $x \in \{0, 1\}^d$ be such a vector. Then we generate the corresponding copy C_x of the Pareto set $\mathcal{P}$ by shifting it by $a + \varepsilon$ in every dimension i with $x_i = 1$. If all solutions in these copies have higher weights than the solutions in the initial Pareto set $\mathcal{P}$, then the initial Pareto set stays Pareto optimal. Furthermore, for each pair of copies C_x and C_y, there is one index i with $x_i = 1$ and $y_i = 0$. Hence, solutions from C_y cannot dominate solutions from C_x. Similarly, one can argue that no solution in the initial copy can dominate any solution from C_x. This shows that all solutions in copy C_x are Pareto optimal. All the copies (including the initial one) have profits in $[0, 2a + \varepsilon]^d$ and together $|\mathcal{P}| \cdot \left(1 + \binom{d}{d_h}\right) \geq |\mathcal{P}| \cdot 2^d/d$ solutions.

We start with an initial Pareto set of a single solution with profit in $[0, 1/\phi]^d$, and hence we can make $\Theta(\log(n\phi))$ copy steps before the hypercube $[0, n]^d$ is filled. In each of these steps the number of Pareto optimal solutions increases by a factor of at least $2^d/d$, yielding a total number of at least $(2^d/d)^{\Theta(\log(n\phi))} = (n\phi)^{\Theta(d-\log(d))}$ Pareto optimal solutions.

In the following, we describe how these copy steps can be realized in the restricted knapsack problem. Again, we have to make a split step because the profit of every object must be in $[0, 1]^d$. Due to such technicalities, the actual bound we prove looks slightly different than the one above. It turns out that we need (before splitting) d new objects $b_1, \ldots, b_d$ for each copy step in contrast to the bi-criteria case, where (before splitting) a single object b was enough.

Let $n_q \geq 1$ be an arbitrary positive integer and let $\phi \geq 2d$ be a real. We consider objects $b_{i,j}$ with weights $2^i/d_h$ and profit vectors

$$q_{i,j} \in Q_{i,j} := \prod_{k=1}^{j-1} \left[0, \frac{\lceil m_i \rceil}{\phi}\right] \times \left(m_i - \frac{\lceil m_i \rceil}{\phi}, m_i\right] \times \prod_{k=j+1}^{d} \left[0, \frac{\lceil m_i \rceil}{\phi}\right],$$

where m_i is recursively defined as

$$m_0 := 0 \quad \text{and} \quad m_i := \frac{1}{\phi - d} \cdot \left(\sum_{l=0}^{i-1} (m_l \cdot (\phi + d) + d)\right), \quad i = 1, \ldots, n_q. \tag{1}$$

The explicit formula for this recurrence is

$$m_i = \frac{d}{\phi + d} \cdot \left(\left(\frac{2\phi}{\phi - d}\right)^i - 1\right), \quad i = 1, \ldots, n_q.$$

The d-dimensional interval $Q_{i,j}$ is of the form that the jth profit of object $b_{i,j}$ is large and all the other profits are small. By using object $b_{i,j}$ the copy of the Pareto set is shifted in direction of the jth unit vector. As mentioned in the motivation we will choose exactly d_h such objects to create additional copies. To give a better intuition for the form of the single intervals the d-dimensional interval $Q_{i,j}$ is constructed of we refer the reader to the explanation in the bi-criteria case.

Let $H(x)$ be the *Hamming weight* of a 0-1-vector x, i.e., the number of ones in x, and let $\hat{S} := \{x \in \{0,1\}^d : H(x) \in \{0, d_h\}\}$ denote the set of all 0-1-vectors of length d with 0 or d_h ones. As set S of solutions we consider $S := \hat{S}^{n_q}$.

Lemma 9. *Let the set S of solutions and the objects $b_{i,j}$ be as above. Then, each solution $s \in S$ is Pareto optimal for $K_S(\{b_{i,j} : i = 1, \ldots, n_q, j = 1, \ldots, d\})$.*

Proof. We show the statement by induction over n_q and discuss the base case and the inductive step simultaneously because of similar arguments. Let $S' := \hat{S}^{n_q - 1}$ and let $(s, s_{n_q}) \in S' \times \hat{S}$ be an arbitrary solution from S. Note that for $n_q = 1$ we get $s = \lambda$, the 0-1-vector of length 0. First we show that there is no domination within one copy, i.e., there is no solution of type $(s', s_{n_q}) \in S$ that dominates (s, s_{n_q}). For $n_q = 1$ this is obviously true. For $n_q \geq 2$ the existence of such a solution would imply that s' dominates s in the knapsack problem $K_{S'}(\{b_{i,j} : i = 1, \ldots, n_q - 1, j = 1, \ldots, d\})$. This contradicts the inductive hypothesis.

Now we prove that there is no domination between solutions from different copies, i.e., there is no solution of type $(s', s'_{n_q}) \in S$ with $s'_{n_q} \neq s_{n_q}$ that dominates (s, s_{n_q}). If $s_{n_q} = \mathbf{0}$, then the total weight of the solution (s, s_{n_q}) is at most $\sum_{i=1}^{n_q - 1} 2^i < 2^{n_q}$. The right side of this inequality is a lower bound for the weight of solution (s', s'_{n_q}) because $s'_{n_q} \neq s_{n_q}$. Hence, (s', s'_{n_q}) does not dominate (s, s_{n_q}). Finally, let us consider the case $s_{n_q} \neq \mathbf{0}$. There must be an index $j \in [d]$

where $(s_{n_q})_j = 1$ and $(s'_{n_q})_j = 0$. We show that the jth total profit of (s, s_{n_q}) is higher than the jth profit of (s', s'_{n_q}). The former one is strictly bounded from below by $m_{n_q} - \lceil m_{n_q} \rceil / \phi$, whereas the latter one is bounded from above by

$$\sum_{i=1}^{n_q-1} \left((d_h - 1) \cdot \frac{\lceil m_i \rceil}{\phi} + \max\left\{ \frac{\lceil m_i \rceil}{\phi}, m_i \right\} \right) + d_h \cdot \frac{\lceil m_{n_q} \rceil}{\phi}.$$

Solution (s', s'_{n_q}) can use at most d_h objects of each group $b_{i,1}, \ldots, b_{i,d}$. Each of them, except one, can contribute at most $\frac{\lceil m_i \rceil}{\phi}$ to the jth total profit. One can contribute either at most $\frac{\lceil m_i \rceil}{\phi}$ or at most m_i. This argument also holds for the n_q^{th} group, but by the choice of index j we know that each object chosen by s'_{n_q} contributes at most $\frac{\lceil m_i \rceil}{\phi}$ to the jth total profit. It is easy to see that $\lceil m_i \rceil / \phi \leq m_i$ because of $\phi > d \geq 1$. Hence, our bound simplifies to

$$\sum_{i=1}^{n_q-1} \left((d_h - 1) \cdot \frac{\lceil m_i \rceil}{\phi} + m_i \right) + d_h \cdot \frac{\lceil m_{n_q} \rceil}{\phi}$$

$$\leq \sum_{i=1}^{n_q-1} \left(d \cdot \frac{m_i + 1}{\phi} + m_i \right) + (d - 1) \cdot \frac{m_{n_q} + 1}{\phi} \qquad (d \geq 2)$$

$$= \frac{1}{\phi} \cdot \left(\sum_{i=1}^{n_q-1} (m_i \cdot (\phi + d) + d) + d \cdot (m_{n_q} + 1) \right) - \frac{m_{n_q} + 1}{\phi}$$

$$= \frac{1}{\phi} \cdot \left(\sum_{i=0}^{n_q-1} (m_i \cdot (\phi + d) + d) + d \cdot m_{n_q} \right) - \frac{m_{n_q} + 1}{\phi} \qquad (m_0 = 0)$$

$$= \frac{1}{\phi} \cdot ((\phi - d) \cdot m_{n_q} + d \cdot m_{n_q}) - \frac{m_{n_q} + 1}{\phi} \qquad (\text{Equ. } (1))$$

$$\leq m_{n_q} - \frac{\lceil m_{n_q} \rceil}{\phi}.$$

This implies that (s', s'_{n_q}) does not dominate (s, s_{n_q}).   □

Immediately, we get a statement about the expected number of Pareto optimal solutions if we randomize.

Corollary 10. *Let $\mathcal{S}$ and $b_{i,j}$ be as above, but the profit vectors $q_{i,j}$ are arbitrarily drawn from $Q_{i,j}$. Then, the expected number of Pareto optimal solutions for $K_{\mathcal{S}}(\{b_{i,j} : i = 1, \ldots, n_q, \ j = 1, \ldots, d\})$ is at least $(2^d/d)^{n_q}$.*

Proof. This result follows from Lemma 9 and $|\hat{S}| = 1 + \binom{d}{d_h} = 1 + \max_{i=1,\ldots,d} \binom{d}{i} \geq 1 + (\sum_{i=1}^{d} \binom{d}{i})/d = 1 + (2^d - 1)/d \geq 2^d/d$.   □

As in the bi-criteria case we now split each object $b_{i,j}$ into $k_i := \lceil m_i \rceil$ objects $b_{i,j}^{(1)}, \ldots, b_{i,j}^{(k_i)}$ with weights $2^i/(k_i \cdot d_h)$ and with profit vectors

$$q_{i,j}^{(l)} \in Q_{i,j}/k_i := \prod_{k=1}^{j-1} \left[0, \frac{1}{\phi}\right] \times \left(\frac{m_i}{k_i} - \frac{1}{\phi}, \frac{m_i}{k_i}\right] \times \prod_{k=j+1}^{d} \left[0, \frac{1}{\phi}\right].$$

Then, we adapt our set $\mathcal{S}$ of solutions such that for any fixed indices i and j either all objects $b_{i,j}^{(1)}, \ldots, b_{i,j}^{(k_i)}$ are put into the knapsack or none of them. Corollary 10 yields the following result.

Corollary 11. *Let $\mathcal{S}$ and $b_{i,j}^{(l)}$ be as described above, but let the profit vectors $p_{i,j}^{(1)}, \ldots, p_{i,j}^{(k_i)}$ be chosen uniformly from $Q_{i,j}/k_i$. Then, the expected number of Pareto optimal solutions of $K_{\mathcal{S}}(\{b_{i,j}^{(l)} : i = 1, \ldots, n_q, \ j = 1, \ldots, d, \ l = 1, \ldots, k_i\})$ is at least $(2^d/d)^{n_q}$.*

Still, the lower bound is expressed in n_q and not in the number of objects used. So the next step is to analyze the number of objects.

Lemma 12. *The number of objects $b_{i,j}^{(l)}$ is upper bounded by $d \cdot n_q + \frac{2d^2}{\phi-d} \cdot \left(\frac{2\phi}{\phi-d}\right)^{n_q}$.*

Proof. The number of objects $b_{i,j}^{(l)}$ is $\sum_{i=1}^{n_q}(d \cdot k_i) = d \cdot \sum_{i=1}^{n_q} \lceil m_i \rceil \leq d \cdot n_q + d \cdot \sum_{i=1}^{n_q} m_i$, and

$$\sum_{i=1}^{n_q} m_i \leq \frac{d}{\phi+d} \cdot \sum_{i=1}^{n_q}\left(\frac{2\phi}{\phi-d}\right)^i \leq \frac{d}{\phi+d} \cdot \frac{\left(\frac{2\phi}{\phi-d}\right)^{n_q+1}}{\left(\frac{2\phi}{\phi-d}\right)-1}$$

$$\leq \frac{d}{\phi} \cdot \left(\frac{2\phi}{\phi-d}\right) \cdot \left(\frac{2\phi}{\phi-d}\right)^{n_q} = \frac{2d}{\phi-d} \cdot \left(\frac{2\phi}{\phi-d}\right)^{n_q}. \qquad \square$$

Now we can prove Theorem 8.

Proof. Without loss of generality let $n \geq 16d$ and $\phi \geq 2d$. For the moment let us assume $\phi - d \leq \frac{4d^2}{n} \cdot \left(\frac{2\phi}{\phi-d}\right)^{\frac{n}{2d}}$. This is the interesting case leading to the first term in the minimum in Theorem 8. We set $\hat{n}_q := \frac{\log\left((\phi-d) \cdot \frac{n}{4d^2}\right)}{\log\left(\frac{2\phi}{\phi-d}\right)} \in \left[1, \frac{n}{2d}\right]$ and obtain $n_q := \lfloor \hat{n}_q \rfloor \geq 1$ by rounding. All inequalities hold because of the bounds on n and ϕ. Now we consider objects $b_{i,j}^{(l)}$, $i = 1, \ldots, n_q$, $j = 1, \ldots, d$, $l = 1, \ldots, k_i$, with weights $2^i/(k_i \cdot d)$ and profit vectors $q_{i,j}$ chosen uniformly from $Q_{i,j}/k_i$. All these intervals have length $1/\phi$ and hence all densities are bounded by ϕ. Let N be the number of objects. By Lemma 12, this number is bounded by

$$N \leq d \cdot n_q + \frac{2d^2}{\phi-d} \cdot \left(\frac{2\phi}{\phi-d}\right)^{n_q} \leq d \cdot \hat{n}_q + \frac{2d^2}{\phi-d} \cdot \left(\frac{2\phi}{\phi-d}\right)^{\hat{n}_q}$$

$$\leq d \cdot \hat{n}_q + \frac{2d^2}{\phi-d} \cdot (\phi-d) \cdot \frac{n}{4d^2} \leq n.$$

Hence, the number N of objects we actually use is at most n, as required. As set $\mathcal{S}$ of solutions we use the set described above, encoding the copy step and the split step. Due to Corollary 11, for fixed $d \geq 2$ the expected number of Pareto optimal solutions of $K_{\mathcal{S}}(\{b_{i,j}^{(l)} : i = 1, \ldots, n_q, \ j = 1, \ldots, d, \ l = 1, \ldots, k_i\})$ is

$$\Omega\left(\left(\frac{2^d}{d}\right)^{n_q}\right) = \Omega\left(\left(\frac{2^d}{d}\right)^{\hat{n}_q}\right) = \Omega\left(\left(\frac{2^d}{d}\right)^{\frac{\log\left((\phi-d)\cdot\frac{n}{4d^2}\right)}{\log\left(\frac{2\phi}{\phi-d}\right)}}\right)$$

$$= \Omega\left(\left((\phi-d)\cdot\frac{n}{4d^2}\right)^{\frac{\log\left(\frac{2^d}{d}\right)}{\log\left(\frac{2\phi}{\phi-d}\right)}}\right) = \Omega\left((\phi\cdot n)^{\frac{d-\log(d)}{\log\left(\frac{2\phi}{\phi-d}\right)}}\right)$$

$$= \Omega\left((\phi\cdot n)^{(d-\log(d))\cdot(1-\Theta(1/\phi))}\right),$$

where the last step holds because of the same reason as in the proof of Theorem 1.

In the case $\phi - d > \frac{4d^2}{n}\cdot\left(\frac{2\phi}{\phi-d}\right)^{\frac{n}{2d}}$ we construct the same instance above, but for a maximum density $\phi' > d$ where $\phi' - d = \frac{4d^2}{n}\cdot\left(\frac{2\phi'}{\phi'-d}\right)^{\frac{n}{2d}}$. Since $n \geq 16d$, the value ϕ' exists, is unique and $\phi' \in [65d, \phi)$. Futhermore, we get $\hat{n}_q = \frac{n}{2d}$. As above, the expected size of the Pareto set is $\Omega\left((2^d/d)^{\hat{n}_q}\right) = \Omega\left((2^d/d)^{n/(2d)}\right) = \Omega\left(2^{\Theta(n)}\right)$. $\square$

References

[BRV07]   Beier, R., Röglin, H., Vöcking, B.: The smoothed number of pareto optimal solutions in bicriteria integer optimization. In: Fischetti, M., Williamson, D.P. (eds.) IPCO 2007. LNCS, vol. 4513, pp. 53–67. Springer, Heidelberg (2007)

[BV04]   Beier, R., Vöcking, B.: Random knapsack in expected polynomial time. Journal of Computer and System Sciences 69(3), 306–329 (2004)

[Ehr05]   Ehrgott, M.: Multicriteria Optimization, 2nd edn. Springer, Heidelberg (2005)

[MO10]   Moitra, A., O'Donnell, R.: Pareto optimal solutions for smoothed analysts. Technical report, CoRR, abs/1011.2249 (2010), http://arxiv.org/abs/1011.2249; To appear in Proc. of the 43rd Annual ACM Symposium on Theory of Computing (STOC) (2011)

[RT09]   Röglin, H., Teng, S.-H.: Smoothed analysis of multiobjective optimization. In: Proc. of the 50th Ann. IEEE Symp. on Foundations of Computer Science (FOCS), pp. 681–690 (2009)

[ST04]   Spielman, D.A., Teng, S.-H.: Smoothed analysis of algorithms: Why the simplex algorithm usually takes polynomial time. Journal of the ACM 51(3), 385–463 (2004)

[ST09]   Spielman, D.A., Teng, S.-H.: Smoothed analysis: an attempt to explain the behavior of algorithms in practice. Communications of the ACM 52(10), 76–84 (2009)

Approximating Minimum Cost Source Location Problems with Local Vertex-Connectivity Demands

Takuro Fukunaga

Department of Applied Mathematics and Physics,
Graduate School of Informatics, Kyoto University, Japan
takuro@amp.i.kyoto-u.ac.jp

Abstract. The source location problem is a problem of computing a minimum cost source set in an undirected graph so that the connectivity between the source set and each vertex is at least the demand of the vertex. In this paper, the connectivity between a source set S and a vertex v is defined as the maximum number of paths between v and S no two of which have common vertices except v. We propose an $O(d^* \log d^*)$-approximation algorithm for the problem with maximum demand d^*. We also define a variant of the source location problem and propose an approximation algorithm for it.

1 Introduction

Nowadays we enjoy many services through the Internet. As our life heavily depends on these services, network failure may make enormous damage to our society. Accordingly service providers are required to locate their servers efficiently to guarantee that their services are always available. The *source location problem* is a problem motivated by this situation.

In the source location problem, it is supposed that each vertex $v \in V$ in an undirected graph $G = (V, E)$ has a demand $d(v) \in \mathbb{Z}_+$, where $\mathbb{Z}_+$ denotes the sets of non-negative integers. With regards to a certain measurement $\psi : 2^V \times V \to \mathbb{Z}_+$ on connectivity between a subset of V and a vertex in G, a set $S \subseteq V$ is called a *source set* when $\psi(S, v) \geq d(v)$ for every $v \in V \setminus S$. The source location problem is a problem of finding a minimum cost source set, which is defined formally as follows.

Source location problem with connectivity measurement ψ
Input: An undirected graph $G = (V, E)$, a demand $d : V \to \mathbb{Z}_+$, and a cost $c : V \to \mathbb{Q}_+$, where $\mathbb{Q}_+$ denotes the set of non-negative rationals.
Output: A source set $S \subseteq V$ (i.e., $\psi(S, v) \geq d(v)$ for each $v \in V \setminus S$) that minimizes $\sum_{v \in S} c(v)$.

Throughout the paper, we represent $|V|$ by n and $|E|$ by m. Moreover, the maximum value $\max_{v \in V} d(v)$ of demands is denoted by d^*. We call the problem

M. Ogihara and J. Tarui (Eds.): TAMC 2011, LNCS 6648, pp. 428–439, 2011.

with measurement ψ by ψ-SLP for short. In previous works on the source location problem, three connectivity measurements λ, κ and $\hat{\kappa}$ are mainly discussed. Their definitions are given below, where a path connecting $v \in V$ and a vertex in S is called (S, v)-path, and a singleton $\{v\}$ is written simply by v.

- Edge-connectivity λ: $\lambda(S, v)$ is defined as the maximum number of (S, v)-paths no two of which have a common edge. This notion represents the tolerance to connection failures on edges.
- Vertex-connectivity κ: $\kappa(S, v)$ is defined as the maximum number of (S, v)-paths no two of which have a common vertex in $V \setminus (S \cup v)$. This notion represents the tolerance to connection failures on vertices in $V \setminus (S \cup v)$. In other word, it is supposed that located servers will not be broken.
- Vertex-connectivity $\hat{\kappa}$: $\hat{\kappa}(S, v)$ is defined as the maximum number of (S, v)-paths no two of which have a common vertex in $V \setminus v$. This notion represents the tolerance to connection failures on vertices in $V \setminus v$. In contrast with κ, this considers a situation where located servers are possibly broken.

Although λ-SLP and κ-SLP are important, we only refer to several previous works [1,7] because this paper focuses on $\hat{\kappa}$-SLP. For $\hat{\kappa}$-SLP with uniform demand (i.e., $d(v) = d^*$ for $v \in V$), Nagamochi, Ishii and Ito [12] presented an $O(\min\{d^*, \sqrt{n}\} d^* n^2)$-time algorithm. For the general case, Sakashita, Makino and Fujishige [15] presented a $(1 + \ln \sum_{v \in V} d(v))$-approximation algorithm. This result is almost best because they also proved that the problem admits no $C \ln \sum_{v \in V} d(v)$-approximation algorithm for some constant C unless every problem in NP has an $O(N^{\log \log N})$-time algorithm. On the other hand, Ishii, Fujita and Nagamochi [4,5] discussed the case where d^* is bounded from the above. In [4], they presented a linear-time algorithm for the case with $d^* \leq 3$ and uniform costs, and in [5], they presented a polynomial-time algorithm for the case with $d^* \leq 3$ and general costs. As a negative result, Ishii, Fujita and Nagamochi [4] showed that the problem is NP-hard even if the cost is uniform and $d^* = 4$. This NP-hardness is strengthen by Ishii [6], who proved that the problem is APX-hard even if the cost is uniform and $d^* = 4$, and presented a $\max\{d^*, 2d^* - 6\}$-approximation algorithm for the uniform cost. His algorithm achieves approximation factor 3 especially when the cost is uniform and $d^* \leq 4$. Table 1 summarizes these results.

Considering these, an interesting question is whether $\hat{\kappa}$-SLP with arbitrary costs is approximable within a factor depending on d^* only. This paper answers to this question by presenting an $O(d^* \log d^*)$-approximation algorithm for $\hat{\kappa}$-SLP with arbitrary costs.

In addition, this paper considers a variant of the source location problem. In the above definition of the problem, a subset S of V is a source set if $\psi(S, v) \geq d(v)$ for each $v \in V \setminus S$. This means that once a source is located on a vertex v, the demand of v is satisfied completely. However it is sometimes natural to suppose that if a source is located on v, then only one unit of the demand of v is satisfied. Moreover, in such a situation, it may be reasonable to consider the case where more than one source can be located on a vertex. This motivates us to define

Table 1. Results for $\hat{\kappa}$-SLP with bounded d^*

uniform costs	general costs
linear-time exact algorithm ($d^* \leq 3$) [4] APX-hardness ($d^* = 4$) [6] 3-approximation ($d^* \leq 4$) [6] max$\{d^*, 2d^* - 6\}$-approximation [6]	exact algorithm ($d^* \leq 3$) [5] $O(d^* \log d^*)$-approximation [This paper]

a new vertex-connectivity measurement $\tilde{\kappa}$ and a new formulation of the source location problem. In the new problem, we represent the numbers of sources located on vertices by $x : V \to \mathbb{Z}_+$. For $x : V \to \mathbb{Z}_+$, let $S_x = \{v \in V \mid x(v) > 0\}$. We define the node-connectivity $\tilde{\kappa}(x, v)$ as $x(v) + \hat{\kappa}(S_x \setminus v, v)$. x is a source set when $\tilde{\kappa}(x, v)$ is not smaller than the demand of each $v \in V$. We assume that a non-decreasing convex function c_v is given for each vertex $v \in V$, and the cost of x is defined as $\sum_{v \in V} c_v(x(v))$.

We can observe that this new problem is contained by $\hat{\kappa}$-SLP with a graph the size of which depends on the total demand. Simultaneously it is a special case of the survivable network design problem with root vertex-connectivity demand. We explain these relationship in Section 3. As mentioned above, an $O(d^* \log d^*)$-approximation algorithm will be presented for $\hat{\kappa}$-SLP in this paper. For the survivable network design problem with root vertex-connectivity demand, there is an $O(d^* \log d^*)$-approximation algorithm presented by Nutov [13]. Either way, we have an $O(d^* \log d^*)$-algorithm for the new problem. The other contribution of this paper is to present a $(2d^* - 1)$-approximation algorithm for the problem, described in Section 4. Our algorithm is based on the iterative rounding method, which has been applied to the network design problem so far (e.g., [3,8,10,17]).

As a related work, we would like to mention the results due to Sakashita, Makino and Fujishige [14,15]. They extended the demands on the connectivities λ and κ to the demands on the flow values in edge-capacitated networks. A solution in their problem is a supply on each vertex, and the cost is defined by a monotone concave function given on each vertex. Refer to [14,15] for the detail.

The rest of this paper is organized as follows. Section 2 presents an $O(d^* \log d^*)$-approximation algorithm for $\hat{\kappa}$-SLP. Section 3 describes the relationship between the new variant of the source location problem and other problems. Section 4 gives a $(2d^* - 1)$-approximation algorithm for it. Section 5 concludes the paper. Due to the space constraints, we cannot present all proofs. Refer to the full version for the omitted proofs.

2 $O(d^* \log d^*)$-Approximation Algorithm for $\hat{\kappa}$-SLP

In this section, we discuss $\hat{\kappa}$-SLP. For $X \subseteq V$, let $d^*(X)$ denote $\max_{v \in X} d(v)$, and $N(X)$ be the set of neighbors of X, i.e., vertices in $V \setminus X$ which are adjacent to some in X. We write $|N(X)|$ simply by $n(X)$. This function n satisfies both of

$$n(X) + n(Y) \geq n(X \cap Y) + n(X \cup Y) \tag{1}$$

and
$$n(X) + n(Y) \geq n(X \setminus (Y \cup N(Y))) + n(Y \setminus (X \cup N(X))) \qquad (2)$$

for any $X, Y \subseteq V$, which can be proven by counting the contribution of each vertex to the both sides. A subset X of V is called *deficient set* if and only if $n(X) < d^*(X)$. We let $\mathcal{X}$ denote the family of all minimal deficient sets.

The next observation derived from Menger's theorem is often observed in the source location problem (its proof can be found in [6] for example).

Theorem 1. $S \subseteq V$ *is a source set if and only if* $S \cap X \neq \emptyset$ *for each* $X \in \mathcal{X}$.

Given a subset family $\mathcal{A} \subseteq 2^V$ of a finite set V, $B \subseteq V$ is called a *transversal* of $\mathcal{A}$ if and only if $B \cap A \neq \emptyset$ for each $A \in \mathcal{A}$. The *frequency* $f_{\mathcal{A}}(v)$ of $v \in V$ in the family $\mathcal{A}$ is defined as $|\{A \in \mathcal{A} \mid v \in A\}|$. Since finding a minimum cost transversal is equivalent to the set cover problem, it is known that a greedy algorithm [2,9,11] achieves approximation factor $1 + \log f^*$ for computing a minimum cost transversal of $\mathcal{A}$ where $f^* = \max_{v \in V} f_{\mathcal{A}}(v)$.

Theorem 1 implies that $\hat{\kappa}$-SLP is equivalent to the problem of computing a minimum cost transversal of $\mathcal{X}$. Applying the greedy algorithm to $\mathcal{X}$ directly does not result in a good approximation factor because we cannot bound the maximum frequency in $\mathcal{X}$. However we see that the greedy algorithm provides an $O(d^* \log d^*)$-approximation algorithm.

Call a minimal deficient set $X \in \mathcal{X}$ by k-*deficient set* if $n(X) = k$. Let S be a transversal of the family of k'-deficient sets for all $k' < k$. $X \in \mathcal{X}$ is called S-*avoiding* if $S \cap X = \emptyset$. Let $\mathcal{X}(k, S)$ denote the family of all S-avoiding k-deficient sets. We show that the maximum frequency in $\mathcal{X}(k, S)$ is small.

Lemma 1. *Let* $X, Y \in \mathcal{X}(k, S)$ *be distinct S-avoiding k-deficient sets where* $\max_{v \in X} d(v)$ *and* $\max_{v \in Y} d(v)$ *are respectively attained by* $x \in X$ *and* $y \in Y$. *If* $X \cap Y \neq \emptyset$, *then* $x \in N(Y)$ *or* $y \in N(X)$ *holds.*

Notice that Lemma 1 tells that each vertex $v \in V$ attains $\max_{v' \in X} d(v')$ for at most one k-deficient set $X \in \mathcal{X}(k, s)$.

Lemma 2. $\max_{v \in V \setminus S} f_{\mathcal{X}(k,S)}(v) \leq 2k + 1$.

Proof. Let $v \in V \setminus S$. We show that $f_{\mathcal{X}(k,S)}(v) \leq 2k + 1$.

Let us consider a digraph $D_v = (U, A)$ in which each vertex $u_X \in U$ corresponds to a member X of $\mathcal{X}(k, S)$ such that $v \in X$. Notice that $f_{\mathcal{X}(k,S)}(v) = |U|$. The arc set A of D_v is defined so that it contains an arc from $u_X \in U$ to $u_Y \in U$ if and only if a vertex $x \in X$ attaining $\max_{v \in X} d(v)$ is in $N(Y)$.

Lemma 1 tells that each two vertices u_X and u_Y are joined by an arc in D_v. Thus D_v has at least $|U|(|U| - 1)/2$ arcs. On the other hand, since $n(X) = k$ for all $X \in \mathcal{X}(k, S)$, the in-degree of each vertex in D_v is at most k. Thus D_v has at most $k|U|$ arcs. Consequently $k|U| \geq |U|(|U| - 1)/2$ holds, implying that $2k + 1 \geq |U|$. $\qquad \square$

Lemma 2 implies an $O(d^* \log d^*)$-approximation algorithm to $\hat{\kappa}$-SLP.

Theorem 2. *The source location problem is approximable within a factor of* $O(d^* \log d^*)$

Proof. Our algorithm begins with $S = \emptyset$ and $k = 1$. It repeats the following operations until S becomes a source set; Compute an approximate solution $S' \subseteq V \setminus S$ of the minimum cost transversal of $\mathcal{X}(k, S)$ by the greedy algorithm, which achieves approximation factor $1 + \log(2k + 1)$ by Lemma 2; Add S' to S and increase k by 1. For any (k, S) constructed in the above algorithm, the minimum cost of transversals of $\mathcal{X}(k, S)$ is at most the minimum cost of source sets. Hence this algorithm achieves approximation factor $\sum_{k=1}^{d^*}(1 + \log(2k+1)) = O(d^* \log d^*)$.

Let us observe that $\mathcal{X}(k, S)$ has a polynomial size and can be computed in polynomial time for each (k, S) appeared in the above algorithm. This fact indicates that the greedy algorithm computes a transversal of $\mathcal{X}(k, S)$ in polynomial time. Lemma 1 tells that each vertex $x \in V \setminus S$ attains $\max_{v \in X} d(v)$ for at most one S-avoiding k-deficient set $X \in \mathcal{X}(k, S)$. Hence $|\mathcal{X}(k, S)| \leq |V \setminus S|$. Let $G' = (V \cup s, E \cup E')$ be the graph obtained from G by adding a new vertex s and the set E' of edges joining s and all $v \in S$. In the above algorithm, S is always a transversal of the family of all k'-deficient sets for $k' < k$. Hence if $X \subseteq V \setminus S$ contains a vertex $x \in X$ with $d^*(x) > k$, then $n(X) \geq k$. This means that $X \subseteq V \setminus S$ is an S-avoiding k-deficient set in G if and only if $n(X) = k$ and X is a minimal set such that $N(X)$ is a minimum vertex-cut separating s and some vertex $x \in X$ with $d^*(x) > k$ in G'. Thereby $\mathcal{X}(k, S)$ can be computed by applying a max-flow algorithm at most $|V \setminus S|$ times. □

3   Formulation of the Source Location Problem with Vertex-Connectivity $\tilde{\kappa}$

In Section 1, we have defined the vertex-connectivity $\tilde{\kappa}$ and the source location problem with $\tilde{\kappa}$. We formulate this problem as a special case of the following survivable network design problem.

Root vertex-connectivity survivable network design (RVSND)
Input: An undirected graph $G = (V \cup r, E)$ with the root r, a demand $d : V \rightarrow \mathbb{Z}_+$, a cost $c : E \rightarrow \mathbb{Q}_+$ and a capacity $u : E \rightarrow \mathbb{Z}_+$.
Output: $x : E \rightarrow \mathbb{Z}_+$ minimizing $\sum_{e \in E} c(e)x(e)$ such that $x(e) \leq u(e)$ for each $e \in E$ and $\kappa(r, v) \geq d(v)$ for each $v \in V$ in the graph which consists of the vertex set $V \cup r$ and the edge set containing $x(e)$ copies of each edge $e \in E$.

For $X \subseteq V$ and $F \subseteq E$ in a graph $G = (V \cup r, E)$, let $\delta_F(X)$ denote the set of edges in F joining vertices in X and those in $(V \cup r) \setminus X$. The source location problem with $\tilde{\kappa}$ is equivalent to the set of instances of RVSND such that $c(e) = 0$ for each edge $e \in E \setminus \delta_E(r)$. In such instances, we can let a solution x satisfy $x(e) = u(e)$ for each zero-cost edge $e \in E$ because choosing these edges does not increase the objective value. Hence the task in such instances is to decide $x(e)$ for positive-cost edge $e \in E$. We call this special case of RVSND *root vertex-connectivity star-augmentation problem* (RVSAP). Let us reformulate the problem as follows.

Root vertex-connectivity star-augmentation problem (RVSAP)
Input: An undirected multigraph $G = (V, E)$, a root r, a set F of edges joining r with vertices in V, a demand $d : V \rightarrow \mathbb{Z}_+$, a cost $c : F \rightarrow \mathbb{Q}_+$ and a capacity $u : F \rightarrow \mathbb{Z}_+$. Note that r is not contained by V.
Output: For $x : F \rightarrow \mathbb{Z}_+$, let $G + x$ denote the graph obtained by adding r and $x(e)$ copies of $e \in F$ to G. The output is $x : F \rightarrow \mathbb{Z}_+$ minimizing $\sum_{e \in F} c(e)x(e)$ such that $x(e) \le u(e)$ for each $e \in F$ and $\kappa(r, v) \ge d(v)$ for each $v \in V$ in the graph $G + x$.

From a solution x to RVSAP, define $x' \in V \rightarrow \mathbb{Z}_+$ by $x'(v) = \sum_{e \in \delta_F(v)} x(e)$. Then $\tilde{\kappa}(x', v)$ in G is equal to $\kappa(r, v)$ in $G + x$. Order the edges $e_1, e_2, \ldots, e_k$ in $\delta_F(v)$ so that $c(e_1) \le c(e_2) \le \cdots \le c(e_k)$. Since $e_i, e_j \in \delta_F(v)$ are parallel, if $x(e_i) > 0$ and $x(e_j) < u(e_j)$, then decreasing $x(e_i)$ and increasing $x(e_j)$ by small amount keeps the feasibility of x in RVSAP. Hence without loss of generality, a solution x to RVSAP satisfies $x(e_i) > 0$ whenever $x(e_j) = u(e_j)$ for all $j < i$. Define the cost function $c_v : \mathbb{Z}_+ \rightarrow \mathbb{Q}_+$ by $c_v(0) = 0$ and $c_v(i) = c_v(i - 1) + c(e_j)$ for $i > 0$ such that $\sum_{j'=1}^{j-1} u(e_{j'}) < i \le \sum_{j'=1}^{j} u(e_{j'})$ where we let $\sum_{j'=1}^{j-1} u(e_{j'}) = 0$ if $j = 1$. Then c_v is non-decreasing convex and $\sum_{e \in F} c(e)x(e) = \sum_{v \in V} c_v(x'(v))$. On the contrary, an instance of RVSAP can be constructed from each instance of the source location problem with $\tilde{\kappa}$. Therefore RVSAP is equivalent to the source location problem with $\tilde{\kappa}$.

Whereas RVSAP is a special case of RVSND, it is simultaneously contained by $\hat{\kappa}$-SLP with a graph the size of which depends on $\sum_{e \in F} u(e)$. Let (G, r, F, d, c, u) be an instance of RVSAP. For each $v \in V$ and $e \in \delta_F(v)$, add new vertices $v_e^1, v_e^2, \ldots, v_e^{u(e)}$ and edges $vv_e^1, vv_e^2, vv_e^{u(e)}$ to G. We call the obtained graph $G' = (V', E')$, and the vertex v_e^i a *copy of* $e \in F$. See Figure 1 for an example of this construction. Moreover define the demand $d' : V' \rightarrow \mathbb{Z}_+$ and the cost $c' : V' \rightarrow \mathbb{Q}_+$ by

$$d'(v) = \begin{cases} d(v) & v \in V, \\ 0 & v \in V' \setminus V, \end{cases}$$

and

$$c'(v) = \begin{cases} +\infty & v \in V, \\ c(e) & v \in V' \text{ is a copy of } e \in F. \end{cases}$$

By the setting of c', we can assume without loss of generality that a solution S to the instance (G', d', c') of $\hat{\kappa}$-SLP consists of only vertices in $V' \setminus V$. From such S, define $x(e)$ as the number of copies of e contained in S for each $e \in F$. Then S is feasible to the instance (G', d', c') of $\hat{\kappa}$-SLP if and only if x is feasible to the instance (G, r, F, d, c, u) of RVSAP. Moreover $\sum_{v \in S} c'(v) = \sum_{e \in F} c(e)x(e)$.

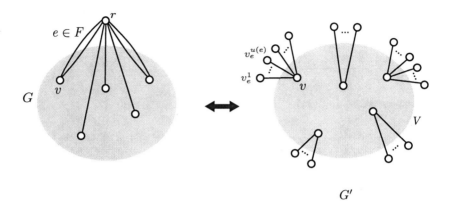

Fig. 1. (G, r, F) in RVSAP and G' in $\hat{\kappa}$-SLP

4 $(2d^* - 1)$-Approximation Algorithm to RVSAP

Let $(G = (V, E), r, F, d, c, u)$ be an instance of RVSAP. Throughout this section, for $X \subseteq V$, $N(X)$ denotes the set of neighbors of X in G, and $n(X)$ denotes $|N(X)|$. Notice that r is not contained by $N(X)$. By the Menger's Theorem, $x : F \to \mathbb{Z}_+$ is feasible to the instance (G, r, F, d, c, u) of RVSAP if and only if $\sum_{e \in \delta_F(X)} x(e) \geq d^*(X) - n(X)$ for every $X \subseteq V$. Therefore RVSAP can be formulated by the next integer programming:

$$
\begin{aligned}
\text{minimize} \quad & \sum_{e \in F} c(e) x(e) \\
\text{subject to} \quad & \sum_{e \in \delta_F(X)} x(v) \geq d^*(X) - n(X) \quad \text{for all } X \subseteq V, \\
& x(e) \in \{0, 1, \dots, u(e)\} \quad \text{for all } e \in F.
\end{aligned}
$$

Our algorithm is based on the iterative rounding method. Following the fashion of the iterative rounding method, we consider a linear programming relaxation of the problem and repeatedly fix the values of variables to integers. The relaxation we use is obtained by replacing the constraint $x(e) \in \{0, 1, \dots, u(e)\}$ by $0 \leq x(e) \leq u(e)$. We let $F' \subseteq F$ be the set of edges whose corresponding variables have not been fixed yet, and $\bar{x} : F \setminus F' \to \mathbb{Z}_+$ represent the values to which variables of $F \setminus F'$ have been fixed. Then the relaxation is represented as follows:

$$
\begin{aligned}
\text{minimize} \quad & \sum_{e \in F'} c(e) x(e) \\
\text{subject to} \quad & \sum_{e \in \delta_{F'}(X)} x(e) \geq d^*(X) - n(X) - \sum_{e \in \delta_{F \setminus F'}(X)} \bar{x}(e) \quad \text{for all } X \subseteq V, \\
& 0 \leq x(e) \leq u(e) \quad \text{for all } e \in F'.
\end{aligned}
$$

$$(3)$$

The basic optimal solution x^* to (3) is the optimal solution which is not represented by a linear combination of the other feasible solutions. The key of our result is to show that the basic optimal solution to (3) has a good property.

Theorem 3. *Let x^* be the basic optimal solution to (3). Then there exists $e \in F'$ such that $x^*(e) = 0$ or $x^*(e) \geq 1/(2d^* - 1)$.*

Our algorithm is a basic iterative rounding algorithm defined as follows.

Iterative rounding algorithm to RVSAP
Input: An instance $(G = (V, E), r, F, d, c, u)$ of RVSAP
Output: $\bar{x} : F \to \mathbb{Z}_+$
Step 1: Let $F' := F$.
Step 2: Compute the basic optimal solution x^* to (3) with F'.
Step 3: If there exists $e \in F'$ such that $x^*(e) = 0$, then remove e from F' and set $\bar{x}(e) := 0$.
Step 4: If there exists $e \in F'$ such that $x^*(e) \geq 1/(2d^* - 1)$, then remove it from F' and set $\bar{x}(e) := \lceil x^*(e) \rceil$.
Step 5: If $F' = \emptyset$, then output $\bar{x}$. Otherwise, return to Step 2.

An optimal solution of (3) can be computed in polynomial time by the ellipsoid method because its separation can be reduced to the max-flow computation. In another way, (3) can be transformed into a polynomial-size formulation by adding variables to represent flow values. It is known that the basic optimal solution of a linear programming can be computed in polynomial time if an optimal solution of the linear programming can be computed in polynomial time (see [16] for example). Hence Step 1 of the above algorithm can be implemented in polynomial time. Alternatively, solving (3) with additional constraint $x(e) = 0$ or $x(e) \geq 1/(2d^* - 1)$ for each $e \in F$, we can judge which edge satisfies the conditions in Steps 2 and 3 without computing x^* explicitly.

It can be proven similarly with other iterative rounding algorithms that the above algorithm achieves $(2d^* - 1)$-approximation if it outputs a solution. We do not present the proof for it here. See [3,8] for example. If Theorem 3 holds, then the above algorithm outputs a feasible solution. Hence in the rest of this section, we concentrate on proving Theorem 3.

We have to introduce some notations. For $X \subseteq V$, let $x^*(X)$ stand for $\sum_{e \in \delta_{F'}(X)} x^*(e)$, and $\bar{x}(X)$ denote $\sum_{e \in \delta_{F \setminus F'}(X)} \bar{x}(e)$. By counting contributions of each edge, we can prove that

$$x^*(X) + x^*(Y) \geq x^*(X \cap Y) + x^*(X \cup Y) \tag{4}$$

and

$$x^*(X) + x^*(Y) \geq x^*(X \setminus (Y \cup N(Y))) + x^*(Y \setminus (X \cup N(X))) + 2x^*(X \cap Y) \tag{5}$$

hold for any $X, Y \subseteq V$. In the same way, it can be proven that

$$\bar{x}(X) + \bar{x}(Y) \geq \bar{x}(X \cap Y) + \bar{x}(X \cup Y) \tag{6}$$

and

$$\bar{x}(X) + \bar{x}(Y) \geq \bar{x}(X \setminus (Y \cup N(Y))) + \bar{x}(Y \setminus (X \cup N(X))) \tag{7}$$

hold for any $X, Y \subseteq V$. From the basic optimal solution x^* to (3), define $f_{x^*}(X) = x^*(X) + \bar{x}(X) + n(X)$ for $X \subseteq V$. (1), (4) and (6) show that f_{x^*} satisfies

$$f_{x^*}(X) + f_{x^*}(Y) \geq f_{x^*}(X \cap Y) + f_{x^*}(X \cup Y) \tag{8}$$

for any $X, Y \subseteq V$. Moreover (2), (5) and (7) show that f_{x^*} satisfies

$$f_{x^*}(X) + f_{x^*}(Y) \geq f_{x^*}(X \setminus (Y \cup N(Y))) + f_{x^*}(Y \setminus (X \cup N(X))) + 2x^*(X \cap Y) \tag{9}$$

for any $X, Y \subseteq V$.

We let $P = \{v \in V \mid \sum_{e \in \delta_{F'}(v)} x^*(e) > 0\}$, and $\mathcal{T} = \{X \subseteq V \mid f_{x^*}(X) = d^*(X) > 0, X \cap P \neq \emptyset\}$. Notice that each $X \in \mathcal{T}$ satisfies $n(X) < d^*(X)$ because $d^*(X) - n(X) = f_{x^*}(X) - n(X) \geq x^*(X) > 0$.

Lemma 3. *Let $v \in P$. The number of minimal members containing v in $\mathcal{T}$ is at most $2d^* - 1$.*

Proof. Let $X_1, X_2, \ldots, X_p$ be the minimal members containing v in $\mathcal{T}$. We suppose that $d^*(X_i)$ is attained by a vertex $v_i \in X_i$ for $i \in \{1, 2, \ldots, p\}$.

First, assume that $X_i \cap X_j \cap \{v_i, v_j\} \neq \emptyset$ holds for some $i, j \in \{1, 2, \ldots, p\}$. Then $d^*(X_i \cap X_j) \geq \min\{d^*(X_i), d^*(X_j)\}$. Since $X_i \cup X_j$ contains both v_i and v_j, $d^*(X_i \cup X_j) \geq \max\{d^*(X_i), d^*(X_j)\}$ also holds. Hence

$$d^*(X_i) + d^*(X_j) \leq d^*(X_i \cap X_j) + d^*(X_i \cup X_j).$$

This and the property (8) of f_{x^*} implies that

$$\begin{aligned} f_{x^*}(X_i \cap X_j) + f_{x^*}(X_i \cup X_j) &\leq f_{x^*}(X_i) + f_{x^*}(X_j) \\ &= d^*(X_i) + d^*(X_j) \\ &\leq d^*(X_i \cap X_j) + d^*(X_i \cup X_j). \end{aligned}$$

Since x^* is feasible to (3), $f_{x^*}(X_i \cap X_j) \geq d^*(X_i \cap X_j)$ and $f_{x^*}(X_i \cup X_j) \geq d^*(X_i \cup X_j)$. These imply that $X_i \cap X_j \in \mathcal{T}$ and $X_i \cup X_j \in \mathcal{T}$, but the former fact contradicts the minimality of X_i or X_j.

Next, assume that $v_i \in X_i \setminus (X_j \cup N(X_j))$ and $v_j \in X_j \setminus (X_i \cup N(X_i))$ hold for some $i, j \in \{1, 2, \ldots, p\}$. In this case, $d^*(X_i \setminus (X_j \cup N(X_j))) = d^*(X_i)$ and $d^*(X_j \setminus (X_i \cup N(X_i))) = d^*(X_j)$ hold. Thus

$$d^*(X_i) + d^*(X_j) = d^*(X_i \setminus (X_j \cup N(X_j))) + d^*(X_j \setminus (X_i \cup N(X_i))).$$

Hence with (9), it shows that

$$\begin{aligned} f_{x^*}(X_i \setminus (X_j \cup N(X_j))) + f_{x^*}(X_j \setminus (X_i \cup N(X_i))) &+ 2x^*(X_i \cap X_j) \\ &\leq f_{x^*}(X_i) + f_{x^*}(X_j) \\ &= d^*(X_i) + d^*(X_j) \\ &= d^*(X_i \setminus (X_j \cup N(X_j))) + d^*(X_j \setminus (X_i \cup N(X_i))). \end{aligned}$$

On the other hand, $f_{x^*}(X_i \setminus (X_j \cup N(X_j))) \geq d^*(X_i \setminus (X_j \cup N(X_j)))$ and $f_{x^*}(X_j \setminus (X_i \cup N(X_i))) \geq d^*(X_j \setminus (X_i \cup N(X_i)))$ because x^* is feasible to (3). These imply that $x^*(X_i \cap X_j) = 0$. However, from $v \in X_i \cap X_j$, it follows that $x^*(X_i \cap X_j) \geq \sum_{e \in \delta_{F'}(v)} x^*(e) > 0$, a contradiction.

By these discussions, we see that at least one of $v_i \in N(X_j)$ and $v_j \in N(X_i)$ holds for any $1 \leq i < j \leq p$. Define a digraph D consisting of p vertices $u_1, u_2, \ldots, u_p$ so that D has an arc from u_i to u_j whenever $v_i \in N(X_j)$. The above fact implies that the underlying undirected graph of D is a clique. Hence D has at least $p(p-1)/2$ arcs. On the other hand, since $n(X_i) < d^*$ for each $i \in \{1, 2, \ldots, p\}$, the in-degree of each vertex in D is less than d^*. Hence D has at most $p(d^* - 1)$ arcs. Therefore $p(p-1)/2 \leq p(d^* - 1)$, implying that $p \leq 2d^* - 1$.   □

For $X \subseteq V$, let $\chi(X)$ denote the incidence vector of $\delta_{F'}(X)$, i.e., $|F'|$-dimensional vector a component of which takes 1 if it is indexed by $e \in \delta_{F'}(X)$, and 0 otherwise. We call $\mathcal{T}' \subseteq \mathcal{T}$ *independent* if for any $X \in \mathcal{T}'$, $\chi(X)$ is not represented by a linear combination of $\chi(Y)$, $Y \in \mathcal{T}' \setminus \{X\}$. By the linear algebra, the basic optimal solution for a linear programming problem with n variables satisfies n linearly independent constraints with equality. In our linear programming (3), this indicates that there exists an independent subset $\mathcal{T}^*$ of $\mathcal{T}$ such that $|\mathcal{T}^*|$ is equal to the number of positive variables in x^*.

Proof of Theorem 3. Suppose that $0 < x^*(e) < 1/(2d^* - 1)$ holds for each $e \in F'$. If F' contains two parallel edges e_1 and e_2, then $x_\epsilon : F' \to \mathbb{Q}_+$ defined by

$$x_\epsilon(e) = \begin{cases} x^*(e) & e \notin \{e_1, e_2\} \\ x^*(e_1) + \epsilon & e = e_1 \\ x^*(e_2) - \epsilon & e = e_2 \end{cases}$$

is also a feasible solution for any ϵ such that $|\epsilon| \leq \min\{x^*(e_1), x^*(e_2), u(e_1) - x^*(e_1), u(e_2) - x^*(e_2)\}$. The existence of such feasible solutions contradicts the definition of x^*. Thus F' contains no parallel edges, and hence $|F'| = |P|$. Moreover there exists an independent subset $\mathcal{T}^* \subseteq \mathcal{T}$ such that $|\mathcal{T}^*| = |F'|$. We assume without loss of generality that $\mathcal{T}^*$ minimizes the number of crossing pairs (i.e., $(X, Y) \in \mathcal{T}^* \times \mathcal{T}^*$ such that all of $X \cap Y$, $X \setminus Y$, $Y \setminus X$ are not empty) among such subsets. Below we show that $|\mathcal{T}^*| < |P|$ to derive a contradiction.

Suppose that $v \in P$ distributes $\sum_{e \in \delta_{F'}(v)} x^*(e)$ tokens to each of minimal members containing v in $\mathcal{T}^*$. By Lemma 3, at most $2d^* - 1$ members of $\mathcal{T}^*$ receive the tokens from $v \in P$. Moreover $\sum_{e \in \delta_{F'}(v)} x^*(e) < 1/(2d^* - 1)$ because $\delta_{F'}(v)$ contains at most one edge. In consequence, the total amount of distributed tokens is less than $|P|$. Below we prove that each member of $\mathcal{T}^*$ obtains at least 1 tokens. This means that $|\mathcal{T}^*|$ is at most the total amount of tokens.

Let $X \in \mathcal{T}^*$, and $Y_1, Y_2, \ldots, Y_p$ be the subsets of X in $\mathcal{T}^*$. Obviously $x^*(X) \geq x^*(Y_1) + x^*(Y_2) + \cdots + x^*(Y_p)$. Since $\{X, Y_1, Y_2, \ldots, Y_p\}$ is independent and $x^*(e) > 0$ for all $e \in F'$, this inequality holds strictly, i.e., $x^*(X) > x^*(Y_1) + x^*(Y_2) + \cdots + x^*(Y_p)$. Since $d^*(X') - n(X') - \bar{x}(X')$ is integer for any $X' \subseteq V$,

$x^*(X)$ and all of $x^*(Y_i)$, $i = 1, 2, \ldots, p$ are integers. Hence $x^*(X) \geq x^*(Y_1) + x^*(Y_2) + \cdots + x^*(Y_p) + 1$. The amount of tokens X obtains is $x^*(X) - (x^*(Y_1) + x^*(Y_2) + \cdots + x^*(Y_p)) \geq 1$.   □

5   Conclusion

We have proposed a new approximation algorithm for $\hat{\kappa}$-SLP. We have also considered a new formulation of the source location problem and have given an approximation algorithm to it. Our algorithms are designed in different frameworks; The former one is a greedy algorithm, and the latter one is based on the iterative rounding method. However the key observations for their analyses are proven in a unified way.

An interesting future work is to design $O(d^*)$-approximation algorithms for $\hat{\kappa}$-SLP and for RVSND. For $\hat{\kappa}$-SLP, Ishii [6] also said that it is open whether the problem is approximable within a constant independent from d^*. For RVSND, Nutov [13] mentioned the existence of $O(d^*)$-approximation algorithms as an open question. We believe that the discussion in Section 4 is useful for solving these questions.

Acknowledgments

This research was partially supported by the Scientific Grant-in-Aid from Ministry of Education, Science, Sports and Culture of Japan.

References

1. Arata, K., Iwata, S., Makino, K., Fujishige, S.: Locating sources to meet flow demands in undirected networks. Journal of Algorithms 42, 54–68 (2002)
2. Chvátal, V.: A greedy heuristic for the set covering problem. Mathematics of Operations Research 4, 233–235 (1979)
3. Fleischer, L., Jain, K., Williamson, D.P.: Iterative rounding 2-approximation algorithms for minimum-cost vertex connectivity problems. Journal of Computer and System Sciences 72, 838–867 (2006)
4. Ishii, T., Fujita, H., Nagamochi, H.: The source location problem with local 3-vertex-connectivity requirements. Discrete Applied Mathematics 155, 2523–2538 (2007)
5. Ishii, T., Fujita, H., Nagamochi, H.: Minimum cost source location problem with local 3-vertex-connectivity requirements. Theoretical Computer Science 372, 81–93 (2007)
6. Ishii, T.: Greedy approximation for the source location problem with vertex-connectivity requirements in undirected graphs. Journal of Discrete Algorithms 7, 570–578 (2009)
7. Ito, H., Ito, M., Itatsu, Y., Uehara, H., Yokoyama, M.: Source location problems considering vertex-connectivity and edge-connectivity simultaneously. Networks 40, 63–70 (2002)

8. Jain, K.: Factor 2 approximation algorithm for the generalized Steiner network problem. Combinatorica 21, 39–60 (2001)

9. Johnson, D.S.: Approximation algorithms for combinatorial problems. Journal of Computer and System Sciences 9, 256–278 (1974)

10. Lau, L.C., Singh, M.: Additive approximation for bounded degree survivable network design. In: 40th ACM Symposium on Theory of Computing, pp. 759–768 (2008)

11. Lovász, L.: On the ratio of optimal integral and fractional covers. Discrete Mathematics 13, 383–390 (1975)

12. Nagamochi, H., Ishii, T., Ito, H.: Minimum cost source location problem with vertex-connectivity requirements in digraphs. Information Processing Letters 80, 287–294 (2001)

13. Nutov, Z.: Approximating minimum cost connectivity problems via uncrossable bifamilies and spider-cover decompositions. In: 50th Annual IEEE Symposium on Foundations of Computer Science, pp. 417–426 (2009)

14. Sakashita, M., Makino, K., Fujishige, S.: Minimizing a monotone concave function with laminar covering constraints. Discrete Applied Mathematics 156, 2004–2019 (2008)

15. Sakashita, M., Makino, K., Fujishige, S.: Minimum cost source location problems with flow requirements. Algorithmica 50, 555–583 (2008)

16. Schrijver, A.: Theory of Linear and Integer Programming. John Wiley & Sons, Chichester (1986)

17. Singh, M., Lau, L.C.: Approximating minimum bounded degree spanning trees to within one of optimal. In: 39th ACM Symposium on Theory of Computing, pp. 661–670 (2007)

Improved Approximation Bounds for the Student-Project Allocation Problem with Preferences over Projects

Kazuo Iwama[1], Shuichi Miyazaki[2], and Hiroki Yanagisawa[3]

[1] Graduate School of Informatics, Kyoto University
iwama@kuis.kyoto-u.ac.jp
[2] Academic Center for Computing and Media Studies, Kyoto University
shuichi@media.kyoto-u.ac.jp
[3] IBM Research - Tokyo
yanagis@jp.ibm.com

Abstract. Manlove and O'Malley [9] proposed the Student-Project Allocation Problem with Preferences over Projects (SPA-P). They proved that the problem of finding a maximum stable matching in SPA-P is APX-hard and gave a polynomial-time 2-approximation algorithm. In this paper, we give an improved upper bound of 1.5 and a lower bound of $21/19$ (> 1.1052).

1 Introduction

Assignment problems based on the preferences of participants, which originated from the famous *Hospitals/Residents problem (HR)* [4], are important almost everywhere, such as in education systems where students must be allocated to elementary schools or university students to projects. In the university case, each student may have preferences over certain research projects supervised by professors and usually there is an upper bound on the number of students each project can accept. Our basic goal is to find a "stable" allocation where no students (or projects or professors if they also have preferences over students) can complain of unfairness. This notion of stability was first introduced by Gale and Shapley in the context of the famous *Stable Marriage problem* in 1962 [3].

The *Student-Project Allocation problem (SPA)* is a typical formulation of this kind of problem originally described by Abraham, Irving, and Manlove [1]. The participants here are *students*, *projects*, and *lecturers*. Each project is offered by a single lecturer, though one lecturer may offer multiple projects. Each project and each lecturer has a *capacity* (called a quota in the original HR). Students have preferences over projects, and lecturers have preferences over students. Our goal is to find a stable matching between students and projects satisfying all of the capacity constraints for projects and lecturers. They proved that all stable matchings for a single instance have the same size, and proposed linear-time algorithms to find one [1].

Manlove and O'Malley [9] proposed a variant of SPA, called *SPA with Preferences over Projects (SPA-P)*, where lecturers have preferences over projects they

M. Ogihara and J. Tarui (Eds.): TAMC 2011, LNCS 6648, pp. 440–451, 2011.

offer rather than preferences over students. In contrast to SPA, they pointed out
that the sizes of stable matchings may differ, and proved that the problem of finding a maximum stable matching in SPA-P, denoted *MAX-SPA-P*, is APX-hard.
They also presented a polynomial-time 2-approximation algorithm. Specifically,
they provided a polynomial-time algorithm that finds a stable matching, and
proved that any two stable matchings differ in size by at most a factor of two.

Our Contributions. In this paper, we improve both the upper and lower
bounds on the approximation ratio for MAX-SPA-P. We give an upper bound of
1.5 and a lower bound of $21/19$ (> 1.1052) (assuming P $\neq$ NP). For the upper
bound, we modify Manlove and O'Malley's algorithm SPA-P-APPROX [9] using
Király's idea [7] for the approximation algorithm to find a maximum stable
matching in a variant of the stable marriage problem (*MAX-SMTI*). We also
show that our analysis is tight. For the lower bound, we give a gap-preserving
reduction from the Minimum Vertex Cover problem, which is similar to the one
used in [5] to prove the approximation lower bound for MAX-SMTI.

2   Preliminaries

Here we give a formal definition of SPA-P and MAX-SPA-P, derived directly
from the literature [9]. An instance I of SPA-P consists of a set S of *students*, a
set P of *projects*, and a set L of *lecturers*. Each lecturer $\ell_k \in L$ offers a subset
P_k of projects. Each project is offered by exactly one lecturer, i.e., $P_{k_1} \cap P_{k_2} = \emptyset$
if $k_1 \neq k_2$. Each student $s_i \in S$ has an *acceptable* set of projects, denoted A_i,
and has a strict order on A_i according to preferences. Each lecturer ℓ_k also has
a strict order on P_k according to preferences. Also, each project p_j and each
lecturer ℓ_k has a positive integer, called a *capacity*, c_j and d_k, respectively.

An *assignment* M is a subset of $S \times P$ where $(s_i, p_j) \in M$ implies $p_j \in A_i$.
Let $(s_i, p_j) \in M$ and ℓ_k be the lecturer who offers p_j. Then we say that s_i *is
assigned to p_j* in M, and p_j *is assigned s_i* in M. We also say that s_i *is assigned
to ℓ_k* in M and ℓ_k *is assigned s_i* in M.

For $r \in S \cup P \cup L$, let $M(r)$ be the set of assignees of r in M. If $M(s_i) = \emptyset$, we
say that the student s_i is *unassigned* in M, otherwise s_i is *assigned* in M. We say
that the project p_j is *under-subscribed*, *full*, or *over-subscribed* with respect to M
according to whether $|M(p_j)| < c_j$, $|M(p_j)| = c_j$, or $|M(p_j)| > c_j$, respectively,
under M. If $|M(p_j)| > 0$, we say that p_j is *non-empty*, otherwise, it is *empty*.
Corresponding definitions apply to each lecturer ℓ.

A *matching* M is an assignment such that $|M(s_i)| \leq 1$ for each s_i, $|M(p_j)| \leq c_j$
for each p_j, and $|M(\ell_k)| \leq d_k$ for each ℓ_k. For a matching M, if $|M(s_i)| = 1$, we
may use $M(s_i)$ to denote the unique project which s_i is assigned to. The *size* of
a matching M, denoted $|M|$, is the number of students assigned in M.

Given a matching M, a (student, project) pair (s_i, p_j) *blocks* M, or is a *blocking
pair* for M, if the following three conditions are met:

1. $p_j \in A_i$.
2. Either s_i is unassigned or s_i prefers p_j to $M(s_i)$.
3. p_j is under-subscribed and either

(a) $s_i \in M(\ell_k)$ and ℓ_k prefers p_j to $M(s_i)$, or

(b) $s_i \notin M(\ell_k)$ and ℓ_k is under-subscribed, or

(c) $s_i \notin M(\ell_k)$, ℓ_k is full, and ℓ_k prefers p_j to the worst non-empty project,

where ℓ_k is the lecturer who offers p_j.

Given a matching M, a *coalition* is a set of students $\{s_{i_0}, s_{i_1}, \dots, s_{i_{r-1}}\}$ for some $r \geq 2$ such that each s_{i_j} is assigned in M and prefers $M(s_{i_{j+1}})$ to $M(s_{i_j})$, where $j+1$ is taken modulo r. A matching that has no blocking pair nor coalition is *stable*. Refer to [9] for the validity of this definition of stability. SPA-P is the problem of finding a stable matching, and MAX-SPA-P is the problem of finding a maximum stable matching.

We say that A is an r-approximation algorithm if it satisfies $\max\{opt(x)/A(x)\} \leq r$ over all instances x, where $opt(x)$ and $A(x)$ are the sizes of the optimal and the algorithm's solutions, respectively.

3   Approximability

3.1   Algorithm SPA-P-APPROX-PROMOTION

Manlove and O'Malley's algorithm SPA-P-APPROX [9] proceeds as follows. First, all students are unassigned. Any student (s) who has non-empty preference list applies to the top project (p) on the current list of s. If the lecturer (ℓ) who offers p has no incentive to accept s for p, then s is rejected. When rejected, s deletes p from the list. Otherwise, (s, p) is added to the current matching. If, as a result, ℓ becomes over-subscribed, ℓ rejects a student from ℓ's worst non-empty project to satisfy the capacity constraint. This continues until there is no unassigned student whose preference list is non-empty. Manlove and O'Malley proved that the obtained matching is stable.

We extend SPA-P-APPROX using Király's idea [7]. During the execution of our algorithm SPA-P-APPROX-PROMOTION, each student has one of two states, "unpromoted" or "promoted". At the beginning, all of the students are unpromoted. The application sequence is unchanged. When a student (s) becomes unassigned with her preference list exhausted, s is promoted. When promoted, s returns to her original preference list (i.e., all of the previous deletions are canceled) and starts a second sequence of applications from the top of her list. For the decision rule for acceptance or rejection by the lecturers, they will prefer promoted students to unpromoted students within the same project. The formal description of SPA-P-APPROX-PROMOTION is given as Algorithm 1.

3.2   Correctness

It is straightforward to show that SPA-P-APPROX-PROMOTION outputs a matching in polynomial time. We will now show that the output matching M is stable. We first prove two useful lemmas:

Algorithm 1. SPA-P-APPROX-PROMOTION

1: $M := \emptyset$.
2: Let all students be unpromoted.
3: **while** (there exists an unassigned student s_i such that s_i's list is non-empty or s_i is unpromoted) **do**
4:      **if** (s_i's list is empty and s_i is unpromoted) **then**
5:          Promote s_i.
6:      **end if**
7:      $p_j :=$ first project on s_i's list.
8:      $\ell_k :=$ lecturer who offers p_j.
9:      /* s_i applies to p_j */
10:     **if** (A. (p_j is full) or (ℓ_k is full and p_j is ℓ_k's worst non-empty project)) **then**
11:         **if** ((s_i is unpromoted) or (there is no unpromoted student in $M(p_j)$)) **then**
12:             Reject s_i.
13:         **else**
14:             Reject an arbitrary unpromoted student in $M(p_j)$ and add (s_i, p_j) to M.
15:         **end if**
16:     **else if** (B. ℓ_k is full and prefers s_i's worst non-empty project to p_j) **then**
17:         Reject s_i.
18:     **else if** (C. Otherwise) **then**
19:         Add (s_i, p_j) to M.
20:         **if** (ℓ_k is over-subscribed) **then**
21:             $p_z := \ell_k$'s worst non-empty project. (Note that $p_z \neq p_j$.)
22:             **if** ($M(p_z)$ contains an unpromoted student) **then**
23:                 Reject an arbitrary unpromoted student in $M(p_z)$.
24:             **else**
25:                 Reject an arbitrary student in $M(p_z)$.
26:             **end if**
27:         **end if**
28:     **end if**
29: **end while**
30: Return M.

Lemma 1. *Suppose that, during the execution of* SPA-P-APPROX-PROMOTION, *a project p_a rejected a promoted student. Then (i) after that point, no new student can be accepted to p_a, and (ii) no unpromoted student can be assigned to p_a in M.*

Proof. Suppose that a promoted student s is rejected by p_a. Let ℓ_k be the lecturer who offers p_a. It is easy to see that just after this rejection, no unpromoted student can be assigned to p_a. We show that after that point, if a student s' applies to p_a when there is no unpromoted student assigned to p_a, then s' must be rejected. It is easy to see that the lemma follows by using this fact inductively.

Note that just after this rejection, either (1) p_a is full or (2) p_a is under-subscribed and ℓ_k is full. We consider Case (2) first. Since p_a is under-subscribed but s was rejected from p_a, p_a must be ℓ_k's worst non-empty project before the rejection. Then after this rejection, p_a is still ℓ_k's worst non-empty project or p_a

becomes worse than it was (if s was the only student assigned to p_a). Note that now ℓ_k remains full until the end of the execution. Then after this point, when any student applies to p_a, only Cases A (line 10) or B (line 16) of the algorithm can apply. Since there is no unpromoted student in $M(p_a)$, s' must be rejected.

In Case (1), if p_a is still full when s' applies to p_a, Case A of the algorithm applies and hence s' must be rejected since $M(p_a)$ contains no unpromoted student. If p_a is under-subscribed when s' applies to p_a, then some student was already rejected from p_a. At that time, ℓ_k must have been full and p_a was ℓ_k's worst non-empty project. Therefore, ℓ_k is still full and p_a is ℓ_k's worst non-empty project or worse than it was. Then we can apply the same argument as in Case (2).   □

The proof of the following lemma is basically similar and is omitted.

Lemma 2. *Suppose that, during the execution of* SPA-P-APPROX-PROMOTION, *a project p_a has rejected a student. Then after that point, no new unpromoted student can be accepted for p_a.*

To prove the stability, we need to prove that there is no coalition or blocking pair.

Lemma 3. *The output matching M is coalition-free.*

Proof. Suppose that there is a coalition $\{s_{i_0}, s_{i_1}, \ldots, s_{i_{r-1}}\}$ for some $r \geq 2$. Let $p_{i_j} = M(s_{i_j})$ for each j ($0 \leq j \leq r-1$). Thus s_{i_j} prefers $p_{i_{j+1}}$ to p_{i_j} (where $j+1$ is taken modulo r). Therefore, at some point of the execution, $p_{i_{j+1}}$ was deleted from s_{i_j}'s list. Note that during the execution of the algorithm, one project may be deleted from a student's list twice (because of a promotion). Hereafter, a "deletion" means the final deletion unless otherwise stated.

Now suppose that among such deletions, the first occurrence was the deletion of p_{i_1} from s_{i_0}'s list. First, suppose that s_{i_0} is eventually unpromoted. Note that s_{i_1} applied to and was accepted by p_{i_1} after s_{i_0} was rejected by p_{i_1}. Therefore s_{i_1} is eventually promoted by Lemma 2. Then s_{i_1} was rejected from p_{i_2} when s_{i_1} was promoted. This means that s_{i_2} is eventually promoted by Lemma 1(ii). Repeating this argument, we can conclude that $s_{i_{r-1}}$ is eventually promoted. Then this contradicts Lemma 1(ii) since p_{i_0} rejected the promoted student $s_{i_{r-1}}$ but is assigned an unpromoted student s_{i_0} in M.

Next suppose that s_{i_0} is eventually promoted. Then since p_{i_1} rejected a promoted student, p_{i_1} accepts no new students by Lemma 1(i). This contradicts the fact that s_{i_1} was accepted to p_{i_1} later.   □

Lemma 4. *The output matching M has no blocking pair.*

Proof. Assume that there exists a blocking pair (s_r, p_t) for M. Then it is clear that s_r was rejected from p_t during the execution (recall that this rejection is the second one if s_r was eventually promoted). Let ℓ_k be the lecturer who offers p_t. Rejections occur at lines 12, 14, 17, 23, and 25. If this rejection occurred at line 17, 23, or 25, then p_t was already ℓ_k's worst non-empty project or worse than that, and this is also the case in M. We know that ℓ_k was full at this rejection point, and remains full in M. Therefore, (s_r, p_t) cannot block M. If this rejection occurred at line 12 or 14 as a result of ℓ_k being full and p_t being ℓ_k's worst non-empty project, then the same argument holds. Therefore suppose that this rejection occurred at line 12 or 14 as a result of p_t being full. Since (s_r, p_t) blocks M, p_t is under-subscribed in M. Then p_t changed from being full to being under-subscribed at some point. This can happen only when ℓ_k is full and p_t is ℓ_k's worst non-empty project. Again, we can use the same argument to show that (s_r, p_t) cannot block M, a contradiction.   □

The following lemma follows immediately from Lemmas 3 and 4.

Lemma 5. SPA-P-APPROX-PROMOTION *returns a stable matching.*

3.3 Analysis of the Approximation Ratio

For a given instance I, let M be a matching output from SPA-P-APPROX-PROMOTION, and let M_{opt} be a largest stable matching for I.

Lemma 6. $|M_{opt}| \leq \frac{3}{2}|M|$.

Proof. Based on M and M_{opt}, we define a bipartite graph $G_{M,M_{opt}} = (U, V, E)$ as follows: Each vertex in U corresponds to a student in I, and each vertex in V corresponds to a position of a project in I. Precisely speaking, for each project p_j whose capacity is c_j, we create c_j "positions" of p_j, each of which can accept at most one student, and each vertex in V corresponds to each such position. We use s_i to denote the vertex in U corresponding to a student s_i and $p_{j,1}, p_{j,2}, \ldots, p_{j,c_j}$ to denote the vertices in V corresponding to a project p_j.

If a student s_i is assigned to a project p_j in M (M_{opt}, respectively), we include an edge $(s_i, p_{j,t})$ for some t ($1 \leq t \leq c_j$), called an M-edge (M_{opt}-edge, respectively), in E. If s_i is assigned to the same project p_j both in M and M_{opt}, then M- and M_{opt}-edges corresponding to this assignment include the same position of p_j, which means we give parallel edges $(s_i, p_{j,t})$ for some t. We also ensure that there are no two vertices p_{j,t_1} and p_{j,t_2} such that p_{j,t_1} is matched in M but not in M_{opt}, and p_{j,t_2} is matched in M_{opt} but not in M. In such a case, there will be M-edge (s_{i_1}, p_{j,t_1}) and M_{opt}-edge (s_{i_2}, p_{j,t_2}). Then we can remove (s_{i_1}, p_{j,t_1}) and add (s_{i_1}, p_{j,t_2}) instead.

Note that each vertex of $G_{M,M_{opt}}$ has degree at most two. Therefore its connected components are alternating paths or alternating cycles. Now we will modify $G_{M,M_{opt}}$ while retaining this property and keeping the numbers of M-edges and M_{opt}-edges unchanged. Note that the resulting graph may not correspond to a feasible solution for I. We use this modification only for the purpose of comparing the sizes of M and M_{opt}.

A connected component consisting of only one M_{opt}-edge is called a *Type-I component*. A connected component which is a length-three alternating path consisting of two M_{opt}-edges and one M-edge in the middle is called a *Type-II component*. We show that there are no Type-I or Type-II components in the resulting bipartite graph. If this is true, the connected component having the largest ratio of the number of M_{opt}-edges to that of M-edges is a length-five alternating path with three M_{opt}-edges and two M-edges, which has the ratio of 1.5. This proves the lemma.

Consider a Type-I component $(s_i, p_{j,t})$. Let ℓ_k be the lecturer who offers p_j. Since $p_{j,t}$ is not matched in M, p_j is under-subscribed in M. Then ℓ_k must be full in M since otherwise (s_i, p_j) blocks M. Therefore, we can find a vertex $p_{a,x}$ in V which is matched in M but not in M_{opt}, where p_a is offered by ℓ_k. We can remove $(s_i, p_{j,t})$ and add $(s_i, p_{a,x})$ to remove this Type-I component.

Consider a Type-II component $s_i - p_{a,x} - s_j - p_{b,y}$. Note that $p_a \neq p_b$ due to the construction of $G_{M,M_{opt}}$. Since s_i is unassigned in M, s_i is promoted. Then s_i applied to p_a when promoted, but was rejected. Therefore s_j must be promoted by Lemma 1(ii). This means that s_j applied to p_b at least once, but was rejected. Let ℓ_k be the lecturer who offers p_b. As mentioned several times before, this rejection can happen only when (1) p_b is full or (2) ℓ_k is full and p_b is ℓ_k's worst non-empty project or worse than that, and either (1) or (2) also holds for the output matching M. However $p_{b,y}$ is unassigned in M, so only (2) is possible. Since ℓ_k is full in M, there must be a vertex $p_{c,z}$ in V which is matched in M but not in M_{opt}, where p_c is offered by ℓ_k. We can remove the edge $(s_j, p_{b,y})$ and add $(s_j, p_{c,z})$ to remove this Type-II component.

Note that in both of these cases, we used the property that ℓ_k is full in M. This implies that for each Type-I or Type-II component, we can find a distinct vertex in V which is matched only in M to perform the above mentioned replacement. We do this replacement for all Type-I and Type-II components in $G_{M,M_{opt}}$. This operation does not change any M-edges, so the number of students assigned to each lecturer or project in M is unchanged. In particular, a lecturer or a project full in M is still full in the modified graph.

As a result of these operations, we may still have a Type-II component. This can happen only when we removed a Type-I component, such as $(s_i, p_{j,t})$, using a length-two path, such as $p_{a,x} - s_r - p_{b,y}$, where $(s_r, p_{a,x})$ is an M-edge and $(s_r, p_{b,y})$ is an M_{opt}-edge. In this example, we removed $(s_i, p_{j,t})$ and added $(s_i, p_{a,x})$. Note that p_a and p_j must be offered by the same lecturer, such as ℓ_k, because of the definition of the operation for Type-I components. Also, by the construction of $G_{M,M_{opt}}$, p_a and p_j must be different projects because $p_{j,t}$ is matched only in M_{opt} and $p_{a,x}$ is matched only in M.

If p_b is also offered by ℓ_k, then corresponding to the M_{opt}-edge $(s_r, p_{b,y})$, we can find a vertex $p_{c,z}$ in V which is matched in M but not in M_{opt}, where p_c is offered by ℓ_k, since ℓ_k is full in M. Then we can remove this Type-II component by replacing $(s_r, p_{b,y})$ with $(s_r, p_{c,z})$. Otherwise, let $\ell_{k'}(\neq \ell_k)$ be the lecturer who offers p_b. Suppose that s_r prefers p_b to p_a. Since p_b is under-subscribed in M, $\ell_{k'}$ must be full in M, since otherwise (s_r, p_b) blocks M. Then we can use the same argument as before to show the existence of a vertex $p_{c,z}$ which is matched in M but not in M_{opt}, where p_c is offered by $\ell_{k'}$. Suppose that s_r prefers p_a to p_b. If ℓ_k prefers p_a to p_j, then (s_r, p_a) blocks M_{opt}, a contradiction (note that $p_{a,x}$ is not matched and hence p_a is under-subscribed in M_{opt}). If ℓ_k prefers p_j to

p_a, then (s_i, p_j) blocks M, a contradiction. We have exhausted all of the cases, and have shown that all Type-I and Type-II components can be removed. This completes the proof.   □

The following theorem follows immediately from Lemmas 5 and 6.

Theorem 1. SPA-P-APPROX-PROMOTION *is a 1.5-approximation algorithm for MAX-SPA-P.*

3.4 Tightness of the Analysis

We give an instance to show that our analysis of the approximation ratio is tight. There are three students s_1, s_2, and s_3 and one lecturer ℓ_1 with $d_1 = 3$ who offers three projects p_1, p_2, and p_3 with $c_1 = c_2 = c_3 = 1$. The preferences of the students and the lecturer are as follows:

$$s_1: p_1 \qquad\qquad \ell_1: p_3 \;\; p_2 \;\; p_1$$
$$s_2: p_1 \;\; p_2$$
$$s_3: p_2 \;\; p_3$$

Note that the matching $\{(s_1, p_1), (s_2, p_2), (s_3, p_3)\}$ of size three is stable, but the following execution of SPA-P-APPROX-PROMOTION yields a stable matching of size two $\{(s_2, p_1), (s_3, p_2)\}$: (1) s_1 applies to p_1 and is accepted. (2) s_3 applies to p_2 and is accepted. (3) s_2 applies to p_1 and is rejected. (4) s_2 applies to p_2 and is rejected. (5) s_2 is promoted. (6) s_2 applies to p_1 and is accepted; s_1 is rejected. (7) s_1 is promoted. (8) s_1 applies to p_1 and is rejected.

4   Inapproximability

The stable marriage problem (SM) [3,4] is the problem of finding a stable matching, given sets of men and women and each person's preference list over the members of the opposite gender. If ties are allowed in the preference lists and if the preference lists may be incomplete (i.e., unacceptable persons may be dropped from the lists), then the problem of finding a maximum stable matching (MAX-SMTI) is NP-hard even if ties appear on only one side (e.g., the men's lists must be totally ordered) [8]. We call this restricted problem *MAX-SMTI-1T*.

There is a similarity between MAX-SMTI-1T and MAX-SPA-P, so we can define the following natural reduction from MAX-SMTI-1T to MAX-SPA-P: Suppose that in the MAX-SMTI-1T instance I, the men's lists are strict and the women's lists may contain ties. Then in the MAX-SPA-P instance I', the students and lecturers correspond to men and women in I, respectively. For each woman w's list, we create a project for each tie in the list, where a man not in a tie is considered as a tie of size one. These projects are offered by the lecturer ℓ_w corresponding to the woman w, and the order of projects in ℓ_w's list is consistent with w's list in I. Each project p is acceptable to the students corresponding to the men in the tie associated with this project p. The order of projects in the preference list of a student is naturally generated from corresponding man's (strictly ordered) list in I. The capacity of each lecturer and each project is one.

Using the above reduction, we can prove that the sizes of a maximum stable matching of I and a maximum blocking-pair-free matching of I' coincide. The only problem is that there is a coalition-freeness condition in the stability definition of SPA-P. Therefore a reduction from the general instances of MAX-SMTI-1T to MAX-SPA-P cannot be applied. However, it turns out that if we use only the instances generated by the reduction in [5], then this problem can be resolved and the sizes of the optimal solutions for MAX-SMTI-1T and MAX-SPA-P coincide, so that the approximation lower bound of $21/19$ for MAX-SMTI-1T proved in [5] applies to MAX-SPA-P. For the completeness of this article, however, we give a direct reduction from the *Minimum Vertex Cover* problem (*MVC*) to MAX-SPA-P.

For a graph $G = (V, E)$, a subset $C \subseteq V$ of vertices is called a *vertex cover* for G if for any edge, at least one of its endpoints is in C. MVC is the problem of finding a vertex cover of minimum size for a given graph. Let $OPT(G)$ be the size of a minimum vertex cover for G. We can now use the well-known Proposition 1.

Proposition 1. *[2] For any $\epsilon > 0$ and $p < \frac{3-\sqrt{5}}{2}$, if there is a polynomial-time algorithm that, given a graph $G = (V, E)$, distinguishes between these two cases, then P=NP.*

   (1) $OPT(G) \leq (1 - p + \epsilon)|V|$.
   (2) $OPT(G) > (1 - \max\{p^2, 4p^3 - 3p^4\} - \epsilon)|V|$.

For an instance I of MAX-SPA-P, let $OPT(I)$ be the size of a maximum stable matching for I. Then we can prove Theorem 2.

Theorem 2. *For any $\epsilon > 0$ and $p < \frac{3-\sqrt{5}}{2}$, if there is a polynomial-time algorithm that, given a MAX-SPA-P instance I of N students, distinguishes between these two cases, then P=NP.*

   (1) $OPT(I) \geq \frac{2+p-\epsilon}{3}N$.
   (2) $OPT(I) < \frac{2+\max\{p^2, 4p^3-3p^4\}+\epsilon}{3}N$.

Proof. Given a graph $G = (V, E)$, we will construct, in polynomial time, an instance I_G of MAX-SPA-P with N students. Our reduction satisfies conditions (i) $N = 3|V|$ and (ii) $OPT(I_G) = 3|V| - OPT(G)$. Then it is not hard to see that Proposition 1 implies Theorem 2.

Now we show the reduction. For each vertex v_i of G, we construct three students a_i, b_i, and c_i and three lecturers x_i, y_i, and z_i. Suppose that v_i is adjacent to k vertices $v_{i_1}, v_{i_2}, \cdots, v_{i_k}$ ($i_1 < i_2 < \cdots < i_k$). Then we construct $k + 4$ projects X_i, Y_i, $Z_{i,-}$, Z_{i,i_1}, $\cdots$, Z_{i,i_k} and $Z_{i,+}$, where X_i is offered by x_i, Y_i by y_i, and $Z_{i,-}$, Z_{i,i_1}, $\cdots$, Z_{i,i_k}, $Z_{i,+}$ by z_i. The capacity of each project and each lecturer is one.

Next, we define the acceptability of projects to students. The project X_i is acceptable to only one student a_i. The project Y_i is acceptable to two students

a_i and b_i. $Z_{i,-}$ is acceptable to only b_i, and $Z_{i,+}$ is acceptable to only c_i. For each $j = 1, 2, \ldots, k$, the project Z_{i,i_j} is acceptable to only one student a_{i_j} (corresponding to the adjacent vertex v_{i_j}). Finally, we define preference lists of the students and lecturers corresponding to v_i as:

$$a_i\colon Y_i \ Z_{i_1,i} \ Z_{i_2,i} \ \cdots \ Z_{i_k,i} \ X_i \qquad x_i\colon X_i$$

$$b_i\colon Y_i \ Z_{i,-} \qquad\qquad\qquad\qquad y_i\colon Y_i$$

$$c_i\colon Z_{i,+} \qquad\qquad\qquad\qquad z_i\colon Z_{i,-} \ Z_{i,i_1} \ \cdots \ Z_{i,i_k} \ Z_{i,+}$$

Obviously, this reduction can be performed in polynomial time. Since the capacities of all of the projects and lecturers are one, for a project or a lecturer r assigned in M, we may use $M(r)$ to denote the unique student assigned to r. Clearly condition (i) holds. In the rest of the proof, we show that condition (ii) holds. To see this, we show that (A) if there is a vertex cover C of G, then there is a stable matching M of I_G such that $|M| = 3|V| - |C|$, and (B) if there is a stable matching M of I_G, then there is a vertex cover C of G such that $|C| = 3|V| - |M|$. The statement (A) implies $OPT(I_G) \geq 3|V| - OPT(G)$ and (B) implies $OPT(G) \leq 3|V| - OPT(I_G)$, which together implies condition (ii).

We show (A) first. Given a vertex cover C for G, we construct a stable matching M for I_G as follows: For each vertex v_i, if $v_i \in C$, let $M(a_i) = Y_i$, $M(b_i) = Z_{i,-}$, and leave c_i unassigned. If $v_i \notin C$, let $M(a_i) = X_i$, $M(b_i) = Y_i$, and $M(c_i) = Z_{i,+}$. Since the capacity of each lecturer is one, we can regard M as a matching between students and lecturers. Fig. 1 shows a part of M corresponding to v_i. By an easy calculation, we can see that $|M| = 2|C| + 3(|V| - |C|) = 3|V| - |C|$ as required.

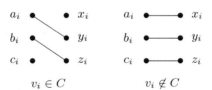

$v_i \in C$ $v_i \notin C$

Fig. 1. A part of matching M

We will show that M is stable. We first show that there is no blocking pair. For $v_i \in C$, a_i is assigned to the top project, so that a_i cannot be part of a blocking pair. The student b_i is assigned to the second project, but the first project Y_i and the lecturer y_i who offers Y_i are both full and hence b_i cannot form a blocking pair. Student c_i is unassigned but the lecturer z_i, who offers c_i's only acceptable project $Z_{i,+}$, is full and prefers c_i's assigned project $Z_{i,-}$ to $Z_{i,+}$, so that c_i cannot be part of a blocking pair. For $v_i \notin C$, b_i and c_i are assigned to the top projects respectively. The only possibility is that a_i forms a blocking pair with some project among $Y_i, Z_{i_1,i}, Z_{i_2,i}, \cdots, Z_{i_k,i}$, but it is easy to see that Y_i is excluded. Therefore, suppose that a_i forms a blocking pair with $Z_{i_j,i}$ for some j. Then by construction there is an edge between v_i and v_{i_j}, and the lecturer z_{i_j} is assigned the student c_{i_j} for the project $Z_{i_j,+}$ (since in the other case, z_{i_j} receives a student for the most preferred project and hence $(a_i, Z_{i_j,i})$ cannot be a blocking pair). This means that $v_{i_j} \notin C$ by the construction of M. Then this contradicts the assumption that C is a vertex cover for G. We then show that M admits no coalition. Note that in M, each student corresponding to the vertex v_i of G is assigned to a project corresponding to v_i. This implies that any coalition must consist of students and projects corresponding to the same vertex. However we can easily verify that there is no coalition in either the case of $v_i \in C$ or $v_i \notin C$, which completes the stability proof.

Next we show (B). Let M be a stable matching for I_G. First, if the project Y_i is unassigned, then both (a_i, Y_i) and (b_i, Y_i) block M, which is a contradiction. Therefore either $M(Y_i) = a_i$ or $M(Y_i) = b_i$.

First, suppose that $M(Y_i) = a_i$. Then $M(b_i) = Z_{i,-}$ since otherwise, $(b_i, Z_{i,-})$ blocks M. Then c_i is unassigned and x_i and X_i are empty in M. In this case, we say that v_i causes *Pattern 1*. A diagrammatic representation of Pattern 1 is given in Fig. 2.

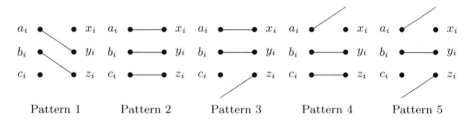

Pattern 1   Pattern 2   Pattern 3   Pattern 4   Pattern 5

Fig. 2. Five patterns caused by v_i

Next, suppose that $M(Y_i) = b_i$. Then a_i is assigned in M, since otherwise (a_i, X_i) blocks M. Since Y_i is already taken by b_i, there remain two cases: (a) $M(a_i) = X_i$ and (b) $M(a_i) = Z_{i_j,i}$ for some j. Similarly, if z_i is empty in M, then $(c_i, Z_{i,+})$ blocks M. This means either (c) $M(z_i) = c_i$ or (d) $M(z_i) = a_{i_j}$ for some j. Hence, we have a total of four cases. These cases are referred to as Patterns 2 through 5 (see Fig. 2). For example, a combination of cases (b) and (c) corresponds to Pattern 4. Lemma 7, whose proof is omitted by the space restriction, excludes the possibility of Patterns 3 or 4.

Lemma 7. *Each vertex causes Pattern 1, 2 or 5.*

By Lemma 7, each vertex v_i will lead to Pattern 1, 2, or 5. We construct the subset C of vertices in this way: If v_i causes Pattern 1 or 5, then let $v_i \in C$, otherwise, let $v_i \notin C$.

We show that C is a vertex cover for G. Assume that C is not a vertex cover for G. Then there are two vertices v_i and v_j in $V \setminus C$ such that $(v_i, v_j) \in E$ and both of them cause Pattern 2. Then both $(a_i, Z_{j,i})$ and $(a_j, Z_{i,j})$ block M, contradicting the stability of M. Hence, C is a vertex cover for G. It is obvious that $|M| = 2|C| + 3(|V| - |C|) = 3|V| - |C|$. Hence, statement (B) holds. This completes the proof of Theorem 2. □

By letting $p = \frac{1}{3}$ in Theorem 2, we have Corollary 1.

Corollary 1. *Assume that P$\neq$NP. Then for any constant $\delta > 0$, there is no polynomial-time $(21/19 - \delta)$-approximation algorithm for MAX-SPA-P.*

Remark. Using the same argument as Remark 3.6 of [5], we can claim that MAX-SPA-P is hard to approximate within $1.25 - \delta$ if MVC is hard to approximate within $2 - \epsilon$ (where δ and ϵ are arbitrary positive constants).

5   Conclusions

In this paper, we improved the upper and lower bounds on the approximation ratio for MAX-SPA-P. One research direction is to further improve the upper bound. For example, a recent approximation algorithm for MAX-SMTI-1T [6] generalizes Király's idea [7] using Linear Programming approach. Its approximation ratio of $25/17 (\simeq 1.4706)$ is slightly better than 1.5. One possible next step is to verify whether this idea can be applied to SPA-P-APPROX-PROMOTION.

Acknowledgments. The authors would like to thank anonymous reviewers for their valuable comments. This work was supported by KAKENHI 22240001 and 20700009.

References

1. Abraham, D.J., Irving, R.W., Manlove, D.F.: Two algorithms for the Student-Project Allocation problem. J. Discrete Algorithms 5(1), 73–90 (2007)
2. Dinur, I., Safra, S.: On the hardness of approximating minimum vertex-cover. Annals of Mathematics 162(1), 439–485 (2005)
3. Gale, D., Shapley, L.S.: College admissions and the stability of marriage. Amer. Math. Monthly 69, 9–15 (1962)
4. Gusfield, D., Irving, R.W.: The Stable Marriage Problem: Structure and Algorithms. MIT Press, Boston (1989)
5. Halldórsson, M.M., Iwama, K., Miyazaki, S., Yanagisawa, H.: Improved approximation results for the stable marriage problem. ACM Transactions on Algorithms 3(3), article no. 30 (2007)
6. Iwama, K., Miyazaki, S., Yanagisawa, H.: A 25/17-approximation algorithm for the stable marriage problem with one-sided ties. In: de Berg, M., Meyer, U. (eds.) ESA 2010. LNCS, vol. 6347, pp. 135–146. Springer, Heidelberg (2010)
7. Király, Z.: Better and simpler approximation algorithms for the stable marriage problem. Algorithmica (2009), doi: 10.1007/s00453-009-9371-7
8. Manlove, D.F., Irving, R.W., Iwama, K., Miyazaki, S., Morita, Y.: Hard variants of stable marriage. Theoretical Computer Science 276(1-2), 261–279 (2002)
9. Manlove, D.F., O'Malley, G.: Student-project allocation with preferences over projects. Journal of Discrete Algorithms 6, 553–560 (2008)

Hardness Results and an Exact Exponential Algorithm for the Spanning Tree Congestion Problem

Yoshio Okamoto[1,*], Yota Otachi[2], Ryuhei Uehara[3], and Takeaki Uno[4]

[1] Center for Graduate Education Initiative,
JAIST, Asahidai 1–1, Nomi, Ishikawa 923–1292, Japan
okamotoy@jaist.ac.jp
[2] Graduate School of Information Sciences, Tohoku University, Sendai 980–8579, Japan
JSPS Research Fellow
otachi@dais.is.tohoku.ac.jp
[3] School of Information Science, JAIST, Asahidai 1–1, Nomi, Ishikawa 923–1292, Japan
uehara@jaist.ac.jp
[4] National Institute of Informatics, 2–1–2 Hitotsubashi, Chiyoda-ku, Tokyo, 101–8430, Japan
uno@nii.ac.jp

Abstract. Spanning tree congestion is a relatively new graph parameter, which has been studied intensively. This paper studies the complexity of the problem to determine the spanning tree congestion for non-sparse graph classes, while it was investigated for some sparse graph classes before. We prove that the problem is NP-hard even for chain graphs and split graphs. To cope with the hardness of the problem, we present a fast (exponential-time) exact algorithm that runs in $O^*(2^n)$ time, where n denotes the number of vertices. Additionally, we provide a constant-factor approximation algorithm for cographs, and a linear-time algorithm for chordal cographs.

1 Introduction

Spanning tree congestion is a graph parameter defined by Ostrovskii [18] in 2004. Simonson [21] also studied the same parameter under a different name as a variant of cutwidth. After Ostrovskii [18], several graph-theoretic results have been presented [2, 6, 12–17, 19], and very recently the complexity of the problem for determining the parameter has been studied [3, 20]. The parameter is defined as follows. Let G be a connected graph and T be a spanning tree of G. The *detour* for an edge $\{u, v\} \in E(G)$ is a unique u–v path in T. We define the *congestion* of $e \in E(T)$, denoted by $cng_{G,T}(e)$, as the number of edges in G whose detours contain e. The *congestion of G in T*, denoted by $cng_G(T)$, is the maximum congestion over all edges in T. The *spanning tree congestion* of G, denoted by $stc(G)$, is the minimum congestion over all spanning trees of G. We denote by STC the problem of determining whether a given graph has spanning tree congestion at most given k. If k is fixed, then we denote the problem by k-STC.

---

* Supported by Grant-in-Aid for Scientific Research from Ministry of Education, Science and Culture, Japan, and Japan Society for the Promotion of Science.

M. Ogihara and J. Tarui (Eds.): TAMC 2011, LNCS 6648, pp. 452–462, 2011.

Bodlaender, Fomin, Golovach, Otachi, and van Leeuwen [3, 20] studied the complexity of STC and k-STC. They showed that k-STC is linear-time solvable for apex-minor-free graphs and bounded-degree graphs, while k-STC is NP-complete even for K_6-minor-free graphs with only one vertex of unbounded degree if $k \geq 8$. They also showed that STC is NP-complete for planar graphs. Bodlaender, Kozawa, Matsushima, and Otachi [2] showed that the spanning tree congestion can be determined in linear time for outerplanar graphs. Although several complexity results are known as mentioned above, they are restricted to sparse graphs. The complexity for non-sparse graphs such as chordal graphs and chordal bipartite graphs were unknown.

In this paper, we show that STC is NP-complete for these important non-sparse graph classes. More precisely, we show that STC is NP-complete even for chain graphs and split graphs. It is known that every chain graph is chordal bipartite, and every split graph is chordal. The hardness for chain graphs is quite unexpected, since there is no other natural graph parameter that is known to be NP-hard for chain graphs, to the best of our knowledge. The hardness for chain graphs also implies the hardness for graphs of clique-width at most three. To cope with the hardness of the problem, we present a fast exponential-time exact algorithm. Our algorithm runs in $O^*(2^n)$ time, while a naive algorithm that examines all spanning trees runs in $O^*(2^m)$ or $O^*(n^n)$ time, where n and m denote the number of vertices and the number of edges. Note that $O^*(f(n)) = O(f(n) \cdot \text{poly}(n))$. The idea, which allows us to achieve this running time, is to enumerate all possible combinations of cuts instead of all spanning trees. Using this idea, we can design a dynamic-programming-based algorithm that runs in $O^*(3^n)$ time. Then, by carefully applying the fast subset convolution method developed by Björklund, Husfeldt, Kaski, and Koivisto [1], we finally get the running time $O^*(2^n)$. We also study the problem on cographs. It is known that cographs are precisely the graphs of clique-width at most two. For some cographs such as complete graphs and complete p-partite graphs, the closed formulas for the spanning tree congestion are known [12, 14, 18]. Although the complexity of STC for cographs remains unsettled, we provide a constant-factor approximation algorithm for them. Furthermore, we present a linear-time algorithm for chordal cographs.

Due to space limitation all proofs are omitted.

2 Preliminaries

Graphs in this paper are finite, simple, and connected, if not explicitly stated otherwise. We deal with edge-weighted graphs in Subsections 2.2 and 3.1. Our exponential-time exact algorithm runs in $O^*(2^n)$ time for edge-weighted graphs, too.

2.1 Graphs

Let G be a connected graph. For $S \subseteq V(G)$, we denote by $G[S]$ the subgraph induced by S. For an edge $e \in E(G)$, we denote by $G - e$ the graph obtained from G by the deletion of e. Similarly, for a vertex $v \in V(G)$, we denote by $G - v$ the graph obtained from G by the deletion of v and its incident edges. By $N_G(v)$, we denote the (*open*) *neighborhood* of v in G; that is, $N_G(v)$ is the set of vertices adjacent to v in G. For $S \subseteq V(G)$, we denote $\bigcup_{v \in S} N_G(v)$ by $N_G(S)$. We define the *degree* of v in G as $deg_G(v) = |N_G(v)|$. If $deg_G(v) = |V(G)| - 1$, then v is a *universal vertex* of G.

Let G and H be graphs. We say that G and H are *isomorphic*, and denote it by $G \simeq H$, if there is a bijection $f: V(G) \to V(H)$ such that $\{u, v\} \in E(G)$ if and only if $\{f(u), f(v)\} \in E(H)$. Now assume $V(G) \cap V(H) = \emptyset$. Then the *disjoint union* of G and H, denoted by $G \cup H$, is the graph with the vertex set $V(G) \cup V(H)$ and the edge set $E(G) \cup E(H)$. The *join* of G and H, denoted by $G \oplus H$, is the graph with the vertex set $V(G) \cup V(H)$ and the edge set $E(G) \cup E(H) \cup \{\{u, v\} \mid u \in V(G), v \in V(H)\}$.

For $A, B \subseteq V(G)$, we define $E_G(A, B) = \{\{u, v\} \in E(G) \mid u \in A, v \in B\}$. For $S \subseteq V(G)$, we define the *boundary edges* of S, denoted by $\theta_G(S)$, as $\theta_G(S) = E_G(S, V(G) \setminus S)$. Note that $\theta_G(\emptyset) = \theta_G(V(G)) = \emptyset$. The congestion $cng_{G,T}(e)$ of an edge $e \in E(T)$ satisfies $cng_{G,T}(e) = |\theta_G(A_e)|$, where A_e is the vertex set of one of the two components of $T - e$. For an edge e in a tree T, we say that e *separates* A and B if $A \subseteq A_e$ and $B \subseteq B_e$, where A_e and B_e are the vertex sets of the two components of $T - e$. Clearly, if T is a spanning tree of G and $e \in E(T)$ separates A and B, then $cng_{G,T}(e) \geq |E(A, B)|$. If e separates A and B, we also say that e *divides* $A \cup B$ into A and B.

Let T be a tree rooted at $r \in V(T)$. Then we denote by $prt_T(v)$ the parent of $v \in V(T)$ in T. The parent of the root r is not defined. We denote by $Chd_T(v)$ the children of $v \in V(T)$ in T. Clearly, $N_T(v) = \{prt_T(v)\} \cup Chd_T(v)$ for every non-root vertex v.

2.2   Spanning Tree Congestion of Weighted Graphs

A graph G may be associated with an edge-weight function $wei: E(G) \to \mathbb{Z}^+$. If a graph has such a function, then we call it an *edge-weighted graph* or just a *weighted graph*. Note that unweighted graphs can be considered as weighted graphs by setting $wei(e) = 1$ for each edge e. For an edge-weighted graph G and $F \subseteq E(G)$, we define $wei(F) = \sum_{f \in F} wei(f)$ for $F \subseteq E(G)$. We extend the notion of spanning tree congestion to edge-weighted graphs by defining the congestion of an edge e as the sum of the weights of edges whose detours pass through the edge e. If $e \in E(T)$ separates vertex sets A and B, then $cng_{G,T}(e) \geq wei(E(A, B))$.

For a weighted graph G, we define the *weighted degree* of v in G as $wdeg_G(v) = wei(\theta_G(\{v\}))$. It is not difficult to see that the following fact holds.

Proposition 2.1. *Let G be a weighted graph, and let $S \subseteq V(G)$. Then*

$$wei(\theta_G(S)) = \sum_{v \in S} wdeg_G(v) - 2wei(E(G[S])).$$

It is known that STC for weighted graphs is equivalent to STC for unweighted graphs in the following sense.

Lemma 2.2 ([3, 20]). *Let G be a weighted graph and let $e \in E(G)$. Let G' be the graph obtained from G by removing the edge e and adding $wei(e)$ internally disjoint paths of arbitrary lengths between the ends of e, where each edge in the added paths is of unit weight. Then, $stc(G) = stc(G')$.*

2.3   Graph Classes

A graph is *chordal* if it has no induced cycle of length greater than three. A graph G is a *split graph* if its vertex set $V(G)$ can be partitioned into two sets C and I so that C is a clique of G and I is an independent set of G. Clearly, every split graph is a chordal graph (see [10]). A *cograph* (or *complement-reducible graph*) is a graph that can be constructed recursively by the following rules:

1. K_1 is a cograph;
2. if G and H are cographs, then so is $G \cup H$;
3. if G and H are cographs, then so is $G \oplus H$.

Note that if G is a connected cograph with at least two vertices, then G can be expressed as $G_1 \oplus G_2$ for some nonempty cographs G_1 and G_2. A cograph is a *chordal cograph* if it is also a chordal graph. Chordal cographs are also known as *trivially perfect graphs* [4, 10] and *quasi-threshold graphs* [22]. It is known that in the construction of a chordal cograph by the above rules, we can assume one of two operands of $\oplus$ is K_1 [22].

Analogous to chordal graphs, *chordal bipartite graphs* are defined as the bipartite graphs without induced cycle of length greater than four. A bipartite graph $G = (X, Y; E)$ is a *chain graph* if there is an ordering $<$ on X such that $u < v$ implies $N_G(u) \subseteq N_G(v)$. It is known that every chain graph is $2K_2$-free [23], and thus chordal bipartite. It is also known that every chain graph has clique-width at most three [5].

Clique-width is a graph parameter which generalizes treewidth in some sense. Many hard problems can be solved efficiently for graphs of bounded clique-width. For the definition and further information of clique-width, see a recent survey by Hliněný, Oum, Seese, and Gottlob [11].

3   Hardness for Split Graphs and Chain Graphs

This section presents our hardness results for split graphs and chain graphs. Namely, we prove the following theorems.

Theorem 3.1. *STC is NP-complete for split graphs.*

Theorem 3.2. *STC is NP-complete for chain graphs.*

Since every chain graph has clique-width at most three, we have the following corollary.

Corollary 3.3. *STC is NP-complete for graphs of clique-width at most three.*

The weighted edge argument [3, 20] allows us to present a simple proof for split graphs. However, we are unable to present a simple proof based on the weighted edge argument for chain graphs. This is because, in the process of modifying a weighted graph to an unweighted graph, we may introduce many independent edges (see Lemma 2.2). Although we need somewhat involved arguments for chain graphs, the proofs are based on essentially the same idea.

Clearly, STC is in NP. The proofs of NP-hardness are done by reducing the following well-known NP-complete problem to STC for both graph classes.

Problem: 3-Partition [9, SP15]

Instance: A multi-set $A = \{a_1, a_2, \ldots, a_{3m}\}$ of $3m$ positive integers and a bound $B \in \mathbb{Z}^+$ such that $\sum_{a_i \in A} a_i = mB$, $a_1 \le a_2 \le \cdots \le a_{3m}$, and $B/4 < a_i < B/2$ for each $a_i \in A$.

Question: Can A be partitioned into m disjoint sets $A_1, A_2, \ldots, A_m$ such that, for $1 \le i \le m$, $\sum_{a \in A_i} a = B$? (Thus each A_i must contain exactly three elements from A.)

It is known that 3-Partition is NP-complete in the strong sense [9]. Thus we assume $a_{3m} \le \text{poly}(m)$, where $\text{poly}(m)$ is some polynomial on m. By scaling each $a \in A$, we can also assume that $a_1 \ge 3m + 2$, $m \ge 3$, $B \ge 8$, and $B/4 + 1 \le a_i \le B/2 - 1$.

3.1   Hardness for Split Graphs

In this subsection, we prove that STC is NP-hard for split graphs. We first show that STC is NP-hard for edge-weighted split graphs with weighted edges only in the maximum clique, by reducing an instance A of 3-Partition to an edge-weighted split graph G_A such that A is a yes instance if and only if $stc(G_A) \le k$ for some k. We then show that G_A can be modified to an unweighted split graph G'_A in polynomial time so that $stc(G_A) = stc(G'_A)$. This proves Theorem 3.1.

Let A be an instance of 3-Partition. We now construct G_A from A in polynomial time. Let $I = \{u_i \mid 1 \le i \le 3m\}$ and $C = \{x\} \cup V \cup W$, where $V = \{v_i \mid 1 \le i \le m\}$ and $W = \{w_i \mid m + 1 \le i \le a_{3m}\}$. The graph G_A has vertex set $I \cup C$. The sets I and C are independent set and a clique of G_A, respectively. Each $u_i \in I$ is adjacent to all vertices in V and vertices $w_1, w_2, \ldots, w_{a_i}$. More formally, $E(G_A)$ is defined as follows:

$$E(G_A) = \{\{c, c'\} \mid c, c' \in C\} \cup \{\{u, v\} \mid u \in I, v \in V\} \cup \{\{u_i, w_j\} \mid u_i \in I, m + 1 \le j \le a_i\}.$$

Recall that $a_i > m$ for any $i \ge 1$. The degrees of vertices in G_A can be determined as follows: $deg_{G_A}(u_i) = a_i$, $deg_{G_A}(v_i) = |C| + |I| - 1$, and $deg_{G_A}(w_i) = |C| + |\{j \mid a_j \ge i\}| - 1$. Some edges of G_A have heavy weights. Let $k = 2B + 2|C| + 2|I| - 15$. Then

$$wei(e) = \begin{cases} \alpha := (k + 1)/2 & \text{if } e = \{x, v_i\}, \\ \beta_i := k - deg_{G_A}(w_i) + 1 & \text{if } e = \{x, w_i\}, \\ 1 & \text{otherwise.} \end{cases}$$

Clearly, G_A is a split graph with weighted edges only in the clique C. The weighted degrees of vertices in G_A is as follows: $wdeg_{G_A}(u_i) = a_i$, $wdeg_{G_A}(v_i) = \alpha + |C| + |I| - 2 = k - B + 6$, and $wdeg_{G_A}(w_i) = k$.

Lemma 3.4. Let $k = 2B + 2|C| + 2|I| - 15$. Then A is a yes instance if and only if $stc(G_A) \le k$.

Proof (Sketch). ($\implies$) Let $A_1, \ldots, A_m$ be a partition of A such that $\sum_{a \in A_i} a = B$ for $1 \le i \le m$. Let U_i denote the set $\{u_j \mid a_j \in A_i\}$. The desired spanning tree T of G_A can be obtained from the partition $A_1, \ldots, A_m$ as follows: $E(T) = \{\{x, c\} \mid c \in C \setminus \{x\}\} \cup \bigcup_{1 \le i \le m} \{\{v_i, u_j\} \mid u_j \in U_i\}$. We can show $cng_{G_A}(T) \le k$. ($\impliedby$) Omitted. $\square$

Now we prove the NP-hardness of STC for unweighted split graphs. To this end, we first reduce an instance A of 3-Partition to a weighted split graph G_A as stated above. Recall that all weighted edges of G_A are in $G_A[C]$. We need the following lemma.

Lemma 3.5. *Let G be an edge-weighted split graph with a partition (C, I) of $V(G)$, where C and I are a clique and an independent set of G, respectively. If the weighted edges are only in $G[C]$ and the maximum edge weight is w_{max}, then an edge-unweighted split graph G' satisfying $stc(G) = stc(G')$ can be obtained from G in $O(w_{max} \cdot |E(G)|)$ time.*

Observe that the maximum edge-weight in G_A is bounded by a polynomial function on B and m. Thus the above lemma implies that from an instance A of 3-PARTITION, we can construct in polynomial time an unweighted split graph G'_A and $k \in \mathbb{Z}^+$ such that A is a yes instance if and only if $stc(G'_A) \leq k$. This proves Theorem 3.1.

3.2 Hardness for Chain Graphs

Next we prove the NP-hardness for chain graphs. Given an instance A of 3-PARTITION, we construct the graph $G_A = (P, Q; E)$. For convenience, let $M = B + 3m - 4$ and $\gamma_i = |\{a_j \in A \mid a_j \geq i\}|$. Note that $0 < \gamma_i \leq 3m$ for $m + 1 \leq i \leq a_{3m}$. In particular, $\gamma_{m+1} = 3m$ and $\gamma_{a_{3m}} > 0$. First we define the vertex sets $P = U \cup V \cup W$ and $Q = X \cup Y \cup Z$ as follows:

$$U = \{u_i \mid 1 \leq i \leq m\}, \quad V = \{v_i \mid m + 1 \leq i \leq a_{3m}\}, \quad W = \{w_i \mid 1 \leq i \leq M - a_{3m}\},$$
$$X = \{x_i \mid 1 \leq i \leq 3m\}, \quad Y = \{y_i \mid m + 1 \leq i \leq a_{3m}\}, \quad Z = \{z_i \mid 1 \leq i \leq M - a_{3m}\}.$$

Next we define the edge set as follows:[1]

$$E = (X \times U) \cup (Y \times (U \cup V)) \cup (Z \times (U \cup V \cup W))$$
$$\cup \{\{x_i, v_j\} \mid x_i \in X, m + 1 \leq j \leq a_i\}$$
$$\cup \{\{y_i, w_j\} \mid y_i \in Y, 1 \leq j \leq M - a_{3m} - \gamma_i\}.$$

See Fig. 1 for a simplified illustration of G_A.

Let G_0 and G_1 be two disjoint copies of G_A. That is, $G_A \simeq G_0 \simeq G_1$ and $V(G_0) \cap V(G_1) = \emptyset$. By P_i, Q_i, U_i, V_i, W_i, X_i, Y_i, and Z_i, we denote the vertex sets of G_i, $i \in \{0, 1\}$, that correspond to the vertex sets P, Q, U, V, W, X, Y, and Z of G_A, respectively. Similarly, we denote the vertices of G_i, $i \in \{0, 1\}$, that correspond to vertices u_j, v_j, w_j, x_j, y_j, and z_j of G_A by u^i_j, v^i_j, w^i_j, x^i_j, y^i_j, and z^i_j, respectively. We define the graph H_A as follows (see Fig. 1): $V(H_A) = V(G_0) \cup V(G_1)$ and $E(H_A) = E(G_0) \cup E(G_1) \cup (P_0 \times P_1)$.

Lemma 3.6. *The graph H_A is a chain graph.*

Lemma 3.7. *The degrees of vertices in H_A satisfy the following relations: $deg_{H_A}(u^i_j) = 2M + 2m$, $deg_{H_A}(v^i_j) = 2M - m + \gamma_j > 2M - m$, $2M - a_{3m} \leq deg_{H_A}(w^i_j) \leq 2M - m$, $deg_{H_A}(x^i_j) = a_i$, $deg_{H_A}(y^i_j) = M - \gamma_j < M$, and $deg_{H_A}(z^i_j) = M$. Moreover, $\Delta(H_A) = 2M + 2m$ and $\delta(H_A) = a_1$.*

Now we prove that A is a yes instance of 3-PARTITION if and only if $stc(H_A) \leq k$. We divide the proof into two only-if part (Lemma 3.8) and if part (Lemma 3.9).

---

[1] For simplicity, we denote by $S \times T$ the set of unordered pairs $\{\{s, t\} \mid s \in S, t \in T\}$.

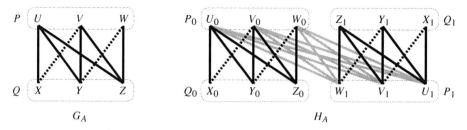

Fig. 1. Graphs G_A and H_A. A solid line between two sets implies that the two sets induce a complete bipartite graph, and a dotted line between two sets implies that there are some (but not all) edges between the two sets. Two color classes of H_A are $P_0 \cup Q_1$ and $Q_0 \cup P_1$.

Lemma 3.8. *Let $k = 3M - m - 2$. If A is a yes instance, then $stc(H_A) \le k$.*

Proof (Sketch). Let T be a spanning tree of H_A with the edge set as follows:

$$E(T) = \{\{u_1^0, v\} \mid v \in Q_0 \cup P_1\} \cup \{\{u_j^1, x_h^1\} \mid a_h \in A_j, \ 1 \le j \le m\} \cup \{\{u_j^0, x_j^0\} \mid 2 \le j \le m\}$$
$$\cup \{\{v_j^i, y_j^i\} \mid i \in \{0, 1\}, \ m + 1 \le j \le a_{3m}\} \cup \{\{w_j^i, z_j^i\} \mid i \in \{0, 1\}, \ 1 \le j \le M - a_{3m}\}.$$

Then, we can prove that $cng_{H_A}(T) \le k$. □

Lemma 3.9. *Let $k = 3M - m - 2$. If $stc(H_A) \le k$, then A is a yes instance.*

4   Exponential-Time Exact Algorithm

We have shown that STC is NP-complete even for very simple graphs. It is widely believed that NP-hard problems cannot be solved in polynomial time. Thus we need *fast* exponential-time (or sub-exponential-time) algorithms for these problems. Nowadays, designing fast exponential-time exact algorithms becomes an important topic in theoretical computer science. See a recent textbook of exponential-time exact algorithms by Fomin and Kratsch [8]. For STC, we can easily design an $O^*(2^m)$- or $O^*(n^n)$-time algorithm that examine all spanning trees of input graphs, where n and m denote the number of vertices and the number of edges, respectively. In this section, we describe an algorithm for STC that runs in $O^*(2^n)$ time. Although it is still an exponential-time algorithm, it is significantly faster than a naive algorithm.

Let $G = (V, E)$ be a given undirected graph. For convenience, we denote $|\theta_G(X)|$ by $c(X)$. Note that $c(\emptyset) = c(V) = 0$. Consider a spanning tree T with congestion at most k. We regard T as a rooted tree with root $r \in V$. We denote this rooted tree by (T, r). Let $e = \{u, v\} \in E(T)$ be an edge of T, and without loss of generality, let u be the parent of v. Then, the congestion of e in T is equal to $c(D_{T,r}(v))$, where $D_{T,r}(v)$ denotes the set of descendants of v in (T, r). Since the congestion of T is at most k, we see that $c(D_{T,r}(v)) \le k$. See Fig. 2. Conversely, if $c(D_{T,r}(v)) \le k$ for all $v \in V \setminus \{r\}$, then the congestion of T is at most k. This is because there exists a one-to-one correspondence between the edges e of T and the vertices v in $V \setminus \{r\}$ so that v is a deeper endpoint of e. We summarize this observation in the following lemma.

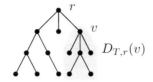

Fig. 2. The definition of $D_{T,r}(v)$

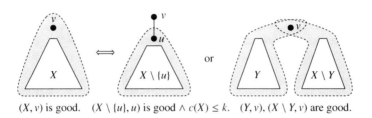

(X, v) is good. $(X \setminus \{u\}, u)$ is good $\wedge c(X) \le k$. $(Y, v), (X \setminus Y, v)$ are good.

Fig. 3. An illustration of Lemma 4.2

Lemma 4.1. *The congestion of a rooted tree* (T, r) *is at most* k *if and only if* $c(D_{T,r}(v)) \le k$ *for every vertex* $v \in V \setminus \{r\}$.

The lemma above suggests the following dynamic-programming approach. We call a pair (X, v) of a subset $X \subseteq V$ and a vertex $v \notin X$ a *rooted subset* of V. By definition, $X \ne V$ for a rooted subset (X, v) of V. A rooted subset (X, v) of V is *good* if there exists a rooted spanning tree (T, v) of $G[X \cup \{v\}]$ such that $c(D_{T,v}(u)) \le k$ for all $u \in X$. Here, c is a cut function of G, not of $G[X \cup \{v\}]$. By definition (X, v) is good when $X = \emptyset$. Note that there exists a rooted spanning tree (T, r) of G with congestion at most k if and only if the rooted set $(V \setminus \{r\}, r)$ is good.

The following lemma provides a recursive formula that forms a basis of our algorithm (see Fig. 3).

Lemma 4.2. *Let* (X, v) *be a rooted subset of* V *with* $|X| \ge 1$. *Then*, (X, v) *is good if and only if at least one of the following holds.*

1. *There exists a vertex* $u \in X \cap N_G(v)$ *such that* $c(X) \le k$ *and* $(X \setminus \{u\}, u)$ *is good.*
2. *There exists a non-empty proper subset* $Y \subseteq X$ *such that both of* $(Y, v), (X \setminus Y, v)$ *are good.*

Lemmas 4.1 and 4.2 above readily give an $O^*(3^n)$-time dynamic programming algorithm. However, the fast subset convolution method enables us to solve the problem in $O^*(2^n)$ time. We give a more detail below.

Let S be a finite set. For two functions $f, g: 2^S \to \mathbb{R}$, their *subset convolution* is a function $f * g: 2^S \to \mathbb{R}$ defined as

$$(f * g)(X) = \sum_{Y \subseteq X} f(Y)g(X \setminus Y)$$

for every $X \subseteq S$. Given $f(X), g(X)$ for all $X \subseteq S$, we can compute $(f * g)(X)$ for all $X \subseteq S$ in $O^*(2^n)$ total time, where $n = |S|$ [1].

Back to the spanning tree congestion problem, let $v \in V$ be an arbitrary vertex. We define the function $f_v \colon 2^{V \setminus \{v\}} \to \mathbb{R}$ by the following recursion: $f_v(X) = 1$ if $X = \emptyset$; otherwise,

$$f_v(X) = \sum_{u \in X \cap N_G(v)} f_u(X \setminus \{u\}) \max\{0, k - c(X) + 1\} + \sum_{\emptyset \neq Y \subsetneq X} f_v(Y) f_v(X \setminus Y),$$

where the empty sum is defined to be 0. It is easy to verify that $f_v(X)$ is non-negative for every $v \in V$ and every $X \subseteq V \setminus \{v\}$.

The following lemma connects the functions f_v, $v \in V$ and good rooted sets.

Lemma 4.3. *Let (X, v) be a pair of a subset $X \subseteq V \setminus \{v\}$ and a vertex $v \in V$. Then, $f_v(X) > 0$ if and only if (X, v) is a good rooted subset of V.*

To apply the subset convolution method, we use the following functions. For each $i \in \{0, 1, \dots, n-1\}$, where $n = |V|$, and $v \in V$, let $f_v^i \colon 2^{V \setminus \{v\}} \to \mathbb{R}$ be defined by

$$f_v^i(X) = \begin{cases} f_v(X) & \text{if } |X| \leq i, \\ 0 & \text{if } |X| > i, \end{cases}$$

for all $X \subseteq V \setminus \{v\}$. Then, it is not difficult to see the following.

1. For all $v \in X$ and $X \subseteq V \setminus \{v\}$, $f_v^{n-1}(X) = f_v(X)$.

$$f_v^{n-1}(X) = f_v(X).$$

2. For all $v \in V$ and $X \subseteq V \setminus \{v\}$,

$$f_v^0(X) = \begin{cases} 1 & \text{if } X = \emptyset, \\ 0 & \text{otherwise.} \end{cases}$$

3. For all $i \in \{1, \dots, n-1\}$, $v \in V$, and $X \subseteq V \setminus \{v\}$

$$f_v^i(X) = \sum_{u \in X \cap N_G(v)} f_u^{i-1}(X \setminus \{u\}) \max\{0, k - c(X) + 1\}$$

$$+ \sum_{\emptyset \neq Y \subsetneq X} f_v^{i-1}(Y) f_v^{i-1}(X \setminus Y)$$

$$= \sum_{u \in X \cap N_G(v)} f_u^{i-1}(X \setminus \{u\}) \max\{0, k - c(X) + 1\}$$

$$+ \sum_{Y \subseteq X} f_v^{i-1}(Y) f_v^{i-1}(X \setminus Y) - 2 f_v^{i-1}(\emptyset) f_v^{i-1}(X)$$

$$= \sum_{u \in X \cap N_G(v)} f_u^{i-1}(X \setminus \{u\}) \max\{0, k - c(X) + 1\}$$

$$+ (f_v^{i-1} * f_v^{i-1})(X) - 2 f_v^{i-1}(\emptyset) f_v^{i-1}(X).$$

Our algorithm is based on these formulas.

Step 1. For all $v \in V$ and $X \subseteq V \setminus \{v\}$, compute $f_v^0(X)$ based on the formulas above.

Step 2. For each $i = 1, \ldots, n - 1$ in the ascending order, do the following.

 Step 2-1. For all $v \in V$, compute the subset convolution $f_v^{i-1} * f_v^{i-1}$.

 Step 2-2. For all $v \in V$ and all $X \subseteq V \setminus \{v\}$, compute $f_v^i(X)$ based on the formula above.

Step 3. If $f_v^{n-1}(V) > 0$, then output Yes. Otherwise, output No.

The correctness is immediate from the discussion so far. The running time is $O^*(2^n)$ since the running time of each step is bounded by $O^*(2^n)$. This is an algorithm for solving the decision problem, but a simple binary search on $k \in \{1, \ldots, |E|\}$ can provide the spanning tree congestion. Thus, we obtain the following theorem.

Theorem 4.4. *The spanning tree congestion of a given undirected graph can be computed in $O^*(2^n)$ time.*

Note that the algorithm also works for the weighted case with the $O(n)$-factor increase of the running time, since the number of distinct cut values $c(X)$ is bounded by 2^n and so the binary search over the all possible values of $c(X)$ takes at most $O(\log(2^n)) = O(n)$ iterations. This is possible if we compute $c(X)$ for all $X \subseteq V$ beforehand, which only takes $O^*(2^n)$ time.

5 Remarks on Cographs

We showed NP-completeness of STC for graphs of clique-width at most three. Therefore, it is quite natural to ask whether or not STC is NP-complete for graphs of clique-width at most two; that is, for cographs [7]. Although the complexity of STC for cographs remains unsettled, we have the following results.

Theorem 5.1. *The spanning tree congestion of cographs can be approximated within a factor four in polynomial time. Furthermore, the spanning tree congestion of chordal cographs can be determined in linear time.*

References

1. Björklund, A., Husfeldt, T., Kaski, P., Koivisto, M.: Fourier meets Möbius: Fast subset convolution. In: Proceedings of the 39th Annual ACM Symposium on Theory of Computing (STOC 2007), pp. 67–74 (2007)
2. Bodlaender, H.L., Kozawa, K., Matsushima, T., Otachi, Y.: Spanning tree congestion of k-outerplanar graphs. In: WAAC 2010, pp. 34–39 (2010)
3. Bodlaender, H.L., Fomin, F.V., Golovach, P.A., Otachi, Y., van Leeuwen, E.J.: Parameterized complexity of the spanning tree congestion problem (submitted)
4. Brandstädt, A., Le, V.B., Spinrad, J.P.: Graph Classes: A Survey. SIAM, Philadelphia (1999)
5. Brandstädt, A., Lozin, V.V.: On the linear structure and clique-width of bipartite permutation graphs. Ars Combin. 67, 273–281 (2003)
6. Castejón, A., Ostrovskii, M.I.: Minimum congestion spanning trees of grids and discrete toruses. Discuss. Math. Graph Theory 29, 511–519 (2009)
7. Courcelle, B., Olariu, S.: Upper bounds to the clique width of graphs. Discrete Appl. Math. 101, 77–114 (2000)

8. Fomin, F.V., Kratsch, D.: Exact Exponential Algorithms. Springer, Heidelberg (2010)
9. Garey, M.R., Johnson, D.S.: Computers and Intractability: A Guide to the Theory of NP-Completeness. Freeman, New York (1979)
10. Golumbic, M.C.: Algorithmic Graph Theory and Perfect Graphs, 2nd edn. Annals of Discrete Mathematics, vol. 57. North Holland, Amsterdam (2004)
11. Hliněný, P., Oum, S., Seese, D., Gottlob, G.: Width parameters beyond tree-width and their applications. Comput. J. 51, 326–362 (2008)
12. Hruska, S.W.: On tree congestion of graphs. Discrete Math. 308, 1801–1809 (2008)
13. Kozawa, K., Otachi, Y.: Spanning tree congestion of rook's graphs. Discuss. Math. Graph Theory (to appear)
14. Kozawa, K., Otachi, Y., Yamazaki, K.: On spanning tree congestion of graphs. Discrete Math. 309, 4215–4224 (2009)
15. Law, H.F.: Spanning tree congestion of the hypercube. Discrete Math. 309, 6644–6648 (2009)
16. Law, H.F., Ostrovskii, M.I.: Spanning tree congestion of some product graphs. Indian J. Math. (to appear)
17. Löwenstein, C., Rautenbach, D., Regen, F.: On spanning tree congestion. Discrete Math. 309, 4653–4655 (2009)
18. Ostrovskii, M.I.: Minimal congestion trees. Discrete Math. 285, 219–226 (2004)
19. Ostrovskii, M.I.: Minimum congestion spanning trees in planar graphs. Discrete Math. 310, 1204–1209 (2010)
20. Otachi, Y., Bodlaender, H.L., van Leeuwen, E.J.: Complexity results for the spanning tree congestion problem. In: Thilikos, D.M. (ed.) WG 2010. LNCS, vol. 6410, pp. 3–14. Springer, Heidelberg (2010)
21. Simonson, S.: A variation on the min cut linear arrangement problem. Math. Syst. Theory 20, 235–252 (1987)
22. Yan, J.H., Chen, J.J., Chang, G.J.: Quasi-threshold graphs. Discrete Appl. Math. 69, 247–255 (1996)
23. Yannakakis, M.: Computing the minimum fill-in is NP-complete. SIAM J. Alg. Disc. Mech. 2, 77–79 (1981)

Switching to Hedgehog-Free Graphs Is NP-Complete

Eva Jelínková

Department of Applied Mathematics*
Charles University
Malostranskénám. 25, 118 00 Praha, Czech Republic
eva@kam.mff.cuni.cz

Abstract. We study the problem of deciding if, for a fixed graph H, a given graph is switching-equivalent to an H-free graph. In all cases of H that have been solved so far, the problem is decidable in polynomial time. We give infinitely many graphs H such that the problem is NP-complete, thus solving an open problem [Kratochvíl, Nešetřil and Zýka, Ann. Discrete Math. **51** (1992)].

Keywords: Seidel's switching, H-free graph, NP-completeness.

1 Introduction

Seidel's switching is a graph operation which makes a given vertex adjacent to precisely those vertices to which it was non-adjacent before, while keeping the rest of the graph unchanged. Two graphs are called switching-equivalent if one can be made isomorphic to the other one by a sequence of switches.

The concept of Seidel's switching was introduced by the Dutch mathematician J. J. Seidel in connection with algebraic structures, such as systems of equiangular lines, strongly regular graphs, or the so-called two-graphs, see [12,13,14].

Applications of Seidel's switching can be found in various algorithms; for example, it plays an important role in Hayward's polynomial-time algorithm for solving the P_3-structure recognition problem [5]. The algorithm of Rotem and Urrutia [11] to recognize circular permutation graphs is based on switching, as well as its linear-time variant found later by Sritharan [15]. Switching also plays part in the algorithm for bi-join decomposition of graphs by de Montgolfier and Rao [10].

Several authors have addressed the problem of deciding if a given graph is switching-equivalent to a graph having a certain desired property. As observed by Kratochvíl et al. [8] and also by Ehrenfeucht at al. [2], there is no obvious correlation between the computational complexity of this problem and the computational complexity of deciding if a graph itself has the property.

For example, the problems of deciding if a graph contains a Hamiltonian path or cycle are well known to be NP-complete [3]. However, Kratochvíl et al. [8]

---

* Supported by project 1M0021620838 of the Czech Ministry of Education.

M. Ogihara and J. Tarui (Eds.): TAMC 2011, LNCS 6648, pp. 463–470, 2011.
© Springer-Verlag Berlin Heidelberg 2011

proved that any graph is switching-equivalent to a graph containing a Hamiltonian path, and it is polynomial to decide if a graph is switching-equivalent to a graph containing a Hamiltonian cycle. These results have been extended to graph pancyclicity by Ehrenfeucht et al. [1].

On the other hand, the problem of deciding switching-equivalence to a regular graph was proven NP-complete by Kratochvíl [9], and switching-equivalence to a k-regular graph for a fixed k is polynomial, while both the regularity and k-regularity of a graph can be tested in polynomial time. Three-colorability and switching-equivalence to a three-colorable graph are both NP-complete [2].

We focus on the problem of deciding switching equivalence to an H-free graph, i. e., to a graph which does not contain an induced copy of H, where H is a certain fixed graph. Polynomial-time algorithms for some graphs H have already been known.

Let K_n denote the complete graph on n vertices, let $K_{n,m}$ be the complete bipartite graph with partities of size n and m, and let P_n be the path with n vertices. The algorithm for K_2 is simple (see [4]), the one for $K_{1,2}$ is due to Kratochvíl et al. [8]. Hayward [5] and independently Hage at al. [4] found an algorithm for K_3; the result is a core of the polynomial-time algorithm for recognizing P_3-structures of graphs. The case of P_4 has been solved by Hertz [6] in connection to the so called perfect switching classes. The algorithm for $K_{1,3}$ is due to Jelínková and Kratochvíl [7].

It can be observed that an algorithm for H, when run on a complement of the input graph, gives an algorithm for the complement $\overline{H}$. (The switching-equivalence of H and H', however, does not yield any obvious relation of algorithms for H and H'.)

Thus, in the cases that have been solved so far, there are polynomial-time algorithms for all graphs H with at most three vertices, and for some graphs H with four vertices. No hardness result has been known.

In Section 3, we give infinitely many examples of graphs H for which the problem is NP-complete. Thus, we answer a question first mentioned by Kratochvíl et al. [8].

2   Preliminaries

All graphs considered are finite, undirected, and without loops or multiple edges. The set of vertices of a graph G is denoted by V_G, and the set of edges of G is denoted by E_G.

We say that the graph H is an *induced subgraph* of G, written $H \leq G$, if $V_H \subseteq V_G$ and $E_H = \binom{V_H}{2} \cap E_G$. For a set $A \subseteq V_G$ we call the graph $(A, \binom{A}{2} \cap E_G)$ the *subgraph of G induced by A*. If an isomorphic copy of H is an induced subgraph of G, we shall say that G *contains H as an induced subgraph* or just that G *contains H*. We say that a graph G is H-*free* if it does not contain H.

A graph $G = (V, \binom{V}{2})$ is called a *complete graph*; a complete graph with n vertices is denoted by K_n. A complete subgraph of a graph is also called a *clique*.

Definition 1. *Let G be a graph. Seidel's switch of a vertex $v \in V_G$ results in a graph called $S(G, v)$ whose vertex set is the same as of G and the edge set is the symmetric difference of E_G and the full star centered in v, i. e.,*

$$V_{S(G,v)} = V_G$$

$$E_{S(G,v)} = E_G \setminus \{xv : x \in V_G, \ xv \in E_G\}) \cup \{xv : x \in V_G, \ x \neq v, \ xv \notin E_G\}.$$

It is easy to observe that the result of a sequence of vertex switches in G depends only on the parity of the number of times each vertex is switched. This allows generalizing switching to vertex subsets of G.

Definition 2. *Let G be a graph. Then the Seidel's switch of a vertex subset $A \subseteq V_G$ is called $S(G, A)$ and*

$$S(G, A) = (V_G, E_G \bigtriangleup \{xy : x \in A, \ y \in V_G \setminus A\}),$$

where the symbol $\bigtriangleup$ denotes the symmetric difference of sets.

Definition 3. *We say that two graphs G and G' are switching equivalent (denoted by $G \sim G'$) if there is a set $A \subseteq V_G$ such that $S(G, A)$ is isomorphic to G'.*

We remark that in our case of switching to an H-free graph, the fact that the above definition deals with isomorphism is not crucial; if $S(G, A)$ is isomorphic to an H-free graph, then $S(G, A)$ is H-free, too. Hence, no isomorphism needs to be explicitly tested.

3 Switching to Hedgehog-Free Graphs

In this Section, we provide infinitely many graphs H such that switching to an H-free graph is NP-complete.

For each $k \geq 3$ we define a graph called *hedgehog* on $2k + 3$ vertices, denoted by H_k. It is composed of a clique on $2k$ vertices called the *body* of the hedgehog, and three other vertices called *pins*. Exactly k vertices of the body are adjacent to all pins, and the remaining k vertices of the body are non-adjacent to any pin (see Fig. 1).

Theorem 1. *For each $k \geq 3$ and each hedgehog H_k, it is NP-complete to decide if a given graph can be switched to an H_k-free graph.*

Proof. Consider a fixed $k \geq 3$. We proceed by reduction of the problem MONO-TONE NOT ALL EQUAL 3-SAT. An instance of MONOTONE NOT ALL EQUAL 3-SAT is a formula in CNF, in which every clause contains exactly three variables and there are no negations. The formula is satisfiable if and only if there exists a truth assignment so that in each clause, at least one variable is true and at least one is false.

Fig. 1. Hedgehogs H_3 and H_4

The problem is known to be NP-complete [3]. Clearly, we may assume that the input formulas contain at least eight distinct variables.

Let φ be an instance of MONOTONE NOT ALL EQUAL 3-SAT. We transform φ into a graph $G = G_\varphi$ as follows. For each variable v, there is a vertex x_v. A clause $c = (v_i \vee v_j \vee v_l)$ is represented by two disjoint cliques K_c and K_c', each having $4k - 2$ vertices. Exactly $2k - 1$ vertices of K_c and $2k - 1$ vertices of K_c' are adjacent to x_{v_i}, x_{v_j} and x_{v_l}, and are called *outer vertices*. Other vertices of K_c and K_c' are called *inner*. See Fig. 2 for illustration.

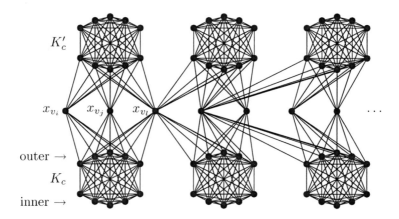

Fig. 2. An example of the graph G_φ (for $k = 3$). The cliques K_c and K_c' represent a clause $c = (v_i \vee v_j \vee v_l)$.

Suppose that there is a satisfying valuation μ of φ. Let A be the subset of vertices corresponding to true variables, i. e., $A = \{x_v : \mu(v) = \text{true}\}$. We prove that $S(G, A)$ is H_k-free.

We use the following terminology: vertices in A are called *switched*, other vertices of G are called *non-switched*. Vertices of G that correspond to a variable of φ are called *variable vertices*, others are called *non-variable vertices*.

Assume that there is a hedgehog H_k in $S(G, A)$.

As none of the non-variable vertices is switched, it is clear that they induce the same cliques in $S(G, A)$ as in G, and that vertices in distinct cliques are not adjacent. Hence, the hedgehog body contains vertices of at most one such clique. At the same time, it contains vertices of at least one such clique, because variable vertices form a complete bipartite graph in $S(G, A)$, so at most two variable vertices can be in the body. We denote this clique by K, the set of inner vertices of K by I, and the set of outer vertices of K by O.

Let c be the clause represented by K. We further denote the set of non-switched variable vertices representing the variables (literals) of c as L, the set of other non-switched variable vertices as N (non-literals). L_A and N_A denote the corresponding sets of switched variable vertices.

Since variable vertices induce a complete bipartite subgraph in $S(G, A)$, the hedgehog body contains at most two variable vertices, namely, one in N_A and one in L. We denote them by x_{N_A} and x_L, respectively. The remaining body vertices are elements of $O \cup I$.

Now consider pins of the hedgehog; they are adjacent to exactly k body vertices out of the total $2k$. The adjacency of vertices of $S(G, A)$ to the body is as follows.

- Vertices in L are adjacent to vertices in O and to x_{N_A},
- vertices in L_A are adjacent to vertices in I and to x_L,
- vertices in N are adjacent to x_{N_A},
- vertices in N_A are adjacent to vertices in O, I, and to x_L,
- vertices of other cliques are adjacent only to x_{N_A} and/or x_L.

We observe that in the latter three cases, vertices have either too few or too many neighbors in the body, as $k > 2$. A vertex of the first case can be a pin, so can a vertex of the second case, but not at the same time, as their neighborhoods in the body are disjoint.

Then, the pins are either three vertices of L or three vertices of L_A. In both cases, the variables of the clause c are all assigned the same value, and c is unsatisfied, which is a contradiction.

Now assume that there is a set A such that $S(G, A)$ is H_k-free. We define a valuation μ of variables so that $\mu(v)$ is true if and only if $x_v \in A$ (there may also be non-variable vertices in A, but these do not influence μ).

Assume that a clause $c = (v_i \vee v_j \vee v_l)$ is unsatisfied in μ. We prove that there is a hedgehog H_k in $S(G, A)$ whose body is in the union of vertices of K_c and K'_c, and whose pins are variable vertices.

K_c and K'_c are two cliques in G and regardless of A, their vertices induce two cliques in $S(G, A)$ as well. One of these cliques contains at least $4k - 2$ vertices. We denote this clique by K.

Consider the role of a vertex of K in G and its relation to the set A. It is either an outer vertex of a clique and not an element of A, or an inner vertex and not in A, or outer in A, or inner in A. We denote the sets of these vertices of K by O, I, O_A, and I_A, respectively.

Similarly, we divide all variable vertices according to whether they are literals or non-literals of the clause c, and whether they belong to A, into sets L, N, L_A, and N_A.

In $S(G, A)$, the adjacency between vertices of K and variable vertices is as follows ("a" meaning adjacent and "n" meaning non-adjacent).

	L	N	N_A	L_A
O	a	n	a	n
I	n	n	a	a
I_A	a	a	n	n
O_A	n	a	n	a

To find a hedgehog, we discuss the sizes of the sets above. As we assume that the clause c is unsatisfied, the three vertices x_{v_i}, x_{v_j} and x_{v_l} are either all in L, or all in L_A. We consider the first case, the second being symmetrical. Then the size of L is three.

Vertices in O and I_A are adjacent to vertices in L, vertices in I and O_A are not. Hence, if there are at least k vertices in $O \cup I_A$, and at least k vertices in $I \cup O_A$, then these $k + k$ vertices form a hedgehog with pins in L.

Otherwise, assume that $|O \cup I_A|$ is at most $k - 1$. Then $|I \cup O_A|$ is at least $3k - 1$. By the construction, each of I and O_A contains at most $2k - 1$ vertices. Then the sizes of O_A and I are at least k, and these vertices form the body of a hedgehog. The pins are either in N or in N_A; we have assumed that there are at most eight distinct variables in φ, and at most three of them occur in the clause c, hence either N or N_A contains at least three vertices. Fig. 3 shows an example of a hedgehog with pins in N.

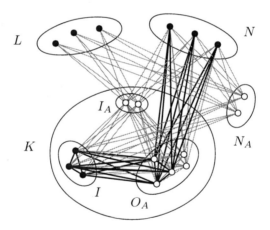

Fig. 3. A hedgehog in $I \cup O_A \cup N$ in $S(G, A)$. The sets O and L_A are empty. Vertices in A are white.

If, on the other hand, $|I \cup O_A|$ is at most $k - 1$, we proceed symmetrically with the role of outer and inner vertices interchanged. Again, we find a hedgehog with pins in N or in N_A.

Hence, as $S(G, A)$ is H_k-free, no clause can be unsatisfied. Then μ is a satisfying valuation, which we wanted to prove.

The transformation can clearly be done in polynomial time.

It is easy to see that the problem is in NP. A certificate containing the set A can be checked in polynomial time, we simply test each induced subgraph of $S(G, A)$ having $2k + 3$ vertices, whether it is a hedgehog.

4   Concluding Remarks

We have examined the problem of switching-equivalence to an H-free graph, and we have given infinitely many graphs H for which the problem is NP-complete. Thus, in the solved cases there are polynomial-time algorithms for K_2 (this is easy to prove, see also [4]), K_3 ([5,4]), $K_{1,2}$ ([8]), P_4 ([6]), $K_{1,3}$ ([7]), and a NP-hardness result for hedgehogs.

However, we are still very far from a complete characterization of graphs H. Hence, the following problem remains open.

Problem 1. For which graphs H is the problem of deciding switching-equivalence to an H-free graph polynomial? For which graphs H is it NP-complete? (Are there some graphs for which it is neither?)

The most symmetrical case is the complete graph K_k. So far we have not been able to solve this special case for any $k \geq 4$, hence we find it worth interest.

Problem 2. What is the computational complexity of switching-equivalence to a K_k-free graph, for any fixed $k \geq 4$?

References

1. Ehrenfeucht, A., Hage, J., Harju, T., Rozenberg, G.: Pancyclicity in switching classes. Inform. Process. Letters 73, 153–156 (2000)
2. Ehrenfeucht, A., Hage, J., Harju, T., Rozenberg, G.: Complexity issues in switching of graphs. In: Ehrig, H., Engels, G., Kreowski, H.-J., Rozenberg, G. (eds.) TAGT 1998. LNCS, vol. 1764, pp. 59–70. Springer, Heidelberg (2000)
3. Garey, M.R., Johnson, D.S.: Computers and Intractability: A Guide to the Theory of NP-Completeness. W. H. Freeman and Company, New York (1979)
4. Hage, J., Harju, T., Welzl, E.: Euler graphs, triangle-free graphs and bipartite graphs in switching classes. In: Corradini, A., Ehrig, H., Kreowski, H.-J., Rozenberg, G. (eds.) ICGT 2002. LNCS, vol. 2505, pp. 148–160. Springer, Heidelberg (2002)
5. Hayward, R.B.: Recognizing P_3-structure: A switching approach. J. Combin. Th. Ser. B 66, 247–262 (1996)
6. Hertz, A.: On perfect switching classes. Discrete Appl. Math. 94, 3–7 (1999)
7. Jelínková, E., Kratochvíl, J.: On Switching to H-Free Graphs. In: Ehrig, H., Heckel, R., Rozenberg, G., Taentzer, G. (eds.) ICGT 2008. LNCS, vol. 5214, pp. 379–395. Springer, Heidelberg (2008)
8. Kratochvíl, J., Nešetřil, J., Zýka, O.: On the computational complexity of Seidel's switching. In: Proc. 4th Czech. Symp., Prachatice (1990); Ann. Discrete Math. 51, 161–166 (1992)

9. Kratochvíl, J.: Complexity of hypergraph coloring and Seidel's switching. In: Bodlaender, H. (ed.) WG 2003. LNCS, vol. 2880, pp. 297–308. Springer, Heidelberg (2003)

10. De Montgolfier, F., Rao, M.: The bi-join decomposition. Electronic Notes in Discrete Mathematics 22, 173–177 (2005)

11. Rotem, D., Urrutia, J.: Circular permutation graphs. Networks 12, 429–437 (1982)

12. Seidel, J.J.: Graphs and two-graphs. In: Proc. 5th. Southeastern Conf. on Combinatorics, Graph Theory, and Computing, Winnipeg, Canada (1974)

13. Seidel, J.J.: A survey of two-graphs. Teorie combinatorie, Atti Conv. Lincei, vol. 17, pp. 481–511. Academia Nazionale dei Lincei, Rome (1973)

14. Seidel, J.J., Taylor, D.E.: Two-graphs, a second survey. Algebraic Methods in Graph Theory, Conf. Szeged, vol. II (1978); Colloq. Math. Janos Bolyai 25, 689–711 (1981)

15. Sritharan, R.: A linear time algorithm to recognize circular permutation graphs. Networks 27, 171–174 (1996)

Locally Injective Homomorphism to the Simple Weight Graphs*

Ondřej Bílka, Bernard Lidický, and Marek Tesař

Department of Applied Mathematics , Charles University,
Malostranské nám. 25, 118 00 Prague, Czech Republic
neleai@seznam.cz, {bernard,tesar}@kam.mff.cuni.cz

Abstract. A Weight graph is a connected (multi)graph with two vertices u and v of degree at least three and other vertices of degree two. Moreover, if any of these two vertices is removed, the remaining graph contains a cycle. A Weight graph is called simple if the degree of u and v is three. We show full computational complexity characterization of the problem of deciding the existence of a locally injective homomorphism from an input graph G to any fixed simple Weight graph by identifying some polynomial cases and some NP-complete cases.

Keywords: computational complexity, locally injective homomorphism, Weight graph.

1   Introduction

Graphs in this paper are generally simple. The only graphs with allowed loops and parallel edges are the Weight graphs. We denote the set of vertices of a graph G by $V(G)$ and the set of edges by $E(G)$. The degree of a vertex v in a graph G is denoted by $\deg_G(v)$ (recall that in multigraphs, degree of a vertex v is defined as the number of edges going to other vertices plus twice the number of loops at v) and the set of all neighbors of v by $N_G(v)$. We omit G in the subscript if G is clear from the context. By $[n]$ we denote the set of integers $\{1, \ldots, n\}$.

A connected (multi)graph H is a *Weight graph* (or sometimes known also as a dumbbell graph) if it contains two vertices u and v of degree at least three and all other vertices are of degree two. Moreover, both $H - u$ and $H - v$ contain a cycle or a loop. The Weight graph H is *simple* if both u and v have degree 3. Note that a simple Weight graph consists of a path connecting u and v and two cycles. A simple Weight graph H is denoted by $\mathcal{W}(a, b, c)$ if H is a union of cycles of lengths a and b and a path of length c. We call $\mathcal{W}(a, b, c)$ *reduced* if the greatest common divisor of a, b and c is one.

Let G and H be graphs. A *homomorphism* $f : G \to H$ is an edge preserving mapping from $V(G)$ to $V(H)$. If H is not a multigraph then homomorphism is *locally injective (resp. surjective, bijective)* if $N_G(v)$ is mapped to $N_H(f(v))$

---

* Supported by Charles University as GAUK 95710 and by the grant SVV-2010-261313 (Discrete Methods and Algorithms).

M. Ogihara and J. Tarui (Eds.): TAMC 2011, LNCS 6648, pp. 471–482, 2011.
© Springer-Verlag Berlin Heidelberg 2011

injectively (resp. surjectively, bijectively) for every $v \in V(G)$. Locally bijective homomorphism is also known as a *covering projection* or simply a *cover*. Similarly, locally injective homomorphism is known as a *partial covering projection* or a *partial cover*. In this paper we denote locally injective homomorphism as *LI-homomorphism*. In case that H is multigraph (and G is simple) LI-homomorphism generalizes as a mapping $f : V(G) \cup E(G) \rightarrow V(H) \cup E(H)$ such that:

i) for every $u \in V(G)$: $f(u) \in V(H)$
ii) for every $e = \{u, v\} \in E(G)$: $f(e) \in E(H)$ and $f(e) = \{f(u), f(v)\}$
iii) for every $u \in V(G)$ and every non loop edge $e \in E(H)$ such that $f(u) \in e$, there is at most one edge $e' \in E(G)$ such that $u \in e'$ and $f(e') = e$
iv) for every $u \in V(G)$ and every loop edge $e \in E(H)$ on vertex $f(u)$, there are at most two edges $e', e'' \in E(G)$ such that $u \in e'$, $u \in e''$ and $f(e') = f(e'') = e$

We can generalize the definition of locally surjective, resp. locally bijective homomorphism from simple graph to multigraph by simply changing the phrase "at most" by "at least", resp. "exactly" in iii) and iv). In the following text we denote homomorphism f from G to H simply as $f : G \rightarrow H$.

We consider the following decision problem. Let H be a fixed graph and G be an input graph. Determine the existence of a locally injective (surjective, bijective) homomorphism $f : G \rightarrow H$. We denote the problem by H-LIHom (resp. H-LSHom, H-LBHom). If there is no local restriction on the homomorphism, the problem is called H-Hom.

In this paper we focus on the H-LIHom problem.

Problem: H-LIHom
Input: Graph G
Task: Determine the existence of a LI-homomorphism $f : G \rightarrow H$.

Locally injective homomorphisms are closely related to $H(2, 1)$-labelings, which have an applications in frequency assignment. Let H be a graph. An $H(2, 1)$-*labeling* of a graph G is a mapping $f : V(G) \rightarrow V(H)$ such that image of every pair of adjacent vertices is two distinct and nonadjacent vertices. Moreover, image of every pair of vertices in distance two is two distinct vertices. For simple graphs the mapping f corresponds to a LI-homomorphism from G to the complement of H.

The computational complexity of H-Hom was fully determined by Hell and Nešetřil [11]. They showed that the problem is solvable in polynomial time if H is bipartite and it is $\mathcal{NP}$-complete otherwise.

The study of H-LSHom was initiated by Kristiansen and Telle [15] and completed by Fiala and Paulusma [10] who gave a full characterization by showing that H-LSHom is $\mathcal{NP}$-complete for every connected graph on at least three vertices.

The computational complexity of locally bijective homomorphisms was first studied by Bodlaender [2] and Abello et al. [1]. Despite of the effort [12,13,14] the complete characterization is not known.

Similarly, for the locally injective homomorphism the dichotomy for the computational complexity is not known. Some partial results can be found in [6,7,9]. Fiala and Kratochvíl [8] also considered a list version of the problem and showed dichotomy. Note that no direct consequences of complexity of H-HOM or H-LSHOM to H-LIHOM are known.

The authors of [7] show that H-LBHOM is reducible in polynomial time to H-LIHOM. Hence it makes sense to study the complexity of H-LIHOM where H-LBHOM is solvable in polynomial time.

In this paper we focus on graphs with at most two vertices of degree three. Restriction on just two vertices of degree more than two is a parallel to boolean logic where two values CSP is fully characterized from the complexity point of view, but multi-value CSP is still open [3]. A complete dichotomy for locally bijective homomorphism of graphs with at most two vertices of degree more than two was shown in [12].

In case of LI-homomorphism this problem was solved for *Theta graphs*. Theta graph is a graph with exactly two vertices u and v of degree at least three and several paths connecting them. Note that u and v may be connected by parallel edges. Study of LI-homomorphisms to the Theta graphs started by a work of Fiala and Kratochvíl [6], continued in [9] and was finished by showing a complete dichotomy by Lidický and Tesař [16].

Also Weight graphs, considered in this paper, have at most two vertices of degree more than two. Study of Weight graphs was initiated by Fiala [5] by showing the following theorems.

Theorem 1. $\mathcal{W}(a, a, a)$-LIHOM *is $\mathcal{NP}$-complete.*

Theorem 2. $\mathcal{W}(a, a, b)$-LIHOM *is polynomial time solvable, whenever a is divisible by strictly higher power of two than b, and is $\mathcal{NP}$-complete otherwise.*

Fiala and Kratochvíl [6] observed that it is sufficient to consider only reduced Weight graphs.

Proposition 1. *Let W be a Weight graph and d be the greatest common divisor of lengths of simple paths in W. Let W' be obtained from W by shortening each of its simple path by a factor of d. Then W-LIHOM is reducible to W'-LIHOM in polynomial time.*

In this paper we continue the study of Weight graphs by showing complete dichotomy for simple Weight graphs.

Theorem 3. *Let H be a bipartite simple reduced Weight graph. Then H-LIHOM problem is solvable in polynomial time.*

Theorem 4. *Let H be a non-bipartite simple reduced Weight graph. Then H-LIHOM problem is $\mathcal{NP}$-complete.*

In the following section, we introduce several definitions and observations. Then we give the proof of Theorem 3 and finish with Section 4 containing the proof of Theorem 4. Some of the proofs are ommited due to page limit.

2   Preliminaries

Let G be a graph and H be a spanning subgraph of G. We say that H is a *2-factor* of G if for all $v \in V(G) : \deg_H(v) = 2$. Let C be a set of colors. A mapping $\varphi : E(G) \to C$ is called an *edge coloring* if for every $e_1, e_2 \in E(G)$ that share a common vertex holds $\varphi(e_1) \neq \varphi(e_2)$.

Let G be a graph and $v_0 v_1 v_2 \ldots v_n$ be a path P of length n in G. The path is *simple* if v_0 and v_n are vertices of degree at least three and all inner vertices of P have degree two. We denote a simple path of length n by SP_n.

Let H be a Weight graph with vertices w_A and w_B of degree at least three. Let $f : G \to H$ be a LI-homomorphism. Note that f must map all vertices of degree at least three to w_A or w_B. Hence every end vertex of every simple path of G must be mapped to w_A or w_B. We call a vertex *big* if it has degree at least three. We denote the set of big vertices of G by $B(G)$. Note that w_A and w_B are also big vertices.

We need to control what are the possible LI-homomorphism of simple paths. Hence we define a function $g_f^P(v_0, v_n) = x$ if the edge $v_0 v_1$ is mapped by a LI-homomorphism f to an edge of SP_x in H. We also use notation $st_f^P(v_0, v_n)$ to denote $f(v_0 v_1)$. We omit the superscript P if there is only one simple path containing v_0 and v_n and the subscript f if it is clear from the context.

We say that SP_n allows *decomposition* $x-y$ if there exists a graph G containing a simple path P of length n with end vertices v_0 and v_n and a LI-homomorphism $f : G \to H$ such that $g_f^P(v_0, v_n) = x$ and $g_f^P(v_n, v_0) = y$. We denote the decomposition by $x -_k y$ (resp. $x -_c y$) if it forces that $f(v_0) = f(v_n)$ (resp. $f(v_0) \neq f(v_n)$).

In the case of decomposition $x-_k y$ (resp. $x-_c y$) we say, that the decomposition keeps (resp. changes) the parity.

Let n be a positive integer and $\mathcal{E} \subseteq \{a_1, a_2, \ldots a_n\}$. The following notation

$$n_{\mathcal{E}}^H : \quad x_1 - y_1, x_2 - y_2, \ldots, x_s - y_s, (z_1 - w_1), (z_2 - w_2), \ldots, (z_t - w_t)$$

describes the list of all decompositions $x - y$ of SP_n where $x, y \in \mathcal{E}$. Decompositions $x_i - y_i$ must be possible and decompositions $z_j - w_j$ are optional for all $i \in [s]$ and $j \in [t]$. Moreover, $-_k$ and $-_c$ can be used instead of just $-$.

We call an edge e of H *bridge edge* if it is on a simple path with distinct end vertices and *loop edge* otherwise. If an edge e' is mapped by a LI-homomorphism f to a bridge (loop) edge, we call e' also a bridge (resp. loop) edge.

We denote the greatest common divisor of $x_1, \ldots x_n$ by $GCD(x_1, \ldots, x_n)$.

Proposition 2. *Let $a, b, d \in \mathbb{N}$ such that $GCD(a, b) = d$. Then for every $z \in \mathbb{Z}$ there exist $s, t \in \mathbb{Z}$ such that $as + bt = zd$.*

3   Proof of Theorem 3 (Polynomial Case)

In this section we give the proof of Theorem 3.

First, we define $\mathcal{W}(1^{n_a}, 1^{n_b}, 1^{n_c})$ to be a Weight graph with n_a, resp. n_b loops on vertices w_A, resp. w_B and n_c parallel edges between w_A and w_B. Next we define that Weight graph $\mathcal{W}(a^{n_a}, b^{n_b}, c^{n_c})$ is obtained from $\mathcal{W}(1^{n_a}, 1^{n_b}, 1^{n_c})$ by subdividing all loops on vertex w_A, resp. w_B $a - 1$, resp. $b - 1$ times and all parallel edges between w_A and w_B $c - 1$ times. Theorem 3 is a special case of the following Theorem 5 for $n_a = n_b = n_c = 1$.

Theorem 5. *Let* $H = \mathcal{W}(a^{n_a}, b^{n_b}, c^{n_c})$, *where* $a, b, c, n_a, n_b, n_c \in \mathbb{N}$ *such that* $GCD(a, b, c) = 1$, *be bipartite. Then* H-LIHOM *problem is solvable in polynomial time.*

Proof. Let H be a fixed Weight graph from the statement of the theorem. Let $w_A, w_B \in V(H)$ be of degrees $2n_a + n_c$ and $2n_b + n_c$ respectively. As H is bipartite and $GCD(a, b, c) = 1$, we conclude that c is odd and a and b are (not necessarily distinct) even numbers.

Let G be an input graph. We assume that G is bipartite. If not, there is no LI-homomorphism from G to H and the algorithm returns NO answer instantly.

First we partition big vertices of $V(G)$ to two sets of the bipartition A and B. Note that all vertices of A must be mapped to w_A and all vertices of B to w_B or vice versa. We try both possibilities and without loss of generality we assume that vertices of A are mapped to w_A and vertices of B are mapped to w_B.

We reduce H-LIHOM problem to a *flag factor* problem of an auxiliary graph G' (of size polynomial in size of G).

Problem: Flag factor
Input: graph G' and functions $f_l : V(G') \to \mathbb{N}_0$ and $f_u : V(G') \to \mathbb{N}$
Output: spanning subgraph F of G' satisfying $f_l(v) \leq deg_F(v) \leq f_u(v)$ for all $v \in V(G')$.

We call the edges of F *matched edges*. The flag factor problem is solvable in polynomial time [6]. We use the flag factor to identify edges of G, which should be mapped to bridge edges of H. Let us now describe G'. The auxiliary graph G' contains two sets of vertices A' and B' corresponding to A and B. We define $f_u(v) = n_c$ for all $v \in B' \cup A'$, $f_l(v) = deg_G(v) - 2n_a$ for all $v \in A'$ and $f_l(v) = deg_G(v) - 2n_b$ for all $v \in B'$.

For every (simple) path between big vertices $u, v \in B(G)$ we construct a list L of all possible mappings under some LI-homomorphism (since H is bipartite, all these decompositions either change or keep parity) and we only distinguish if these decompositions begin, resp. end by loop edge or bridge edge. In list L we denote loop edge as "$\sim$" and bridge edge as "$-$". We join corresponding $u' \in V(G')$ and $v' \in V(G')$ by gadget according to L as depicted in Figure 1 (note that corresponding lists for vertices u, v and v, u are symmetric and so corresponding gadgets are symmetric as well).

We show that there exists a LI-homomorphism $f : G \to H$ if and only if there exists a flag factor F of G'.

If f exists, we can get F in a following way: for every edge $\{u, w\} \in E(G)$ such that $u \in B(G)$ and $\{u, w\}$ is mapped to the bridge edge, we take an edge incident to u' in a gadget corresponding to a simple path containing an edge $\{u, w\}$ to F. For such F for every $u \in B(G)$ holds that $f_l(u') \leq deg_F(u') \leq f_u(u')$ and

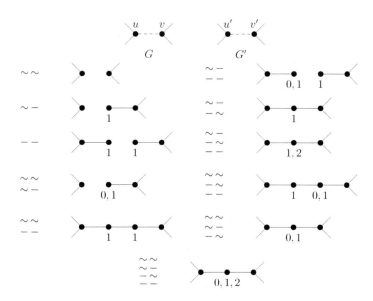

Fig. 1. Gadget joining vertices u' and v' in G'. The right gadget is chosen according to possible decompositions of a simple path joining big vertices u and v in G. The numbers below the vertices indicate the intervals given by f_l and f_u.

it is not hard to prove (because of choosing of appropriate gadgets in construction of G') that if F is not a flag factor then we can add to F some edges which are not incident to any u' for $u \in B(G)$. So if a flag factor F does not exist, neither does a LI-homomorphism f.

If a flag factor F exists, we show that a LI-homomorphism $f : G \to H$ also exists. The choice of f_l and f_u assures that each vertex of A', resp. B' has at most n_c incident matched edges and at most $2n_a$, resp. $2n_b$ incident non-matched edges.

Let $v \in V(G)$ be big and v' be the corresponding vertex in G'. Then if $v \in A$ we prescribe $f(v) = w_A$ and otherwise $f(v) = w_B$. Let P be a simple path beginning with v and g be the gadget corresponding to P in G'. If there is a matched edge in g incident to v', then f will map the beginning of P to some bridge edge and to some loop edge incident to vertex $f(v)$ otherwise. We do not define f on P yet. We only prescribe one of decompositions $\{a-a, a-b, a-c, b-c, b-b, c-c\}$, which splits P into paths of lengths a, b and c and it prescribes which internal vertices of P are mapped to w_A and w_B. So, we have fixed which vertices of G are mapped to w_A and w_B.

Let T be vertices of G which are mapped to big vertices of H. Note that also vertices of degree less than three may be in T. Also note that a path P where $T \cap P$ are the only endpoints of P has length in $\{a, b, c\}$. We just need to decide to which loop or bridge of H the path P will be mapped.

We create auxiliary graphs U_a, U_b, resp. U_c with vertex set $\{u \in T \mid f(u) = w_A\}$, $\{u \in T \mid f(u) = w_B\}$, resp. T. For all $x \in \{a, b, c\}$ two vertices of U_x are joined by an edge if they are connected by a path of length x which is internally disjoint with T and U_x contains a loop at vertex v if there is a cycle C of length x where $v = T \cap C$ for $x \in \{a, b\}$. The graph U_a can be made $2n_a$-regular by adding vertices and edges. It is well known that any $2n_a$-regular graph can be partitioned to n_a 2-factors in polynomial time. For each 2-factor Z of U_a we use one cycle C_a of length a of H and map vertices in cycles of Z to C_a. We treat U_b in analogous way. Note that there are no loops in U_c. Graph U_c is bipartite with maximum degree at most n_c and by König's theorem there exists an edge coloring $\varphi : E(U_c) \rightarrow [n_c]$. We assign one simple path of length c of H to each color class of φ. So f can be constructed from F.

All steps of the reduction can be computed in polynomial time and the flag factor problem is also solvable in polynomial time. Hence we conclude that f can be computed in polynomial time if exists or detect that it does not exist.   □

4   Proof of Theorem 4 ($\mathcal{NP}$-Complete Case)

The goal of this section is to give a proof of Theorem 4 by showing a reduction from 1-IN-3 SAT or NAE-3-SAT. These problems ask for existence of an evaluation of formula in CNF with clauses of size exactly 3 such that in every clause there is exactly one positive literal, resp. in every clause there exist at least one positive as well as at least one negative literal. Both problems are $\mathcal{NP}$-complete by Schaefer [17]. Note that H-LIHOM is in $\mathcal{NP}$ as a description of the desired homomorphism is of linear size and can be verified in polynomial time. Hence we only need to proof $\mathcal{NP}$-hardness.

We use the same basic notation as in the previous section. Let H be a fixed non-bipartite Weight graph $\mathcal{W}(a, b, c)$ with big vertices w_A and w_B. Recall that Theorem 2 implies $a \neq b$.

We start by restricting a, b and c by the following corollary of Theorem 1.

Corollary 1. *If there exist $x, y \in \mathbb{N}$ such that $c = ax = by$, then H-LIHOM is $\mathcal{NP}$-complete.*
(proof is ommited)

Lemma 1. *Let $a, b, c \in \mathbb{N}$ such that $GCD(a, b, 2c) = 1$. Then exist $s, t, z \in \mathbb{N}$ such that $as = bt + 2cz + c$ and $t > z$.*
(proof is in ommited)

We use the notation $u \sim n - v$, where $u, v \in \{w_A, w_B\}$ and $n \in \mathbb{N}$ to denote existence of a LI-homomorphism f from a simple path $u = v_0 v_1 \ldots v_n = v$ to H such that $g_f(v_0, v_1)$ is a loop edge and $g_f(v_n, v_{n-1})$ is a bridge edge. Variants where $\sim$ and $-$ are combined differently are defined similarly. Also $\simeq$ is used instead of $\sim$ and $-$ if the exact mapping is not known. We say that f is *beginning* with $g_f(v_0, v_1)$ and f is *ending* with $g_f(v_n, v_{n-1})$ on path $v_0 v_1 \ldots v_n$.

The following corollary connects Lemma 1 and mappings of simple paths to H.

Corollary 2. *There exists $k \in \mathbb{N}$ such that there exist LI-homomorphisms f_1 and f_2 where $f_1 = w_A \sim k \sim x$, $f_2 = w_B \sim k - x$ and $x \in \{w_A, w_B\}$.*

Proof. As H is non-bipartite, a, b and c satisfy the assumptions of Lemma 1. Let $k = as + az = bt + 2cz + az + c$ and $x = w_A$. As k is divisible by a, f_1 can just use the the cycle of length a to achieve $w_A \sim k \sim w_A$. We construct the mapping f_2 starting from w_A by z repetitions of the pattern $cbca$ and then by adding c and finally $(t - z)$ times b. As $t > z$, we have that $g_{f_2}(v_k, v_{k-1}) = b$. Hence we constructed $w_A - k \sim w_B$.   □

In the following, we assume that k is the smallest possible number, whose existence is guaranteed by the previous Corollary, such that there exists mappings $f_1 = w_A \sim k \simeq x$ and $f_2 = w_B \sim k \simeq x$. We use y for the vertex of H such that $\{x, y\} = \{w_A, w_B\}$.

Lemma 2. *There do not exist mappings*

1. *both $w_A \sim k - x$ and $w_B \sim k - x$*
2. *both $w_A \sim k \sim x$ and $w_B \sim k \sim x$*

Proof. In the first case we observe that there exist mappings $f_1' = w_A \sim (k-c) \simeq y$ and $f_2' = w_B \sim (k-c) \simeq y$ contradicting the choice of k. In the second case we consider mappings $f_1'' = w_A \sim (k - d) \simeq x$ and $f_2'' = w_B \sim (k - d) \simeq x$, where d is the length of the loop at x in H. This also contradicts the choice of k.   □

Without loss of generality, in what follows we assume that $f_1 = w_A \sim k \sim x$ and $f_2 = w_B \sim k - x$ (since Lemma 2 and the fact that $\mathcal{W}(a, b, c)$ is isomorphic to $\mathcal{W}(b, a, c)$).

Lemma 3. *There do not exist mappings*

1. $w_A \sim k \sim y$
2. $w_B \sim k - y$

(proof is ommited)

Let us now summarize the results of previous lemmas:

- mappings $w_A \sim k \sim x$ and $w_B \sim k - x$ exist
- mappings $w_A \sim k - y$ and $w_B \sim k \sim y$ may or may not exist
- mappings $w_A \sim k - x$, $w_A \sim k \sim y$, $w_B \sim k \sim x$, and $w_B \sim k - y$ do not exist

Next we introduce several gadgets. Let $z \in \{w_A, w_B\}$ then Z-gadget is a graph containing a vertex v_z of degree one such that any LI-homomorphism from Z-gadget to H maps v_z to z and the edge incident with v_z is mapped to a loop edge. We call v_z a z-vertex.

Lemma 4. *For every $z \in \{w_A, w_B\}$ there exists an Z-gadget.*
(proof is ommited)

If z is w_A, w_B, resp. x then we denote appropriate Z-gadget as A-gadget, B-gadget, resp. X-gadget and appropriate z-vertex as a-vertex, b-vertex, resp. x-vertex.

A *variable gadget* $VG(i)$ for $i \in \mathbb{N}$ is a graph containing two subsets of vertices A and B such that $|A| = |B| = i$ and vertices of $A \cup B$ have degree two. Moreover, if a graph G contains a copy of $VG(i)$ and all vertices of $A \cup B$ are big, then for every LI-homomorphism $f : G \to H$ holds that edges adjacent to vertices of $A \cup B$ not contained in $VG(i)$ are mapped to loop edges and either $\forall v \in A : f(v) = w_A$ and $\forall v \in B : f(v) = w_B$ or in the other case $\forall v \in A : f(v) = w_B$ and $\forall v \in B : f(v) = w_A$.

Lemma 5. *If c is not divisible neither by a nor b then there exists a variable gadget $VG(i)$ for all $i \in \mathbb{N}$.*

Proof. We first describe a construction of $VG(i)$ and then argue that it is indeed a variable gadget.

We start the construction by taking i copies of the X-gadget $X_1, \dots, X_i$ with x-vertices $x_1, \dots, x_i$ (see Figure 2). We continue by adding paths $x_j a_j b_j x_{j+1}$ of length three for all $j \in [i]$ where x_{i+1} is x_1. Finally, we subdivide both edges incident to x_j that are not contained in gadget X_j $k-1$ times and the edge $a_j b_j$ $c-1$ times for all $j \in [i]$. The resulting graph is $VG(i)$ and $A = \{a_j : 1 \le j \le i\}$ and $B = \{b_j : 1 \le j \le i\}$.

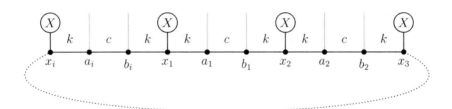

Fig. 2. Gadget $VG(i)$ from Lemma 5. Light lines indicate edges which will be present in a graph containing the gadget.

Let G be a graph containing a copy of $VG(i)$ where $A \cup B \subseteq B(G)$ and $f : G \to H$ be LI-homomorphism. Observe that $f(x_j) = x$ because it belongs to a copy of X-gadget and one of the two edges incident to x_j not contained in the X-gadget is mapped to a loop edge and the other one is mapped to a bridge edge for all $j \in [i]$. Since neither a nor b divides c, the only decomposition of a simple path of length c is $c -_c c$ and so $g_f(a_j, b_j) = c = g_f(b_j, a_j)$ and $f(a_j) \neq f(b_j)$ for every $j \in [i]$.

Suppose that $st(x_1, a_1)$ is a loop edge. We know that $st(a_1, x_1)$ is a loop edge (because $st(a_1, b_1)$ is a bridge edge) and so $f(a_1) = w_A$ and $g_f(a_1, x_1) = a$ (because $x \sim k \sim w_A$ and Lemma 2). It means that $f(b_1) = w_B$ and since $st(b_1, x_2)$ is a loop edge and $f(x_2) = x$ we have that $st(x_2, b_1)$ is a bridge edge and consequently $st(x_2, a_2)$ is a loop edge. Now we can continue in the same way in gadget $VG(i)$ and we get that $st(a_j, b_j)$ as well as $st(b_j, a_j)$ is a bridge edge, $f(a_j) = w_A$ and $f(b_j) = w_B$ for all $j \in [i]$, what proves that $VG(i)$ is a variable gadget.

If $st(x_1, a_1)$ is a bridge edge then $st(x_1, b_i)$ is a loop edge and we can continue in a similar way (but in an opposite direction) as in the previous paragraph. We get that $st(a_j, b_j)$ as well as $st(b_j, a_j)$ is a bridge edge, $f(a_j) = w_B$ and $f(b_j) = w_A$ for all $j \in [i]$, what proves that $VG(i)$ is a variable gadget. □

Lemma 6. *If c is divisible by exactly one of the numbers a or b then there exists a variable gadget $VG(i)$ for all $i \in \mathbb{N}$.*
(proof is ommited)

Recall that for every $i \in \mathbb{N}$ there exists variable gadget $VG(i)$ such that there exist LI-homomorphisms $f_1, f_2 : VG(i) \to H$ such that for every vertex $v \in A \cup B$ one edge incident to v is mapped to loop edge and one is mapped to bridge edge (for both f_1 and f_2) and $\forall u \in A, \forall v \in B : f_1(u) = f_2(v) = w_A$ and $f_1(v) = f_2(u) = w_B$. In the following text we use only such variable gadgets.

The last gadget is a gadget for representing clauses, so called *CL-gadget*. It contains three vertices of degree one z_1, z_2 and z_3 which are connected by three internally disjoint paths of length k to a vertex z_4 of degree three.

Lemma 7. *Let f be a locally injective homomorphism from CL-gadget to H such that $f(z_1), f(z_2), f(z_3) \in \{w_A, w_B\}$ and $st(z_1, z_4), st(z_2, z_4)$ and $st(z_3, z_4)$ are loop edges. Then $\{f(z_1), f(z_2), f(z_3)\} = \{w_A, w_B\}$.*

Proof. Suppose for contradiction that $|\{f(z_1), f(z_2), f(z_3)\}| = 1$. First, let $f(z_1) = f(z_2) = f(z_3) = w_A$. If $f(z_4) = x$, then we get a contradiction with Lemma 2 as f implies existence of a decomposition $w_A \sim k - x$. If $f(z_4) = y$, then we get existence of $w_A \sim k \sim y$ contradicting Lemma 3.

The case $f(z_1) = f(z_2) = f(z_3) = w_B$ is analogous. Both cases $f(z_4) = x$ and $f(z_4) = y$ contradict Lemma 2 or Lemma 3. □

Let ϱ be a formula in CNF with clauses $C_1, \ldots, C_m$ and variables $p_1, \ldots, p_n$ where all clauses contain exactly three literals. We denote the number of occurrences of a variable p in ϱ by $occ(p)$.

Now describe a construction of a graph G_ϱ. First take a copy CL_i of CL-gadget for every $i \in [m]$ and then take a copy VG_j of $VG(occ(p_j))$-gadget for every $j \in [n]$ (see Figure 3 for an example). We denote A and B of VG_j by A_j and B_j for every $j \in [n]$ and $\{z_1, z_2, z_3\}$ of CL_i by Z_i and z_j by z_j^i for every $j \in \{1, 2, 3, 4\}$ and $i \in [m]$.

Next we identify some vertices. If p_j occurs as a positive literal in C_i, we identify one vertex of Z_i with one vertex of A_j and if the occurrence is negative, we identify one of Z_i with one of B_j. The identification can be done such that every vertex is identified at most once as $occ(p_j) \leq |A_j| = |B_j|$. Finally, for every $w \in A_j \cup B_j$ of degree 2, we add a new vertex of degree one adjacent to w (so w is big vertex).

We prove Theorem 4 as a consequence of the following Lemmas 8 and 9.

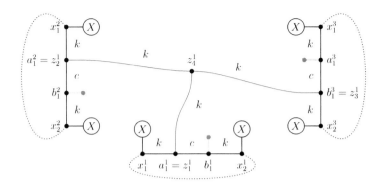

Fig. 3. Clause $C_1 = (p_1 \lor p_2 \lor \neg p_3)$

Lemma 8. *If there exist both decompositions $w_A \sim k - y$ and $w_B \sim k \sim y$, then H-LIHOM is $\mathcal{NP}$-hard.*

Proof. Let ϱ be an instance of NAE-3-SAT.

If there exists a LI-homomorphism $f : G_\varrho \to H$, then we evaluate p_j true if $f(A_j) = w_B$ and false otherwise for all $j \in [n]$ (this evaluation is well defined because of definition of variable gadget). Lemma 7 implies that there is no clause with all literals equal.

On the other hand let φ be an NAE evaluation of ϱ. We predefine a LI-homomorphism $f : G_\varrho \to H$ by mapping VG_j while requiring $f(A_j) = w_B$ if $\varrho(p_j)$ is true and $f(A_j) = w_A$ otherwise. This can be done as VG_j's are disjoint. Observe that $Z_i = \{w_A, w_B\}$. Hence f can be also defined on CL_i for all $i \in [m]$.   □

Lemma 9. *If at least one of decompositions $w_A \sim k - y$ and $w_B \sim k \sim y$ does not exist, then H-LIHOM is $\mathcal{NP}$-hard.*

Proof. Let ϱ be an instance of 1-IN-3-SAT.

If there exists a LI-homomorphism $f : G_\varrho \to H$, then we evaluate p_j true if $f(A_j) = w_B$ and false otherwise for all $j \in [n]$. The assumptions imply that $f(z_4^i) = x$ for all $i \in [n]$, hence exactly one vertex of $f(z_1^i)$, $f(z_2^i)$ and $f(z_3^i)$ is w_B and so every clause has exactly one literal evaluated as true.

On the other hand let φ be an 1-IN-3 evaluation of ϱ. We predefine a LI-homomorphism $f : G_\varrho \to H$ by mapping VG_j while requiring $f(A_j) = w_B$ if $\varrho(p_j)$ is true and $f(A_j) = w_A$ otherwise. This can be done as VG_j's are disjoint. Next we map $f(z_4^i)$ to x. As exactly one literal in every clause is true, exactly one of $f(z_1^i)$, $f(z_2^i)$ and $f(z_3^i)$ is w_B. Hence f can be extended to a locally injective homomorphism to H.   □

Recall that predefined mappings from proofs of Lemmas 8 and 9 can be easily extended to edges and so to LI-homomorphisms (similarly as in the proof of Theorem 5).

Acknowledgment

We would like to thank Jiří Fiala and Jan Kratochvíl for interesting discussions and fruitful comments.

References

1. Abello, J., Fellows, M.R., Stillwell, J.C.: On the complexity and combinatorics of covering finite complexes. Australian Journal of Combinatorics 4, 103–112 (1991)
2. Bodlaender, H.L.: The classification of coverings of processor networks. Journal of Parallel Distributed Computing 6, 166–182 (1989)
3. Feder, T., Vardi, M.Y.: Monotone monadic SNP and constraint satisfaction. In: Proceedings of the Tenth Annual ACM Symposium on Theory of Computing, STOC 1993, pp. 612–622 (1993)
4. Fiala, J.: NP completeness of the edge precoloring extension problem on bipartite graphs. Journal of Graph Theory 43, 156–160 (2003)
5. Fiala, J.: Locally injective homomorphisms, disertation thesis (2000)
6. Fiala, J., Kratochvíl, J.: Complexity of partial covers of graphs. In: Eades, P., Takaoka, T. (eds.) ISAAC 2001. LNCS, vol. 2223, pp. 537–549. Springer, Heidelberg (2001)
7. Fiala, J., Kratochvíl, J.: Partial covers of graphs. Discussiones Mathematicae Graph Theory 22, 89–99 (2002)
8. Fiala, J., Kratochvíl, J.: Locally injective graph homomorphism: Lists guarantee dichotomy. In: Fomin, F.V. (ed.) WG 2006. LNCS, vol. 4271, pp. 15–26. Springer, Heidelberg (2006)
9. Fiala, J., Kratochvíl, J., Pór, A.: On the computational complexity of partial covers of Theta graphs. Discrete Applied Mathematics 156, 1143–1149 (2008)
10. Fiala, J., Paulusma, D.: The computational complexity of the role assignment problem. In: Baeten, J.C.M., Lenstra, J.K., Parrow, J., Woeginger, G.J. (eds.) ICALP 2003. LNCS, vol. 2719, pp. 817–828. Springer, Heidelberg (2003)
11. Hell, P., Nešetřil, J.: On the complexity of H-colouring. Journal of Combinatorial Theory, Series B 48, 92–110 (1990)
12. Kratochvíl, J., Proskurowski, A., Telle, J.A.: Covering regular graphs. Journal of Combinatorial Theory B 71, 1–16 (1997)
13. Kratochvíl, J., Proskurowski, A., Telle, J.A.: Covering directed multigraphs I. colored directed multigraphs. In: Möhring, R.H. (ed.) WG 1997. LNCS, vol. 1335, pp. 242–257. Springer, Heidelberg (1997)
14. Kratochvíl, J., Proskurowski, A., Telle, J.A.: Complexity of graph covering problems. Nordic Journal of Computing 5, 173–195 (1998)
15. Kristiansen, P., Telle, J.A.: Generalized H-coloring of graphs. In: Lee, D.T., Teng, S.-H. (eds.) ISAAC 2000. LNCS, vol. 1969, pp. 456–466. Springer, Heidelberg (2000)
16. Lidický, B., Tesař, M.: Locally injective homomorphism to the Theta graphs. In: Iliopoulos, C.S., Smyth, W.F. (eds.) IWOCA 2010. LNCS, vol. 6460, pp. 326–336. Springer, Heidelberg (2011)
17. Schaefer, T.J.: The complexity of satisfiability problems. In: Proceedings of the Tenth Annual ACM Symposium on Theory of Computing, STOC 1978, pp. 216–226 (1978)

Maximal Matching and Path Matching Counting in Polynomial Time for Graphs of Bounded Clique Width

Benjamin Hellouin de Menibus[1] and Takeaki Uno[2]

[1] ENS Lyon, France
benjamin.hellouin_de_menibus@ens-lyon.fr
[2] National Institute of Informatics, Tokyo 101–8430, Japan
uno@nii.jp

Abstract. In this paper, we provide polynomial-time algorithms for different extensions of the matching counting problem, namely maximal matchings, path matchings (linear forest) and paths, on graph classes of bounded clique-width. For maximal matchings, we introduce matching-cover pairs to efficiently handle maximality in the local structure, and develop a polynomial time algorithm. For path matchings, we develop a way to classify the path matchings in a polynomial number of equivalent classes. Using these, we develop dynamic programing algorithms that run in polynomial time of the graph size, but in exponential time of the clique-width. In particular, we show that for a graph G of n vertices and clique-width k, these problems can be solved in $O(n^{f(k)})$ time where f is exponential in k or in $O(n^{g(l)})$ time where g is linear or quadratic in l if an l-expression for G is given as input.

Keywords: maximal matching, path matching, counting, clique-width.

1 Introduction

Counting problems in graphs can be very difficult, i.e. #P-hard in the general case, even for simple objects such as trees and independent sets. Research on graph classes has been motivated by such "hard" decision or optimization problems, and restricting the input to given graph classes has led to numerous polynomial-time algorithms. Despite this, only a few useful algorithms for counting problems exist, and these are relatively recent.

In this paper, we focus on counting maximal matching and path matching (linear forest). Matching counting and all extensions considered in this paper have been proved #P-complete in the general case. Some sparse graph classes such as planar graphs or graphs of bounded tree-width allow polynomial-time algorithms for perfect matching counting (see [13] and [1]); on the negative side, Valiant, when introducing the class #P, proved that counting perfect matchings as well as general matchings in bipartite graphs is #P-complete [19,20]. Valiant's proof concerning matchings has since been extended to 3-regular bipartite graphs [8], bipartite graphs of maximum degree 4 and bipartite planar graphs of maximum degree 6 [18].

M. Ogihara and J. Tarui (Eds.): TAMC 2011, LNCS 6648, pp. 483–494, 2011.
© Springer-Verlag Berlin Heidelberg 2011

The problem of counting perfect matchings in chordal and chordal bipartite graphs is also #P-complete [16], but good results on independent sets [15] give the impression that the chordal structure could nevertheless be interesting regarding matching counting. This led us to focus on a related graph class, the $(5, 2)$-crossing-chordal graphs. We especially make use of the bounded clique-width of this graph class.

Courcelle et al. introduced clique-width in [5] as a generalization of tree-width, and it attracted attention mainly for two reasons. On the one hand, in a similar fashion as the tree width, putting a bound on the clique-width makes many difficult problems decidable in polynomial time (see for example [6]). On the other hand, this class contains dense graphs as well as sparse graphs, which leads to more general results.

Makowsky et al. already proved as a consequence of a result in [14] that matching counting on graphs of bounded clique-width is polynomial in the size of the input graph. In this paper, we will extend this result by adapting their method to maximal matchings and path matchings. Our algorithms are polynomial of the graph size, but exponential of the clique-width k, i.e., $O(n^{poly(k)})$ time. It might be hard to develop a fixed parameter tractable algorithm such as an $O(c^{poly(k)}poly(n))$ time algorithm, since many graph algorithms, e.g. vertex coloring, have to spend $O(n^{poly(k)})$ time unless FPT $\neq W[1]$ [10].

The existing matching counting algorithms can not be used to count maximal matchings directly. The algorithms in [14] classify matchings of local graphs according to their sizes and the colors of the endpoints, and then get information about larger graphs by merging the matchings. However, in this way, each classified group may contain both matchings included in maximal matchings and those not included in any maximal matching. Actually, it seems to be difficult to characterize the number of matchings included in some maximal matching, by using only their sizes and their endpoints. In this paper, we introduce matching-cover pairs for this task. When we restrict a maximal matching to a subgraph, it can be decomposed into the matching edges belonging to the subgraph and end vertices of matching edges not included in the subgraph. From the maximality, the end vertices form a vertex cover of the edges of the subgraph. Thus, we count such pairs of matching and vertex cover according to their sizes and colors, and obtain a polynomial time algorithm for the problem.

For the problem of counting paths and path matchings, we must have some way to handle the connectivity of edge sets. Actually, connectivity is not easy to handle; for example, checking for the existence of Hamiltonian path is equivalent to checking whether the number of paths of length $n-1$ is larger than zero or not. Gimenez et al. devised an algorithm based on Tutte polynomial computation to count the number of forests in bounded-clique-width graphs in sub-exponential time, running in $2^{O(n^c)}$ time for constant $c < 1$ [11]. We use the properties of bounded-clique-width graphs so that we can classify the path matchings in a polynomial number of groups of equivalent path matchings, and thereby compute the number of paths and path matchings in polynomial time.

2   Clique Width

We shall introduce clique-width on undirected, non-empty labeled graphs by a construction method. Let G_i be the subgraph of vertices labeled i in a graph G. We define the singleton S_i as the labeled graph with one vertex of label i and no edge, and the following construction operations:

- Renaming : $\rho_{i \rightarrow j}(G)$ is G where all labels i are replaced by labels j;
- Disjoint union : $(V_1, E_1) \oplus (V_2, E_2) = (V_1 \cup V_2, E_1 \cup E_2)$;
- Edge creation : $\eta_{i,j}((V,E)) = (V, E \cup \{(v_1, v_2) \mid v_1 \in G_i, v_2 \in G_j\})$.

The class of graphs with clique-width $\leq k$ is the smallest class containing the singletons S_i, closed under $\rho_{i \rightarrow j}, \oplus$ and $\eta_{i,j}$ $(1 \leq i, j \leq k)$. In other words, the *clique-width* of a graph G, denoted as $cwd(G)$, is the minimal number of labels necessary to construct G by using singletons and renaming, disjoint union and edge creation operations.

For an unlabeled graph G, we define its clique-width by labeling all vertices with label 1. This is necessarily the best labeling, since any labeling can be renamed to a monochromatic labeling. Note that the clique-width of a graph of order n is at most n.

$(5,2)$-crossing-chordal graphs are known to have clique-width ≤ 3 [3] (we recall that a $(5,2)$-crossing-chordal graph is a graph where any cycle of length ≥ 5 has a pair of crossing diagonals). Other interesting results include: cographs are exactly the graphs with $cwd(G) \leq 2$, planar graphs of bounded diameter have bounded clique-widths, and any graph class of treewidth $\leq k$ also has a bounded clique-width of $\leq 3.2^{k-1}$ [4]. A complete review can be found in [12].

An l-*expression* is a term using $S_i, \rho_{i \rightarrow j}, \eta_{i,j}$ and $\oplus$ (with $i, j \leq l$) that respects the arity of each operation. It can be represented more conveniently in a tree structure, and we can inductively associate the current state of the construction with each node. If G is the graph associated with the root, we say that this term is an l-expression for G, and it is a certificate that G is of clique-width $\leq l$. An example is given in Fig.1.

Fellows et al. proved the NP-hardness of computing the minimum clique-width for general graphs [9]. The current best approximation is due to Oum and Seymour [17], who provided a linear time algorithm that, given a graph G and an integer c as input, returns an 2^{3c+2}-expression for G or certifies that the graph has a clique-width larger than c.

This implies that we can compute in quadratic time a 2^{3k+2}-expression for a graph of clique-width k by applying this algorithm for $c = 1, 2 \ldots$. As the bound is independent of n, algorithms requiring expressions as input will still work in polynomial time, although the time complexity will usually be extremely poor. For $(5,2)$-crossing-chordal graphs, though, this is not a concern since it is possible to compute a 3-expression in linear time [3].

An l-expression is called *irredundant* if every edge-creation operation $\eta_{i,j}$ is applied to a graph where no two vertices in G_i and G_j are adjacent. Any l-expression can be turned into an l-irredundant expression in linear time [7]. Therefore, we assume w.l.o.g. that the input expression is irredundant.

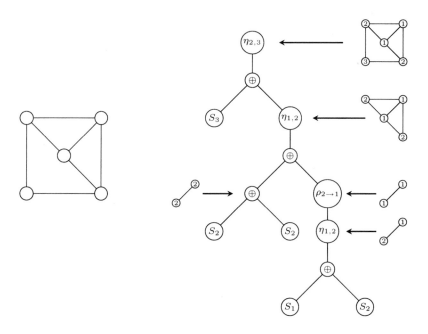

Fig. 1. Graph of clique-width 3, and a possible 3-expression tree (the last renaming operations are omitted)

3   Framework of Our Algorithms

The input of our algorithms is a graph G on n vertices and an l-expression for G, and the output is the number of objects (ex. matchings, paths) in G. The procedure works by counting these objects at each step of the construction, by using the expression tree : we start from the leaves and process a node once all its children have been processed. Finally, the value at the root of the tree is the output of the algorithm. Instead of doing it directly with the considered object, we introduce appropriate intermediate objects, and we compute tables of values at each step.

To avoid tedious case studies, we shall assume that requesting the value of any vector outside of the range $\{0 \ldots n\}$ returns the value 0. Also, $\Delta_r(l)$ is the vector $(\delta_{i,r})_{1 \leq i \leq l}$, and $\Delta_{r,s}(l)$ is the vector $(\delta_{i,r} \cdot \delta_{j,s})_{\substack{0 \leq i \leq j \leq l \\ (i,j) \neq (0,0)}}$, where $\delta_{i,j}$ is the

Kronecker delta:

$$\delta_{i,j} = \begin{cases} 1 \text{ if } i = j; \\ 0 \text{ otherwise.} \end{cases} \tag{1}$$

We will omit the l when it is obvious from the context.

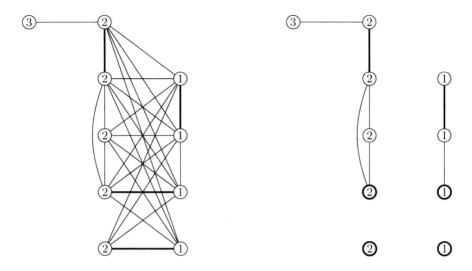

Fig. 2. Maximal matching of $\eta_{1,2}(G')$, and the corresponding matching-cover pair of G'

4 Counting Maximal Matchings

Theorem 1. *Computing the number of maximal matchings of a graph with n vertices with a corresponding l-expression can be done in polynomial time in n (but exponential w.r.t l).*

We cannot directly use the previous framework on maximal matchings. Indeed, consider M a maximal matching of $G = \eta_{i,j}(G')$ and M' the induced matching in G': M' is not necessarily maximal. However, we can keep track of the vertices of G' that are covered in M, and those vertices must form a vertex cover of the subgraph left uncovered by M'. See Fig.2 for an example.

A *matching-cover* pair of a graph $G = (V, E)$ is a pair (m, c) such that:

- $m \subseteq E$ is a matching of G (i.e. no vertex is covered more than once);
- $c \subseteq V$ is a vertex cover of the subgraph left uncovered by m (i.e. every *edge* is covered at least once).

We show that computing the number of matching-cover pairs of a graph with n vertices with a corresponding l-expression can be done in polynomial time in n.

Let $M = (m_i)_{1 \leq i \leq l}$ and $C = (c_i)_{1 \leq i \leq l}$ be two vectors of non-negative integers. For a graph G, we say that a pair (m, c) satisfies the condition $\varphi_{M,C}(G)$ if m covers m_i vertices in G_i and c uses c_i vertices in G_i for all i, and we denote by $mc_{M,C}(G)$ the number of pairs that satisfy $\varphi_{M,C}(G)$. Note that maximal matchings are exactly pairs with an empty cover; therefore, the number of maximal matchings of G is $\sum_{k \leq n} mc_{k \cdot \Delta_1, 0}(G)$.

Now we will follow the framework described above and compute $mc_{M,C}$ for all possible M and C, at each step of the construction. We associate to each node of the tree a table of size n^{2l} corresponding to the values of $mc_{M,C}$ on this graph for M and C ranging from $(0, .., 0)$ to $(n, .., n)$. For a singleton S_i, we can easily see that:

$$mc_{M,C}(S_i) = \begin{cases} 1 \text{ if } M = 0 \text{ and } C = 0 \text{ or } \Delta_i; \\ 0 \text{ otherwise.} \end{cases} \tag{2}$$

For the renaming operation $G = \rho_{i \to j}(G')$, the graph is not modified, but all vertices of label i are set to label j. Hence, we modify the entries i and j accordingly.

$$mc_{M,C}(G) = \sum_{\substack{M':(M,M') \vdash \phi_{i,j} \\ C':(C,C') \vdash \phi_{i,j}}} mc_{M',C'}(G') \tag{3}$$

$$\text{where } (X, X') \vdash \phi_{i,j} \Leftrightarrow \begin{pmatrix} x_j = x_i' + x_j' \\ x_i = 0 \\ \forall k \notin \{i, j\}, x_k = x_k' \end{pmatrix} \tag{4}$$

For the disjoint union of two graphs $G = G_1 \oplus G_2$, we have a bijection between matching-cover pairs (m, c) in G and pairs $(m_1, c_1), (m_2, c_2)$ of matching-cover pairs in G_1 and G_2, respectively. Moreover, if (m, c) satisfies $\varphi_{M,C}$, (m_1, c_1) satisfies φ_{M_1,C_1} and (m_2, c_2) satisfies φ_{M_2,C_2}, we have $M = M_1 + M_2$ and $C = C_1 + C_2$.

$$mc_{M,C}(G) = \sum_{\substack{M_1+M_2=M \\ C_1+C_2=C}} mc_{M_1,C_1}(G_1) \cdot mc_{M_2,C_2}(G_2) \tag{5}$$

For the edge creation operation $G = \eta_{i,j}(G')$, we have to choose the extremities of the edges added to the matching among the vertices in the vertex cover. If q is the number of new edges, we have:

$$mc_{M,C}(G) = \sum_{q=0}^{n} mc_{M',C'}(G') \cdot \binom{c_i'}{q} \cdot \binom{c_j'}{q} \cdot q! \tag{6}$$

$$\text{where } M' = M - q\Delta_i - q\Delta_j, \ C' = C + q\Delta_i + q\Delta_j \tag{7}$$

Once the maximal matchings of all sizes are computed, it is straightforward to count the number of perfect matchings and the number of minimum maximal matchings in polynomial time. Note that counting perfect matchings can be achieved in $O(n^{2l+1})$ time by adapting the matching counting algorithm presented in [14] in a similar fashion.

Complexity study: Obviously, there are exactly n singleton operations, and each operation requires a constant amount of time. Every other operation requires one to compute n^{2l} values. As the expression is irredundant, every edge creation operation adds at least one edge, so there are at most n^2 edge creation operations, processed in linear time. As a disjoint union operation has two children in the tree, and there are n leaves, there are $n - 1$ disjoint union operations, and they require $O(n^{2l})$ time.

For the renaming operation, consider the number of different labels at each step of the construction. This number is one for a singleton, the edge creation operation has no effect, the disjoint union is an addition in the worst case (no shared label) and the renaming operation diminishes this number by one. Therefore, there are at most n renaming operations, and they are done in $O(n^4)$ time. The final sum requires $O(n^l)$ operations.

Therefore, the overall complexity of the algorithm is

$$O(n) + O(n^{2l}) \cdot \left(O(n^5) + O(n^{2l+1}) + O(n^3)\right) + O(n^l) = O(n^{4l+1}) \; (l \geq 2). \quad (8)$$

For $(5,2)$-crossing-chordal graphs, we can compute an expression of width $l = 3$ in linear time and the algorithm runs in time $O(n^{13})$.

5   Counting Paths and Path Matchings

A *path matching* (or *linear forest*) is a disjoint union of paths, in other words, a cycle-free set of edges such that no vertex is covered more than twice.

Theorem 2. *Computing the number of paths $pth(G)$ and the number of path matchings $pm(G)$ of a graph of clique-width $\leq k$ can be done in polynomial time (but exponential w.r.t. k).*

Proof. Let $K = (k_{i,j})_{\substack{0 \leq i \leq j \leq l \\ (i,j) \neq (0,0)}}$ be a vector of non-negative integers. We say that a path matching P of G satisfies the condition ψ_K if:

- $\forall i > 0, k_{0,i}$ vertices in G_i are left uncovered by P;
- $\forall (i,j), i \leq j, k_{i,j}$ paths in P have extremities in G_i and G_j.

We denote the number of path matchings in G satisfying ψ_K by $pm_K(G)$. If $i > j$, we denote $k_{i,j} = k_{j,i}$. As K is of size $\frac{l(l+3)}{2}$, we compute tables of size $n^{\frac{l(l+3)}{2}}$ at each step.

For a singleton S_i, the only possible path matching is empty and leave the vertex uncovered.

$$\forall K, pm_K(S_i) = \begin{cases} 1 & \text{if } K = \Delta_{0,i}; \\ 0 & \text{otherwise.} \end{cases} \quad (9)$$

For the renaming operation $G = \rho_{i \to j}(G')$, the method is the same as for maximal matchings.

$$pm_K(G) = \sum_{K':(K,K') \vdash \phi} pm_{K'}(G') \quad (10)$$

$$\text{where } (K, K') \vdash \phi \Leftrightarrow \begin{pmatrix} k_{j,j} = k'_{j,j} + k'_{i,j} + k'_{i,i} \\ \forall a \notin \{i,j\}), k_{a,j} = k'_{a,i} + k'_{a,j} \\ \forall a, k_{a,i} = 0 \\ \forall a \notin \{i,j\}, b \notin \{0,i,j\}, k_{a,b} = k'_{a,b} \end{pmatrix} \quad (11)$$

For the disjoint union operation $G = G_1 \oplus G_2$, we have a bijection between path matchings p in G and pairs (p_1, p_2) of path matchings in G_1 and G_2, respectively. Plus, if p_1 satisfies ψ_{K_1}, p_2 satisfies ψ_{K_2} and p satisfies ψ_K, we have $K = K_1 + K_2$.

$$pm_K(G) = \sum_{K_1 + K_2 = K} pm_{K_1}(G_1) \cdot pm_{K_2}(G_2) \tag{12}$$

Consider now the edge creation operation $G = \eta_{i,j}(G')$. We say a path matching P in G is an *extension* of a path matching P' in G' if $P \cap G' = P'$, so that $P = P' \cup E_{i,j}$ where $E_{i,j}$ is a subset of the edges added by the operation. Now, if we consider a path matching P' in G' that satisfies $\psi_{K'}$, we claim that the number of extensions of P' in G that satisfy ψ_K depends only on i, j, K' and K (and not on P' or G'), and we represent it as $N_{i,j}(K', K)$. Since every path matching of G is an extension of an unique path matching of G', we have:

$$pm_K(G) = \sum_{K'} pm_{K'}(G') \cdot N_{i,j}(K', K) \tag{13}$$

Moreover, we can compute all the $N_{i,j}(K', K)$ beforehand in $O(n^{l(l+4)})$ time. The proof of these claims is given in the appendix.

We can then compute the number of paths $pth(G)$ and the number of path matchings $pm(G)$ with the formulas:

$$
\begin{aligned}
pth(G) &= \sum_{0 \le a \le n} pm_{K(a)}(G) \quad \text{where } K(a) = a \cdot \Delta_{0,1} + \Delta_{1,1} \\
pm(G) &= \sum_{1 \le a + 2b \le n} pm_{K(a,b)}(G) \text{ where } K(a,b) = a \cdot \Delta_{0,1} + b \cdot \Delta_{1,1}
\end{aligned}
\tag{14}
$$

Complexity study: A singleton operation requires constant time. Every other operation requires us to compute $n^{\frac{l(l+3)}{2}}$ values. For each value, the renaming operation in processed in linear time, the disjoint union operation in $O(n^{l^2})$ time and the edge creation operation in $O(n^{\frac{l(l+3)}{2}})$ time.

The overall complexity of the algorithm is:

$$
\begin{cases}
O(n^{l^2 + 4l}) \text{ for } l \le 5; \\
O(n^{\frac{3}{2}(l^2 + l) + 1}) \text{ for } l > 5.
\end{cases}
\tag{15}
$$

For $(5, 2)$-crossing-chordal graphs, we can compute in linear time an expression of width $l = 3$ and we have an algorithm running in $O(n^{21})$ time.

6   Conclusion

These results seem to confirm the intuition that bounding clique-width is an efficient restriction on the input of #P-hard problems in order to allow the use of polynomial algorithms. Notably, being able to count paths and path matchings in connected structures are usually very difficult to count. In that sense, the next logical step was to study the tree (or, equivalently, forest) counting problem. However, our attempts to do so by using a method similar to the one we used in the paper, only produced algorithms running in exponential time. Our feeling is that the tree counting problem remains #P-complete for graphs of bounded clique-width, as this intuitive method keeps giving bad results. It remains an open problem for now.

References

1. Arnborg, S., Lagergren, J., Seese, D.: Easy Problems for Tree-Decomposable Graphs. Journal of Algorithms 12, 308–340 (1991)
2. Bandelt, H.J., Mulder, H.M.: Distance-hereditary Graphs. Journal of Combinatorial Theory, Series B 41(2), 182–208 (1986)
3. Corneil, D.G., Habib, M., Lanlignel, J.-M., Reed, B., Rotics, U.: Polynomial Time Recognition of Clique-Width ≤ 3 Graphs. In: Gonnet, G.H., Viola, A. (eds.) LATIN 2000. LNCS, vol. 1776, pp. 126–134. Springer, Heidelberg (2000)
4. Corneil, D.G., Rotics, U.: On the Relationship Between Clique-Width and Treewidth. SIAM Journal of Computing 34(4), 825–847 (2005)
5. Courcelle, B., Engelfriet, J., Rozenberg, G.: Handle-Rewriting Hypergraph Grammars. J. Comput. Syst. Sciences 46, 218–270 (1993)
6. Courcelle, B., Makowsky, J.A., Rotics, U.: Linear Time Solvable Optimization Problems on Graphs of Bounded Clique Width. Theory of Computing Systems 33, 125–150 (2000)
7. Courcelle, B., Olariu, S.: Upper Bounds to the Clique-Width of Graphs. Discrete Applied Mathematics 101, 77–114 (2000)
8. Dagum, P., Luby, M.: Approximating the Permanent of Graphs with Large Factors. Theoretical Computer Science 102, 283–305 (1992)
9. Fellows, M.R., Rosamond, F.A., Rotics, U., Szeider, S.: Clique width Minimization is NP-hard. Proceedings of the 38th annual ACM Symposium on Theory of Computing, session 8B, pp. 354–362 (2006)
10. Fomin, F.V., Golovach, P.A., Lokshtanov, D., Saurabh, S.: Algorithmic Lower Bounds for Problems Parameterized by Clique-Width. In: Proceedings of the 21th ACM-SIAM Symposium on Discrete Algorithms (SODA 2010), pp. 493–502. ACM and SIAM (2010)
11. Giménez, O., Hliněný, P., Noy, M.: Computing the Tutte Polynomial on Graphs of Bounded Clique-Width. In: Kratsch, D. (ed.) WG 2005. LNCS, vol. 3787, pp. 59–68. Springer, Heidelberg (2005)
12. Golumbic, M.C., Rotics, U.: On the Clique-Width of Some Perfect Graph Classes. International Journal of Foundations of Computer Science 11(3), 423–443 (2000)
13. Kasteleyn, P.W.: Dimer Statistics and Phase Transitions. Journal of Mathematical Physics 4, 287–293 (1963)
14. Makowsky, J.A., Rotics, U., Averbouch, I., Godlin, B.: Computing Graph Polynomials on Graphs of Bounded Clique-Width. In: Fomin, F.V. (ed.) WG 2006. LNCS, vol. 4271, pp. 191–204. Springer, Heidelberg (2006)
15. Okamoto, Y., Uno, T., Uehara, R.: Counting the Independent Sets in a Chordal Graph. Journal of Discrete Algorithms 6(2), 229–242 (2008)
16. Okamoto, Y., Uehara, R., Uno, T.: Counting the Number of Matchings in Chordal and Chordal Bipartite Graph Classes. In: Paul, C., Habib, M. (eds.) WG 2009. LNCS, vol. 5911, pp. 296–307. Springer, Heidelberg (2010)

17. Oum, S., Seymour, P.: Approximating Clique-Width and Branch-Width. Journal of Combinatorial Theory, Series B 96(4), 514–528 (2006)
18. Vadhan, S.P.: The Complexity of Counting in Sparse, Regular, and Planar Graphs. SIAM Journal on Computing 31, 398–427 (2001)
19. Valiant, L.G.: The Complexity of Computing the Permanent. Theoretical Computer Science 8, 189–201 (1979)
20. Valiant, L.G.: The Complexity of Enumeration and Reliability Problems. SIAM Journal on Computing 8, 410–421 (1979)

Appendix

We now prove the case of Thm. 2 we have omitted. Let G and G' be two labeled graphs such that $G = \eta_{i,j}(G')$ (for some $i < j$) and P' a path matching of G' satisfying $\psi_{K'}$ for some K'. For any K, we want to compute the number of extensions of P' in G satisfying ψ_K.

Definitions. For any path matching satisfying ψ_K, a path with two extremities $x \in G_i$ and $y \in G_j$ is called an (i,j)-*path*, and x and y are called *partners*. We denote by $V_a(b)$ the vertices of G_a whose partner is in G_b, and by $V_a(0)$ the uncovered vertices in G_a. We also note $v_{a,b} = \#V_a(b)$, which means that $v_{a,b} = k_{a,b}$, except for $v_{a,a} = 2k_{a,a}$ (note that $v_{a,b}$ depend only on K). An edge which extremities are in $V_i(a)$ and in $V_j(b)$, respectively, is called an (a,b)-*edge*.

We use a dynamic programming technique to build all possible extensions of P' by considering each vertex of G_i' one by one in the order $V_i(j), V_i(0), .., V_i(l)$ (the reason for this order will be explained later). If $X = (x_0, .., x_l)$ is a vector of non-negative integers and P_1 a path matching in G_1 that satisfies ψ_{K_1}, $T_{i,j}(G_1, P_1, K_2, X)$ stands for the number of extensions of P_1 in $\eta_{i,j}(G_1)$ that satisfy ψ_{K_2} and that uses only the x_k last vertices of $V_i(k)$ for every k.

At each step of the computation, the equations show that knowing G_1 and P_1 is not necessary as long as ψ_{K_1} is satisfied: this proves our first claim, and we write $T_{i,j}(K_1, K_2, X)$ for $T_{i,j}(G_1, P_1, K_2, X)$. Also, since i, j and K_2 are not modified during the computation, we write $T(K_1, X)$ for $T_{i,j}(K_1, K_2, X)$.

We now detail the different steps by increasing difficulty (instead of the actual order of the algorithm). First, assume that $x_j = x_0 = .. = x_{k-1} = 0$ and $x_k \neq 0$ (for some $k \neq i$). We consider the first vertex in $V_i(k)$ that has not been considered yet in the computation. We have only two possibilities:

- No new edge adjacent to this vertex is added to the path matching.
- One new (k, a)-edge (possibly $a = 0$) is added to the path matching: we have $v_{j,a}$ choices for the edge. A (i, k)-path and a (j, a)-path are transformed into a (k, a)-path.

In each case, the value of the current K is updated accordingly and the vertex is deleted from X. Next to each term is the set that contains the other extremity of the edge being considered.

$$T(K", X) = T(K", X - \Delta_k) \qquad\qquad\qquad \emptyset$$
$$+ \sum_{1 \le a \le l} v_{j,a}.T(K" - \Delta_{i,k} - \Delta_{j,a} + \Delta_{k,a}, X - \Delta_k)\, V_j(a) \qquad (16)$$
$$+v_{0,j}.T(K" - \Delta_{i,k} - \Delta_{0,j} + \Delta_{j,k}, X - \Delta_k) \qquad V_j(0)$$

Note that if a (k, i)-edge is added, the partner of another vertex of $V_i(j)$ is also modified: this is why $V_i(j)$ is considered first in the computation, so that it does not appear in X anymore at this step. This remark holds for all the other cases except for $k = j$.

Now, we consider the first step $(k = j)$. The situation is similar, but the vertex cannot be linked to its own partner when $k' = i$. Note that adding a (j, i)-edge changes the partner of another vertex of $V_i(j)$, but the new partner is still in G_j, so doing this brings no modification to X.

$$T(K", X) = T(K", X - \Delta_j) \qquad\qquad\qquad \emptyset$$
$$+ \sum_{\substack{1 \le a \le l \\ a \ne i}} v_{j,a} T(K" - \Delta_{i,j}, X - \Delta_j) \qquad V_j(a)$$
$$\qquad\qquad\qquad\qquad (17)$$
$$+(v_{i,j} - 1)T(K" - \Delta_{i,j}, X - \Delta_j) \qquad V_j(i)$$
$$+v_{0,j}.T(K" - \Delta_{i,j} - \Delta_{0,j} + \Delta_{j,j}, X - \Delta_j)\, V_j(0)$$

For the uncovered vertices $(k = 0)$, up to two edges can be added to the path matching. The possibilities are:

- No new edge adjacent to this vertex is added to the matching.
- One new (k, k')-edge is added to the matching: we have $v_{j,k'}$ choices for the edge. An uncovered vertex and a (j, k')-path are transformed into a (x, k')-path.
- Two new (k, k') and $(k, k")$-edges are added to the matching: we have $v_{j,k'}.v_{j,k"}$ choices for the two edges (only half of those when $k' = k"$). An uncovered vertex, a (j, k')-path and a $(j, k")$-path are transformed into a $(k', k")$-path.

Then $T(K'', X) =$

$$T(K", X - \Delta_0) \qquad\qquad\qquad\qquad\qquad\qquad\qquad \emptyset$$
$$+ \sum_{1 \le a \le l} v_{j,a}.T(K" - \Delta_{0,i} - \Delta_{j,a} + \Delta_{i,a}, X - \Delta_0) \qquad V_j(a)$$
$$+ v_{0,j}.T(K" - \Delta_{0,i} - \Delta_{0,j} + \Delta_{i,j}, X - \Delta_0) \qquad V_j(0)$$
$$+ \sum_{1 \le a < b \le l} v_{j,a}.v_{j,b}.T(K" - \Delta_{0,i} - \Delta_{j,a} - \Delta_{j,b} + \Delta_{a,b}, X - \Delta_0)\, V_j(a) \mid V_j(b)$$
$$+ \sum_{1 \le a \le l} v_{j,a}.v_{0,j}.T(K" - \Delta_{0,i} - \Delta_{0,j}, X - \Delta_0) \qquad V_j(a) \mid V_j(0)$$
$$+ \sum_{\substack{1 \le a \le l \\ a \ne j}} \frac{v_{j,a}.(v_{j,a} - 1)}{2}.T(K" - \Delta_{0,i} - 2\Delta_{j,a} + \Delta_{a,a}, X - \Delta_0)\, V_j(a) \mid V_j(a)$$
$$+ \frac{v_{0,j}.(v_{0,j} - 1)}{2}.T(K" - \Delta_{0,i} - 2\Delta_{0,j} + \Delta_{j,j}, X - \Delta_0) \qquad V_j(0) \mid V_j(0)$$
$$+ \frac{v_{j,j}.(v_{j,j} - 2)}{2}.T(K" - \Delta_{0,i} - \Delta_{j,j}, X - \Delta_0) \qquad V_j(j) \mid V_j(j)$$
$$\qquad\qquad\qquad\qquad\qquad\qquad\qquad\qquad\qquad\qquad (18)$$

For $k = i$, we consider the two extremities of the (i, i)-path at the same time. Therefore, this situation is similar to the previous one, except that we have to choose one of the extremities in each case. There are twice as many possibilities as in the previous case. We have $T(K", X) =$

$$
\begin{array}{ll}
T(K", X - 2\Delta_i) & \emptyset \\
+ \displaystyle\sum_{1 \le a \le l} 2v_{j,a}.T(K" - \Delta_{i,i} - \Delta_{j,a} + \Delta_{i,a}, X - 2\Delta_i) & V_j(a) \\
+ 2v_{0,j}.T(K" - \Delta_{i,i} - \Delta_{0,j} + \Delta_{i,j}, X - 2\Delta_i) & V_j(0) \\
+ \displaystyle\sum_{1 \le a < b \le l} 2v_{j,a}.v_{j,b}.T(K" - \Delta_{i,i} - \Delta_{j,a} - \Delta_{j,b} + \Delta_{a,b}, X - 2\Delta_i) & V_j(a) \mid V_j(b) \\
+ \displaystyle\sum_{1 \le a \le l} 2v_{j,a}.v_{0,j}.T(K" - \Delta_{i,i} - \Delta_{0,j}, X - 2\Delta_i) & V_j(a) \mid V_j(0) \\
+ \displaystyle\sum_{\substack{1 \le a \le l \\ a \ne j}} v_{j,a}.(v_{j,a} - 1).T(K" - \Delta_{i,i} - 2\Delta_{j,a} + \Delta_{a,a}, X - 2\Delta_i) & V_j(a) \mid V_j(a) \\
+ v_{0,j}.(v_{0,j} - 1).T(K" - \Delta_{i,i} - 2\Delta_{0,j} + \Delta_{j,j}, X - 2\Delta_i) & V_j(0) \mid V_j(0) \\
+ v_{j,j}.(v_{j,j} - 2).T(K" - \Delta_{i,i} - \Delta_{j,j}, X - 2\Delta_i) & V_j(j) \mid V_j(j)
\end{array}
$$
$$(19)$$

Now, if we have $\forall k, x_k = 0$, then all the vertices have been considered and:

$$
T_{i,j}(G_1, K_1, K_2, 0) = \begin{cases} 1 \text{ if } K_1 = K_2 \\ 0 \text{ otherwise} \end{cases}
\tag{20}
$$

The table of all possible $T_{i,j}(K_1, K_2, X)$ is of size $l^2.n^{l(l+4)}$. Using the previous equations, we can compute the table by increasing X in $O(n^{l(l+4)})$ operations (individual equations are independent of n). We now have $N_{i,j}(K', K) = T_{i,j}(K', K, X)$ where $\forall k \ne i, x_k = k'_{i,k}$ and $x_i = 2k'_{i,i}$.

Hide-and-Seek: Algorithms for Polygon Walk Problems[*]

Atlas F. Cook IV[1], Chenglin Fan[2], and Jun Luo[2,**]

[1] Dept. of Computing Sciences, Utrecht University, Netherlands
atlas@cs.uu.nl
[2] Shenzhen Institutes of Advanced Technology, Chinese Academy of Sciences, China
{cl.fan,jun.luo}@siat.ac.cn

Abstract. Jack and Jill want to play hide-and-seek on the boundary of a simple polygon. Given arbitrary paths for the two children along this boundary, our goal is to determine whether Jack can walk along his path without ever being seen by Jill. To solve this problem, we use a linear-sized *skeleton invisibility diagram* to implicitly represent invisibility information between pairs of points on the boundary of the simple polygon. This structure has additional applications for any *polygon walk problem* where one entity wishes to remain hidden throughout a traversal of some path. We also show how Jack can avoid being seen not just by one moving child but by an arbitrary number of moving children.

1 Introduction

Our problem is to determine whether two children can play hide-and-seek along the boundary of a simple polygon. Specifically, if Jill moves with constant speed along a given path on this boundary, then we want to determine whether Jack can control his speed so that he reaches the end of his given path without being seen by Jill. Both children have excellent eyesight and can see infinitely far inside the simple polygon. However, neither child can see past any opaque segment that bounds the simple polygon.

We refer to a *polygon walk problem* as any situation where one or more entities traverse the boundary of a simple polygon while maintaining certain properties. These types of problems have been extensively studied. For example, Suzuki et al. [8] and Tan [9] show how to search a simple polygon office space for an intruder by using a mobile camera that is mounted on a cyclic track along the walls.

Related work by Icking and Klein [6] permits two guards to patrol the boundary of a simple polygon art gallery. Both guards begin patrolling from a single entrance point, and the two guards walk along the boundary. For security purposes, the guards always maintain mutual visibility. The goal is to determine whether the two guards can always detect an intruder if they begin their patrol at a given entrance point. Their solution to this problem requires $O(n \log n)$ time, where n is the number of simple polygon vertices. Heffernan [5] has improved this algorithm to optimal $O(n)$ time. Additionally, Zhang and Kameda [11] have given an $O(n)$ time algorithm that detects all possible

---

[*] This work is supported by Shenzhen Key Laboratory of High Performance Data Mining (grant no. CXB201005250021A).
[**] A previous version of this work has appeared as [2].

M. Ogihara and J. Tarui (Eds.): TAMC 2011, LNCS 6648, pp. 495–504, 2011.

entrance points on the boundary that permit the two guards to successfully detect an intruder.

Polygon walk problems with varying numbers of guards have also been studied. For example, LaValle et al. [7] show how to search a polygonal region with a single guard in $O(n^2)$ time. Efrat et al. [3] sweep a polygonal chain of guards along the boundary of a simple polygon in $O(n^3)$ time. Each consecutive pair of guards in their chain are always mutually visible, and the goal is to detect an intruder using the minimum number of guards.

Our problem is different from previous approaches because we do not try to maintain mutual visibility between moving entities. Instead, we seek to maintain mutual *invisibility* as two entities traverse paths on the boundary of a simple polygon. Note that our problem could be applied to any scenario where an individual needs to reach some destination without being seen. For example, a rabbit may want to reach its home without being seen by one or more wolves, a person who is late for work may wish to travel along roads without being seen by a police officer, and resources may need to be covertly delivered through hostile territory.

Results. Let P be a simple polygon with n vertices, and let π_{Jack} and π_{Jill} be two arbitrary paths on the boundary of P. Assume that Jill walks with constant speed from the start point of π_{Jill} to the end point of π_{Jill}. This paper describes an $O(n \log n)$ time and $O(n)$ space algorithm that determines whether it is possible for Jack to control his speed so that he walks from the start point to the end point of π_{Jack} without ever being seen by Jill. The algorithm can additionally report any existing traversal for Jack in output-sensitive time.

We also consider the following more general problem. Let π_{Jack} and $\pi_{c_1}, ..., \pi_{c_m}$ be arbitrary paths along the boundary of P. Assume that m children $c_1, ..., c_m$ all walk with constant speed along their respective paths. This paper describes an algorithm that determines whether Jack can control his speed so that he reaches the end point of his path without being seen by any child.

2   Preliminaries

Given a simple polygon P, two points p and q are *mutually visible* if the line segment $\overline{pq}$ is completely contained inside P; otherwise, p and q are *mutually invisible*. As in [7,11], we define a *configuration* $\langle p, q \rangle \in \partial P \times \partial P$ by a pair of points on the boundary of P. We call such a configuration *safe* when p and q are mutually invisible and *unsafe* otherwise. We pick an arbitrary point on the boundary ∂P as the origin, and we measure all distances along ∂P in clockwise order from this origin. Let $|\partial P|$ denote the total length of ∂P.

Let x_1, x_2, y_1, y_2 be points on ∂P such that Jack travels from x_1 to x_2 and Jill travels from y_1 to y_2 (see Figure 1a). Define a two-dimensional configuration space such that the x-axis represents the position of Jack on ∂P and the y-axis represents the position of Jill on ∂P (see Figure 1b). Observe that if Jack and Jill are located at the same point on ∂P, then they must be mutually visible. Consequently, every point on the lines $y = x$ and $y = x - |\partial P|$ in this configuration space is always unsafe. This permits us to prune our search space to the area between the lines $y = x$ and $y = x - |\partial P|$. We shade

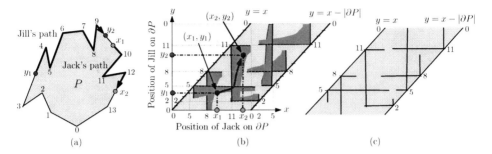

Fig. 1. (a) Simple polygon P, (b) Invisibility diagram of P, (c) Skeleton invisibility diagram of P. The numbers $0, 1, ..., 13$ represent the 14 vertices of P

all safe configurations in this search space a gray color and refer to this augmented space as an *invisibility diagram*. See Figure 1b. To speedup our algorithm, we will use an implicit *skeleton invisibility diagram* that represents all safe configurations as a discrete set of line segments. Although Zhang and Kameda [11] have also explored these diagrams, they were interested in searching a simple polygon while maintaining mutual visibility between a pair of guards. By contrast, we are interested in maintaining mutual invisibility.

2.1 Cells and Skeletons

We now describe all safe configurations as a discrete set of shapes in the invisibility diagram. To do this, it will be important to consider each reflex vertex of P. A *reflex vertex* is a vertex $r \in P$ whose interior angle inside P is at least π. For example, vertices $2, 5, 8$, and 11 in Figure 1a are reflex vertices. Each reflex vertex of P has the capability of blocking visibility inside P and will consequently be associated with a collection of safe configurations that we will refer to as a *cell*.

As illustrated in Figure 1b, each cell is bounded by a horizontal segment and a vertical segment that intersect on either $y = x$ or $y = x - |\partial P|$. We define the *skeleton* of a cell as the union of these horizontal and vertical segments. The *skeleton invisibility diagram* is simply the union of all skeletons in the invisibility diagram (see Figure 1c).

2.2 The Boundary of a Cell

This section shows that a cell in the invisibility diagram is always bounded by two horizontal segments, two vertical segments, and a monotone curve. In order to describe this boundary, it will be useful to think of the boundary of P as a clockwise sequence of vertices. The vertices immediately preceding and succeeding a vertex r on ∂P are denoted $Pred(r)$ and $Succ(r)$, respectively. Let $B(r) \in \partial P$ be the *backward extension point* where the extension of the edge from $Succ(r)$ to r leaves P for the first time. Let $F(r)$ be the *forward extension point* where the extension of the edge from $Pred(r)$ to r leaves P for the first time.

Figure 2 illustrates how to calculate the boundary of the cell for a reflex vertex $r \in P$. In Figure 2a, Jack moves from r to $Succ(r)$ while Jill stands still at $B(r)$.

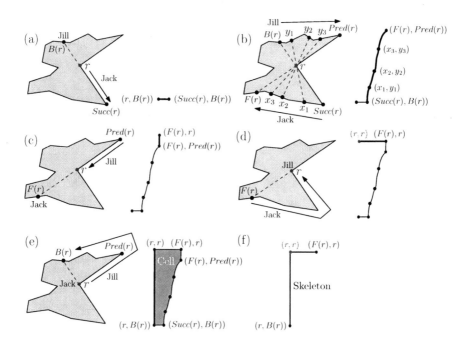

Fig. 2. A reflex vertex $r \in P$ defines one cell in the invisibility diagram

This scenario corresponds to a horizontal line segment in the invisibility diagram that has endpoints $(r, B(r))$ and $(Succ(r), B(r))$. The reason this line segment bounds a cell is because Jill clearly sees Jack throughout this walk; however, if Jill moves infinitesimally clockwise on ∂P, then she no longer sees Jack at any point except r.

In Figure 2b, Jack moves along multiple line segments from $Succ(r)$ to $F(r)$ while Jill moves from $B(r)$ to $Pred(r)$. Notice that all lines-of-sight from Jack to Jill that pass through r will define a boundary of the cell for r because an infinitesimal motion of either child can break visibility between the two children. Since all of these lines-of-sight can be obtained by rotating a line through r about the point r, the portion of the cell boundary from $(Succ(r), B(r))$ to $(F(r), Pred(r))$ will be a piecewise monotone curve with $O(n)$ vertices.

In Figure 2c, Jack stands still at $F(r)$ while Jill moves from $Pred(r)$ to r. This defines a vertical boundary segment for the cell because an infinitesimal movement of Jack can break visibility. In Figure 2d, Jack moves from $F(r)$ to r while Jill stands still at r. This defines a horizontal boundary segment for the cell. Figure 2e shows Jack standing still at r while Jill moves from r to $B(r)$. This defines a vertical boundary segment for the cell. Notice that the cell boundary is now a closed curve that is composed of two horizontal segments, two vertical segments, and a monotone set of curves. All points inside this cell represent safe points where Jack and Jill cannot see each other because the reflex vertex r blocks visibility between the two children.

Although the boundary of a cell has a computable structure, we will show that an implicit representation of a cell that we call a *skeleton* is often sufficient. Figure 2f illustrates that the *skeleton* of a cell consists of just two segments: the vertical boundary segment from $(r, B(r))$ to (r, r) and the horizontal boundary segment from (r, r) to $(F(r), r)$. We will sometimes refer to (r, r) as a *corner* point.

3   Properties of the Skeleton Invisibility Diagram

This section proves that a pair of cells intersect if and only if their skeletons intersect. This nice property will permit us to compute paths through the invisibility diagram without explicitly computing any cells.

The following notation will be useful. For any two points $a, b \in \partial P$, the open and closed portions of ∂P from a to b in clockwise order are, respectively, denoted by $\partial P(a, b)$ and $\partial P[a, b]$. We now show that two skeletons can intersect in at most one point.

Lemma 1. *A pair of skeletons have at most one intersection point.*

Proof. Let S and S' be a pair of skeletons with respective corner points (r, r) and (r', r') in the invisibility diagram. If both of these corners intersect the line $y = x$ (or both corners intersect the line $y = x - |\partial P|$), then one skeleton's vertical line segment must lie entirely to the left of the other skeleton. This scenario guarantees that there is at most one intersection point between S and S'. Please refer to Figure 3a.

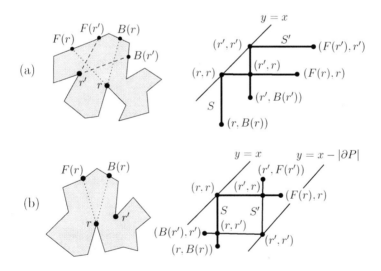

Fig. 3. A pair of skeletons have at most one intersection point

Suppose now that the corner of S lies on the line $y = x$ and the corner of S' lies on the line $y = x - |\partial P|$. Assume that the two skeletons S and S' could have two intersection points (r', r) and (r, r') as illustrated in Figure 3b. Notice that the y-coordinates in the skeletons appear from bottom-to-top as the sequence $B(r)$, r', r, $F(r')$. This means that these four points must appear in clockwise order along ∂P. In particular, r' must precede r in the clockwise ordering around ∂P. Furthermore, the x-coordinates in the skeletons appear from left-to-right as the sequence $B(r')$, r, r', $F(r)$. However, this implies that r must precede r' in the clockwise ordering around ∂P. This is a contradiction. Thus, a pair of skeletons have at most one intersection point.   □

We show next that a pair of cells intersect if and only if their skeletons intersect.

Lemma 2. *Two cells intersect if and only if their skeletons intersect.*

Proof. If two skeletons intersect, then the cells for these skeletons must also intersect. This follows because a cell is always a superset of its skeleton. We will now prove that if two cells intersect, then their skeletons must also intersect.

Recall that the corner point of every cell in the invisibility diagram must lie on either the line $y = x$ or the line $y = x - |\partial P|$. Let c_1 and c_2 be cells in the invisibility diagram, and assume that the boundaries of c_1 and c_2 intersect in at least one point (a, b). Let r_1 and r_2 be the reflex vertices of P that define these cells, and without loss of generality, assume that r_1 precedes r_2 in the clockwise ordering on ∂P. By construction, the point (a, b) in the invisibility diagram corresponds to a pair of points on the boundary of the simple polygon P. Both r_1 and r_2 must affect the visibility between a and b; therefore, all four points a, r_1, r_2, and b are collinear in P.

Assume for a moment that the two points $B(r_1) \in \partial P$ and $F(r_2) \in \partial P$ are on the same side of the line segment $\overline{ab}$ (see Figure 4a). This scenario ensures that no skeleton endpoint can appear between r_1 and r_2 in the clockwise ordering on ∂P. Thus, no skeleton endpoint can appear in the invisibility diagram for any $r_1 < x < r_2$ or $r_1 < y < r_2$. Consequently, the horizontal segment of the skeleton for r_1 must intersect the vertical segment of the skeleton for c_2.

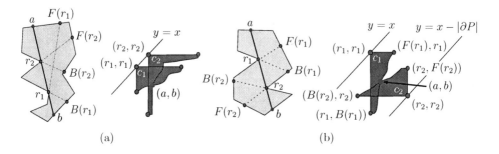

Fig. 4. Two cells intersect if and only if their skeletons intersect

Now suppose that the two points $B(r_1) \in \partial P$ and $F(r_2) \in \partial P$ are on opposite sides of the line segment $\overline{ab}$ (see Figure 4b). Since r_1 precedes r_2 in the clockwise ordering on ∂P, the point $(r, B(r_1)) \in c_1$ in the invisibility diagram is below the point $(r_2, r_2) \in c_2$ and the point $(B(r_2), r_2) \in c_2$ is to the left of the point $(r_1, r_1) \in c_1$. Consequently, the vertical segment of c_1's skeleton must intersect the horizontal segment of c_2's skeleton. This means that whenever two cells intersect, their skeletons must also intersect.   □

Recall that the start and end points of Jack are $x_1, x_2 \in \partial P$, and the start and end points of Jill are $y_1, y_2 \in \partial P$ (see Figure 1). Let $s = (x_1, y_1)$ and $t = (x_2, y_2)$ be points in the invisibility diagram. We define a *safe path* $\pi_{\text{safe}}(s,t)$ as any path that connects the points s and t in the invisibility diagram such that Jack and Jill do not see each other at any point along this path. This means that a safe path is entirely contained in the gray area that represents the union of the cells (see Figure 1b). Note that Jack can traverse his path without being seen by Jill if and only if a safe path $\pi_{\text{safe}}(s,t)$ exists in the invisibility diagram. Lemmas 1 and 2 permit us to determine whether a safe path exists without explicitly computing any cells. Instead, we only need to compute connected components of skeletons.

4   Computing a Safe Path for Jack and Jill

A safe path $\pi_{\text{safe}}(s,t)$ exists if and only if s and t lie in the same connected component of gray points in the invisibility diagram. To decide whether the points s and t are in the same connected component, we determine the skeleton that contains s and the skeleton that contains t. We then decide whether these two skeletons are in the same connected component of gray points.

4.1   Determining Skeletons for s and t

Recall that two points $x, y \in \partial P$ define a point (x, y) in the invisibility diagram. Our task is to determine one skeleton whose cell contains (x, y). To simplify this process, we define the visibility polygon $Vis(P, x)$ as the set of all points in P that are directly visible from x (see Figure 5).

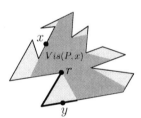

Fig. 5. Determining the skeleton of a cell that contains a point (x, y) in the invisibility diagram

Lemma 3. *Given any fixed point (x, y) in the invisibility diagram, the skeleton of a cell that contains (x, y) can be determined in $O(n)$ time.*

Proof. Compute the visibility polygon $Vis(P, x)$ in $O(n)$ time [4]. Traverse the boundary of P and determine the maximally connected interval of ∂P that contains y and is *not* directly visible from x (see Figure 5). If no such interval exists, then no cell contains (x, y), so we report that the point (x, y) is unsafe. Otherwise, one endpoint of this interval (e.g., r) must be a reflex vertex of P, and we associate (x, y) with the skeleton for this reflex vertex.   □

Lemma 3 implies that we can calculate in $O(n)$ time the skeleton of a cell that contains $s = (x_1, y_1)$ and the skeleton of a cell that contains $t = (x_2, y_2)$.

4.2   Determining Connected Components of Skeletons

Given the skeleton of a cell that contains s and the skeleton of a cell that contains t, we now wish to determine whether there exists a safe path between these two skeletons. Given the horizontal and vertical line segments defining all $O(n)$ skeletons in the skeleton invisibility diagram, we can compute the connected components of these orthogonal line segments in $O(n \log n)$ time and $O(n)$ space [10]. These connected components are useful because Lemma 2 ensures that a pair of cells intersect if and only if the skeletons for these cells intersect. Hence, a safe path will only exist from s to t when the skeleton for s is in the same connected component as the skeleton for t.

We can now state one of our main results.

Theorem 1. *Suppose Jack and Jill walk along fixed paths on the boundary of a simple polygon that has n vertices. We can decide whether Jack and Jill can simultaneously walk from their respective start points to their respective endpoints without ever seeing each other in $O(n \log n)$ time and $O(n)$ space.*

Proof. Compute the skeletons of the invisibility diagram in a preprocessing step as in [11], and compute the connected components for these skeletons [10]. Using Lemma 3, associate the start point s of Jack and Jill with the connected component of its associated skeleton. Now associate the end point t of Jack and Jill with the connected component of its associated skeleton. If these two connected components are the same, then a rectilinear path can be returned through the skeletons in this connected component. Otherwise, no safe path exists.   □

5   Hide-and-Seek with Multiple Children

This section describes an algorithm that determines whether Jack can traverse a path on the boundary of a simple polygon without being seen by any of m children $c_1, ..., c_m$. Recall that P is a simple polygon with n vertices and that ∂P is the boundary of P. Both Jack and the m children $c_1, ..., c_m$ will be traversing given paths along ∂P. Assume that each child c_i travels along his or her path with a constant velocity v_i and that all children (including Jack) begin moving at the same instant. The problem is to determine whether Jack can control his speed so that he reaches the end of his path without being seen by any child $c_1, ..., c_m$.

As in the previous sections, each pair of Jack with a child c_i defines a unique start point s_i in the invisibility diagram for P. Let L_e be the vertical line that is associated with the endpoint of Jack's path in the invisibility diagram.

In order to determine whether Jack can traverse his path without being seen by any child $c_1, ..., c_m$, we need to determine if a safe path exists from each start point s_i to the vertical line L_e. However, these safe paths cannot be computed independently because Jack's fixed velocity at any given x-coordinate affects the slopes of all m of these paths.

To keep things simple, assume for just a moment that each child moves with the same velocity (i.e., $v_1 = ... = v_m$) and that each child moves in a clockwise manner around ∂P. Create a copy of the invisibility diagram for each child $c_1, ..., c_m$. Denote these invisibility diagrams as $D_1, ..., D_m$, and associate the start point s_i with the invisibility diagram D_i. Translate each of the m invisibility diagrams so that all start points $s_1, ..., s_m$ lie at a common point $s \in \mathbb{R}^2$ after performing the translations. Let $\mathcal{I} = D_1 \cap ... \cap D_m$ be the intersection of the safe points in these translated invisibility diagrams. Note that Jack can traverse his path without being seen by any child $c_1, ..., c_m$ if and only if some path exists in $\mathcal{I}$ that connects s and the vertical line L_e that is associated with the endpoint of Jack's path in the invisibility diagram.

We now relax our assumptions so that each child is permitted to travel with a unique constant velocity in either the clockwise or counter-clockwise direction around ∂P. To permit children to move at different velocities, we vertically scale each invisibility diagram D_i by a factor of $\frac{1}{v_i}$, where v_i is the velocity of child c_i. Furthermore, we can vertically invert the invisibility diagram of any child that moves counter-clockwise along ∂P. This process of vertically inverting, vertically scaling, and translating each invisibility diagram has the effect of normalizing the invisibility diagrams $D_1, ..., D_m$. After this normalization, we can once again let $\mathcal{I}$ be the intersection of the safe points in these normalized invisibility diagrams. We can now state the following result.

Theorem 2. *Suppose Jack and m children walk along fixed paths on the boundary of a simple polygon that has n vertices. We can determine whether Jack can traverse his path without being seen by any child in $O(m^2 n^3 \log mn)$ time and $O(m^2 n^3)$ space.*

Proof. The invisibility diagram for a simple polygon with n vertices always has $O(n)$ cells, and each cell has $O(n)$ complexity. Since each cell is a monotone shape, a pair of cells intersect at most $O(n)$ times. More generally, the arrangement of m normalized invisibility diagrams $D_1, ..., D_m$ involves $O(mn)$ cells. Each of the $O(m^2 n^2)$ pairs of cells intersect $O(n)$ times, so the intersection $\mathcal{I}$ of the safe points in these normalized invisibility diagrams has $O(m^2 n^3)$ complexity. Any standard arrangement algorithm such as [1] can be used to construct the intersection $\mathcal{I}$ in $O(m^2 n^3 \log(mn))$ time.

Recall that L_e is the vertical line that is associated with the end point of Jack's path. To determine whether Jack can traverse his path without being seen by any child, we need to determine whether any path exists from the start point of Jack $s \in \mathcal{I}$ to any point on the vertical line L_e. This implies that a simple scan through the faces of $\mathcal{I}$ is sufficient to return a safe path for Jack or to report that no safe path exists.   □

6 Conclusion

This paper has shown how to determine whether two children can possibly traverse paths along the boundary of a simple polygon without ever seeing each other. We also explore the scenario where one child should avoid being seen by multiple children. Linear-sized structures are used to implicitly represent invisibility information between pairs of points on the boundary of a simple polygon. These structures have applications for any polygon walk problem where one entity wishes to remain hidden at all times during a traversal of a path. For example, resources may need to be covertly delivered to some destination in hostile territory. In the future, it might be interesting to consider the problem where the movements of the children along the boundary of the simple polygon are constrained by speed limits.

References

1. Agarwal, P.K., Sharir, M.: Davenport–Schinzel Sequences and Their Geometric Applications. In: Handbook of Computational Geometry, pp. 1–47. Elsevier, Amsterdam (2000)
2. Cook IV, A.F., Fan, C., Luo, J.: Hide-and-seek: A linear time algorithm for polygon walk problems. In: 26th European Workshop on Computational Geometry (EuroCG), pp. 121–124 (2010)
3. Efrat, A., Guibas, L.J., Har-Peled, S., Lin, D.C., Mitchell, J.S.B., Murali, T.M.: Sweeping simple polygons with a chain of guards. In: 11th Symposium on Discrete Algorithms (SODA), pp. 927–936 (2000)
4. Guibas, L.J., Hershberger, J., Leven, D., Sharir, M., Tarjan, R.E.: Linear time algorithms for visibility and shortest path problems inside simple polygons. In: 2nd Symposium on Computational Geometry (SoCG), pp. 1–13 (1986)
5. Heffernan, P.J.: An optimal algorithm for the two-guard problem. In: 9th Symposium on Computaitonal Geometry (SoCG), pp. 348–358 (1993)
6. Icking, C., Klein, R.: The two guards problem. In: 7th Symposium on Computational Geometry (SoCG), pp. 166–175 (1991)
7. LaValle, S.M., Simov, B.H., Slutzki, G.: An algorithm for searching a polygonal region with a flashlight. In: 16th Symposium on Computational Geometry (SoCG), pp. 260–269 (2000)
8. Suzuki, I., Tazoe, Y., Yamashita, M., Kameda, T.: Searching a polygonal region from the boundary. International Journal of Computational Geometry and Applications 11, 529–553 (2000)
9. Tan, X.: A characterization of polygonal regions searchable from the boundary. In: Akiyama, J., Baskoro, E.T., Kano, M. (eds.) IJCCGGT 2003. LNCS, vol. 3330, pp. 200–215. Springer, Heidelberg (2005)
10. Thurimella, R., Lopez, M.A.: On computing connected components of line segments. IEEE Transactions on Computers 44, 597–601 (1995)
11. Zhang, J.Z., Kameda, T.: A linear-time algorithm for finding all door locations that make a room searchable. In: Agrawal, M., Du, D.-Z., Duan, Z., Li, A. (eds.) TAMC 2008. LNCS, vol. 4978, pp. 502–513. Springer, Heidelberg (2008)

Linear-Time Algorithms for Graphs of Bounded Rankwidth: A Fresh Look Using Game Theory
(Extended Abstract)[*]

Alexander Langer, Peter Rossmanith, and Somnath Sikdar

RWTH Aachen University, 52074 Aachen, Germany

Abstract. We present an alternative proof of a theorem by Courcelle, Makowski and Rotics [6] which states that problems expressible in MSO_1 are solvable in linear time for graphs of bounded rankwidth. Our proof uses a game-theoretic approach and has the advantage of being self-contained. In particular, our presentation does not assume any background in logic or automata theory. Moreover our approach can be generalized to prove other results of a similar flavor, for example, that of Courcelle's Theorem for treewidth [3,19].

1 Introduction

In this paper we give an alternate proof of the theorem by Courcelle, Makowski and Rotics [6]: *Every decision or optimization problem expressible in MSO_1 is linear time solvable on graphs of bounded cliquewidth.* We prove the same theorem for graphs of bounded rankwidth. Since rankwidth and cliquewidth are equivalent width measures in the sense that a graph has bounded rankwidth iff it has bounded cliquewidth, it does not matter which of these width measures is used to state the theorem [21].

The proof by Courcelle et al. [6,7] makes use of the Feferman-Vaught Theorem [10] adapted to MSO (cf. [14,15]) and MSO transductions (cf., [4]). Understanding this proof requires a reasonable background in logic and as such this proof is out of reach of many practicing algorithmists. An alternative proof of this theorem has been recently published by Ganian and Hliněný [11] who use an automata-theoretic approach to prove the theorem. Our approach to proving this theorem is game-theoretic, an outline of which follows.

It is known that any graph of rankwidth t can be represented by a t-labeled parse tree [11]. Given any integer q, one can define an equivalence relation on the class of all t-labeled graphs as follows: t-labeled graphs G_1 and G_2 are equivalent, denoted $G_1 \equiv_q^{MSO} G_2$, iff for every MSO_1-formula of quantifier rank at most q $G_1 \models \varphi$ iff $G_2 \models \varphi$, i.e., no formula with at most q nested quantifiers can distinguish them. The number of equivalence classes depends on the quantifier rank q and the number of labels t and each equivalence class can be represented

---

[*] This work is supported by the Deutsche Forschungsgemeinschaft (DFG) under grant RO 927/8. A full version is available at http://arxiv.org/abs/1102.0908.

M. Ogihara and J. Tarui (Eds.): TAMC 2011, LNCS 6648, pp. 505–516, 2011.

by a tree-like structure of size $f(q, t)$, where f is a computable function of q and t only.

This tree-like representative of an equivalence class, called a *reduced characteristic tree of depth q* and denoted by $\mathrm{RC}_q(G)$, captures all model-checking games that can be played on graphs in that equivalence class and formulas of quantifier rank at most q (see Section 3). One can construct a reduced characteristic tree of depth q given a t-labeled parse tree of an n-vertex graph in time $O(f'(q, t) \cdot n)$ (Section 4). Finally to decide whether $G \models \varphi$, for some MSO_1-formula φ of quantifier rank at most q, we simply simulate the model checking game on φ and G using $\mathrm{RC}_q(G)$. This takes an additional $O(f(q, t))$ time and shows that one can decide whether $G \models \varphi$ in time $O(f''(q, t) \cdot n)$ proving the main theorem:

The Main Theorem ([6,11]). *Let φ be an MSO_1-formula with $\mathrm{qr}(\varphi) \leq q$. There is an algorithm that takes as input a t-labeled parse tree decomposition T of a graph G and decides whether $G \models \varphi$ in time $O(f(q, t) \cdot |T|)$, where f is some computable function and $|T|$ is the number of nodes in T.*

The notions of q-equivalence $\equiv_q^{\mathrm{MSO}}$ and related two-player pebble games (such as the Ehrenfeucht-Fraïssé game) are fundamental to finite model theory and can be found in any book on the subject (cf. [9]). However for understanding this paper, one does not need any prior knowledge of these concepts.

2   Preliminaries

Rankwidth was originally defined by Oum and Seymour in terms of *branchwidth* [22]. However this definition is not very useful from an algorithmic point-of-view and this prompted Courcelle and Kanté [5] to introduce the notion of bilinear products of multi-colored graphs and algebraic expressions over these products as an equivalent description of rankwidth. Ganian and Hliněný [11] formulated the same ideas in terms of labeling joins and parse trees which we briefly describe here.

t-labeled graphs. A *t-labeling* of a graph G is a mapping $lab \colon V(G) \to 2^{[t]}$ which assigns to each vertex of G a subset of $[t] = \{1, \ldots, t\}$. A *t-labeled graph* is a pair (G, lab), where lab is a labeling of G and is denoted by $\bar{G}$. Since a t-labeling function may assign the empty label to each vertex, an unlabeled graph is considered to be a t-labeled graph for all $t \geq 1$. A t-labeling of G may also be interpreted as a mapping from $V(G)$ to the t-dimensional binary vector space $\mathrm{GF}(2^t)$ by associating the subset $X \subseteq [t]$ with the t-bit vector $\mathbf{x} = x_1 \ldots x_t$, where $x_i = 1$ if and only if $i \in X$. Thus one can represent a t-labeling lab of an n-vertex graph as an $n \times t$ binary matrix.

A *t-relabeling* is a mapping $f \colon [t] \to 2^{[t]}$. One can also view a t-relabeling as a linear transformation from the space $\mathrm{GF}(2^t)$ to itself and one can therefore represent a t-relabeling by a $t \times t$ binary matrix T_f. For a t-labeled graph $\bar{G} = (G, lab)$, we define $f(\bar{G})$ to be the t-labeled graph $(G, f \circ lab)$, where $(f \circ lab)(v)$ is the vector in $\mathrm{GF}(2^t)$ obtained by applying the linear transformation f to the vector $lab(v)$. It is easy to see that the labeling $lab' = f \circ lab$ is the matrix product $lab \times T_f$. Informally, to calculate $(f \circ lab)(v)$, apply the map f to each element of $lab(v)$ and "sum the elements modulo 2".

We now define three operators on t-labeled graphs that will be used to define parse tree decompositions of t-labeled graphs. These operators were first described by Ganian and Hliněný in [11]. The first operator is denoted $\odot$ and represents a nullary operator that creates a new graph vertex with the label 1. The second operator is the t-labeled join and is defined as follows. Let $\bar{G}_1 = (G_1, lab_1)$ and $\bar{G}_2 = (G_2, lab_2)$ be t-labeled graphs. The t-labeled join of $\bar{G}_1$ and $\bar{G}_2$, denoted $\bar{G}_1 \otimes \bar{G}_2$, is defined as taking the disjoint union of G_1 and G_2 and adding all edges between vertices $u \in V(G_1)$ and $v \in V(G_2)$ such that $|lab_1(u) \cap lab_2(v)|$ is odd. The resulting graph is unlabeled.

Note that $|lab_1(u) \cap lab_2(v)|$ is odd if and only if the scalar product $lab_1(u) \bullet lab_2(v) = 1$, that is, the vectors $lab_1(u)$ and $lab_2(v)$ are *not orthogonal* in the space $\mathrm{GF}(2^t)$. For $X \subseteq V(G_1)$, the set of vectors $\gamma(\bar{G}_1, X) = \{ lab_1(u) \mid u \in X \}$ generates a subspace $\langle \gamma(\bar{G}_1, X) \rangle$ of $\mathrm{GF}(2^t)$. The following result shows which pair of vertex subsets do not generate edges in a t-labeled join operation.

Proposition 1 ([12]). *Let $X \subseteq V(G_1)$ and $Y \subseteq V(G_2)$ be arbitrary nonempty subsets of t-labeled graphs $\bar{G}_1$ and $\bar{G}_2$. In the join graph $\bar{G}_1 \otimes \bar{G}_2$ there is no edge between any vertex of X and a vertex of Y if and only if the subspaces $\langle \gamma(\bar{G}_1, X) \rangle$ and $\langle \gamma(\bar{G}_2, Y) \rangle$ are orthogonal in the vector space $\mathrm{GF}(2^t)$.*

The third operator is called the t-labeled composition and is defined using the t-labeled join and t-relabelings. Given three t-relabelings $g, f_1, f_2 \colon [t] \to 2^{[t]}$, the t-labeled composition $\otimes[g|f_1, f_2]$ is defined on a pair of t-labeled graphs $\bar{G}_1 = (G_1, lab_1)$ and $\bar{G}_2 = (G_2, lab_2)$ as follows:

$$\bar{G}_1 \otimes[g|f_1, f_2] \bar{G}_2 := \bar{H} = (\bar{G}_1 \otimes g(\bar{G}_2), lab),$$

where $lab(v) = f_i \circ lab_i(v)$ for $v \in V(G_i)$ and $i \in \{1, 2\}$. Thus the t-labeled composition first performs a t-labeling join of $\bar{G}_1$ and $g(\bar{G}_2)$ and then relabels the vertices of G_1 using f_1 and the vertices of G_2 with f_2. Note that a t-labeling composition is not commutative and that $\{u, v\}$ is an edge of $\bar{H}$ if and only if $lab_1(u) \bullet (lab_2(v) \times T_g) = 1$, where T_g is the matrix representing the linear transformation g.

Definition 1 (t-labeled Parse Trees). A t-labeled parse tree T is a finite, ordered, rooted subcubic tree (with the root of degree at most two) such that

1. all leaves of T are labeled with the $\odot$ symbol, and
2. all internal nodes of T are labeled with a t-labeled composition symbol.

A parse tree T generates the graph G that is obtained by the successive leaves-to-root application of the operators that label the nodes of T.

It is known that rankwidth can be defined using t-labeled parse trees.

Theorem 1 (The Rankwidth Parsing Theorem [5,11]). *A graph G has rankwidth at most t if and only if some labeling of G can be generated by a t-labeled parse tree. Moreover, a width-k rank-decomposition of an n-vertex graph can be transformed into a t-labeled parse tree on $\Theta(n)$ nodes in time $O(t^2 \cdot n^2)$.*

Monadic second-order logic (MSO) is an extension of first-order logic which allows quantification over sets of objects. We briefly fix notation, for details please refer to [9]. A *vocabulary* τ is a finite set of relation symbols $P, Q, R, \ldots$ each associated with a natural number known as its *arity*. A τ-*structure* $\mathscr{A}$ consists of a set A called the *universe* of $\mathscr{A}$ and a p-ary relation $R^{\mathscr{A}} \subseteq A^p$ for every p-ary relation symbol R in τ. Graphs can be expressed in a natural way as relational structures with universe the vertex set and a vocabulary consisting of a single binary (edge) relation symbol. To express a t-labeled graph G, we may use a vocabulary τ consisting of the binary relation symbol E (representing, as usual, the edge relation) and t unary relation symbols $L_1, \ldots, L_t$, where L_i represents the set of vertices labeled i.

The *quantifier rank* $\mathrm{qr}(\varphi)$ of a *formula* φ is the maximum number of nested quantifiers occurring in it. A variable in a formula is *free* if it is not within the scope of a quantifier. By $\mathrm{free}(\varphi)$ we denote the set of free variables of φ. A formula without free variables is called a *sentence*.

An *assignment* in $\mathscr{A}$ is a function α that assigns *values* to the free variables of φ. For a variable x and an assignment α, we let $\alpha[x/a]$ denote an assignment that agrees with α except that it assigns the value a to x. We write $\mathscr{A} \models \varphi[\alpha]$ if φ *holds* in $\mathscr{A}$, when the free variables of φ have been assigned the values given by α.

3   The $\equiv_q^{\mathrm{MSO}}$-Relation and Its Characterization

Given a vocabulary τ and a natural number q, one can define an equivalence relation on the class of τ-structures as follows. For τ-structures $\mathscr{A}$ and $\mathscr{B}$ and $q \in \mathbf{N}$, define $\mathscr{A} \equiv_q^{\mathrm{MSO}} \mathscr{B}$ (q-equivalence) if and only if $\mathscr{A} \models \varphi \Longleftrightarrow \mathscr{B} \models \varphi$ for all MSO sentences φ of quantifier rank at most q. In other words, two structures are q-equivalent if and only if no sentence of quantifier rank at most q can distinguish them. We provide a characterization of the relation $\equiv_q^{\mathrm{MSO}}$ using objects called characteristic trees of depth q. We show that two τ-structures $\mathscr{A}$ and $\mathscr{B}$ have identical characteristic trees of depth q if and only if $\mathscr{A} \equiv_q^{\mathrm{MSO}} \mathscr{B}$. We shall see that characteristic trees are specially useful because their size is "small" and for graphs of bounded rankwidth can be constructed efficiently given their parse tree decomposition. However before we can do that, we need a few definitions.

Definition 2 (Induced Structure and Sequence). Let $\mathscr{A}$ a τ-structure with universe A and let $\bar{c} = c_1, \ldots, c_m \in A^m$. The structure $\mathscr{A}' = \mathscr{A}[\bar{c}] = \mathscr{A}[\{c_1, \ldots, c_m\}]$ induced by $\bar{c}$ is a τ-structure with universe $A' = \{c_1, \ldots, c_m\}$ and interpretations $P^{\mathscr{A}'} := P^{\mathscr{A}} \cap \{c_1, \ldots, c_m\}^r$ for every relation symbol $P \in \tau$ of arity r. For an arbitrary sequence of objects $\bar{c} = c_1, \ldots, c_m$ and a set U, we let $\bar{c}[U]$ be the subsequence of $\bar{c}$ that contains only objects in U. For a sequence of sets $\bar{C} = C_1, \ldots, C_p$ we let $\bar{C} \cap A$ denote the sequence $C_1 \cap U, \ldots, C_p \cap U$ and write $\bar{C} \cap \bar{c}$ for $C_1 \cap \{c_1, \ldots, c_m\}, \ldots, C_p \cap \{c_1, \ldots, c_m\}$.

Definition 3 (Partial Isomorphism). Let $\mathscr{A}$ and $\mathscr{B}$ be structures over the vocabulary τ with universes A and B, respectively, and let π be a map such that $\mathrm{domain}(\pi) \subseteq A$ and $\mathrm{range}(\pi) \subseteq B$. The map π is said to be a *partial isomorphism* from $\mathscr{A}$ to $\mathscr{B}$ if (1) π is one-to-one and onto; and (2) for every p-ary relation symbol $R \in \tau$ and all $a_1, \ldots, a_p \in \mathrm{domain}(\pi)$, $R^{\mathscr{A}} a_1, \ldots, a_p$ iff $R^{\mathscr{B}} \pi(a_1), \ldots, \pi(a_p)$. If $\mathrm{domain}(\pi) = A$ and $\mathrm{range}(\pi) = B$, then π is an *isomorphism* between $\mathscr{A}$ and $\mathscr{B}$ and $\mathscr{A}$ and $\mathscr{B}$ are *isomorphic*.

Let $(\mathscr{A}, \bar{A})$ and $(\mathscr{B}, \bar{B})$ be tuples, where $\bar{A} = A_1, \ldots, A_s$ and $\bar{B} = B_1, \ldots, B_s$, $s \geq 0$, such that for all $1 \leq i \leq s$, we have $A_i \subseteq A$ and $B_i \subseteq B$. We say that π is a partial isomorphism between $(\mathscr{A}, \bar{A})$ and $(\mathscr{B}, \bar{B})$ if (1) π is a partial isomorphism between $\mathscr{A}$ and $\mathscr{B}$, and (2) for each $a \in \mathrm{domain}(\pi)$ and all $1 \leq i \leq s$, it holds that $a \in A_i$ iff $\pi(a) \in B_i$. The tuples $(\mathscr{A}, \bar{A})$ and $(\mathscr{B}, \bar{B})$ are *isomorphic* if π is an isomorphism between $\mathscr{A}$ and $\mathscr{B}$.

In Definition 2 of an induced structure we ignore the order of the elements in $\bar{c}$. For the purposes in this paper, the order in which the elements are chosen is important because it is used to map variables in the formula to elements in the structure. Moreover, elements could repeat in the vector $\bar{c}$ and this fact is lost when we consider the induced structure $\mathscr{A}[\bar{c}]$. To capture both the order and the multiplicity of the elements in vector $\bar{c}$ in the structure $\mathscr{A}[\bar{c}]$, we introduce the notion of an *ordered induced structure*.

Let U be a set and $\equiv$ be an equivalence relation on U. For $u \in U$, we let $[u]_\equiv = \{ u' \in U \mid u \equiv u' \}$ be the *equivalence class* of u under $\equiv$, and $U/\equiv = \{ [u]_\equiv \mid u \in U \}$ be the *quotient space* of U under $\equiv$. A vector $\bar{c} = c_1, \ldots, c_m \in A^m$ defines a natural equivalence relation $\equiv_{\bar{c}}$ on the set $[m] = \{1, \ldots, m\}$: for $i, j \in [m]$, we have $i \equiv_{\bar{c}} j$ if and only if $c_i = c_j$. For simplicity, we write $[i]_{\bar{c}}$ for $[i]_{\equiv_{\bar{c}}}$.

Definition 4 (Ordered Induced Structure). Let $\mathscr{A}$ be a τ-structure and $\bar{c} = c_1, \ldots, c_m \in A^m$. The *ordered structure induced by* $\bar{c}$ is the τ-structure $\mathscr{H} = \mathrm{Ord}(\mathscr{A}, \bar{c})$ with universe $H = [m]/\equiv_{\bar{c}}$ such that the map $h \colon c_i \mapsto [i]_{\bar{c}}$, $1 \leq i \leq m$, is an isomorphism between $\mathscr{A}[\bar{c}]$ and $\mathscr{H}$. Thus $\mathrm{Ord}(\mathscr{A}, \bar{c})$ is simply the structure $\mathscr{A}[\bar{c}]$ with element c_i being called $[i]_{\bar{c}}$. Let $\bar{C} = C_1, \ldots, C_p$ with $C_i \subseteq A$, $1 \leq i \leq p$. Then we let $\mathrm{Ord}(\mathscr{A}, \bar{c}, \bar{C}) := \big(\mathrm{Ord}(\mathscr{A}, \bar{c}), \bar{h}, h(\bar{C} \cap \bar{c}) \big)$, where $h \colon c_i \mapsto [i]_{\bar{c}}$, $1 \leq i \leq m$, $\bar{h} = h(1), \ldots, h(m)$ and $h(\bar{C} \cap \bar{c}) = h(C_1 \cap \bar{c}), \ldots, h(C_p \cap \bar{c})$. See Figure 1.

3.1   Model Checking Games and Characteristic Trees

Testing whether a non-empty structure models a formula can be specified by a *model checking game* (also known as *Hintikka game*, see [16,13]). Let $\mathscr{A}$ be a τ-structure with universe A. Let φ be a formula and α be an assignment to the free variables of φ. The game is played between two players called the *verifier* and the *falsifier*. The verifier tries to prove that $\mathscr{A} \models \varphi[\alpha]$ whereas the falsifier tries to disprove this. We assume without loss of generality that φ is in negation normal form, i.e., negations in φ appear only at the atomic level. This can always be achieved by applying simple rewriting rules such as $\neg \forall x \varphi(x) \rightsquigarrow \exists x \neg \varphi(x)$. The model checking game $\mathcal{MC}(\mathscr{A}, \varphi, \alpha)$ is positional with positions (ψ, β), where ψ is a subformula of φ and β is an assignment to the free variables of ψ. The game starts at position (φ, α). At a position $(\forall X \psi(X), \beta)$, the falsifier chooses

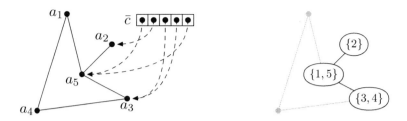

Fig. 1. The vector $\bar{c} = a_5 a_2 a_3 a_3 a_5$ lists vertices in the graph $\mathscr{G}$ on the left. The resulting ordered induced structure $\mathrm{Ord}(\mathscr{G}, \bar{c})$ is depicted in black on the right.

a subset $D \subseteq A$, and the game continues at position $(\psi, \beta[X/D])$. Similarly, at a position $(\forall x \psi(x), \beta)$ or $(\psi_1 \wedge \psi_2, \beta)$, the falsifier chooses an element $d \in A$ or some $\psi := \psi_i$ for some $1 \leq i \leq 2$ and the game then continues at position $(\psi, \beta[x/d])$ or (ψ, β), respectively. The verifier moves analogously at existential formulas. If an element is chosen then the move is called a *point move*; if a set is chosen then the move is a *set move*. The game ends once a position (ψ, β) is reached, such that ψ is an atomic or negated formula. The verifier *wins* if and only if $\mathscr{A} \models \psi[\beta]$. We say that the verifier has a *winning strategy* if they win every play of the game irrespective of the choices made by the falsifier. It is well known that the model checking game characterizes the satisfaction relation $\models$. The following lemma can easily be shown by induction over the structure of φ.

Lemma 1 (cf., [13]). *Let $\mathscr{A}$ be a τ-structure, let φ be an MSO formula, and let α be an assignment to the free variables of φ. Then $\mathscr{A} \models \varphi[\alpha]$ if and only if the verifier has a winning strategy on the model checking game on $\mathscr{A}$, φ, and α.*

A model checking game on a τ-structure $\mathscr{A}$ and a formula φ with quantifier rank q can be represented by a tree of depth q in which the nodes represent positions in the game and the edges represent point and set moves made by the players. Such a tree is called a *game tree* and is used in combinatorial game theory for analyzing games (see [2], for instance). For our purposes, we define a notion related to game trees called *full characteristic trees* which are finite rooted trees, where the nodes represent positions and edges represent moves of the game. A node is a tuple that represents the sets and elements that have been chosen thus far. The node can be thought of as a succinct representation of the state of the game played till the position represented by that node. However, note that a full characteristic tree depends on the quantifier rank q and *not* on a particular formula.

Definition 5 (Full Characteristic Trees). Let $\mathscr{A}$ be a τ-structure with universe A and let $q \in \mathbf{N}$. For elements $\bar{c} = c_1, \ldots, c_m \in A^m$, sets $\bar{C} = C_1, \ldots, C_p$ with $C_i \subseteq A$, $1 \leq i \leq p$, let $T = \mathrm{FC}_q(\mathscr{A}, \bar{c}, \bar{C})$ be a finite rooted tree such that (1) $\mathrm{root}(T) = (\mathscr{A}[\bar{c}], \bar{c}, \bar{C} \cap \bar{c})$, and (2) if $m + p + 1 \leq q$ then the subtrees of the root of $\mathrm{FC}_q(\mathscr{A}, \bar{c}, \bar{C})$ is the set $\{ \mathrm{FC}_q(\mathscr{A}, \bar{c}d, \bar{C}) \mid d \in A \} \cup \{ \mathrm{FC}_q(\mathscr{A}, \bar{c}, \bar{C}D) \mid D \subseteq A \}$.

The *full characteristic tree of depth q* for $\mathscr{A}$, denoted by $\mathrm{FC}_q(\mathscr{A})$, is defined as $\mathrm{FC}_q(\mathscr{A}, \varepsilon, \varepsilon)$, where ε is the empty sequence.

Let $T = (V, E)$ be a rooted tree. We let $\mathrm{root}(T)$ be the root of T and for $u \in V$ we let $\mathrm{children}_T(u) = \{ v \in V \mid (u, v) \in E \}$ and $\mathrm{subtree}_T(u)$ be a subtree of T rooted at u, and $\mathrm{subtrees}(T) = \{ \mathrm{subtree}_T(u) \mid u \in \mathrm{children}_T(\mathrm{root}(T)) \}$.

We now define a model checking game $\mathcal{MC}(F, \varphi, \bar{x}, \bar{X})$ on full characteristic trees $F = \mathrm{FC}_q(\mathscr{A}, \bar{c}, \bar{C})$ and formulas φ with $\mathrm{qr}(\varphi) \leq q$, where $\bar{x} = x_1, \ldots, x_m$ are the free object variables of φ, $\bar{X} = X_1, \ldots, X_p$ are the free set variables of φ, $\bar{c} = c_1, \ldots, c_m \in A^m$, and $\bar{C} = C_1, \ldots, C_p$ with $C_i \subseteq A$, $1 \leq i \leq p$. The rules are similar to the classical model checking game $\mathcal{MC}(\mathscr{A}, \varphi, \alpha)$. The game is positional and played by two players called the *verifier* and the *falsifier* and is defined over subformulas ψ of φ. However instead of choosing sets and elements explicitly, the tree F is traversed top-down. At the same time, we "collect" the list of variables the players encountered, such that we can make the assignment explicit once the game ends. The game starts at the position $(\varphi, \bar{x}, \bar{X}, \mathrm{root}(F))$. Let $(\psi, \bar{y}, \bar{Y}, v)$ be the position at which the game is being played, where $v = (\mathscr{H}, \bar{d}, \bar{D})$ is a node of $\mathrm{FC}_q(\mathscr{A}, \bar{c}, \bar{C})$, and ψ is a subformula of φ with $\mathrm{free}(\psi) = \bar{y} \cup \bar{Y}$. At a position $(\forall X \vartheta(X), \bar{y}, \bar{Y}, v)$ the falsifier chooses a child $u = (\mathscr{H}, \bar{d}, \bar{D}D)$ of v, where $D \subseteq A$, and the game continues at position $(\vartheta, \bar{y}, \bar{Y}X, u)$. Similarly, at a position $(\forall x \vartheta(x), \bar{y}, \bar{Y}, v)$ the falsifier chooses a child $u = (\mathscr{H}', \bar{d}d, \bar{D})$, where $d \in A$, and the game continues in $(\vartheta, \bar{y}x, \bar{Y}, u)$, and at a position $(\vartheta_1 \wedge \vartheta_2, \bar{y}, \bar{Y}, v)$, the falsifier chooses some $1 \leq i \leq 2$, and the game continues at position $(\vartheta_i, \bar{y}, \bar{Y}, v)$. The verifier moves analogously at existential formulas.

The game stops once an atomic or negated formula has been reached. Suppose that a particular play of the game ends at a position $(\psi, \bar{y}, \bar{Y}, v)$, where ψ is a negated atomic or atomic formula with $\mathrm{free}(\psi) = \{y_1, \ldots, y_s, Y_1, \ldots, Y_t\}$ and $v = (\mathscr{H}, \bar{d}, \bar{D})$ some node of F, where $\bar{d} = d_1, \ldots, d_s$ and $\bar{D} = D_1, \ldots, D_t$. Let α be an assignment to the free variables of φ, such that $\alpha(y_i) = d_i$, $1 \leq i \leq s$, and $\alpha(Y_i) = D_i$, $1 \leq i \leq t$. The verifier *wins* the game if and only if $\mathscr{H} \models \psi[\alpha]$. The verifier has a *winning strategy* if and only if they can win every play of the game irrespective of the choices made by the falsifier. In what follows, we identify a position $(\psi, \bar{y}, \bar{Y}, v)$ of the game $\mathcal{MC}(\mathrm{FC}_q(\mathscr{A}, \bar{c}, \bar{C}), \varphi, \bar{x}, \bar{X})$, where $v = (\mathscr{H}, \bar{d}, \bar{D})$, with the game $\mathcal{MC}(\mathrm{FC}_q(\mathscr{A}, \bar{d}, \bar{D}), \psi, \bar{y}, \bar{Y})$.

Lemma 2. *Let $\mathscr{A}$ be a τ-structure and let φ be an MSO formula with $\mathrm{qr}(\varphi) \leq q$ and free variables $\{x_1, \ldots, x_m, X_1, \ldots, X_m\}$. Let α be an assignment to the free variables of φ. Then the verifier has a winning strategy in the model checking game $\mathcal{MC}(\mathscr{A}, \varphi, \alpha)$ if and only if the verifier has a winning strategy in the model checking game $\mathcal{MC}(\mathrm{FC}_q(\mathscr{A}, \bar{c}, \bar{C}), \varphi, \bar{x}, \bar{X})$, where $\bar{c} = \alpha(x_1), \ldots, \alpha(x_m)$ and $\bar{C} = \alpha(X_1), \ldots, \alpha(X_p)$.*

Lemma 2 is shown by simulating each play of the model checking game $\mathcal{MC}(\mathscr{A}, \varphi, \alpha)$ in $\mathcal{MC}(\mathrm{FC}_q(\mathscr{A}, \bar{c}, \bar{C}), \varphi, \bar{x}, \bar{X})$ and vice versa. Therefore, a full characteristic tree of depth q for a structure $\mathscr{A}$ can be used to simulate the model checking game on $\mathscr{A}$ and any formula φ of quantifier rank at most q. However the size of such a tree is of the order $(2^n + n)^q$, where n is the number of elements in the universe of $\mathscr{A}$. We now show that one can "collapse" equivalent branches of a full characteristic tree to obtain a much smaller labeled tree (called a reduced characteristic tree) that is in some sense equivalent to the original (full) tree. We will then show that for a graph G of rankwidth at most t, the reduced characteristic tree of G is efficiently computable given a t-labeled parse tree decomposition of G. We achieve this collapse by replacing the induced structures $\mathscr{A}[\bar{c}]$ in the full characteristic tree by a more generic, implicit representation — that of their ordered induced substructures $\mathrm{Ord}(\mathscr{A}, \bar{c})$.

Definition 6 (Reduced Characteristic Trees). Let $\mathscr{A}$ be a τ-structure and let $q \in \mathbf{N}$. For elements $\bar{c} = c_1, \ldots, c_m \in A^m$, sets $\bar{C} = C_1, \ldots, C_p$ with $C_i \subseteq A$, $1 \leq i \leq p$, we let $\mathrm{RC}_q(\mathscr{A}, \bar{c}, \bar{C})$ be a finite rooted tree such that (1) the root of $\mathrm{RC}_q(\mathscr{A}, \bar{c}, \bar{C})$ is $\mathrm{Ord}(\mathscr{A}, \bar{c}, \bar{C})$, and (2) if $m + p + 1 \leq q$ then the subtrees of the root of $\mathrm{RC}_q(\mathscr{A}, \bar{c}, \bar{C})$ is the set $\{\, \mathrm{RC}_q(\mathscr{A}, \bar{c}d, \bar{C}) \mid d \in A \,\} \cup \{\, \mathrm{RC}_q(\mathscr{A}, \bar{c}, \bar{C}D) \mid D \subseteq A \,\}$. The *reduced characteristic tree of depth q* for the structure $\mathscr{A}$, denoted by $\mathrm{RC}_q(\mathscr{A})$, is defined to be $\mathrm{RC}_q(\mathscr{A}, \varepsilon, \varepsilon)$, where ε is the empty sequence.

One can define the model checking game $\mathcal{MC}(R, \varphi, \bar{x}, \bar{X})$ on a tree $R = \mathrm{RC}_q(\mathscr{A}, \bar{c}, \bar{C})$ in exactly the same manner as $\mathcal{MC}(\mathrm{FC}_q(\mathscr{A}, \bar{c}, \bar{C}), \varphi, \bar{x}, \bar{X})$. As mentioned before, our interest in $\mathrm{RC}_q(\mathscr{A}, \bar{c}, \bar{C})$ lies in that: (1) they are equivalent to $\mathrm{FC}_q(\mathscr{A}, \bar{c}, \bar{C})$; (2) they are "small"; and, (3) they are efficiently computable if $\mathscr{A}$ is a graph of rankwidth at most t. We first show that $\mathrm{RC}_q(\mathscr{A}, \bar{c}, \bar{C})$ is equivalent to its full counterpart $\mathrm{FC}_q(\mathscr{A}, \bar{c}, \bar{C})$.

Lemma 3. *Let $\mathscr{A}$ be a τ-structure and let $q \in \mathbf{N}$. Let $\bar{c} = c_1, \ldots, c_m \in A^m$ and $\bar{C} = C_1, \ldots, C_p$ with $C_i \subseteq A$, $1 \leq i \leq p$. Let $F = \mathrm{FC}_q(\mathscr{A}, \bar{c}, \bar{C})$ and $R = \mathrm{RC}_q(\mathscr{A}, \bar{c}, \bar{C})$. Then the verifier has a winning strategy in the model checking game $\mathcal{MC}(F, \varphi, \bar{x}, \bar{X})$ if and only if the verifier has a winning strategy in the game $\mathcal{MC}(R, \varphi, \bar{x}, \bar{X})$, where $\varphi \in \mathrm{MSO}(\tau)$ with $\mathrm{qr}(\varphi) \leq q$ with free object variables $\bar{x} = x_1, \ldots, x_m$ and free set variables $\bar{X} = X_1, \ldots, X_p$.*

From Lemmas 1, 2, and 3, we obtain the important fact that reduced characteristic trees are in fact equivalent to their full counterparts and characterize the equivalence relation $\equiv_q^{\mathrm{MSO}}$.

Corollary 1. *Let $\mathscr{A}$ and $\mathscr{B}$ be τ-structures and $q \in \mathbf{N}$. Then $\mathrm{RC}_q(\mathscr{A}) = \mathrm{RC}_q(\mathscr{B})$ iff $\mathscr{A} \equiv_q^{\mathrm{MSO}} \mathscr{B}$.*

The next lemma shows that reduced characteristic trees have small size. For $i \in \mathbf{N}$, we define $\exp^{(i)}(\cdot)$ as: $\exp^{(0)}(x) = x$, $\exp^{(1)}(x) = 2^x$ and $\exp^{(i)}(x) = 2^{2\exp^{(i-1)}(x)}$ for $i \geq 2$.

Lemma 4. *Let $\mathscr{A}$ be a τ-structure with universe A such that each relation symbol in τ has arity at most r, and $q \in \mathbf{N}$. Then the number of reduced characteristic trees $\mathrm{RC}_q(\mathscr{A}, \bar{c}, \bar{C})$ for all possible choices of $\bar{c}, \bar{C}$ is at most $\exp^{(q+1)}(|\tau| \cdot q^r + q \log q + q^2)$. The size of a reduced characteristic tree $\mathrm{RC}_q(\mathscr{A}, \bar{c}, \bar{C})$ is at most $(\exp^{(q)}(|\tau| \cdot q^r + q \log q + q^2))^4$.*

4   Constructing Characteristic Trees

In this section, we show how to construct reduced characteristic trees of depth q for a graph G of rankwidth t when given a t-labeled parse tree decomposition of G. A t-labeled graph may be represented as τ-structure where $\tau = \{E, L_1, \ldots, L_t\}$. The symbol E is a binary relation symbol representing the edge relation and L_i for $1 \leq i \leq t$ is a unary relation symbol representing the set of vertices with label i. In what follows, whenever we talk about a τ-structure $\mathscr{A}$, we mean a graph viewed as a structure over the vocabulary $\{E, L_1, \ldots, L_t\}$.

Lemma 5. *Let $\mathscr{A}$ be a τ-structure with $|A| = 1$. Let $q \geq 0$ and $\bar{c} \in A^m$ and $\bar{C} = C_1, \ldots, C_p$ with $C_i \subseteq A$, $1 \leq i \leq p$. Then $\mathrm{RC}_q(\mathscr{A}, \bar{c}, \bar{C})$ can be constructed in constant time for each fixed q.*

In what follows, we let $\mathscr{A}_1, \mathscr{A}_2$ and $\mathscr{A} = \mathscr{A}_1 \otimes \mathscr{A}_2$ be τ-structures, where $\otimes = \otimes[g|f_1, f_2]$ for t-relabelings g, f_1, and f_2. Recall that if $\mathscr{A} = \mathscr{A}_1 \otimes \mathscr{A}_2$, then we assume that A_1 and A_2 (the universes of $\mathscr{A}_1$ and $\mathscr{A}_2$, respectively) are disjoint. Furthermore for a fixed constant $q \geq 0$, let m and p be nonnegative integers such that $m + p \leq q$, $\bar{c} = c_1, \ldots, c_m \in (A_1 \cup A_2)^m$ and $\bar{C} = C_1, \ldots, C_p$, where $C_j \subseteq A_1 \cup A_2$, $1 \leq j \leq p$. For $i \in \{1, 2\}$, we let $\bar{c}_i = c_{i,1}, \ldots, c_{i,m_i} = \bar{c}[A_i]$.

In the remainder of this section, we show how to construct $\mathrm{RC}_q(\mathscr{A}, \bar{c}, \bar{C})$ given $\mathrm{RC}_q(\mathscr{A}_1, \bar{c}_1, \bar{C} \cap \bar{c}_1)$ and $\mathrm{RC}_q(\mathscr{A}_2, \bar{c}_2, \bar{C} \cap \bar{c}_2)$. For the construction, we need to know the order in which the elements in $\bar{c}_1$ and $\bar{c}_2$ appear in $\bar{c}$. This motivates us to define the notion of an *indicator vector* $\mathrm{ind}(A_1, A_2, \bar{c})$.

Definition 7. The *indicator vector* of $\bar{c} = c_1, \ldots, c_m$, denoted $\mathrm{ind}(A_1, A_2, \bar{c})$, is the vector $\bar{d} = d_1, \ldots, d_m$, such that for $i \in \{1, 2\}$ and all $1 \leq j \leq m$ it holds that $d_j = (i, k)$ iff $c_j = c_{i,k}$. That is, $d_j = (i, k)$ iff c_j is the kth element in the vector $\bar{c}_i = \bar{c}[A_i]$. If $\bar{d} = d_1, \ldots, d_m$ and $(i, k) \in \{1, 2\} \times [m + 1]$, then we use $\bar{d}(i, k)$ to denote the vector $d_1, \ldots, d_{m+1}$, where $d_{m+1} = (i, k)$.

Constructing $R = \mathrm{RC}_q(\mathscr{A}, \bar{c}, \bar{C})$ when given $R_1 = \mathrm{RC}_q(\mathscr{A}_1, \bar{c}_1, \bar{C} \cap \bar{c}_1)$, $R_2 = \mathrm{RC}_q(\mathscr{A}_2, \bar{c}_2, \bar{C} \cap \bar{c}_2)$, and $\bar{d} = \mathrm{ind}(A_1, A_2, \bar{c})$ consists of the following two steps: First construct the label for $\mathrm{root}(R) = \mathrm{Ord}(\mathscr{A}, \bar{c}, \bar{C})$, and then recursively construct its subtrees. Since $\mathrm{Ord}(\mathscr{A}, \bar{c}) \cong \mathscr{A}[\bar{c}]$ and $\mathscr{A}_i[\bar{c}_i] \cong \mathrm{Ord}(\mathscr{A}_i, \bar{c}_i)$, one easily sees that $\mathrm{Ord}(\mathscr{A}, \bar{c}) \cong \mathrm{Ord}(\mathscr{A}_1, \bar{c}_1) \otimes \mathrm{Ord}(\mathscr{A}_2, \bar{c}_2)$. For the first step, we therefore just need to rename elements in $\mathrm{Ord}(\mathscr{A}_1, \bar{c}_1) \otimes \mathrm{Ord}(\mathscr{A}_2, \bar{c}_2)$ in an appropriate way. The information on how elements are to be renamed is stored in the indicator vector $\bar{d}$ of $\bar{c}$. The formal definition of the renaming operator $\otimes_{\bar{d}}$ and Lemma 6 are technical and may be skipped if the reader believes that one can construct $\mathrm{Ord}(\mathscr{A}, \bar{c})$ from $\mathrm{Ord}(\mathscr{A}_1, \bar{c}_1)$ and $\mathrm{Ord}(\mathscr{A}_2, \bar{c}_2)$ using $\bar{d}$.

Definition 8. For $i \in \{1, 2\}$, let $\mathrm{Ord}(A_i, \bar{c}_i, \bar{C} \cap A_i) = (\mathscr{H}_i, \bar{c}'_i, \bar{C}'_i)$. Define a map $f \colon [m] \to H_1 \uplus H_2$ as follows: for all $1 \leq j \leq m$, let $f(j) = [k]_{\bar{c}_i}$ iff $d_j = (i, k)$. Then we define $\mathrm{Ord}(\mathscr{A}_1, \bar{c}[A_1], \bar{C} \cap A_1) \otimes_{\bar{d}} \mathrm{Ord}(\mathscr{A}_2, \bar{c}[A_2], \bar{C} \cap A_2)$ as $\mathrm{Ord}(\mathscr{H}_1 \otimes \mathscr{H}_2, f(1) \ldots f(m), \bar{C}'_1 \cup \bar{C}'_2)$.

Lemma 6. *Let $\mathscr{A}_1$ and $\mathscr{A}_2$ be τ-structures and let $\otimes = \otimes[g|f_1, f_2]$ for some t-relabelings g, f_1, f_2. Let $\bar{c} = c_1, \ldots, c_m \in (A_1 \cup A_2)^m$ and $\bar{C} = C_1, \ldots, C_p$, where $C_j \subseteq A_1 \cup A_2$ for $1 \leq j \leq p$. Also let $\bar{d} = \mathrm{ind}(A_1, A_2, \bar{c})$. Then $\mathrm{Ord}(\mathscr{A}_1 \otimes \mathscr{A}_2, \bar{c}, \bar{C}) = \mathrm{Ord}(\mathscr{A}_1, \bar{c}[A_1], \bar{C} \cap A_1) \otimes_{\bar{d}} \mathrm{Ord}(\mathscr{A}_2, \bar{c}[A_2], \bar{C} \cap A_2)$.*

We now define the *tree cross product* $R_1 \times (q, \otimes, \bar{d}) R_2$ of R_1 and R_2 and then show that in fact $R = R_1 \times (q, \otimes, \bar{d}) R_2$. As motivated before, the root of the tree cross product is simply $\mathrm{root}(R_1) \otimes_{\bar{d}} \mathrm{root}(R_2)$. For the construction of the subtrees, recall that each subtree of R corresponds to either a set move $U \subseteq A$ or a point move $a \in A$. Here, $\{U \subseteq A\} = \{U_1 \uplus U_2 \mid U_1 \subseteq A_1, U_2 \subseteq A_2\}$ and $A = A_1 \uplus A_2$. We can therefore reconstruct the subtrees of R by recursively combining each subtree for a set $U_1 \subseteq A_1$ with a subtree for a set $U_2 \subseteq A_2$ (the set S_2 in the following definition), and by choosing subtrees of R_1 for point moves in A_1, and choosing subtrees of R_2 for point moves in A_2 (the set S_1 in the following definition).

Definition 9 (Tree Cross Product). Let $\mathscr{A}_1$ and $\mathscr{A}_2$ be τ-structures and let $\otimes = \otimes[g|f_1, f_2]$ for some t-relabelings g, f_1, f_2. For a fixed constant $q \geq 0$, let m and p be nonnegative integers such that $m + p \leq q$. Let $\bar{c} = c_1, \ldots, c_m \in (A_1 \cup A_2)^m$ and $\bar{C} = C_1, \ldots, C_p$, where $C_j \subseteq A_1 \cup A_2$, $1 \leq j \leq p$. For $i \in \{1, 2\}$, let $\bar{c}_i = c_{i,1}, \ldots, c_{i,m_i} = \bar{c}[A_i]$, $q_i \geq q - m - p$, and $R_i = \mathrm{RC}_{q_i}(\mathscr{A}_i, \bar{c}_i, \bar{C} \cap A_i)$ with $\mathrm{root}(R_i) = (\mathscr{H}_i, \bar{c}_i', \bar{C}_i') = \mathrm{Ord}(A_i, \bar{c}_i, \bar{C} \cap A_i)$. We define the *tree cross product* of R_1 and R_2, $R = R_1 \times (q, \otimes, \bar{d}) R_2$, be a finite, rooted tree such that (1) $\mathrm{root}(R) = \mathrm{root}(R_1) \otimes_{\bar{d}} \mathrm{root}(R_2)$, and (2) if $m + p + 1 \leq q$, then $\mathrm{subtrees}(R) = S_1 \cup S_2$, where

$$S_1 = \left\{ \mathrm{subtree}_{R_1}(u_1) \times (q, \otimes, \bar{d}(1, m_1 + 1)) R_2 \mid \right.$$
$$u_1 = (\mathscr{H}_1', \bar{c}_1'c, \bar{C}_1') \in \mathrm{children}_{R_1}(\mathrm{root}(R_1)) \left. \right\} \cup$$
$$\left\{ R_1 \times (q, \otimes, \bar{d}(2, m_2 + 1)) \mathrm{subtree}_{R_2}(u_2) \mid \right.$$
$$u_2 = (\mathscr{H}_2', \bar{c}_2'c, \bar{C}_2') \in \mathrm{children}_{R_2}(\mathrm{root}(R_2)) \left. \right\},$$
$$S_2 = \left\{ \mathrm{subtree}_{R_1}(u_1) \times (q, \otimes, \bar{d}) \mathrm{subtree}_{R_2}(u_2) \mid \right.$$
$$u_i = (\mathscr{H}_i', \bar{c}_i', \bar{C}_i'D_i) \in \mathrm{children}_{R_i}(\mathrm{root}(R_i)), 1 \leq i \leq 2 \left. \right\}.$$

Lemma 7. *Let $\mathscr{A}_1$ and $\mathscr{A}_2$ be τ-structures and let $\otimes = \otimes[g|f_1, f_2]$ for some t-relabelings g, f_1, f_2. For nonnegative integers q, m, p with $m + p \leq q$, let $\bar{c} = c_1, \ldots, c_m \in (A_1 \cup A_2)^m$ and $\bar{C} = C_1, \ldots, C_p$, where $C_j \subseteq A_1 \cup A_2$ for $1 \leq j \leq p$. Also let $\bar{d} = \mathrm{ind}(A_1, A_2, \bar{c})$ and for $1 \leq i \leq 2$ let $q_i \geq q - m - p$. Then $\mathrm{RC}_q(\mathscr{A}_1 \otimes \mathscr{A}_2, \bar{c}, \bar{C}) = \mathrm{RC}_{q_1}(\mathscr{A}_1, \bar{c}_1, \bar{C} \cap A_1) \times (q, \otimes, \bar{d}) \mathrm{RC}_{q_2}(\mathscr{A}_2, \bar{c}_2, \bar{C} \cap A_2)$.*

Lemma 8. *Given R_1 and R_2, the tree cross product $R_1 \times (q, \otimes, \bar{d}) R_2$ can be computed time $\mathrm{poly}(|R_1|, |R_2|)$, where $|R_i|$ denotes the number of nodes in R_i.*

We can now finally prove the Main Theorem.

Proof (Main Theorem). It is no loss of generality to assume that G has at least one vertex. Otherwise deciding whether $G \models \varphi$ takes constant time. By Lemmas 1, 2 and 3, to prove that $G \models \varphi$ it is sufficient to show that the verifier has a winning strategy in the model checking game $\mathcal{MC}(\mathrm{RC}_q(G), \varphi, \epsilon, \epsilon)$.

By Lemma 4, the size of the reduced characteristic tree $\mathrm{RC}_q(G)$ of a t-labeled graph is at most $f_1(q,t)$ for some computable function f_1 of q and t alone. By Lemma 8, the time taken to combine two reduced characteristic trees of size $f_1(q,t)$ is $f(q,t) = \mathrm{poly}(f_1(q,t))$.

We claim that the total time taken to construct $\mathrm{RC}_q(G)$ from its parse tree decomposition T is $O(f(q,t) \cdot |T|)$. This is where we use the fact that the graph G has rankwidth at most t. The proof is by induction on $|T|$. By Lemma 5, the claim holds when $|T| = 1$. Suppose that $\bar{G} = \bar{G}_1 \otimes [g|h_1, h_2] \bar{G}_2$, where g, h_1, h_2 are t-relabelings and let T_1 and T_2 be parse trees of $\bar{G}_1$ and $\bar{G}_2$, respectively. Then $|T| = |T_1| + |T_2| + 1$, where T is a parse tree of $\bar{G}$. By induction hypothesis, one can construct the reduced characteristic trees $\mathrm{RC}_q(G_1)$ and $\mathrm{RC}_q(G_2)$ in times $O(f(q,t) \cdot |T_1|)$ and $O(f(q,t) \cdot |T_2|)$, respectively. By Lemma 7, one can indeed construct $\mathrm{RC}_q(G)$ given $\mathrm{RC}_q(G_1)$, $\mathrm{RC}_q(G_2)$ and $\bar{d} = \varepsilon$. By using Lemma 8, the time taken to construct $\mathrm{RC}_q(G)$ is $O(f(q,t) + f(q,t) \cdot |T_1| + f(q,t) \cdot |T_2|) = O(f(q,t) \cdot |T|)$, proving the claim.

In order to check whether the verifier has a winning strategy in the model checking game $\mathcal{MC}(\mathrm{RC}_q(G), \varphi, \epsilon, \epsilon)$, one can use a very simple recursive algorithm (see also [13]). A position $p = (\psi, \bar{x}, \bar{X}, u)$ of the model checking game can be identified with a call of the algorithm with arguments p. If ψ is universal, then the algorithm recursively checks whether the verifier has a winning strategy from all positions u' that are reachable from u in the model checking game. If otherwise ψ is existential, then the algorithm checks whether there is one subsequent position in the game from which the verifier has a winning strategy. This algorithm visits each node of the reduced characteristic tree $\mathrm{RC}_q(G)$ at most once. Therefore the time taken to decide whether $G \models \varphi$ is $O(f_1(q,t) + f(q,t) \cdot |T|) = O(f(q,t) \cdot |T|)$, as claimed.   □

5   Discussion and Conclusion

With some additional effort the proof of the Main Theorem can be extended to linear optimization problems expressible in MSO_1 (the *LinMSO*-framework). Moreover the results of this paper naturally extend to directed graphs and bi-rankwidth. This allows us to conclude that any decision or optimization problem on directed graphs expressible in MSO_1 is linear-time solvable on graphs of bounded birankwidth [6,18]. Finally, the game-theoretic approach has already been used to prove Courcelle's result for treewidth [3,1,8] with an emphasis on practical implementability [19].

References

1. Arnborg, S., Lagergren, J., Seese, D.: Easy problems for tree-decomposable graphs. J. Algorithms 12(2), 308–340 (1991)
2. Berlekamp, E.R., Conway, J.H., Guy, R.K.: Winning Ways for Your Mathematical Plays. A.K. Peters, Wellesley (1982)
3. Courcelle, B.: The monadic second order theory of Graphs I: Recognisable sets of finite graphs. Information and Computation 85, 12–75 (1990)

4. Courcelle, B.: Monadic second-order definable graph transductions: A survey. Theor. Comput. Sci. 126(1), 53–75 (1994)
5. Courcelle, B., Kanté, M.M.: Graph operations characterizing rank-width and balanced graph expressions. In: Brandstädt, A., Kratsch, D., Müller, H. (eds.) WG 2007. LNCS, vol. 4769, pp. 66–75. Springer, Heidelberg (2007)
6. Courcelle, B., Makowsky, J.A., Rotics, U.: Linear Time Solvable Optimization Problems on Graphs of Bounded Clique Width. Theory Comput. Syst. 33, 125–150 (2000)
7. Courcelle, B., Makowsky, J.A., Rotics, U.: On the fixed parameter complexity of graph enumeration problems definable in monadic second-order logic. Discrete Applied Mathematics 108(1-2), 23–52 (2001)
8. Courcelle, B., Mosbah, M.: Monadic second-order evaluations on tree-decomposable graphs. Theor. Comput. Sci. 109(1-2), 49–82 (1993)
9. Ebbinghaus, H.-D., Flum, J.: Finite Model Theory. Springer, Heidelberg (1999)
10. Feferman, S., Vaught, R.: The first order properties of algebraic systems. Fund. Math. 47, 57–103 (1959)
11. Ganian, R., Hliněený, P.: On parse trees and Myhill–Nerode–type tools for handling graphs of bounded rank-width. Disc. App. Math. 158(7), 851–867 (2010)
12. Ganian, R., Hliněný, P., Obdržálek, J.: Unified approach to polynomial algorithms on graphs of bounded (bi-)rank-width (2009) (submitted)
13. Grädel, E.: Finite model theory and descriptive complexity. In: Finite Model Theory and Its Applications, pp. 125–230. Springer, Heidelberg (2007)
14. Gurevich, Y.: Modest Theory of Short Chains. I. J. Symb. Log. 44(4), 481–490 (1979)
15. Gurevich, Y.: Monadic second-order theories. In: Jon Barwise, S.F. (ed.) Model-Theoretic Logics, pp. 479–506. Springer, Heidelberg (1985)
16. Hintikka, J.: Logic, Language-Games and Information: Kantian Themes in the Philosophy of Logic. Clarendon Press, Oxford (1973)
17. Hliněný, P., Oum, S.: Finding branch-decomposition and rank-decomposition. SIAM Journal on Computing 38, 1012–1032 (2008)
18. Kante, M.M.: The rankwidth of directed graphs (2007) (preprint), http://arxiv.org/abs/0709.1433
19. Kneis, J., Langer, A., Rossmanith, P.: Courcelle's Theorem – a game-theoretic approach (2010) (submitted)
20. Oum, S.: Graphs of Bounded Rankwidth. PhD thesis, Princeton University (2005)
21. Oum, S., Seymour, P.D.: Approximating clique-width and branch-width. Journal of Combinatorial Theory Series B 96(4), 514–528 (2006)
22. Øverlier, L., Syverson, P.: Locating hidden servers. In: Proceedings of the 2006 IEEE Symposium on Security and Privacy. IEEE CS, Los Alamitos (May 2006)

On the Polynomial Depth of Various Sets of Random Strings

Philippe Moser

Department of Computer Science,
NUI Maynooth, Maynooth, Co. Kildare, Ireland
pmoser@cs.nuim.ie

Abstract. This paper proposes new notions of polynomial depth (called monotone poly depth), based on a polynomial version of monotone Kolmogorov complexity. We show that monotone poly depth satisfies all desirable properties of depth notions i.e., both trivial and random sequences are not monotone poly deep, monotone poly depth satisfies the slow growth law i.e., no simple process can transform a non deep sequence into a deep one, and monotone poly deep sequences exist (unconditionally).

We give two natural examples of deep sets, by showing that both the set of Levin-random strings and the set of Kolmogorov random strings are monotone poly deep.

1 Introduction

From the observation that nature contains both very simple and highly complex structures, Bennett introduced the profound concept of logical depth [7], as a formal definition of *useful* information, as opposed to (random) information in the traditional algorithmic information theory sense. Bennett's original idea is to categorize structures in three groups: trivial, random and the remaining ones; with the idea that trivial structures being completely predictable contain no useful information; random ones, being completely unpredictable, do not contain any useful information either; both (trivial and random) being therefore shallow objects. On the other hand, structures that are neither random nor trivial i.e., that contain intricate patterns that are neither fully predictable nor completely unpredictable, contain useful information; they are called deep structures. Although random sequences contain a lot of information (in the sense of algorithmic information theory), this information is not of much value, and such sequences are shallow.

Bennett observed that deep objects, because they contain complex well-hidden patterns, cannot be created by easy processes. This observation was formalized in the so-called *slow growth law*, which states that if a simple process (a truth table reduction) transforms some (source) sequence into an (image) sequence that is deep, then the source sequence it started from must be deep i.e., no easy process can transform a shallow sequence into a deep one.

M. Ogihara and J. Tarui (Eds.): TAMC 2011, LNCS 6648, pp. 517–527, 2011.
© Springer-Verlag Berlin Heidelberg 2011

Bennett's logical depth is based on Kolmogorov complexity. Intuitively, a binary sequence is deep, if the more time an algorithm is given, the better it can compress the sequence. Although Bennett's formulation is theoretically very elegant, it is uncomputable, due the uncomputability of Kolmogorov complexity.

To overcome the uncomputability of logical depth, several notions of feasible depth have been proposed so far [11,5,9]. In [5] Antunes, Fortnow, van Melkebeek, and Vinodchandran studied several polynomial-time formulations of depth, with connections to average-case complexity, nonuniform circuit complexity, and efficient search for satisfying assignments to Boolean formulas. In [9], both a notion of finite-state and polynomial depth were investigated, and the depth of polynomial weakly useful languages was shown.

Unfortunately, the feasible notions proposed so far suffer some limitations, e.g. a notion in [5] requires a complexity assumptions to prove the existence of deep sequences; and the polynomial depth of [9] is based on polynomial time predictors that cannot read their input (predictors must predict the nth bit of a sequence without access to the history, i.e. bits $1, 2, \ldots, n-1$).

As noticed in [14], depth is not an absolute concept, but depends on the power of two competing group of observers Δ and Δ'. Informally a sequence is (Δ, Δ')-deep if for any observer O from Δ there is an observer O' in Δ' such that O' performs (e.g. compresses, predicts, etc.) better than O on the sequence. Δ and Δ' can be the same class e.g. for recursive depth [11], $\Delta = \Delta'$ are recursive time bounds, or different classes e.g. for Bennett's depth [7], Δ are recursive time bounds but Δ' is unbounded Kolmogorov complexity.

In this paper, we use the idea of competing observers from [14] to construct new notions of polynomial depth (called monotone-polynomial depth), aiming at notions that satisfy the slow growth law, and for which deep objects can be proved to exist unconditionally. The classes of observers (the classes Δ and Δ') we consider are based on the notion of monotone polynomial time compression [8], which is a polynomial version of monotone Kolmogorov complexity, with the advantage that unlike polynomial predictors [9], monotone polynomial compressors can read their input. We show that our notions of monotone polynomial depth have all the desired properties of a depth notion, i.e. both trivial and random sequences are shallow, they satisfy a slow-growth law, and deep objects can be shown to exist unconditionally.

Although logical depth is a very profound concept, there have not been many examples of natural deep sequences in the literature so far. Bennett [7] showed that the halting language is deep. Lathrop and Lutz [10] generalized Bennett's result by showing that every weakly useful sequence (i.e. a sequence such that the set of languages that can be reduced to it has measure non-zero) is deep, a result that was shown to hold in the context of polynomial depth [9]. In this paper, we give two natural examples of deep languages, in the context of monotone poly depth, namely the set of Levin-random strings and the set of Kolmogorov random strings. Levin randomness is a standard randomness notion due to Levin [12]; it is a computable approximation of Kolmogorov complexity, that enjoys many useful properties, among others it provides a search strategy for finding solutions of NP problems, that is optimal up to a multiplicative constant (see [13]). Curiously although random sequences are shallow, our result shows that the set of Levin-random strings is not. This shows that in the context of polynomial monotone depth, having a test that detects randomness (i.e the set of Levin-random strings), is more useful than having access to randomness (a random sequence).

Several authors [2,4,1,3] showed the computational power of the set of Kolmogorov random strings by reducing (using several types of reduction) a broad range of complexity classes to it. Our observation that the set of Kolmogorov random strings is monotone-poly deep is consistent with the results by these authors [2,4,1,3] whose results intuitively show that this set contains a lot of useful information.

Due to lack of space, some proofs are postponed to the full version of this paper.

2   Preliminaries

We write $\mathbb{N}$ for the set of all nonnegative integers. Let us fix some notations for strings and languages. A *string* is an element of $\{0,1\}^n$ for some integer n. We denote by $s_0, s_1, \ldots, s_n$ the standard enumeration of strings in lexicographic order. For a string x, its length is denoted by $|x|$. The empty string is denoted by λ. We say string y is a prefix of string x, denoted $y \sqsubset x$ (also $y \sqsubseteq x$), if there exists a string a such that $x = ya$. We write $x \sim y$ if x is a prefix of y or vice-versa. For a string x, $\mathrm{dbl}(x)$ is x with every bit doubled.

A sequence is an infinite binary string, i.e. an element of $\{0,1\}^\infty$. For $S \in \{0,1\}^\infty$ and $i, j \in \mathbb{N}$, we write $S[i..j]$ for the string consisting of the ith through j^{th} bits of S, with the convention that $S[i..j] = \lambda$ if $i > j$, and $S[1]$ is the leftmost bit of S. We write $S[i]$ for $S[i..i]$ (the ith bit of S). For a sequence S divided into blocks $S = S_1 S_2 S_3 \ldots$, where S_i are strings, $S \upharpoonright S_i$ (resp $S \uparrow S_i$) denotes $S_1 \ldots S_i$ (resp. $S_1 \ldots S_{i-1}$). For $w \in \{0,1\}^*$ and $S \in \{0,1\}^\infty$, we write $w \sqsubseteq S$ if w is a prefix of S, i.e., if $w = S[1..|w|]$. Unless otherwise specified, logarithms are taken in base 2.

A *language* is a set of strings. The characteristic sequence of a language L is the sequence $\chi_L \in \{0,1\}^\infty$, whose nth bit is one iff $s_n \in L$. We will often use the notation L for χ_L.

TM stands for Turing machine. A monotone TM is a TM such that for any strings x, y, $M(xy) \sqsupseteq M(x)$.

E denotes the standard linear exponential time complexity class $\mathsf{E} = \cup_{c \in \mathbb{N}} \mathsf{DTIME}(2^{cn})$. A time bound is a monotone time constructible function $t : \mathbb{N} \to \mathbb{N}$, i.e. there is a TM that on input any string of length n halts in exactly $t(n)$ steps. We will consider the following standard time bound families: $\mathrm{Poly} = \cup_{k \in \mathbb{N}} \{t(n) = kn^k\}$, $\mathrm{Lin} = \cup_{k \in \mathbb{N}} \{t(n) = kn\}$, $\mathrm{Polylog} = \cup_{k \in \mathbb{N}} \{t(n) = \log^k n\}$ and $\mathrm{Rec} = \{t \mid t \text{ is a time bound}\}$.

3   Polynomial Depth

Our polynomial depth notions are based on polynomial monotone compression from [8].

Definition 1. *Let Δ be a family of (at least linear) time bounds (e.g. Poly, Lin, etc) and $S \in \{0,1\}^\infty$. A Δ-compression of S is a 3-tuple (C, D, p) where C, D are TMs and $p \in \{0,1\}^\infty$ such that there exists a time bound $t \in \Delta$ such that*

1. *Decompression: For all $j \in \mathbb{N}$, $D(p[1..j])$ outputs $S[1..i_{D,j}]$ in time $t(i+j)$, where $i_{D,j}$ is a monotone sequence of integers.*
2. *Compression: For all $i \in \mathbb{N}$, $C(S[1..i])$ outputs several strings in time $t(i)$, one of which is a prefix p' of p, such that $D(p') \sqsupseteq S[1..i]$.*

The integer $i_{D,j}$ is the number of bits the decompressor D can output given i bits of input i.e., the larger the difference $i_{D,j} - j$ the greater the compression.

When $\Delta = \text{Rec}$, we drop the compression requirement, i.e. a Rec-compression is a 2-tuple (D, p). This is because the compressor C may be uncomputable. When $\Delta = \text{Rec}$ we are in the realms of Kolmogorov complexity, where similarly there is no (computable) compressor but only a computable decompressor (the universal TM U).

To avoid extreme compressions of the form "On input n, output $2^{2^{\cdot^{\cdot^{\cdot 2^n}}}}$ ze-roes", we fix the maximal compression factor we allow i.e., let MD (maximal decompression) be a function such that $\text{MD}(j)$ is computable in $O(\text{MD}(j))$ time for any integer j (e.g. $\text{MD}(j) = 2^{2^{2^j}}$). We require that for any Δ-compression (C, D, p), and for every integer j,

$$i_{D,j} \leq \text{MD}(j)$$

i.e. MD is the same for all compressors.

Let us introduce our (Δ, Δ')-depth notion, based on competing observers' classes Δ and Δ'.

Definition 2. *$S \in \{0,1\}^\infty$ is a.e. (resp i.o.) (Δ, Δ')-deep if for every Δ-compression (C, D, p) of S and any $a > 0$, there exists a Δ'-compression (C', D', p') of S such that for almost every (resp. infinitely many) $j \in \mathbb{N}$*

$$i_{D',j} - i_{D,j} \geq a \log i_{D',j}. \tag{1}$$

A sequence is (Δ, Δ')-shallow if it is not (Δ, Δ')-deep.

The choice of the log function in Equation 1 is arbitrary. In Bennett's original notion [7], it was only required the difference be unbounded and the rate was not specified, but Bennett's notion would also work with a log rate function. Most feasible depth notions published after Bennett's paper [5,9] used a logarithmic rate function. We choose to do the same.

As noticed in [14], the notion of depth is a relative notion, that depends on the power of the observers. Our goal is to study polynomial versions of Bennett's original depth notion [7], and its recursive version called recursive depth [11]. Recursive depth [11] is defined in terms of recursive observers competing against recursive observers, i.e. Δ and Δ' have the same power. The natural polynomial version is to choose $\Delta = \Delta' = \text{Poly}$. We call this notion monotone-Poly-depth.

Bennett's depth [7] on the other hand is based on observers of different strength i.e., recursive observers competing against noncomputable Kolmogorov complexity. For Bennett's notion, there is no unique translation into the polynomial world. We propose to study ($\Delta = \text{Lin}, \Delta' = \text{Poly}$) as a polynomial version of Bennett's depth (called monotone-Lin-depth), which encompasses the idea of observers of different strength (Lin vs Poly), but keeping both in the polynomial setting. The choice ($\Delta = \text{Lin}, \Delta' = \text{Poly}$) is actually flexible, and Poly (resp. Lin) could be replaced by anything strictly stronger (resp. weaker) than Lin, e.g. $O(n^2)$ (resp. Polylog), without modifying our results on monotone-Lin-depth from Section 5 (the choice Δ =Polylog, would require a modification of the notion of Δ-compression [8] to allow for sublinear running time, in the same way as martingales where modified to allow sublinear time bounds in [15]. We defer this generalization to the full version of this paper). The choice Lin vs Poly somehow reflects the difference in power of complexity classes E and EXP, which are the complexity classes on which Δ-compression was first introduced [8], to define a measure notion.

In [5] Antunes et al. proposed another resource-bounded version of Bennett's depth [7] called basic computational depth, by looking at bounded (sublinear or polynomial) Kolmogorov complexity vs unbounded Kolmogorov complexity. We introduce a translation of basic computational depth [5] in the setting of polynomial monotone compressors, by choosing ($\Delta = \text{Poly}, \Delta' = \text{Rec}$). We call this notion basic-monotone-Poly-depth (bm-Poly-depth). bm-Poly-depth captures the idea behind basic computational depth [5] but with Kolmogorov complexity replaced by monotone compressors.

The difference between a.e. and i.o. depth is similar to the difference between (resource-bounded)-packing dimension and (resource-bounded)-dimension (see e.g. [6]), where a compressor is required to compress infinitely many prefixes, or almost all prefixes. Bennett's depth [7] is an a.e. notion. Sometimes when the observers are very weak e.g. finite-state, i.o. is the best achievable (e.g. see [9]). All our results use the stronger formulation i.e. a.e. (which implies an i.o. result), except Theorem 4.

4 Basic Properties of Monotone-Poly-Depth

In the next section we study the basic properties of monotone-Poly-depth. All results remain true for both monotone-Lin-depth and bm-Poly-depth.

It is a key feature of logical depth [7] that both trivial (recursive) and random sequences are shallow. In this section we show that a similar result holds in the context of monotone-Poly-depth. Let us define what is meant by trivial sequences in the context of polynomial depth. Informally a sequence is trivial if its prefixes can be maximally compressed.

Definition 3. *Let $S \in \{0,1\}^\infty$. S is Poly-optimally-compressible if there exists a Poly-compression (C, D, p) of S, such that $i_{D,j} = MD(j)$ for almost every $j \in \mathbb{N}$.*

As an example, it is easy to check that the characteristic sequences of languages in E are Poly-optimally-compressible. The following result shows that optimally-compressible sequences are shallow.

Theorem 1. *Every Poly-optimally-compressible sequence is a.e. Poly-shallow.*

On the other extremity of the scale of randomness, we have random sequences. Here is a definition in the context of polynomial depth.

Definition 4. *Let $S \in \{0,1\}^\infty$. S is Poly-random if for every Poly-compression (C, D, p) of S, there exists $c \in \mathbb{N}$ such that for almost every $j \in \mathbb{N}$*

$$i_{D,j} \leq j + c.$$

The following result shows that random sequences are shallow.

Theorem 2. *Every Poly-random sequence is a.e. Poly-shallow.*

4.1   Slow Growth Law

A key property of logical depth [7], is that depth cannot be easily created. The formalization of this idea is known as the slow-growth law. It states that if a simple process transforms some (source) sequence into an (image) sequence that is deep, then the source sequence it started from must be deep i.e., no easy process can transform a shallow sequence into a deep one. Bennett proved a slow growth law for truth-table reductions (i.e. in the context of logical depth, simple process corresponds to truth-table reductions).

In the following section, we prove a slow growth law in the context of monotone-Poly-depth. As the power of polynomial monotone compressors is much smaller than the unbounded time case considered for Bennett's logical depth, we need to reduce the power of "simple processes" accordingly, by choosing weaker reductions. Here is a definition.

Definition 5. *Let $S, T \in \{0,1\}^\infty$. S is Poly-monotone reducible to T, if there exists a Poly-time monotone TM M such that*

1. *Reduction: for every $n \in \mathbb{N}$, $M(T[1..n]) \sqsubseteq S$.*
2. *Honesty: There exists $a > 0$ such that for every $n \in \mathbb{N}$*

$$n - a \log n \leq |M(T[1..n])| \leq n + a \log n$$

3. *Monotone injectivity: If $M(x) \sim M(y)$ then $x \sim y$.*

The following result is a slow-growth law for monotone-Poly-depth. A similar result holds for both monotone-Lin-depth and bm-Poly-depth (provided the reduction is linear-time bounded for monotone-Lin-depth).

Theorem 3. *Let $S, T \in \{0,1\}^\infty$, such that S is a.e. monotone-Poly-deep and Poly-monotone reducible to T. Then T is a.e. Poly-deep.*

A similar proof shows that the result holds for both monotone-Lin-depth and bm-Poly-depth (provided the reduction is linear-time bounded for monotone-Lin-depth).

4.2 A Poly Deep Sequence

Some previous polynomial depth notions (e.g. distinguishing complexity from [5]) require complexity assumption to prove the existence of deep sequences. The following result shows that our notion is unconditional. Similarly to other feasible depth notions with restricted power [9], our result is an i.o. result.

The proof uses the equivalence between compressors and martingales from [8]. A direct proof can be given without martingales, but using martingales makes the proof easier to read since it is easier to sum and diagonalize against martingales than it is against compressors directly. It is also interesting to see the correspondence martingales-compressor in the context of depth.

Theorem 4. *There exists an i.o. monotone-Poly-deep sequence.*

5   The Set of Levin Random Strings Is Deep

Whereas random sequences are shallow, we show that the characteristic sequence of the set of random strings is deep. Our result holds for the standard randomness notion due to Levin [12]; Levin's notion is a computable approximation of Kolmogorov complexity, that enjoys many useful properties, among others it provides a search strategy for finding solutions of NP problems, that is optimal up to a multiplicative constant (see [13]). Here is a definition.

Definition 6. *Fix a prefix-free universal Turing machine U. The Levin complexity of a string x is*

$$\mathrm{Kt}(x) = \min\{|p| + \log t : \ U(p) = x \ \text{in at most } t \ \text{steps}\}.$$

The definition of Kt does not depend on the choice of the universal TM U, up to an additive constant (see [13]).

The set of Levin random strings is

$$R_{\mathrm{Kt}} = \{x \in \{0,1\}^* : \ \mathrm{Kt}(x) \geq |x| + \log|x|\}. \tag{2}$$

By a standard program counting argument, it is easy to see that $R_{\mathrm{Kt}} \neq \emptyset$. Although the strings in R_{Kt} are shallow, the characteristic sequence of R_{Kt} contains useful information, i.e. is monotone-Lin-deep, as the following result shows.

Theorem 5. *R_{Kt} is a.e. monotone-Lin-deep.*

Proof. We need the following lemma.

Lemma 1. *For every Lin-compression* (C, D, p) *and for almost every* $j \in \mathbb{N}$

$$i_{D,j} < 2^{2^{3j}+1}.$$

Let us prove the lemma by contradiction. Suppose there is an infinite set J of integers j such that $i_{D,j} \geq 2^{2^{3j}+1}$; in particular for every $j \in J$, $i_{D,\frac{1}{3}\log j} > 2^{j+1}$. Thus

$$R_{\mathrm{Kt}} \sqsupseteq D(p[1..\frac{1}{3}\log j]) \sqsupseteq R_{\mathrm{Kt}}[1..2^{j+1}].$$

Let $j \in J$ be large (to be determined later). Letting $d = p[1..\frac{1}{3}\log j]$ yields a string with high Kt complexity: from $\pi = \langle D, d \rangle$ recover $R_{\mathrm{Kt}}[1..2^{j+1}]$ and j (from the length of d). Output the first y with $|y| = j$ and $R_{\mathrm{Kt}}(y) = 1$, i.e.

$$\mathrm{Kt}(y) \geq j + \log j.$$

By encoding π the standard way, i.e. $\pi = \mathrm{dbl}(\langle D \rangle)01d$

$$|\pi| \leq |d| + O(1) \leq \frac{1}{3}\log j + O(1).$$

The time to construct j is the time to recover $R_{\mathrm{Kt}}[1..2^{j+1}]$ (less than $O(2^{j+1})$) and the time to find y in $R_{\mathrm{Kt}}[1..2^{j+1}]$ (less than $O(2^{j+1})$ steps), i.e. a total of at most $O(2^{j+1})$ steps. Therefore

$$\mathrm{Kt}(y) \leq |\pi| + \log O(2^{j+1}) \leq \frac{1}{3}\log j + j + O(1) < j + \log j$$

for j large enough, which contradicts $R_{\mathrm{Kt}}(y) = 1$; thus ending the proof of the lemma. □

Lemma 2. *There exists a Poly-compression* (C, D, p) *of* R_{Kt} *such that for almost every* $j \in \mathbb{N}$

$$i_{D,j} = \mathrm{MD}(j).$$

Let $p = 0^\infty$. D on input $p[1..j]$ computes $i_{D,j} := \mathrm{MD}(j)$. D constructs $R_{\mathrm{Kt}}[1..i_{D,j}]$ by simulating the universal TM on all programs π_l of size at most $\log i_{D,j} + \log\log i_{D,j}$ during t_l steps ($t_l \leq 2^{\log i_{D,j} + \log\log i_{D,j}}$), the results string of such a simulation is denoted x_l. All strings x_l with $|x_l| \leq \log i_{D,j}$, for which $|\pi_l| + \log t_l \leq |x_l| + \log|x_l|$ have membership bit 0 in the characteristic sequence $R_{\mathrm{Kt}}[1..i_{D,j}]$. All remaining bits in $R_{\mathrm{Kt}}[1..i_{D,j}]$ are 1s. The running time of D is less than

$$O(2^{\log i_{D,j} + \log\log i_{D,j}}) \cdot 2^{\log i_{D,j} + \log\log i_{D,j}} \leq (i_{D,j})^c$$

for some $c \in \mathbb{N}$.

The compressor C on input $R_{\mathrm{Kt}}[1..i]$ finds the smallest j such that $\mathrm{MD}(j) \geq i$, and outputs 0^j. C runs in time polynomial in i. This ends the proof of the lemma. □

Let us show that R_{Kt} is monotone-Lin-deep. Let $a > 0$ and (C, D, p) be a Lin-compression of R_{Kt}, and let (C', D', p') be the Poly-compression from Lemma 2. We have

$$
\begin{aligned}
i_{D',j} - i_{D,j} &= \mathrm{MD}(j) - i_{D,j} && \text{by Lemma 2} \\
&\geq \mathrm{MD}(j) - 2^{2^{3j}+1} && \text{by Lemma 2} \\
&\geq \frac{1}{2}\mathrm{MD}(j) && \text{by definition of MD} \\
&= \frac{1}{2}i_{D',j} && \text{by definition of } i_{D',j} \\
&\geq a \log i_{D',j}
\end{aligned}
$$

for almost every j i.e. R_{Kt} is monotone-Lin-deep.

6   The Set of Kolmogorov-Random Strings Is Deep

The next result shows that the set of Kolmogorov random strings is bm-Poly-deep.

Definition 7. *Fix a prefix-free universal Turing machine U. The Kolmogorov complexity of x is the length of the shortest program that outputs x.*

$$K(x) = \min\{|p| : \ U(p) = x\}.$$

The definition of K does not depend on the choice of the universal TM U, up to an additive constant (see [13]).

For a time bound t, the t-bounded Kolmogorov complexity of x is

$$K^t(x) = \min\{|p| : \ U(p) = x, \text{ and } U \text{ halts in at most } t(|x|) \text{ steps}\}.$$

Let $0 < \epsilon < 1$. The set of Kolmogorov random string is

$$R_{K,\epsilon} = \{x \in \{0,1\}^* : \ K(x) \geq \epsilon|x|\}. \tag{3}$$

Theorem 6. *Let $0 < \epsilon < 1$. $R_{K,\epsilon}$ is a.e. bm-Poly-deep.*

Proof. Let $0 < \epsilon < 1$ We need the following lemma.

Lemma 3. *For every Poly-compression (C, D, p) of $R_{K,\epsilon}$ and for almost every $j \in \mathbb{N}$*

$$i_{D,j} < 2^{j+1}.$$

Let us prove the lemma by contradiction. Suppose there is a Poly-compression (C, D, p) of $R_{K,\epsilon}$ and an infinite set N of integers j such that $i_{D,j} \geq 2^{j+1}$. Let $c = 4/(1 - \epsilon)$ and $j \in N$.

Let $y_1, \ldots, y_c \in \{0,1\}^j$ such that $K^{2^{n^2}}(\langle y_1, \ldots, y_c \rangle) \geq cj - O(\log j)$ but $K(\langle y_1, \ldots, y_c \rangle) \leq O(\log j)$. Such a c-tuple can be found by simulating U on all programs of appropriate size running in at most 2^{n^2} steps. We have $R_{K,\epsilon}(y_t) = 0$ for every $t = 1, \ldots, c$.

Consider $L = \{(l_1, \cdots, l_c) | \ 1 \leq l_t \leq 2^{\epsilon(j+1)}, \ t = 1, \ldots, c\}$. Let $Q = \{q_l | l \in L\}$ with $q_l = \langle \text{code}, p[1..j], l \rangle$ be the set of programs such that U on input q_l simulates $D(p[1..j])$ to reconstruct $R_{K,\epsilon}[1..i_{D,j}] \sqsupseteq R_{K,\epsilon}[1..2^{j+1}]$, which takes time less than $O(2^j)$ (U stops once D already output the 2^{j+1} first bits of $R_{K,\epsilon}$). U constructs

$$R_0 = \{r_1 < r_2 < \ldots | \ r_t \in \{0,1\}^{\leq j}, R_{K,\epsilon}(r_t) = 0\}$$

the lexicographical ordered set of all strings of length at most j whose characteristic bit in $R_{K,\epsilon}$ is 0, which takes time $O(2^j)$. If $r_{l_1}, \cdots, r_{l_c} \in R_0$ then output $\langle r_{l_1}, \cdots, r_{l_c} \rangle$ else halt, which takes time $O(2^j)$.

On any program $q_l \in Q$, U runs in less than $2^{O(j)}$ steps. Moreover all l_t ($t = 1, \ldots, c$) can be encoded in at most $\epsilon(j+1)$ bits i.e., all programs $q_l \in Q$ have size bounded by

$$|q_l| \leq c\epsilon(j+1) + j + O(\log j) \leq (c\epsilon + 1)j + O(\log j).$$

Because $R_{K,\epsilon}(y_t) = 0$ for every $t = 1, \ldots, c$, let $v = (v_1, \ldots, v_c) \in L$ be the vector of the positions of $y_1, \ldots, y_c$ in $R_{K,\epsilon}$ i.e., $r_{v_t} = y_t$ for every $t = 1, \ldots, c$. Thus U on input q_v outputs $\langle y_1, \ldots, y_c \rangle$ i.e., q_v is a program for $\langle y_1, \ldots, y_c \rangle$ that runs in less than $2^{O(j)}$ steps. Thus we have

$$K^{2^{n^2}}(\langle y_1, \ldots, y_c \rangle) \leq (c\epsilon + 1)j + O(\log j) \quad \text{which implies}$$

$$cj - O(\log j) \leq (c\epsilon + 1)j + O(\log j) \quad \text{i.e.,}$$

$$cj \leq (c\epsilon + 1)j + O(\log j) \leq (c\epsilon + 2)j$$

thus $c(1 - \epsilon) \leq 2$ which is a contradiction. □

Lemma 4. *There exists a Rec-compression (D, p) of $R_{K,\epsilon}$ such that for almost every $j \in \mathbb{N}$*

$$2^{j+1} \geq i_{D,j} \geq 2^{j/\epsilon}.$$

Let $p = \Omega[1..n]$ be the halting probability $\Omega = \sum_{p:U(p)\downarrow} 2^{-|p|}$. D on input $p[1..\epsilon j]$ can compute using standard Dove-tailing (see [13]) whether $U(p) \downarrow$ for all programs p with $|p| \leq \epsilon j$ i.e., it can reconstruct $R_{K,\epsilon}[1..2^{j+1} - 1]$. We have $i_{D,\epsilon j} \geq 2^j$ i.e., $i_{D,j} \geq 2^{j/\epsilon}$. By construction $2^{j+1} \geq i_{D,j}$. □

Let us show that $R_{K,\epsilon}$ is bm-Poly-deep. Let $a > 0$ and (C, D, p) be a Poly-compression of $R_{K,\epsilon}$, and let (D', p') be the Rec-compression from Lemma 4. We have

$$
\begin{aligned}
i_{D',j} - i_{D,j} &\geq 2^{j/\epsilon} - i_{D,j} && \text{by Lemma 4} \\
&\geq 2^{j/\epsilon} - 2^{j+1} && \text{by Lemma 3} \\
&= 2^j(2^{(1/\epsilon - 1)j} - 2) && \\
&> 2^{j+1} && \text{for } j \text{ large enough} \\
&\geq i_{D',j} && \text{by Lemma 4} \\
&\geq a \log i_{D',j}
\end{aligned}
$$

for almost every j i.e., $R_{K,\epsilon}$ is a.e. bm-Poly-deep. □

References

1. Allender, E., Buhrman, H., Koucký, M.: What can be efficiently reduced to the K-random strings? In: Diekert, V., Habib, M. (eds.) STACS 2004. LNCS, vol. 2996, pp. 584–595. Springer, Heidelberg (2004)
2. Allender, E., Buhrman, H., Koucký, M.: What can be efficiently reduced to the Kolmogorov-random strings? Ann. Pure Appl. Logic 138(1-3), 2–19 (2006)
3. Allender, E., Buhrman, H., Koucký, M., van Melkebeek, D., Ronneburger, D.: Power from random strings. In: FOCS, pp. 669–678. IEEE Computer Society, Los Alamitos (2002)
4. Allender, E., Buhrman, H., Koucký, M., van Melkebeek, D., Ronneburger, D.: Power from random strings. SIAM J. Comput. 35(6), 1467–1493 (2006)
5. Antunes, L., Fortnow, L., van Melkebeek, D., Vinodchandran, N.: Computational depth: Concept and applications. Theoretical Computer Science 354, 391–404 (2006)
6. Athreya, K.B., Hitchcock, J.M., Lutz, J.H., Mayordomo, E.: Effective strong dimension in algorithmic information and computational complexity. SIAM J. Comput. 37(3), 671–705 (2007)
7. Bennett, C.H.: Logical depth and physical complexity. The Universal Turing Machine, A Half-Century Survey, pp. 227–257 (1988)
8. Buhrman, H., Longpré, L.: Compressibility and resource bounded measure. SIAM J. Comput. 31(3), 876–886 (2001)
9. Doty, D., Moser, P.: Feasible depth. In: Cooper, S.B., Löwe, B., Sorbi, A. (eds.) CiE 2007. LNCS, vol. 4497, pp. 228–237. Springer, Heidelberg (2007)
10. Juedes, D.W., Lathrop, J.I., Lutz, J.H.: Computational depth and reducibility. Theor. Comput. Sci. 132(2), 37–70 (1994)
11. Lathrop, J.I., Lutz, J.H.: Recursive computational depth. Inf. Comput. 153(1), 139–172 (1999)
12. Levin, L.A.: Randomness conservation inequalities; information and independence in mathematical theories. Information and Control 61(1), 15–37 (1984)
13. Li, M., Vitanyi, P.: Introduction to Kolmogorov complexity and its applications. Springer, Heidelberg (1993)
14. Moser, P.: A general notion of useful information. In: Neary, T., Woods, D., Seda, A.K., Murphy, N. (eds.) CSP. EPTCS, vol. 1, pp. 164–171 (2008)
15. Moser, P.: Martingale families and dimension in P. Theor. Comput. Sci. 400(1-3), 46–61 (2008)

Edge Contractions in Subclasses
of Chordal Graphs*

Rémy Belmonte, Pinar Heggernes, and Pim van 't Hof

Department of Informatics, University of Bergen,
P.O. Box 7803, N-5020 Bergen, Norway
{remy.belmonte,pinar.heggernes,pim.vanthof}@ii.uib.no

Abstract. Modifying a given graph to obtain another graph is a well-studied problem with applications in many fields. Given two input graphs G and H, the CONTRACTIBILITY problem is to decide whether H can be obtained from G by a sequence of edge contractions. This problem is known to be NP-complete already when both input graphs are trees of bounded diameter. We prove that CONTRACTIBILITY can be solved in polynomial time when G is a trivially perfect graph and H is a threshold graph, thereby giving the first classes of graphs of unbounded treewidth and unbounded degree on which the problem can be solved in polynomial time. We show that this polynomial-time result is in a sense tight, by proving that CONTRACTIBILITY is NP-complete when G and H are both trivially perfect graphs, and when G is a split graph and H is a threshold graph. If the graph H is fixed and only G is given as input, then the problem is called H-CONTRACTIBILITY. This problem is known to be NP-complete on general graphs already when H is a path on four vertices. We show that, for any fixed graph H, the H-CONTRACTIBILITY problem can be solved in polynomial time if the input graph G is a split graph.

1   Introduction

The problem of deciding whether a given graph can be obtained from another given graph by contracting edges is motivated from Hamiltonian graph theory and graph minor theory, and it has applications in computer graphics and cluster analysis [11]. This problem has recently attracted increasing interest, in particular when restrictions are imposed on the input graphs [2,9,10,11,12]. We continue this line of research with new polynomial-time and NP-completeness results.

For a fixed graph H, the H-CONTRACTIBILITY problem is to decide whether H can be obtained from an input graph G by a sequence of edge contractions. This problem is closely related to the well-known H-MINOR CONTAINMENT problem, which is the problem of deciding whether H can be obtained from a *subgraph* of G by contracting edges. A celebrated result by Robertson and Seymour [16] shows that H-MINOR CONTAINMENT can be solved in polynomial time on general graphs for any fixed H. As a contrast, H-CONTRACTIBILITY is

---

* This work has been supported by the Research Council of Norway.

M. Ogihara and J. Tarui (Eds.): TAMC 2011, LNCS 6648, pp. 528–539, 2011.
© Springer-Verlag Berlin Heidelberg 2011

NP-complete already for very simple fixed graphs H, such as a path or a cycle on four vertices [2]. The version of the problem where both graphs are given as input, called CONTRACTIBILITY, is NP-complete on trees of bounded diameter, as well as on trees all whose vertices but one have degree at most 5 [15].

Given these hardness results, it is perhaps not surprising that hardly any positive results are known on the CONTRACTIBILITY problem. So far, CON-TRACTIBILITY is known to be solvable in polynomial time only when G has bounded treewidth and H has bounded degree [15]. A few more positive results are known on the H-CONTRACTIBILITY problem. For example, for every fixed graph H on at most 5 vertices, H-CONTRACTIBILITY can be solved on general graphs in polynomial time when H has a dominating vertex, and it is NP-complete otherwise [11,12]. However, it is known that for larger fixed graphs H, the presence of a dominating vertex in H is not a guarantee for polynomial-time solvability of the problem [9]. Very recently, Kamiński, Paulusma and Thilikos [10] showed that H-CONTRACTIBILITY can be solved in polynomial time on planar input graphs for every fixed H.

In this paper, we study the CONTRACTIBILITY and H-CONTRACTIBILITY problems on subclasses of chordal graphs. Chordal graphs constitute one of the most famous graph classes, with a large number of practical applications (see e.g., [5,6,17]). Edge contractions preserve the property of being chordal. Since trees are chordal graphs, it follows from the above-mentioned hardness result on trees that CONTRACTIBILITY is NP-complete when G and H are both chordal. We show that the problem remains NP-complete even when G and H are both trivially perfect graphs or both split graphs. Note that trees are neither trivially perfect nor split. Trivially perfect graphs and split graphs are two unrelated subclasses of chordal graphs, and both classes are well-studied with several theoretical applications [1,6]. These two classes share a common subclass called threshold graphs, which is another well-known subclass of chordal graphs [14]. We prove that CONTRACTIBILITY remains NP-complete even when G is split and H is threshold. On the positive side, we show that CONTRACTIBILITY can be solved in polynomial time when G is trivially perfect and H is threshold. This result can be considered tight by the above-mentioned hardness results. For H-CONTRACTIBILITY, we give a polynomial-time algorithm when G is a split graph and H is an arbitrary fixed graph. The results of this paper are summarized in Table 1.

On the way to obtain our results, we show that the problems CONTRACTIBIL-ITY and INDUCED SUBGRAPH ISOMORPHISM are equivalent on connected trivially perfect graphs. Hence our results imply that the latter problem is NP-complete on connected trivially perfect graphs, and that this problem can be solved in polynomial time when G is trivially perfect and H is threshold. We would like to mention that INDUCED SUBGRAPH ISOMORPHISM is known to be NP-complete on split graphs and on cographs [3]. Trivially perfect graphs constitute a subclass of cographs, and threshold graphs are both cographs and split graphs. Hence our results tighten previously known hardness results on INDUCED SUBGRAPH ISOMORPHISM. The relationships between the graph classes mentioned in this paper are given in Figure 1.

Table 1. The complexity of deciding whether G can be contracted to H, according to our results; (i) stands for "part of the input", (f) stands for "fixed"

G	H	Complexity
Trivially perfect (i)	Trivially perfect (i)	NP-complete
Trivially perfect (i)	Threshold (i)	Polynomial
Trivially perfect (i)	Trivially perfect (f)	Polynomial
Threshold (i)	Arbitrary (i)	Linear
Split (i)	Threshold (i)	NP-complete
Split (i)	Arbitrary (f)	Polynomial

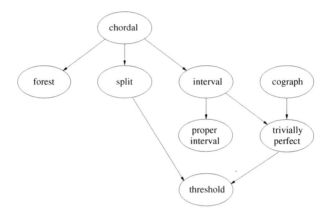

Fig. 1. The graph classes mentioned in this paper, where $\rightarrow$ represents the $\supset$ relation

2   Preliminaries

All graphs considered in this paper are undirected, finite and simple. A vertex $v \in V(G)$ is called *universal* if it is adjacent to every other vertex of G. For any set $S \subseteq V(G)$, we write $G[S]$ to denote the subgraph of G *induced* by S. We write $G - v$ to denote the graph $G[V(G) \setminus \{v\}]$. The set S is said to be *connected* if $G[S]$ is connected. We say that two disjoint sets $S, S' \subseteq V(G)$ are *adjacent* if there exist vertices $s \in S$ and $s' \in S'$ that are adjacent. A connected component of a graph is called *nontrivial* if it contains at least one edge. An ordering $\alpha = (v_1, v_2, \ldots, v_n)$ of the vertices of a graph G is called a *non-increasing degree ordering* of G if $d(v_1) \geq d(v_2) \geq \cdots \geq d(v_n)$.

The *contraction* of edge uv in G removes u and v from G, and replaces them by a new vertex, which is made adjacent to precisely those vertices that were adjacent to at least one of the vertices u and v. Instead of speaking of the contraction of edge uv, we sometimes say that a vertex u is *contracted onto* v if the new vertex resulting from the contraction is still called v. We write G/uv to denote the graph obtained from G by contracting the edge uv. We say that a graph G can be *contracted* to a graph H, or is *H-contractible*, if H is isomorphic to a graph that can be obtained from G by a sequence of edge contractions.

Let $S \subseteq V(G)$ be a connected set. If we repeatedly contract edges in $G[S]$ until only one vertex of $G[S]$ remains, we say that we *contract S into a single vertex*. Let H be a graph with vertex set $\{h_1, \ldots, h_{|V(H)|}\}$. Saying that a graph G can be contracted to H is equivalent to saying that G has a so-called H-*witness structure* $\mathcal{W}$, which is a partition of $V(G)$ into *witness sets* $W(h_1), \ldots, W(h_{|V(H)|})$, such that each witness set induces a connected subgraph of G, and such that for every two vertices $h_i, h_j \in V(H)$, the corresponding witness sets $W(h_i)$ and $W(h_j)$ are adjacent in G if and only if h_i and h_j are adjacent in H. By contracting each of the witness sets into a single vertex, we obtain a graph which is isomorphic to H. See Figure 2 for an example that shows that, in general, an H-witness structure of G is not uniquely defined. For any subset $S \subseteq V(H)$, we write $W(S)$ to denote the set of vertices of G that are contained in a witness set $W(v)$ for some $v \in S$, i.e., $W(S) = \cup_{v \in S} W(v)$.

Fig. 2. Three different H-witness structures of a threshold graph

Cographs are the graphs that do not contain a path on four vertices as an induced subgraph. *Interval graphs* are the intersection graphs of intervals of a line, and they form a subclass of chordal graphs. *Chordal graphs* are the graphs without induced cycles of length more than 3.

Trivially perfect graphs have various characterizations [1,6,7,20]. For our purposes, it is convenient to use the following characterization as a definition. A graph G is *trivially perfect* if and only if each connected induced subgraph of G contains a universal vertex [18,19]. Let $\alpha = (v_1, v_2, \ldots, v_n)$ be an ordering of the vertices of a trivially perfect graph G. If α has the property that v_i is universal in a connected component of $G[\{v_i, , v_{i+1}, \ldots, v_n\}]$ for $i = 1, \ldots, n$, then α is called a *universal-in-a-component ordering (uco)*. A graph is trivially perfect if and only if it has a uco, and if and only if every non-increasing degree ordering is a uco [7,20]. Consequently, for every edge uv in a trivially perfect graph, either $N[u] \subseteq N[v]$ or $N[v] \subseteq N[u]$ [20].

Every rooted tree T defines a connected trivially perfect graph, which is obtained by adding edges to T so that every path between the root and a leaf becomes a clique. In fact, all connected trivially perfect graphs can be created this way, and there is a bijection between rooted trees and connected trivially perfect graphs [20]. Given a connected trivially perfect graph G, a rooted tree T_G corresponding to G, which we call a *uco-tree* of G, can be obtained in the following way. If G is a single vertex, then T_G is this vertex. Otherwise, take a universal vertex v of G, make it the root of T_G, and delete it from G. In the remaining graph, for each connected component G', build a uco-tree $T_{G'}$ of G' recursively and make v the parent of the root of $T_{G'}$. All rooted trees that can be obtained from a connected trivially perfect graph in this way are isomorphic, and hence T_G is unique for every connected trivially perfect graph G. If G is disconnected, then it has a *uco-forest*, which is the disjoint union of the uco-trees of the connected components of G.

A graph G is a *split graph* if its vertex set can be partitioned into a clique C and an independent set I, where (C, I) is called a *split partition* of G. If C is not a maximum clique, then there is a vertex $v \in I$ that is adjacent to every vertex of C. In this case, $C' = C \cup \{v\}$ is a maximum clique, and $(C', I \setminus \{v\})$ is also a split partition of G. In this paper, unless otherwise stated, we assume that the clique C of a split partition (C, I) is maximum. This implies that none of the vertices in I is adjacent to every vertex of C. Split graphs form a subclass of chordal graphs.

Threshold graphs constitute a subclass of both trivially perfect graphs and split graphs. Threshold graphs have several characterizations [1,6,14], and we use the following one as a definition. A graph G is a *threshold graph* if and only if it is a split graph and, for any split partition (C, I) of G, there is an ordering $(v_1, v_2, \ldots, v_k)$ of the vertices of C such that $N[v_1] \supseteq N[v_2] \supseteq \ldots \supseteq N[v_k]$, and there is an ordering $(u_1, u_2, \ldots, u_\ell)$ of the vertices of I such that $N(u_1) \subseteq N(u_2) \subseteq \ldots \subseteq N(u_\ell)$ [14]. In that case, $(v_1, v_2, \ldots, v_k, u_\ell, \ldots, u_2, u_1)$ is a non-increasing degree ordering, and hence a uco, of G. Every connected threshold graph has a universal vertex, e.g., vertex v_1 in the ordering given above. Since we assume the clique of any split partition to be maximum, a vertex of C of smallest degree, e.g., vertex v_k in the ordering given above, has no neighbors in I. If a threshold graph is disconnected, then it has at most one nontrivial connected component; all other connected components are isolated vertices.

Split graphs, trivially perfect graphs, and threshold graphs are *hereditary* graph classes, meaning that the property of belonging to each of these classes is closed under taking induced subgraphs. These graph classes can be recognized in linear time; split partitions and uco-trees can also be obtained in linear time [1,6,7,20].

3   Contractions and Induced Subgraph Isomorphisms of Trivially Perfect Graphs

In this section, we will give results on the computational complexity of CONTRACTIBILITY on trivially perfect graphs, corresponding to the first four rows of Table 1. The first theorem reveals the equivalence of the problems CONTRACTIBILITY and INDUCED SUBGRAPH ISOMORPHISM on the class of connected trivially perfect graphs.

Theorem 1. *For any two connected trivially perfect graphs G and H, the following three statements are equivalent:*

(i) G can be contracted to H;
(ii) G contains an induced subgraph isomorphic to H;
(iii) T_G can be contracted to T_H.

Proof. First we prove the equivalence between (i) and (ii). Suppose G is H-contractible, and let uv be one of the edges of G that were contracted to obtain a graph isomorphic to H. Since G is trivially perfect, we have either $N_G[u] \subseteq N_G[v]$ or $N_G[v] \subseteq N_G[u]$. Without loss of generality, assume that $N_G[u] \subseteq N_G[v]$. Then contracting edge uv in G is equivalent to deleting vertex u from G. We can repeat this argument for every edge that was contracted, and conclude that G has an induced subgraph isomorphic to H.

For the opposite direction, suppose G' is an induced subgraph of G isomorphic to H. Let x be a universal vertex of G. We claim that G has an induced subgraph G'' isomorphic to H such that G'' contains x. If G' already contains x, then we can take $G'' = G'$. Suppose $x \notin V(G')$. Since G' is a connected trivially perfect graph, it has a universal vertex x'. Since x is a universal vertex in G, we have $N_G(x') \subseteq N_G[x]$. Hence the graph $G'' = G[(V(G') \setminus \{x'\}) \cup \{x\}]$ is isomorphic to G', and is therefore also isomorphic to H. Now let $y \neq x$ be one of the vertices that has to be deleted from G to obtain its induced subgraph G'', i.e., $y \in V(G) \setminus V(G'')$. Since x is a universal vertex, we know that $N_G(y) \subseteq N_G[x]$. Then deleting vertex y from G is equivalent to contracting edge xy in G. Since $x \in V(G'')$, we can repeat this argument for every vertex of $V(G) \setminus V(G'')$, and conclude that G is H-contractible.

Next we prove the equivalence between (ii) and (iii). Suppose G contains an induced subgraph G' isomorphic to H, and let y be one of the vertices of G that has to be deleted to obtain G'. As argued above, we can assume that G' contains a universal vertex $x \neq y$ of G, which we can assume to be the root of T_G. This means in particular that $G - y$ is connected. Let x be the parent of y in T_G, and let T' be the tree obtained from T_G by contracting y onto x. This makes x the parent in T' of all children of y in T_G. Other than this, all parent-children relations are the same in T' as they were in T_G. Since x was already adjacent in G to all the vertices in the subtree of T_G rooted at y, we see that T' is indeed a uco-tree of $G - y$, and hence T' is isomorphic to T_{G-y}. Now we can repeat this argument for every vertex of $V(G) \setminus V(G')$, and conclude that T_G is T_H-contractible.

For the opposite direction, suppose T_G is T_H-contractible, and let xy be one of the edges of T_G that were contracted to obtain a tree isomorphic to T_H. Let $T' = T_G/xy$, and assume without loss of generality that x is the parent of y in T_G and that y is contracted onto x. Let G' be the trivially perfect graph having T' as a uco-tree. Note that a vertex u belongs to the subtree rooted at a vertex v in T' if and only if it already belonged to the subtree of T_G rooted at v. Therefore, for every pair of vertices $u, v \in V(G) \setminus \{y\}$, $uv \in E(G')$ if and only if $uv \in E(G)$, and hence G' is isomorphic to $G - y$. Now we can repeat this argument for every edge of T_G that was contracted, and conclude that G has an induced subgraph isomorphic to H.   □

Theorem 1 immediately gives us the result mentioned in the third row of Table 1, since checking whether a *fixed* graph H appears as an induced subgraph of an input graph G can be done trivially in polynomial time. Since Matoušek and Thomas [15] implicitly proved CONTRACTIBILITY to be NP-complete on rooted trees, Theorem 1 also implies the following result.

Corollary 1. *Both* CONTRACTIBILITY *and* INDUCED SUBGRAPH ISOMORPHISM *are NP-complete on connected trivially perfect graphs.*

The results below show that both problems can be solved in polynomial time when G is a trivially perfect graph and H is a threshold graph, even if both G and H are disconnected. Observe that these results are tight in light of Corollary 1.

Theorem 2. *Given a threshold graph G and an arbitrary graph H, it can be decided in linear time whether G can be contracted to H.*

Proof. (Sketch.) It is not hard to see that we may assume that both G and H are connected threshold graphs. We first compute the uco-trees T_G and T_H in linear time. By Theorem 1, G is H-contractible if and only if T_G is T_H-contractible. Since G is a threshold graph, T_G contains a unique maximal path $S = s_1 \cdots s_g$, starting from the root, such that each vertex on the path has at least one child. Let $T = t_1 \cdots t_h$ be the analogous path in T_H. If T_G is T_H-contractible, then a T_H-witness structure $\mathcal{W}$ of T_G can be generated as follows. For increasing i from 1 to h, we build the witness set $W(t_i)$ by repeatedly adding vertices from S to $W(t_i)$, each time picking the vertex of S with the lowest possible index that has not yet been assigned to a witness set, until the vertices in $W(t_i)$ together have at least as many neighbors in T_G as vertex t_i has neighbors in T_H. After the witness sets $W(t_1) \ldots , W(t_h)$ have been generated, witness sets for the p_i leafs adjacent to vertex t_i in T_H are formed by arbitrarily picking p_i leaves in T_G that are adjacent to the set $W(t_i)$, for every $i \in \{1, \ldots, h\}$. Greedily assigning the remaining vertices of T_G to appropriate witness sets yields $\mathcal{W}$. It can be shown that if the procedure described above does not yield a T_H-witness structure of T_G, then T_G is not T_H-contractible.   □

Theorem 3. *Given a trivially perfect graph G and a threshold graph H, it can be decided in polynomial time whether G can be contracted to H.*

Proof. (Sketch.) Using the fact that a threshold graph has at most one nontrivial connected component, it can be argued that we may assume both G and H to be connected graphs. Let T_G and T_H be the uco-trees of G and H, respectively. For every path S in T_G from the root to a leaf, we define $C(S)$ to be the graph obtained by contracting every edge of T_G, apart from those edges that have both endpoints in S or that are incident to a leaf. By Theorem 1, $C(S)$ is the uco-tree of an induced subgraph G_S of G, and it is not hard to see that G_S is a threshold graph. Note also that $C(S)$ has as many leaves as G. It can be shown that G is H-contractible if and only if there is a path S in T_G such that $C(S)$ is T_H-contractible. Hence, in order to prove Theorem 3, we do as follows. For each distinct maximal path S of T_G from the root containing only vertices that have at least one child, we check whether $C(S)$ is contractible to T_H using the linear-time procedure described in the proof of Theorem 2. Since the number of distinct paths S is $O(|V(G)|)$, the total running time is polynomial.   □

By Theorem 1, INDUCED SUBGRAPH ISOMORPHISM is equivalent to CONTRACTIBILITY on connected trivially perfect graphs. Hence the only difference between the proofs of the following results and those of the two previous theorems is in the connectivity arguments.

Theorem 4. *Given a trivially perfect graph G and a threshold graph H, it can be decided in polynomial time whether G contains an induced subgraph isomorphic to H.*

Theorem 5. *Given a threshold graph G and an arbitrary graph H, it can be decided in linear time whether G has an induced subgraph isomorphic to H.*

4   Contracting Split Graphs

In the previous section, we showed that deciding whether a threshold graph G can be contracted to an arbitrary graph H can be done in linear time. The next theorem shows that this result is not likely to be extendable to split graphs. A *hypergraph* F is a pair $(Q, \mathcal{S})$ consisting of a set $Q = \{q_1, \ldots, q_k\}$, called the *vertices* of F, and a set $\mathcal{S} = \{S_1, \ldots, S_\ell\}$ of nonempty subsets of Q, called the *hyperedges* of F. A *2-coloring* of a hypergraph $F = (Q, \mathcal{S})$ is a partition (Q_1, Q_2) of Q such that $Q_1 \cap S_j \neq \emptyset$ and $Q_2 \cap S_j \neq \emptyset$ for $j = 1, \ldots, \ell$.

Theorem 6. CONTRACTIBILITY *is NP-complete on input pairs (G, H) where G is a connected split graph and H is a connected threshold graph.*

Proof. We use a reduction from HYPERGRAPH 2-COLORABILITY, which is the problem of deciding whether a given hypergraph has a 2-coloring. This problem, also known as SET SPLITTING, is NP-complete [13]. The problem remains NP-complete when restricted to hypergraphs in which every vertex is contained in at least two hyperedges.

Let $F = (Q, \mathcal{S})$ be a hypergraph with $Q = \{q_1, \ldots, q_k\}$ and $\mathcal{S} = \{S_1, \ldots, S_\ell\}$ such that every vertex of Q appears in at least two hyperedges. We construct a split graph G as follows. We start with a clique $A = \{a_1, \ldots, a_k\}$, where the vertex $a_i \in A$ corresponds to the vertex $q_i \in Q$ for $i = 1, \ldots, k$. We add an independent set $B = \{b_1, \ldots, b_\ell\}$, where the vertex $b_i \in B$ corresponds to the hyperedge $S_i \in \mathcal{S}$ for $i = 1, \ldots, \ell$. Finally, for $i = 1, \ldots, k$ and $j = 1, \ldots, \ell$, we add an edge between a_i and b_j in G if and only if $q_i \in S_j$. We also construct a threshold graph H from a single edge $x_1 x_2$ by adding an independent set $Y = \{y_1, \ldots, y_\ell\}$ on ℓ vertices, and making each vertex of Y adjacent to both x_1 and x_2. We claim that G can be contracted to H if and only if F has a 2-coloring.

Suppose F has a 2-coloring, and let (Q_1, Q_2) be a 2-coloring of F. Let (A_1, A_2) be the partition of A corresponding to this 2-coloring of F. Note that A_1 and A_2 both form a connected set in G, since the vertices of A form a clique in G. We contract A_1 into a single vertex p_1, and we contract A_2 into a single vertex p_2. Let G' denote the resulting graph. Since (Q_1, Q_2) is a 2-coloring of F, every vertex in B is adjacent to at least one vertex of A_1 and at least one vertex of A_2

in the graph G. As a result, every vertex in B is adjacent to both p_1 and p_2 in G'. Hence G' is isomorphic to H, which means that G can be contracted to H.

Now suppose G can be contracted to H, and let $\mathcal{W}$ be an H-witness structure of G. Since we assumed that every vertex of F appears in at least two hyperedges, every vertex in A has at least two neighbors in B. This means that B is the only independent set of size ℓ in G. Since Y is an independent set of size ℓ in H, the witness sets $W(y_1), \ldots, W(y_\ell)$ each must contain exactly one vertex of B. In fact, since every vertex of A has at least two neighbors in B, we have $W(Y) = B$. This means that the two witness sets $W(x_1)$ and $W(x_2)$ form a partition of the vertices of A. By the definition of an H-witness structure and the construction of H, each witness set $W(y_i)$ is adjacent to both $W(x_1)$ and $W(x_2)$. Hence the partition $(W(x_1), W(x_2))$ of A corresponds to a 2-coloring of F.   □

Although the problem of deciding whether a split graph G can be contracted to a split graph H is NP-complete when both G and H are given as input, we will show in the remainder of this section that the problem can be solved in polynomial time when H is fixed.

Definition 1. *Let G and H be two split graphs with split partitions (C_G, I_G) and (C_H, I_H), respectively. A set $U \subseteq I_G$ with $|U| = |I_H|$ is called H-compatible if G has an H-witness structure $\mathcal{W}$ such that $W(I_H) = U$.*

Lemma 1. *Let G and H be two split graphs. Then G is H-contractible if and only if G contains an H-compatible set.*

If U is an H-compatible set of G, then, by Definition 1, G has an H-witness structure $\mathcal{W}$ such that $W(I_H) = U$. The next technical lemma shows that each of the witness sets of $\mathcal{W}$ contains a small subset, bounded in size by a function of $|V(H)|$ only, such that the collection of these subsets provide all the necessary adjacencies between the witness sets of $\mathcal{W}$.

Lemma 2. *Let G and H be two connected split graphs with split partitions (C_G, I_G) and (C_H, I_H), respectively. Let $C_H = \{x_1, \ldots, x_k\}$. A set $U \subseteq I_G$ with $|U| = |I_H|$ is H-compatible if and only if there exists a collection $\mathcal{M}$ of pairwise disjoint subsets $M(x_1), \ldots, M(x_k)$ of $V(G) \setminus U$ satisfying the following properties:*

(i) *at most one set of $\mathcal{M}$ contains a vertex of I_G, and such a set has cardinality 1 if it exists;*

(ii) *for every subset $X \subseteq U$, $M(x_i)$ contains at most two vertices a and b such that $N_G(a) \cap U = N_G(b) \cap U = X$, for $i = 1, \ldots, k$;*

(iii) $\bigcup_{i=1}^{k} |M(x_i)| \leq |C_H| \cdot 2^{|I_H|+1}$;

(iv) *for every $v \in V(G) \setminus (U \cup \bigcup_{i=1}^{k} M(x_i))$, there is a set in $\mathcal{M}$ that is adjacent to every vertex in $N_G(v) \cap U$;*

(v) *the graph $G' = G[U \cup \bigcup_{i=1}^{k} M(x_i)]$ has an H-witness structure $\mathcal{W}'$ such that $W'(I_H) = U$ and $W'(x_i) = M(x_i)$ for $i = 1, \ldots, k$.*

We call the collection $\mathcal{M}$ in Lemma 2 an *essential collection* for U, and the sets $M(x_i)$ are called *essential sets*. The fact that the total size of an essential collection does not depend on the size of G plays a crucial role in the proof of the following lemma.

Lemma 3. *Let G and H be two split graphs with split partitions (C_G, I_G) and (C_H, I_H), respectively. Given a set $U \subseteq I_G$ with $|U| = |I_H|$, it can be decided in $f(|V(H)|) \cdot |V(G)|^3$ time whether U is H-compatible, where the function f depends only on H and not on G.*

Proof. (Sketch.) Let U be a subset of I_G with $|U| = |I_H|$, and let $C_H = \{x_1, \ldots, x_k\}$. We present an algorithm that checks whether or not there exists an essential collection for U. By Lemma 2, it can be shown that U is H-compatible if and only if such a collection exists. We distinguish two cases, depending on whether or not every vertex of C_H has at least one neighbor in I_H. Here we only describe the case where every vertex of C_H has at least one neighbor in I_H. The other case, as well as the correctness proof and running time analysis, have been omitted due to page restrictions.

For every subset $X \subseteq U$, we define the set $Z_X = \{v \in V(G) \setminus U \mid N_G(v) \cap U = X\}$. Note that there are at most $2^{|U|}$ non-empty sets Z_X, and that these sets form a partition of $V(G) \setminus U$. Let $\mathcal{Z} = \{Z_X \mid X \subseteq U\}$ be the collection of these sets Z_X. Let $\mathcal{A}$ be the power set of $\mathcal{Z}$, i.e., $\mathcal{A}$ is the set consisting of all possible subsets of $\mathcal{Z}$. For every element $A \in \mathcal{A}$, we have $A = \{Z_{X_1}, \ldots, Z_{X_\ell}\}$ for some $1 \leq \ell \leq 2^{|U|}$, where $X_i \subseteq U$ for $i = 1, \ldots, \ell$ and $X_i \neq X_j$ whenever $i \neq j$. Finally, let $\mathcal{B}$ be the set of all ordered k-tuples of elements in $\mathcal{A}$, where elements of $\mathcal{A}$ may appear more than once in an element $B \in \mathcal{B}$. For any element $B \in \mathcal{B}$, we have $B = (A_1, A_2, \ldots, A_k)$, where $A_i \in \mathcal{A}$ for $i = 1, \ldots, k$.

For every $B = (A_1, A_2, \ldots, A_k) \in \mathcal{B}$, we generate a "candidate" essential set $M(x_i)$ for every vertex $x_i \in C_H$ as follows. At the start, all the vertices of C_G are unmarked, and all the vertices of $I_G \setminus U$ are marked. Of every set in A_1 that contains at least one unmarked vertex, we add one unmarked vertex to $M(x_1)$. We mark all the vertices that are added to $M(x_1)$. We then generate a candidate essential set $M(x_2)$ as before, adding an unmarked vertex from every set in A_2 that contains such a vertex to $M(x_2)$, and marking all the vertices added to $M(x_2)$. After we have generated a candidate essential set $M(x_i)$ for every vertex $x_i \in C_H$ in the way described, we define $M = \bigcup_{i=1}^{k} M(x_i)$, i.e., M is the set of marked vertices of C_G. Let $\mathcal{M}$ denote the collection of all candidate essential sets $M(x_i)$. Note that the sets of $\mathcal{M}$ are pairwise disjoint subsets of C_G. It is clear that, by construction, $\mathcal{M}$ satisfies properties (i), (ii), and (iii) of Lemma 2.

We now check whether $\mathcal{M}$ satisfies properties (iv) and (v). In order to check property (iv), we determine for every vertex $v \in V(G) \setminus (U \cup M)$ whether $\mathcal{M}$ contains a candidate essential set that is adjacent to every vertex in $N_G(v) \cap U$. $\mathcal{M}$ satisfies property (iv) if and only if such a set exists for every vertex of $V(G) \setminus (U \cup M)$. In order to check property (v), we first delete all the vertices in $V(G) \setminus (U \cup M)$, and then contract each of the candidate essential sets $M(x_i)$ into a single vertex. $\mathcal{M}$ satisfies property (v) if and only if the obtained graph is isomorphic to H. If $\mathcal{M}$ satisfies properties (iv) and (v), then $\mathcal{M}$ is an essential collection for U, and the algorithm concludes that U is H-compatible. If $\mathcal{M}$ does not satisfy properties (iv) and (v), then we unmark all vertices of C_G (the vertices of $I_G \setminus U$ remain marked) and repeat the procedure on the next element of $\mathcal{B}$. If we have processed all elements of $\mathcal{B}$ without finding an essential collection for U, then we conclude that U is not H-compatible due to Lemma 2.   □

Theorem 7. *For every fixed graph H, the problem of deciding whether a given split graph G can be contracted to H can be solved in polynomial time.*

Proof. Let G be a split graph with split partition (C_G, I_G), and suppose G is an input graph of H-CONTRACTIBILITY. Observe that contracting any edge of a split graph yields another split graph. Hence G can be contracted to a graph H only if H is a split graph with $|V(G)| \geq |V(H)|$. We can check this in time linear in the size of H, which is constant. Suppose H is a split graph, and let (C_H, I_H) be a split partition of H. By Lemma 1, G can be contracted to H if and only if G contains an H-compatible set. The number of different subsets of I_G of cardinality $|I_H|$ is $\binom{|I_G|}{|I_H|} \leq |V(G)|^{|I_H|}$. For each of those sets, we can test in $f(|V(H)|) \cdot |V(G)|^3$ time whether it is H-compatible by Lemma 3. Since the graph H is fixed, it can be decided in time polynomial in $|V(G)|$ whether G can be contracted to H.   □

5   Concluding Remarks

It is known that INDUCED SUBGRAPH ISOMORPHISM is NP-complete on cographs and on interval graphs [3,4]. Hence Corollary 1 strengthens these existing NP-completeness results. The INDUCED SUBGRAPH ISOMORPHISM problem is also known to be NP-complete on another subclass of interval graphs, called proper interval graphs [3,4], but only if the input graph H is disconnected [8]. Thus we find it interesting that this problem is NP-complete on *connected* trivially perfect graphs.

The positive results on H-CONTRACTIBILITY given in this paper and in [10] give rise to other interesting questions as well. Is H-CONTRACTIBILITY solvable in polynomial time when G is a chordal graph? Is CONTRACTIBILITY fixed parameter tractable, parameterized by the size of H, when G is a split graph or a planar graph, i.e., is there an algorithm with running time $f(|V(H)|) \cdot |V(G)|^{O(1)}$, where the function f only depends on H and not on G?

References

1. Brandstädt, A., Le, V. B., Spinrad, J.: Graph Classes: A Survey. SIAM Monographs on Discrete Mathematics and Applications (1999)
2. Brouwer, A.E., Veldman, H.J.: Contractibility and NP-completeness. Journal of Graph Theory 11, 71–79 (1987)

3. Damaschke, P.: Induced subgraph isomorphism for cographs is NP-complete. In: Möhring, R.H. (ed.) WG 1990. LNCS, vol. 484, pp. 72–78. Springer, Heidelberg (1991)
4. Garey, M.R., Johnson, D.S.: Computers and Intractability. W.H. Freeman and Co., New York (1979)
5. George, A., Liu, J.W.: Computer Solution of Large Sparse Positive Definite. Prentice Hall Professional Technical Reference (1981)
6. Golumbic, M.C.: Algorithmic Graph Theory and Perfect Graphs. Annals of Discrete Mathematics, vol. 57. Elsevier, Amsterdam (2004)
7. Heggernes, P., Kratsch, D.: Linear-time certifying recognition algorithms and forbidden induced subgraphs. Nordic Journal of Computing 14, 87–108 (2007)
8. Heggernes, P., Meister, D., Villanger, Y.: Induced Subgraph Isomorphism on Interval and Proper Interval Graphs. In: Cheong, O., Chwa, K.-Y., Park, K. (eds.) ISAAC 2010, Part II. LNCS, vol. 6507, pp. 399–409. Springer, Heidelberg (2010)
9. van 't Hof, P., Kamiński, M., Paulusma, D., Szeider, S., Thilikos, D.M.: On Contracting Graphs to Fixed Pattern Graphs. In: van Leeuwen, J., Muscholl, A., Peleg, D., Pokorný, J., Rumpe, B. (eds.) SOFSEM 2010. LNCS, vol. 5901, pp. 503–514. Springer, Heidelberg (2010)
10. Kamiński, M., Paulusma, D., Thilikos, D.M.: Contractions of planar graphs in polynomial time. In: de Berg, M., Meyer, U. (eds.) ESA 2010. LNCS, vol. 6346, pp. 122–133. Springer, Heidelberg (2010)
11. Levin, A., Paulusma, D., Woeginger, G.J.: The computational complexity of graph contractions I: polynomially solvable and NP-complete cases. Networks 51, 178–189 (2008)
12. Levin, A., Paulusma, D., Woeginger, G.J.: The computational complexity of graph contractions II: two tough polynomially solvable cases. Networks 52, 32–56 (2008)
13. Lovász, L.: Coverings and colorings of hypergraphs. In: Proceedings of the 4th Southeastern Conference on Combinatorics, Graphs Theory, and Computing, pp. 3–12. Utilitas Mathematica Publishing, Winnipeg (1973)
14. Mahadev, N., Peled, U.: Threshold graphs and related topics. Annals of Discrete Mathematics, vol. 56. North Holland, Amsterdam (1995)
15. Matoušek, J., Thomas, R.: On the complexity of finding iso- and other morphisms for partial k-trees. Discrete Mathematics 108(1-3), 343–364 (1992)
16. Robertson, N., Seymour, P.D.: Graph Minors.XIII. The Disjoint Paths Problem. Journal of Combinatorial Theory, Series B 63(1) (1995)
17. Semple, C., Steel, M.: Phylogenetics. Graduate series Mathematics and its Applications. Oxford University Press, Oxford (2003)
18. Wolk, E.S.: The comparability graph of a tree. Proceedings of the American Mathematical Society 13, 789–795 (1962)
19. Wolk, E.S.: A note on "The comparability graph of a tree". Proceedings of the American Mathematical Society 16, 17–20 (1965)
20. Yan, J.-H., Chen, J.-J., Chang, G.J.: Quasi-threshold graphs. Discrete Applied Mathematics 69, 247–255 (1996)

Planarity Testing Revisited

Samir Datta and Gautam Prakriya

Chennai Mathematical Institute
India
{sdatta,gautam}@cmi.ac.in

Abstract. Planarity Testing is the problem of determining whether a given graph is planar while planar embedding is the corresponding construction problem. The bounded space complexity of these problems has been determined to be exactly deterministic logarithmic space by Allender and Mahajan [AM00] with the aid of Reingold's result [Rei08]. Unfortunately, the algorithm is quite daunting and generalizing it to, say the bounded genus case, seems a tall order.

We present a simple planar embedding algorithm running in logspace. The algorithm uses the unique embedding of 3-connected planar graphs, a variant of Tutte's criterion on the conflict graphs of cycles and an explicit change of basis for the cycle space.

We also present a logspace algorithm to find an obstacle to planarity, viz. a Kuratowski minor, for non-planar graphs. To the best of our knowledge this is the first logspace algorithm for this problem.

1 Introduction

Planarity Testing is a fundamental problem in algorithmic graph theory. Along with the problem of actually obtaining a planar embedding, it is a prerequisite for algorithms designed to work specifically for planar graphs.

Our focus is on the bounded space complexity of the planar embedding problem because we know that many graph theoretic problems like reachability [BTV09], perfect matching [DKR10, DKT10], and even isomorphism [DLN08, DLN+09] have efficient bounded space algorithms when provided graphs embedded on the plane.

Almost a decade ago, building on previous work by Ramachandran and Reif [RR94], Allender and Mahajan [AM00, AM04], proved that Planarity Testing is contained in SL and is L-hard. With Reingold's result [Rei08] proving SL = L, this gave a tight complexity theoretic classification for the problem. This seemed to be the end of the story as far as the problem of Planarity Testing vis-a-vis the logspace world was concerned.

The only catch was, the algorithm described in the paper was quite complicated - in fact a simpler SL algorithm was listed as one of the open questions in [AM00]. We would be satisfied with a complicated algorithm if all we were concerned with was pigeonholing the complexity of the problem. Planarity testing, however, happens to be a fundamental task in Topological Graph theory and

M. Ogihara and J. Tarui (Eds.): TAMC 2011, LNCS 6648, pp. 540–551, 2011.
© Springer-Verlag Berlin Heidelberg 2011

a first step towards problems like toroidicity testing and bounded genus testing (also listed as open problems in [AM00]). Therefore, if we are to make any progress towards a tighter classification of these problems (for which no efficient bounded space algorithm is known), especially, through non-blackbox use of a planarity algorithm, it is advisable to search for a less daunting algorithm. This work is the result of this search.

In addition to its simplicity, the interplay between the properties of 3-connected planar graphs lends a certain elegance to the algorithm, at least in our, necessarily, biased eyes. We also describe a logspace algorithm to identify a Kuratowski minor if the graph is non-planar.

2   Related Work

Here we survey the related work very briefly - see the paper by Allender and Mahajan [AM00] for a detailed survey. The algorithmic aspects of Planarity Testing have been studied since the inception of Computer Science. It is clear that as far as sequential computation is concerned, a linear algorithm is optimal. Such algorithms include the one by Hopcroft and Tarjan [HT74]. The next result concerns parallel models of computation. Ramachandran and Reif proposed a complicated algorithm which worked in logarithmic time and performed almost linear work and can be interpreted as placing the problem in the complexity class AC^1. The final frontier was bounded space computation and initial sorties had already taken place in the early eighties. Reif [Rei84] proved that planarity testing for degree 3 graphs is in the SL-hierarchy while JáJá and Simon [JS82a, JS82b] proved that planarity testing is in the NL-hirerarchy. After Nisan and TaShma's result [NTS95] and the Immerman-Szelpcsenyi Theorem [Imm88, Sze88] these bounds become SL and NL respectively. The paper by Allender and Mahajan completed the campaign when they proved that Planarity Testing is SL-complete and Reingold's result [Rei08] set the final seal by proving SL = L.

2.1   Comparison with [AM04]

[AM04] works with biconnected graphs. The main steps of [AM04] are given below. Given biconnected graph G as input

1. Construct the open ear decomposition.
2. Direct the ears to get an acyclic graph G_{st} also known as the (s,t)-numbering graph.
3. Construct a local replacement graph G_l.
4. G_{st} and G_l uniquely specify a spanning tree T_{st}.
5. Construct conflict graphs w.r.t each fundamental cycle. For each cycle obtain the constraint graph G^* by adding some additional constraints.
6. This gives an embedding of G_l from which we obtain an embedding of G.

Our algorithm deals with 3-connected components of the graph. This significantly simplifies the subsequent steps primarily because 3-connected planar graphs have a unique planar embedding. In other words, the conflict graph is connected, therefore it has a unique 2-coloring (if any). Given a 3-connected graph G, our algorithm constructs the conflict graphs w.r.t the fundamental cycles of some spanning tree of G, obtains the unique 2-coloring of each of these conflict graphs and extracts the embedding of G by a change of cycle basis. Putting together the (unique)embeddings of the triconnected components of a graph is not difficult. Therefore we side-step the more complicated steps of [AM04] including the algorithm to obtain an open-ear decomposition.

3   Definitions and Preliminaries

We assume familiarity with basic complexity theory, in general, and bounded space classes, in particular, as described in any standard text e.g [AB09]. We will also assume familiarity with graph theory as described in texts like [Die05, Wes00]. Below, we explicate all non-standard material we will have occasion to use.

Definition 1. *The bridges of a cycle C consist of:*

- *For every connected component X of $G \setminus C$, the induced graph $G[X \cup A_X]$ where $A_X \subseteq C$ are the vertices of C adjacent to some vertex of X (the vertices in A_X are also called points of attachment).*
- *The chords of C - here the endpoints of the chord are its points of attachment*

Definition 2. *Two bridges B_1, B_2 of a cycle C conflict iff either of the following conditions hold:*

1. *a_i, a'_i are two points of attachment of B_i on C for $i \in \{1, 2\}$ such that they occur in the order a_1, a_2, a'_1, a'_2 along the cycle C.*
2. *B_1, B_2 have three common points of attachment on the cycle C.*

Definition 3. *The conflict graph $H_C(G)$ of a graph w.r.t. a cycle C is the graph obtained by taking the bridges of C as vertices and joining two vertices by an edge iff the corresponding bridges conflict.*

Definition 4. *Given a spanning tree T of a biconnected graph G, and an edge $e \in E(G) \setminus E(T)$, the subgraph $T \cup e$ contains a unique cycle $C(e)$ called the fundamental cycle of e. We say that a face of an embedded planar graph is fundamental if it is a fundamental cycle w.r.t some fixed spanning tree.*

The following is an easy consequence of Reingold's result [Rei08] We single it out of a set of similar elementary graph computations summarized in Section 3.2 of [AM00] which can be done in logspace because it is of special significance for us.

Fact 1. *The list of edges in each fundamental cycle of G w.r.t. a spanning tree T can be obtained by a logspace transducer.*

Fact 2. *(Proposition 4.2.10 and Theorem 4.5.2 [Die05]) The faces of a 3-connected planar graph G are exactly the induced non-separating cycles of G. Further, a 3-connected graph is planar iff every edge lies on exactly two induced, non-separating cycles.*

Definition 5. *A Combinatorial Embedding of a graph is a set of cyclic orders of the edges around each of the vertices.*

We omit the proof of the following proposition.

Proposition 1. *Given the faces of a 3-connected embedded graph, it is possible to construct in logspace, a combinatorial embedding of the graph.*

4   Outline of Algorithms

4.1   Outline of Planar Embedding Algorithm

We now motivate and informally describe the planarity Testing Algorithm. It is easy to to see that the conflict graph w.r.t any cycle in a planar graph G is bipartite .

Conversely, Tutte [Tut58] has shown that a graph is planar iff the conflict graph with respect to *every* cycle is bipartite. But since a graph potentially has exponentially many cycles a direct application of this result does not yield even a polynomial time algorithm. Thus we ought to seek a small set of cycles such that restricting our attention to the bipartiteness of their conflict graphs suffices to yield a planarity test. The set of fundamental cycles w.r.t. some spanning tree seems to be a natural candidate. Unfortunately, there are non-planar graphs which have a spanning tree such that the conflict graph w.r.t. each of the fundamental cycles is bipartite. For instance, in a K_5 with the spanning tree being a star centered at some vertex, the conflict graph w.r.t. every fundamental cycle has a single vertex.

Suppose, however that we are able to find a deterministic algorithm that constructs a valid planar embedding whenever its input is a planar graph and either fails when supplied with a non-planar graph or outputs a non-planar embedding.

Theorem 1 describes how we can eliminate any non-planar graphs at this stage.

Thus it suffices to focus on finding an embedding algorithm that works correctly under the promise that the given graph is planar. We can without much loss of generality, strengthen the promise and assume that the graph is, in addition to being planar, also 3-connected. This is because finding the triconnected components is known to be in logspace (see Lemma 1 and so is patching together the given planar embeddings of triconnected components of a graph. (see Lemma 2).

Concentrating on 3-connected planar graphs we observe that:

- The graph has a unique planar embedding.
- The conflict graphs w.r.t. any cycle is connected (see Section 5.2), which enables us to bipartition the bridges in a unique way so that bridges in a partition lie on one side of the cycle in question.
- Ideally we would like to pick a small set of cycles and determine which edges lie inside them and somehow piece together the combinatorial embedding from this information. A natural choice for such a set is the set of fundamental cycles w.r.t. an arbitrary spanning tree. But the following two problems crop up:
  - Though we have a bipartition of bridges for each fundamental cycle , it is not clear which partition is mapped inside and which outside.
  - How do we actually piece together a combinatorial embedding once we know this information?
- The first problem can be solved if we knew at least one face of the unique embedding, because then we can just think of the face as the external face and stipulate that every edge (not part of the face) is contained inside it. Thus for every fundamental cycle, any bridge containing edges of the external face would lie outside the cycle - this fixes which bridges lie inside and which outside. But Corollary 1 tells us that for the graphs in question there is always a fundamental cycle which is a face. This, combined with Fact 2 ensures that we can find such a face in logspace.
- We solve the second problem in Section 5.4, where we first show that finding a solution is equivalent to a change of the cycle basis from fundamental cycles to faces. Then we show an explicit way to perform this change of basis in logspace.

4.2   Outline of Algorithm to Find Kuratowski Minors

In Section 6 we describe an algorithm to find a Kuratowski (i.e. a K_5 or $K_{3,3}$) minor in a non-planar graph. The algorithm indentifies a cycle with a non-bipartite conflict graph. It then finds an induced odd cycle in this conflict graph. Finally, it contracts/ deletes some of the bridges and edges of attachment of these bridges to yield a Kuratowski minor.

5   Planarity Testing

5.1   Reduction to the Triconnected case

It is a well known fact that if a graph is non-planar then one of it's 3-connected components is non-planar. [Wes00]

Lemma 1. *(Lemma 3.3 [DNTW09]) The triconnected components of a graph can be obtained in logpace.*

We omit the following proof due to space constraints.

Lemma 2. *Given a combinatorial planar embedding of the triconnected components of a graph, it is possible to obtain the planar embedding of the graph in logspace.*

See [DP11] for the proof.

5.2   The Conflict Graph

Lemma 3. *Given a 3-connected graph G and an arbitrary cycle C in the graph the conflict graph $H_C(G)$ is connected.*

Proof. It is easy to see that for an embedded planar graph G and any cycle C contained in it, the conflict graph, $H_C(G)$ is bipartite (with the bipartition being given by whether a bridge is mapped "inside" or "outside" the cycle). We now show that $H_C(G)$ is connected.

For every pair of points $x, y \in C$ there are two paths from x to y along C. We will call these paths $P_{x,y}$ and $P'_{x,y}$. Let $\mathcal{B}$ be a set of bridges that correspond to one of the connected components of $H_C(G)$.

Pick a pair u, v such that:(a) u and v are points of attachment of bridges in $\mathcal{B}$. (b) u and v are not adjacent. (c) The points of attachment of all bridges in $\mathcal{B}$ lie in either $P_{u,v}$ or $P'_{u,v}$. (Assume WLOG that they lie in $P_{u,v}$).

We now show that one can always find such a pair u, v. Pick a bridge $B' \notin \mathcal{B}$. The points of attachment of B' divide C into a number of segments .(atleast 3 since every bridge has 3 or more points of attachment.) Since no bridge in $\mathcal{B}$ conflicts with B' and the vertices corresponding to $\mathcal{B}$ in $H_C(G)$ form a connected component, it follows easily that all points of attachments of $\mathcal{B}$ must lie in one of the segments(as $B' \notin \mathcal{B}$. This implies that there is a point w in C(a point of attachment of B') that is not a point of attachment of any bridge in $\mathcal{B}$. Therefore we can find a pair u, v with the properties listed above because we can pick u, v to be points of attachment of $\mathcal{B}$ on either side of w.

Now, for every point $x \notin \{u, v\}$ on $P_{u,v}$, $\exists$ a bridge $B \in \mathcal{B}$ with points of attachment b_1, b_2 s.t b_1 precedes x and b_2 succeeds x in $P_{u,v}$. If not then $\forall B \in \mathcal{B}$ either all points of attachment of B lie between u and x or all of them lie between x and v. This implies that in $H_C(G)$ vertices corresponding to bridges in $\mathcal{B}$ with u as a point of attachment are not connected to vertices corresponding to bridges with v as a point of attachment. This is a contradiction since $\mathcal{B}$ corresponds to a connected component in $H_C(G)$.

Now, if $\{u, v\}$ is not a separating pair then $\exists$ a bridge $B' \notin \mathcal{B}$ with points of attachment in both $P_{u,v} \setminus \{u, v\}$ and $P'_{u,v} \setminus \{u, v\}$. From the above it follows that B' conflicts with a bridge in $\mathcal{B}$ which is not possible. Therefore, $\{u, v\}$ is a separating pair.   □

Fact 3. *In a planar graph G the conflict graph with respect to every cycle is bipartite.*

See [Tut58].

5.3   Inside and Outside a Fundamental Cycle

Next we determine which bridges w.r.t. each of the fundamental cycles are mapped inside and which are mapped outside the concerned cycle. Basically we find one *face* of the graph and call it the external face. Thus w.r.t every cycle bridges containing edges from this face are mapped outside. This completely fixes the 2-coloring of conflict graphs w.r.t all fundamental cycles.

Proposition 2. *Any biconnected embedded planar graph has a fundamental* face *(i.e. a fundamental cycle which is also a face) w.r.t. each of its spanning trees.*

Proof. Given a biconnected planar graph and a spanning tree in the graph, it is easy to see that the set of dual edges corresponding to the non-tree edges of the biconnected planar graph form a tree (after reducing all multi-edges to a single edge) - the so called dual tree. This follows from observing that cycles in the primal correspond to cuts in the dual (Proposition 4.6.1 [Die05]) and therefore the set of edges dual to the non-tree edges are connected in the dual graph, the proof being completed by applying Euler's relation to verify that the number of such edges is exactly one less than the number of faces(vertices in the dual). It is easy to see that the face of the original graph corresponding to a leaf of the dual tree is a fundamental face.   □

A direct consequence of Fact 1, Proposition 2, Fact 2 is the following (since finding whether a cycle is induced and non-separating is a logspace predicate given the list of edges in the cycle on the input tape).

Corollary 1. *Every 3-connected planar graph has at least one fundamental face and this can be found by a Logspace transducer.*

Knowing which edges of the graph map to the region enclosed by each of the fundamental cycles, we proceed to construct a purported embedding of the given graph. If the graph is indeed planar we will obtain a valid planar embedding, else we will either fail to obtain an embedding or obtain a non-planar embedding. Theorem 1 describes how we eliminate any non-planar graphs.

5.4   Obtaining a Planar Embedding

Here we exhibit a way to identify the edges in each face of the given planar 3-connected graph. We also describe how to reject non-planar graphs. Let $C(e_0)$ be the fundamental face which is picked to be the external face.

At the heart of the proof is a change of basis in the cycle space over $\mathbb{Z}_2$. Notice that we have an "implicit" representation of the faces (see below) and we are seeking an explicit representation in terms of which edges constitute a face. Since fundamental cycles and internal faces both form a basis of the cycle space over $\mathbb{Z}_2$, we just need to invert the matrix that expresses the fundamental cycles as a linear combination of faces. Though seemingly this would place the problem in $\oplus\mathsf{L}$ we give an explicit way to perform this inversion which yields a L upper bound.

Returning to the "implicit" representation of faces, given an embedding of the graph, there is a natural bijection between non-tree edges and faces(excluding the external face) viz. one that maps a non-tree edge e to that face $f(e)$ adjacent to e which lies *inside* the fundamental cycle $C(e)$ (w.r.t. the external face $C(e_0)$). With faces labeled in this way, the 2-coloring of the conflict graph described in the preceding section tells us which faces are contained within the fundamental cycle $C(e)$.

Let us start by fixing some notation. For distinct non-tree edges e_1, e_2, define $e_1 \prec e_2$ iff in a 2-coloring of the conflict graph $H_{C(e_2)}(G)$, the colors of the vertices corresponding to bridges containing e_0, e_1 get different colors. Intuitively, this means that (if G is planar) $C(e_2)$ separates e_1 from e_0 in the unique planar embedding of G.

For each $e \in E(G) \setminus \{e_0\}$, let $P(e)$ denote:

$$\{e' \in E(G) \setminus (E(T) \cup \{e\})|e' \prec e \wedge \nexists e'' : e' \prec e'' \prec e\},$$

i.e, $P(e)$ is the set of maximal elements w.r.t $\prec$ in $\{e'|e' \prec e\}$ in other words it is the set of immediate predecessors of e w.r.t $\prec$. Further, let $F(e)$ denote

$$\bigoplus_{e' \in P(e) \cup \{e\}} C(e')$$

Note that for sets A, B the set $A \oplus B$ denotes the symmetric difference of the two sets and the notation above refers to an iteration of this operation.

If the graph is 3-connected planar then in its unique planar embedding with $C(e_0)$ as the external face, $P(e)$ consists of the set of non-tree edges that are enclosed by $C(e)$ but not by any $C(e'')$ which is also enclosed by e. Intuitively, it clear that these are exactly the non-tree edges occurring on the face $f(e)$ (see description in the preceding paragraph). We will make this intuition precise and in fact, show the following:

Lemma 4. *For each non-tree edge e, the face $f(e)$ consists exactly of the edges in the set $F(e)$.*

Proof. Given a fundamental cycle C let $\phi(C)$ represent the number of faces that lie inside C and $\psi(C)$ represent the number of fundamental cycles that lie inside C including C. Notice that for fundamental cycles which are also faces, $\phi(C) = \psi(C) = 1$. In fact for every fundamental cycle C, as an easy consequence of Euler's formula it follows that $\phi(C) = \psi(C)$.

Let $e \in E(G) \setminus E(T)$. We will first show that $F(e)$ is a face. Since each $e' \in P(e)$ lies inside $C(e)$ and for every non-tree $e'' \notin P(e)$, such that $e'' \prec e$, there is exactly one $e' \in P(e)$ such that $e'' \prec e'$. There is one such e' because we know that every non-tree edge enclosed by $C(e)$ is either enclosed by some other fundamental cycle $C(e_1)$ or otherwise is in $P(e)$. Thus by induction on the "enclosure depth" of an edge we get an $e' \in P(e)$ which encloses it. The uniqueness follows from observing that if both $e \prec e' \wedge e \prec e'_1$ then either $e'_1 \prec e'$ or $e' \prec e'_1$. - this is due to planarity - therefore both e', e'_1 cannot be in $P(e)$. Thus we get,

$$\psi(C(e)) = 1 + \sum_{e' \in P(e)} \psi(C(e'))$$

therefore:

$$\phi(C(e)) = 1 + \sum_{e' \in P(e)} \phi(C(e'))$$

Now since every fundamental cycle can be written as a sum of internal faces of G and $F(e)$ is a sum of a set of fundamental cycles (where all computation is over

$\mathbb{Z}_2$), it is also a sum of some internal faces. Since only faces contained within $C(e)$ figure in this sum, $F(e)$ must be a disjoint sum of some cycles enclosed by $C(e)$. We can be more specific, the sum:

$$\bigoplus_{e' \in P(e)} C(e')$$

includes exactly the faces inside $C(e)$ which are also faces inside some $C(e')$ for $e' \in P(e)$. Thus $F(e)$ includes exactly the faces contained inside $C(e)$ but not in any $C(e')$ for $e' \in P(e)$. Now, it is easy to see from the expression for $\phi(C(e))$ that there is exactly one such face, so it follows from the linear independence of the faces that this must be $f(e)$.   □

Thus, we can complete the proof of the following:

Theorem 1. *Given a graph G, constructing a planar embedding for G if it is planar and otherwise rejecting it, can be done in logspace.*

Proof. Given a graph we obtain its 3-connected components using Lemma 1. If each triconnected compoenent is planar, we will successfully obtain a planar embedding for each component and then obtain the proof with the aid of Lemma 2. If some triconnected component is not planar, we will either obtain an $F(e)$ which is not induced or is separating or we will obtain an edge lying on at least three $F(e)$'s and therefore reject.   □

6   Finding Kuratowski Subgraphs

We describe an an algorithm to obtain a Kuratowski subgraph given a cycle with a non-bipartite conflict graph. As a consequence, to obtain a Kuratowski subgraph, it is sufficient to find such a cycle in a subgraph G' of G. We do this in a series of lemmas:

Lemma 5. *Given a non-planar graph G, we can, in logspace, find a non-planar subgraph G' and a cycle $C \subseteq G'$ s.t. $H_C(G')$ is non-bipartite.*

Proof. Given access to a routine for planarity checking, and a non-planar graph G, we can find a minimal non-planar subgraph G' of g in the following sense. Order the edges of G arbitrarily as $e_1, \ldots, e_m$. Now consider the smallest i such that the graph G' formed by the union of the first i edges $e_1, \ldots, e_i$ is non-planar. Notice that this means that $G' \setminus e_i$ is necessarily planar.

Now, we show how one can find a cycle in this graph G' such that the conflict graph $H_C(G')$ is non-bipartite. Construct a planar embedding of $G' \setminus \{e\}$. The endpoints, say x, y, of e_i must lie on different faces of this embedding because G' is non-planar.

Find a path in the dual graph of the embedding between any two faces F_x, F_y incident respectively on x, y. (this path must avoid any other faces incident on x or y.) The symmetric difference of the faces in this path is a cycle C one of whose bridges is e_i.

We claim that $H_C(G')$ is non-bipartite. If it were bipartite then it would be possible to give an orientation to each of it's bridges such that conflicting bridges got opposing orientations. But since the bridges of C are all planar, this would imply that G' is planar . Therefore, $H_C(G')$ is non-bipartite.   □

Lemma 6. *Given a non-planar graph G' and a cycle C witnessing this via the non-bipartiteness of $H_C(G')$, we can, in logspace, find an induced odd cycle $\bar{C}$ in $H_C(G')$.*

Proof. Since the conflict graph $H_C(G')$ is non-bipartite, it contains an odd cycle. We aim to find an induced odd cycle in this graph. For this consider a spanning tree of $H_C(G')$. 2-color the spanning tree. Notice that there must exist a non-tree edge between two vertices with the same color (else the graph would have been bipartite). Find a non-tree edge e such that all the chords of the fundamental cycle $C(e)$ get opposite colors - a simple induction on the number of chords that a fundamental cycle has, shows that such an edge must exist and locating it in logspace is easy.

We will construct a chordless cycle from $C(e)$ by replacing some tree paths by chords of $C(e)$. To do this, let $e = (u_1, u_k)$ and let the vertices of the tree path from u_1 to u_k be $u_2, \ldots, u_{k-1}$ in order. Call the point u_i of a chord (u_i, u_j) for $i < j$ as the origin of the chord and u_j the end of the chord. Now start walking from u_1 to u_k along the tree path, outputting tree edges till the origin of a chord is encountered. Output the chord and move on to the end of the chord and repeat. It is easy to see that the edges output by this procedure along with the non-tree edge (u_1, u_k) form an induced cycle because either the origin or endpoint of any other chord of $C(e)$ is not on this cycle. Also, because the endpoints of all edges on this cycle are oppositely colored except for (u_1, u_k), this is an odd cycle.   □

At this point we have a cycle C in G with a non-bipartite conflict graph $H_C(G')$ and an induced odd cycle $\bar{C}$ in $H_C(G')$ witnessing this. Now we prove the following,

Lemma 7. *Given an induced odd cycle $\bar{C}$ in $H_C(G')$, we can, in logspace, find a Kuratowski subgraph.*

Proof. First, suppose that two conflicting bridges B_1 and B_2 in the odd cycle share 3 points of attachment a_1, a_2 and a_3. If either of the bridges(B_1 say.) has another point of attachment a_4 then, clearly the points of attachment a_1, a_3 of B_2 alternate with the points of attachment a_2, a_4 of B_1. Therefore, it is sufficient to deal with the case when the case where both B_1 and B_2 have just 3 points of attachment. In this case it is not difficult to see that any bridge that conflicts with B_1 also conflicts with B_2 and vice-versa. Hence $\bar{C}$ must be a 3-cycle. We deal with this case in lemma 8.

Next, consider the case where $\bar{C}$ is an odd cycle of size greater than 3 and the conflict of a bridge B in the odd cycle with any other bridge in the odd-cycle is witnessed by 2 points of attachment of B i.e we exclude the possibility that 2 bridges share 3 points of attachment.

It is easy to see that $\exists$ 2 points of attachment for every bridge that witness both it's conflicts. Thus by contracting all bridges (excluding points of attachment) to single points and removing edges of attachment, one can reduce all bridges to paths while maintaining all conflicts. This can be done in logspace since for each bridge one only needs to remember the witnesses of conflict of the bridge with its neighbours (in $\bar{C}$).(atmost 8 points.)

If we label the vertices of $\bar{C}$ as $v_{B_0}, v_{B_1}, \ldots, v_{B_{2k}}$ $(k > 1)$ for some positive integer k, and the points of attachment of bridge B_i as u_i, v_i then the points occur along C in the order: $u_0, v_{2k}, u_1, v_0, u_2, v_1, \ldots, u_{2k-1}, v_{2k-2}, u_{2k}, v_{2k-1}$.

Now, consider the cycle $u_{2k}, u_0, v_{2k}, v_0, U'$ (Where $U' = \{\{v_1, u_3\} \cup B_3 \cup \{v_3\} \cup \ldots \cup \{u_{2k-1}\}\}$) along with the bridges B_{2k}, B_0 the paths connecting u_0 and U', v_{2k} and U' and the path $\{v_0, u_2\} \cup B_2 \ldots \{v_{2k-2}, u_{2k}\}$ Clearly, this is a K_5 minor.(see appendix for an example). □

Finally we are left with the following case:

Lemma 8. *Given a cycle in G' with three mutually conflicting bridges, it is possible to extract a Kuratowski minor of G' in logspace.*

Proof. Clearly, it is sufficient to consider 4 points of attachment for each bridge since 4 points of attachment suffice to witness conflict with two other bridges. We can contract the bridges (excluding the points of attachment) to single points and reduce the problem of finding a Kuratowski subgraph in the above graph to that of finding a Kuratowski minor in a non-planar graph with atmost 15 vertices and 24 edges. Since this is a constant sized graph we can find a Kuratowski minor by brute-force. □

Acknowledgements

The first author would like to acknowledge discussions with Eric Allender, Abhishek Bhrushundi, Ashish Kabra, Raghav Kulkarni, Nutan Limaye, Meena Mahajan, and Prajakta Nimbhorkar on this topic.

References

[AB09]   Arora, S., Barak, B.: Computational Complexity: A Modern Approach, 1st edn. Cambridge University Press, New York (2009)

[AM00]   Allender, E., Mahajan, M.: The complexity of planarity testing. In: Reichel, H., Tison, S. (eds.) STACS 2000. LNCS, vol. 1770, pp. 87–98. Springer, Heidelberg (2000)

[AM04]   Allender, E., Mahajan, M.: The complexity of planarity testing. Inf. Comput. 189(1), 117–134 (2004)

[BTV09]   Bourke, C., Tewari, R., Vinodchandran, N.V.: Directed planar reachability is in unambiguous log-space. TOCT 1(1) (2009)

[Die05]   Diestel, R.: Graph Theory. Graduate Texts in Mathematics. Springer, Heidelberg (2005)

[DKR10]   Datta, S., Kulkarni, R., Roy, S.: Deterministically isolating a perfect matching in bipartite planar graphs. Theory Comput. Syst. 47(3), 737–757 (2010)

[DKT10]   Datta, S., Kulkarni, R., Tewari, R.: Perfect matching in bipartite planar graphs is in ul. Electronic Colloquium on Computational Complexity (ECCC) 17, 201 (2010)

[DLN08]   Datta, S., Limaye, N., Nimbhorkar, P.: -Connected planar graph isomorphism is in log-space3. In: Proceedings of the 28th Annual Conference on Foundations of Software Technology and Theoretical Computer Science (FSTTCS), pp. 153–162 (2008)

[DLN$^+$09]   Datta, S., Limaye, N., Nimbhorkar, P., Thierauf, T., Wagner, F.: Planar graph isomorphism is in log-space. In: IEEE Conference on Computational Complexity, pp. 203–214 (2009)

[DNTW09]   Datta, S., Nimbhorkar, P., Thierauf, T., Wagner, F.: Graph isomorphism for $K_{3,3}$-free and K_5-free graphs is in log-space. In: FSTTCS, pp. 145–156 (2009)

[DP11]   Datta, S., Prakriya, G.: Planarity testing revisited. Electronic Colloquium on Computational Complexity (ECCC) 18, 9 (2011)

[HT74]   Hopcroft, J.E., Tarjan, R.E.: Efficient planarity testing. Journal of the ACM 21(4), 549–568 (1974)

[Imm88]   Immerman, N.: Nondeterministic space is closed under complementation. SIAM J. Comput. 17(5), 935–938 (1988)

[JS82a]   JáJá, J., Simon, J.: Parallel algorithms in graph theory: Planarity testing. SIAM J. Comput. 11(2), 314–328 (1982)

[JS82b]   JáJá, J., Simon, J.: Space efficient algorithms for some graph theoretical problems. Acta Inf. 17, 411–423 (1982)

[NTS95]   Nisan, N., Ta-Shma, A.: Symmetric logspace is closed under complement. Chicago J. Theor. Comput. Sci. (1995)

[Rei84]   Reif, J.H.: Symmetric complementation. J. ACM 31(2), 401–421 (1984)

[Rei08]   Reingold, O.: Undirected connectivity in log-space. J. ACM 55(4) (2008)

[RR94]   Ramachandran, V., Reif, J.H.: Planarity testing in parallel. Journal of Computer and System Sciences 49, 517–561 (1994)

[Sze88]   Szelepcsényi, R.: The method of forced enumeration for nondeterministic automata. Acta Inf. 26(3), 279–284 (1988)

[Tut58]   Tutte, W.T.: A homotopy theorem for matroids, I, II. Trans. AMS 98, 144–174 (1958)

[Wes00]   West, D.B.: Introduction to Graph Theory, 2nd edn. Prentice-Hall, Englewood Cliffs (August 2000)

Generalized Satisfiability
for the Description Logic $\mathcal{ALC}$
(Extended Abstract)

Arne Meier[1] and Thomas Schneider[2]

[1] Leibniz Universität Hannover, Germany
meier@thi.uni-hannover.de
[2] University of Bremen, Germany
tschneider@informatik.uni-bremen.de

Abstract. The standard reasoning problem, concept satisfiability, in the basic description logic $\mathcal{ALC}$ is PSPACE-complete, and it is EXPTIME-complete in the presence of unrestricted axioms. Several fragments of $\mathcal{ALC}$, notably logics in the $\mathcal{FL}$, $\mathcal{EL}$, and DL-Lite families, have an easier satisfiability problem; sometimes it is even tractable. We classify the complexity of the standard satisfiability problems for all possible Boolean and quantifier fragments of $\mathcal{ALC}$ in the presence of general axioms.

1 Introduction

Standard reasoning problems of description logics, such as satisfiability or subsumption, have been studied extensively. Depending on the expressivity of the logic, the complexity of reasoning for DLs between fragments of the basic DL $\mathcal{ALC}$ and the OWL 2 standard $\mathcal{SROIQ}$ is between trivial and NEXPTIME.

For $\mathcal{ALC}$, concept satisfiability is PSPACE-complete [32]. In the presence of unrestricted axioms, it is EXPTIME-complete due to the correspondence with propositional dynamic logic [30, 34, 17]. Since the standard reasoning tasks are interreducible, subsumption has the same complexity.

Several fragments of $\mathcal{ALC}$, such as logics in the $\mathcal{FL}$, $\mathcal{EL}$ or DL-Lite families, are well-understood. They usually restrict the use of Boolean operators and of quantifiers, and it is known that their reasoning problems are often easier than for $\mathcal{ALC}$. We now need to distinguish between satisfiability and subsumption because they are no longer obviously interreducible if certain Boolean operators are missing. Concept subsumption with respect to acyclic and cyclic terminologies, and even with general axioms, is tractable in the logic $\mathcal{EL}$, which allows only conjunctions and existential restrictions, [4, 11], and it remains tractable under a variety of extensions such as nominals, concrete domains, role chain inclusions, and domain and range restrictions [5, 6]. Satisfiability for $\mathcal{EL}$, in contrast, is trivial, i.e., every $\mathcal{EL}$-ontology is satisfiable. However, the presence of universal quantifiers usually breaks tractability: Subsumption in $\mathcal{FL}_0$, which allows only conjunction and universal restrictions, is coNP-complete [27] and increases to PSPACE-complete with respect to cyclic terminologies [3, 21] and to

M. Ogihara and J. Tarui (Eds.): TAMC 2011, LNCS 6648, pp. 552–562, 2011.
© Springer-Verlag Berlin Heidelberg 2011

EXPTIME-complete with general axioms [5, 20]. In [15, 16], concept satisfiability and subsumption for several logics below and above $\mathcal{ALC}$ that extend $\mathcal{FL}_0$ with disjunction, negation and existential restrictions and other features, is shown to be tractable, NP-complete, coNP-complete or PSPACE-complete. Subsumption in the presence of general axioms is EXPTIME-complete in logics containing both existential and universal restrictions plus conjunction or disjunction [18], as well as in $\mathcal{AL}$, where only conjunction, universal restrictions and unqualified existential restrictions are allowed [14]. In DL-Lite, where atomic negation, unqualified existential and universal restrictions, conjunctions and inverse roles are allowed, satisfiability of ontologies is tractable [13]. Several extensions of DL-Lite are shown to have tractable and NP-complete satisfiability problems in [1, 2]. The logics in the $\mathcal{EL}$ and DL-Lite families are so important for (medical and database) applications that OWL 2 has two profiles that correspond to logics in these families.

This paper revisits restrictions to the Boolean operators in $\mathcal{ALC}$. Instead of looking at one particular subset of $\{\sqcap, \sqcup, \neg\}$, we are considering all possible sets of Boolean operators, and therefore our analysis includes less commonly used operators such as the binary exclusive *or* $\oplus$. Our aim is to find for *every* possible combination of Boolean operators whether it makes satisfiability of the corresponding restriction of $\mathcal{ALC}$ hard or easy. Since each Boolean operator corresponds to a Boolean function—i.e., an n-ary function whose arguments and values are in $\{0, 1\}$—there are infinitely many sets of Boolean operators that determine fragments of $\mathcal{ALC}$. The complexity of the corresponding concept satisfiability problems without theories has already been classified in [19] between being PSPACE-complete, coNP-complete, tractable and trivial for all combinations of Boolean operators and quantifiers.

The tool used in [19] for classifying the infinitely many satisfiability problems was Post's lattice [29], which consists of all sets of Boolean functions closed under superposition. These sets directly correspond to all sets of Boolean operators closed under composition. Similar classifications have been achieved for satisfiability for classical propositional logic [22], Linear Temporal Logic [8], hybrid logic [24], and for constraint satisfaction problems [31, 33].

In this paper, we classify the concept satisfiability problems with respect to theories for $\mathcal{ALC}$ fragments obtained by arbitrary sets of Boolean operators and quantifiers. We separate these problems into EXPTIME-complete, NP-complete, P-complete and NL-complete, leaving only two single cases with non-matching upper and lower bound. We will also put these results into the context of the above listed results for $\mathcal{ALC}$ fragments.

This study extends our previous work in [25] by matching upper and lower bounds and considering restricted use of quantifiers.

2   Preliminaries

Description Logic. We use the standard syntax and semantics of $\mathcal{ALC}$ [7], with the Boolean operators $\sqcap, \sqcup, \neg, \top, \bot$ replaced by arbitrary operators $\circ_f$ that correspond to Boolean functions $f : \{0,1\}^n \to \{0,1\}$ of arbitrary arity n. Let N_C, N_R and N_I be sets of atomic concepts, roles and individuals. Then the set of *concept descriptions*, for short *concepts*, is defined by

$$C := A \mid \circ_f(C, \ldots, C) \mid \exists R.C \mid \forall R.C,$$

where $A \in N_C$, $R \in N_R$, and $\circ_f$ is a Boolean operator. For a given set B of Boolean operators, a *B-concept* is a concept that uses only operators from B. A *general concept inclusion (GCI)* is an axiom of the form $C \sqsubseteq D$ where C, D are concepts. We use "$C \equiv D$" as the usual syntactic sugar for "$C \sqsubseteq D$ and $D \sqsubseteq C$". A *TBox* is a finite set of GCIs without restrictions. An *ABox* is a finite set of axioms of the form $C(x)$ or $R(x,y)$, where C is a concept, $R \in N_R$ and $x, y \in N_I$. An *ontology* is the union of a TBox and an ABox. This simplified view suffices for our purposes.

An *interpretation* is a pair $\mathcal{I} = (\Delta^{\mathcal{I}}, \cdot^{\mathcal{I}})$, where $\Delta^{\mathcal{I}}$ is a nonempty set and $\cdot^{\mathcal{I}}$ is a mapping from N_C to $\mathfrak{P}(\Delta^{\mathcal{I}})$, from N_R to $\mathfrak{P}(\Delta^{\mathcal{I}} \times \Delta^{\mathcal{I}})$ and from N_I to $\Delta^{\mathcal{I}}$ that is extended to arbitrary concepts as follows:

$$\circ_f(C_1, \ldots, C_n)^{\mathcal{I}} = \{x \in \Delta^{\mathcal{I}} \mid f(\|x \in C_1^{\mathcal{I}}\|, \ldots, \|x \in C_n^{\mathcal{I}}\|) = 1\},$$

$$\text{where } \|x \in C_1^{\mathcal{I}}\| = 1 \text{ if } x \in C_1^{\mathcal{I}} \text{ and } \|x \in C_1^{\mathcal{I}}\| = 0 \text{ if } x \notin C_1^{\mathcal{I}},$$

$$\exists R.C^{\mathcal{I}} = \{x \in \Delta^{\mathcal{I}} \mid \{y \in C^{\mathcal{I}} \mid (x,y) \in R^{\mathcal{I}}\} \neq \emptyset\},$$

$$\forall R.C^{\mathcal{I}} = \{x \in \Delta^{\mathcal{I}} \mid \{y \in C^{\mathcal{I}} \mid (x,y) \notin R^{\mathcal{I}}\} = \emptyset\}.$$

An interpretation $\mathcal{I}$ *satisfies* the axiom $C \sqsubseteq D$, written $\mathcal{I} \models C \sqsubseteq D$, if $C^{\mathcal{I}} \subseteq D^{\mathcal{I}}$. Furthermore, $\mathcal{I}$ satisfies $C(x)$ or $R(x,y)$ if $x^{\mathcal{I}} \in C^{\mathcal{I}}$ or $(x^{\mathcal{I}}, y^{\mathcal{I}}) \in R^{\mathcal{I}}$. An interpretation $\mathcal{I}$ satisfies a TBox (ABox, ontology) if it satisfies every axiom therein. It is then called a *model* of this set of axioms.

Let B be a finite set of Boolean operators and $\mathcal{Q} \subseteq \{\exists, \forall\}$. We use $\mathsf{Con}_{\mathcal{Q}}(B)$, $\mathfrak{T}_{\mathcal{Q}}(B)$ and $\mathfrak{D}_{\mathcal{Q}}(B)$ to denote the set of all concepts, TBoxes and ontologies that use operators in B only and quantifiers from $\mathcal{Q}$ only. The following decision problems are of interest for this paper.

Concept satisfiability $\mathrm{CSAT}_{\mathcal{Q}}(B)$:
   Given a concept $C \in \mathsf{Con}_{\mathcal{Q}}(B)$, is there an interpretation $\mathcal{I}$ s.t. $C^{\mathcal{I}} \neq \emptyset$?

TBox satisfiability $\mathrm{TSAT}_{\mathcal{Q}}(B)$:
   Given a TBox $\mathcal{T} \subseteq \mathfrak{T}_{\mathcal{Q}}(B)$, is there an interpretation $\mathcal{I}$ s.t. $\mathcal{I} \models \mathcal{T}$?

TBox-concept satisfiability $\mathrm{TCSAT}_{\mathcal{Q}}(B)$:
   Given $\mathcal{T} \subseteq \mathfrak{T}_{\mathcal{Q}}(B)$ and $C \in \mathsf{Con}_{\mathcal{Q}}(B)$, is there an $\mathcal{I}$ s.t. $\mathcal{I} \models \mathcal{T}$ and $C^{\mathcal{I}} \neq \emptyset$?

Ontology satisfiability $\mathrm{OSAT}_{\mathcal{Q}}(B)$:
   Given an ontology $\mathcal{O} \subseteq \mathfrak{D}_{\mathcal{Q}}(B)$, is there an interpretation $\mathcal{I}$ s.t. $\mathcal{I} \models \mathcal{O}$?

Ontology-concept satisfiability $\mathrm{OCSAT}_{\mathcal{Q}}(B)$:
   Given $\mathcal{O} \subseteq \mathfrak{D}_{\mathcal{Q}}(B)$ and $C \in \mathsf{Con}_{\mathcal{Q}}(B)$, is there an $\mathcal{I}$ s.t. $\mathcal{I} \models \mathcal{O}$ and $C^{\mathcal{I}} \neq \emptyset$?

By abuse of notation, we will omit set parentheses and commas when stating $\mathcal{Q}$ explicitly, as in $\mathrm{TSAT}_{\exists\forall}(B)$. The above listed decision problems are interreducible independently of B and $\mathcal{Q}$ in the following way:

$$\mathrm{CSAT}_{\mathcal{Q}}(B) \leq^{\log}_{\mathrm{m}} \mathrm{OSAT}_{\mathcal{Q}}(B)$$

$$\mathrm{TSAT}_{\mathcal{Q}}(B) \leq^{\log}_{\mathrm{m}} \mathrm{TCSAT}_{\mathcal{Q}}(B) \leq^{\log}_{\mathrm{m}} \mathrm{OSAT}_{\mathcal{Q}}(B) \equiv^{\log}_{\mathrm{m}} \mathrm{OCSAT}_{\mathcal{Q}}(B)$$

Some reductions in the main part of the paper consider another decision problem which is called *subsumption* (SUBS) and is defined as follows: Given a TBox $\mathcal{T}$ and two atomic concepts A, B, does every model of $\mathcal{T}$ satisfy $A \sqsubseteq B$?

Complexity Theory. We assume familiarity with the standard notions of complexity theory as, e.g., defined in [28]. In particular, we will make use of the classes NL, P, NP, coNP, and EXPTIME, as well as logspace reductions $\leq^{\log}_{\mathrm{m}}$.

Boolean operators. This study is complete with respect to Boolean operators, which correspond to Boolean functions. The table below lists all Boolean functions that we will mention, together with the associated DL operator where applicable.

Table 1. Boolean functions with description and corresponding DL operator symbol

Function symbol	Description	DL operator symbol
0, 1	constant 0, 1	$\bot, \top$
and, or	binary conjunction/disjunction $\wedge, \vee$	$\sqcap, \sqcup$
neg	unary negation $\overline{\cdot}$	$\neg$
xor	binary exclusive *or* $\oplus$	$\boxplus$
andor	$x \wedge (y \vee z)$	
sd	$(x \wedge \overline{y}) \vee (x \wedge \overline{z}) \vee (\overline{y} \wedge \overline{z})$	
equiv	binary equivalence function	

A set of Boolean functions is called a *clone* if it contains all projections (also known as identity functions, the eponym of the I-clones below) and is closed under composition (also referred to as superposition). The lattice of all clones has been established in [29], see [10] for a more succinct but complete presentation. Via the inclusion structure, lower and upper complexity bounds can be carried over to higher and lower clones under certain conditions. We will therefore state our results for minimal and maximal clones only, together with those conditions.

Given a finite set B of functions, the smallest clone containing B is denoted by $[B]$. The set B is called a *base* of $[B]$, but $[B]$ often has other bases as well. For example, nesting of binary conjunction yields conjunctions of arbitrary arity. The table below lists all clones that we will refer to, using the following definitions. A Boolean function f is called *self-dual* if $f(\overline{x_1}, \ldots, \overline{x_n}) = \overline{f(x_1, \ldots, x_n)}$, *c-reproducing* if $f(c, \ldots, c) = c$ for $c \in \{0, 1\}$, and *c-separating* if there is an $1 \leq i \leq n$ s.t. for each $(b_1, \ldots, b_n) \in f^{-1}(c)$, it holds that $b_i = c$.

From now on, we will use B to denote a finite set of Boolean operators. Hence, $[B]$ consists of all operators obtained by nesting operators from B. By abuse of notation, we will denote operator sets with the above clone names when this is not ambiguous. Furthermore, we call a Boolean operator corresponding to

Table 2. List of all relevant clones in this paper with their standard bases

Clone	Description	Base
BF	all Boolean functions	$\{\text{and}, \text{neg}\}$
R_0, R_1	0-, 1-reproducing functions	$\{\text{and}, \text{xor}\}$, $\{\text{or}, \text{equiv}\}$
M	all monotone functions	$\{\text{and}, \text{or}, 0, 1\}$
S_1	1-separating functions	$\{x \wedge \overline{y}\}$
S_{11}	1-separating, monotone functions	$\{\text{andor}, 0\}$
D	self-dual functions	$\{\text{sd}\}$
L	affine functions	$\{\text{xor}, 1\}$
L_0	affine, 0-reproducing functions	$\{\text{xor}\}$
L_3	affine, 0- and 1-reproducing functions	$\{x \text{ xor } y \text{ xor } z \text{ xor } 1\}$
E_0, E	conjunctions and 0 (and 1)	$\{\text{and}, 0\}$, $\{\text{and}, 0, 1\}$
V_0, V	disjunctions and 0 (and 1)	$\{\text{or}, 0\}$, $\{\text{or}, 0, 1\}$
N_2, N	negation (and 1)	$\{\text{neg}\}$, $\{\text{neg}, 1\}$
I_0, I	0 (and 1)	$\{0\}$, $\{0, 1\}$

a monotone (self-dual, 0-reproducing, 1-reproducing, 1-separating) function a monotone (self-dual, $\perp$-reproducing, $\top$-reproducing, $\top$-separating) operator.

Known complexity results for CSAT. In [19], the complexity of concept satisfiability has been classified for modal logics corresponding to all fragments of $\mathcal{ALC}$ with arbitrary combinations of Boolean operators and quantifiers: $\text{CSAT}_{\mathcal{Q}}(B)$ with $\mathcal{Q} \subseteq \{\exists, \forall\}$ is either PSPACE-complete, coNP-complete, or in P. Some of the latter cases are trivial, i.e., every concept in such a fragment is satisfiable. These results generalize known complexity results for $\mathcal{ALE}$ and the $\mathcal{EL}$ and $\mathcal{FL}$ families. On the other hand, results for $\mathcal{ALU}$ and the DL-Lite family cannot be put into this context because they only allow unqualified existential restrictions. See [25] for a more detailed discussion.

3 Complexity Results for TSAT, TCSAT, OSAT, OCSAT

In this section we will almost completely classify the above mentioned satisfiability problems for their tractability with respect to sub-Boolean fragments and put them into context with existing results for fragments of $\mathcal{ALC}$. Full proofs of every theorem and auxiliary lemmata are given in [26].

We use $\star\text{SAT}_{\mathcal{Q}}(B)$ to speak about any of the four satisfiability problems $\text{TSAT}_{\mathcal{Q}}(B), \text{TCSAT}_{\mathcal{Q}}(B), \text{OSAT}_{\mathcal{Q}}(B)$ and $\text{OCSAT}_{\mathcal{Q}}(B)$ introduced above; for the three problems having the power to speak about a single individual, we abuse this notion and write $\star\text{SAT}_{\tilde{\mathcal{Q}}}(B)$ for the problems $\star\text{SAT}_{\mathcal{Q}}(B)$ without $\text{TSAT}_{\mathcal{Q}}(B)$.

3.1 Both Quantifiers

Theorem 1 ([30, 34, 17]). $\text{OCSAT}_{\exists\forall}(\text{BF}) \in \text{EXPTIME}$.

Due to the interreducibilities stated in Section 2, it suffices to show lower bounds for TSAT and upper bounds for OCSAT. Moreover one can show that a base independence result holds which enables us to restrict the proofs to the standard basis of each clone for stating general results. Several proof sketches involve the ability to express the constant $\top$ through a fresh concept. This technique goes back to Lewis 1979 [22] and often will be referred to as $\top$-knack.

The following theorem improves [25] by stating completeness results.

Theorem 2. *Let B be a finite set of Boolean operators.*

1. *If $\mathsf{I} \subseteq [B]$ or $\mathsf{N}_2 \subseteq [B]$, then $\mathrm{TSAT}_{\exists\forall}(B)$ is EXPTIME-complete.*
2. *If $\mathsf{I}_0 \subseteq [B]$ or $\mathsf{N}_2 \subseteq [B]$, then $\star\mathrm{SAT}^{\sim}_{\exists\forall}(B)$ is EXPTIME-complete.*
3. *If $[B] \subseteq \mathsf{R}_0$, then $\mathrm{TSAT}_{\exists\forall}(B)$ is trivial.*
4. *If $[B] \subseteq \mathsf{R}_1$, then $\star\mathrm{SAT}_{\exists\forall}(B)$ is trivial.*

Proof sketch. 1. Membership is immediate from Theorem 1. Hardness can be shown in two steps: via a reduction from the positive entailment problem for Tarskian set constraints (cf. [18]) to $\mathrm{TSAT}_{\exists\forall}(\mathsf{E})$ and eliminating in this reduction the conjunction operator. The latter is achieved by an extended version of the normalization algorithm in [12]. The case $\mathsf{N}_2 \subseteq [B]$ then follows directly from Lemma 1 in [25]. 2. follows from 1. by simulating the constant $\top$ with a fresh concept. 3. and 4. follow from [25]. □

Part (2) for I_0 generalizes the EXPTIME-hardness of subsumption for $\mathcal{FL}_0$ and $\mathcal{AL}$ with respect to GCIs [18, 14, 5, 20]. The contrast to the tractability of subsumption with respect to GCIs in $\mathcal{EL}$, which uses only existential quantifiers, undermines the observation that, for negation-free fragments, the choice of the quantifier affects tractability and not the choice between conjunction and disjunction. DL-Lite and $\mathcal{ALU}$ cannot be put into this context because they use unqualified restrictions.

Parts (1) and (2) show that satisfiability with respect to theories is already intractable for even smaller sets of Boolean operators. One reason is that sets of axioms already contain limited forms of implication and conjunction. This also causes the results of this analysis to differ from similar analyses for sub-Boolean modal logics in that hardness already holds for bases of clones that are comparatively low in Post's lattice.

Part (3) reflects the fact that TSAT is less expressive than the other three decision problems: it cannot speak about one single individual.

3.2 Restricted Quantifiers

In this section we investigate the complexity of the problems $\mathrm{OCSAT}_{\mathcal{Q}}$, $\mathrm{OSAT}_{\mathcal{Q}}$, $\mathrm{TCSAT}_{\mathcal{Q}}$, and $\mathrm{TSAT}_{\mathcal{Q}}$, where $\mathcal{Q}$ contains at most one of the quantifiers $\exists$ or $\forall$. Even the case $\mathcal{Q} = \emptyset$ is nontrivial: for example, $\mathrm{TSAT}_{\mathcal{Q}}(B)$ does not reduce to propositional satisfiability for B because restricted use of implication and conjunction is implicit in sets of axioms.

TSAT-Results

Theorem 3. *Let B be a finite set of Boolean operators.*

1. *If $\mathsf{L}_3 \subseteq [B]$ or $\mathsf{M} \subseteq [B]$, then $\mathrm{TSAT}_\emptyset(B)$ is NP-complete.*
2. *If $\mathsf{E} = [B]$ or $\mathsf{V} = [B]$, then $\mathrm{TSAT}_\emptyset(B)$ is P-complete.*
3. *If $[B] \in \{\mathsf{I}, \mathsf{N}_2, \mathsf{N}\}$, then $\mathrm{TSAT}_\emptyset(B)$ is NL-complete.*
4. *Otherwise (if $[B] \subseteq \mathsf{R}_1$ or $[B] \subseteq \mathsf{R}_0$), then $\mathrm{TSAT}_\emptyset(B)$ is trivial.*

Proof sketch. For the monotone case in *1.* it holds that $\overline{\mathrm{IMP}(\mathsf{M})} \leq_m^{\log} \mathrm{TSAT}_\emptyset(\mathsf{M})$ where $\mathrm{IMP}(\mathsf{M})$ being coNP-complete is shown in [9]. For L_3 using a knack from Theorem 2 lets us easily state a reduction from the NP-complete problem 1-in-3-SAT involving the binary exclusive-or. Hardness in the E-case of *2.* is achieved via the hypergraph accessibility problem HGAP; for $\mathsf{V} = [B]$ we argue via contraposition. Membership comes from containment in $\mathrm{OCSAT}_\exists(\mathsf{E})$. In case *3.* hardness is shown for $\mathsf{I} = [B]$ by reducing from the graph inaccessibility problem $\overline{\mathrm{GAP}}$. Membership is entailed by $\mathrm{TCSAT}_\emptyset(\mathsf{N})$. *4.* follows from Theorem 2.   □

Theorem 4. *Let B be a finite set of Boolean operators and $\mathcal{Q} \in \{\forall, \exists\}$.*

1. *If $\mathsf{M} \subseteq [B]$ or $\mathsf{N}_2 \subseteq [B]$, then $\mathrm{TSAT}_\mathcal{Q}(B)$ is EXPTIME-complete.*
2. *If $\mathsf{E} = [B]$, $\mathsf{V} = [B]$, or $\mathsf{I} = [B]$, then $\mathrm{TSAT}_\mathcal{Q}(B)$ is P-complete.*
3. *Otherwise (if $[B] \subseteq \mathsf{R}_1$ or $[B] \subseteq \mathsf{R}_0$), then $\mathrm{TSAT}_\mathcal{Q}(B)$ is trivial.*

Proof sketch. In case *1.* with $\mathsf{M} \subseteq [B]$ and $\mathcal{Q} = \exists$ we reduce from $\overline{\mathcal{ELU}\text{-SUBS}}$, the subsumption problem of the logic $\mathcal{ELU}$, whose EXPTIME-completeness has been proven in [5]. For $\mathcal{Q} = \forall$ we reduce from $\mathrm{TSAT}_\exists(B)$ in combination with a contraposition argument. For N_2, negation can simulate both constants which leads to a simple reduction from $\mathrm{TSAT}_{\exists\forall}(\mathsf{I})$. The hardness results of item *3.* follow from $\mathrm{TSAT}_\exists(\mathsf{I})$ whose hardness is via a reduction from the word problem for a particular Turing machine model of the class P. The P-algorithm for the case $\forall$ and $\exists$ extends the algorithm in [11]. The upper bound for $\mathrm{TSAT}_\exists(\mathsf{E})$ results from $\mathrm{OCSAT}_\exists(\mathsf{E})$; for the remainder we use a contraposition argument.   □

Part (3) generalizes the fact that every $\mathcal{EL}$- and $\mathcal{FL}_0$-TBox is satisfiable, and the whole theorem shows that separating either conjunction and disjunction, or the constants is the only way to achieve tractability for TSAT.

TCSAT-, OSAT-, OCSAT-Results.

Theorem 5. *Let B be a finite set of Boolean operators.*

1. *If $\mathsf{S}_{11} \subseteq [B]$ or $\mathsf{L}_3 \subseteq [B]$ or $\mathsf{L}_0 \subseteq [B]$, then $\star\mathrm{SAT}_\emptyset^\sim(B)$ is NP-complete.*
2. *If $[B] \in \{\mathsf{E}_0, \mathsf{E}, \mathsf{V}_0, \mathsf{V}\}$, then $\star\mathrm{SAT}_\emptyset^\sim(B)$ is P-complete.*
3. *If $[B] \in \{\mathsf{I}_0, \mathsf{I}, \mathsf{N}_2, \mathsf{N}\}$, then $\star\mathrm{SAT}_\emptyset^\sim(B)$ is NL-complete.*
4. *Otherwise (if $[B] \subseteq \mathsf{R}_1$), then $\star\mathrm{SAT}_\emptyset^\sim(B)$ is trivial.*

Proof sketch. Hardness in *1.* follows from the respective $\text{TSAT}_\emptyset(B)$ case together with the fact that these fragments can simulate $\top$. Membership is via a reduction to SAT involving a construction using implication for the terminology. In case *2.* we can adjust the lower bounds from $\text{TSAT}_\emptyset(B)$ and $[B] \in \{V, E\}$ by simulating $\top$ again. Membership is due to $\text{OCSAT}_\exists(E)$. In *3.*, membership for N is via an algorithm that searches for cycles containing a concept and its negation in the directed graph induced by the terminology. Hardness results from the case $\text{TSAT}_\emptyset(I)$ plus the $\top$-knack. The last item is due to Theorem 1. $\qquad\square$

Theorem 6. *Let B be a finite set of Boolean operators, and $\mathcal{Q} \in \{\forall, \exists\}$.*

1. *If $S_{11} \subseteq [B]$, $N_2 \subseteq [B]$, or $L_0 \subseteq [B]$ then $\star\text{SAT}_{\mathcal{Q}}^{\sim}(B)$ is EXPTIME-complete.*
2. *If $I_0 \subseteq [B] \subseteq V$, then $\text{TCSAT}_\exists(B)$ and $\star\text{SAT}_\forall^{\sim}(B)$ are P-complete[1].*
3. *If $[B] \in \{E_0, E\}$, then $\star\text{SAT}_\forall^{\sim}(B)$ is EXPTIME-complete,*
   and $\star\text{SAT}_\exists^{\sim}(B)$ is P-complete.
4. *If $[B] \subseteq R_1$, then $\star\text{SAT}_{\mathcal{Q}}^{\sim}(B)$ is trivial.*

Proof sketch. For *1.*, combine EXPTIME-completeness of $\text{TSAT}_{\mathcal{Q}}(B)$ with the $\top$-knack. EXPTIME-hardness of $\text{TCSAT}_\forall(B)$ in case *3.* is achieved via a reduction from $\overline{\mathcal{FL}_0\text{-SUBS}}$ which is EXPTIME-complete [5, 20]. P-hardness in *2.* and *3.* result from $\text{TSAT}_{\mathcal{Q}}(I)$ again with the help of the $\top$-knack. P-membership of $\text{OCSAT}_\exists(E)$ is accomplished through a reduction to the subsumption problem for the logic $\mathcal{EL}^{++}$ [6], and a contraposition argument is used to reduce $\text{OCSAT}_\forall(V)$ to $\text{OCSAT}_\exists(E)$. *4.* is due to Theorem 2. $\qquad\square$

Theorem 6 shows one reason why the logics in the $\mathcal{EL}$ family have been much more successful as "small" logics with efficient reasoning methods than the $\mathcal{FL}$ family: the combination of the $\forall$ with conjunction is intractable, while $\exists$ and conjunction are still in polynomial time. Again, separating either conjunction and disjunction, or the constants is crucial for tractability.

Table 3. Complexity overview for all Boolean function and quantifier fragments. All results are completeness results for the given complexity class, except for the case marked §: here, OCSAT and OSAT are in EXPTIME and P-hard.

$\text{TSAT}_{\mathcal{Q}}(B)$	I	V	E	N/N_2	M	L_3 to BF	otherwise
$\mathcal{Q} = \emptyset$	NL	P		NL	NP		trivial
$\|\mathcal{Q}\| = 1$	P			EXPTIME			trivial
$\mathcal{Q} = \{\exists, \forall\}$	EXPTIME						trivial

$\star\text{SAT}_{\mathcal{Q}}^{\sim}(B)$	I/I_0	V/V_0	E/E_0	N/N_2	S_{11} to M	L_3/L_0 to BF	otherwise
$\mathcal{Q} = \emptyset$	NL	P		NL	NP		trivial
$\mathcal{Q} = \{\exists\}$	P	P§	P	EXPTIME			trivial
$\mathcal{Q} = \{\forall\}$	P			EXPTIME			trivial
$\mathcal{Q} = \{\exists, \forall\}$	EXPTIME						trivial

---

[1] $\text{OSAT}_\exists(B)$ and $\text{OCSAT}_\exists(B)$ are P-hard for $[B] \in \{V_0, V\}$ and in EXPTIME.

Table 3 gives an overview of our results. [26] contains figures showing how the results arrange in Post's lattice.

4   Conclusion

With Theorems 2 to 6, we have completely classified the satisfiability problems connected to arbitrary terminologies and concepts for $\mathcal{ALC}$ fragments obtained by arbitrary sets of Boolean operators and quantifiers—only the fragments emerging around ontologies with existential quantifier and disjunction as only allowed connective resisted a full classification. In particular we improved and finished the study of [25]. In more detail we achieved a dichotomy for all problems using both quantifiers (EXPTIME-complete vs. trivial fragments), a trichotomy when only one quantifier is allowed (trivial, EXPTIME-, and P-complete fragments), and a quartering for no allowed quantifiers ranging from trivial, NL-complete, P-complete, and NP-complete fragments.

Furthermore the connection to well-known logic fragments of $\mathcal{ALC}$, e.g., $\mathcal{FL}$ and $\mathcal{EL}$ now enriches the landscape of complexity by a generalization of these results. These improve the overall understanding of where the tractability border lies. The most important lesson learnt is that the separation of quantifiers together with the separation of either conjunction and disjunction, or the constants, is the only way to achieve tractability in our setting.

Especially in contrast to similar analyses of logics using *Post's lattice*, this study shows intractable fragments quite at the bottom of the lattice. This illustrates how expressive the concept of terminologies and assertional boxes is: restricted to only the Boolean function *false* besides both quantifiers we are still able to encode EXPTIME-hard problems into the decision problems that have a TBox and a concept as input. Thus perhaps the strongest source of intractability can be found in the fact that unrestricted theories already express limited implication and disjunction, and not in the set of allowed Boolean functions alone.

For future work, it would be interesting to see whether the picture changes if the use of general axioms is restricted, for example to cyclic terminologies—theories where axioms are cycle-free definitions $A \equiv C$ with A being atomic. Theories so restricted are sufficient for establishing taxonomies. Concept satisfiability for $\mathcal{ALC}$ w.r.t acyclic terminologies is still PSPACE-complete [23]. Is the tractability border the same under this restriction? One could also look at fragments with unqualified quantifiers, e.g., $\mathcal{ALU}$ or the DL-lite family, which are not covered by the current analysis. Furthermore, since the standard reasoning tasks are not always interreducible under restricted Boolean operators, a similar classification for other decision problems such as concept subsumption is pending.

Acknowledgements

We thank Peter Lohmann and the anonymous referees for helpful comments and suggestions.

References

[1] Artale, A., Calvanese, D., Kontchakov, R., Zakharyaschev, M.: DL-Lite in the light of first-order logic. In: Proc. AAAI, pp. 361–366 (2007)

[2] Artale, A., Calvanese, D., Kontchakov, R., Zakharyaschev, M.: Adding weight to DL-Lite. In: Proc. DL (2009), `http://CEUR-WS.org`

[3] Baader, F.: Using automata theory for characterizing the semantics of terminological cycles. Ann. Math. Artif. Intell. 18(2-4), 175–219 (1996)

[4] Baader, F.: Terminological cycles in a description logic with existential restrictions. In: Proc. IJCAI, pp. 325–330 (2003)

[5] Baader, F., Brandt, S., Lutz, C.: Pushing the $\mathcal{EL}$ envelope. In: Proc. IJCAI, pp. 364–369 (2005)

[6] Baader, F., Brandt, S., Lutz, C.: Pushing the $\mathcal{EL}$ envelope further. In: Proc. OWLED DC (2008)

[7] Baader, F., Calvanese, D., McGuinness, D.L., Nardi, D., Patel-Schneider, P.F.: The Description Logic Handbook: Theory, Implementation, and Applications. Cambridge University Press, Cambridge (2003)

[8] Bauland, M., Schneider, T., Schnoor, H., Schnoor, I., Vollmer, H.: The complexity of generalized satisfiability for Linear Temporal Logic. LMCS 5(1) (2009)

[9] Beyersdorff, O., Meier, A., Thomas, M., Vollmer, H.: The Complexity of Propositional Implication. IPL 109(18), 1071–1077 (2009)

[10] Böhler, E., Creignou, N., Reith, S., Vollmer, H.: Playing with Boolean blocks, part I: Post's lattice with applications to complexity theory. ACM-SIGACT Newsletter 34(4), 38–52 (2003)

[11] Brandt, S.: Polynomial time reasoning in a description logic with existential restrictions, GCI axioms, and—what else? In: Proc. ECAI, pp. 298–302 (2004)

[12] Brandt, S.: Reasoning in $\mathcal{ELH}$ w.r.t. general concept inclusion axioms. LTCS-Report LTCS-04-03, Dresden University of Technology, Germany (2004)

[13] Calvanese, D., De Giacomo, G., Lembo, D., Lenzerini, M., Rosati, R.: DL-Lite: Tractable description logics for ontologies. In: Proc. AAAI, pp. 602–607 (2005)

[14] Donini, F.M.: Complexity of reasoning. In: Description Logic Handbook [7], pp. 96–136

[15] Donini, F.M., Lenzerini, M., Nardi, D., Hollunder, B., Nutt, W., Marchetti-Spaccamela, A.: The complexity of existential quantification in concept languages. AI 53(2-3), 309–327 (1992)

[16] Donini, F.M., Lenzerini, M., Nardi, D., Nutt, W.: The complexity of concept languages. Inf. Comput. 134(1), 1–58 (1997)

[17] Donini, F.M., Massacci, F.: EXPTIME tableaux for $\mathcal{ALC}$. AI 124(1), 87–138 (2000)

[18] Givan, R., McAllester, D., Wittny, C., Kozen, D.: Tarskian set constraints. Information and Computation 174, 105–131 (2002)

[19] Hemaspaandra, E., Schnoor, H., Schnoor, I.: Generalized modal satisfiability. CoRR, abs/0804.2729 (2008)

[20] Hofmann, M.: Proof-theoretic approach to description-logic. In: Proc. LICS, pp. 229–237 (2005)

[21] Kazakov, Y., de Nivelle, H.: Subsumption of concepts in $\mathcal{FL}_0$ for (cyclic) terminologies with respect to descriptive semantics is PSPACE-complete. In: Proc. DL (2003), `http://www.CEUR-WS.org`

[22] Lewis, H.: Satisfiability problems for propositional calculi. Math. Sys. Theory 13, 45–53 (1979)

[23] Lutz, C.: Complexity of terminological reasoning revisited. In: Ganzinger, H., McAllester, D., Voronkov, A. (eds.) LPAR 1999. LNCS, vol. 1705, pp. 181–200. Springer, Heidelberg (1999)

[24] Meier, A., Mundhenk, M., Schneider, T., Thomas, M., Weber, V., Weiss, F.: The complexity of satisfiability for fragments of hybrid logic—part I. In: Královič, R., Niwiński, D. (eds.) MFCS 2009. LNCS, vol. 5734, pp. 587–599. Springer, Heidelberg (2009)

[25] Meier, A., Schneider, T.: The complexity of satisfiability for sub-Boolean fragments of $\mathcal{ALC}$. In: Proc. of DL 2010. CEUR-WS.org (2010)

[26] Meier, A., Schneider, T.: Generalized satisfiability for the description logic $\mathcal{ALC}$. CoRR (2011), http://arxiv.org/abs/1103.0853

[27] Nebel, B.: Terminological reasoning is inherently intractable. AI 43(2), 235–249 (1990)

[28] Papadimitriou, C.H.: Computational Complexity. Addison-Wesley, Reading (1994)

[29] Post, E.: The two-valued iterative systems of mathematical logic. Ann. Math. Studies 5, 1–122 (1941)

[30] Pratt, V.R.: A practical decision method for propositional dynamic logic: Preliminary report. In: STOC, pp. 326–337. ACM, New York (1978)

[31] Schaefer, T.J.: The complexity of satisfiability problems. In: Proc. STOC, pp. 216–226. ACM Press, New York (1978)

[32] Schmidt-Schauß, M., Smolka, G.: Attributive concept descriptions with complements. AI 48(1), 1–26 (1991)

[33] Schnoor, H.: Algebraic Techniques for Satisfiability Problems. PhD thesis, Leibniz University of Hannover (2007)

[34] Vardi, M.Y., Wolper, P.: Automata-theoretic techniques for modal logics of programs. JCSS 32(2), 183–221 (1986)

Author Index